ANNUAL REVIEW OF ASTRONOMY AND ASTROPHYSICS

EDITORIAL COMMITTEE (1984)

ANNUAL REVIEW OF ASTRONOMY AND ASTROPHYSICS

VOLUME 22, 1984

GEOFFREY BURBIDGE, *Editor*
Kitt Peak National Observatory

DAVID LAYZER, *Associate Editor*
Harvard College Observatory

JOHN G. PHILLIPS, *Associate Editor*
University of California, Berkeley

ANNUAL REVIEWS INC.   4139 EL CAMINO WAY   PALO ALTO, CALIFORNIA 94306 USA

International Standard Serial Number : 0066-4146
International Standard Book Number : 0-8243-0922-7
Library of Congress Catalog Card Number : 63-8846

TYPESET BY AUP TYPESETTERS (GLASGOW) LTD., SCOTLAND
PRINTED AND BOUND IN THE UNITED STATES OF AMERICA

PREFACE

This volume of the Annual Review of Astronomy and Astrophysics was planned at a meeting of the Editorial Committee held on May 8, 1982, in San Francisco, California. Those present were Editor Geoffrey Burbidge; Associate Editors David Layzer and John Phillips; Committee Members Ann Boesgaard, Marshall Cohen, Sandra Faber, Herbert Gursky, and Jack Harvey; and Production Editor Keith Dodson. Barbara Jones attended as a guest and helped with the choice of articles. Once again I would like to thank the Associate Editors for carrying out an efficient job of scientific editing, and Keith Dodson for doing an excellent job in Palo Alto.

THE EDITOR

SOME RELATED ARTICLES IN OTHER *ANNUAL REVIEWS*

From the *Annual Review of Nuclear and Particle Science*, Volume 34 (1984):

Low-Energy Neutrino Physics and Neutrino Mass, F. Boehm and P. Vogel

Nucleosynthesis, James W. Truran

Annual Review of Astronomy and Astrophysics
Volume 22, 1984

CONTENTS

Jesse L. Greenstein

Ann. Rev. Astron. Astrophys. 1984. 22 : 1–35

AN ASTRONOMICAL LIFE

Jesse L. Greenstein

Department of Astronomy and Palomar Observatory,
California Institute of Technology, Pasadena, California 91125

PERSONAL

People write history although never certain what the future could learn from the past. Professional historians recreate a possible past with emphasis on what documentary evidence exists; they reimage it, conditioned by their own world view. Events remembered by participants may differ from history so much as to be nearly unrecognizable. My prefatory chapter will be quite personal and anecdotal; it is only one possible account of institutions and events of the over 50 years through which I have lived as a scientist. The mental landscape recreated is in part memory, in part illusion, but not necessarily deceptive. It naturally puts me too much at the center of events. Another landscape could be found in the roughly 600 pages of transcribed, personal oral history, and in many shelves of archives. Which is the true picture? I would have liked, sometime, to describe objectively the growth and maturity of the institutions where I have been and to help document the explosive growth of the knowledge and funding of astronomy in the United States. But a personal approach should give readers a feeling for the startling change in style of research, dramatic even though spread over 50 years. Ten or more years spent for an experiment in space, multiauthor papers, and computer-generated theory are quite alien to me. Here I limit myself to my own activities and interests, involvements with government, and the characteristics of a few of those leaders in astronomy with whom I have worked. By chance I am the first US-born astronomer to write a prefatory chapter for this series. My work is rooted in personality; my public activities reflect my private world. I was born in the Year of the Comet (1909), a comet that appears again on a Palomar CCD image (1982); 1909 is not a lost world, for me.

I lack the often-quoted advantages of an impoverished and embittered childhood; my earliest memories are of being indulged. My paternal

1

0066–4146/84/0915–0001$02.00

grandfather had come to America in 1888; in the usual miraculous American way, he and my father prospered. When I was eight, my grandfather gave me a brass telescope on a tripod. With it I lectured my friends on the planets, stars, and nebulae. I founded an "Interesting Topics Club" to describe the miracles in the sky. As in many Jewish families, there was an excellent and varied library. At home I read C. Flammarion, J. Verne, S. Newcomb, and the *Splendors of the Heavens*. In a basement laboratory, I used a prism spectroscope (from Gaertner Scientific), an arc, a rotary spark, a rectifier, and a radio transmitter. From Kayser's *Handbook of Spectroscopy*, with spectroscope and arc, I tried to identify atoms; line series were simple, but not so the new multiplets and energy-level diagrams. I read contemporary literature as well as science as I skipped through the New York public schools, until I entered Horace Mann School for Boys (a private high school) when I was eleven. My first hero was my Latin teacher. He threw me out of the classroom regularly until I caught up with my classmates, three years older, who had already studied Latin for two years. (I still can read Latin.) Mr. Nagle introduced me to the "ideal" of hard work. No longer a child's oyster, the world was a fascinating sea, navigable by work. I studied and enjoyed chemistry, then the most appealing experimental science; physics, with levers, pulleys and magnets, was dull.

In 1926, when 16 and at Harvard, I met Naomi Kitay, whom I was fortunate enough to marry in 1934, after she graduated from the Horace Mann School for Girls and Mount Holyoke College in 1933. I had one younger brother; she, however, had a brother and three older, educated, and talkative sisters! Our youths were spent in comfortable, patriarchal ways no longer imaginable, dependent on stability and middle-class good manners. Perhaps the important feature of this prospersous, decent world was that it was an easy one to rebel against and eventually to leave. It was naturally expected that I should continue the family business and prosper. When I chose not to do so, later, and still in the depths of the Great Depression (1934), it was the radical nature of the change toward an academic career that convinced our two families that we really wanted that world. No one believed that there was such a paid profession as being an astronomer. To teach meant to be poor. This story will move back and forth in time frequently, since my career did not move in a straight line. The period I describe (and only through its impact on myself) is covered in a more complete and unbiased fashion in Struve & Zebergs (71).

After my Harvard AB, October 1929 saw the decisive collapse of the US and world economy. I was faced with a "duty" to help my family survive the business crash. I worked through disasters to our manufacturing and real-estate business. People did jump from skyscrapers. The banks closed, business stopped, the streets filled, and strangers spoke to each other. The actions of President Roosevelt in responding with Federal intervention shaped the future; they were resisted but inevitable. During four years I learned a great deal: how to talk to different kinds of people; how to handle money; what poverty did to good people; the unattractive nature of the politics of extremists.

I might have become some type of theorist if an originally planned stay at Oxford had happened in 1929. Certainly I would have been less of a manager and leader. The Depression changed attitudes toward government involvement in education and research. European university life had been part of the State's responsibility, sometimes represented at a high level by a Minister of Education. Technical schools were often linked to the modernization of industry. But in the United States, research groups were largely within the government departments, not independent agencies, and provided no university support. The list is small although honorable. The Geological Survey, the Bureau of Mines, Coast and Geodetic Survey, the Weather Service, the Navy and its Observatory, the Bureau of Standards, and the National Advisory Committee for Aeronautics were among the few successful groups in the physical sciences. An established profession in research, where a job would carry financial support, was rarely part of a budding scientist's expectation. Astronomy was particularly small. Teaching was the usual outlet, normally incompatible with doing much research. Thus I have lived through a complete change of attitude—to the complete Federal support of graduate and postdoctoral education in the 1980s. The present scheme has an emotional unreality to me, possibly to others who lived through the Depression and the wars. It still seems always precarious. Much of my activity in planning for funding of science had its origin in this sense of danger to a delicately balanced enterprise. Only the stronger feeling for the giant opportunity provided by new technologies helped balance my apprehension. But unlike many colleagues, I always looked for and welcomed private support; G. E. Hale had already shown how well such riches could be harnessed. All institutions that have paid me or provided me extraordinary research opportunities were founded or touched by his genius—the National Research Council, Yerkes Observatory, Mount Wilson Observatory, Palomar Observatory, and Caltech. Unfortunately, I never met him.

HARVARD

> *God offers for every mind the choice between truth and repose.*
> Ralph Waldo Emerson

For undergraduates, astronomy was taught in a building on Jarvis Street (now buried under the Law School), with associated transit circles, clocks, and courses in practical astronomy and navigation. The main text was

Russell, Dugan & Stewart; the professor (H. T. Stetson) went on two eclipse expeditions, and so I did some tutoring. At first there was no contact with the Harvard College Observatory (HCO) staff or knowledge of the rapid growth of astrophysics at HCO. In 1928–29, I attended a lecture course by the HCO staff that covered stellar astronomy, astrophysics, and some extragalactic research. There were a few graduate students at HCO. A young theorist, H. H. Plaskett, had recently come from Canada and impressed me strongly; he arranged that I meet E. A. Milne that summer at Michigan to plan for study at Oxford in 1929. The interests of C. Payne (stellar atmospheres and their composition) and of D. H. Menzel (solar astrophysics and gaseous nebulae) were part of a modernization program to have been lead by Plaskett, but Plaskett did not stay at Harvard. I became closest to Miss Payne, a person of wide culture and astronomical knowledge. The obvious discrimination against her as a woman scientist worthy of normal academic recognition exacerbated the stressful life she led. She was unhappy, emotional, in rivalry with Menzel and Plaskett. But with me, she was charming and humorous as we exchanged quotations from T. S. Eliot, Shakespeare, the Bible, Gilbert and Sullivan, and Wordsworth. Her *Stellar Atmospheres* (61) is one of the great theses in astronomy.

My undergraduate education was diffuse, covering a variety of topics. The first I heard of quantum mechanics was in lectures by E. C. Slater on visits from MIT. My advisor suggested that quantum mechanics was only one more fad and that I should instead take more classical physics! There was no physics experimental lab, and in fact I had lost my early gadgeteering interests. I listened to A. N. Whitehead on the philosophy of science and the foundations of mathematics. An illness prevented my trip to Oxford, so I stayed for the AM (1930), working at the Observatory. My first research was on the temperature scale for B and O stars. I planned to use the mean "color equivalents," c_2/T, tabulated by E. Hertzsprung (and others) from visual and photographic magnitudes, to determine a main-sequence temperature scale. Miss Payne had noted the abnormally low color temperatures of O and B stars, in contrast with those indicated by their lines of high ionization and excitation potential (using Saha and Boltzmann theory). I found their mean color temperatures to be lowest at right ascensions of 3 to 7 hr (notably in Orion) and again at 17 to 21 hr. What I had found was the general interstellar reddening of B stars by dust in the galactic plane. Instead, I explained away the reddening as a seasonal systematic error in atmospheric extinction. Why? First, the concept of air mass had just been developed by the meteorologist C. G. Rossby and was fashionable. Second, H. Shapley claimed to have disproved interstellar reddening, since he had found blue faint stars in the Milky Way using

Selected Area magnitude sequences. In addition, reddening by Rayleigh scattering would have severely distorted energy distributions (λ^{-4} rather than the λ^{-1} observed). Both ideas, although fashionable, were irrelevant; Shapley's observations were incorrect. I lacked the independent wisdom to establish the existence of interstellar reddening in my first paper (15). I remember the excitement at HCO when R. J. Trumpler's paper on galactic clusters (73) arrived, showing that a general interstellar absorption existed. Trumpler records his own unwillingness to accept general reddening at first, although he had proved absorption to be real. He notes that Wallenquist (77) had also detected reddening in a galactic cluster. There was a lesson to be learned, but science has always provided an ample lifetime quota of such shocks for would-be pioneers.

During the depression years in New York City (1930–34), I luckily met the physicist I. I. Rabi; in his usual no-nonsense style, he asked me what a bright boy was doing in the real-estate business. He offered to let me do volunteer work in his laboratory at Columbia, designing and winding deflection coils and computing particle trajectories. Since his work for the Nobel Prize was done in 1937, my fear of electromagnetic theory may have been unfortunate. Rabi introduced me to J. Schilt, who also offered volunteer work. He had 75 plates of the globular cluster Messier 3, taken with the Mount Wilson 60-inch in 1926; I searched for new RR Lyrae variables and gave periods (16) and light curves for 199 stars. W. J. Eckert at Columbia was using the first large array of IBM mechanical calculators for celestial mechanics. My work on Messier 3 was monotonous, carried out by eye and using hand calculators, but it convinced me that I had more love for astronomy than for money. I visited Shapley at Harvard, who told me that science had advanced too rapidly for me to hope to catch up. I persisted, and returned for the summer school of 1934. The total financial support was a $400 scholarship in my second graduate year and $700 in my last.

During my four-year absence, HCO had changed radically; groups of graduate students were associated either with Menzel or B. J. Bok. A detailed history of the years 1930–39 at HCO would be a fascinating study in the growth of US astronomy and in the training of a generation of leaders. In retrospect, HCO now appears to have been unfortunately isolated from the great observational efforts at Mount Wilson and Lick, possibly because of some past hostility to Shapley. HCO students were expected to learn a little about everything. Some strengths of the Observatory were in its patrol-plate collection covering both hemispheres, the Henry Draper objective-prism spectra, and the photographic survey of the Magellanic Clouds. Few students were involved with Shapley's extragalactic programs. In the liveliness of the HCO approach to a wide variety of astronomical topics, and its inheritance from sky surveys, one could disregard the modest quality of the telescopes. H. N. Russell, active in laboratory spectroscopy, analysis of complex spectra, and links with the new atomic physics, was a frequent visitor from Princeton. Among the summer visitors, O. Struve had the sharpest impact on me; he showed what seemed then an almost excessive regard for astrophysical depth and details of the interpretation of binaries. He also used atomic physics; with him, stellar astronomy was clearly part of current physics. From Menzel I learned the physics of ionized gases and recombination theory, together with more atomic spectroscopy. Many of Menzel's students became leaders of American astronomy, e.g. L. H. Aller, J. G. Baker, L. Goldberg. With Bok were E. Lindsay, S. W. McCuskey, F. D. Miller, and C. K. Seyfert. F. L. Whipple, with a West Coast background, was then a young instructor. He already had a devouring interest in planets, comets, and meteors. Shapley, along with other senior US astronomers, had begun to take part in rescuing scientists from Europe; some of these became our close friends.

I was briefly in Bok's star-counting circus but counted only a few stars. I shared his interest in the absorbing dust clouds and undertook observations to determine the interstellar-reddening law and the physical theory of reddening. Three papers (17–19) came from my thesis. I applied the classical theories (dating from 1908) of G. Mie and P. Debye to compute the extinction of light by dust particles smaller than or near the wavelength of light. Such computations had been carried out by C. Schalén (in Sweden) in 1929 and by E. Schönberg and B. Jung (in Germany). It was a mathematical boundary-value problem, which except for small particles requires computation of a series of multipole amplitudes using special functions (which were often not fully tabulated). I computed the integrated extinction by a power-law frequency distribution of particle sizes. The observed reddening law was derived from the calibrated photographic spectrophotometry of objective-prism spectra of 38 B stars, obtained with the 24-inch reflector at Agassiz station on a few, cold and fortunately clear nights. The ratios of fluxes between reddened and nearly unreddened B's gave an extinction law near $\lambda^{-0.7}$. In my thesis, I mention photoelectric photometry by J. Stebbins and C. M. Huffer; later this method was used, with six photoelectric colors, by Stebbins and A. E. Whitford. I studied the radiation pressure on grains, important for the interstellar medium as a whole only if interstellar hydrogen is neglected. Star counts by Bok's group had established the existence of dense dark nebulae, which I grouped into five large "cloud complexes." These are essentially groups of giant molecular clouds, within which much of the general absorption arises. The grains for which I did computations included metals, silicates, and frozen water (the latter two having a high ratio of scattering to extinction, i.e. high albedo) important for reflection nebulae. In 1937 the gas producing then-known interstellar lines seemed only a trace constituent. At my oral examination, Shapley asked me how to find interstellar hydrogen, rather than dust. I suggested recombination lines (eventually seen in the H II regions) and subordinate lines of H I (such as Struve had found in circumstellar shells). I did not mention Lα which seemed hopelessly unobservable; it was 20 years before Sputnik.

A new venture closes my Harvard adventures; I had a long interest in amateur radio. A flurry of publicity (May 1933) marked K. Jansky's discovery of "cosmic static"—but I didn't notice it. His other papers, published in engineering journals, had little effect on astronomers. I have written elsewhere (for the fiftieth anniversary celebration of radio astronomy) about my involvement. Fred Whipple and I attempted to explain Jansky's radio signals by thermal emission from dust in the galactic center; we assumed that the space densities of stars and dust grains increased inward by a steep power law. We (78) used S. Chandrasekhar's radiation-transfer formulation for spherical atmospheres to derive the maximum dust temperatures in the Galaxy's central blaze of stars and dust, but we could not reach antenna temperatures above 30 K, even with 10,000 times the radiative energy density in our part of the Galaxy. Our purely thermal explanation failed to explain Jansky's observed fluxes by a factor of 10,000. At that time, neither relativistic particles nor magnetic fields were conceivable parts of the astronomical repertory. A polite editorial in the *Boston Evening Transcript* gives us credit for a very large failure, but it was a failure that persisted for 15 years.

EARLY YEARS AT THE YERKES OBSERVATORY

I was fortunate to obtain a National Research Council Fellowship for 1937–39, one of the few available in the physical sciences. The stipend was $2200 and permitted a choice of where to work. I went to the Yerkes Observatory of the University of Chicago at Williams Bay, Wisconsin. It was indeed a change of scene. When my wife and I drove from Harvard, we reached—at 20 miles past the Hudson River—the farthest point west we had traveled, although we had traveled extensively in Europe. Williams Bay was a town of 600 people on a beautiful glacial lake in farming country. Yerkes and Harvard differed as much as did their landscapes, and I have benefited much from both. Yerkes was entering into its great period, with an expanding faculty and instrumentation plans for the McDonald Observatory near Fort Davis, Texas. Yerkes (1897) was the first observatory built by Hale; his 40-inch refractor was its main instrument, and his taste dictated its elaborate architecture. My fellowship was from an organization born from his revitalization of the National Academy of

Sciences after World War I. But Yerkes was fully reshaped under Struve's leadership. I believe it was the first US working observatory to have theoretical astrophysicists on its staff. Struve also imported onto the Yerkes staff leading astronomers from abroad such as Chandrasekhar, G. P. Kuiper, B. Strömgren, P. Swings, and visitors like A. Unsöld and K. Wurm. Elsewhere, among observing astronomers, xenophobia was not uncommon. Another innovation was the construction of the 82-inch reflector at McDonald as a joint project of the Universities of Chicago and Texas. Completion of the 82-inch in 1939 marked the first major telescope construction since the Mount Wilson 100-inch in 1918 and the beginning of the migration to the good observing climates of the Southwest.

Yerkes was 80 miles from the University of Chicago and 60 miles from the University of Wisconsin. Graduate students lived in the Observatory and thus lacked exposure to contemporary physics. Courses were given by Yerkes staff and visitors; research was at first only with the 40-inch, later with the 82-inch. The use of the 40-inch was romantic, exhausting, and often cold. Having new optics in development for the 82-inch (with F. Ross as advisor) made it more common to experiment with new instruments than at Harvard. Struve and Kuiper were interested in state-of-the-art developments, although at first neither had the experimental skills common on the West Coast. Struve believed in quick responses to new opportunities; he had been interested in the airglow and in faint surface photometry and spectroscopy. Some new advances were the McDonald coudé, the nebular spectrograph at Yerkes, a Fabry photometer on the 40-inch, some photoelectric photometry, near-infrared photography, and later Kuiper's near-infrared spectroscopy. Important scientific neighbors were Stebbins and Whitford, the first of the converted physicist-astronomers I met. Young scientists at Yerkes included W. W. Morgan, an artist in spectral classification (whose work could be done with the 40-inch), and L. G. Henyey, a theorist with whom I became closely involved. Our first paper (44) was based on the newly invented nebular spectrograph. We collaborated on eight observational and theoretical papers and five (classified) reports on optical design in a few years. His perfectionism blended with my somewhat coarser energy into confidence that we could finish anything we tried. I have been lucky to enjoy such well-matched collaboration often. Henyey's life was unfortunately shortened by an illness he had suffered from since childhood. His Yerkes thesis was on reflection nebulae and showed elegant mathematical skill. The observed sizes and surface brightnesses of the nebulae required high albedo; my thesis suggested that the dust could be ice or silicate glass. We solved many difficult radiation-transfer problems analytically, some with methods related to V. A. Ambartsumian's and

Chandrasekhar's invariance and reciprocity theorems. We used the Fabry-lens photometer at the focus of the 40-inch refractor, setting on empty space between the visible stars in the Milky Way, to measure the diffuse galactic light (46). This was stimulated by Struve and C. T. Elvey's discovery of the high surface brightness of a dark nebula. We found the dust to have a high albedo and a forward-throwing phase function. The birth of the Yerkes nebular spectrograph may illustrate the style of Struve's leadership. On a cloudy night when I was assigned the 40-inch, I found Struve and Henyey in the library. Struve posed the question, "In principle, what is the most efficient possible spectrograph?" We decided it was one with the fewest possible components, ultraviolet optics, and the fastest focal ratio camera. We set a slit on top of the far end of the 40-inch tube, omitted the collimator, and put a wooden box with the McDonald quartz prisms and f/1 Schmidt at the eye end, 69 feet away. I used it two nights later on another cloudy night; a long exposure at the zenith gave a magnificent spectrum of an aurora. Spectra obtained of emission and reflection nebulae were also exciting (44, 45). $H\alpha$ was nearly everywhere in the Milky Way, with known emission nebulae often only brighter patches. Henyey's spectrum of Comet Encke shows NH and OH strongly, now first made visible by the UV optics. An improved 150-foot-long nebular spectrograph was built at the unfinished McDonald Observatory site; with it, Struve and Elvey completed the discovery of H II regions, soon after explained by Strömgren. The nebular spectrograph permitted my brash first venture into extragalactic astronomy and cosmology in a paper (20) written only 13 months after I had left Harvard, early enough to be only the tenth McDonald Contribution. It used the energy distribution of Messier 31, measured on 7 spectra, to determine temperature from the spectrophotometric gradient, c_2/T, on the Greenwich system. The galaxy had a roughly blackbody energy distribution, at 4200 K, with some ultraviolet deficiency. Comparing this distribution with photoelectric colors by Stebbins and Whitford gave me the range of color temperatures for E to Sc galaxies as 4900–6200 K. A serious cosmological problem then arose in reanalyzing Hubble's work on the effect of redshift on galaxy magnitudes, since these temperatures were much lower than those that Hubble had used. Galaxies therefore dimmed too rapidly with increasing redshift to be compatible with the Hubble counts as a function of magnitude. I determined the redshift corrections as a function of $d\lambda/\lambda$. When I later met Hubble, I was at my boldest. In 1930, I had failed to use color temperatures to discover interstellar reddening; by 1939, I had to believe in good observational data. This was a most valuable lesson: believe in data, improve data, try new experimental techniques, and (at least for me) try to do a first theoretical interpretation. While still limited to using photographic techniques, I could see that a "new" astronomy existed that had always to be renewed. Important new directions with a different instrument soon opened for me.

The McDonald 82-inch (now renamed the Struve telescope) was dedicated in May 1939. Under the shadow of World War II, Europeans enjoyed a Texan barbecue and rodeo. I met Milne again; for the first time, my wife and I met W. S. Adams, W. Baade, E. P. Hubble, R. McMath, J. Oort, and A. Unsöld, all of whom played an important part in the future of astronomy. On the first 82-inch observing run, I helped Struve take coudé spectra of τ Scorpii for Unsöld to analyze in Germany; these became testing grounds for improved composition analyses by successive generations of Unsöld students.

That year effectively ended my youth—I passed 30, my interests changed, and the war altered my life. Struve said that high-dispersion spectroscopy was exciting and possible and that υ Sagittarii was interesting. The 82-inch coudé prism spectrograph rivaled that which T. Dunham was developing on the Mount Wilson 100-inch. I obtained both coudé and cassegrain spectra to determine abundances over a range of temperatures, gravities, and apparent composition anomalies. Most important was the analysis (21) of υ Sagittarii. This hydrogen-poor object provided the second quantitative analysis of a star of abnormal composition; L. Berman had recently analyzed a carbon-rich star, R Coronae Borealis, in a pioneer study. My first analysis at coudé resolution (22) was of α Carinae—a star far south, even for McDonald. Its composition proved normal, although it was a supergiant; that paper outlines the practical method of differential-curve-of-growth analysis, what Unsöld called "grobanalyse." Since few atomic parameters were then available, the transition probabilities were derived by hook or by crook, using solar gf-values obtained by Menzel and Goldberg, strengths from supermultiplets or transition arrays, and opacity theory to set the atmospheric mass. No models existed—the ionization and excitation temperatures had to be estimated from metallic lines. Reliable spectrophotometry of the continuum, which would be used nowadays to determine the effective temperature, did not exist. It was υ Sgr, with a dominantly helium atmosphere, that introduced me to complex problems in stellar spectroscopy and atmospheres. Plate IV of the υ Sgr paper (21) is still one of the most dramatic illustrations of an important composition abnormality; it was so used by Burbidge, Burbidge, Fowler & Hoyle in 1957 (3). The curve of growth was made more sophisticated by Unsöld's methods of weighting functions and thermal stratification. The book *Spectroscopic Astrophysics* (47) recapitulates Struve's early work in stellar spectroscopy, as well as the later research stimulated by his work.

On 7 September 1941, the fiftieth anniversary of the University of

Chicago and three months before Pearl Harbor, the American Astronomical Society (J. Stebbins, President) met in Williams Bay during a 5-inch rainstorm. The dedication of Yerkes Observatory had occurred in 1897, leading to the founding of the American Astronomical and Astrophysical Society (S. Newcomb, President) in 1899. For the meeting, I arranged an exhibit of 50 years of astronomical photography, including 30-foot-long blowups of coudé spectra from both Mount Wilson and McDonald that covered walls of the 40-inch dome. The group photograph (4) brings back pleasant memories. There are past and present leaders; friends there included G. Randers (later head of the Norwegian atomic-energy project), J. S. Hall, W. A. Hiltner, and L. Spitzer. These names bring hints of the future in plasmas, magnetic fields, and electronics. At the meeting, the AAS took responsibility for publications; the *Astronomical Journal* was transferred from Dudley Observatory to Yale (with D. Brouwer as editor), the *Astrophysical Journal* to the University of Chicago Press (with editor Struve, later Chandrasekhar). Photoelectric photometry had spread; Stebbins and Whitford had improved the interstellar reddening law over my photographic determination. Eight years later, Hall and Hiltner independently discovered interstellar polarization. Before the meeting, I had become a friend of G. Reber as I revived my interest in radio astronomy. The technology-centered growth of modern observational astronomy was starting.

In the front row of this group is my wife, Naomi, fully occupied both as manager and actress at a summer community theater and as mother of a one-year-old son (George; now professor of astrophysics at Amherst College). Our second son, Peter, was born in 1946; he is now active in music and drives a mobile library near Oakland. But informative and nostalgic as it may be, the picture is a prelude to large, and sad, changes that followed the 1941 meeting of the AAS.

YERKES: THE WAR YEARS

Struve was strongly oriented toward international problems of astronomy and had close ties with Europe. He saw war as inevitable and was concerned that his staff would join large laboratory engineering groups elsewhere. Several young astronomers had left Yerkes and Harvard for defense well before Pearl Harbor. He felt that the future of astronomy was threatened unless astronomers stayed together, proffering talents in research groups. The military draft also threatened to produce an unfortunate loss of scientific talent. A discussion of Struve's efforts, and his correspondence with Shapley and Russell, is in De Vorkin's (9) study of the Yerkes Optical Bureau. A more general discussion of astronomers involved with military optics is in Dunham (10), who headed the section of the Office of Scientific Research and Development (OSRD) with which I became involved. De Vorkin quotes extensively Struve's (70) pessimistic, but farsighted, thoughts on problems concerning the intellectual survival of astronomical research. Until I read De Vorkin's account, I did not know how intensely Struve had struggled to keep the telescopes in operation and to retain a group at Yerkes. By mid-1942, Henyey and I (with D. Popper, briefly, and G. Van Biesbroeck) were designing lenses and optical systems. Later F. Pearson, under Hiltner, provided a working optical shop. I negotiated with Dunham, the OSRD, or potential military users to clarify an idea for a needed instrument. The OSRD had other, larger groups at Harvard, Mount Wilson, and Rochester. No satisfactory texts on optics existed; optical aberration theory was taken from an old National Bureau of Standards (NBS) handbook. US industry had longstanding proprietary relations with Germany. Most of our effort was spent fitting requirements on space, weight, and materials; much was wasted due to rivalry under conditions of secrecy. I retain little affection for the optical industry.

We had much to do. Henyey could outline the mathematical theory of a required lens system, often overnight; within limits, optimum lens systems could be quickly defined. But the tedium of ray tracing with hand-cranked calculators is incredible; in 90 seconds we could trace a single ray through an optical surface with six-digit accuracy, interpolating in trigonometric tables. Two independent computations were needed. We soon developed a more efficient ray-tracing method. I struggled with priorities for rare glasses and machinery. I carried one-of-a-kind optics to a proving ground and remember sleeping on a misshapen box containing a gunsight. We designed and built extraordinarily complex lenses—including one for photography through a periscope. But radar was taking over, making optics a secondary solution. As during the depression, I learned much about new types of people, in this case the military, and benefited. Scientific spinoffs were the development of ultrafast lens and mirror systems and the use of unusual materials. Before Henyey left for Berkeley in 1947, we designed a fast system for X-ray fluoroscopy of the digestive system for the Billings Hospital in Chicago. We obtained patents, one of which helped to lead, after several changes, to wide-screen movie projection. (I still own 7500 shares of worthless stock in such an enterprise.) Several of our devices are in museums, including the $f/2$ $140°$ wide-field camera. With this, our Galaxy was photographed as an edge-on spiral with a central dust lane.

The Yerkes staff was severely depleted by the war, but telescopes were fully operated. I had several observing runs at McDonald but no time to think. I was not drafted; I got to know some senior officers. Like other scientists who "had won the war," I was unconsciously being prepared for major changes in my scientific career and outside activities. At Yerkes, before 1941, the younger staff had not been given planning or operating responsibilities—that was not Struve's style. Yet for reasons of personality, availability, and some outside friendships, I found myself involved in new organizational structures for astronomy. By 1947, the Office of Naval Research (ONR) had instituted a small grants program, and I was on that committee. Somehow, before the birth of the National Science Foundation (NSF), colleagues and I were involved in its plans, and I became the first chairman of its astronomy advisory committee. Scientists returned to Yerkes with new gadgets. Kuiper learned to use a PbS photoconductive Cashman cell, leading him to infrared planetary spectroscopy at McDonald by 1946.

Because of his work, Reber was familiar with radar. He and I wrote the first résumé of the rapid pace of discovery in high-frequency electronics and its applications in radio astronomy (62). W. T. Sullivan's (72) history of that field includes my account. Henyey, P. C. Keenan, and I computed the free-free radiation of H II regions and failed, again, to account for the intense low-frequency power in the Milky Way. A book edited by Kuiper (49) records an epochal symposium marking another new technique for astronomy: research from space. It includes a brief account of the traumatic, unsuccessful flight, in 1947, of a high-resolution solar spectrograph I built for a V2 rocket. Funded at $7000 by J. Van Allen's Applied Physics Laboratory at Johns Hopkins, it made me the first, but certainly not the last, conventionally trained astronomer to have an experiment fail in space! I predicted the continuous ultraviolet spectrum of the Sun as depressed by absorption lines and continua, with quite low boundary temperature (49). The book in which this paper appears includes many new names: Van Allen, R. Tousey, E. Durand, J. J. Hopfield, H. Tatel, H. Friedman. The sponsorship of astronomy by the NACA, the predecessor of NASA, opened a new road. Rocket scientists of the Naval Research Laboratory, the Bureau of Ordnance, together with old hands in astronomy like Goldberg and Whipple, created an advisory apparatus. The space program thus started with scientific goals in which some astrophysical, as well as solar-system, problems were recognized, although it was a long and difficult struggle that still continues.

An important feature of the success of astrophysics since 1945 has been its assimilation of physicists and engineers with other formal backgrounds. It succeeded in converting them into hyphenated-astronomers. Fortunately (and it was an actively discussed problem), the AAS and the *Astrophysical Journal* co-opted such people to present and publish their work. New areas of technology and relevant physics had to be understood by older astronomers; they were, but with omissions. The new fields were co-opted. I had been anxious to see radio observations part of astronomy; a letter from me to Struve (1946) explains some practical difficulties facing us were we to undertake the new venture and make a radio-astronomer faculty appointment at Yerkes. They involved military and industrial secrecy, overhead, engineering-level salaries, and operations at McDonald with Navy funds. The project fell through, but a few years later, ONR and NSF funds were available, and Harvard, Michigan, Cornell, and the Carnegie Institution of Washington entered that field.

Theoretical astrophysics was less directly affected, but new concepts entered this field, notably from hydrodynamics, turbulence, shocks, and convection. Theoretical problems arising from observation are seldom at a level of abstraction requisite for mathematical solution. But observations of the solar flux and limb darkening confirmed theoretical distributions of temperature with depth (6). Theoretical astrophysicists became increasingly confident and broad in fields such as atmospheres, galactic dynamics, the patchy interstellar medium, and the internal structure, energy generation, and evolution of stars.

Before modern computers, a model atmosphere remained a serious task. For stars, the empirical approach is exemplified in papers (23, 24, 38) based on McDonald coudé spectra. The compositions of Am (metallic-line) and normal F and G stars were determined by the differential curve of growth. The method was used for 15 years until model atmospheres became available and "coarse analysis" was replaced by "fine analysis." I found abundances relative to the Sun for a dozen stars, with an accuracy of ± 0.3 dex (rarely better, often worse). The Am stars, however, had apparent deficiencies from $+0.5$ to -1.5 dex. Morgan (55) had discovered many complex peculiarities among the A stars, which are now generally called "chemically peculiar" stars. His results should have suggested that we would probably soon exhaust theoretical explanations of such anomalies if they were taken as deep-seated. Most F and G stars had nearly solar composition; I attempted to explain away the Am phenomenon by anomalous charge-transfer ionization, rather than composition. The current explanation is a competition between gravitational diffusion and selective radiation pressure in nonconvective atmospheres. The important fact is that Morgan's chemically peculiar stars were all in the nonconvective range of surface temperatures, although the explanation did not come for 20 years.

TRANSITION

> *...but something ere the end,*
> *some work of noble note, may yet be done,*
> Alfred Lord Tennyson, *Ulysses*

The uniformity of stellar composition, coupled with some variety (as in R CrB, v Sgr), led to exploration of the effects of nuclear physics on composition. Magnetic fields in space were used by E. Fermi to accelerate cosmic rays. Spitzer enlivened an AAS meeting by introducing plasma physics and doing an illustrative "dance of an electron in a magnetic field." Among World War II advances was the RCA photomultiplier tube, the 1P 21, which made everyone a master of precision photometry. In this epoch of new ideas I was faced with a personal decision; Struve had relinquished the Yerkes directorship, Henyey had left for Berkeley, and I had received several tempting offers involving leadership. My family had lived 11 years in a country village; we missed the intellectual and cultural activities and the more active life of the city. I was asked to come to Caltech to help it prepare for operation of the Palomar Observatory, which it owned, and to create a graduate school and gather the scientific staff for Palomar. We left Yerkes with genuine regrets. I arrived in Pasadena in June 1948 for the dedication of the 200-inch, prepared to become involved in enormous changes. I entered a world in which I had to become two persons—scientist and organizer. Struve ran and financed Yerkes. I had not wanted to be involved, nor was I consulted, in organizational matters at Yerkes; I had not raised money, nor had I been an active committee member. Now, however, organization, administration, and relations with the university, the government, and the public became at least half my life. But I decided, firmly, not to abandon research, which also took more than half my life. In 1949, at age 40, the bibliography of my published papers numbered 73; in 1959, there were 158; in 1969, 261; and in 1979, 353. A few are wrong. Perhaps 20 percent are ephemera or related to the public side of science, including encyclopedia contributions, review articles, and published committee reports. The numbers show that I remained an active scientist, good or bad, but prolific.

I no longer remember how the balancing act was possible; in a running biography, there are about 75 named lectures and memberships on 50 major committees and study or advisory groups. At one time I held over 20 simultaneous committee memberships. Had I stayed at Yerkes, I would possibly have been a better astronomer, but I had already traversed most of the road from near-theorist to observer. My experience was far from unique in my generation. Historians will need to study the broadening of technology, the changes of institutional arrangements, and the daily novelty (which was somehow to be fully funded) that became an expected feature of scientific life. Contemporaries managed in different ways to remain active scientists under overwhelming pressures. In Europe, university life and research had been part of the government apparatus. In the US, in 30 years we lived through a culturally diverse transition to a pluralistic arrangement in which some older, independent units flourished, while completely new institutions, observatories, and consortia appeared. In optical astronomy, one major given fact was that large telescopes existed before Federal support was available. In radio and space astronomy, the Federal role immediately had to become central.

CALTECH AND THE MOUNT WILSON AND PALOMAR OBSERVATORIES: ORGANIZATION

> (1) *What means were there to examine what it was like before heaven and earth had taken shape?*
> (2) *Who planned and measured out the round shape and nine old gates of heaven?*
> *Anthology of Heavenly Questions* (China, 4th c. B.C.)

The California Institute of Technology (CIT) is a young institution with a historic devotion to research, tracing back to R. A. Millikan, G. E. Hale, A. A. Noyes, and T. H. Morgan. The post-WWII growth of research in other universities makes it now seem less exceptional than in the past. Fine, and few, students and a relatively large faculty are fortunate characteristics. Another major factor in its development as a center of astronomical research is the 200-inch, built on Hale's persuasion and given to CIT by the Rockefeller General Education Boards. The success of the 60- and 100-inch reflectors at Mount Wilson had led to planning for the 200-inch by astronomers of the Carnegie Institution of Washington (CIW). CIT and CIW joined in a full partnership agreement for the joint operation of the Mount Wilson and Palomar observatories (MWP) the year I arrived. The agreement was negotiated during the Mount Wilson directorships of W. S. Adams and I. S. Bowen. Bowen, who had been a CIT professor of physics, became director of the combined observatories. Each institution fully supported its own mountain and staff; the staff were to use the instruments of either mountain as their science required, as were CIT graduate students. The director had his offices at Santa Barbara Street in Pasadena, two miles from CIT. Management of the final Palomar construction and instrumentation was under the direction of B. Rule (a CIT engineer) in the Robinson Laboratory at CIT, where I established the new teaching and research faculty. Bowen did much of the optical design and supervised the final refiguring of the 200-inch mirror. The CIW staff was large but aging; it included outstanding observers of whom I name only a few: W. Baade, E. Hubble, M. Humason (comprising the nebular group); A. H. Joy, P. Merrill, R. Sanford, O. C. Wilson and R. E. Wilson, H. D. Babcock and, soon, H. W. Babcock (spectroscopists); S. Nicholson, R. Richardson (the solar group). CIT had only F. Zwicky and J. Anderson (who had run the 200-inch construction project after Max Mason). I was responsible through the dean of the faculty (E. C. Watson) to Millikan and soon to CIT's new president, L. A. DuBridge, with whom I had excellent rapport. The organizational structure was, and remained, mysteriously complex, but it worked. The Observatory Committee (of which I was a permanent member) advised the director on MWP matters. Faculty appointments at CIT needed approval by the normal chain: approval by the Observatory Committee and by the presidents and trustees of both CIT and CIW. The budgets had two fully independent sources, and both needed joint approval; CIT paid the operating expenses of Palomar and its faculty salaries. Later, with Federal assistance, CIT established the Big Bear Solar Observatory and the Owens Valley Radio Observatory, the latter independent of MWP and reporting through the Caltech division chairman (R. F. Bacher for many years). Astronomy was one of three "options" within the Division of Physics, Mathematics, and Astronomy. Much astronomical research was conducted within Physics: cosmic ray, X-ray, radio, and infrared astronomy, theoretical astrophysics. In Planetary Sciences were geochemistry and the advanced camera developments, such as the CCD for the Space Telescope. Astronomy spread over five other buildings and to the Jet Propulsion Lab. Some teaching was to be done, at first by volunteers from the CIW staff; some theses were done under their guidance. This management plan was hopelessly complicated; it worked for a long time because of loyal devotion to science and the marvelous telescopes available. My major task was to suggest and carry through (although with difficulty) the new appointments. In the Robinson Lab I inherited a five-story building designed by Hale, filled with a 15-year accumulation of squatters from other departments and with laboratory equipment from another epoch. I threw out mountains of junk and evicted tenants, including the Dean of Graduate Studies and the Department of Mathematics. There was an uncompleted solar tower; now we have the Big Bear Solar Observatory. Robinson has since overflowed physically, with computers (three VAX's and the VLBI processors on two floors outside the old building frame), laboratories, and shops. An ambitious plan for two new large buildings, one for CIT and one for CIW, foundered, unfortunately, when CIW decided to build (with private funds) the Las Campanas Observatory in Chile, which has proved to be an outstanding success.

On arrival, I taught the graduate (full-year) courses on stellar atmospheres and on stellar interiors; some appealed to graduate students in physics, an advantage of a divisional over a departmental system. These classes were small; later I taught a one-term introductory course for sophomores, usually 50 brilliant boys, from whom I recruited several future astronomers. The backbone of CIT graduate instruction was in mathematical physics, also required of astronomy students; the courses used problems of numbing difficulty to strain all minds and, it was hoped, to train the best. The survivors knew an order-of-magnitude more physics than my generation had been required to learn. In the first ten years we were fortunate to have some of the best students in astronomy and to graduate many of its present leaders. I took more than paternal interest in them all, although few theses were actually done with me, since I was spread far too thinly. The department was rated highest in several evaluations by the Council on Graduate Education. The early faculty choices included several from Yerkes, largely those with theoretical abilities. I had found that a reformed theorist may become the most intelligent observer. Another factor was a change in the type of person appointed at CIW and to the joint MWP staff. Some were Caltech PhD's, others had substantial theoretical as well as instrumental abilities. The CIW staff had shrunk through retirements, and for financial reasons its earlier size could not be maintained. The size of the total CIT and CIW staff barely approached that of Mt. Wilson in the 1930s, in spite of the doubling of large telescopes available. One effect of the joint operation was that CIT astronomers were precluded from receiving government money for research, since CIW would not, on principle, do so for many years. The combined largest observatory in the world was fully privately supported and was always short of money for new instrumentation and for postdoctoral fellows. Beginning in 1957 (until 1970), after something of a crisis, I arranged substantial funding (by the Air Force Office of Scientific Research) of an "*Abundance Project.*" Caltech is currently one of the largest university recipients of Federal funding for its radio and solar observatories, infrared experiments (including IRAS), the submillimeter-wave interferometer, and Palomar advanced instrumentation. But Palomar is still mostly privately funded. (A recent example of such funding is the upgrading of the 48-inch Schmidt for the repetition of an improved Sky Survey.) I am now in principle retired (since 1979), after having resigned my management tasks in 1972. I cannot here describe the work of my younger colleagues, whom I admire, because of too close personal involvement. Many were students or postdoctoral fellows. For the same reason of closeness, I should not assess the contributions to astronomy by the Mount Wilson and Palomar Observatories. I believe them to have been very significant, and hope that they still are. They fulfill many of Hale's dreams, even if in fields he could not have foreseen. They combine development of the best possible instruments with extended programs of difficult observations and interpretation.

EARLY RESEARCH IN PASADENA

The emphasis in the 1940s on plasma and magnetic fields in space helped lead to an understanding of interstellar polarization and its implications for the magnetic field in the Galaxy. Knowing how to compute extinction by interstellar grains, I found a perfect collaborator in a younger physicist, L. Davis, Jr. He displayed the merits of the Caltech emphasis on classical mechanics and electromagnetism. His example persuaded me to force later generations of graduate students to take such courses. We enjoyed a lively controversy with L. Spitzer. In Davis & Greenstein (8), the alignment mechanism for rapidly spinning, elongated grains is thoroughly discussed. The fields deduced lay (correctly) along galactic spiral arms. Davis later studied the trajectories of cosmic rays in fields with small-scale deviations from parallelism and the morphology of the fields as deduced from maps of polarization. The Spitzer-Tukey suggestion of a ferromagnetic contribution to the relaxation process and, much later, Purcell's ideas may save the theory when the composition of the grains is better known. Our treatment seems to be one of the few in astrophysics relevant after 30 years.

The next learning experience was in nuclear physics, with W. A. Fowler as guide. The Kellogg Lab had been studying low-energy cross sections relevant to stellar energy production. My interest in anomalies of stellar composition meshed with the work in Kellogg. With various collaborators, using Mt. Wilson, Palomar, and solar spectra, I studied (1950–56) the abundances of $^{13}C/^{12}C$, $^3He/^4He$, ^6Li, ^7Li, Be, and Tc. The $^{13}C/^{12}C$ ratio in most stars, and even in a comet, seemed close to that in the Sun. The Li/H ratio in young stars was found elevated, and the $^3He/^4He$ ratio was high in some chemically peculiar stars. In much of this work, in retrospect, the hope of linking a surface abundance anomaly to specific, exoergic nuclear reactions in the interior seems naive. Early reviews (25, 26) describe the subject when hopeful and new. In a lecture (26) I noted that the exoergic $^{13}C(\alpha, n)^{16}O$ reaction was a possible neutron source. A. G. W. Cameron explored chains of neutron captures in detail that explain the heavy-element stars and S-type giants containing technetium, which P. Merrill had found. I developed a somewhat philosophical, vintage-1952 plan for studying energy-generating reactions as the source of peculiar stellar composition. From 1957 to 1970 I published (with many collaborators) 60 papers in this field, largely with AFOSR support. Many US and foreign colleagues came to Pasadena, observed at Mt. Wilson and Palomar, or used my available spectra. There was also much theory, some agony, and much fun; in January 1970, a committee (G. Wallerstein, W. L. W. Sargent, and L. Searle) organized a surprise symposium on the "Chemical History of the Galaxy" for my 60th birthday. Pagel (60) has given a partial résumé, but unrecorded are glimpses of old friends convulsed with laughter at some particularly outrageous remark (usually by Geoff Burbidge, Willy Fowler, or Jerry Wasserburg). I have a pleasant letter from Unsöld (who really did not believe much in stellar nucleosynthesis), with his watercolor of Robinson Lab. There is an unpublished paper by P. Conti and A. Schadee on "The Presence of Greenstones in White Dwarf Stars," with abstract as follows:

The fact that greenstones have a half life of 60 years suggests that nucleosynthesis from "teeny-tiny" bangs (Waggoner 1970) has occurred recently in white dwarfs.

I am proud to count the many postdocs who came through the Abundance Project, with the early graduate students at Caltech, as among my valued scientific foster-children. The quotation from Conti & Schadee correctly foreshadows the end of my work on stellar composition and the subsequent shift into white dwarfs, where, alas, greenstones have not yet been identified. The general theory of nucleosynthesis in stars (3) was based on what is really quite early knowledge. Abundances have since been determined from steadily improved models (made possible by the computer) and spectral synthesis (e.g. by R. A. Bell) and depend on electro-optical, higher-resolution data (e.g. by D. Lambert). Composition analyses of stranger and fainter stars and abundance gradients in the composite spectra of galaxies completely depend on new technology.

The metal-poor subdwarf G stars studied by L. H. Aller and myself in 1960 were of 9th magnitude; the 14th magnitude globular-cluster giants studied at 18 A mm^{-1} dispersion by L. Helfer, Wallerstein and myself in 1959 took me longer than a full night to obtain exposures at the coudé, which illustrates the limit of photographic spectroscopy. The 200-inch coudé was designed by Bowen, with a series of Schmidt cameras made by D. Hendrix and a mosaic of original gratings by H. W. Babcock. When it began operation in 1952, it was a superb, ultimate instrument made inefficient by photographic plates and by slit-losses when used at high resolution. Work on faint stars that no one else could observe seemed a proper use of the 200-inch. The prime-focus spectrograph used dark-of-the-moon time, competing with the high-priority programs of Baade, Humason, Minkowski, and Sandage on galaxies and clusters. But it opened up work on faint stars of low luminosity and on quasars. This change in subject matter reflected a pattern in my research. I enjoyed exploring a new speciality in order to learn a new area of astronomy and its related physics; I tended to leave a field once it was well established as a result of my low threshold for boredom and my inability to resist use of newly available equipment on a new type of object. In 1957, for example, I took the first high-resolution spectra of comets at the Palomar coudé. I wrote a half-dozen papers with collaborators and found the rotation of a comet by a generalization of the Swings fluorescence mechanism. I built a high-resolution image-tube camera (0.2 Å) for Comet Kohoutek, but the latter's brightness failed to justify early expectations and I also encountered foul weather. I never returned to comets. Such incursions and retreats from

future projects from which they will never benefit. Others choose whole-hearted devotion to their own scientific work, knowing that they will face frustrations with time's passage. I enjoyed running my department and doing battle in committee rooms, while still working with the telescope. I tried to stay in competition for scientific novelty and insight. Several occurrences drove me further into the nonresearch world, even as available facilities were made more tempting by technological innovation. A personal bias preexisted; Hitler, WWII, and Korea had frightened me. From 1950 onward the international scene appeared ominous, and it has not changed. The reputation of Caltech and the Observatory made me an easily targeted and persuadable spokesman. I found that some national advisory activities required technical knowledge, while others needed mainly general "wisdom." I learned some of the limitations of a purely rational scientific approach. I am grateful that some officials and industrialists accepted my limited horizons of advice when for ten years during and after the Korean War I worked in major, nonscience areas.

It was a very hectic life. Some committees of which I was a member carry only initials that I cannot recognize. In one year I gave three dedicatory addresses for new buildings. I was on the National Academy of Sciences' Committee on Science and Public Policy (COSPUP) and then the NAS Council, which led to my chairing the study *Astronomy and Astrophysics for the 1970's.* I served as a special advisor to NASA under N. Ramsey, and in a smaller group directly advisory to the Administrator, J. Webb. The Ramsey committee was an early link in the planning of the Large Space Telescope. Several ad hoc NASA committees helped bridge gaps between scientists and NASA management. One led to the decision to locate the NASA Infrared Telescope Facility on Mauna Kea; my last service for NASA was on the Source Evaluation Board for the Science Institute for the Space Telescope. I received the title of Lee A. DuBridge Professor of Astrophysics (1970 till retirement), which honored me and a man I both liked and admired. I was particularly honored to serve as Chairman of Caltech's Faculty Board at a time when important changes (1965–67) were occurring in university life. In spite of committee and administrative distractions lasting many years, I was honored for scientific work: the California Scientist of the Year Award for 1964, the Russell Lectureship of the American Astronomical Society (1970), the Bruce Medal of the Astronomical Society of the Pacific (1971), the NASA Distinguished Public Service Medal (1974), and the Gold Medal of the Royal Astronomical Society (1975).

I returned to Harvard as a member of its Board of Overseers (1965–71; successive chairmen, D. Dillon and D. Rockefeller). After not having visited Harvard since 1939, I attended 40 meetings of the Overseers and related committees. It was a time of student unrest and violence. The contrast was dramatic between young near-revolutionaries and the devoted judges, publishers, lawyers, industrialists, and academics. Harvard's long survival as a national resource required vigorous money raising, as it does at all universities. Financially generous Overseers and their friends really enjoyed the opportunity of participating in novel academic adventures. The challenge for me was to explain what academic life was for; it was not easy with such intelligent people, since academic goals were abstruse and diverse. The renewed contacts with old friends still at Harvard were stimulating. At the same time, I remained active in fund raising and planning for Caltech.

My central activity from 1947 to 1981 was in work with Federal agencies involved in funding science, especially astronomy. I came slightly late; many leaders of WWII science had tried to slow the dismantling of some of the applied-science organizations created for WWII, or to transfer to universities some support for basic research. Active on the larger scene were W. Baker, D. Bronk, V. Bush, J. Conant, L. DuBridge, J. Killian, etc. Others had special enthusiasms, like L. Berkner (of Associated Universities, Inc.) for radio astronomy, R. McMath (of the University of Michigan) for astronomy in the about-to-be-born National Science Foundation (NSF), and C. C. Lauritsen (Caltech), who was instrumental in establishing the Office of Naval Research (ONR). I served (1947) on the first ONR grants committee for astronomy. It was ONR that funded Caltech's radio observatory beginning in 1954. My involvement with the NSF was long; I was member and chairman of its first astronomy advisory committee (1952–55) when it considered its first grants, including the early approaches for the national optical observatory (2). In 1954, I was secretary of the group organizing a conference on radio astronomy (sponsored by CIT, CIW, and NSF). The conference led to planning for the National Radio Astronomy Observatory. It seems that committee memberships or leadership positions are addictive. This brings to mind a *New Yorker* cartoon that shows a father and child viewing a statue of a group of business-suited figures, with the caption "There are no great men, only great committees." Oral history programs at the American Institute of Physics attempt to reconstruct these hectic years when astronomy made its quantum jump in size. Our first, privately supported, big-science became one of the most successful in (per capita) Federal funding and, we believe, in achievement.

Most circumstances are irresistible, and no individual plays a decisive role. The cost effectiveness of being an advisor cannot be objectively measured. The tasks that had to be done were well done by US astronomers, effectively and honestly. A worthwhile historical background is a study of the President's Science Advisory Committee, edited by my good friend W. T. Golden (14), who helped found this committee. The study illuminates developments at the highest Federal level. The seamy side may be found in S. Hersh or H. Kissinger. I emphasized a technique that suited my times and personality, namely, to avoid confrontation and to stress rational compromise. Initiating a new program is a long task, and many people of diverse talents and styles are involved. After the scientific need and a consensus are established, scientific leaders must be matched with responsive individuals in Federal agencies, in a creative, symbiotic relation. Lobbying is not enough, and I am unsure of the benefits of letter-writing pressure. Scientists tend to get bored too soon. Someone within the bureaucracy must carry the burden of internal persuasion, of interminable briefings and hearings, with little personal reward.

My personal copy of the *Greenstein Report* (56) is inscribed by G. Kistiakowsky (science advisor to the President for the years 1959–61), on whom I relied heavily. It reads, "Sorry that in my ignorance I started all of this." But I am not sorry. In 1964, the NAS had published the Whitford report on ground-based astronomy; there were two studies of radio-astronomy facilities by panels headed by R. Dicke; there were also ongoing studies by the Space Science Board for NASA. Attitudes within the government had changed by 1965. The leveling off (and therefore real decline) in funding went with an impatience with "shopping lists." Thus, H. Brooks (in COSPUP) emphasized that the scientists must set the priorities, recommend that the obsolete facilities be closed, and even reject some new proposals. The *Greenstein Report* (56) had to include all astronomy (except planetary missions), to establish priorities, and to give reasonably accurate prices. Ten years later, the *Field Report* (57) was to function in an even more difficult environment; the astronomical community had become less cohesive, and major space projects more costly. A large portion of the Federal funds available were required to operate the national observatories. Costs in astronomy rival, per capita, those in high-energy physics. The *Field Report* recommends programs costing $1.9 billion (1980 dollars); it states that the *Greenstein Report* recommended programs costing $844 million (in 1970 dollars, equivalent to $1.7 billion in 1980), of which most have been implemented.

The growth of knowledge and understanding in the last few decades are, for me, sufficient moral, aesthetic, and scientific justifications for such levels of expenditure. Human illness and poverty cannot be cured merely by spreading money, but the human condition can be ennobled by spending money wisely. Rising levels of health, education, and prosperity within the US are based on our technological revolution, which affords us the luxury of basic research from which technology springs. The growth of particle physics, with its ever more expensive, rapidly obsolescent facilities, parallels ours. After decades of leadership in that field, the US has fallen behind the orderly developments in Europe, at CERN and in Germany. I hope I am not being merely nationalistic. The competitive spirit in science seems to me to be necessary. Astronomy suited the American genius in its mixture of romantic subject matter and sophisticated tinkering. Observations, data, suit American empiricism. In Pasadena, a concern had existed in 1948 that I would bring in too much "theory" with the new staff and disregard the traditions of obtaining good data. But I feel that there is never enough good data. What if Grand Unified Theory does link cosmology and particle physics; what if the Universe is closed by invisible matter? Is astronomy to reach a dead end? Or will its practitioners readapt once more to produce ideas and new instruments to make visible the invisible? I hope so, and there are many past examples. The magnitude limit for photography was near 22 in 1952; it is now near 26 with the CCD at the 200-inch reflector. An excellent improvement, for half a percent the cost of a space experiment. Prolonged involvement in plans gave me opinions (but not conclusions) about the best strategy in the balance between the private and public sectors, on the role of the national observatories, and on the importance of the original, talented individual. The unique instrument may grow from the ideas of an individual, in response to either a scientific goal or to an irresistible urge for the ultimate technology. Groups or individuals plan and use experiments on the Space Telescope, the Advanced X-Ray Astronomy Facility, or the Space Shuttle; no individual can take full credit. But a ten-year delay makes success feel humanly remote. An excellent history of government and of external advisory committees preceding the 1972 funding of the Very Large Array was prepared by G. Lubkin (51). A Nieman Fellow (journalism) at Harvard, Lubkin bases her report on correspondence in the NAS (Brooks, COSPUP, my NAS Survey) and in the National Science Foundation, and supplements it by interviews with the protagonists. The National Radio Astronomy Observatory planning for a high-resolution system started in 1962, in many studies. External groups, the Whitford Committee, two Dicke Panels, and my NAS Survey gave it highest priority. Over 60 individuals are mentioned (51). The Bureau of the Budget and the NSF were convinced by 1969; so were the Office of Science and Technology (OST), the PSAC, and the congressional committee staffs involved. The procedure was slow and incredibly complex. It must be studied to be understood. It succeeded, and the VLA is successful; its story is a useful one for aspiring promoters of further large projects.

The issue of balance between individual and national goals is indirectly addressed in the Academy study (56); during our final discussions, it caused me intense discomfort. I resigned (51) for a brief time as chairman, since I was uncertain that I could fully support all the recommendations. That survey report is schizoid as published—it says build large, new national instruments (but please do not neglect to support university scientists and their new instruments). The parenthetical phrase may be intellectually correct but it is impotent, with no political or budgetary clout. The individualistic style of my own research was possible at institutions founded to pursue new, unplanned, and often changing goals. That system was good; I remain skeptical that it is completely outmoded.

LATER RESEARCH ON FAINT OBJECTS

Life has a limit, knowledge none.
Chung-Tzu

The decisions on such arguments lie in the future; I wish good fortune to those who must answer such questions. I did not disappear from the national scene in 1972, but I had lost my optimism about large further contributions and resigned from many outside activities. Yet the busy, double life continued; I was on the Associated Universities for Research in Astronomy (AURA) Board (serving as its chairman from 1974 to 1977); I was a visiting professor at the Princeton Institute for Advanced Study, at NORDITA, at the Bohr Institute in Copenhagen, and at the Institute for Astronomy in Hawaii. I resigned from heading astronomy at Caltech in 1972 after 24 years. I had been assigned 20 to 30 nights a year (1952–79) at the 200-inch, and I could profitably use new instrumentation as it was developed. The prime-focus nebular spectrograph, designed for galaxy redshifts, was an exciting instrument and permitted work on faint objects, quasars, and white dwarfs. The first papers by Eggen & Greenstein (11–13) contain photographic spectral classifications, photoelectric colors, luminosities, and space motions of over 200 white dwarfs. One of these, Tonantzintla 202, has an interesting story. Called a white dwarf (40 pc distant), its prime-focus spectrum is probably the first taken (1960) of a quasar. Ton 202 appeared to be a nearly featureless, DC star, with flat energy distribution and possible emission lines; with electronic detectors, these would have been obvious. In a paper by Greenstein & Oke (40) it is shown to have a redshift of 37%, i.e. 2×10^9 pc distant. If one must be wrong, be so on a grand scale!

My general plan was to explore the lower left-hand quadrant of the HR diagram containing the hot subluminous stars, a complex and then nearly unknown group. Only a few, brighter members were accessible to detailed analysis at the coudé. The nebular spectrograph permitted classification of horizontal-branch and subdwarf stars (sdB, sdO) and some quantitative work. I first categorized the subluminous stars in two earlier papers (27, 30); the end results are in Greenstein & Sargent (41). Only recently have sdO's received the full attention they deserve, by D. Schoenberner, D. Koester, and R. P. Kudritzki, in work at ESO, and with Kiel models. They range from nuclei of planetary nebulae, above 150,000 K, down to the predecessors of hot white dwarfs. Humason & Zwicky (48) had found faint blue stars at the galactic pole, and Zwicky believed them to be intergalactic wanderers. While some apparently normal B stars do exist at surprisingly large distances from the galactic plane, most faint blue stars prove to be highly evolved, low-mass remnants. They outline a horizontal branch extending to higher temperatures than found in globular clusters. Their low luminosity was established from line profiles and models, as well as from proper motions. They provided a new, broad insight into a wide variety of the nearly terminal stages of evolution. The *Strasbourg Conference Proceedings* (52) reviews the early stages of our recognition of questions raised by these stars.

The luminosity of a normal B star is 100,000 times that of a hot white dwarf; and it is another step of nearly 100,000 from the hottest to the coolest known degenerate. A broad range of phenomena awaited exploration. At first, white dwarfs seemed attractive since they were supposed to form a simple group; this proved far from the truth. Lacking an energy source other than cooling of the nucleons in their cores, they travel down a straight line in the $\log L$, $\log T$ plane at a radius fixed by their mass. Chandrasekhar (5) and Hamada & Salpeter (43) had given the zero-temperature mass-radius diagram, so that, unique in astrophysics, a structure existed, depending upon a single variable. An IAU Symposium (53) documents the appearance of a variety of new problems. The observational data available in 1965 was severely limited; eventually I gave 12 lists of white dwarfs observed spectroscopically or spectrophotometrically (over 550 stars). The last list (34) was published in 1980. The number of presently well-observed degenerates exceeds 2000, many from the work of R. F. Green, Schmidt and Liebert (in preparation), G. Wegner, and others. A computer-stored list, with tables of most data, is being prepared by E. M. Sion and collaborators at Villanova. In 1965, only 17 white dwarfs had parallaxes; now over 100 precise values exist, coming largely from the US Naval Observatory (USNO) program initiated by K. Aa. Strand. The subject now seems inexhaustible. The IUE added ultraviolet fluxes for hot stars; HZ 43 is one of the hottest, and also is detected in the EUV. The IUE data checked the effective-temperature scale for hot degenerates. It provided the exciting detection of C I and permitted composition determinations of trace elements, like metals, in a few yellow degenerates (7, 68).

My collaboration with Eggen (a mine of knowledge on binaries, clusters, motions, and photoelectric photometry) was exciting. On the 200-inch, I graduated from photography to the image-tube spectrograph, and then to

Oke's remarkable multichannel spectrophotometer (59). This instrument provides absolute fluxes for comparison with computed models. Shipman (65–67) and Greenstein (32, 33, 35) established the relation between theoretical fluxes and temperature using Oke's (59) and my own multichannel data (32, 37). The Oke-Gunn double CCD spectrograph has now become the ultimate instrument. In 1982, long after retirement, I observed 200 spectra in 8 nights at 5 Å resolution, which approaches that of prime-focus spectra. Spectra with signal/noise over 100 were obtainable in 10 minutes at 16th magnitude. I am grateful to Oke for his unselfish development of this advanced instrumentation for the 200-inch. My results spread over 60 papers, exploring such topics as mass, luminosity, temperature, surface composition, energy distribution, line profiles, magnetic field, the lack of rotation, gravitational redshift, motions, gravitational diffusion, and cooling theory. Some of the papers give general treatments of properties of white dwarfs with hydrogen-dominated atmospheres (33, 37) and with helium atmospheres (31, 37).

It was a pleasure to have to learn a new part of physics (solid state) for application within the theory of cooling and internal structure. The termination of nuclear burning in the parent red giant leaves a core of carbon (plus helium), and possibly a skin of an unknown, but small, mass of hydrogen. Interstellar matter may add more hydrogen by accretion. Thus, the internal composition of a low-mass white dwarf is almost certainly carbon, the envelope helium, and a skin of hydrogen may or may not exist. Observations probe only the thin atmosphere, which for 70% of the degenerates is hydrogen (H/He > 100), and for the balance, helium plus trace elements (He/H > 10,000). The white dwarf spectra divide observationally into the DA's and non-DA's, with the latter commonly having carbon and metals in concentrations far below their solar values. Only about 20 stars have had atmospheric composition analyses performed for carbon/helium or metal/helium ratios. The spectral class of a white dwarf should now include both a composition parameter and a temperature estimate. The latter is easy, since modern detectors give colors quantitatively. All white dwarfs have nearly the same radius; a new spectral classification system has been proposed by a group of workers active in the field, in which the type symbol indicates the dominant composition, the temperature, and thus the luminosity (69). This replaces my classification system (27) of 1960.

In their enormous gravitational fields, the composition of white dwarf atmospheres will be altered by differential diffusion of the heavy elements. They are also subject to accretion from interstellar space (1, 35), a process providing hydrogen, helium, and heavy elements. Accretion would contaminate the non-DA's with hydrogen; only rarely would He and metals complicate the DA's, in which rapid diffusion of heavy elements purifies very thin atmospheres. Composition-stratification clearly exists (from the theory of nonradial vibrations of ZZ Ceti stars). A competition exists between convective instability and radiation pressure on trace elements (75, 76). The known complexity in composition of the thin surface layers of white dwarfs grows with each new quantitative analysis. Accretion theory is not certain; the observations currently suggest that some new mechanism is needed to inhibit hydrogen accretion. An interesting fact is that no white dwarf is known with high O/He (in fact O has never been detected), while relatively high C/He is common. If dredge-up mechanisms are involved in the appearance of C and metals in some He-atmosphere white dwarfs, it seems likely that nuclear burning ends at C, rather than at O in red giants that survive to become white dwarfs. Alternatively, gravitational diffusion may have buried the O so that the cores are also compositionally stratified. Burnt-out cores of red giants of initial masses up to $5M_\odot$ or even higher become white dwarfs without explosion, indicating how high is the fraction of mass lost.

If we know the core composition, cooling theory predicts the luminosity function. An excellent review is given by Liebert (50), who is now pressing forward the search for cool, red degenerates of very low luminosity. The first cooling stages, to luminosity L, predict a lifetime of $t_0 L^{-5/7}$; the constant t_0 depends on the mean atomic weight $\langle A \rangle$ in the core. Then, as new degenerates appear, in a steady-streaming model, the number of degenerates per unit volume should vary as $n_0 L^{-5/7}$, i.e. as $n_0 T^{-20/7}$. Below a certain core temperature $T_c(Z, A)$, the core solidifies; below the Debye temperature $T_D(Z, A)$, the available heat content is low from quantum effects. If there is a simple dependence of effective temperature on core temperature, the number of white dwarfs above T_D should and does increase steeply with increasing bolometric magnitude. Below T_D, with plausible simplifications, the number per bolometric magnitude interval is constant [see Greenstein (29)]. The increasing frequency of cool degenerates is established down to $M_V = +15$, but no very low-luminosity degenerates have yet been found. I was convinced in 1969 that cool degenerates were rare; the maximum cooling age found was near 7×10^9 yr. By 1979 I had spectrophotometry of 25 red degenerates with $\langle M_V \rangle = +15.3$. There are no known red degenerates fainter than $+16.1$. Liebert, and workers at USNO (notably C. Dahn), are pressing this search. In the oldest globular clusters, such red degenerates would be fainter than 29th magnitude. The search for new, low-luminosity red degenerates is a challenging technical problem; the total mass contributed by white dwarfs in our neighborhood depends critically on whether the observed luminosity function should be extrapolated as $L^{-5/7}$ or as L^0. The rapid increase of bolometric correction below 4000 K suggests proper-motion surveys in the red for stars that are not M dwarfs. Binaries containing faint old dM stars have been searched for even fainter red degenerate companions, without success. A whole-sky search to 20th magnitude would probably reveal only 30 red degenerates of $M_V = +17.5$. None are now known. Liebert (private communication) has discussed the detailed prospects for such a search.

Other physical studies concern the properties of individual degenerates and the statistical value of the mean gravitational redshift. The latter is $+50$ km s^{-1}, from photographic low-resolution spectra, giving a reasonable value of $\langle M/R \rangle$ and a reasonable mean mass near $0.7 M_\odot$. The detailed analysis by V. Weidemann and collaborators has shown a remarkably narrow spread of masses, as deduced from accurate surface gravities and temperatures, near $0.55 \pm 0.10 M_\odot$ for H- and He-atmosphere degenerates and ZZ Ceti variables. Another aspect of stellar evolution is the remarkably slow rotation of single white dwarfs, as illuminated by the discovery of sharp cores at the center of Hα and Hβ in a few bright objects. This discovery was made possible by visits of A. Boksenberg's IPCS to Palomar. The resolution, at the coudé, was 25 km s^{-1}; the deduced rotational velocities were from 35 to 100 km s^{-1}. This would be unexpectedly low for a star that has shrunk by a factor of 100 in radius, if angular momentum were conserved. Like the pulsars, single white dwarfs are, in fact, slow rotators. Both the specific angular momentum and magnetic field must be drastically reduced during their prior evolution, presumably by the stellar winds that are reducing their mass by about 80%.

For many, the white dwarfs in close binaries are fascinating, and their accretion disks are more easily studied than those near neutron stars. For me, there are still mysteries in the single white dwarfs; as a spectroscopist I find it exciting that we have not yet identified broad absorption features in some magnetic white dwarfs. White-dwarf fields lie halfway between normal and neutron stars, on a logarithmic scale. The cataclysmic variables have matter falling into relatively shallow potential wells, barely deep enough to produce X rays. In AM Her stars a single magnetic funnel guides matter from the main-sequence star to a surface where the white-dwarf magnetic field can sometimes be directly observed. The binaries containing white dwarfs are useful guides, on a smaller scale, to some of the phenomena in X-ray binaries and X-ray pulsars.

If one likes puzzles, take the example of an enigmatic white dwarf GD 356. This star had intrigued me for years because its continuum appeared rough with the low resolution of the multichannel. With the high signal/ noise of the CCD, I discovered triple, Zeeman-resolved emission lines of Hα and Hβ split by ± 400 Å, at a contrast of only a few percent with the continuum (36). The splitting suggests a 10–20 MG field, presumably in a thin chromosphere. More detailed study provides new problems; $H\beta$ is stronger than $H\alpha$, which is possible when Lyman lines and continuum are optically thick. In a dipole spread over an appreciable volume, the lines become too broad. The magnetic energy density controls the emitting region. No trace of a companion has been found, and GD 356 is not an X-ray source nor is it polarized or variable. It may be demonstrating hydromagnetic heating.

Since white dwarfs provide such a cornucopia of new and unexpected phenomena, I am delighted at the continued observations at Arizona, Kitt Peak, Texas, and Australia and at the theoretical studies at Delaware, Kiel, Montreal, Rochester, and elsewhere. The field is healthy and in good hands. I was indeed honored that colleagues working in the field dedicated to me the volume resulting from the IAU Colloquium No. 53 (74). The rapid intrinsic variability of the ZZ Ceti white dwarfs (not binaries) provides a seismic probe of the stratification of composition in their atmospheres and envelopes; the driving-pump mechanism is a hydrogen-ionization zone for the DA's. This was generalized, and rapid variability predicted (79, 80) and found, in a hot DB, GD 358, where it is driven by the helium-ionization zone. The computer is as necessary a tool for study of white dwarfs as a large telescope. Even I have a graphics terminal in my office.

SPECULATIONS ON THE FUTURE

> It is not for you to finish the work,
> But neither may you exempt yourself from it.
> Ethics of the Fathers (1st c. B.C.)

This prefatory chapter has oscillated between my views on trends in the organization and funding of astronomy and a personal account of research interests. Astronomy is now too complex for even a well-informed generalist to make significant comments about its future. One obvious remark is that overspecialization is easy and common. In the 1950s an attempt was made to keep meetings of the American Astronomical Society to single sessions so as to induce all members to hear about all subdisciplines. While astronomy remains a romantic, all-embracing view of the Universe, it now employs too many types of eyes for easy communication. Struve's (70) rather pessimistic thoughts about the future of astronomy were written in the face of a war following a depression. His concern was whether his staff would remain interested in astronomy as it had been constituted. We now face financial crises, for example, in the undersupport of planetary missions and X-ray astronomy, but a new concern is that the nation's astronomical staff comprises such an enormous diversity of scientists, working with many different tools. The adaptability of our institutions and publications has made the transition apparently painless, but there are still problems. I have difficulties in reading, let alone understanding, five pounds of *Astrophysical Journal* each month. Scientists with training as chemists, computer engineers, electronics engineers, geochemists, and physicists are now found under our broadened umbrella. While they often rediscover what was obvious to earlier generations, they also find new things that an older generation could only have imagined. Can the new, broadened family learn to speak a mutually intelligible language? Can we retain a sense of daily excitement and of pride and interest in each other's achievements?

Historically, optical astronomers have needed more detected photons. We have done well along one road, where electro-optical sensors approach limiting quantum efficiency. Availability of large collecting areas is more general, with the provision of telescopes of 4-m aperture for the astronomical community at the national observatories in both hemispheres. Much larger telescopes are planned with radically new designs. While radio astronomy long suffered an enormous deficit of capital expenditures, in the VLA it now has an extraordinarily successful instrument with millisecond resolution on galaxies and stars, better than any optical instrument. The very-long-baseline array should follow (57) the very-long-baseline-interferometry, working to microsecond resolution. The Space Telescope will provide many more photons for the ultraviolet (since the IUE had only an 18-inch mirror) and much higher resolution against a darker sky in the optical wavelengths. Infrared progress, including work done using the IRTF on Mauna Kea and the IRAS in space, has been remarkable. There seem to be no more wavelength regions to be opened. Easily crossed frontiers have vanished.

Gravitational-wave detection remains open ended in that it now gives information only on upper limits. Particle detection has remained peripheral to astronomy, even though some of the data obtained by cosmic-ray physicists are published in the *Astrophysical Journal*. But consider what happens if the "missing-mass" (which I hope does not exist) consists of magnetic monopoles, gravitinos, or neutrinos. (For a brief period, neutrinos were supposed to oscillate and have mass, but they now don't.) In such problems, how will astronomers (broadly defined) be involved, except in a speculative way? In particle physics our education and skills seem irrelevant. But there is a lesson to be learned from our past—that of resourcefulness. Many discoveries are part of what now seems obvious, but they cost a great deal to obtain. Some new results could have been guessed at or predicted. Astronomers have always had to be resourceful in deductions concerning unobservable wavelength regions. The Zanstra mechanism (81), where far-ultraviolet photons are counted from the visible line emission of gaseous nebulae, is an example of such a resourceful technique. The X-ray emission from solar flares was deduced from ionospheric phenomena. In the 1960s, the high luminosity of quasars called for an unknown energy source, magnetic fields, and relativistic electrons. That energy source, now known to be gravitational collapse, is merely an extension of the idea of disks in deep potential wells. The nearly universal existence of magnetic fields is plausible from interstellar polarization and the isotropy of cosmic rays. It was noted that relativistic electrons would suffer inverse Compton collisions, turning radio-frequency photons into hard photons. A first response to that prediction might be "implausible." But the X-ray sky is now full of quasars. Had we been sufficiently brave, we would have predicted the X-ray sky with its binaries and quasars. Is it possible that we were forced to be more clever when severely limited in what we could hope to observe? I doubt that, and believe that this cleverness will reappear as scientists interpret the dramatic, unexpected revelations of the future, using typical astronomical ingenuity. I applaud their success in advance.

Literature Cited

1. Alcock, C., Illarionov, A. 1980. *Ap. J.* 235:541
2. Association of Universities for Research in Astronomy (AURA). 1983. *The First Twenty Five Years.* Tucson, Ariz: AURA
3. Burbidge, E. M., Burbidge, G. R., Fowler, W. A., Hoyle, F. 1957. *Rev. Mod. Phys.* 29:547
4. Calvert, M. 1941. *Sky* 5(5):12
5. Chandrasekhar, S. 1939. *An Introduction to the Study of Stellar Structure.* Chicago: Univ. Chicago Press
6. Chandrasekhar, S. 1950. *Radiative Transfer.* 1950. Oxford: Clarendon
7. Cottrell, P., Greenstein, J. L. 1980. *Ap. J.* 238:941
8. Davis, L. Jr., Greenstein, J. L. 1951. *Ap. J.* 114:206
9. De Vorkin, D. 1980. *Minerva* 18:595
10. Dunham, T. Jr., ed. 1946. *Optical Instruments: Summ. Tech. Rpt. Div. 16, Natl. Def. Res. Comm.* Washington, DC: Natl. Def. Res. Comm.
11. Eggen, O. J., Greenstein, J. L. 1965. *Ap. J.* 141:83
12. Eggen, O. J., Greenstein, J. L. 1965. *Ap. J.* 142:925
13. Eggen, O. J., Greenstein, J. L. 1967. *Ap. J.* 150:927
14. Golden, W. T., ed. 1980. *Science Advice to the President. Technology and Society,* Vol. 2, Nos. 1, 2. New York: Pergamon
15. Greenstein, J. L. 1930. *Harvard Coll. Obs.*
Bull. No. 876, p. 32
16. Greenstein, J. L. 1935. *Astron. Nachr.* 257:302
17. Greenstein, J. L. 1936. *Ann. Harvard Coll. Obs.* 105:359
18. Greenstein, J. L. 1937. *Ap. J.* 85:242
19. Greenstein, J. L. 1938. *Ap. J.* 87:151
20. Greenstein, J. L. 1938. *Ap. J.* 88:605
21. Greenstein, J. L. 1940. *Ap. J.* 91:438
22. Greenstein, J. L. 1942. *Ap. J.* 95:161
23. Greenstein, J. L. 1947. *Ap. J.* 107:151
24. Greenstein, J. L. 1949. *Ap. J.* 109:121
25. Greenstein, J. L. 1953. *Mem. Soc. R. Liège* 14:307
26. Greenstein, J. L. 1954. In *Modern Physics for the Engineer,* ed. L. Ridenour, pp. 235–71. New York: McGraw-Hill
27. Greenstein, J. L. 1960. In *Stellar Atmospheres,* ed. J. L. Greenstein, Chap. 19. Chicago: Univ. Chicago Press
28. Greenstein, J. L. 1964. *Ap. J.* 140:666
29. Greenstein, J. L. 1969. *Comments Astron. Astrophys.* 1:62
30. Greenstein, J. L. 1976. *Mem. Soc. R. Liège* 9:246
31. Greenstein, J. L. 1976. *Ap. J.* 210:524
32. Greenstein, J. L. 1976. *Astron. J.* 81:323
33. Greenstein, J. L. 1979. *Ap. J.* 233:239
34. Greenstein, J. L. 1980. *Ap. J.* 242:738
35. Greenstein, J. L. 1982. *Ap. J.* 258:661
36. Greenstein, J. L. 1983. *IAU Circ. No. 3823*

37. Greenstein, J. L. 1984. *Ap. J.* 276: 602
38. Greenstein, J. L., Hiltner, W. A. 1949. *Ap. J.* 109: 265
39. Greenstein, J. L., Matthews, T. A. 1963. *Nature* 197: 1041
40. Greenstein, J. L., Oke, J. B. 1970. *Publ. Astron. Soc. Pac.* 82: 898
41. Greenstein, J. L., Sargent, A. I. 1974. *Ap. J. Suppl.* 28: 157
42. Greenstein, J. L., Schmidt, M. 1964. *Ap. J.* 140: 1
43. Hamada, T., Salpeter, E. E. 1961. *Ap. J.* 134: 683
44. Henyey, L. G., Greenstein, J. L. 1937. *Ap. J.* 86: 619
45. Henyey, L. G., Greenstein, J. L. 1938. *Ap. J.* 87: 79
46. Henyey, L. G., Greenstein, J. L. 1941. *Ap. J.* 93: 70
47. Herbig, G., ed. 1970. *Spectroscopic Astrophysics.* Berkeley: Univ. Calif. Press
48. Humason, M. L., Zwicky, F. 1947. *Ap. J.* 105: 85
49. Kuiper, G. P., ed. 1952. *The Atmospheres of the Earth and Planets.* Chicago: Univ. Chicago Press. 2nd ed.
50. Liebert, J. 1980. *Ann. Rev. Astron. Astrophys.* 18: 363
51. Lubkin, G. 1975. *The Decision to Build the Very Large Array.* Unpublished study for the Natl. Acad. Sci.
52. Luyten, W. J., ed. 1965. *First Conference on Faint Blue Stars.* Minneapolis: Obs. Univ. Minn.
53. Luyten, W. J., ed. 1971. *IAU Symp. No. 42.* Dordrecht: Reidel
54. Matthews, T. A., Sandage, A. R. 1963. *Ap. J.* 138: 30
55. Morgan, W. W. 1935. *Publ. Yerkes Obs.* 7: 133
56. National Academy of Sciences. 1972. *Astronomy and Astrophysics for the 1970's,* Vols. 1, 2. Washington, DC: Natl. Acad. Sci. (Greenstein Report)
57. National Academy of Sciences. 1982. *Astronomy and Astrophysics for the 1980's,* Vols. 1, 2. Washington, DC: Natl. Acad. Sci. (Field Report)
58. Oke, J. B. 1963. *Nature* 197: 1040
59. Oke, J. B. 1974. *Ap. J. Suppl.* 27: 21
60. Pagel, B. E. J. 1970. *Q. J. R. Astron. Soc.* 11: 172
61. Payne, C. 1925. *Stellar Atmospheres.* Cambridge, Mass: Harvard Univ. Press
62. Reber, G., Greenstein, J. L. 1947. *Observatory* 67: 115
63. Robinson, I., Schild, A., Schucking, E. L., eds. 1964. *Quasi-Stellar Sources and Gravitational Collapse.* Chicago: Univ. Chicago Press
64. Schmidt, M. 1963. *Nature* 197: 1040
65. Shipman, H. L. 1972. *Ap. J.* 177: 723
66. Shipman, H. L. 1977. *Ap. J.* 213: 138
67. Shipman, H. L. 1979. *Ap. J.* 228: 240
68. Shipman, H. L., Greenstein, J. L. 1983. *Ap. J.* 266: 761
69. Sion, E. M., Greenstein, J. L., Landstreet, J. D., Liebert, J., Shipman, H. L., Wegner, G. 1983. *Ap. J.* 269: 253
70. Struve, O. 1942. *Pop. Astron.* 50: 465
71. Struve, O., Zebergs, V. 1962. *Astronomy of the 20th Century.* New York: MacMillan
72. Sullivan, W. T. III. 1984. *The Early History of Radio Astronomy.* Cambridge: Cambridge Univ. Press. In press
73. Trumpler, R. J. 1930. *Lick Obs. Bull. No. 420*
74. Van Horn, H. M., Weidemann, V., eds. 1979. *White Dwarfs and Variable Degenerate Stars, IAU Colloq. No. 53.* Rochester, NY: Univ. Rochester
75. Vauclair, G., Reisse, C. 1977. *Astron. Astrophys.* 61: 415
76. Vauclair, G., Vauclair, S., Greenstein, J. L. 1979. *Astron. Astrophys.* 80: 79
77. Wallenquist, A. 1929. *Medd. Uppsala. No. 42*
78. Whipple, F. L., Greenstein, J. L. 1937. *Proc. Natl. Acad. Sci. USA* 23: 177
79. Winget, D. E., Robinson, E. L., Nather, R. N., Fontaine, G. 1982. *Ap. J. Lett.* 262: L11
80. Winget, D. E., Van Horn, H. M., Hansen, C. T., Fontaine, G. 1983. *Ap. J. Lett.* 268: L33
81. Zanstra, H. 1931. *Z. Ap.* 2: 1; 2: 329

Ann. Rev. Astron. Astrophys. 1984. 22:37–74

STRUCTURE AND EVOLUTION OF IRREGULAR GALAXIES

John S. Gallagher, III[1,2]

Department of Astronomy, University of Illinois, Urbana, Illinois 61801

Deidre A. Hunter

Kitt Peak National Observatory,[3] Tucson, Arizona 85726

INTRODUCTION

Irregular galaxies (Irrs) usually are smaller, less massive, and optically dimmer than commonly studied giant spirals, S0s, and classical ellipticals. At first glance they also appear to be rare objects that make up only a few percent of the major bright galaxy catalogs (87, 300). Is there then much reward in pursuing such faint and elusive galaxies? The answer to this question turns out to be a surprisingly strong "yes." An examination of more nearly complete samples of galaxies (213, 341, 366) reveals that Irrs account for a substantial (1/3–1/2) fraction of *all* galaxies, and that they certainly are dominant by number density among actively star-forming galaxies. A diversity of nearby examples therefore abound [including, of course, the Large and Small Magellanic Clouds (LMC and SMC)], making Irrs prime targets for detailed investigations of galactic stellar content and star formation processes. Recognition of their structural simplicity has provided an additional stimulus for recent interest in Irrs as tests for theories of galaxy structure and evolution. These galaxies also are comparatively unevolved, and thus they may yield further rewards by allowing conditions to be defined in the poorly understood realm of low-density extragalactic systems, which retain considerable information about galaxy formation (101, 327).

[1] Supported in part by the National Science Foundation through grant AST 82-14127.

[2] Now at Kitt Peak National Observatory.

[3] Operated by Association of Universities for Research in Astronomy, Inc. under contract from the National Science Foundation.

37

In the broadest sense, the irregular galaxy class is loosely defined (e.g. see illustrations in 9, 23, 378, 379). Hubble (175, 176) originally built on earlier nebular classification schemes (e.g. 153) and considered galaxies to be "irregular" if they showed chaotic, nonsymmetrical blue-light distributions, in contrast with the axial symmetry of normal "regular" systems. Later classification systems subdivide irregular galaxies into two major groups: Magellanic systems (Irr I, Im), which resemble the Magellanic Clouds; and peculiar, often amorphous galaxies, which are classified as Irr II or I0 systems (72, 174, 290, 292).

Unfortunately, irregular structures may arise from a variety of physical causes, and as a result a wide range of physical types of irregular galaxies are known to exist: (a) There may be substantial chaos in the projected stellar mass distributions in galaxies, although such nonequilibrium structures are unlikely to survive for more than a few rotation periods ($\sim 10^9$ yr). Most currently known galaxies in disturbed states seem to be involved in galaxy-galaxy interactions (311, 358, 359, 384, 392, 393), although newly formed galaxies could also find themselves in this situation (45). Some galaxies classified as Irr II or I0 also belong in the category of interacting galaxies. (b) Similarly, unusual distributions of dense interstellar gas across the face of a galaxy may produce an optical image that is mottled by dark lanes. The dusty Irr II galaxies discussed by Krienke & Hodge (214) belong in this group. As these authors note, extended interstellar matter also can lead to emission filaments and reflection nebulae on galactic scales that give rise to abnormal optical morphologies (e.g. M82; 34, 67, 281, 332). The origins of peculiar global distributions of interstellar matter are not well understood, but some cases are caused by interactions between galaxies (65, 136, 373). (c) Young, massive stars have very low mass-to-light ratios, and thus sites of recent star formation stand out against even moderately high density projected stellar backgrounds. Furthermore, OB stars tend to form in spatially localized groups that are seen as OB associations or perhaps larger units with dimensions of up to 1 kpc ("constellations," 253; "star complexes," 91). Thus, in galaxies with low background stellar density levels and spatially incoherent star formation patterns, patches of young stars stand out against symmetrically distributed older stars and give rise to irregular optical brightness structures. This effect is beautifully illustrated in UV photographs of the Large Magellanic Cloud that were obtained from the lunar surface during the Apollo program (263, 264).

Most irregular galaxies belong in the third category, which includes Magellanic-type irregulars and spirals (70, 72, 78, 233, 290, 292, 365, 366, 368) as well as a smattering of genetically related systems such as intergalactic H II regions (304, 312), amorphous galaxies (defined by 294, 300; blue galaxies with relatively smooth optical light distributions; a subclass of Irr IIs or I0s: 174, 321; also "blue Es": 104), or luminous, clumpy Irr galaxies (50, 51, 149, 150, 302). Hereafter, we limit our scope to the Magellanic galaxy family (and mainly systems that are not involved in obvious galaxy-galaxy interactions), which we refer to simply as Irrs.

The Irrs blend from lower luminosities smoothly into the spirals, as emphasized by both G. de Vaucouleurs and A. Sandage, and thus can be viewed as an extension of actively star-forming disk galaxies to lower densities and luminosities. Luminous members of the Irr family overlap with spirals in optical luminosity and are often preferentially selected in surveys for blue or emission-line galaxies (e.g. the G. Haro and B. Markarian surveys; see 208). Thus Irrs probably now comprise the bulk of known galaxies with dominant hot stellar populations. We close this section with Figure 1, which illustrates NGC 4449, a classical nearby giant Irr. In the remainder of this article, we discuss specific physical characteristics of Irrs that are structurally related to NGC 4449, with an emphasis on exploiting the Irrs as probes of evolutionary processes in galaxies.

BASIC PROPERTIES

Light

Irrs exist over an extreme range in optical luminosity extending from clumpy irregulars with $M_B \lesssim -20$ to intrinsically faint dwarfs with $M_B > -13$. Usually, luminous systems have moderate-to-high optical surface brightnesses (blue SB $\sim 100\ L_\odot\ pc^{-2}$ in an effective radius), while most dwarfs (i.e. galaxies with $M_B > -16$) are barely detectable above the night-sky background (blue SB $\sim 10\ L_\odot\ pc^{-2}$). There are exceptions; for example, dwarf "blue compacts" can have high surface brightness (361), and luminous examples of low-surface-brightness Irrs also exist (231, 352). Most Irrs, however, are low-surface-brightness dwarfs, which accounts for their rarity in catalogs of bright galaxies.

Optical surface photometry is available for a number of Irrs, most of which are dwarfs (e.g. 2, 3, 32, 71, 73, 74, 79, 160, 165, 166, 323). The mean radial brightness profiles of these galaxies are well represented by exponential intensity distributions $I(R) = I(0) \exp(-\alpha R)$ with scale lengths $\alpha^{-1} \sim 1\text{--}3$ kpc. The light distributions in Irrs are therefore similar in form, but of lower surface brightness and shorter scale length, to those of spiral galaxy disks (117, 310). We thus infer that Irrs have lower mean projected stellar densities than typical spirals. The smooth profiles of Irrs are often disturbed by (a) star formation activity, which produces islands of high surface brightness, especially in the inner regions; (b) by bars, which are common in Irrs; and (c) by the effects of individual luminous stars in nearer systems (79). Deep images (172, 199) sometimes reveal that the actively star-forming cores of Irrs are embedded in smooth halos that qualitatively resemble diffuse dwarf elliptical galaxies. This has led several authors to suggest a close structural relationship between elliptical and Irr dwarfs (39, 125, 229, 386).

In terms of integrated optical colors, the Irrs are the bluest of the "normal" galaxy classes, with $(B-V) \sim 0.4$ and $(U-B) \sim -0.3$ (77, 177, 178). Standard models of stellar populations can yield such blue colors for galaxies of normal cosmological age only with some difficulty, which has

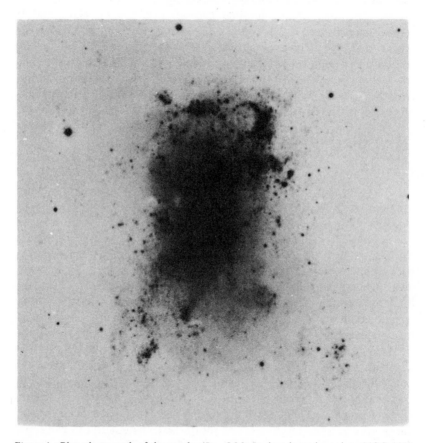

Figure 1 Blue photograph of the nearby $(D \sim 5$ Mpc), giant irregular galaxy NGC 4449, taken for the authors by G. Lelievre with the prime-focus camera on the Canada-France-Hawaii 3.6-m telescope. The scale of this print is approximately 3″ mm⁻¹; north is to the upper right corner, and east is counterclockwise. The chaotic structure of this galaxy largely results from many bright star-forming complexes superimposed on a strongly barred, amorphous, older stellar background. A few dark nebulae (light regions on this negative copy) are also visible.

fostered the idea that many Irrs have been detected in a stage of heightened star formation activity, i.e. very blue galaxies may be caused by star formation bursts (177, 178, 313). It also is possible that the colors could arise if the initial mass function (IMF) differed significantly from that deduced for stars in the Milky Way. Among intrinsically brighter Irrs, there is no strong trend between color and luminosity, but the fainter dwarfs in the David Dunlap Observatory (DDO) catalog of low surface brightness galaxies (365, 366) are systematically blue (82). This may be indicative of preferential selection of galaxies currently experiencing major star-forming events in samples of intrinsically small, faint galaxies. An example of the importance of star formation on the detectability of dwarf Irrs can be found by comparing VII Zw 403, a high surface brightness Local Group dwarf (361), with the low surface brightness Local Group Irr LSG 3 (363); the differences between these two galaxies stem from the presence of a small OB stellar complex in VII Zw 403.

Unlike spirals, Irrs often show a central bluing of optical color (81, 83). This implies a concentration of star-forming activity to the inner regions of the galaxy, which is consistent with observed steep radial falloffs in distributions of H II regions, supergiant stars, and other Pop I star indicators in the LMC (171) and other Irrs (165, 166, 188, 189, 291). Many years ago, de Vaucouleurs (72) remarked on the smooth transitions from the pure irregulars to true spiral galaxies. This in part involves a change in styles of OB star formation. Unlike Irrs, *relative* star formation rates in spirals are highest (and colors bluest) in the mid-to-outer optical disk. Thus, in spirals the young stellar component often has a *flatter* radial distribution than the overall light. The Sdm-Sm galaxies (the morphological transition between spiral galaxies and pure Magellanic irregulars) are intermediate cases in this regard, having extensive OB stellar components that extend well beyond the obvious older stellar amorphous backgrounds. This phenomenon does not seem to correlate with the form of the outer low-density H I distribution; for example, NGC 4449 has a very extensive H I envelope (376) but tight OB star distribution, while in NGC 4214 the H I profile is of more normal dimensions (7), even though far-flung OB stars abound.

Irr galaxies are only beginning to be extensively studied in the non-traditional X-ray, rocket ultraviolet, and infrared spectral regions, but interesting results have already appeared. Irrs may have high X-ray to optical flux ratios compared with nonactive spiral galaxies. This is consistent with the presence of binary X-ray sources in their large, young stellar population fractions (99, 100, 126, 333). Early OAO rocket-UV photometry revealed that Irr galaxies have integrated ultraviolet energy distributions somewhat like late B stars (59), a result that now has received support from a variety of UV observations (48, 49, 61, 239, 263). These energy distributions underscore the high-visibility, dominant role played by luminous OB stars in Irr galaxies, but they also clearly show the composite nature of the stellar populations in Irrs, since $m_{1550} - V \sim -1.5$ is redder than B stars.

Infrared JHK photometry of Irrs has been obtained by Aaronson (1), Thuan (349), and Hunter & Gallagher (in preparation). Most Irrs fall near globular clusters and star-burst nuclei in their JHK colors, and thus they contain cool, presumably evolved stars. Interpretation of these IR data is complicated by the important role of asymptotic giant branch stars in stellar populations of intermediate age and by uncertainties in red supergiant populations (193, 272, 279), but typical Irrs have IR colors near those expected on the basis of constant star formation rates (335; B. Tinsley, private communication, 1978). As emphasized by Thuan (349) and Huchra et al. (179), the blue $V - K$ colors ($\lesssim 1$) of some galaxies, however, could indicate a very large proportion of young stars or large-amplitude star formation bursts. Problems here include the sensitivity of young red star populations to low metallicities (24, 44) and the need to observe the entire old stellar component, which may be distributed across an underlying galaxy of low surface brightness and of larger angular size than the prominent young stellar complexes.

IR photometry is thus a potentially powerful tool in understanding recent star formation histories of galaxies (122, 123). Longer wavelength, thermal infrared emission from dust in Irrs has been detected in only a very few cases, but since large infrared luminosities (at wavelengths of $> 10 \ \mu m$ and especially $\sim 100 \ \mu m$) are characteristic of high star formation rates in a variety of galactic environments, we can expect Irrs to begin to show up in this category as improvements are made in far-IR sensitivity (282, 345, 382). By providing information on IR luminosities and spectral characteristics as functions of readily determined gas metallicities and star formation rates, the Irrs play an important role in calibrating models for thermal IR emission from dust in "young," metal-poor galaxies.

Masses and Kinematics

Global dynamics for large samples of Irrs have been derived from H I 21-cm line surveys (53, 109, 110, 135, 231, 241, 350–352). These observations provide integrated Doppler line widths, and usually the full width at 20% of peak intensity is interpreted as a good estimate of $2V'_c \sin i$, where V'_c is a characteristic circular gas velocity in a galaxy disk of inclination i. The distribution of H I line velocity widths shown in Figure 2 demonstrates that the majority of Magellanic Irrs have peak orbital velocities of $V_c <$ 100 km s^{-1}, with $V_c \sim 50$–70 km s^{-1} a representative value, in agreement with Brosche (41). Compared with spirals, for which $V_c \gtrsim 200$ km s^{-1} is typical (41, 102, 287, 374), Magellanic Irrs are slow rotators and have low specific angular momenta ($\lesssim 0.1$ of the solar neighborhood value).

Dynamical masses within the optical radii R_{opt} of galaxies are normally calculated on the basis of spherical mass distributions $M \approx (V_c^2 R_{opt}/G)$. This may lead to mass overestimates by a factor of two for Irr galaxies, which are disk dominated with low central mass concentrations. Since Magellanic Irrs have both smaller V_c and R_{opt} values than archetypal spirals, their masses are considerably smaller, with values of 10^8–10^{10} $M_\odot$ typical of survey results. Dynamical masses for individual galaxies are subject to a variety of uncertainties [e.g. inclinations are difficult to estimate for barred, chaotic galaxies (351, 352), velocity fields may be complex (376), and warps in the gas disk may not be uncommon (326)]. Even so, the mass range (but perhaps not the mass distribution function) derived from surveys of Magellanic Irrs should be reliable. Mass-to-blue-light ratios found from the H I surveys mainly scatter between 2 and 10 for $H_0 = 100$ km s^{-1} Mpc^{-1}. As a fiducial point, the LMC has a mass of $5 \pm 1 \times 10^9$ $M_\odot$ (105), which yields a value of $M/L_B = 1.6 \pm 0.2$. There is then some justification for believing that the M/L_B values based on H I survey results are systematically too high.

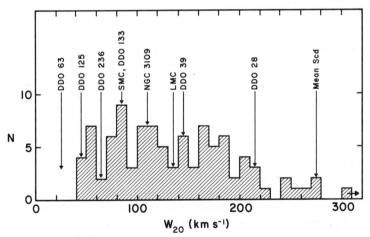

Figure 2 Distribution of H I line Doppler full widths at 20% of peak intensity for Im and Sm galaxies in the Fisher & Tully (110) survey that meet the conditions of (*a*) redshift velocity less than 1200 km s^{-1} and (*b*) estimated inclination of 50° or larger. Also shown are positions of several Irrs with known rotation curves based on references in the text (which are plotted at twice the maximum circular velocity) and the mean location of Scd galaxies [from Faber & Gallagher (102)]. Irr galaxies rotate more slowly than spirals, and virtually all systems with $W_{20} \gtrsim 180$ km s^{-1} are transition spirals, while for $W_{20} < 100$ km s^{-1} pure irregular morphologies greatly predominate.

Internal kinematics have been measured in several nearer Irrs using accurate H I velocity maps derived from pencil-beam observations (181–183, 286; earlier references in 374; rotation curve data are summarized in 18) and from H I aperture synthesis studies (5–7, 21, 67, 136, 137, 227, 303, 326, 360a, 373, 377). Both of these methods suffer to some degree from modest angular resolution. Velocities from optical emission lines can provide good spatial resolution and, in some cases, good velocity resolution but are sensitive to motions induced in the ionized gas by young stars (54, 78, 80, 105, 134, 188, 189, 236).

From these observations we can identify an underlying unity to the properties of Magellanic Irrs. Cool gas is located in uniformly rotating disks. The dispersion velocity of H I in the disks is typically $\sim 10 \pm 2$ km s^{-1}, a value that appears to be universal in H I disks of all galaxies (7, 182, 183, 375) and must therefore be an intrinsic property of interstellar matter rather than a property of specific galaxies. Rotation velocities within the gas disks, on the other hand, are quite sensitive to galaxy structure, with V_c declining from ~ 100 km s^{-1} in giant Irrs to virtually undetectable levels in extreme dwarfs and intergalactic H II regions. Among the very luminous clumpy Irrs and related blue compact galaxies, rotation velocities approach those of normal spirals (12, 53, 135), and thus these systems originate from a rather different state than the common Magellanic Irrs.

Mean rotation curves of Irrs also differ in form from those of spirals. Near-rigid-body rotation extends over most of the optical dimensions, with shallow velocity gradients of $dV/dr \sim 5$–20 km s^{-1} kpc^{-1} that reach peak velocities near the optical peripheries of the systems. In spirals the rigid-body rotation region is often quite limited in radius and has a steep gradient of $dV/dr > 20$ km s^{-1} kpc^{-1} (134, 287) [often larger than 50–100 km s^{-1} kpc^{-1} in more massive spirals (62, 133, 212)]. The peak velocities in spirals are attained in a nearly flat rotation curve that extends over most of the optical galaxy (37, 38, 102). Thus Irrs exist in a state with minimal differential rotation in star-forming regions, while star formation in spiral disks suffers strong shear due to differential rotation.

The forms of the rotation curves thus strongly affect the optical appearances of galaxies and insure that low-density disk systems will be morphologically distinct from spirals. Strom (334) demonstrated that the degree of gas compression during passage of interstellar matter through a density-wave spiral arm depends upon the maximum V_c value, and that for typical arm inclinations, only galaxies with $V_c \gtrsim 50$–100 km s^{-1} will produce spiral arm shocks. Thus, slowly rotating galaxies should not be capable of arm shock-induced star formation and thus will not appear as spirals. Similarly, patterns produced by propagating star formation will not distort into spirals in the absence of differential rotation (116, 315, 316), nor will shear amplification of small inhomogeneities lead to the creation of spiral armlets (357). So whatever the reader's favorite spiral arm theory may be, it probably will not directly stimulate star formation in slowly rotating Irrs (cf. 255). Star formation processes in Irrs are free from internal dynamical forcing and therefore appear in a free-wheeling, chaotic natural state. From this perspective, the luminous clumpy Irrs with their spirallike rotation properties will be more difficult to produce, which may explain their rarity and tendency to be found in interacting systems where star formation can be externally stimulated (50, 177, 201).

The kinematics of Irrs are qualitatively consistent with their exponential brightness distributions. Mass models based on exponential disks (117) or low central concentration, Gaussian density distributions (360a) reproduce observed Irr rotation curves for $M/L_B \lesssim 5$ in the central regions but fail in spirals (16, 17). A recent study of star cluster kinematics in the LMC similarly finds no evidence for a spheroidal component, even among old globular clusters (119). The Irrs thus have the expected character of pure disk galaxies containing (nearly) the observed amount of luminous mass, while spirals must be pinned on high-density central cores and also embedded in extensive stellar (and nonstellar?) halos. The lack of dense cores explains why Irrs rarely have optically identifiable nuclei and are not commonly sites for violent activity (20, 72; but see also 148a, 151). Evidently, a deep central gravitational potential well is a necessary ingredient for formation of massive, dense galactic nuclei and their associated fireworks.

When velocity fields in Irrs are observed with sufficient spatial and velocity resolution, numerous complexities appear. (a) H I is usually clumped into large clouds, which may not lie on the smooth rotation curves (as in the LMC; 241). (b) Irr galaxies are preferentially barred, and the bars primarily result from an enhanced density of older stars (75, 105, 188, 189), and usually do not lie on center [i.e. Irrs strangely prefer to be asymmetric rotators (78, 105)]. The presence of bars is to be expected if the Irrs are indeed dynamically cold systems that do not contain disk-stabilizing stellar or dark halos (93, 262, 318, 371). The combined impact of asymmetries and bars introduces significant perturbations into the velocity field (cf. 7, 236). Off-center bars remain theoretically poorly understood (63, 78, 106), which presents an impediment to construction of dynamical models for Irrs. (c) Injection of energy into the interstellar medium by massive stars can produce relatively large kinematic effects in slowly rotating Irrs. For example, giant H II regions expanding with velocities of ~ 20 km s^{-1} (124, 188, 347, 348) can significantly distort rotation properties as deduced from ionized gas. Blowouts in the cool interstellar medium due to stellar winds and supernovae lead to large-scale expanding gas shells (141, 243, 383) that are seen as "holes" in the H I distributions and can cause H I maps of Irrs to resemble Swiss cheese in both velocity and physical spaces (e.g. in the SMC; 19, 154, 331).

If one takes only optical light into account, then Irrs, like spirals (102, 255), must contain "dark" matter, but much of this is in the form of gas, including both easily detectable H I (and its associated helium) and elusive molecular material that nonetheless must be present at some level (197, 227, 390). Based on detections of CO emission in NGC 1569 and the LMC, the total mass of gas in inner regions of some Irrs may be as much as twice the H I value. The situation regarding dark mass, possibly nonnucleonic, that does not radiate detectable electromagnetic radiation is at present extremely unclear. Tinsley (354) and Lin & Faber (229) applied indirect arguments to conclude that dwarf Irrs probably have spirallike dark envelopes, but in both cases gas content, stellar population properties, and accuracy of dynamical masses introduce uncertainties. Feitzinger's (105) investigation of LMC dynamics, on the other hand, suggests there is little or no unaccounted mass, and a similar result was obtained by Gallagher et al. (126) for normal-mass Irrs ($M \lesssim 10^{10}\ M_\odot$) based on constant star formation rate models. Blue Irr galaxies of higher inferred dynamical mass, however, are found by Gallagher et al. to have excess mass for their optical luminosities; thus, two dynamical classes of Irr galaxies may exist, although the uncertainties are large.

The issue of spirallike dark envelopes in Irrs is relevant to fundamental points such as the necessity of dark matter for galaxy formation and whether the luminous baryon mass to total mass ratio could be nearly constant in all galaxies (cf. 101, 103, 229). Furthermore, the presence of dark envelopes in small systems would place useful constraints on relic elementary particle interpretations of dark matter [e.g. hot particles such as light neutrinos are unable to cluster on such small scales (68, 276, 360)]. We have seen that evidence for significant amounts of dark matter within Irrs is at best ambiguous. As in spirals, kinematic observations of Irrs at large galactocentric radii therefore must eventually play a crucial role, i.e. we should seek to identify dark envelopes from flat or rising rotation curves in regions exterior to most of the luminous mass. It does seem clear from the available data that single Irrs are unlikely to have rising rotation curves (but see 218), although the current observations of the gas kinematics of Irrs do not extend far enough or have sufficient spatial resolution to distinguish between flat and falling rotation curves. A new systematic high-resolution study of H I kinematics in inclined, noninteracting dwarf Irrs is needed to clarify the issue.

Abundances

The metallicities of the ionized gas (i.e. abundances of O, Ne, and N; for a review, see 268) have been determined from the emission-line ratios under the assumption of photoionization (4, 30) for a large number of high surface brightness Irrs (120, 191, 209, 216a, 226, 340, 381) and low surface brightness, dwarf Irrs (cf. 189, 226, 267, 270, 329, 340, 361). The abundances generally range from SMC-like to LMC-like, with no distinction between the high and low surface brightness systems. Only some of the dwarf blue compacts or "intergalactic H II regions" seem to be much more metal poor than the SMC (4, 120, 209, 216a, 226, 377), and no systems are known with emission spectra consistent with extreme metal-poor galactic Pop II abundances. The lower half of Figure 3 shows the number distribution of Irrs as a function of O/H for galaxies in the above references (including blue compact systems and intergalactic H II regions). Generally, O/H is $1-4 \times 10^{-4}$, where the solar value is $\sim 7.6 \times 10^{-4}$. Interestingly, several very luminous intergalactic H II regions have been observed at moderate redshifts and are found to have normal metallicities for Irrs; for example, B234 at redshift 0.06 (121) and B272 at redshift 0.04 (88) both have O/H $\sim 2 \times 10^{-4}$. Some very luminous blue compact galaxies and the unusual M82 system, however, may approach or even exceed solar metallicity levels in the gas (36, 132, 200, 257).

There are difficulties in measuring abundances of Irrs from emission lines. Often [O III] $\lambda 4363$ is not strong enough to be accurately measured, so some other means must be found for determining the electron temperature, such as the empirical relationship between forbidden oxygen and hydrogen Balmer emission intensities developed by Pagel et al. (266). This can lead to an ambiguity in the derivation of O/H for low-abundance objects (189). For example, consider the very metal-poor object I Zw 18 (120). The [O III] $\lambda 4363$ ratio gives a temperature of 16,300 K, whereas simple application of the relationship determined by Pagel et al. gives 6400 K. The former results in an O/H value of 0.2×10^{-4} and the latter in a value of 5×10^{-4}. A plot of [O II]/[O III] intensity vs O/H (189) appears to be able to resolve this ambiguity, since ionization levels and abundances are well correlated in H II regions; however, one must keep in mind that the errors in even routine emission-line abundance determinations can be large. For example, of the 45 overlapping observations in the references cited at the beginning of this section, 15 pairs agree within 20%, 5 within 50%, while 6 are off by factors of two or more.

A further complication in using emission lines as abundance indicators in more distant galaxies where individual H II regions are not resolved stems from anomalous inter-H II emission-line ratios found by Hunter (190). In nearby Irrs, the [O II]/[O III] intensity ratio from diffuse emission considerably exceeds the ratio for H II regions within the same galaxies. Thus, integral emission-line flux ratios will differ from those of typical H II regions. The origin of anomalous line ratios is unclear and could involve either shocks or some type of photoionization process. It is also uncertain whether these anomalous emission regions are similar to the diffuse Hα emission seen in late-type spirals (246) and the Milky Way (280).

Since Irrs do cover a range in gas metallicity, they have proven valuable in efforts to determine the pregalactic helium abundance level, which is an

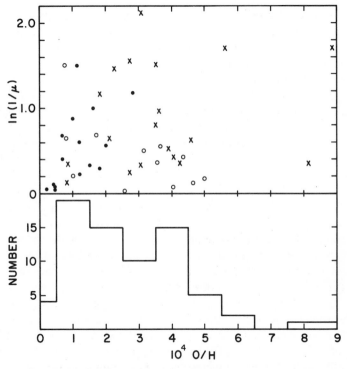

Figure 3 (*lower panel*) The distribution of Irr galaxies as a function of the oxygen abundance in H II regions is displayed. This sample includes low and high surface brightness Magellanic Irrs, luminous Irrs, blue compact systems, and intergalactic H II regions (see text for references). Gas abundances in Irrs, with few exceptions, are similar to the Magellanic Clouds. (*upper panel*) Dependence of oxygen H II abundances on gas mass fraction is shown in terms of $\mu = M_{gas} (M_{gas} + M_{stars})^{-1}$. Simple chemical evolution models predict a linear relationship between O/H and $\ln (1/\mu)$, which is not found in this sample or in more traditional plots where $\mu = M_{gas}/M_{dyn}$ (238). All galaxies from the lower panel with sufficient data to enable μ to be estimated are included. Symbols are as follows: ×, high surface brightness normal Irrs; ○, low surface brightness dwarfs; ●, luminous and blue compact Irrs.

important constraint on cosmological models (see 14, 259, 268, 389). For example, an upper bound on the primordial helium abundance is provided by the most metal-poor intergalactic H II regions, and studies of larger samples of Irrs can yield insight into the variation of helium enrichment by stars as a function of metallicity, which allows an extrapolation to be made to zero metallicity (for excellent reviews, see 210, 216a, 265). Considerable effort therefore has been devoted to enlarging the sample of very metal deficient emission-line galaxies (cf. 215), but only limited success has been achieved thus far in that I Zw 18 remains the most metal-poor emission-line galaxy.

The metallicities of stars in Irrs seem to be consistent with the moderately metal-poor nature of the gas. Star clusters in the LMC have been found to have metallicities of 1/2 to 1/40 of the Sun (cf. 60, 168). The SMC, while on the average more metal poor than the LMC, is not known to contain objects as metal poor as the most metal-poor objects in the LMC or the Galaxy (127, 128), and thus it evidently has a smaller spread in abundances. Finally, *JHK* colors of Irrs are near those of several-billion-year-old intermediate metallicity systems (349). If gas captures from external sources are a common occurrence, e.g. among star-burst Irrs, then it is possible that gas and stellar metallicity levels may not mesh in some systems. This phenomenon has not to our knowledge been observed, but data on stellar abundances are still very sketchy.

The relatively low abundances of the Irrs tell us that these systems are less evolved than most spirals, in the sense that the interstellar gas is less processed. Less processing, however, does not mean that the returning metals are not mixed throughout the system. Spectrophotometry of individual H II regions (cf. 191, 266, 340, 381) shows that the O/H and N/S abundance ratios are remarkably constant throughout the disks of Irrs. By whatever process, the metals in these galaxies are fairly well mixed over at least the optically prominent regions.

Correlations between the gas abundances and other global parameters would be expected to give us some clue as to why this particular morphological type of galaxy (the Irrs) would be less evolved and better mixed than most spirals. As already mentioned, except for the intergalactic H II regions, which are the most metal poor, the abundances do not separate according to the luminosity of the systems. There also do not seem to be any compelling relationships between metallicity and current stellar birthrate or galaxian spatial dimensions, and thus our hopes for an obvious clue to evolutionary processes go unrewarded.

The closed-system model with the instantaneous recycling approximation provides the simplest and most widely used galaxy chemical evolution model (268, 312, 339, 353). In this model, the metallicity of the gas is given by

$Z_g = y \ln(1/\mu)$, where y is the stellar heavy element yield and μ is the gas fraction, i.e., mass in gas per unit total mass (stars plus gas); thus a correlation between Z_g and gas fraction should exist. Since the Irrs are evidently well approximated by a single zone, they provide a good test for the applicability of the basic chemical evolution model to real galaxies. The upper half of Figure 3 is a plot of $\ln(1/\mu)$ against the abundance parameter O/H for high surface brightness Irrs (189, 191, 226, 340), for low surface brightness Irrs (189, 226, 340), and for luminous Irrs and blue compact systems (4, 120, 209). The parameter μ is computed for all objects in the manner described by Hunter et al. (191), where the gas mass is taken to be 1.34 times the total hydrogen mass and the mass in stars is estimated using the UBV colors to determine the M_{star}/L_V ratio from Larson & Tinsley's (220) stellar population models. Thus, the total mass is the mass in gas plus stars, rather than the often-used dynamical mass, which could include contributions from matter that does not partake in the chemical evolution. Clearly, no correlation is evident in Figure 3, in agreement with the similar study by Matteucci & Chiosi (238). Thus we see that while simple chemical evolution models provide useful qualitative insights, they do not quantitatively fit the observations.

There are several possible ways to modify the basic model. For example, in using the closed-system model one must consider over what region does the galaxy really behave as a closed system. Complications along these lines could include gas infall from the outer parts of the galaxy (55, 126, 238), gas that does not actively participate in the galaxy's chemical evolution (188), or ejection of metals from star-forming regions (146). In loosely bound galaxies, material ejected due to supernovae could be lost from the system entirely or rain back down on other regions of the galaxy (115, 346). These types of processes may allow us to understand the empirical correlation that exists between metallicity and dynamical mass in dwarf Irrs and extragalactic H II regions (191, 226, 238, 340). Thus it may well be that total mass is equally or more important than relative gas fraction in controlling Z_g. In this case the simplest closed-system models would not apply, and more specific models will be needed instead.

Stellar Populations

The blue colors of Irrs are generally taken to mean that a proportionally larger component of the stellar population consists of early-type stars (247, 248). This is consistent with the fact that the high surface brightness Irrs are endowed with numerous H II regions and are actively forming stars. Bagnuolo (15) and Huchra (178), for example, have fit the colors of Irrs with a composite between old and young stellar populations, while continuous star formation models have been computed by Searle et al. (313), by Huchra (178) and by Code & Welch (59).

Specifics of the OB star populations have been explored through IUE spectra that are now available for a variety of high surface brightness Irr galaxy family members (27–29, 179, 180, 219, 377; see also 47, 196). Generally the $\lambda\lambda1150$–2000 Å UV spectra of Irrs show features consistent with rich OB star clusters having near normal IMFs, but there are important exceptions. On the basis of UV spectra, a case was made for a very massive superstar ($\sim 2 \times 10^3\ M_\odot$) in the core of the giant 30 Doradus H II complex in the LMC (52, 107, 305) and more recently for the somewhat similar NGC 604 giant H II region in the nearby spiral M33 (237). Even if we do not accept the presence of superstars in these systems (245a), it is clear that extraordinary concentrations of high-mass stars ($\sim 10^2\ M_\odot$) must be present to meet the ionization requirements and to fit the observed spectral characteristics. Giant H II regions are common in Irrs (169), and their possible relationship to very massive stars is now receiving careful scrutiny.

The low surface brightness dwarf Irrs present more of a problem with regard to stellar content, since they are also fairly blue but do not have the many obvious star-forming regions that characterize the high surface brightness systems. In general, the dwarfs seem to have stellar population mixes similar to those of the larger Irrs (81, 82, 160, 189), but the numbers of luminous stars are down in a manner qualitatively consistent with a lower total star formation rate (173, 185, 228, 295). Some dwarf systems have extremely blue colors and high surface brightnesses, perhaps indicating that bursts of star formation have recently been completed (158, 177, 178, 232, 313); detailed population studies of a few resolved galaxies support these viewpoints (57, 288).

The Irrs, therefore, are correctly noted for their young stellar population, but they also contain older stars. Only a few of the extreme "intergalactic H II regions" seem to be without a possible older stellar component (cf. 179, 312, 349, 377), but the nature of this older population and the number of previous generations of stars are not so clear. In the dwarf Irr VII Zw 403, for example, the metallicity of the gas is sufficiently low and the level of current star formation sufficiently vigorous that one is forced to doubt whether the galaxy could have formed stars in this vigorous way at an earlier stage (361). Nevertheless, the existence of a diffuse stellar component implies that an older population is in fact present. (Star formation histories are discussed in a later section.)

Many Irr galaxies are close enough that they can be resolved into individual stars. The LMC and SMC are, in fact, the best systems outside our own for studying stellar populations and support the concept of a nearly constant IMF in disk galaxies (see below). Beyond the MC, only intermediate- to high-mass stars can be individually observed at present, and aside from very luminous supergiants, only colors and magnitudes are available (187, and references therein). In more luminous Irr dwarfs, observed color-magnitude diagrams are similar in form, with pronounced blue and red supergiant branches well separated by a Hertzsprung gap (184–186, 202, 295, 296, 301). Massive stars evidently are present with relatively constant properties in Irrs, and thus the door is open to modeling the light from young stellar populations in terms of fairly standard components (as in 76, 180, 189, 191, 225).

Differences in OB stellar content between galaxies are largely explainable in terms of statistical effects, which can be quite severe in faint dwarfs, where OB star formation probably involves a series of time-disconnected discrete events (172, 288, 313). At the upper extremes of stellar mass, the situation is, as we have seen, less well defined; for example, the presence of many Wolf-Rayet stars in a small galaxy like Tololo 3 (216) might be due either to statistics or to special processes (e.g. very massive stars) in very large star formation events that are seen as giant and supergiant H II regions (169, 237). Finally, we point out that intermediate-mass stars become very luminous during the AGB evolutionary phase (see 193) and may be seen as resolvable stars in the diffuse light of Irrs (140), as long-period variables of interest to the extragalactic distance scale (387), and as major contributors to the infrared luminosity (272).

Star clusters provide further important clues to stellar populations in galaxies (e.g. 56), and currently they can be detected to distances of several Mpc in Irrs as a result of the open structures of the parent galaxies (155, 188). While numbers, sizes, and richnesses of star clusters vary from galaxy to galaxy (160, 167, 369), the clusters themselves are found to be remarkably similar in their integral optical stellar properties within such diverse Irr systems as the LMC (370a), M82 (257), and NGC 6822 (165, 372). The main variables affecting integral cluster observables are well known to be IMF, age, and chemical composition, although stellar richness can also be a significant factor (279). From color-magnitude diagrams of individual Magellanic Cloud star clusters it has been possible to calibrate approximately variations in global cluster parameters as a function of age for metallicity levels appropriate to most Irrs. Unfortunately, some disturbing inconsistencies remain in the details of the age scales (170, 249), and subtle differences exist between clusters and stellar evolution model predictions (26, 112, 113). Still, the analysis of LMC cluster photometry in the classic work of Searle et al. (314), as well as Rabin's (277) investigations of individual cluster spectra, assures us that among younger clusters age is the major determinant of spectral properties, while in very old clusters metallicity is a primary factor. Star clusters thus are a comparatively reliable means for unraveling the stellar age/metallicity strata that hold the histories of galaxies.

Studies of the Magellanic Clouds and other nearby Irrs reveal star clusters covering a full range of age classes, and thus these galaxies have been actively producing stars for at least several billion years (162, 167). Recently photometry has been obtained by Stryker (336, see also 338) down to the main sequence turnoff in the red LMC halo cluster NGC 2257. As this cluster contains a well-defined horizontal branch and RR Lyrae variables, the preferred age calibration method, developed by Rood & Iben (284; see Iben 192) and based on the distance-independent luminosity difference between main sequence turnoff and horizontal branch, can be applied to show that NGC 2257 is as old as Galactic globulars. Evidently the LMC produced or obtained star clusters from the same early epoch as the Milky Way. In contrast to the Milky Way, however, the Magellanic Clouds are still making globularlike star clusters, the "populous blue clusters" or "blue globular clusters" (156). Although there has been some resistance to considering these as total parallels to young globular clusters, LMC blue globular cluster masses lie in the range of 10^4–10^5 $M_\odot$ and therefore overlap with true globulars (58, 114, 118, 148, 254). These clusters are not unique to the Magellanic Clouds, and the luminous, near-stellar knot seen in actively star-forming regions of Irrs such as NGC 1569 or NGC 5253 (2, 84, 188, 189, 367, 370) may be populous stellar clusters in early evolutionary phases when OB stars and circumstellar gas are still present. It is therefore not appropriate to attribute the production of globular star clusters only to unique conditions in the early Universe (e.g. 85), but rather there may be a variety of channels for the formation of dense, spheroidal star clusters.

Not all Irrs, however, display the same small-scale spatial patterns of star-forming activity. At one extreme, the low astration rate dwarfs often lack rich clusters and pronounced OB associations, even when massive stars are present (e.g. 288, 301, 369). At the other extreme, some rapidly star-forming amorphous Irrs are also quite smooth in their optical appearances (125, 294, 300) and are pervaded by diffuse optical emission lines from ionized gas (84, 157, 188). It is quite clear that these galaxies may contain large complements of massive young stars, as evidenced by their high Hα luminosities and hot IUE ultraviolet spectra (219). The optically distinct, large star-forming complexes (OB associations, H II regions, etc.), which are the hallmark of most Irrs, are, however, missing. Perhaps in these systems the individual star-forming sites are overlapping or the stars are forming via a different mechanism than in most Irrs. But in either case, the amorphous Irrs illustrate that kinematically similar galaxies do not necessarily follow identical evolutionary paths.

STAR FORMATION PROPERTIES

Current Global Star Formation Rates

Measures of Lyman continuum photon luminosities, e.g. as determined from the Hβ or Hα emission lines or radio thermal fluxes, can be used to estimate the current rate of formation of massive stars. When coupled with an IMF such as from Salpeter (289) or Miller & Scalo (245), this information can yield the total rate at which gas is condensing into stars of all masses (143a, 319). If the emission flux is that for the entire galaxy, then the global star formation rate is known. Galaxy-wide star formation rates for Irr galaxies have been estimated in this fashion by extrapolating from large-aperture spectrophotometry (191), from flux-calibrated Hα images (126), from Hα photometry (205, 206), from radio continuum observations (195), and from UV luminosities (29, 86). For one galaxy, NGC 1569, the number of Lyman continuum photons determined from the radio observations is about 3 times higher than the number found from Hα emission. This is probably due to extinction within H II complexes, and thus optical measurements will usually yield underestimates of star formation rates (205).

The star formation rates determined in this manner are found to cover a wide range, but the average rate per unit area in high surface brightness Irrs is comparable to the Milky Way disk ($\sim 5 \times 10^{-9} \, M_\odot \, \text{yr}^{-1} \, \text{pc}^{-2}$; 330, 353, 364). If we consider the total rate of star formation per unit gas mass, the average high surface brightness Irr is again comparable to a typical spiral, although some Irrs are overachievers in this regard. (NGC 1569, for example, has a rate 30 times that of our Galaxy; 390.) It is important to keep in mind that the samples chosen for star formation rate studies have been intentionally biased toward high surface brightness, rapidly star-forming galaxies, and that there do exist low surface brightness dwarf Irrs that have very little current star formation activity (e.g. 57, 189, 288). Nevertheless, these studies show that Irrs exist that are uninfluenced by interactions or other outside perturbations and yet are actively forming stars. We must conclude, therefore, that *spiral density waves are not necessary to a vigorous production of stars.*

In order to understand the star formation mechanisms and the galactic characteristics that govern them, a search has been made for correlations between star formation rates and various global parameters. One might expect, for example, a correspondence between stellar birthrate and gas density such as in the Schmidt (308) empirical model in which star formation rate varies as the gas density squared. The higher the mean volume density in the interstellar medium, the higher the star formation rate would be. In practice it is difficult to ascertain from the observed projected density of interstellar matter the fraction that is in a proper physical state to produce stars. Furthermore, it is both the fluctuations and mean gas density that are important in producing stars, so we may expect some problems with this type of approach. From a study of spiral and Irr galaxies, Lequeux (222) concluded that the star formation rate actually decreases with increasing average gas density, while Guibert (142) found that if the star formation rate is proportional to a power of the gas density, the proportionality constant decreases with increasing gas fraction. Young & Scoville (391), on the other hand, suggest that in spirals the production rate of stars per gas nucleon is constant. In their study of Irrs, Hunter et al. (191) found no relationship between star formation rates and average H I gas density or gas mass, although Donas & Deharveng (86) do find a correlation with gas mass. Low CO fluxes from actively star-forming Irrs (95, 390) further indicate that no simple relationship exists between measurable *global* gas characteristics and current stellar production rates.

A few studies (86, 191) have also searched for relationships between star formation rates and other integral galactic quantities such as metallicity, total mass, the ratio of mass in stars to dynamical mass, and the H I mass per unit luminosity. No convincing correlations have been found. From this it is concluded that local parameters are probably more important than global ones in determining star formation patterns in noninteracting Irrs.

The manner in which these local processes interact, then, must set the global states of Irrs. Unfortunately, the mechanisms that provide coupling between star formation sites are not known and may not include the traditional local moderators of star formation processes: gravitational and magnetic instabilities. Galactic-scale magnetic fields probably are present in Irrs, as evidenced by organized interstellar polarization in the Magellanic Clouds (198, 309), and potentially play an important, although as yet poorly defined, role in large-scale star formation processes (96, 250, 251). Gravitational instabilities against axisymmetric perturbations have also been proposed as drivers of global star formation in disk galaxies (see 143, 255). Both gas and stars in Irrs, however, have typical velocity dispersions of $\sigma \sim 10$ km s^{-1} (105, 119, 182, 183) and therefore by the usual criteria (356, 357) are safely stable unless the dispersions are highly anisotropic. The most probable mode of interaction in star-forming processes within Irrs is thus through modifications of conditions in the interstellar medium, which can be induced by the young stars themselves.

Distributions of Star-Forming Regions

Irregular galaxies, particularly high surface brightness Irrs, are noted for and defined by the rather chaotic spatial distributions of their star-forming regions (cf. 164). Correlations in the distributions and spacings of H II regions seem to be lacking, but there do seem to be a few significant chains of H II regions that are probably coeval (188). This indicates, as does the presence of ill-defined "spiral arms" in some systems, that the star formation mechanism is not entirely random. In Irrs with bars, for example, large star-forming regions often occur at one end of the bar, posing the possibility that gas flows due to the bar may be important there (98, 106).

In spite of the apparent chaos, star-forming regions are not uniformly distributed over the disks of the Irr galaxies; instead, they seem to be asymmetrical and clumped on large scales. Hodge (159) interpreted the clumping in terms of localized star formation bursts on a scale of ~ 1 kpc, and the major star-forming regions then must migrate around a galaxy with time. This effect can also be seen from the distributions of star clusters in NGC 6822 and IC 1613 (167) and of the Cepheids in the LMC (92, 269); the main star-forming centers of the recent past ($\sim 10^7$–10^8 yr) are in different locations than currently active regions. Large-aperture spectrophotometry centered on the most active areas of Irrs also shows that the star formation rates are too high and the metallicity and gas content too low for star formation to have always continued *in that region* at the current rate (191). Furthermore, in galaxies such as NGC 3738, NGC 4214, and NGC 4449, one can identify large complexes that obviously recently supported star formation but that are now in decline (188). In fact, the H II complexes typically cover less than 4% of the optical areas of Irrs; thus each position in a galaxy probably experiences a major star formation event once every 10^8–10^9 yr. This shifting of the major centers of star formation seems physically reasonable, since the star formation process must certainly deplete the local cool gas. But why star formation would necessarily clump on the observed scales and whether it migrates in a systematic manner are not known.

In addition, we do not find bright H II regions out to the optical "edges" of Irrs. That is, typically the current activity is within the inner 60% of the blue optical dimensions given in the Uppsala General Catalogue (UGC, 254a). A few extreme galaxies, such as NGC 5253, seem to be forming stars mostly in their central cores, although star formation obviously did occur in the outer parts of NGC 5253 at some time in the past (370). It is possible that areas outside of the active central zones no longer have gas above the critical density necessary for star formation, and so these galaxies will never form stars in their outer regions again (see below). However, an alternative possibility is that star formation in these less dense outer regions of Irrs is continuing but in a more diffuse manner. This is analogous to the problem of the dwarf Irrs, which also lack the giant H II complexes and large OB associations: Is star formation temporarily absent, or instead is the process of star formation different (i.e. higher mass stars and/or clusters are not formed)?

Star-Forming Complexes and the Interstellar Medium

The great star-forming complexes found in Irrs are very similar to those associated with giant H II regions in spirals, as in M33 or M101 (188,

189, 377). This includes their optical properties (225), ultraviolet spectra (180), sizes (188, 203, 204, 297–299), kinematics (compare 188 with 285), and morphology (Hunter & Gallagher, in preparation). These similarities indicate that once a gas cloud complex is stimulated or naturally reaches a stage where stars will condense, the region forgets what kind of galaxy it is in, and that the upper limit to star-forming cloud complex sizes is similar in spirals and Irrs. In addition, Hodge (169) has demonstrated that the distribution of sizes of all H II regions in a galaxy can be fit by an exponential law, although this fit steepens for less luminous parent galaxies. The giant H II regions in some Irrs are then found to be anomalously large relative to this distribution, and therefore they could result from special (but common) cloud formation processes (98).

There are also some differences between the interstellar mediums of Irrs and spirals. First, the Irrs appear to be deficient in large, dense interstellar clouds as judged from the low optical visibility of dark clouds, although a few dark nebulae are clearly present, especially on smaller spatial scales (161, 163, 165, 166, 188). Star-forming regions in the Galaxy and in spirals are most often adjacent to dusty areas that mark molecular clouds (97), so a dearth of dark clouds in star-forming galaxies is unexpected. An examination of blue and red passband images, Balmer decrements, and optical- and radio-determined star formation rates shows that even in and around star-forming sites, optical interstellar extinction in Irrs tends to be low [maximum $E(B-V) \lesssim 0.5$ (188, 191, 340); for the Magellanic Clouds, see, for example, 89, 111, 184, 186, 234, 271]. This may be in part a reflection of the underabundance of heavy elements, which certainly results in some modifications of dust properties (e.g. 283a, 306). In addition the low surface brightnesses and irregular light distributions could hinder the detection of dust in Irrs. A lack of extensive regions of high column density gas, however, is probably the main factor. The notable exception among Irrs is the peculiar galaxy M82, which is loaded with dust even as compared with spirals, although M82 is also relatively metal rich (257). Why some galaxies have optically obvious dust clouds and others do not and how the presence or lack of dark clouds affects the star formation processes are not known, since high and low star formation rate examples of both extremes are known to exist.

Furthermore, the high surface brightness Irrs are underluminous in CO molecular microwave emission relative to regions in spirals with similar luminosities and stellar content (95, 135a, 390). Although the global quantity of molecular matter evidently is down, the presence of vigorous star-forming regions in Irrs argues that on local scales the molecular clouds are normal. It is possible that the current star formation activity has disrupted most of the parent clouds, but then we should find systems with a reduced star formation rate that are full of CO clouds. An alternative is for diffuse molecular clouds to be comparatively short lived in Irrs, which is possibly related to the low specific angular momentum of Irr disks. If molecular clouds formed in low angular momentum environments in fact collapse more readily than in spirals, then in a *local* sense Irrs may be more efficient star-formers than spirals, despite the absence of dynamical forcing by arms..This could explain the anomalous presence of spectacular star-forming complexes (e.g. giant H II regions) in otherwise undistinguished galaxies (but see 35). Refueling from outer gaseous reservoirs (disks, halos; 126) or differences in the thermal structure of the interstellar medium are also important factors in determining the state of interstellar matter that potentially could differentiate Irrs from spirals.

The Initial Mass Function

The lack of understanding about the IMF (initial mass function) is a serious stumbling block in the study of galactic star formation histories (see 307 for a comprehensive review). The Magellanic Clouds are the only systems that are sufficiently near to check the IMF outside of the Galaxy, although as we noted earlier, massive star populations are relatively similar in most nearby Irrs. An inventory of Magellanic Cloud supergiants by Dennefeld & Tammann (69), for example, indicates that the stellar component of masses $\gtrsim 9\ M_\odot$ is not radically different from the Galaxy (see also 223, 224, 296). Deep luminosity functions measured in the LMC by Butcher (46) and Stryker & Butcher (337) extend this conclusion to $\sim 1\text{--}3\ M_\odot$ stars. The presence of classical Cepheids (269) and RR Lyrae variables (138, 139) in the Magellanic Clouds further shows that stars covering a range in mass from $1\text{--}10\ M_\odot$ have evolved off the main sequence. Cepheids also have been observed in several other Local Group and nearby Irrs, but only in IC 1613 are the numbers sufficient to probe the stellar content (291), which is found to be like the Magellanic Clouds (25, 240). Thus the stellar mixes of the Magellanic Clouds are surprisingly similar to those of the disks of spiral Local Group members (144, 145), although there are hints from the luminosity functions that the star formation rates are not smooth functions of time in either the LMC or SMC (147, 337).

If the IMF is to be different in the Irrs, we require that high surface brightness systems with extreme blue colors and strong emission lines be overabundant in high-mass stars, while the low surface brightness Irrs have fewer massive stars. In making such comparisons, however, one must distinguish between changes in the form (e.g. slope) of the IMF and statistical effects, i.e. very large star-forming events naturally will produce many massive stars. The similarities between the large H II complexes in Irrs and those in other types of galaxies give no reason to expect that local environments are sufficiently diverse that a different IMF would result. Hence, we are reluctant to invoke an unusual IMF when there is no compelling evidence for it. Terlevich & Melnick (347, 348) have argued that there is a systematic variation of the IMF with metal abundance, such that more metal-poor systems have flatter IMF slopes, but their approach to this problem is still controversial (cf. 124). Some effect of this type may, however, be necessary to explain the correlation between ionization level and metallicity that exists in H II regions. Below, we discuss further evidence that a normal IMF is at least consistent with the evolutionary histories of typical Irrs. However, we do not know over what time or spatial scales it is necessary to average the stellar populations in order to obtain the "normal" IMF. It is possible that during galaxy-wide bursts of star formation, conditions may be such that this spatial- or temporal-averaging process is disrupted, and a peculiar IMF results.

Usable Gas

In measuring star formation rates and other parameters that depend on the gas content, the gas mass used is the total mass *detected*. The proper value to consider is really the mass that has the potential to engage in star formation. Irrs often have halos of neutral hydrogen extending to several optical diameters. An extreme example is the dwarf Irr IC 10, which has a hydrogen envelope 20 times its optical diameter (326); however, this galaxy suffers heavy Galactic extinction, so it is possible that the optical size has been underestimated. It is not clear that this outer gas necessarily participates in the star formation process (an idea suggested for M33 in 235). If such is the case, our concept of the Irrs as systems that homogeneously evolve as a single spatial unit must be altered.

In a sample of 21 Irrs (mostly low surface brightness), Huchtmeier et al. (183) found that the FWHM of the H I distribution occurred at the Holmberg radius but that H I usually extended to several Holmberg radii at a density of $\geq 10^{19}$ atoms cm^{-2}. Spirals, on the other hand, can have H I values to 1.5 times the Holmberg dimension at 10^{20} atoms cm^{-2} (37, 38). However, there are important exceptions. NGC 4449 still has a column density $\sim 10^{20}$ atoms cm^{-2} at 4 times its Holmberg radius (376) and IC 10 at 2–3 times (326). Yet, in neither galaxy is there direct evidence for ongoing star formation at such large radii. All of this suggests that much of the gas in the outer parts of many Irrs may not be able to contribute to the star formation as effectively as the inner gas. This provides the empirical basis for our earlier assumption that condensation of at least detectably young OB stars is a threshold phenomenon that does not occur (or takes place with much lower frequency) in gas below a critical density (94, 244). Based on the data for spirals given by Bosma (37, 38) and the properties of Irrs, a rough empirical guess is that gas in H I disks is below the OB star formation threshold density when $\sigma_{HI} \lesssim 5 \times 10^{20}$ atoms cm^{-2}.

STAR-FORMING HISTORIES

Constant Star Formation Rates

The localized bursts of star formation that characterize Irrs do not necessarily imply galaxy-wide bursts of activity; the global mean rates could still be constant while local variations are large (cf. 223). Gallagher et al. (126) have explored the star formation histories of a sample of high surface brightness Irrs chosen for their blue colors by considering parameters that measure the stellar birthrate over different time scales: The galaxian mass is a clue to the astration rates integrated over the galaxy's lifetime, the blue luminosity is dominated by stars formed over the last few billion years, and the ionizing photons give the current rate. They found that for most Irrs these parameters are consistent with a constant mean rate of star formation and IMF over the galaxy's lifetime, in agreement with results for the Magellanic Clouds (e.g. 283).

Few Irr systems seem to be undergoing honest global star formation bursts, i.e. only in unusual circumstances does the current global stellar production rate exceed the lifetime average rate (126, 191, 223, 283). In addition, Hunter (189) showed that the time scales to exhaust the present gas content at current astration rates and the metallicities were also consistent with constant stellar birthrates *if* all of the detected gas readily participates in the system's evolution. These results stand in contrast to studies based on colors and emission-line properties in which bursts and peculiar IMFs are found to be a normal feature of blue galaxies (see the following section). We are not currently able to reconcile these conclusions, although galaxy sample selections, data characteristics (i.e. local vs global measurements), and possible problems with stellar population models are all factors that may lead to differences.

If stars are formed at constant rates in Irrs, what does this imply about these galaxies? As we have seen, there is empirical evidence that gas densities must exceed some critical value for star formation to proceed at normal levels. A constant stellar birthrate then implies that a galaxy must maintain a constant amount of gas above this critical density, in a state suitable for starbirth, even as gas is continually being locked into new stars. How then does a galaxy manage to keep shuffling the same amount of gas into stars? We would expect that as more and more gas was turned into stars, the overall gas density would drop and the star formation rate would decrease. Thus the standard models predict that star formation should steeply decline with time in all galaxies (94, 126). This difficulty can be avoided either by postulating a continuous resupply of gas to star-forming regions, i.e. by maintaining constant average gas density (126, 221, 253), or by presuming the star formation rate is not solely dependent on mean gas density. Models of the former type require gas inflow from a halo or outer disk, which has never been confirmed by direct observation. In the latter class of models, the star formation rate must not depend on the total amount of gas. For example, the production of dense interstellar clouds from which stars are born could be the controlling factor and could be set by the OB stars themselves (e.g. 324).

A complementary question concerns the early stages of formation of an Irr. Given that the global star formation rates in the past might not have exceeded current levels by much, and that most Irrs probably formed as gravitationally bound entities at approximately the same time as other types of galaxies, then it seems that these systems managed to collapse into rotating "puddles of gas" before beginning serious star formation. Star formation in the earliest phases of galaxy evolution is still a heated issue with regard to all types of galaxies, and as Irrs represent the extreme end of the "late-bloomers," they should provide an important point for empirical comparisons with galaxy formation models.

Bursts of Star Formation

In spite of the previous section's tone, it is clear that a constant stellar production rate does not fit the evolution of all Irrs. Repeatedly in the literature, people find themselves forced to conclude that the current global astration rate in a system is much higher than it used to be. Bursts have been suggested for various high surface brightness Irrs from a comparison of evolutionary models of stellar populations with observed broadband colors for late-type galaxies (15, 220, 313, 333, 349), from colors and emission-line strengths in Markarian galaxies (177, 178), from population studies for the Magellanic Clouds (8, 43, 122, 168) and M82 (257), from metallicity enrichment rate arguments (4, 225, 226, 258, 361), from radio observations of Markarian galaxies (33, 148a), and from star cluster studies of NGC 5253 (370), to name a few examples. Similarly, amorphous galaxies such as NGC 1705, which appear to be involved in OB star formation at high rates over most of their optical dimensions, are likely to be in burst phases (22, 219), as are clumpy Irrs (66). It is also evident from the prevalence of very blue galaxies in binary systems that interactions affect star formation processes and may stimulate bursts (10, 11, 31, 201, 220).

It is not immediately obvious, however, how seriously one should take the evidence for star formation bursts as a *general* evolutionary feature of noninteracting Irrs. The formation of OB stars, after all, occurs via gravitational collapse in interstellar cloud complexes, a process that produces spatially compact OB associations and star clusters. Thus, as Searle et al. (313) explicitly recognized, star formation is an intrinsically grainy process, and in small galaxies the normal evolution should proceed as a series of "bursts" associated with the appearance and decay of individual star-forming complexes (cf. 86). These statistical effects will be most important in small galaxies, since the blue luminosities of single star-forming complexes probably do not much exceed $M_B \sim -15$ (388) and spatial sizes of ~ 1 kpc (170, 188). Star formation bursts in small galaxies therefore do not necessarily indicate any evolutionary anomalies. There are also some difficulties in interpreting the empirical evidence for bursts in any galaxies: With metallicity enrichment arguments, one can raise the questions of whether the system is closed and what volumes of gas must be considered (see previous discussions); and with colors, one can say that the evolutionary path to any set of optical colors is not unique. Despite these problems, it is still clear that even some nearby, noninteracting Irrs cannot be explained without global star formation bursts [e.g. NGC 2915 (320), Haro 22 (126)]. In NGC 1569, for example, OB stars are spread over an area of many kpc (84, 188, 191), the current star formation rate is 10 times higher than its average past rate (126), and the optical luminosity is more than 10 times the single event maximum. Alternatively, if galaxies like NGC 1569 are not bursting, then they must be peculiar in other ways, i.e. they must be young or have an unusual IMF that favors production of OB stars.

The existence of Irrs currently undergoing global bursts of star formation implies that (a) some mechanism must exist to organize star formation on large scales, and (b) there must be Irrs of the same basic types as bursters that are not now active. In fact, we should see systems in all phases of postburst decays. The low surface brightness, low star formation rate dwarf Irrs come immediately to mind as postburst candidates. These systems may have had higher star formation rates in the past, but spectrophotometric data indicate that the metallicities and the stellar populations are not consistent with the picture of their being the low star formation rate states of high surface brightness Irr galaxies (189).

High and low surface brightness Irrs, in fact, seem to have experienced parallel recent evolutionary histories that have produced similar integral properties, such as gas metallicities and stellar population mixes. These galaxies thus primarily differ as a result of stellar surface density, but it is also noteworthy that lower surface brightness Irrs rarely contain luminous star-forming complexes (and when exceptions occur, as in NGC 2366, they are obvious but do not change the overall surface brightness). Perhaps factors such as gas density prevent the low surface brightness dwarfs from forming gas clouds in the size and quantity necessary for large star-forming complexes, which are typical of high surface brightness Irrs. In NGC 6822, for example, many H II regions are small (211), and only a few H II regions require more than a single O star for their ionization (137). Based on the Hodge (170) study of H II region sizes in Irrs, this situation seems to be typical. Star formation in dwarfs evidently proceeds in an unspectacular manner due to the lack of giant star-forming complexes. This results in low surface brightness dwarfs having followed approximately the same evolutionary paths as giant Irrs, in agreement with the observations, but leaves the issue of descendants of global burst Irrs unresolved.

SSPSF: A Possible Model

Much work has been done on the theory of density-wave-induced star formation in spiral galaxies, but it is only recently that the first major theory of global star formation processes applicable to Irrs was developed. The stochastic self-propagating star formation (SSPSF) model is one in which star formation is continued through the energy dumped back into the interstellar medium by evolving massive stars through H II region expansions, stellar winds, and supernovae (for a review, see 316). Theoretical models of galaxian evolution with this mechanism were first computed by Mueller & Arnett (252) and have been developed extensively by Gerola, Seiden, and collaborators (130, 131, 315) as well as others (64, 116).

In the SSPSF models of Gerola & Seiden, a two-dimensional galaxy is divided into cells that represent the average size of distinct star-forming complexes; Comins (64) has extended this approach to three dimensions. The probability that a cell will form stars is greatly increased if an adjacent region formed stars in a previous time step, i.e. star formation stimulates further star formation as suggested by Öpik (256). Once a cell has formed stars, it is initially unable to produce any further stars, and the probability of further star formation thereafter increases as a function of time corresponding to the replenishment of cool gas supplies.

Models of low-mass galaxies with little or no differential rotation (64, 108, 131) produce systems with chaotic morphologies and properties (metallicity, stellar content, etc.) that depend primarily on the ratio of galaxy size to the size of star-forming cells. A smaller galaxy has a low average star formation rate, but the rate fluctuates widely between bursts and quiescent levels as cells individually and in small groups experience star formation. Larger systems exhibit a more constant and higher mean star formation rate and display more morphological structure as sites of current star formation activity move around the disk. Because of their higher mean star formation rates, the larger galaxies will be more evolved and hence have higher metallicities and lower fractional gas masses. Even in larger galaxies, however, collective effects can lead to significant nonrandom time variations in total stellar birthrates (317), which could explain the existence of luminous systems in apparent burst states.

Quantitative comparisons between the SSPSF models and observations are not easy. Measurements of the sizes of Hα complexes in nearby (< 10 Mpc) high surface brightness Irrs indicate that the concept of an average "cell" size may not be entirely arbitrary (169, 188, 203, 204, 297). Therefore, the models predict that for a fixed cell size the gas fraction and average metallicity should correlate with galaxy size. In fact, no clean relationships are seen, although very small galaxies do tend to be the most metal poor and gas rich (189, 190, 238). Similarities between the mean colors of galaxies covering a considerable range in optical size also present a problem for basic SSPSF models. *Morphologically* the models reproduce reasonable Irrs. Star formation in many Irrs is *not* observed to be purely random, and any successful model must account for this fact. But SSPSF is not a unique answer. Ultimately the local gas characteristics, such as density, will determine when and where the star formation occurs. The origin of these density fluctuations could be associated with a variety of other mechanisms, such as the global velocity field, gas infall, magnetic field structures, etc., which could also act to organize star formation.

Direct evidence for propagation of star formation from one cloud complex or cell to a neighboring cell is lacking. The simplest version of SSPSF, in which star-forming events directly stimulate preexisting analogues to interstellar clouds, therefore may not apply to real galaxies. Sequential star formation within a cloud complex has been observed in the Galaxy (cf. 217), which *could* be interpreted as star-induced star formation, but again this is *within* a cell rather than between cells. Measurements of the energetics of H II regions show that considerable amounts of energy from massive stars are being dumped back into the interstellar medium (cf. 188). And we can see in nearby galaxies that the process of star formation does have a large effect on the interstellar medium; H I holes centered on NGC 206 in the M31 spiral (40) and the LMC's Constellation III (243, 383), as well as the supershells in our Galaxy (152), are examples. However, we still do not know what ultimate effects star formation may have on interstellar gas and whether these are sufficient to allow star formation to propagate on galactic scales. Until this problem is overcome, SSPSF will not be on a physically sound basis, although it should be emphasized that this theory provides a very general framework (316, 324, 325) in which the role of feedback (either positive or negative) in star formation processes can be readily examined.

EVOLUTIONARY STATUS

Even though Irrs are less evolved than most types of luminous galaxies, they are normally far from being in primordial states. Intermediate metallicity levels, modest gas-to-stellar-mass ratios, and strongly composite stellar populations found in the average Irr are indicative of maturity levels that have required billions of years to achieve. Similarly, stars and gas in all but the smallest dwarfs are in flattened disks and are largely supported by rotation. Thus considerable dissipation of energy has likely occurred since a purely gaseous protogalactic state, and sizes of present-day Irrs therefore reflect the angular momentum and density distributions as well as the extent and mean density of their pregalactic progenitors (101, 117, 327, 328). The smooth rotation properties in the outer disks of *many* (but not all) Irrs further indicate that a few rotation periods (which are in some cases $\sim 10^9$ yr) have elapsed since outer H I disk formation. Most irregulars therefore are probably old systems that formed as gravitationally bound entities $\sim 10^{10}$ yr ago, but we should be aware of possible exceptions. Metal-poor extragalactic H II regions are potential young galaxy candidates, and in some instances they have the disorganized, multiple gas cloud structures that are expected to characterize newly formed galaxies (21, 225, 227, 377). These small systems, however, could be old, but due to their low masses they have simply needed very long times to become dynamically organized and produce stars (cf. the multiple H I cloud structure of the SMC; 242).

Other exceptions to the above "late-bloomer" model are found primarily among the most luminous of the Irr family, i.e. giant blue compacts and clumpy irregulars. These systems have moderate-to-solar metallicities (36, 120, 342; Gallagher, Hunter & Bushouse, in preparation) and rotate at velocities similar to spirals, but they produce OB stars on incredible scales and in chaotic fashions that lead to Irr morphologies (66, 273, 274). The processes that cause such extreme (and probably transitory) evolutionary events in spirallike galaxies are not known, but they could include interactions with other galaxies or external agents (e.g. 220; many luminous Irrs are in binary pairs), major internal instabilities, such as relaxation of nonaligned gas disk components (90, 322), or delayed galaxy formation (e.g. 344). In any case, the existence of luminous Irrs and possibly related bursting Irr galaxies serves to remind us of the point stressed by van den Bergh (367): The evolution of galaxies may not be smooth in time, but instead may proceed by leaps and bounds in the form of postformation eruptions of star formation that dramatically and quickly alter the states of galaxies.

Most Irrs, however, are not in star formation burst phases, but rather they are evolving at nearly constant rates. Even though the mechanisms that produce such equilibrium behavior are not understood, it is clear that *local* processes play an extremely important role in the evolution of Irrs, probably through cumulative effects of many independent star-forming events or collective interactions between star-forming cells as envisioned in SSPSF theories. In the absence of strong differential rotation or dynamical forcing by spiral arms, these processes naturally lead to galaxies with Irr characteristics, i.e. astration is *globally* comparatively inefficient and star-forming complexes are distributed with a high degree of randomness. These features, furthermore, are not sensitive to details of galactic structure; it is then understandable that Irrs are found with remarkably uniform properties among slowly rotating galaxies over a range of $\sim 10^3$ in mass and that they predominate among low-mass disk galaxies.

The common dwarf Irr systems thus ultimately stem from the tendency for density and degree of central concentration to decrease in tandem with mass in disk galaxies, i.e. low-mass disk systems are slow, near-rigid-body rotators. A similar strong correlation between stellar mass and central density is found in diffuse dwarf elliptical galaxies (160, 293). Evidently the link between mass and density is nearly a universal property among galaxies with $M \lesssim 10^{10} \ M_\odot$, independent of morphological type, level of star-forming activity, or environmental factors (e.g. cluster vs field locations). This implies that initial conditions are crucial in determining fundamental properties of less massive galaxies, and that later environmentally induced modifications either have not been very common or have failed to produce major structural modifications.

Furthermore, as dwarf and luminous galaxies apparently have similar, clumpy spatial distributions within the Local Supercluster (362, 380), most regions of space must have produced a large range of galaxy masses, i.e. dwarfs do not originate from special initial conditions. Indeed, some would view the extreme dwarf Irrs as being representative of the types of individual bound fragments from which all galaxies initially arose (see 327). In this regard it is interesting that bound systems consisting only of cool gas (i.e. objects that contain H I but have undergone little or no evolution) are evidently extraordinarily rare (if they exist at all), even among the least massive galaxies (230). The characteristics of extreme dwarf Irrs may also provide useful tests for models in which galaxy formation is explosively induced (194, 260), since this mechanism naturally yields a lower cutoff mass for galaxies, which should be observationally accessible (261).

On the other hand, Irrs are fragile and thus easily influenced by external factors. Indeed, they are often observed to have been affected by close passages near giant galaxies. It is also possible that the seas of diffuse dwarf ellipticals that populate regular clusters of galaxies (278, 293, 385) represent systems that initially formed as dwarf Irrs but were later transformed into stellar fossils as a result of gas removal by the hot intracluster medium via processes such as stripping or thermal evaporation (see 39, 384). Alternatively, initial conditions at or near the time of formation may have separated low-mass galaxies over a spectrum of star formation rates (125, 293, 368), in which case initial conditions were simply not the same in regions destined to become regular clusters and in the much broader general field.

Interactions with the environment, however, need not always be destructive. Violent disruptions suffered by massive galaxies or their companions during deeply interpenetrating collisions or mergers yield sizable fragments that later can become independent dwarf galaxies, perhaps even gas-rich ones of the Irr type (129, 311). Populations and structural-type distributions of small galaxies therefore will be in a constant state of flux, and observational programs directed toward studies of Irrs in a wide range of settings are of considerable interest. The still unsolved problem of the relative importance of environment vs initial conditions ("genetics") in determining traits of galaxies thus extends to the Irrs. Since these galaxies are vulnerable and relatively uniform in their properties when in undisturbed states, comparative studies of Irrs in a range of environments have the potential to allow an empirical solution of this sticky issue.

By combining this information with an improved understanding of internal evolutionary processes, it should also prove possible to develop models for the initial states of Irrs. There are grounds for optimism about this task, since these low-density systems are probably closer to their protogalactic predecessors than are their dense spiral relatives. Thus somewhat paradoxically, Irr galaxies, which are dominated by young stars, are likely to be excellent stepping stones to the epochs of galaxy formation.

ACKNOWLEDGMENTS

We wish to thank our many colleagues who have provided stimulating discussions and preprints of their work. We are grateful to Bill Bagnuolo, Craig Foltz, Paul Hodge, Jim Kaler, and Tom Kinman for reading and commenting on preliminary versions of this manuscript. The task of typing was efficiently dispatched by the office of the Department of Astronomy at the University of Illinois, and we wish to acknowledge Sandie Osterbur, Deana Griffin, Jan Wehmer, and especially Carol Stickrod for their help.

We similarly are indebted to the photographic and drafting personnel at Kitt Peak, who provided their prompt services. DAH acknowledges a research associateship at Kitt Peak National Observatory and JSG research grants from the National Science Foundation.

Literature Cited

1. Aaronson, M. 1977. *Infrared observations of galaxies.* PhD thesis. Harvard Univ., Cambridge, Mass. 300 pp.
2. Ables, H. D. 1971. *Publ. US Nav. Obs.*, 2nd Ser. Vol. 20, Part 4
3. Ables, H. D., Ables, P. N. 1977. *Ap. J. Suppl.* 34:245
4. Alloin, D., Bergeron, J., Pelat, D. 1978. *Astron. Astrophys.* 70:141
5. Allsopp, N. 1978. *MNRAS* 184:397
6. Allsopp, N. 1979. *MNRAS* 188:371
7. Allsopp, N. 1979. *MNRAS* 188:765
8. Ardeberg, A. 1976. *Astron. Astrophys.* 46:87
9. Arp, H. 1966. *Ap. J. Suppl.* 14:1
10. Arp, H. 1981. *Ap. J. Suppl.* 46:75
11. Arp, H. 1982. *Ap. J.* 256:54
12. Arp, H., O'Connell, R. W. 1975. *Ap. J.* 197:291
13. Athanassoula, E., ed. 1983. *Internal Kinematics and Dynamics of Galaxies, IAU Symp. No. 100.* Dordrecht: Reidel
14. Audouze, J., Tinsley, B. M. 1976. *Ann. Rev. Astron. Astrophys.* 14:43
15. Bagnuolo, W. G. 1976. *Stellar composition and evolution of irregular and other late-type galaxies.* PhD thesis. Calif. Inst. Technol., Pasadena. 143 pp.
16. Bahcall, J. N. 1983. *Ap. J.* 267:52
17. Bahcall, J. N., Schmidt, M., Soneira, R. M. 1982. *Ap. J. Lett.* 258:L23
18. Baiesi-Pillastrini, G., Palumbo, G., Vettolani, G. 1983. *Astron. Astrophys. Suppl.* 53:373
19. Bajaja, E., Loiseau, N. 1982. *Astron. Astrophys. Suppl.* 48:71
20. Balick, B., Heckman, T. M. 1982. *Ann. Rev. Astron. Astrophys.* 20:431
21. Balkowski, C., Chamaraux, P., Weliachew, L. 1978. *Astron. Astrophys.* 69:263
22. Barbieri, C., Bonoli, C., Rafanelli, P. 1979. *Astron. Astrophys. Suppl.* 37:541
23. Barbieri, C., Casini, C., Heidmann, J., Serego, S. di, Zambon, M. 1979. *Astron. Astrophys. Suppl.* 37:559
24. Becker, S. A. 1981. *Ap. J. Suppl.* 45:475
25. Becker, S. A., Iben, I. Jr., Tuggle, R. S. 1977. *Ap. J.* 218:633
26. Becker, S. A., Mathews, G. J. 1983. *Ap. J.* 270:155
27. Benacchio, L., Galletta, G. 1981. *Ap. J. Lett.* 243:L65
28. Benvenuti, P., Casini, C., Heidmann, J. 1979. *Nature* 282:272
29. Benvenuti, P., Casini, C., Heidmann, J. 1982. *MNRAS* 198:825
30. Bergeron, J. 1977. *Ap. J.* 211:62
31. Bergvall, N. 1981. *Astron. Astrophys.* 97:302
32. Bertola, F. 1967. *Mem. Soc. Astron. Ital.* 38:417
33. Biermann, P., Fricke, K. 1977. *Astron. Astrophys.* 54:461
34. Blackman, C., Axon, D., Taylor, K. 1979. *MNRAS* 189:751
35. Blitz, L., Glassgold, A. E. 1982. *Ap. J.* 252:481
36. Boesgaard, A. M., Edwards, S., Heidmann, J. 1982. *Ap. J.* 252:487
37. Bosma, A. 1981. *Astron. J.* 86:1721
38. Bosma, A. 1981. *Astron. J.* 86:1825
39. Bothun, G., Caldwell, N. 1984. Preprint
40. Brinks, E. 1981. *Astron. Astrophys.* 95:L1
41. Brosche, P. 1971. *Astron. Astrophys.* 13:293
42. Bruck, H. H., Coyne, G. V., Longair, M. S., eds. 1982. *Astrophysical Cosmology.* Vatican: Pontif. Acad. Sci. Scr. Varia
43. Bruck, M. T. 1980. *Astron. Astrophys.* 87:92
44. Brunish, W., Truran, J. 1982. *Ap. J. Suppl.* 49:447
45. Burbidge, E., Burbidge, G., Hoyle, F. 1963. *Ap. J.* 138:873
46. Butcher, H. 1977. *Ap. J.* 216:372
47. Capaccioli, M. 1982. In *Eur. IUE Conf., 3rd,* pp. 59–68. The Netherlands: ESA Sci. Tech. Publ. Branch
48. Carruthers, G. R., Page, T. 1977. *Ap. J.* 211:728
49. Carruthers, G. R., Page, T. 1984. *Ap. J. Suppl.* 53:623
50. Casini, C., Heidmann, J. 1976. *Astron. Astrophys.* 47:371
51. Casini, C., Heidmann, J. 1978. *Astron. Astrophys. Suppl.* 34:91
52. Cassinelli, J., Mathis, J., Savage, B. 1981. *Science* 212:1497
53. Chamaraux, P. 1977. *Astron. Astrophys.* 60:67
54. Cheriguene, M. F. 1974. In *La Dynamique des Galaxies Spirales,* ed. L. Weliachew, pp. 439–81. Paris: CNRS

55. Chiosi, C., Matteucci, F. 1982. *Astron. Astrophys.* 110:54
56. Christian, C. A., Schommer, R. A. 1982. *Ap. J. Suppl.* 49:405
57. Christian, C. A., Tully, R. B. 1983. *Astron. J.* 88:934
58. Chun, M. S. 1978. *Astron. J.* 83:1062
59. Code, A. D., Welch, G. 1982. *Ap. J.* 256:1
60. Cohen, J. 1981. See Ref. 275, pp. 229–34
61. Coleman, G., Wu, C.-C., Weedman, D. 1981. *Ap. J. Suppl.* 43:393
62. Colin, J., Athanassoula, E. 1981. *Astron. Astrophys.* 97:63
63. Colin, J., Athanassoula, E. 1983. See Ref. 13, pp. 239–40
64. Comins, N. F. 1983. *Ap. J.* 266:543
65. Cottrell, G. A. 1978. *MNRAS* 184:259
66. Coupinot, G., Hecquet, J., Heidmann, J. 1982. *MNRAS* 199:451
67. Crutcher, R. M., Rogstad, O. H., Chu, K. 1978. *Ap. J.* 225:784
68. Davis, M., Lecar, M., Pryor, C., Witten, E. 1981. *Ap. J.* 250:423
69. Dennefeld, M., Tammann, G. 1980. *Astron. Astrophys.* 83:275
70. de Vaucouleurs, G. 1955. *Astron. J.* 60:126
71. de Vaucouleurs, G. 1959. *Ap. J.* 131:265
72. de Vaucouleurs, G. 1959. *Handb. Phys.* 53:275
73. de Vaucouleurs, G. 1960. *Ap. J.* 131:574
74. de Vaucouleurs, G. 1961. *Ap. J.* 133:405
75. de Vaucouleurs, G. 1964. See Ref. 207, p. 269
76. de Vaucouleurs, G., de Vaucouleurs, A. 1959. *Lowell Obs. Bull. No. 92*
77. de Vaucouleurs, G., de Vaucouleurs, A. 1972. *MNRAS* 77:1
78. de Vaucouleurs, G., Freeman, K. 1972. *Vistas Astron.* 14:163
79. de Vaucouleurs, G., Moss, C. 1983. *Ap. J.* 271:123
80. de Vaucouleurs, G., Pence, W. D. 1980. *Ap. J.* 242:18
81. de Vaucouleurs, G., de Vaucouleurs, A., Buta, R. 1981. *Astron. J.* 86:1429
82. de Vaucouleurs, G., de Vaucouleurs, A., Buta, R. 1983. *Astron. J.* 88:764
83. de Vaucouleurs, G., de Vaucouleurs, A., Corwin, H. G. 1976. *Second Reference Catalog of Bright Galaxies.* Austin: Univ. Texas Press
84. de Vaucouleurs, G., de Vaucouleurs, A., Pence, W. 1974. *Ap. J. Lett.* 194:L119
85. Dicke, R. H., Peebles, P. J. E. 1968. *Ap. J.* 154:891
86. Donas, J., Deharveng, J. M. 1984. Preprint
87. Dostal, V. A. 1979. *Sov. Astron. AJ* 23:135
88. Downes, R., Margon, B. 1981. *Astron. J.* 86:19
89. Dufour, R. J., Harlow, W. V. 1977. *Ap. J.* 216:706
90. Durisen, R. H., Tohline, J. E., Burns, J. A., Dobrovoloskis, A. R. 1983. *Ap. J.* 264:392
91. Efremov, Yu. 1982. *Sov. Astron. Lett.* 8:357
92. Efremov, Yu., Pavlovskaya, E. 1982. *Sov. Astron. Lett.* 8(1):4
93. Efstathiou, G., Lake, G., Negroponte, J. 1982. *MNRAS* 199:1069
94. Einasto, J. 1972. *Astrophys. Lett.* 11:195
95. Elmegreen, B., Elmegreen, D., Morris, M. 1980. *Ap. J.* 240:455
96. Elmegreen, B. G. 1982. *Ap. J.* 253:655
97. Elmegreen, D. M. 1980. *Ap. J. Suppl.* 43:37
98. Elmegreen, D., Elmegreen, B. 1980. *Astron. J.* 85:1325
99. Fabbiano, G., Feigelson, E., Zamorani, G. 1982. *Ap. J.* 256:397
100. Fabbiano, G., Panagia, N. 1983. *Ap. J.* 266:568
101. Faber, S. M. 1982. See Ref. 42, pp. 191–217
102. Faber, S. M., Gallagher, J. S. 1979. *Ann. Rev. Astron. Astrophys.* 17:135
103. Faber, S. M., Lin, D. N. C. 1983. *Ap. J. Lett.* 266:L17
104. Fairall, A. P. 1980. *MNRAS* 191:391
105. Feitzinger, J. V. 1980. *Space Sci. Rev.* 27:35 (In German)
106. Feitzinger, J. V. 1983. See Ref. 13, p. 241
107. Feitzinger, J. V., Schlosser, W., Schmidt-Kaler, Th., Winkler, Chr. 1980. *Astron. Astrophys.* 84:50
108. Feitzinger, J. V., Glassgold, A. E., Gerola, H., Seiden, P. 1981. *Astron. Astrophys.* 98:371
109. Fisher, J. R., Tully, R. B. 1975. *Astron. Astrophys.* 44:151
110. Fisher, J. R., Tully, R. B. 1981. *Ap. J. Suppl.* 47:139
111. Fitzpatrick, E. L., Savage, B. D. 1983. *Ap. J.* 267:93
112. Flower, P. 1982. *Publ. Astron. Soc. Pac.* 94:894
113. Flower, P. 1983. *Publ. Astron. Soc. Pac.* 95:122
114. Ford, H. C. 1970. *The kinematics and stellar content of populous clusters in the Magellanic Clouds.* PhD thesis. Univ. Wis., Madison. 148 pp.
115. Fransson, C., Epstein, R. 1982. *MNRAS* 198:1127
116. Freedman, W. L., Madore, B. S. 1983. *Ap. J.* 265:140

117. Freeman, K. 1970. *Ap. J.* 160:811
118. Freeman, K. C. 1980. In *Star Clusters, IAU Symp. No. 85*, ed. J. E. Hesser, pp. 317–23. Dordrecht: Reidel
119. Freeman, K. C., Illingworth, G., Oemler, A. 1983. *Ap. J.* 272:488
120. French, H. 1980. *Ap. J.* 240:41
121. French, H., Miller, J. 1981. *Ap. J.* 248:468
122. Frogel, J., Blanco, V. 1983. *Ap. J. Lett.* 274:L57
123. Frogel, J. 1984. In preparation
124. Gallagher, J., Hunter, D. 1983. *Ap. J.* 274:141
125. Gallagher, J., Hunter, D., Knapp, G. 1981. *Astron. J.* 86:344
126. Gallagher, J., Hunter, D., Tutukov, A. 1984. *Ap. J.* In press
127. Gascoigne, S. C. B. 1980. In *Star Clusters, IAU Symp. No. 85*, ed. J. E. Hesser, pp. 305–15. Dordrecht: Reidel
128. Gascoigne, S., Bessell, M., Norris, J. 1981. See Ref. 275, pp. 223–28
129. Gerola, H., Carnevali, P., Salpeter, E. E. 1983. *Ap. J. Lett.* 268:L75
130. Gerola, H., Seiden, P. 1978. *Ap. J.* 223:129
131. Gerola, H., Seiden, P., Schulman, L. 1980. *Ap. J.* 242:517
132. Gillett, F. C., Kleinmann, D. E., Wright, E. L., Capps, R. W. 1975. *Ap. J. Lett.* 198:L65
133. Goad, J. W., De Veny, J. B., Goad, L. E. 1979. *Ap. J. Suppl.* 39:439
134. Goad, J. W., Roberts, M. S. 1981. *Ap. J.* 250:79
135. Gordon, D., Gottesman, S. T. 1981. *Astron. J.* 86:161
135a. Gordon, M. A., Heidmann, J., Epstein, E. E. 1982. *Publ. Astron. Soc. Pac.* 94:415
136. Gottesman, S., Weliachew, L. 1977. *Ap. J.* 211:47
137. Gottesman, S., Weliachew, L. 1977. *Astron. Astrophys.* 61:523
138. Graham, J. A. 1975. *Publ. Astron. Soc. Pac.* 87:641
139. Graham, J. A. 1977. *Publ. Astron. Soc. Pac.* 89:425
140. Graham, J. A. 1982. *Ap. J.* 252:474
141. Graham, J. A., Lawrie, D. G. 1982. *Ap. J. Lett.* 253:L73
142. Guibert, J. 1979. In *Stars and Star Systems*, p. 85. Boston: Reidel
143. Gunn, J. 1982. See Ref. 42, pp. 233–62
143a. Güsten, R., Mezger, P. 1982. *Vistas Astron.* 26:159
144. Hardy, E. 1977. *Ap. J.* 211:718
145. Hardy, E. 1978. *Publ. Astron. Soc. Pac.* 90:132
146. Hartwick, F. D. A. 1980. *Ap. J.* 236:754
147. Hawkins, M. R. S., Bruck, M. T. 1982. *MNRAS* 198:935
148. Heckman, T. 1974. *Astron. J.* 79:1040
148a. Heeschen, D. S., Heidmann, J., Yin, Q. F. 1983. *Ap. J. Lett.* 267:L73
149. Heidmann, J. 1979. *Ann. Phys. Fr.* 4:205
150. Heidmann, J. 1983. In *Highlights of Astronomy, IAU Gen. Assem., Patras,* 6:611
151. Heidmann, J., Klein, U., Wielebinski, R. 1982. *Astron. Astrophys.* 105:188
152. Heiles, C. 1979. *Ap. J.* 229:533
153. Herschel, J. F. 1846. In *Results of Astronomical Observations Made During the Years 1834–38 at the Cape of Good Hope*, pp. 137–43. London: Smith, Elder & Co.
154. Hindman, J. V., Balnaves, K. V. 1967. *Aust. J. Phys. Astrophys. Suppl. No. 4*
155. Hitchcock, J. L., Hodge, P. W. 1968. *Ap. J.* 152:1067
156. Hodge, P. 1961. *Ap. J.* 133:413
157. Hodge, P. 1966. *Ap. J.* 146:593
158. Hodge, P. 1967. *Ap. J.* 148:719
159. Hodge, P. 1969. *Ap. J.* 156:847
160. Hodge, P. 1971. *Ann. Rev. Astron. Astrophys.* 9:35
161. Hodge, P. 1972. *Publ. Astron. Soc. Pac.* 84:365
162. Hodge, P. 1973. *Astron. J.* 78:807
163. Hodge, P. 1974. *Publ. Astron. Soc. Pac.* 86:263
164. Hodge, P. 1975. *Ap. J.* 201:556
165. Hodge, P. 1977. *Ap. J. Suppl.* 33:69
166. Hodge, P. 1978. *Ap. J. Suppl.* 37:145
167. Hodge, P. 1980. *Ap. J.* 241:125
168. Hodge, P. 1981. See Ref. 275, pp. 205–14
169. Hodge, P. 1983. *Astron. J.* 88:1323
170. Hodge, P. 1983. *Ap. J.* 264:470
171. Hodge, P., Lucke, P. B. 1970. *Astron. J.* 75:933
172. Hoessel, J., Danielson, G. 1983. *Ap. J.* 271:65
173. Hoessel, J., Mould, J. 1982. *Ap. J.* 254:38
174. Holmberg, E. 1958. *Medd. Lunds Astron. Obs., Ser. 2, No. 136*
175. Hubble, E. 1922. *Ap. J.* 56:162
176. Hubble, E. 1926. *Ap. J.* 64:321
177. Huchra, J. 1977. *Ap. J. Suppl.* 35:171
178. Huchra, J. 1977. *Ap. J.* 217:928
179. Huchra, J. P., Geller, M., Willner, S., Hunter, D., Gallagher, J. 1982. In *Advances in Ultraviolet Astronomy: Four Years of IUE Research, NASA Conf. Publ. 2238*, ed. Y. Kondo, J. M. Meade, R. D. Chapman, pp. 151–55
180. Huchra, J., Geller, M., Gallagher, J., Hunter, D., Hartmann, L., et al. 1983. *Ap. J.* 274:125
181. Huchtmeier, W. K. 1979. *Astron. Astrophys.* 75:170

182. Huchtmeier, W. K., Seiradarkis, J. H., Materne, J. 1980. *Astron. Astrophys.* 91:341
183. Huchtmeier, W. K., Seiradarkis, J. H., Materne, J. 1981. *Astron. Astrophys.* 102:134
184. Humphreys, R. M. 1979. *Ap. J. Suppl.* 39:389
185. Humphreys, R. M. 1980. *Ap. J.* 238:65
186. Humphreys, R. M. 1983. *Ap. J.* 269:335
187. Humphreys, R. M. 1983. *Ap. J.* 265:176
188. Hunter, D. 1982. *Ap. J.* 260:81
189. Hunter, D. 1982. *Global and local properties of non-interacting irregular galaxies.* PhD thesis. Univ. Ill., Urbana. 271 pp.
190. Hunter, D. 1984. *Ap. J. Lett.* In press
191. Hunter, D., Gallagher, J., Rautenkranz, D. 1982. *Ap. J. Suppl.* 49:53
192. Iben, I. Jr. 1974. *Ann. Rev. Astron. Astrophys.* 12:215
193. Iben, I. Jr., Renzini, A. 1983. *Ann. Rev. Astron. Astrophys.* 21:271
194. Ikeuchi, S. 1981. *Publ. Astron. Soc. Jpn.* 33:211
195. Israel, F. 1980. *Astron. Astrophys.* 90:246
196. Israel, F. P., Koornneef, J. 1979. *Ap. J.* 230:390
197. Israel, F. P., de Graauw, T., Lidholm, S., van de Stadt, H., de Vries, C. 1982. *Ap. J.* 262:100
198. Isserstedt, J., Reinhardt, M. 1976. *MNRAS* 176:693
199. Johnson, H. M. 1961. *Publ. Astron. Soc. Pac.* 73:20
200. Jura, M., Hobbs, R. W., Maran, S. P. 1978. *Astron. J.* 83:153
201. Karachentsev, I. E. 1981. *Sov. Astron. Lett.* 7:1
202. Kayser, S. 1967. *Astron. J.* 72:134
203. Kennicutt, R. 1979. *Ap. J.* 228:394
204. Kennicutt, R. 1979. *Ap. J.* 228:696
205. Kennicutt, R. 1983. *Ap. J.* 272:54
206. Kennicutt, R. C., Kent, S. M. 1983. *Astron. J.* 88:1094
207. Kerr, F. J., Rodgers, A. W., eds. 1964. *The Galaxy and the Magellanic Clouds, IAU Symp. No. 20.* Canberra: Aust. Natl. Acad. Sci.
208. Kinman, T. 1984. In *Astronomy with Schmidt-Type Telescopes, IAU Colloq. No. 78.* In press
209. Kinman, T., Davidson, K. 1981. *Ap. J.* 243:127
210. Kinman, T. D., Davidson, K. 1982. *Philos. Trans. R. Soc. London Ser. A* 307:37
211. Kinman, T., Green, J., Mahaffey, C. 1979. *Publ. Astron. Soc. Pac.* 91:749
212. Knapp, G. R. 1983. In *Kinematics,* *Dynamics, and Structure of the Milky Way, IAU Symp. No. 101,* ed. W. L. H. Shuter, p. 233. Dordrecht: Reidel
213. Kraan-Korteweg, R., Tammann, G. 1979. *Astron. Nachr.* 300:181
214. Krienke, O. K., Hodge, P. W. 1974. *Astron. J.* 79:1242
215. Kunth, D., Sargent, W. L. W. 1979. *Astron. Astrophys. Suppl.* 36:259
216. Kunth, D., Sargent, W. L. W. 1981. *Astron. Astrophys.* 101:L5
216a. Kunth, D., Sargent, W. L. W. 1983. *Ap. J.* 273:81
217. Lada, C., Blitz, L., Elmegreen, B. 1978. In *Protostars and Planets,* p. 341. Tucson: Univ. Ariz. Press
218. Lake, G., Schommer, R. A. 1984. Preprint
219. Lamb, S., Gallagher, J., Hjellming, M., Hunter, D. 1984. Preprint
220. Larson, R., Tinsley, B. 1978. *Ap. J.* 219:46
221. Larson, R. B., Tinsley, B. M., Caldwell, C. N. 1980. *Ap. J.* 237:692
222. Lequeux, J. 1979. *Rev. Mex. Astron. Astrofis.* 4:325
223. Lequeux, J. 1979. *Astron. Astrophys.* 71:1
224. Lequeux, J. 1979. *Astron. Astrophys.* 80:35
225. Lequeux, J., Maucherat-Joubert, M., Deharveng, J., Kunth, D. 1981. *Astron. Astrophys.* 103:305
226. Lequeux, J., Peimbert, M., Rayo, J., Serrano, A., Torres-Peimbert, S. 1979. *Astron. Astrophys.* 80:155
227. Lequeux, J., Viallefond, F. 1980. *Astron. Astrophys.* 91:269
228. Lequeux, J., West, R. M. 1981. *Astron. Astrophys.* 103:319
229. Lin, D. N. C., Faber, S. 1983. *Ap. J. Lett.* 266:L21
230. Lo, K. Y., Sargent, W. L. W. 1979. *Ap. J.* 227:756
231. Longmore, A., Hawarden, T., Goss, W., Mebold, U., Webster, B. 1982. *MNRAS* 200:325
232. Longmore, A., Hawarden, T., Webster, B., Goss, W., Mebold, U. 1978. *MNRAS* 184:97p
233. Lundmark, K. 1926. *Uppsala Obs. Medd. No. 7*
234. Madore, B. 1982. *Ap. J.* 253:575
235. Madore, B., van den Bergh, S., Rogstad, D. 1974. *Ap. J.* 191:317
236. Marcelin, M., Athanassoula, E. 1982. *Astron. Astrophys.* 105:76
237. Massey, P., Hutchings, J. B. 1983. *Ap. J.* 275:576
238. Matteucci, F., Chiosi, C. 1983. *Astron. Astrophys.* 123:121
239. Maucherat-Joubert, M., Lequeux, J.,

Rocca-Volmerange, B. 1980. *Astron. Astrophys.* 86:299
240. McAlary, C. M., Madore, B. F., Davis, L. E. 1984. Preprint
241. McGee, R. X., Milton, J. A. 1966. *Aust. J. Phys.* 19:343
242. McGee, R. X., Newton, L. M. 1981. *Proc. Astron. Soc. Aust.* 4:189
243. Meaburn, J. 1980. *MNRAS* 192:365
244. Mezger, P. G., Smith, L. F. 1977. In *Star Formation, IAU Symp. No. 75*, ed. T. de Jong, A. Maeder, pp. 133–69. Dordrecht: Reidel
245. Miller, G., Scalo, J. 1979. *Ap. J. Suppl.* 41:513
245a. Moffat, A. F. J., Seggewiss, W. 1983. *Astron. Astrophys.* 125:83
246. Monnet, G. 1971. *Astron. Astrophys.* 12:379
247. Morgan, W. W. 1958. *Publ. Astron. Soc. Pac.* 70:364
248. Morgan, W. W., Osterbrock, D. E. 1969. *Astron. J.* 74:515
249. Mould, J., Aaronson, M. 1982. *Ap. J.* 263:629
250. Mouschovias, T. Ch. 1981. In *Fundamental Problems in the Theory of Stellar Evolution, IAU Symp. No. 93*, ed. D. Q. Lamb, D. N. Schramm, pp. 27–62. Dordrecht: Reidel
251. Mouschovias, T. Ch., Shu, F. H., Woodward, P. R. 1974. *Astron. Astrophys.* 33:73
252. Mueller, M., Arnett, D. 1976. *Ap. J.* 210:670
253. Nail, V., Shapley, H. 1953. *Proc. Natl. Acad. Sci. USA* 39:358
254. Nelson, M., Hodge, P. 1983. *Publ. Astron. Soc. Pac.* 95:5
254a. Nilson, P. 1973. *Uppsala General Catalogue of Galaxies*. Uppsala: Uppsala Offset Cent.
255. Norman, C. 1983. See Ref. 13, p. 163
256. Öpik, E. J. 1953. *Irish Astron. J.* 2:219
257. O'Connell, R., Mangano, J. 1978. *Ap. J.* 221:62
258. O'Connell, R. W., Thuan, T. X., Goldstein, S. J. 1978. *Ap. J. Lett.* 226:L11
259. Olive, K. A., Schramm, D. H., Steigman, G., Turner, M. S., Yang, J. 1981. *Ap. J.* 246:557
260. Ostriker, J. P., Cowie, L. L. 1981. *Ap. J. Lett.* 243:L127
261. Ostriker, J. P., Ikeuchi, S. 1983. *Ap. J. Lett.* 268:L63
262. Ostriker, J. P., Peebles, P. J. E. 1973. *Ap. J.* 186:467
263. Page, T., Carruthers, G. R. 1981. *Ap. J.* 248:906
264. Page, T., Carruthers, G. R., Hill, R. 1978. *S-201 Catalog of Far Ultraviolet Objects, NRL Rep. 8173*
265. Pagel, B. E. J. 1982. *Proc. R. Soc. London Ser. A* 307:19
266. Pagel, B., Edmunds, M., Blackwell, D., Chun, M., Smith, G. 1979. *MNRAS* 189:95
267. Pagel, B., Edmunds, M., Smith, G. 1980. *MNRAS* 193:219
268. Pagel, B. E. J., Edmunds, M. G. 1981. *Ann. Rev. Astron. Astrophys.* 19:77
269. Payne-Gaposhkin, C. 1974. *Smithsonian Contrib. Astrophys. No. 16*
270. Peimbert, M., Spinrad, H. 1970. *Astron. Astrophys.* 7:311
271. Peimbert, M., Torres-Peimbert, S. 1974. *Ap. J.* 193:327
272. Persson, S., Aaronson, M., Cohen, J., Frogel, J., Matthews, K. 1983. *Ap. J.* 266:105
273. Petrosyan, A. R. 1981. *Astrofizika* 17:421
274. Petrosyan, A. R., Saakyan, K., Khachikyan, E. 1979. *Astrofizika* 15:209
275. Philip, A. G. D., Hayes, D. S., eds. 1981. *Astrophysical Parameters for Globular Clusters, IAU Colloq. No. 68*. Schenectady, N.Y: L. Davis
276. Primack, J., Blumenthal, G. 1984. Preprint
277. Rabin, D. 1982. *Ap. J.* 261:85
278. Reaves, G. 1983. *Ap. J. Suppl.* 53:375
279. Renzini, A. 1981. *Ann. Phys. Fr.* 6:87
280. Reynolds, R., Roesler, F., Scherb, F. 1977. *Ap. J.* 211:115
281. Rieke, G. H., Lebofsky, M. J., Thompson, R. I., Low, F. J., Tokunaga, A. T. 1980. *Ap. J.* 238:24
282. Rieke, G. H., Lebofsky, M. J. 1979. *Ann. Rev. Astron. Astrophys.* 17:477
283. Rocca-Volmerange, B., Lequeux, J., Maucherat-Joubert, M. 1981. *Astron. Astrophys.* 104:177
283a. Rocca-Volmerange, B., Prevot, L., Ferlet, R., Lequeux, J., Prevot-Burnichon, M. L. 1981. *Astron. Astrophys.* 99:L5
284. Rood, R. T., Iben, I. Jr. 1968. *Ap. J.* 154:215
285. Rosa, M., Solf, J. 1984. Preprint
286. Rots, A. 1980. *Astron. Astrophys.* 41:189
287. Rubin, V. C., Ford, W. C. Jr., Thonnard, N. 1980. *Ap. J.* 238:471
288. Ruotsalainen, R. W. 1982. *Multicolor photometry of stars in dwarf irregular galaxies.* PhD thesis. Univ. Hawaii, Honolulu. 207 pp.
289. Salpeter, E. E. 1959. *Ap. J.* 129:608
290. Sandage, A. 1961. *Hubble Atlas of Galaxies, Carnegie Inst. Washington Publ. 618*
291. Sandage, A. 1971. *Ap. J.* 166:13
292. Sandage, A. R. 1975. In *Galaxies and the*

Universe, ed. A. Sandage, M. Sandage, J. Kristian, pp. 1–35. Chicago: Univ. Chicago Press

293. Sandage, A. R., Binggeli, B., Tarenghi, M. 1982. *Carnegie Inst. Washington Yearb.* 81:623

294. Sandage, A., Brucato, R. 1979. *Astron. J.* 84:472

295. Sandage, A., Carlson, G. 1982. *Ap. J.* 258:439

296. Sandage, A., Katem, B. 1976. *Astron. J.* 81:743

297. Sandage, A., Tammann, G. 1974. *Ap. J.* 190:525

298. Sandage, A., Tammann, G. 1974. *Ap. J.* 194:223

299. Sandage, A., Tammann, G. 1974. *Ap. J.* 194:559

300. Sandage, A., Tammann, G. 1981. *Revised Shapley-Ames Catalog of Bright Galaxies, Carnegie Inst. Washington Publ. 635.* 157 pp.

301. Sandage, A., Tammann, G. 1982. See Ref. 42, pp. 23–84

302. Sargent, W. L. W. 1970. *Ap. J.* 160:405

303. Sargent, W. L. W., Sancisi, R., Lo, K. V. 1983. *Ap. J.* 265:711

304. Sargent, W. L. W., Searle, L. 1970. *Ap. J. Lett.* 162:L155

305. Savage, B. D., Fitzpatrick, E. L., Cassinelli, J. P., Ebbets, D. B. 1983. *Ap. J.* 273:597

306. Savage, B. D., Mathis, J. S. 1979. *Ann. Rev. Astron. Astrophys.* 17:73

307. Scalo, J. 1984. *Fundam. Cosmic Phys.* In preparation

308. Schmidt, M. 1959. *Ap. J.* 129:243

309. Schmidt, Th. 1976. *Astron. Astrophys. Suppl.* 24:357

310. Schweizer, F. 1976. *Ap. J. Suppl.* 31:313

311. Schweizer, F. 1982. *Ap. J.* 252:455

312. Searle, L., Sargent, W. 1972. *Ap. J.* 173:25

313. Searle, L., Sargent, W., Bagnuolo, W. 1973. *Ap. J.* 179:427

314. Searle, L., Wilkinson, A., Bagnuolo, W. G. 1980. *Ap. J.* 239:803

315. Seiden, P., Gerola, H. 1979. *Ap. J.* 233:56

316. Seiden, P. E., Gerola, H. 1982. *Fundam. Cosmic Phys.* 7:241

317. Seiden, P. E., Schulman, L. S., Feitzinger, J. V. 1982. *Ap. J.* 253:91

318. Sellwood, J. A. 1983. See Ref. 13, pp. 197–202

319. Sérsic, J. L. 1964. See Ref. 207, pp. 208–11

320. Sérsic, J. L., Bajaja, E., Colomb, R. 1977. *Astron. Astrophys.* 59:19

321. Sérsic, J. L., Pastoriza, M. G., Carranza, G. J. 1968. *Astrophys. Lett.* 2:45

322. Sérsic, J. L., Calderón, J. H. 1979. *Astrophys. Space Sci.* 62:211

323. Shapovalova, A. I. 1971. *Vestn. Kiev Univ. Ser. Astron. No. 13*, p. 104

324. Shore, S. N. 1981. *Ap. J.* 249:93

325. Shore, S. N. 1983. *Ap. J.* 265:202

326. Shostak, G., van Woerden, H. 1983. See Ref. 13, pp. 33–34

327. Silk, J. 1983. *Nature* 301:574

328. Silk, J., Norman, C. 1981. *Ap. J.* 247:59

329. Smith, H. E. 1975. *Ap. J.* 199:591

330. Smith, L., Biermann, P., Mezger, P. 1978. *Astron. Astrophys.* 66:65

331. Smith, M. G., Weedman, D. W. 1973. *Ap. J.* 179:461

332. Solinger, A., Morrison, P., Markert, T. 1977. *Ap. J.* 211:707

333. Stewart, G. C., Fabian, A. C., Terlevich, R. J., Hazard, C. 1982. *MNRAS* 200:61p

334. Strom, S. 1980. *Ap. J.* 237:686

335. Struck-Marcell, C., Tinsley, B. 1978. *Ap. J.* 221:562

336. Stryker, L. L. 1983. *Ap. J.* 266:82

337. Stryker, L., Butcher, H. 1982. See Ref. 275, pp. 255–60

338. Stryker, L. L., Butcher, H. R., Jewell, J. L. 1982. See Ref. 275, pp. 267–71

339. Talbot, R. J., Arnett, W. D. 1971. *Ap. J.* 170:409

340. Talent, D. L. 1980. *A spectrophotometric study of H II regions in chemically young galaxies.* PhD thesis. Rice Univ., Houston, Tex. 161 pp.

341. Tammann, G. 1980. See Ref. 343, p. 3

342. Tamura, S., Hasegawa, M. 1979. *Publ. Astron. Soc. Jpn.* 31:329

343. Tarenghi, M., Kjar, K., eds. 1980. *Dwarf Galaxies, ESO/ESA Rep.*

344. Tayler, R. J. 1976. *MNRAS* 177:39

345. Telesco, C., Harper, D. A. 1980. *Ap. J.* 235:392

346. Tenorio-Tagle, G., Yorke, H., Bodenheimer, P. 1982. In *Mechanismes de Production d'Energie dans le Milieu Interstellaire*, ed. J. P. Sivan, pp. 181–99. Marseille: Lab. Astron. Spat. CNRS

347. Terlevich, R., Melnick, J. 1981. *MNRAS* 195:839

348. Terlevich, R., Melnick, J. 1984. *MNRAS.* Submitted for publication (ESO Preprint No. 264)

349. Thuan, T. X. 1983. *Ap. J.* 268:667

350. Thuan, T. X., Martin, G. E. 1981. *Ap. J.* 247:823

351. Thuan, T. X., Seitzer, P. 1979. *Ap. J.* 232:680

352. Thuan, T. X., Seitzer, P. 1979. *Ap. J.* 231:327

353. Tinsley, B. M. 1980. *Fundam. Cosmic Phys.* 5:287

354. Tinsley, B. M. 1981. *MNRAS* 194:63

355. Tinsley, B. M., Larson, R. B., eds. 1977. *The Evolution of Galaxies and Stellar*

Populations. New Haven, Conn: Yale Univ. Obs.
356. Toomre, A. 1964. *Ap. J.* 139:1217
357. Toomre, A. 1977. *Ann. Rev. Astron. Astrophys.* 15:437
358. Toomre, A. 1977. See Ref. 355, pp. 401–16
359. Toomre, A., Toomre, J. 1972. *Ap. J.* 178:628
360. Tremaine, S., Gunn, J. E. 1979. *Phys. Rev. Lett.* 42:407
360a. Tully, R. B., Bottinelli, L., Fisher, J. R., Gouguenheim, L., Sancisi, R., van Woerden, H. 1978. *Astron. Astrophys.* 63:37
361. Tully, R., Boesgaard, A., Dyck, H., Schempp, W. 1981. *Ap. J.* 246:38
362. Tully, R. B. 1982. *Ap. J.* 257:389
363. Tully, R. B., Christian, C. A. 1983. *Astron. J.* 88:934
364. Twarog, B. 1980. *Ap. J.* 242:242
365. van den Bergh, S. 1959. *Publ. Dom. Astrophys. Obs., Victoria, BC* 2:147
366. van den Bergh, S. 1966. *Astron. J.* 71:922
367. van den Bergh, S. 1972. *J. R. Astron. Soc. Can.* 66:237
368. van den Bergh, S. 1977. See Ref. 355, pp. 19–37
369. van den Bergh, S. 1979. *Ap. J.* 230:95
370. van den Bergh, S. 1980. *Publ. Astron. Soc. Pac.* 92:122
370a. van den Bergh, S. 1981. *Astron. Astrophys. Suppl.* 46:79
371. van den Bergh, S. 1982. *Astron. J.* 87:987
372. van den Bergh, S., Humphreys, R. M. 1979. *Astron. J.* 84:604
373. van der Hulst, J. M. 1979. *Astron. Astrophys.* 75:97
374. van der Kruit, P. C., Allen, R. J. 1978. *Ann. Rev. Astron. Astrophys.* 16:103
375. van der Kruit, P. C., Shostak, G. S.

1983. See Ref. 13, pp. 69–76
376. van Woerden, H., Bosma, A., Mebold, U. 1975. In *La Dynamique des Galaxies Spirales*, p. 483. Paris: CNRS
377. Viallefond, F., Thuan, T. X. 1983. *Ap. J.* 269:444
378. Vorontsov-Velyaminov, V. V. 1959. *Atlas and Catalog of Interacting Galaxies.* Moscow: Moscow Univ.
379. Vorontsov-Velyaminov, V. V. 1977. *Astron. Astrophys. Suppl.* 28:1
380. Webster, B. L., Goss, W. M., Hawarden, T. G., Longmore, A. J., Mebold, U. 1979. *MNRAS* 186:31
381. Webster, B. L., Smith, M. G. 1983. *MNRAS* 204:743
382. Werner, M., Becklin, E., Gatley, I., Ellis, M., Hyland, A., et al. 1978. *MNRAS* 184:365
383. Westerlund, B., Mathewson, D. 1966. *MNRAS* 131:371
384. White, S. D. M. 1982. In *Morphology and Dynamics of Galaxies, 12th Saas Fee Course*, ed. L. Martinet, M. Mayor, p. 293. Sauverny: Geneva Obs.
385. Wirth, A., Gallagher, J. S. 1980. *Ap. J.* 242:469
386. Wirth, A., Gallagher, J. S. 1984. *Ap. J.* In press
387. Wood, P. R., Bessell, M. S., Fox, M. W. 1983. *Ap. J.* 272:99
388. Wray, J. D., de Vaucouleurs, G. 1980. *Astron. J.* 85:1
389. Yang, J., Turner, M. S., Steigman, G., Schramm, D. N., Olive, K. A. 1984. *Ap. J.* In press
390. Young, J., Gallagher, J., Hunter, D. 1984. *Ap. J.* 276:476
391. Young, J. S., Scoville, N. 1982. *Ap. J. Lett.* 260:L11
392. Zwicky, F. 1957. *Morphological Astronomy.* Berlin: Springer-Verlag
393. Zwicky, F. 1959. *Handb. Phys.* 53:373

Ann. Rev. Astron. Astrophys. 1984. 22:75–95

OBSERVATIONS OF SUPERNOVA REMNANTS[1]

John C. Raymond

Harvard-Smithsonian Center for Astrophysics, Cambridge,
Massachusetts 02138

INTRODUCTION

Since the reviews of supernova remnant observations by Woltjer (200),
Milne (143), and Gorenstein & Tucker (88), the sensitivity and resolution of
observations at all wavelengths have greatly improved. Some of the most
recent work is available in the proceedings of IAU Symposium 101 (J.
Danziger & P. Gorenstein, eds.) and the NATO Advanced Study Institute
on Supernovae (M. Rees & R. Stoneham, eds.). Excellent reviews on the
interaction of supernovae with the interstellar medium (ISM) [Chevalier
(30)], the theory of interstellar shock waves [McKee & Hollenbach (137)],
and the Crab-like supernova remnants [Weiler (196), Wilson (198)] have
appeared recently. Holt (104) has reviewed the X-ray spectra of supernova
remnants. A catalogue of known remnants is given by van den Bergh (193a).
Here, I avoid duplicating the discussions in these papers as much as is
consistent with clarity.

The basic character of a supernova remnant (SNR) is determined by the
supernova type (I or II), the presence or absence of a pulsar to provide
power, and the age of the remnant. The discussion of galactic SNRs is
organized around these features, and extragalactic remnants are treated
separately. I discuss the prototype of each class in detail and compare the
other members of the class with it, though of course many remnants do not
fit neatly into a single class.

[1] The US Government has the right to retain a nonexclusive royalty-free license in and to
any copyright covering this paper.

75

YOUNG REMNANTS

Metal-Rich SNRs: Cas A

Cas A is the youngest known galactic SNR. It is exceptionally bright at X-ray and radio wavelengths, but its high reddening makes it faint in the optical and invisible in the ultraviolet. At an estimated distance of 3 kpc, its radius is 1.6 pc. It is likely that Flamsteed observed the supernova in 1680 [Ashworth (2)], but the lack of other reports of this supernova suggests an unusually low brightness [Chevalier (29)]. Though the SN type cannot be deduced from seventeenth century observations, the large progenitor mass inferred from optical and X-ray studies has led to the assumption of Type II.

The optical emission of Cas A is dominated by roughly a hundred small emission-line knots that fall into two classes. The fast-moving knots (FMKs) have large proper motions and Doppler shifts, with space motions around 6000 km s^{-1} [van den Bergh & Kamper (194)], while the quasi-stationary flocculi (QSFs) move at about 150 km s^{-1}. The proper motions extrapolate back to an explosion in the year 1658 ± 3, 22 years before the probable Flamsteed observation.

Brecher & Wasserman (21) have attributed the 22-year difference to deceleration of the SN ejecta as a whole. On the other hand, Bell (10) suggested that the FMKs are knots of material originally ejected at modest velocity (6000 km s^{-1} compared with 10,000 km s^{-1} typically observed near SN maximum). The low-density, higher velocity material is decelerated by the interstellar medium, and cold knots, which have been moving in the very low pressure free expansion region, encounter the high pressure in the decelerated ejecta. Shocks propagating into the knots give rise to the optical emission.

If we accept Bell's picture, the present velocity of an FMK (obtained through a Doppler shift and a proper motion) will be the difference between the ejection velocity and the velocity of the shock that illuminates the knot. The apparent 22-year difference between the expansion age and the real age implies that on the average the FMKs are illuminated by 400 km s^{-1} shocks.

The bizarre abundances of the fast-moving knots were discovered by Baade & Minkowski (4) and explored further by Searle (174), Peimbert (151), and Peimbert & van den Bergh (152). Hydrogen and helium are absent from the optical spectra. The most recent work [Kirshner & Chevalier (117), Chevalier & Kirshner (32, 33)] shows a pure oxygen FMK and knots having oxygen mixed with products of oxygen burning, sulfur, argon, and calcium. Neon was detected in one knot by Minkowski (144), but it has not been seen in other spectra. A line at 7376 Å that Kirshner & Chevalier (117) attributed to [Fe II] may be due to [Ni II] [Nussbaumer &

Storey (149)]. Electron densities around 3×10^4 cm^{-3} are inferred from the [S II] lines (32). The relative strengths of the [O I], [O II], and [O III] lines require that the shock temperatures be above 10^5 K. Itoh (109, 110) has constructed models of shocks in pure oxygen, and has found that a preshock density of 30 cm^{-3} and shock velocity of 141 km s^{-1} fit the emission-line strengths of the pure oxygen knot. The [Ar V] line detected in some knots would require faster shocks. Upper limits on the hydrogen and helium content are 2% and 42% by mass (32, 33), respectively, but further theoretical models are badly needed. If the optically observed knots reveal abundances typical of the progenitor, then the pre-SN star was around 20 $M_{\odot}$ judging from the ratio of silicon-group elements to oxygen. However, variations among the knots show that mixing was not complete, and the emitting mass of the observed knots is only about 10^{-4} $M_{\odot}$.

The QSFs differ from the fast knots not only in their much smaller velocities, but also in their abundances. Shock models for the appropriate compositions and densities are not available, but Chevalier & Kirshner (32) estimate that nitrogen is overabundant by a factor of 10, and helium may be similarly overabundant. The standard interpretation (152) is that the QSFs are knots of material ejected at low velocity before the supernova explosion whose densities are high enough that relatively slow shocks propagate into them. Their abundances are typical of planetary nebulae and WN stars. Chevalier & Kirshner (32) find preshock densities of ~ 30 cm^{-3} and shock velocities around 60 km s^{-1}.

Two weaker optical features are present. A faint, diffuse emission is strong in [S II] [van den Bergh (193)], and low surface brightness emission surrounds the nebula (152). The diffuse [S II] emission is at low velocity, and the line ratio indicates low density. This gas and the emission extending beyond the nebula may have been photoionized by the presupernova star or the supernova itself (152), or they could have been ionized by the nonthermal particles accelerated by the SNR [Stewart et al. (183)].

Cas A has been studied extensively at radio wavelengths. Its overall brightness is declining by about 1–2% per year, and its spectrum is growing flatter [Vinyaikin & Razin (195), Dickel & Griesen (64)]. Individual knots of high surface brightness appear or fade over a few years. There is a good overall correlation between the regions bright in radio and the bright optical nebulosity, but only one knot appears at both wavelengths. Bell (10) and Tuffs (188) found that the system of knots is expanding at about one third the expansion rate of the FMKs, but the expansion rate, if any, seems to be very uncertain (64).

The X-ray luminosity of Cas A is $\sim 10^{36}$ erg s^{-1}. X-ray images of Cas A show a clumpy shell of radius 102″ and thickness 17″ [Fabian et al. (81), Dickel et al. (65)] accounting for most of the emission in the *Einstein* energy band. Its mass is estimated at 20 $M_\odot$. The X-ray shell coincides quite well with both the radio and optical shells. The pressures derived from [S II] in the optical knots, the minimum pressure of the magnetic fields and non-thermal particles in the radio knots, and the surface brightness of the X-ray emission are also similar [$\sim 10^{-7}$ dyn cm^{-2} (10, 81)]. A second, fainter shell surrounds the first. Its average radius is 150″, and it is slightly thicker than the bright inner shell (81).

The natural interpretation of the remnant structure is that the outer shell is the shocked interstellar medium, while the inner shell is gas ejected by the supernova that has passed through a "reverse shock" [Gull (93, 94), McKee (135)]. The latter is brighter because it is denser (and cooler to maintain pressure equilibrium with the ISM), and because enhanced abundances of the silicon-group elements increase its emissivity in the *Einstein* band. It is likely that dust forms in Type II supernova ejecta, and this dust should emit strongly in the infrared when it passes through the reverse shock and heats up [Dwek & Werner (80)]. Dinerstein et al. (68, 69) have recently detected 10μ emission from two knots in Cas A.

Estimates of the temperature of the Cas A X-ray emission cover a wide range. Pravdo & Smith (157) report detection of X rays above 10 keV, which, if thermal, require an electron temperature of around 30 keV. Temperatures around 1 keV are needed to fit the lower energy spectra, but nonequilibrium ionization models remove much of the need for a low-temperature component (108, 178). The Kα and Kβ lines of highly ionized iron have been observed with proportional counters and gas scintillation proportional counters [Pravdo et al. (158), Manzo et al. (129)], and their equivalent widths are consistent with approximately cosmic abundances. The energy centroid of the 6.7-keV Kα feature implies that Fe XXV and Fe XXVI dominate the blend, rather than satellite lines due to inner shell transitions in Fe XVIII–XXIV (158). This is surprising in view of the nonequilibrium ionization models [Itoh (108), Hamilton et al. (96), Shull (178), Gronenschild & Mewe (92)] that suggest that Fe XIX–XX should be the dominant ionization stages. Since the age of the remnant is known, either higher densities or higher temperatures are required. *Einstein* SSS spectra reveal very strong K-lines of Si and S [Becker et al. (7)], which require overabundances of these elements by a factor of 3 or so relative to oxygen and lighter elements when compared with the nonequilibrium models. The silicon feature is dominated by Si XIII rather than Si XIV (7), which implies an underionization of the gas that is difficult to reconcile with the presence of Fe XXV. The Focal Plane Crystal Spectrometer on *Einstein* observed several lines at high spectral resolution [Markert et al. (130)]. Doppler shifts of opposite senses were found in the northern and southern parts of the remnant, suggesting a ring of ejecta expanding at a rate of at least 2000 km s^{-1}.

A decade ago Cas A seemed to be unique, but over the last few years five similar remnants have been discovered.

N132D   Danziger & Dennefeld (45, 46) reported high-velocity oxygen-rich knots in the Large Magellanic Cloud (LMC) remnant N132D. The knots are located within a region of diffuse normal abundance gas excited by slow shocks (as inferred from the relative line strengths), which is in turn located in an apparently normal H II region. Lasker (121, 122) found that the slow shock material fills a 11×16 pc disk, while the H II region is 80 pc in diameter. Lines of [Ne III], [O II], and [Ar III], as well as the dominant [O III] lines, were detected, with oxygen at least 100 times overabundant compared with hydrogen. The [O III] knots fall in a 6-pc-radius annulus expanding at 2250 km s^{-1}, implying an age of 1300 years. Clark et al. (39) give a soft X-ray luminosity of 8×10^{37} erg s^{-1}, four times that of any other LMC remnant. The three regions identified in the optical seem to be counterparts of the FMKs, QSFs, and extended H II region seen in Cas A. The main differences are the stronger Ne and weaker S lines in the N132D high-velocity material, which suggest less-advanced nuclear processing, and the lack of N or He enhancement in the low-velocity shocked material.

NGC 4449 supernova remnant   A compact radio source in the irregular galaxy NGC 4449 was found to show high-velocity emission in oxygen lines but no high-velocity Balmer line emission [Balick & Heckman (5)]. Subsequent observations [Blair et al. (18)] detected [O I], [O II], [O III], [Ne III], [S II], and [Ar III] emission with widths corresponding to expansion at 3500 km s^{-1}. The remnant is not only extremely luminous at optical and radio wavelengths, but its 10^{39} erg s^{-1} soft X-ray luminosity makes it the most luminous known SNR in X rays (18). Since the remnant is unresolved (55), little is known about its structure, but its age is estimated to be 120 years. This SNR is relatively free of reddening, and it is the only metal-rich SNR that has been detected in the UV. Blair et al. (19) measured an [O III] $\lambda1666$ luminosity within a factor of two of the predictions of Itoh's (109) models. The [C III] $\lambda1909$ line (which is the brightest UV line in normal abundance remnants) was not detected, a fact that implies a carbon-to-oxygen ratio less than one-third solar and suggests a massive progenitor.

G292.0 + 1.8   Goss et al. (90) discovered emission lines of O and Ne, with Hβ less than 2% as bright as [O III]. Line widths measured by Murdin & Clark (145) correspond to an expansion velocity of 2000 km s^{-1}. Tuohy et al. (189) have interpreted an *Einstein* high-resolution image in terms of an expanding ring of oxygen-rich material within a diffuse ellipsoidal shell. The SSS spectrum reveals line emission suggestive of a modest overabundance of sulfur [Clark et al. (40)].

0540−69.3   The *Einstein* X-ray survey of the LMC [Long et al. (126)] revealed this remnant. It is the third most luminous X-ray SNR in the LMC, and the SSS data fit a nonthermal spectrum, although a very hot oxygen-rich gas could also account for the lack of X-ray emission lines [Clark et al. (39)]. An [O III] image shows a 1.6-pc-diameter ring, and the [O III] velocity width of nearly 3000 km s^{-1} leads to an age estimate of 500 years [Mathewson et al. (132)].

1E0102.2−7219   Seward & Mitchell (176) discovered this luminous soft X-ray source in the SMC. Dopita et al. (76) found a corresponding ring of [O III] emission at least 60 times brighter than Hβ. A nonthermal radio source seems to be present, but the region is confused. Tuohy & Dopita (190) have constructed a velocity map of the [O III] emission and find a distorted ring structure expanding at 3250 km s^{-1}, giving an age of 1000 years. Detection of He II λ4686 emission from the surrounding H II region suggests ionization by the burst of UV from the supernova itself or by emission from the remnant.

Type I Remnants: Tycho's SNR

Roughly half of the galactic SNRs must be remnants of Type I supernovae, but the difference between the types is only apparent for very young remnants. Tycho's supernova remnant (SN 1592) is the prototype. Its X-ray and radio images show it to be very nearly spherical [Henbest (100), Seward et al. (175)], and faint optical filaments lie along the outer edge. Proper motions of both the optical filaments and the radio shell [Kamper & van den Bergh (114), Strom et al. (185)] are about 0.″23 per year. This value is smaller than the average proper motion needed to achieve the observed 220″ radius in the 392-year age of the remnant, implying substantial deceleration.

The optical emission is remarkable for its pure Balmer line spectrum [Kirshner & Chevalier (118)], which is interpreted as emission from neutral hydrogen swept up in the fast shock in the interstellar medium [Chevalier & Raymond (35); see review by McKee & Hollenbach (137)]. The surrounding medium must be largely neutral to produce this spectrum, and a Type II supernova or its precursor is likely to ionize the nearby gas, so a young remnant showing a pure Balmer line emission spectrum is assumed to be Type I. The Hα profiles provide a direct determination of the shock velocity of $\sim$2000 km s^{-1} [Chevalier et al. (34)], which together with the observed proper motion implies a distance of 2–2.6 kpc. Earlier distance estimates based on the radio surface brightness–diameter (Σ-D) relation or on 21-cm absorption profiles [Milne (143), Schwarz et al. (170)] were twice as large. Albinson & Gull (1) obtained a distance of 2–2.5 kpc based on spatially resolved 21-cm absorption profiles and a more detailed model of motions in the direction of the Perseus spiral arm. Absorption-line features at the velocities seen in the 21-cm profiles of Tycho are present in optical and UV spectra of five stars within a few degrees of Tycho at distances of 1.7 to 2.5 kpc [Black & Raymond (15)]. The recent identification of four Type I remnants in the LMC (191) shows that these remnants fall well below the mean Σ-D relation. Thus, all the evidence is consistent with a distance of 2–2.5 kpc.

The radio emission is concentrated in a thin shell, and its polarization shows a generally radial magnetic field [Strom & Duin (184)], often interpreted as evidence that Rayleigh-Taylor instabilities at the shocked ejecta–shocked interstellar medium contact discontinuity (93, 94, 135) have stretched out the field lines. The variations in direction of polarization can also be used to infer that the radio emitting shell is interior to a shell of swept-up interstellar gas.

Tycho is a bright X-ray source. Electron temperatures around 5 keV are inferred from 2–10 keV X-ray observations, with a cooler component having a temperature near 1 keV [Davison et al. (54), Pravdo et al. (158)]. When the *Einstein* SSS spectrum [Becker et al. (8)] is interpreted in terms of a Sedov blast wave model [Shull (178)], modest overabundances of the silicon-group elements are found. While the Sedov model was not intended to describe shocked ejecta before the remnant has evolved to the blast wave phase, Hamilton & Sarazin (95a) have found it to be a reasonable approximation. The energy centroid of the iron 6.7-keV feature shows that Kα from Fe XXII–XXIV dominates the emission (158). Thus iron is farther from coronal ionization equilibrium in Tycho than in Cas A, presumably because of its lower density. A detailed analysis of the *Einstein* image of Tycho [Seward et al. (175)] shows that the X-ray emission is dominated by a clumpy, somewhat spread out shell of shocked SN ejecta, most likely the result of the Rayleigh-Taylor instability of the boundary between shocked ejecta and shocked ISM. A faint outer shell is attributed to the shocked interstellar gas. With this interpretation, the proper motion observed in the radio corresponds to the velocity of ejecta that has passed through the reverse shock, while the velocity of the ISM shock is much higher (~ 6000 km s^{-1}). The lower shock velocity measured for the optical filaments must be due to higher density in the optical regions. HEAO-1 observations of X rays above 10 keV [Pravdo & Smith (157)] require very high electron temperatures if the emission is thermal. This necessitates rapid equilibration of electron and ion temperatures if the shock velocity is that of the optical filaments, or somewhat slower equilibration if the larger velocity is appropriate. A nonthermal interpretation [Reynolds & Chevalier (166)] is also possible.

Two other Type I remnants in the Galaxy are known, and four have been discovered in the LMC.

SN 1006   The remnant of SN 1006 is classified as Type I because of its pure Balmer line optical emission filaments [Schweizer & Lasker (171), Lasker (123)]. No Hα wings were detected, implying a shock velocity above 3000 km s^{-1}. The X-ray emission may be nonthermal, since no emission lines were detected with the SSS, and the slope of the best-fitting power law agrees with that expected [Becker et al. (9)]. Wu et al. (202) discovered broad Fe II absorption in an IUE spectrum of an O subdwarf behind SN 1006, which showed that the large amounts of iron ejected in models of Type I supernova but not detected in X rays may not have reached the reverse shock yet. The IUE spectrum also shows high-velocity silicon absorption. The X-ray appearance of SN 1006 is unusual in that there is a limb-brightened shell and a softer component filling the center [Pye et al. (159)]. The radio map [Caswell et al. (28)] is very similar to the outer X-ray shell, but no interior emission is seen.

Kepler's SNR   Historical records of Kepler's supernova lead to its classification as Type I [Baade (3)], although its optical emission does not resemble that of Tycho. Danziger & Goss (47), Leibowitz & Danziger (124), and Dennefeld (56) have obtained optical and near-infrared spectra that agree with models for a modest velocity shock in fairly dense gas ($\sim$ 120 km s^{-1}, 300 cm^{-3}). Nitrogen is about twice the solar abundance, but helium seems normal. This suggests an analogy with the QSFs in Cas A. The distance to Kepler is very uncertain. Radio and X-ray maps show a very nearly circular shell 2.6 in diameter [Clark & Caswell (36), White & Long (197)]. Either free expansion or Sedov models can account for the surface brightness distribution. The *Einstein* SSS spectrum closely resembles that of Tycho, again requiring modest overabundance of the silicon-group elements [Becker et al. (6)].

LMC Type I remnants   Tuohy et al. (191) identified four Type I remnants from among the SNRs of the *Einstein* survey on the basis of pure or nearly pure Balmer line emission spectra. One remnant shows broad Hα wings indicative of a 2900 km s^{-1} shock and, thus, an age of 500 years. The radio emission of these remnants falls well below the Σ-*D* relation for the LMC, while their X-ray luminosities seem normal [Mathewson et al. (133)].

The Crab

Thorough reviews of observations of the Crab and related remnants by Wilson (198) and Weiler (196) have just appeared. We refer the reader to those papers and limit the discussion here to a few of the most recent results.

Because of its special nature, the Crab has been especially carefully observed. Ultraviolet observations have confirmed the extension of the power law seen in the visual range [Davidson et al. (52)]. Accurate emission-line fluxes in the visible are available for many filaments [Miller (141), Davidson (50, 51), Fesen & Kirshner (86)]. Davidson et al. (52) summed 65 hours of IUE exposures to measure the intensities of [C III] $\lambda1909$, C IV $\lambda1550$, and He II $\lambda1640$ for a particularly bright filament. They obtained a carbon-to-oxygen abundance ratio near unity. If this applies to the nebula as a whole, it implies a progenitor mass of 8 $M_\odot$, but since the helium abundance is known to vary among the filaments, it is possible that the carbon abundance may as well. Recent observations in the near-infrared [Pequinot & Dennefeld (153), Henry et al. (101)] reveal lines of C I, S III, Fe II, and Ni II. The carbon lines provide another estimate of the carbon abundance, although there are still large uncertainties in the relevant atomic rates. The nickel-to-iron abundance ratio obtained from the infrared lines appears to be much larger than the solar one, but it is similar to the ratio obtained from other nebulae. Henry et al. suggest that differing depletions due to grains may be responsible.

Detailed theoretical models incorporating many recent improvements in the atomic rates have been constructed for the interpretation of these observations. The models generally assume photoionization by synchrotron radiation to be the main energy source. Henry & MacAlpine (102) constructed models to fit the Fesen & Kirshner spectra. They confirm the interpretation of variations in the He I $\lambda5876$/Hβ ratio as a reflection of abundance variations and find that the nitrogen abundance varies as well. In addition, they estimate a total filament mass of 1.2 $M_\odot$. Pequinot & Dennefeld (153) present a photoionization model that fits the observed line ratios of Miller's position 2 (141) from the ultraviolet through the near-infrared. They find a substantial helium overabundance, in agreement with other models, but they also find a normal oxygen abundance per nucleon and a modest nitrogen underabundance. In order to obtain enough emission from highly ionized species (C IV and Ne V), a region of relatively low density is required; thus, Pequinot & Dennefeld suggest that the density decreases outward from a peak near 10^3 cm^{-3}.

Clark et al. (37) have recently produced a velocity map of the [O III] emission with 30 km s^{-1} resolution on a $4\overset{''}{.}7 \times 10''$ grid. Since velocity along the line of sight is equivalent to distance along the line of sight in free expansion, they are able to deduce the three-dimensional structure of the emitting filaments. Clark et al. find two distinct shells in the shape of prolate ellipsoids. The inner expands at ~ 720 km s^{-1} and the outer at ~ 1800 km s^{-1}. The thickness of the region between the shells is about 40% of the outer radius, and this region contains a few more or less radial filaments connecting the two shells. The inner shell is the brighter of the two and shows an enhanced helium abundance, while the abundances seem more normal in the outer shell. This is in harmony with the Type II supernova model of Chevalier (31), according to which the inner shell would be material from the outer core that never reached nuclear densities, while the outer shell would be material from the progenitor envelope. The inner shell marks the boundary of the bright synchrotron emission, but fainter synchrotron emission fills the region between the shells. Clark et al. also find weak [O III] emission extending to $+3600$ and -2400 km s^{-1}, providing confirmation of the faint outer emission shell reported by Murdin & Clark (146).

The polarization of the Crab has also recently been mapped with high spatial resolution. McLean et al. (139) present measurements with $1\rlap{.}''5$ resolution. They find no structure on scales below 5″ and confirm the alignment of the magnetic field along the wisps [Schmidt et al. (172)].

Recent observations of the other Crab-like remnants listed by Weiler (196) include the detection of high-velocity emission in 3C 58, supporting its identification with SN 1181 [Fesen (83)]. Optical and X-ray observations of N157B indicate that it is a Crab-like remnant in the LMC [Danziger et al. (48), Clark et al. (39)].

MIDDLE-AGED REMNANTS: THE CYGNUS LOOP

The Cygnus Loop is the best studied of the older remnants as a result of its striking beauty and also, perhaps, its large angular size, its high surface brightness, and its low reddening. It is generally pictured as a 400 km s^{-1} shock in an intercloud gas of density ~ 0.2 cm^{-3} that drives ~ 100 km s^{-1} shocks into the clouds it encounters. The faster (nonradiative) shock produces the observed X rays, while the slower (radiative) cloud shocks produce the bright optical filaments [McKee & Cowie (136), Bychkov & Pikel'ner (25)].

Increasingly detailed maps in optical lines and at X-ray and radio wavelengths are being produced. As is the case for nearly all other remnants, there is an impressive general correlation among the brightnesses at various wavelengths. The northeastern and western edges are bright in X rays, optical emission lines, and radio continuum. There are also many exceptional regions, however. For instance, the southern section of the

Cygnus Loop is bright at radio wavelengths, but the optical filaments are faint. Dickel & Willis (66) present a 1' resolution radio map of the northeastern region. They find that faint emission extends outside the bright filaments to the faint outer Hα filaments reported by Gull et al. (95). The surface brightness and the kink in the radio spectrum [DeNoyer (59)] are in reasonable accord with predictions for compression of ambient nonthermal particles and magnetic field in radiatively cooling shocked gas [Duin & van der Laan (79)]. DeNoyer (60) found that the bright optical filaments correspond to clouds seen just outside the remnant in 21-cm emission with estimated densities of 5–10 cm^{-3}, while Scoville et al. (173) detected CO emission outside the western edge of the Cygnus Loop and inferred densities two orders of magnitude larger.

Low-resolution X-ray maps show substantial variations in the X-ray temperature across the remnant [Gronenschild (91), Kayat et al. (115)]. The X-ray emission and the emission in the [Fe X] and [Fe XIV] optical lines formed at $1–2 \times 10^6$ K generally extend 5 or 10 arcmin beyond the bright optical filaments [Tuohy et al. (192), Woodgate et al. (201), Lucke et al. (127)], apparently to a series of very faint outer filaments at the shock front itself. A HEAO-1 study by Kahn et al. (113) revealed oxygen and iron lines between 0.6 and 0.9 keV. The best single temperature fit to an equilibrium thermal emission spectrum was 3×10^6 K with oxygen and iron depleted, but multitemperature fits indicated that elemental depletion was not necessary.

The most recent X-ray observations are *Einstein* IPC and HRI images and a higher spectral resolution gas scintillation proportional counter image [Ku et al. (119a)]. The Cygnus Loop is more symmetric in X rays than in optical images, but a breakout in the south suggests a shock front encountering a lower density "tunnel." The spectrum is dominated by line emission of the heliumlike and hydrogenic ions of oxygen, carbon, and nitrogen and the lines of Fe XVII. The temperature is 2×10^6 K at the limb, compared with an average temperature of 3×10^6 K for the remnant as a whole, and temperature and surface brightness are anticorrelated, as expected for pressure equilibrium.

Cox (42) found that models of radiative shocks (in which gas suddenly heats up, then radiatively cools and recombines) having velocities near 100 km s^{-1} (heating the gas to $\sim 1.5 \times 10^5$ K) matched the optical emission-line spectra measured by Miller (140). Fesen et al. (84) obtained excellent interference filter images in the lines of [O III], Hα + [N II], and [S II], and they obtained high quality spectra of positions chosen for their unusual line ratios. Many of the spectra do not agree with radiative shock models at all. Some filaments show intensity ratios [O III]/Hβ as high as 40, while shock models with normal abundances do not produce values of this ratio above about 6 (23, 72, 160, 179). One of Miller's positions had already shown embarrassingly strong [O III], leading to a suspicion that the steady flow assumed in the models was inappropriate [Raymond (160), Contini & Shaviv (41)]. The departure from steady flow is characterized as a lack of the low-temperature emitting material, the $T < 10^4$ K recombination zone predicted by the models. This can be attributed to (a) thermal instability [McCray et al. (134)], (b) an inhomogeneous preshock medium in which the intercloud shock has very recently encountered a denser cloud [McKee & Cowie (136)], or (c) a general transition of that section of the SNR from the adiabatic to the "snowplow" phase [Cox (43)].

The new spectra verify that the departure from steady flow, rather than an abundance anomaly, accounts for the strong [O III] filaments, since [O I] and other lines formed at low temperatures are weak in these regions. The new work also makes the problem more acute, however. The [O III] structure is more filamentary, and the [O III] filaments are located toward the outside of the remnant as compared with regions bright in the lower temperature lines. Arcminute separations between the strong [O III] and strong Hα filaments are observed, while the three explanations above would predict separations of a few arcseconds. Fesen et al. (84) suggest one possible explanation. The separation between hot and cool regions in a steady-flow model is given by one fourth of the cloud shock velocity times the cooling time (the factor of four resulting from the compression of the gas by the fast shock). However, if the optical emission originates in a series of tiny clouds being struck by a 400 km s^{-1} intercloud shock, the appropriate velocity is ~300 km s^{-1}, rather than ~25 km s^{-1}, and the predicted separation approaches an arcminute.

Hester et al. (103) have analyzed a series of interference filter photographs of the complex southeastern part of the Cygnus Loop. By normalizing their images to spectrophotometry at a few positions, they obtained the intensities of seven lines in each $3'' \times 3''$ pixel of a $5' \times 10'$ field. The quantitative morphological information makes it possible to compare surface brightness variations with line ratio variations. Hester et al. find several large regions of anomalously strong [O III], similar in line ratios and morphology to those discussed by Fesen et al. They also find a clean feature that seems to be free of line-of-sight confusion with other features, and whose line ratios agree with models for steady-flow shocks. By taking cuts along the shock propagation direction, Hester et al. are able to follow the transition from strong [O III] to steady-flow regions. They are also able to explain the observed stratification in terms of the intercloud shock velocity times the cooling time in the shocked cloud, but they take the cooling time to be the time required for a steady-flow shock to develop within a single cloud. The further inward decline in the ratio of [O III] to

Hα once steady flow has been achieved is attributed to a gradual decline in shock velocity as the cloud shocks move into the clouds. This could result from a density gradient within the cloud or the drop in ram pressure as the intercloud shock passes the cloud (97, 136). The Hester et al. picture agrees better with both direct and indirect (140, 161) estimates of the sizes of clouds that are emitting optically in the Cygnus Loop. It requires that the [O III] bright regions be shocks driven into the sides of the cloud as the intercloud shock passes. Interference filter images have shown that anomalously strong [O III] emission is fairly common in old SNRs. Examples are G65.3 + 5.7, CTA 1, G126.2 + 1.6, and VRO 42.05.01 [Fesen et al. (85), Sitnik et al. (181), Rosado (167)].

Profiles of the Hα, [N II], and [O III] lines have been measured from multislit echelle plates for a large number of knots [Shull et al. (180)]. Since the hydrogen and nitrogen lines are formed at about the same temperature, it is possible to separate thermal and nonthermal line widths. Nonthermal line broadening is seen at all positions, with a typical width of 20–60 km s^{-1}. No detailed model for these widths is yet available, but some combination of thermally unstable cooling and the complex flow associated with the blast wave–cloud interaction [Smith & Dickel (182)] must be responsible.

The Cygnus Loop is the most extensively observed remnant in the ultraviolet because of its low reddening [$E(B - V) = 0.08$; Parker (150)]. Carruthers & Page (26a) obtained broadband images of the northeastern section showing excellent correspondence between optical and UV brightnesses. Spectra of a few 1° square regions with 30-Å resolution were obtained with the ultraviolet spectrometer aboard *Voyager 2* [Shemansky et al. (177)]. Lines of C III and N III lines were reported. A λ1222 line has no obvious identification, and its existence rests upon the accuracy of the subtraction of the strong geocoronal Lyα background. A 1037-Å line seems too bright in comparison with C II λ1335 to be the C II λ1037 line. Shemansky et al. considered O VI λλ1032, 1038 to be an unacceptable identification, since the feature is not as broad as might be expected for such a doublet, but this may still be the most likely identification. Neither the 100 km s^{-1} shocks inferred from the optical spectra nor the 400 km s^{-1} shocks inferred from X rays would produce enough O VI emission to account for the observed feature, but intermediate velocity shocks are bright in O VI, and Hα observations discussed below show that 200–300 km s^{-1} shocks are present. Another interpretation for the Voyager spectra, proposed by Benvenuti et al. (12), is that the wavelength scale is systematically shifted by about 10 Å, and that most of the observed features are Lyman lines.

Several of the bright optical filaments have been observed in IUE spectra [Benvenuti et al. (11, 12), Raymond et al. (161, 162), D'Odorico et al. (70)].

The C IV, N V, and O V lines in these spectra probe higher temperatures than do the optical [O III] lines, and the shock velocities derived from the UV spectra are generally higher than those derived from optical spectra (~ 130 km s^{-1}). The UV spectra also provide lines of carbon and silicon, which have no strong optical lines, so that abundances of these elements can be derived. This is especially interesting, since these elements are liberated from grains destroyed by sputtering or grain-grain collisions in the shocked gas [Draine & Salpeter (78)]. Comparison with shock models suggests that C and Si are depleted by modest factors (0.3 to 1.0). Near-infrared observation of the [C I] lines also suggests depletion of carbon [Dennefeld & Andrillat (57)]. The analysis of the UV spectra is greatly complicated by reductions in the C II λ1335 and C IV λ1550 line intensities by resonant scattering either in the emitting filament itself or in the intervening interstellar gas (70, 162). The two-photon continuum of hydrogen is detected, which confirms Miller's (140) analysis of the optical continuum. The combination of ultraviolet and optical spectra includes lines of up to five ionization stages of nitrogen and five of oxygen, so that powerful tests of shock models are possible. Various discrepancies between models and the observed UV spectra are attributed to either a mixture of shock velocities within the $10'' \times 20''$ IUE aperture (12) or departures from steady flow (161).

Very faint high-velocity Hα emission was detected in Fabry-Perot observations near the center of the Cygnus Loop by Kirshner & Taylor (119) and Doroshenko & Lozinskaya (77), and some of the very faint filaments around the outermost edges of the remnant were observed to be pure Balmer line emission [Raymond et al. (164), Fesen et al. (84)]. This is taken to be emission from neutral hydrogen swept up in a nonradiative shock [Chevalier & Raymond (35), Bychkov & Lebedev (24)]. The velocity shifts and widths [Kirshner & Taylor (119), Treffers (187), Raymond et al. (163)] indicate shock velocities near 200 km s^{-1}. Higher velocity shocks must be present to account for the X-ray observations, but only a small number of the brightest emission regions have been observed. The most detailed study has been done on the brightest of the faint filaments, one located about 5' outside the bright northeastern region [Raymond et al. (163)]. An echelle spectrum shows an Hα line width of 170 km s^{-1} FWHM, and a deep-blue spectrum shows very faint forbidden line emission. The He II λ1640, C IV λ1550, and N V λ1240 lines were detected in an IUE spectrum along with the hydrogen two-photon continuum. Theoretical models of the filament show that the shock velocity is 170 km s^{-1} if the electron and ion temperatures equilibrate on a Coulomb time scale, or 205 km s^{-1} if much more rapid equilibration by means of plasma turbulence is assumed [see McKee & Hollenbach (137) for a discussion of electron-ion equilibration]. The slow equilibration model provides a better fit to the relative line intensities, but the evidence is not conclusive. The data suggest a 30% depletion of carbon and a 30% neutral hydrogen fraction in the preshock gas. The separation between faint and bright filaments is surprisingly large considering the cooling times and velocities involved. Falle & Garlick (82) present a model in which the separation is due to projection effects.

With a few exceptions, the Cygnus Loop observations are typical of the observations of SNRs old enough to have forgotten the details of the supernova explosion. An additional type of observation is available for those SNRs located in front of hot stars. Absorption-line studies of the Vela SNR show high-velocity gas in ionization states as high as O VI [Jenkins et al. (112)], and Phillips & Gondhalekar (155) have detected CO and high-velocity Fe II absorption in S147. Molecular emission has also been detected from SNRs that are older or that are encountering higher density clouds. The most abundant source seems to be IC443. The 2.2-μ line of H_2 and high-velocity components in CO, OH, and HCO^+ have been detected there [Treffers (186), DeNoyer (61, 62), Dickinson et al. (67)].

OLD SUPERNOVA REMNANTS: THE NORTH POLAR SPUR

Very large shells are observed in nonthermal radio emission [Landecker & Wiebelski (120)], H I 21-cm emission [Heiles (98), Hu (105)], Hα emission [Reynolds & Ogden (165), Davies et al. (53)], and soft X rays [e.g. Burstein et al. (22), Cash et al. (27)]. Many of these features are too energetic to be the remnants of a single supernova. They are believed to be bubbles blown by the stellar winds of entire OB associations [e.g. Lockman & Ganzel (125)]. The North Polar Spur is a very large structure, about 55° in radius. Models that portray the neutral hydrogen shell and the 10^6 K soft X-ray emission as the remnants of a single supernova explosion require very large energies, but if the H I shell was created by winds of the Sco-Cen association, a supernova inside the bubble could account for the X rays [Iwan (111), Davelaar et al. (49), Heiles et al. (99)]. Schnopper et al. (169) report a temperature near 4×10^6 K for the shell.

The local soft X-ray and EUV background is brighter than expected from models of the three-phase interstellar medium, so it has been suggested that the Sun is located within an old supernova remnant [Sanders et al. (168), Davelaar et al. (49)], a proposal in harmony with the very low neutral hydrogen densities within 50–100 pc of the Sun [Frisch & York (87)]. X-ray lines of carbon and oxygen have been detected by Inoue et al. (107) and Schnopper et al. (169), and a detailed theoretical model was constructed by Cox & Anderson (44).

The clearest example of an old SNR may be the "Monogem ring," a 20° (50 pc) diameter shell of million degree gas with a density of ~ 0.01 cm^{-3} [Nousek et al. (148)]. A fragmentary shell is also seen in Hα, 21-cm, and radio continuum emission. The Eridanus hot spot [Naranan et al. (147)] is similar in size and age.

TRANSITIONAL SUPERNOVA REMNANTS

The most obvious remnants in between the young SNR categories and the middle-aged remnants are the "combination" type having a central source like the Crab and an outer shell as well. Weiler (196) gives an extensive review of these remnants, of which Vela, CTB 80, and W 50 are the best known examples. MSH15-52 contains two compact X-ray sources, and [Fe II] emission from a bright knot has been detected at 1.6 μm [Seward et al. (175a)]. The radio and X-ray morphologies of G109.1–1.0 suggest excitation by precessing jets [Gregory & Fahlman (90a)].

The most likely case of a transition between oxygen-rich and normal phases is Puppis A. The bright knots show a nitrogen enhancement [Dopita et al. (75)]. The coronal [Fe XIV] line is relatively bright [Clark et al. (38), Lucke et al. (128)], and the X-ray and radio maps indicate interaction of the remnant with a fairly dense cloud [Petre et al. (154)]. High-resolution X-ray spectroscopy shows an overabundance of oxygen, and when the recent atomic rates found by Pradhan et al. (156) are employed, a clear underionization of the X-ray-emitting gas is also seen [Winkler et al. (199), Canizares & Winkler (26)].

SUPERNOVA REMNANTS IN OTHER GALAXIES

Nearby galaxies provide sets of SNRs at known distances and varying stages of evolution. The Magellanic Clouds have been studied most extensively, with early radio and optical surveys [Mathewson & Clark (131), Mills et al. (142)] and the *Einstein* X-ray surveys (126, 176). Dopita et al. (75) and Dopita (73) studied the abundances and expansion velocities of many of these remnants. Mathewson et al. (133) present a sample of LMC remnants believed to be complete for diameters below 40 pc. They find that Type I remnants (identified by pure Balmer line emission spectra) fall below the mean Σ-D relation by factors of 5 to 30. A plot of $N(D)$, the number of remnants having diameters less than D, versus D reveals that $N(D)$ increases as $D^{1.0}$, whereas the Sedov blast wave theory would predict $D^{2.5}$. While there may be important selection effects [Hughes & Helfand (106)], the most straightforward interpretation is that the remnants expand with constant velocity. This would imply that the SNRs have yet to be substantially slowed by interaction with the interstellar gas and would favor the McKee & Ostriker (138) picture of supernova remnant evolution in a cloudy medium.

Abundance gradients and SNR evolution have been systematically studied in M33 [Berkhuijsen (13), Dopita et al (74), Goss et al. (89)] and M31 [D'Odorico et al. (71), Dennefeld & Kunth (58), Blair et al. (16, 17), Dickel et al. (63)]. The nitrogen abundance is found to decline with galactocentric distance, in good agreement with a similar study of Milky Way remnants [Binette et al. (14)] and with the nitrogen abundances inferred from H II region spectra. Bohigas (20) has considered both abundance and evolution effects. A simple model of the SNR-cloud interaction yields an unreasonable correlation between derived supernova energy and remnant radius when applied to the M31 remnants (16). This result can be interpreted as evidence either for magnetic field support of shocked clouds or for an SNR evolution model in which the supernova kinetic energy is fully converted into thermal energy only for SNRs at least 50 pc in diameter. This last explanation would be in harmony with the apparent constant expansion rate of the LMC remnants, but the magnetic field interpretation is also viable.

ACKNOWLEDGMENTS

I would like to thank W. Blair and F. Seward for extensive discussions. This work was partially supported by NASA contracts NAG-117 and NAG5-87 with the Smithsonian Astrophysical Observatory.

Literature Cited

1. Albinson, J. S., Gull, S. F. 1982. In *Regions of Recent Star Formation*, ed. R. S. Roger, P. E. Dewdney, p. 193. Dordrecht: Reidel
2. Ashworth, J. 1980. *J. Hist. Astron.* 11:1
3. Baade, W. 1943. *Ap. J.* 97:119
4. Baade, W., Minkowski, R. 1954. *Ap. J.* 119:206
5. Balick, B., Heckman, T. 1978. *Ap. J. Lett.* 227:L7
6. Becker, R. H., Boldt, E. A., Holt, S. S., Serlemitsos, P. J., McCluskey, G. E. 1980. *Ap. J. Lett.* 237:L77
7. Becker, R. H., Holt, S. S., Smith, B. W., White, N. E., Boldt, E. A., et al. 1979. *Ap. J. Lett.* 234:L73
8. Becker, R. H., Holt, S. S., Smith, B. W., White, N. E., Boldt, E. A., et al. 1980 *Ap. J. Lett.* 235:L5
9. Becker, R. H., Szymkowiak, A. E., Boldt, E. A., Holt, S. S., Serlemitsos, P. J. 1980. *Ap. J. Lett.* 240:L33
10. Bell, A. R. 1977. *MNRAS* 179:573
11. Benvenuti, P., D'Odorico, S., Dopita,
M. 1978. *Nature* 277:99
12. Benvenuti, P., Dopita, M., D'Odorico, S. 1980. *Ap. J.* 238:601
13. Berkhuijsen, E. M. 1983. *Astron. Astrophys* 120:147
14. Binette, L., Dopita, M. A., D'Odorico, S., Benvenuti, P. 1982. *Astron. Astrophys.* 115:315
15. Black, J. H., Raymond, J. C. 1984. *Astron. J.* In press
16. Blair, W. P., Kirshner, R. P., Chevalier, R. A. 1981. *Ap. J.* 247:879
17. Blair, W. P., Kirshner, R. P., Chevalier, R. A. 1984. *Ap. J.* Submitted for publication
18. Blair, W. P., Kirshner, R. P., Winkler, P. F. Jr. 1983. *Ap. J.* 272:84
19. Blair, W. P., Raymond, J. C., Fesen, R. A., Gull, T. R. 1984. *Ap. J.* In press
20. Bohigas, J. 1983. *Rev. Mex. Astron. Astrofis.* 5:271
21. Brecher, K., Wasserman, I. 1980. *Ap. J. Lett.* 240:L105
22. Burstein, P., Borken, R. J., Kraushaar,

W. L., Sanders, W. T. 1977. *Ap. J.* 213:405

23. Butler, S. E., Raymond, J. C. 1980. *Ap. J.* 240:680

24. Bychkov, K. V., Lebedev, V. S. 1979. *Astron. Astrophys.* 80:167

25. Bychkov, K. V., Pikel'ner, S. B. 1975. *Sov. Astron. Lett.* 1:14

26. Canizares, C. R., Winkler, P. F. 1981. *Ap. J. Lett.* 246:L33

26a. Carruthers, G. P., Page, T. 1976. *Ap. J.* 205:397

27. Cash, W., Charles, P., Bowyer, S., Walter, F., Garmire, G., Reigler, G. 1980. *Ap. J. Lett.* 238:L71

28. Caswell, J. L., Haynes, R. F., Milne, D. K., Wellington, K. J. 1983. *MNRAS* 204:921

29. Chevalier, R. A. 1976. *Ap. J.* 208:826

30. Chevalier, R. A. 1977. *Ann. Rev. Astron. Astrophys.* 15:175

31. Chevalier, R. A. 1977. In *Supernovae*, ed. D. Schramm, p. 53. Dordrecht: Reidel

32. Chevalier, R. A., Kirshner, R. P. 1978. *Ap. J.* 219:931

33. Chevalier, R. A., Kirshner, R. P. 1979. *Ap. J.* 233:154

34. Chevalier, R. A., Kirshner, R. P., Raymond, J. C. 1980. *Ap. J.* 235:186

35. Chevalier, R. A., Raymond, J. C. 1978. *Ap. J. Lett.* 225:L27

36. Clark, D. H., Caswell, J. L. 1976. *MNRAS* 174:267

37. Clark, D. H., Murdin, P., Wood, R., Gilmozzi, R., Danziger, J., Furr, A. W. 1983. *MNRAS* 204:415

38. Clark, D. H., Murdin, P., Zarnecki, J. C., Culhane, J. L. 1979. *MNRAS* 188:11p

39. Clark, D. H., Tuohy, I. R., Long, K. S., Szymkowiak, A. E., Dopita, M. A., et al. *Ap. J.* 255:440

40. Clark, D. H., Tuohy, I. R., Becker, R. H. 1980. *MNRAS* 193:129

41. Contini, M., Shaviv, G. 1982. *Astrophys. Space Sci.* 85:203

42. Cox, D. 1972. *Ap. J.* 178:143

43. Cox, D. 1972. *Ap. J.* 178:159

44. Cox, D. P., Anderson, P. R. 1982. *Ap. J.* 253:268

45. Danziger, I. J., Dennefeld, M. 1976. *Publ. Astron. Soc. Pac.* 88:44

46. Danziger, I. J., Dennefeld, M. 1976. *Ap. J.* 207:394

47. Danziger, I. J., Goss, W. M. 1979. *MNRAS* 190:47p

48. Danziger, I. J., Goss, W. M., Murdin, P., Clark, D. H., Boksenberg, A. 1981. *MNRAS* 195:33p

49. Davelaar, J., Bleeker, J. A. M., Deerenberg, A. J. M. 1980. *Astron. Astrophys.* 92:231

50. Davidson, K. 1978. *Ap. J.* 220:177

51. Davidson, K. 1979. *Ap. J.* 228:179

52. Davidson, K., Gull, T. R., Maran, S. P., Stecher, T. P., Fesen, R. A., et al. 1982. *Ap. J.* 253:696

53. Davies, R. D., Elliot, K. H., Meaburn, J. 1976. *MNRAS* 81:89

54. Davison, P. J. N., Culhane, J. L., Mitchell, R. J. 1976. *Ap. J. Lett.* 206:L37

55. deBruyn, A. G. 1983. *Astron. Astrophys.* 119:301

56. Dennefeld, M. 1982. *Astron. Astrophys.* 112:215

57. Dennefeld, M., Andrillat, Y. 1981. *Astron. Astrophys.* 103:44

58. Dennefeld, M., Kunth, D. 1981. *Astron. J.* 86:989

59. DeNoyer, L. K. 1974. *Astron. J.* 79:1253

60. DeNoyer, L. K. 1975. *Ap. J.* 196:479

61. DeNoyer, L. K. 1979. *Ap. J. Lett.* 228:L41

62. DeNoyer, L. K. 1979. *Ap. J. Lett.* 232:L165

63. Dickel, J. R., Felli, M., D'Odorico, S., Dopita, M. A. 1982. *Ap. J.* 252:582

64. Dickel, J. R., Griesen, E. W. 1979. *Astron. Astrophys.* 75:145

65. Dickel, J. R., Murray, S. S., Morris, J., Wells, D. L. 1982. *Ap. J.* 257:145

66. Dickel, J. R., Willis, A. G. 1980. *Astron. Astrophys.* 85:55

67. Dickinson, D. F., Kuiper, E. N. R., St. Clair Dinger, A., Kuiper, T. B. H. 1980. *Ap. J. Lett.* 237:L43

68. Dinerstein, H. L., Werner, M. W., Capps, R. W., Dwek, E. 1982. *Ap. J.* 255:552

69. Dinerstein, H. L., Werner, M. W., Dwek, E., Capps, R. W. 1981. *Bull. Am. Astron. Soc.* 13:895

70. D'Odorico, S., Benvenuti, P., Dennefeld, M., Dopita, M. A., Greve, A. 1980. *Astron. Astrophys.* 92:22

71. D'Odorico, S., Dopita, M. A., Benvenuti, P. 1980. *Astron. Astrophys. Suppl.* 40:67

72. Dopita, M. A. 1977. *Ap. J. Suppl.* 33:437

73. Dopita, M. A. 1979. *Ap. J. Suppl.* 40:455

74. Dopita, M. A., D'Odorico, S., Benvenuti, P. 1980. *Ap. J.* 236:628

75. Dopita, M. A., Mathewson, D. S., Ford, V. L. 1977. *Ap. J.* 214:179

76. Dopita, M. A., Tuohy, I. R., Mathewson, D. S. 1981. *Ap. J. Lett.* 248:L105

77. Doroshenko, V. T., Lozinskaya, T. A. 1977. *Sov. Astron. Lett.* 3:295

78. Draine, B. T., Salpeter, E. E. 1979. *Ap. J.* 231:438

79. Duin, R. M., van der Laan, H. 1975. *Astron. Astrophys.* 40:111

80. Dwek, E., Werner, M. W. 1981. *Ap. J.* 248:138
81. Fabian, A. S., Willingale, R., Pye, J. P., Murray, S. S., Fabbiano, G. 1980. *MNRAS* 193:175
82. Falle, S. A. E. G., Garlick, A. R. 1982. *MNRAS* 201:635
83. Fesen, R. A. 1983. *Ap. J. Lett.* 270:L53
84. Fesen, R. A., Blair, W. P., Kirshner, R. P. 1982. *Ap. J.* 262:171
85. Fesen, R. A., Gull, T. R., Ketelsen, D. A. 1983. *Ap. J. Suppl.* 51:337
86. Fesen, R. A., Kirshner, R. P. 1982. *Ap. J.* 258:1
87. Frisch, P. C., York, D. G. 1983. *Ap. J. Lett.* 271:L59
88. Gorenstein, P., Tucker, W. H. 1976. *Ann. Rev. Astron. Astrophys.* 14:373
89. Goss, W. M., Ekers, R. D., Danziger, I. J., Israel, F. R. 1980. *MNRAS* 193:901
90. Goss, W. M., Shaver, P. A., Zealey, W. J., Murdin, P., Clark, D. H. 1979. *MNRAS* 188:357
90a. Gregory, P. C., Fahlman, G. G. 1983. In *Supernova Remnants and Their X-Ray Emission, IAU Symp. No. 101*, ed. J. Danziger, P. Gorenstein, p. 429. Dordrecht: Reidel
91. Gronenschild, E. H. B. M. 1980. *Astron. Astrophys.* 89:66
92. Gronenschild, E. H. B. M., Mewe, R. 1982. *Astron. Astrophys. Suppl.* 48:305
93. Gull, S. F. 1973. *MNRAS* 162:135
94. Gull, S. F. 1975. *MNRAS* 171:237
95. Gull, T. R., Kirshner, R. P., Parker, R. A. R. 1977. In *Supernovae*, ed. D. Schramm, p. 71. Dordrecht: Reidel
95a. Hamilton, A. J. S., Sarazin, C. L. 1983. In *Supernova Remnants and Their X-Ray Emission, IAU Symp. No. 101*, ed. J. Danziger, P. Gorenstein, p. 113. Dordrecht: Reidel
96. Hamilton, A. J. S., Sarazin, C. L., Chevalier, R. A. 1983. *Ap. J. Suppl.* 51:115
97. Heathcote, S. R., Brand, P. W. J. L. 1983. *MNRAS* 203:67
98. Heiles, C. 1979. *Ap. J.* 229:533
99. Heiles, C., Chu, Y.-H., Reynolds, R. J., Yegingil, I., Troland, T. H. 1980. *Ap. J.* 242:533
100. Henbest, S. N. 1980. *MNRAS* 190:833
101. Henry, R., MacAlpine, G. M., Kirshner, R. P. 1984. Preprint
102. Henry, R., MacAlpine, G. M. 1982. *Ap. J.* 258:11
103. Hester, J. J., Parker, R. A. R., Dufour, R. J. 1983. *Ap. J.* 273:219
104. Holt, S. S. 1983. In *Supernova Remnants and Their X-Ray Emission, IAU Symp. No. 101*, ed. J. Danziger, P. Gorenstein, p. 17. Dordrecht: Reidel
105. Hu, E. M. 1981. *Ap. J.* 248:119
106. Hughes, J., Helfand, D. J. 1983. *Bull. Am. Astrom. Soc.* 14:936
107. Inoue, H., Koyama, K., Matsuoka, M., Ohashi, T., Tanaka, Y., Tsunemi, H. 1980. *Ap. J.* 238:886
108. Itoh, H. 1977. *Publ. Astron. Soc. Jpn.* 29:813
109. Itoh, H. 1981. *Publ. Astron. Soc. Jpn.* 33:1
110. Itoh, H. 1981. *Publ. Astron. Soc. Jpn.* 33:521
111. Iwan, D. 1980. *Ap. J.* 239:316
112. Jenkins, E. B., Silk, J., Wallerstein, G. 1976. *Ap. J. Suppl.* 32:681
113. Kahn, S. M., Charles, P. A., Bowyer, S., Blissett, R. 1980. *Ap. J. Lett.* 242:L19
114. Kamper, K. W., van den Bergh, S. 1978. *Ap. J.* 224:851
115. Kayat, M. A., Rolf, D. P., Smith, G. C., Willingale, R. 1980. *MNRAS* 191:729
116. Deleted in proof
117. Kirshner, R. P., Chevalier, R. A. 1977. *Ap. J.* 218:142
118. Kirshner, R. P., Chevalier, R. A. 1978. *Astron. Astrophys.* 67:267
119. Kirshner, R. P., Taylor, K. 1976. *Ap. J. Lett.* 208:L83
119a. Ku, W. H.-M., Long, K., Pisarski, R., Vartanian, M. 1983. In *Supernova Remnants and Their X-Ray Emission, IAU Symp. No. 101*, ed. J. Danziger, P. Gorenstein, p. 253. Dordrecht: Reidel
120. Landecker, T. L., Wiebelski, R. 1970. *Aust. J. Phys. Astrophys. Suppl.* 16:1
121. Lasker, B. M. 1978. *Ap. J.* 223:109
122. Lasker, B. M. 1980. *Ap. J.* 237:765
123. Lasker, B. M. 1981. *Ap. J.* 244:518
124. Leibowitz, E. M., Danziger, I. J. 1983. *MNRAS* 204:273
125. Lockman, F. J., Ganzel, B. L. 1983. *Ap. J.* 268:117
126. Long, K. S., Helfand, D., Grabelsky, D. A. 1981. *Ap. J.* 248:925
127. Lucke, R. L., Woodgate, B. E., Gull, T. R., Socker, D. G. 1980. *Ap. J.* 235:882
128. Lucke, R. L., Zarnecki, J. C., Woodgate, B. E., Culhane, J. L., Socker, D. G. 1979. *Ap. J.* 228:763
129. Manzo, G., Peacock, A., Taylor, B. G., Andresen, R. D., Culhane, J. L., Catura, R. C. 1983. *Astron. Astrophys.* 122:124
130. Markert, T. H., Canizares, C. R., Clark, G. W., Winkler, P. F. 1983. *Ap. J.* 268:134
131. Mathewson, D. S., Clark, J. N. 1973. *Ap. J.* 180:725
132. Mathewson, D. S., Dopita, M. A., Tuohy, I. R., Ford, V. L. 1980. *Ap. J. Lett.* 242:L73
133. Mathewson, D. S., Ford, V. L., Dopita, M. A., Tuohy, I. R., Long, K. S., Helfand, D. J. 1983. *Ap. J. Suppl.* 51:345

134. McCray, R., Stein, R. F., Kafatos, M. 1975. *Ap. J.* 196:565
135. McKee, C. F. 1974. *Ap. J.* 188:335
136. McKee, C. F., Cowie, L. L. 1975. *Ap. J.* 195:715
137. McKee, C. F., Hollenbach, D. J. 1980. *Ann. Rev. Astron. Astrophys.* 18:219
138. McKee, C. F., Ostriker, J. P. 1978. *Ap. J.* 218:148
139. McLean, I. S., Aspin, C., Reitsema, H. 1983. *Nature* 304:243
140. Miller, J. S. 1974. *Ap. J.* 189:239
141. Miller, J. S. 1978. *Ap. J.* 220:490
142. Mills, B. Y., Little, A. G., Durdin, J. M., Kesteven, M. J. 1982. *MNRAS* 200:1007
143. Milne, D. K. 1979. *Aust. J. Phys.* 32:83
144. Minkowski, R. 1957. In *Radio Astronomy, IAU Symp. No. 4*, ed. H. C. van de Hulst, p. 107. Cambridge: Cambridge Univ. Press
145. Murdin, P., Clark, D. H. 1979. *MNRAS* 189:501
146. Murdin, P., Clark, D. H. 1981. *Nature* 294:543
147. Naranan, S., Shulman, S., Friedman, H., Fritz, G. 1976. *Ap. J.* 208:718
148. Nousek, J. A., Cowie, L. L., Hu, E., Lindblad, C. J., Garmire, G. P. 1981. *Ap. J.* 248:152
149. Nussbaumer, H., Storey, P. J. 1982. *Astron. Astrophys.* 110:295
150. Parker, R. A. R. 1967. *Ap. J.* 149:363
151. Peimbert, M. 1971. *Ap. J.* 170:261
152. Peimbert, M., van den Bergh, S. 1971. *Ap. J.* 167:223
153. Pequinot, D., Dennefeld, M. 1983. *Astron. Astrophys.* 120:249
154. Petre, R. A., Canizares, C. R., Kriss, G. A., Winkler, P. F. 1982. *Ap. J.* 258:22
155. Phillips, A. P., Gondhalekar, P. M. 1983. *MNRAS* 202:483
156. Pradhan, A. K., Norcross, D. W., Hummer, D. G. 1981. *Ap. J.* 246:1031
157. Pravdo, S. H., Smith, B. W. 1979. *Ap. J. Lett.* 108:L83
158. Pravdo, S. H., Smith, B. W., Charles, P. A., Tuohy, I. R. 1980. *Ap. J. Lett.* 235:L9
159. Pye, J. P., Pounds, K. A., Rolf, D. P., Seward, F. D., Smith, A., Willingale, R. 1981. *MNRAS* 194:569
160. Raymond, J. C. 1979. *Ap. J. Suppl.* 39:1
161. Raymond, J. C., Black, J. H., Dupree, A. K., Hartmann, L., Wolff, R. S. 1980. *Ap. J.* 238:881
162. Raymond, J. C., Black, J. H., Dupree, A. K., Hartmann, L., Wolff, R. S. 1981. *Ap. J.* 246:100
163. Raymond, J. C., Blair, W. P., Fesen, R. A., Gull, T. R. 1983. *Ap. J.* 275:636
164. Raymond, J. C., Davis, M., Gull, T. R., Parker, R. A. R. 1980. *Ap. J. Lett.* 238:L21

165. Reynolds, R. J., Ogden, P. M. 1979. *Ap. J.* 229:942
166. Reynolds, S. P., Chevalier, R. A. 1981. *Ap. J.* 245:912
167. Rosado, M. 1982. *Rev. Mex. Astron. Astrofis.* 5:127
168. Sanders, W. T., Kraushaar, W. L., Nousek, J. A., Fried, P. M. 1977. *Ap. J. Lett.* 217:L87
169. Schnopper, H. W., Delvaille, J. P., Rocchia, R., Blondel, C., Cheron, C., et al. 1982. *Ap. J.* 253:131
170. Schwarz, U. J., Arnal, E. M., Goss, W. M. 1980. *MNRAS* 192:67p
171. Schweizer, F., Lasker, B. M. 1978. *Ap. J.* 226:167
172. Schmidt, G. D., Angel, J. R. P., Beaver, E. A. 1979. *Ap. J.* 227:106
173. Scoville, N. Z., Irvine, W. M., Wannier, P. G., Predmore, C. R. 1977. *Ap. J.* 216:320
174. Searle, L. 1971. *Ap. J.* 168:41
175. Seward, F., Gorenstein, P., Tucker, W. 1983. *Ap. J.* 266:287
175a. Seward, F. D., Harnden, F. R. Jr., Murdin, P., Clark, D. H. 1983. *Ap. J.* 267:698
176. Seward, F. D., Mitchell, M. 1981. *Ap. J.* 243:736
177. Shemansky, D. E., Sandel, B. R., Broadfoot, A. L. 1979. *Ap. J.* 231:35
178. Shull, J. M. 1982. *Ap. J.* 262:308
179. Shull, J. M., McKee, C. F. 1979. *Ap. J.* 227:131
180. Shull, P. J. Jr., Parker, R. A. R., Gull, T. R., DuFour, R. J. 1982. *Ap. J.* 253:682
181. Sitnik, T. G., Klementeva, A. Yu., Toropova, M. S. 1983. *Astron. Zh.* 60:503
182. Smith, M. D., Dickel, J. R. 1983. *Ap. J.* 265:272
183. Stewart, G., Fabian, A., Seward, F. 1983. In *Supernova Remnants and Their X-Ray Emission, IAU Symp. No. 101*, ed. J. Danziger, P. Gorenstein, p. 59. Dordrecht: Reidel
184. Strom, R. G., Duin, R. M. 1973. *Astron. Astrophys.* 25:351
185. Strom, R. G., Goss, W. H., Shaver, P. A. 1982. *MNRAS* 200:473
186. Treffers, R. R. 1979. *Ap. J. Lett.* 233:L17
187. Treffers, R. R. 1981. *Ap. J.* 250:213
188. Tuffs, R. J. 1983. In *Supernova Remnants and Their X-Ray Emission, IAU Symp. No. 101*, ed. J. Danziger, P. Gorenstein, p. 49. Dordrecht: Reidel
189. Tuohy, I. R., Clark, D. H., Burton, W. M. 1982. *Ap. J. Lett.* 260:L65
190. Tuohy, I. R., Dopita, M. A. 1983. *Ap. J. Lett.* 260:L65
191. Tuohy, I. R., Dopita, M. A., Mathewson, D. S., Long, K. S., Helfand, D. J. 1982. *Ap. J.* 261:473

192. Tuohy, I. R., Nousek, J. A., Garmire, G. P. 1979. *Ap. J. Lett.* 234:L101

193. van den Bergh, S. 1974. In *Highlights of Astronomy*, ed. G. Contopolous, 3:559. Dordrecht: Reidel

193a. van den Bergh, S. 1983. In *Supernova Remnants and Their X-Ray Emission, IAU Symp. No. 101*, ed. J. Danziger, P. Gorenstein, p. 597. Dordrecht: Reidel

194. van den Bergh, S., Kamper, K. 1983. *Ap. J.* 268:129

195. Vinyaikin, E. N., Razin, V. A. 1979. *Sov. Astron. AJ* 23:515

196. Weiler, K. 1983. *Observatory* 1054:85

197. White, R. L., Long, K. S. 1983. *Ap. J.* 264:196

198. Wilson, A. S. 1983. *Observatory* 1054:73

199. Winkler, P. F., Canizares, C. R., Bromley, B. C. 1983. In *Supernova Remnants and Their X-Ray Emission, IAU Symp. No. 101*, ed. J. Danziger, P. Gorenstein, p. 245. Dordrecht: Reidel

200. Woltjer, L. 1972. *Ann. Rev. Astron. Astrophys.* 10:129

201. Woodgate, B. E., Kirshner, R. P., Balon, R. J. 1977. *Ap. J. Lett.* 218:L129

202. Wu, C.-C., Leventhal, M., Sarazin, C. L., Gull, T. R. 1983. *Ap. J. Lett.* 269:L5

Ann. Rev. Astron. Astrophys. 1984. 22: 97–130

IMAGE FORMATION BY SELF-CALIBRATION IN RADIO ASTRONOMY

T. J. Pearson and A. C. S. Readhead

Owens Valley Radio Observatory, California Institute of Technology, Pasadena, California 91125

1. INTRODUCTION

Since its first application to astronomy in the late 1940s (67, 95, 98), radio interferometry has developed into an imaging tool of unparalleled power. The complex visibility measured by a radio interferometer is a single Fourier component of the sky brightness distribution, so a map of the sky can be made by using Fourier synthesis to combine measurements obtained with different interferometer baselines. This technique, called *aperture synthesis* (97), was the basis of a number of instruments designed in the 1950s, in which several fixed antennas or two or more movable antennas were used to provide the necessary range of baselines. In the 1960s and 1970s, several telescopes were built that made use of the Earth's rotation to change the interferometer baselines, notably the One-Mile and Five-Kilometre telescopes at Cambridge in England (96), and the Westerbork Synthesis Radio Telescope (WSRT) in the Netherlands (3, 56). The most powerful of these is the Very Large Array (VLA) of the National Radio Astronomy Observatory of the United States (52, 54, 112). The VLA, with a maximum baseline of 35 km, routinely makes maps that surpass the resolution of the best optical telescopes. Also during the 1960s, baselines were increased to hundreds of kilometers by using radio links to transmit the signal from one antenna to the other (35, 80), and eventually to thousands of kilometers by the technique of Very Long Baseline Interferometry (VLBI), in which the signals are recorded on magnetic tape at each antenna and later played back for cross-correlation (8, 16, 23, 24).

A common problem in aperture synthesis is that the phase of the complex visibility is badly corrupted by instrumental and propagation effects. Over

97

short baselines, these problems can generally be overcome by careful calibration of the instrument using unresolved sources. For a variety of reasons this solution cannot be applied to VLBI, and thus for some years it was thought that VLBI could never become a true imaging technique. It had been known since the early 1950s that part of the phase information could be recovered from badly corrupted phases when three or more antennas were used simultaneously (59), but methods of applying this information in the construction of images were not explored until the 1970s. The breakthrough came with the realization that this partial phase information could be used, together with the constraint that the source brightness distribution must be positive, to recover the full phase information and hence to make reliable images (86). This knowledge has stimulated the development of methods of mapping that can be used in the presence of large systematic errors in both the amplitude and the phase of the measured visibilities. These methods, now generally known as *self-calibration* or *hybrid mapping*, make use of the fact that most of the errors are associated with individual antennas, and that it is possible to form combinations of the measured visibilities, called *closure phases* and *closure amplitudes*, that are independent of such errors. Although the methods were originally developed for VLBI, where the errors are so large as to prevent conventional mapping, they are now also used in connected-element interferometers such as the VLA and the WSRT, and in the Jodrell Bank MERLIN array in England (33), to correct the much smaller errors that occur in such instruments; their application to VLBI has shown that it is possible to make reliable maps with submilliarcsecond resolution.

This review describes these self-calibration methods and their application to both conventional and very long baseline arrays and includes examples of the high-quality images that they produce. After first defining closure phase and closure amplitude in Section 2, we review conventional aperture synthesis methods in Sections 3 and 4. The new iterative self-calibration algorithms are the subject of Section 5. In Section 6 we consider the capabilities and limitations of these algorithms. Finally, in Section 7 we mention some applications in optical and infrared astronomy and in new radio instruments presently planned or under construction.

2.   CLOSURE QUANTITIES

2.1   *Interferometry*

The theory of a radio interferometer has been described in a number of papers (e.g. 17, 39, 40, 89, 111, 114). In order to clarify the problems involved in calibrating measurements made with an interferometer, we first briefly summarize this theory.

Let $v_m(t)$ and $v_n(t)$ be the analytic signal representation of the voltages received by two antennas m and n, expressed as functions of time t. The interferometer measures the *mutual coherence function* (11) of the voltages received by the two antennas,

$$\Gamma_{mn}(\tau) = \langle v_m(t)v_n^*(t+\tau) \rangle, \qquad\qquad 1.$$

at some time delay τ (which may be adjusted); the angular brackets represent a time average. In continuum observing, the delay τ is adjusted so that $|\tau| \ll 1/\Delta v$, where Δv is the bandwidth of the antenna and associated electronics. In this case the interferometer measures the *mutual intensity* or *complex visibility* $\Gamma_{mn}(0)$. The van Cittert–Zernike theorem (11, 111) shows that $\Gamma_{mn}(0)$ is proportional to a Fourier component of the sky brightness distribution:

$$\Gamma_{mn}(0) \propto \int\int I(\mathbf{s}) \exp(-2\pi i \mathbf{s} \cdot \mathbf{B}_{mn}/\lambda) \, d\Omega, \qquad\qquad 2.$$

where $I(\mathbf{s})$ represents the strength of the signal received from direction $\mathbf{s}$ (a unit vector), $\mathbf{B}_{mn}$ is the vector separation of the two antennas (baseline vector), λ is the wavelength at the center of the received band, and the integration $(d\Omega)$ is over the celestial sphere. In spectral line observing, the delay dependence of $\Gamma_{mn}(\tau)$ is measured, and then the Fourier transform of $\Gamma_{mn}(\tau)$ with respect to τ gives the frequency dependence of $I(\mathbf{s})$. In the technique of aperture synthesis (which we discuss further in Section 3), one seeks to measure Γ_{mn} at a sufficient number of different antenna separations $\mathbf{B}_{mn}$ to form an estimate of $I(\mathbf{s})$ by inversion of Equation 2.

The quantity actually measured in a real interferometer is the complex visibility Γ_{mn} multiplied by a complex gain factor G_{mn}:

$$V_{mn} = G_{mn}\Gamma_{mn} + \text{noise}. \qquad\qquad 3.$$

The gain factor G represents amplitude and phase errors introduced by the interferometer itself and during the propagation of the signals through space and through the Earth's ionosphere and troposphere. If the interferometer electronics are sufficiently stable and well understood, it is possible in principle to calculate their contribution to G; but as there are always additional contributions that cannot be calculated, it is the usual practice to determine the G_{mn} factors from observations of sources of known structure, usually point sources. These calibration observations must be made sufficiently frequently that any time dependence of the G factors can be determined, and the calibration sources must be sufficiently close to the object under study that the propagation effects are indistinguishable.

The phase errors due to refraction in the troposphere become significant for frequencies higher than 2 GHz and baselines longer than 1 km, and they vary on time scales between minutes and hours, owing to changes in water-vapor content of the atmosphere (e.g. 2, 50, 53, 66). As it is not practicable to make calibration observations more frequently than once every few minutes, conventional calibration can be almost impossible for baselines longer than a few kilometers at frequencies higher than 10 GHz. At frequencies below 1 GHz, the effects of tropospheric refraction are small compared with ionospheric refraction, which also makes calibration difficult on long baselines. In VLBI, there is the added problem of phase fluctuations in the local oscillators, which can reduce the coherence time to a minute or less. Amplitude errors are generally less important, and have longer time scales, than the phase errors, but again at high frequencies (>5 GHz) variable absorption by the atmosphere and mispointing of the antennas can introduce large amplitude errors with a time scale of a few minutes that cannot be calibrated.

If the complex gain errors G are varying faster than they can be measured, it would appear that it is impossible to recover the complex visibility Γ from the measurements. Progress can be made, however, when more than two antennas are used simultaneously. This is because the number of unknown gain factors is then less than the total number of measurements. In many cases, the gain factors can be associated with individual antennas:

$$G_{mn} = g_m g_n^*, \qquad\qquad 4.$$

where g_m is a complex gain factor for the individual antenna m. Most of the errors that occur in real interferometers can be factorized in this way; we discuss the origin and effects of errors that cannot be factorized in Section 6. In an array of N antennas, up to $N(N-1)/2$ simultaneous interferometer baselines can be formed. If all the antenna pairs are cross-correlated, this factorization reduces the number of unknown (complex) gain factors that must be determined by calibration from $N(N-1)/2$ to N. When N is less than $N(N-1)/2$, it is thus possible in principle to determine both the gain factors themselves and some parameters of the complex visibility of the source from the observations. This is the basis of the self-calibration methods that are the subject of this review.

In Sections 2.2 and 2.3, we enumerate explicitly two sets of parameters, the *closure phases* and *closure amplitudes*, that can be derived from observations with an N-antenna array even if the antenna gain factors g_m are unknown. It is convenient for this purpose to express the terms in Equation 3 as amplitudes and phases:

$$V_{mn} = |V_{mn}| \exp(i\phi_{mn}),$$

$$G_{mn} = |g_m| |g_n| \exp(i\theta_m) \exp(-i\theta_n), \qquad\qquad 5.$$

$$\Gamma_{mn} = |\Gamma_{mn}| \exp(i\psi_{mn}).$$

Equation 5 defines the observed interferometer phase ϕ_{mn}, the true visibility phase ψ_{mn}, and the phase errors θ_m and θ_n associated with each antenna.

2.2   Closure Phase

From Equations 3 and 5, the visibility phase ψ_{mn} on baseline mn is related to the observed phase ϕ_{mn} by

$$\phi_{mn} = \psi_{mn} + \theta_m - \theta_n + \varepsilon_{mn}, \qquad\qquad 6.$$

where ε_{mn} represents the noise in the measurement. The *closure phase* Ψ_{lmn} is formed by summing the observed phases around a triangle of baselines lm, mn, and nl:

$$\begin{aligned}
\Psi_{lmn} &= \phi_{lm} + \phi_{mn} + \phi_{nl} \\
&= \psi_{lm} + \psi_{mn} + \psi_{nl} + \varepsilon_{lm} + \varepsilon_{mn} + \varepsilon_{nl}.
\end{aligned} \qquad 7.$$

It can be seen that the θ_m terms cancel completely. This is true for any closed loop of baselines, not just a triangle. Thus the observed closure phase is uncorrupted by phase errors that are associated with individual antennas. These errors include those due to propagation effects along the line of sight, oscillator drifts, and uncertainties in the source position and baseline vectors. Certain residual errors that do not cancel remain, but these "closure errors" are usually small; we discuss them in Section 6. Figure 1 shows some measurements of phase on three baselines of the VLA, together with the calculated closure phase: the cancellation of antenna-based errors is dramatic.

The closure phase was first used by Jennison in triple-interferometer experiments at Jodrell Bank in the early 1950s (59, 60). As later interferometers became more phase-stable, the technique was no longer needed, until it was revived by Rogers and coworkers (92) for use in VLBI, where propagation effects and oscillator errors make the direct measurement of visibility phases impossible. The term "closure phase" is due to Rogers.

2.3   Closure Amplitude

With four antennas, k, l, m, n, it is possible to form ratios of the visibility amplitudes that are independent of the antenna gain factors:

$$A_{klmn} = \frac{|V_{kl}| \, |V_{mn}|}{|V_{km}| \, |V_{ln}|} = \frac{|\Gamma_{kl}| \, |\Gamma_{mn}|}{|\Gamma_{km}| \, |\Gamma_{ln}|}. \qquad\qquad 8.$$

Such ratios are called *closure amplitudes*. If all six interferometer pairs formed by the four antennas are correlated, three different closure amplitudes can be calculated (A_{klmn}, A_{klnm}, A_{knml}), but only two of these are independent. Ratios of this type were first measured with a four-antenna interferometer by Twiss et al. (116, 117). Their application to VLBI was first

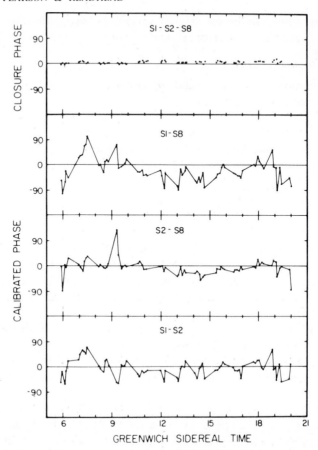

Figure 1   Illustration of the elimination of antenna-based phase errors by the calculation of closure phase. The lower three frames show the visibility phases observed on three baselines of the VLA in bad weather, after conventional calibration on a nearby point source. The upper frame shows the closure phase formed by summing the three visibility phases. The scatter is greatly reduced; the deviations from zero are due to structure in the source (the quasar 3C 147). Reproduced, with permission, from Readhead et al. (84).

discussed by Readhead et al. (85), who introduced the term "closure amplitude" by analogy with "closure phase." Note that at least four antennas are required to measure a quantity that is independent of all the antenna gain factors. With fewer antennas, though, it is still possible to eliminate some of the gain factors; for example, Smith (110), using three antennas and a single correlator, was able to eliminate the unknown gain of one of the antennas. Jennison (59) used three antennas and three correlators, and by monitoring the total power received by one antenna, he could eliminate all the antenna gains; this method is equivalent to using the closure amplitude with one baseline of zero length.

2.4   *Interpretation of Closure Phase and Closure Amplitude*

In an array of N antennas, with $N(N-1)/2$ interferometer pairs, there are N unknown complex antenna gain factors g_m, that is, N unknown amplitude factors and N unknown phases. As only phase differences are important, however, one of the unknown phases can be set to zero, leaving a total of $2N-1$ unknowns. The $N(N-1)/2$ complex visibility measurements thus yield $N(N-1)-(2N-1) = N^2 - 3N + 1$ parameters of the true source visibility. These may be divided into $(N-1)(N-2)/2$ independent closure phases and $N(N-3)/2$ independent closure amplitudes. In an N-antenna array, the fraction of the total visibility information available in the closure quantities is $(N-2)/N$ in the closure phases and $(N-3)/(N-1)$ in the closure amplitudes. These ratios show the rewards to be gained by increasing the number of antennas in the array; with only 4 antennas, 50% of the phase information and 33% of the amplitude information is available, while for 10 antennas (a typical VLBI configuration), these ratios increase to 80% and 78%, respectively.

The interpretation of measurements of closure phase and closure amplitude is difficult, as, unlike the visibilities, they are not related to the sky brightness distribution by a simple Fourier transform relationship. If only closure phases and closure amplitudes are measured, knowledge of both the absolute strength and the absolute position of the source is lost. This can be seen by (*a*) applying a scale factor to $I(\mathbf{s})$ in Equation 2, which merely introduces a scale factor in Γ that cancels when the closure amplitude is calculated; and by (*b*) shifting the function $I(\mathbf{s})$, which introduces phase factors in Γ that cancel when the closure phase is calculated.

In the early work with closure phase and closure amplitude (59, 116), the antennas could be arranged in a redundant configuration, so that all the interferometer baselines were parallel and multiples of a single minimum baseline. In this case it is possible to solve the closure relations to compute the visibilities on all the baselines in terms of that on the minimum baseline, so that the *only* information that is lost is the absolute strength and position of the source. Both the strength and the position of the source can be recovered if it is possible to calibrate the visibility measurement on a single baseline. Rogstad (93) has given an example of a redundant six-antenna array with which it is possible to compute the phases on all multiples of the minimum baseline up to ten, and Rhodes & Goodman (88) have indicated how this technique can be extended to two-dimensional arrays. In VLBI arrays, however, it is not usually possible to arrange the antennas in a redundant configuration, and hence other methods must be used to derive the visibility phases. The methods used, nowadays called either *hybrid mapping* or *self-calibration*, are the subject of Section 5; but we must first digress to discuss the techniques of conventional aperture synthesis, of which the self-calibration methods are an adaptation.

3. APERTURE SYNTHESIS

3.1 *Basic Theory*

Let us rewrite Equation 2 in a coordinate system in which $\mathbf{B}_{mn}/\lambda = (u, v, w)$ and $\mathbf{s} = (x, y, (1 - x^2 - y^2)^{1/2})$:

$$\Gamma(u, v, w) = \int\int I'(\mathbf{s}) \exp[-2\pi i(ux + vy + w(1 - x^2 - y^2)^{1/2}] \, dx \, dy, \qquad 9.$$

where we have written $\Gamma(u, v, w)$ for $\Gamma_{mn}(0)$, and $I'(\mathbf{s}) = I(\mathbf{s})/(1 - x^2 - y^2)^{1/2}$. The sky-coordinates x and y are direction cosines relative to the center of the field of interest. In all the cases that we consider, it is possible to ignore w, and thus Equation 9 reduces to a two-dimensional Fourier transform:

$$\Gamma(u, v) = \int\int I'(\mathbf{s}) \exp[-2\pi i(ux + vy)] \, dx \, dy. \qquad 10.$$

Equation 10 is the fundamental equation of aperture synthesis, and the fundamental problem in aperture synthesis is, given a number of measurements of Γ at different points in the uv-plane, to invert Equation 10 in order to estimate the sky brightness I.

The treatment of w deserves a little discussion. In going from Equation 9 to Equation 10 we have removed a term $-2\pi w$ from the phase, which is usually applied as a phase correction to the measured visibilities, i.e.

$$\Gamma(u, v) = \Gamma(u, v, w) \exp[2\pi i w],$$

and we have ignored higher-order terms in w. When this is not a valid approximation, Equation 9 does not reduce to a Fourier transform, and its inversion is much more difficult (17, 114). Fortunately, w can be ignored in all cases in which all the interferometer baselines lie in a single plane (in which case w may be arranged to be zero by choice of coordinate system) and in cases in which the field of view occupied by the source is sufficiently small $[2\pi w(x^2 + y^2) \ll 1]$, which includes almost all VLBI observations.

The conventional method of inverting Equation 10 is based on the inverse Fourier transform:

$$I'(x, y) = \int\int \Gamma(u, v) \exp[2\pi i(ux + vy)] \, du \, dv. \qquad 11.$$

The measurements are made at a set of only M points (u_j, v_j) in the uv-plane, and the Fourier transform must be approximated by the sum

$$I''(x, y) = \sum_{j=1}^{M} \Gamma(u_j, v_j) W_j \exp[2\pi i(u_j x + v_j y)], \qquad 12.$$

where W_j is a "weight" associated with the jth point. This gives an estimate $I''(x, y)$ of the sky brightness, usually called the "dirty map," which is the convolution of the true sky-brightness $I'(x, y)$ with a point-spread function or "dirty beam":

$$I''(x, y) = I'(x, y) \otimes P(x, y), \qquad 13.$$

where $\otimes$ denotes two-dimensional convolution. The point-spread function $P(x, y)$ is given by the right-hand side of Equation 12, with all the $\Gamma(u_j, v_j)$ terms replaced by 1. The weights may be chosen to adjust the shape of the dirty beam.

In practice, a Fast Fourier Transform (FFT) algorithm is usually used to compute Equation 12, which requires that the data be sampled on a regular rectangular grid. The choice of weights W_j, the techniques for interpolating the measured data onto a rectangular grid, and the distortions that this procedure introduces into the map are described elsewhere (17, 32, 114).

3.2   Image Restoration

The estimate of the sky brightness distribution provided by Equation 12 is a *linear* combination of the measured quantities. It is a convolution of the sky brightness distribution with a beam shape determined by the sampling of the uv-plane and the weights assigned to the measured points. Although this map will be sufficient to show many features of the source in cases where the uv-plane coverage is good, it is not the best possible representation of the sky, and it contains artifacts, most notably the positive and unphysical negative sidelobes around bright peaks. If the sampling of the uv-plane is irregular and uneven, the dirty beam will have large sidelobes that will confuse and obscure the structure of interest in the map. These are due to the erroneous assumption that the visibility is zero at spatial frequencies where it was not measured. The choice of weights W_j is necessarily a compromise between high resolution, low sidelobe level, and good signal-to-noise ratio (78).

A considerable improvement is possible by using *nonlinear* methods of image restoration, and a number of such methods have been introduced into radio astronomy in the last decade. These all attempt to estimate, or "restore," the unmeasured Fourier components in order to produce a more physical map; some of them can actually estimate components at larger spatial frequencies than any that were measured, and thus achieve an increase in resolution (superresolution). These methods also have the advantage over the linear methods that they can take into account the constraint that the map must be everywhere positive or zero (brightness is never negative), and they can also make use of a priori knowledge about the extent of the source and the statistics of the measurement process. Nonlinear methods can always do a better job than any linear technique (121), but the restoration is not unique, because in almost all cases the number N of parameters defining the map (e.g. the brightness at N pixels on a rectangular array) is greater than the number of constraints ($2M$ if complex visibility measurements are made at M points, but less if self-calibration is required). Thus the image reconstruction method must select one of the possible solutions, and the choice of reconstruction method is tantamount to a choice of criteria for selecting the "best" solution. A large number of nonlinear restoration methods have been suggested and experimented with; here, we confine our discussion to three of the most popular: the Högbom CLEAN method, the Gerchberg–Saxton method, and the maximum entropy methods. These methods have matured somewhat since they were last reviewed in this series (13).

CLEANING   The most widely used technique of image reconstruction in radio interferometry is the iterative beam-removing method called CLEAN; as this is a major part of current self-calibration algorithms, we shall discuss it in some detail. CLEAN was developed by Högbom in the early 1970s (55, 94, 107). It starts with a dirty map made by the usual linear Fourier inversion procedure and attempts to decompose this into a number of components, each of which has the shape of the dirty beam. The algorithm searches the dirty map for the pixel with largest absolute value $|I_{max}|$ and subtracts a dirty beam pattern centered on this point with amplitude γI_{max} (the factor γ is called the *loop gain*). Then the residual map is searched for the next largest pixel, and a second beam shape is subtracted, and so on. The iteration is terminated when the maximum residual is consistent with the expected noise level. The result of the iteration is a residual map containing noise and low-level source contributions, plus a list of the amplitudes and positions of the components removed. The list of components, considered as an array of delta functions, is a model of the sky brightness distribution that will reproduce the observed visibilities (within the noise). As this is not convenient for display, the delta functions are usually convolved with a "clean beam" and added back to the residuals to produce a "clean map." The clean beam can be chosen arbitrarily: usually a truncated elliptical Gaussian of about the same size as the central lobe of the dirty beam is used, thus producing a map with the same resolution as the original dirty map but uncontaminated by sidelobes. The Fourier transform of the clean map does *not* reproduce the measured visibilities, but rather the measured visibilities multiplied by a taper function that is the Fourier transform of the clean beam.

It has been shown (105, 106) that CLEAN is equivalent to a least-squares fit of sine functions to the visibility data. There is now considerable experience with CLEAN in practice (e.g. 26). A fast algorithm based on the FFT has been introduced by Clark (21), and a modification of CLEAN with improved behavior has been suggested by Cornwell (28). Perhaps the least-understood aspect of CLEAN is the convolution of the delta-function components with the clean beam: there are no theoretical criteria for choosing the clean beam. Although the normal practice is to use a clean beam comparable to the central lobe of the dirty beam, it is sometimes possible to achieve superresolution by using a smaller clean beam. Experiments have shown, however, that such superresolution is untrustworthy, unless the signal-to-noise ratio is very high and the *uv*-plane coverage is very good. In no case has superresolution by a factor > 2 been demonstrated.

THE GERCHBERG–SAXTON ALGORITHM   The Gerchberg–Saxton or Fienup algorithm was originally introduced for solving the phase problem in electron microscopy (45) and has been adapted for reconstructing an image from measurements of the visibility amplitudes only (37, 38). Although it has not gained a large following in radio astronomy, it is closely related to the iterative self-calibration algorithms that are the subject of Section 5. It is an iterative algorithm involving repeated Fourier transformation between the map (or sky) plane and the *uv*-plane, imposing the known constraints in each domain. In radio astronomy, the constraints are that the map should nowhere be negative and that the *uv*-plane data should agree with the measurements where available; the map can also be constrained to be zero outside one or more selected "windows." Thus, starting with an estimate of the visibilities in the *uv*-plane, as in the standard linear Fourier inversion, we transform to the sky plane, set all negative regions and any regions outside the window to zero, and transform back to the *uv*-plane. To start the next iteration, we replace the visibilities thus derived by the measured visibilities where they are known. The procedure, therefore, allows arbitrary values for the visibility in unsampled regions of the *uv*-plane. Gerchberg (44) has shown that this method is capable of superresolution. Högbom (55) has discussed the use of this method in radio astronomy and pointed out its main drawback (which it shares with the maximum entropy methods)—that the resolution achieved is data dependent and varies across the map. Methods of this type may be regarded as special cases of more general "constrained iterative restoration algorithms" (100).

MAXIMUM ENTROPY AND RELATED METHODS   The limited sampling of the *uv*-plane means that there will be an infinite number of possible maps consistent with the data within the accuracy of the measurements. Not all of these possibilities will be everywhere positive. The maximum entropy methods, which were introduced to radio astronomers by Ables (1), attempt to choose the map that is in some sense the most "likely" or most "reasonable" a priori, and that satisfies the positivity constraint. If we represent the map by the vector $\mathbf{b} = (b_1, b_2, \ldots, b_N)$ formed from the brightnesses of its N pixels and define an "entropy" function $Q(\mathbf{b})$, the methods involve choosing $\mathbf{b}$ to maximize $Q(\mathbf{b})$ subject to the constraint

$$\sum_{j=1}^{M} |\Gamma(u_j, v_j) - V(u_j, v_j)|^2 / \sigma_j^2 \leq M, \qquad 14.$$

where $\Gamma(u_j, v_j)$ is the complex visibility predicted by the map $\mathbf{b}$, $V(u_j, v_j)$ are the visibility measurements, the sum is over the M sampled (u, v) points, and σ_j^2 is the variance of the jth point. The left-hand side of Equation 14 is a χ^2 measure of the disagreement between the visibilities predicted by the map and the measured visibilities, and the constraint ensures that the resulting map will be consistent with the data, taking into account its noise statistics, which are assumed to be Gaussian. Other models of the noise statistics, or other constraints (19), can be incorporated in the maximum entropy method, but this is the most usual. The solution of the constrained maximization of $Q(\mathbf{b})$ is not a trivial matter, but Wernecke (122) and Gull & Daniell (48) have developed iterative numerical techniques that can solve the problem in a reasonable amount of computer time by use of the FFT algorithm.

For the entropy $Q(\mathbf{b})$, two functions have been suggested:

$$Q_1 = \sum_n \log b_n, \qquad 15.$$

$$Q_2 = -\sum_n b_n \log b_n. \qquad 16.$$

There is still some controversy, which we do not enter into here, as to which of these functions should be used and which, if either, deserves to be called entropy. The Q_1 function, favored by Wernecke & D'Addario (122, 123), is a measure of the entropy of the electromagnetic field sampled by the antenna, while Q_2, favored by Frieden (42) and Gull & Daniell (48, 49), is a measure of the configurational entropy or information content of the map itself. In practice, it appears that both functions give similar maps when there are sufficient data. Other "entropy" functions with similar behavior have been suggested (74), corresponding to different a priori assumptions about the sky brightness. A radically different approach is the "maximum sharpness" criterion (4, 71):

$$Q_3 = -\sum_n b_n^2.$$ 17.

The maximum entropy methods have not yet gained a strong following among radio astronomers, probably because the behavior of these methods is not yet fully understood. For example, they can superresolve, in some cases by up to a factor of two, but the reliability of this superresolution has not yet been demonstrated. Nor are all astronomers convinced that the maximum entropy map is the best possible; for example, the maximum of the entropy function may be very broad, and in such cases it is not really appropriate to suggest that any single map is a good estimate of the sky brightness. The maximum entropy methods do, however, have the advantage that they can identify such cases.

4. INTERNAL CALIBRATION METHODS

The calibration problems described above can be overcome if there is a suitable reference source in the same isoplanatic patch as the object. Ideally, such a reference source should be unresolved, and it must be possible to separate its contribution to the visibility measurements from that of the source under study. Such a separation is possible if the reference source is sufficiently far removed from the object, or if the reference source and the object appear in different frequency channels. If the displacement between the reference source and the object is large enough, the interferometer delay (τ in Equation 1) and fringe rate (the rate of change of the visibility phase) will be different for reference source and object, which allows their contributions to the visibility to be separated (82). The power of this technique is illustrated by the VLBI detection of a third compact radio component in the gravitational lens $0957 + 561$ (47). In one particular class of object, the cosmic maser sources, it is possible to separate an internal reference source by frequency. The cosmic masers (both OH and H_2O) generally consist of a number of well-separated components, some of which are unresolved or have very simple structure; the maser lines are narrow, and the different velocities of the components cause them to appear in different frequency channels. In some cases, one frequency channel is found to contain a single unresolved source, which may be used as a reference for calibrating the other channels. Such cases are rare, but if it is possible to obtain the brightness distribution in a particular frequency channel without using phase (e.g. by model-fitting using the amplitude measurements), or if a single channel can be mapped by one of the hybrid-mapping methods of Section 5 (77), the data in that channel can be used as a phase reference. Two methods, "fringe rate mapping" and "phase mapping," have been used to map maser sources.

In fringe rate mapping (69, 70, 119), narrowband spectra of the object are examined, and a reasonably strong unblended spectral feature is chosen as the reference source. The most sensitive interferometer baseline is selected for comparing the fringe rates of all the features. The interferometer phases at each point in the spectrum are adjusted so that the reference source has zero phase at all times, and hence zero fringe rate. The residual fringe rates that remain for the other sources are due to position offsets from the reference source, and the relative positions of the different spectral features can thus be determined from the relative fringe rates. In principle, accuracy of about 10^{-4} arcsec can be achieved by this method (43).

In the phase-mapping (or aperture synthesis) method (87, 120), a reference source in one channel is used to calibrate the phases in all the other channels, which can then be mapped by the usual aperture synthesis methods. The drawback of this method for VLBI observations of masers is that the maser components are spread over a field of view that may be thousands of times larger than the beamwidth, so that the computation involved in aperture synthesis is prohibitive. In some cases it is useful to begin by making a fringe rate map of the entire masing region using the most sensitive interferometer pair, followed by low-resolution phase maps, and finally by phase maps at the full resolution of selected areas encompassing the most interesting groups of masers. The reference features are sometimes spatially resolved, in which case they must be modeled before being used as a reference. The relative phases of the other features depend on baseline vectors, station clocks, and propagation effects, as well as on the positions and the frequencies of these features. The errors are usually dominated by clock and baseline errors, and the positional accuracy achieved is usually about 5×10^{-5} arcsec.

The internal calibration methods are restricted in scope because of the requirement that there be a calibration source close to the object under study. The hybrid-mapping methods, which are the subject of the next section, circumvent this difficulty by using the object itself as the calibration source.

5.   HYBRID-MAPPING METHODS

When insufficient data are available to make a dirty map, the standard mapping and image-restoration procedures cannot be used. The missing data may be, for example, the visibility phases (in which case we want to derive a map from the amplitudes only, and thus implicitly make an estimate of the phases) or the complex antenna gains (in which case we want to estimate the antenna gains as well as the sky brightness distribution from the available data). We might even be faced with the problem of mapping with phase information but no amplitudes.

A number of iterative procedures have been introduced for tackling problems of this type. Although they differ in detail, they all have a similar structure, involving repeated Fourier transformation between the map plane and the *uv*-plane, with adjustments in each domain. They could be regarded as variants of the Gerchberg–Saxton algorithm; but in radio astronomy they are generally termed "hybrid-mapping" methods, following Baldwin & Warner (6) and Readhead & Wilkinson (86).

A single iteration consists of the following steps (Figure 2): (*a*) Starting with a *model* of the sky-brightness distribution, called a *trial map*, obtain *trial visibilities* by Fourier transformation. (*b*) Combine the (incomplete) visibility measurements with sufficient information from the trial visibilities to complete the sampling of the *uv*-plane. (*c*) Compute a *hybrid map* by Fourier inversion of these estimated visibilities. Baldwin & Warner (6) defined a hybrid map to be a map made using amplitudes that are appropriate to one sky brightness distribution and phases that are appropriate to a different one. It seems reasonable to generalize the definition to include maps made from any mixture of measurements and guesses from the trial map. (*d*) Examine the hybrid map and use it to derive a better trial map for the next iteration.

The procedure should be iterated until the hybrid map is indis-

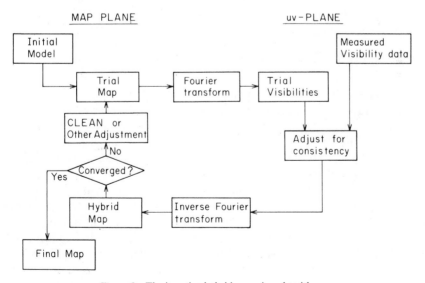

Figure 2   The iterative hybrid-mapping algorithm.

tinguishable from the trial map. There is no guarantee that the procedure will converge—this depends principally on step (d)—but if it does, the final hybrid map is the best estimate of the correct map.

The various hybrid-mapping algorithms in the literature differ in how they combine measured data with trial-map data [step (b)] and in the procedure for deriving the trial map [step (d)]. They all require an initial trial map to start the iteration, which may be obtained in a number of ways. In Sections 5.1, 5.2, and 5.3 we describe first the method of Baldwin & Warner for phaseless aperture synthesis and then discuss two methods for mapping in the presence of large antenna gain and phase errors: the original method of Readhead & Wilkinson, and the more general self-calibration method of Schwab and Cornwell & Wilkinson. Following this discussion, we examine the adaptation of maximum entropy for self-calibration (Section 5.4), as well as two related iterative calibration procedures: the use of redundancy (Section 5.5) and global fringe-fitting in VLBI (Section 5.6).

5.1   Method of Baldwin & Warner

Baldwin & Warner (5–7) were concerned with the problem of phaseless aperture synthesis. The hybrid map in this case is made from the measured amplitudes and trial phases. They considered sky brightness distributions consisting of a number of point sources in an otherwise empty field, and they showed that it is fairly easy in this case to obtain an initial trial map from the map of amplitude squared (the autocorrelation function of the true map) and to identify features that are missing from the trial map by inspection of the hybrid map.

5.2   Method of Readhead & Wilkinson

If we have measurements of amplitude and closure phase only in an array of N antennas, there are $(N-1)$ unknown phase errors that must be estimated by hybrid mapping. Readhead & Wilkinson (86, 125), inspired by the work of Fort & Yee (41), chose to use phases computed from the trial map on $(N-1)$ of the $N(N-1)/2$ baselines, and to use the measured closure phases to deduce phases for the remaining baselines. This requires the observer to specify which visibility phases should be obtained from the trial map; this choice affects the speed of convergence of the algorithm, but it should not change the ultimate solution. Cotton (31) suggested that it would be preferable to choose the $(N-1)$ unknown phase errors to minimize the differences between the hybrid and trial visibilities. Both Readhead & Wilkinson and Cotton used CLEAN on the hybrid map to generate a delta-function model to be used as the next trial map. This has to be done with care, since if CLEAN is allowed to converge, the delta-function trial map would give identical visibilities to the hybrid map, and the iteration would not advance. The solution to this is to not clean too deeply, to omit large negative CLEAN components from the trial map, and to restrict CLEAN to a small "window" (30).

Readhead & Wilkinson made "blind tests" of their method that showed that the procedure usually does converge quite quickly to the correct solution. It has been used widely in VLBI since its first application to observations of the quasar 3C 147 (126) and has been applied to VLA observations made in conditions of poor phase stability (84): Figure 3 shows the great improvement in the map obtained for 3C 147. Readhead et al. (85) extended the method to observations with poor amplitude calibration by utilizing the closure amplitudes in a similar way to the closure phases, using amplitudes from the trial map on some baselines and deducing the remainder from the closure amplitudes. This method has successfully corrected amplitude errors of a factor of 20 (83).

Fort & Yee (41) and Rogers (90) suggested variants of the Readhead–Wilkinson method in which instead of using CLEAN, they obtained a trial map by setting to zero (a) regions of low or negative brightness in the hybrid map, and (b) regions outside an a priori window.

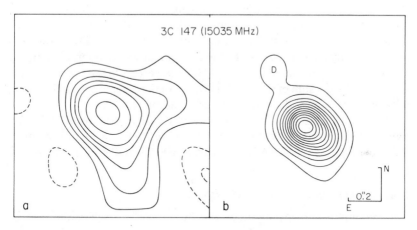

Figure 3   Comparison of maps made with and without self-calibration. (a) VLA map of 3C 147 made with conventionally calibrated phases, some of which are shown in Figure 1. The calibrated phases are severely corrupted owing to bad weather during the observations. All of the extended features in this map are spurious; note the large negative sidelobes. Contour levels: -20, -10, 10, 20, 30,...% of the peak value. (b) Hybrid map made from the same observations, using the closure phases. The extensions seen here have been confirmed by subsequent observations. Contour levels: 4, 12, 20,...% of the peak value. Reproduced, with permission, from Readhead et al. (84).

5.3   *Method of Schwab and Cornwell & Wilkinson*

In the Readhead–Wilkinson method, the closure phases and closure amplitudes are computed explicitly, and the closure relations are solved on a baseline-by-baseline basis. Both Schwab (102) and Cornwell & Wilkinson (29) showed that it was preferable to solve for the unknown antenna gains directly, without explicit calculation of the closure quantities, as this allows a correct treatment of noise in the measurements and is more economical of computer resources. This procedure involves adjusting the estimates of the N complex antenna gain errors to minimize the mean square difference between the measured visibilities and the true visibilities:

$$\sum_{m<n} w_{mn}|V_{mn}-g_m g_n^* \Gamma_{mn}|^2, \qquad\qquad 18.$$

where w_{mn} are weighting factors that can be chosen on the basis of the observed signal-to-noise ratio. This is a nonlinear least-squares problem that must be solved iteratively. When used in hybrid mapping, an estimate of Γ_{mn} is obtained from the trial map, and the estimates of the gain-factors g_m are used with the measured visibilities V_{mn} to form a hybrid map for the next iteration. Both Schwab and Cornwell & Wilkinson used CLEAN to derive a new trial map from the hybrid map. Cornwell & Wilkinson improved on Schwab's method by showing how to take account of prior information about the signal-to-noise ratio of the data and the expected variances and time scales of the gain errors. Least-squares (Equation 18) is not necessarily the best way of choosing the g-factors. An alternative that is less sensitive to occasional large, non-Gaussian errors in the measurements is to minimize the sum of the absolute values of the differences (the L_1 norm) instead of the sum of their squares (the L_2 norm) (14, 103).

Maps made from VLA data (by Schwab's method) and from MERLIN data (by the Cornwell–Wilkinson method) are shown in Figures 4 and 5. Other examples can be found in the article by Bridle & Perley in this volume (15).

5.4   *Maximum Entropy Methods*

The maximum entropy methods can be adapted to self-calibration by changing the constraints on the maximization to ensure that the maximum entropy map is consistent not with the visibilities but with the measured closure quantities. The algorithms that have been used for solving the maximum entropy problem all involve repeated transformation between the sky and the *uv*-plane, so it is a fairly simple matter to incorporate the closure constraints in the *uv*-plane in the same way as is done in the Readhead–Wilkinson or Schwab–Cornwell–Wilkinson methods. We have

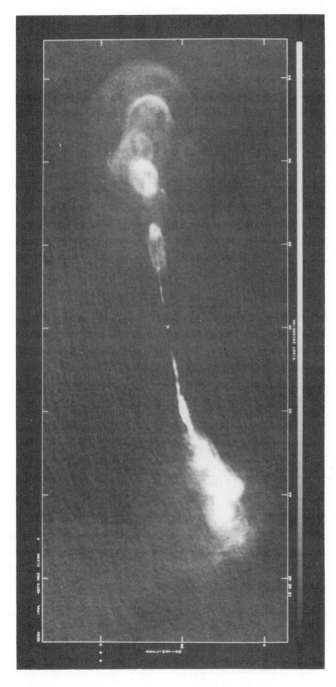

Figure 4   Map of the radio galaxy Her A (3C 348), made using the VLA in three configurations at 5 GHz. The data were self-calibrated using Schwab's method. The resolution is 0″.5, and the radio brightness ranges from 0.05 to 22 mJy per beam area. Reproduced, with permission, from Dreher & Feigelson (34). Copyright © 1984 by Macmillan Journals Limited.

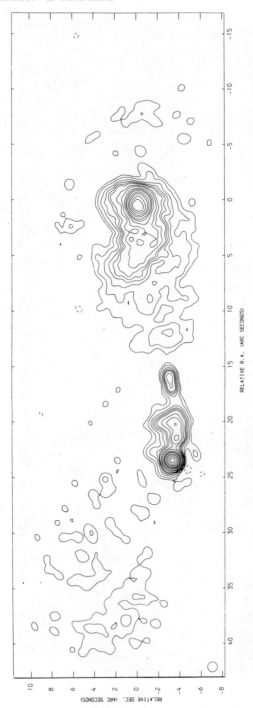

Figure 5   Map of the quasar 3C 249.1 at 408 MHz, made with the six-antenna MERLIN array using the Cornwell–Wilkinson self-calibration method. The lowest contour level is 0.5% of the peak brightness. Reproduced, with permission, from Lonsdale & Morison (65). Copyright © 1983 by the Royal Astronomical Society.

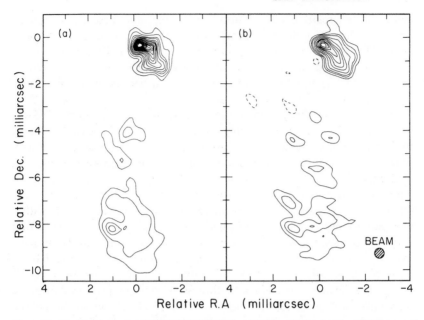

Figure 6   Maps of the nucleus of NGC 1275 (3C 84) at 22 GHz, obtained with a VLBI array of six antennas. (*a*) Map made by the maximum entropy method; (*b*) map made by the Readhead–Wilkinson method. The contour levels are the same in both maps, and the lowest contour is 4% of the peak in (*b*). The maximum entropy map has a higher peak owing to its higher resolution.

had considerable experience with a program of this sort written by S. F. Gull (private communication), and Sanroma & Estalella (99) have described some tests of a similar program. An example of a map made from VLBI data by the maximum entropy method is shown in Figure 6. The only problem with these methods is that the final map is more difficult to interpret than one with the same resolution at all points. It is clear, however, that as maximum entropy methods become better understood in conventional aperture synthesis, they will also be used for self-calibration.

5.5   *Redundancy*

As we mentioned in Section 2.4, if there is sufficient "redundancy" in the interferometer array, it is possible to recover almost all the visibility data from measurements of the closure quantities. A redundant array is one in which there are baselines of the same length and orientation between different pairs of antennas. If there is enough redundancy, the number of independent errors may be reduced to one gain and one phase error, and only the absolute intensity and position of the source are lost. For arrays with few redundant baselines the errors in the relative gains accumulate and make the full relative gain solution unreliable, but these errors can be reduced in heavily redundant arrays if all of the redundant information is used. (This is not, however, a good argument for building a redundant array: with a fixed number of antennas, it is usually best to arrange them to optimize the sampling of the uv-plane, rather than to maximize redundancy.)

Noordam & de Bruyn (75, 76) have used this method on the WSRT, which has many redundant spacings. The WSRT is a linear east-west array that, at any instant, has a one-dimensional fan beam. The beam rotates with time, and the brightness distribution is built up from successive scans, each of duration $\ll 12$ hr. For each scan, all spacings that carry redundant information are used to obtain a weighted least-squares solution of the relative antenna gains. This fixes the relative gains of all antennas within one scan. Ionospheric and tropospheric wedges cause phase gradients that vary between scans, and there are a variety of effects that cause long-term overall gain variations. Successive scans must therefore be corrected for both phase gradients and overall gain variations. This is done by comparing the one-dimensional brightness distribution produced from the scan with a source model, and then correcting the level and position of the brightness distribution. This is nearly the same as the self-calibration procedures described above; the only difference is that the correction is done on a number of one-dimensional scans sequentially instead of on the whole source simultaneously.

When this method is used, the WSRT can achieve a very high dynamic range. The dynamic range is the ratio of the strongest feature to the weakest believable feature in the map. In their map of the active galaxy NGC 1275, Noordam & de Bruyn obtained a dynamic range of 10,000:1, which is the highest that has yet been obtained. NGC 1275 is a very strong source; with weaker sources, the dynamic range will be limited by the noise in the measurements. An example of a high dynamic range map produced by this method is displayed in Figure 7, which shows the faint extended structure in the quasar 3C 345. The lowest contour is a factor of 8000 below the peak brightness of the core.

A similar method has been used to correct phase errors in an interferometer at Toyokawa Observatory (58).

5.6   Global Fringe-Fitting in VLBI

In VLBI, as in other types of interferometry, one measures the amplitude and phase of the complex visibility. Owing to uncertainties in the propagation delay, the baseline parameters, and the oscillator phase, the delay (τ in Equation 1) and the fringe rate (the rate of change of the measured visibility phase) cannot be predicted accurately, so the procedure

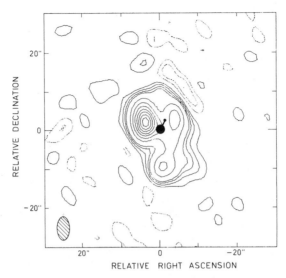

Figure 7   Map of the quasar 3C 345 at 5 GHz, made using the redundancy method on the WSRT. A point source of 10.15 Jy and a jet of 0.35 Jy have been subtracted at the positions indicated by black dots. The lowest contour level is 1/8000 of the peak brightness. Reproduced, with permission, from Schilizzi & de Bruyn (101). Copyright © 1983 by Macmillan Journals Limited.

adopted (70) is to cross-correlate the signals at several different trial delays and to integrate with several trial fringe rates, and then to search for the maximum amplitude as a function of delay and fringe rate. In the past, this search was carried out independently for each baseline. This requires a high signal-to-noise ratio on every baseline in order to obtain reliable detections. A large part of the uncertainty in delay and fringe rates is due to errors that cancel around triangles, like the closure phases. The number of unknown delays or fringe rates is N (the number of antennas), rather than $N(N-1)/2$ (the number of baselines). This constraint can be used to advantage if the data for all $N(N-1)/2$ baselines are used simultaneously in a *global* search instead of a baseline-by-baseline one. For example, if the source is unresolved, the "closure delays" and "closure fringe rates" are exactly zero, so that if fringes can be detected with good signal-to-noise ratio on two baselines, the delay and fringe rate can be predicted exactly for the third baseline of the triangle, which allows the third visibility to be measured even if its signal-to-noise ratio is low.

If the source is resolved, however, the closure delays and closure fringe rates are not zero, and a more complicated procedure must be used. Schwab & Cotton (104) have devised and programmed such a method of performing the global solution for a resolved source. As in the hybrid-

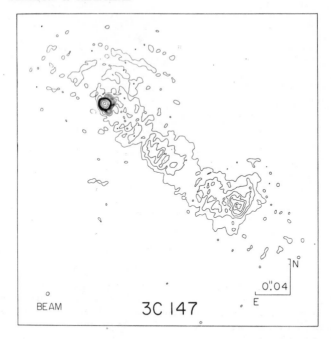

Figure 8 Ten-antenna VLBI map of the quasar 3C 147 at 1.7 GHz, made using the global fringe-fitting technique to obtain high signal-to-noise ratio and high dynamic range. This is a higher-resolution map of the object shown in Figure 3: the resolution is 0″004. The lowest contour level is 0.5% of the peak brightness. Reproduced, with permission, from Simon et al. (109).

mapping methods, a source model is used to predict the structure effects in delay and fringe rate, which are used in the global solution. The visibility measurements obtained are then used to make a map by self-calibration methods, and the map is in turn used as a model for another iteration of global fringe-fitting. Global fringe-fitting is thus a doubly iterative procedure, with an iterative self-calibration nested within each iteration of the fitting procedure. It is expensive in computer time, but the rewards it brings are well worth this investment, as it permits the use of data on baselines that are much too weak for single-baseline fitting and increases the overall signal-to-noise ratio. An example of a map made from VLBI data using this technique is shown in Figure 8.

6.   CAPABILITIES AND LIMITATIONS

The various self-calibration algorithms attempt to estimate the sky brightness distribution from measurements of the closure phases and closure amplitudes. If we knew the sky brightness distribution, we could calculate the closure quantities; thus self-calibration is a member of the general class of *inverse problems* that occur in many fields (e.g. 10, 81). Many techniques have been introduced for solving these problems, and some of them are closely related to the self-calibration algorithms. For any algorithm, we should consider the questions of *uniqueness, construction,* and *stability.* Uniqueness is a mathematical matter: given perfect observations, is there only one solution to the problem? If there is not, there certainly will not be a unique solution with incomplete and noisy data. The question of construction is, will the algorithm actually find the solution in a reasonable number of iterations? Finally, the algorithm is unstable if a small change in the measured quantities leads to a large change in the solution. None of these questions has yet been answered for self-calibration, but in this section we discuss the effects of incomplete *uv*-plane coverage and measurement errors on the convergence of the algorithms used for self-calibration.

6.1   *Uniqueness and uv-Plane Coverage*

One may naturally be worried that there might be several brightness distributions compatible with the closure quantities, and that the final self-calibrated map might not be unique, even if there were no errors in the measurements. We have already seen that if only the closure quantities are measured, we cannot determine the absolute strength and location of the source; that is, if $I(\mathbf{s})$ is a brightness distribution consistent with the closure quantities, then so is $cI(\mathbf{s} + \mathbf{s}')$ for any factor c and displacement $\mathbf{s}'$. Are these the *only* distributions consistent with the closure quantities? The question of uniqueness is tied up with that of the adequacy of the *uv*-plane sampling for mapping the source, and we should distinguish the problems of *uv*-plane coverage that are common to both conventional aperture synthesis and self-calibration from those that occur only with self-calibration.

There are no general rules available for determining whether a given sampling of the *uv*-plane is sufficient to map a source by aperture synthesis. To map a source that is confined within a rectangle of size θ radians, it is sufficient to sample the *uv*-plane on a rectangular grid with spacing $< 1/(2\theta)$ wavelengths (the Nyquist sampling rate); but unless this sampling is extended indefinitely, there will be infinitely many brightness distributions consistent with the measured visibilities (12). If the *uv*-plane coverage extends to a maximum baseline of u_{max}, the resolution of the map is $\sim 1/u_{max}$, which amounts to saying that all the possible brightness distributions will look similar when convolved with a beam of this width. In practice, particularly with VLBI arrays, the sampling of the *uv*-plane can be sparse and irregular, and there are no clear-cut rules for what field of view can be mapped, or with what resolution; it depends very much on the source structure and the signal-to-noise ratio. Because the total number of measurements is less, it is clear that the number of picture elements that can be mapped will be less when closure phases are measured in place of visibility phases, or closure amplitudes in place of visibility amplitudes.

A case that has received considerable attention is that of mapping with amplitudes but no phases. This "phase problem" occurs in X-ray diffraction, electron microscopy, speckle interferometry, and many other fields. In general, the solution is ambiguous. Complete knowledge of the visibility amplitudes is insufficient to define a unique brightness distribution, even with the constraints that the brightness distribution must be positive and confined to a finite solid angle; but it now appears that in *most* cases, the brightness distribution can be reconstructed uniquely (ignoring the "trivial" ambiguities: an arbitrary translation of the image or a rotation through 180° does not change the visibility amplitudes). There is still some controversy, however, as to the correct interpretation of the word "most" (e.g. 9, 18, 36, 57). It is likely that knowledge of the closure phases is sufficient to resolve the ambiguities, but this has yet to be proved.

The converse problem of mapping with phase measurements but no amplitude measurements has received less attention, as, in practice, phases are usually much harder to measure than amplitudes. An interesting review of phase-only mapping has been presented by Oppenheim & Lim (79), who point out that the phase of the visibility function carries more information about the brightness distribution than the amplitude does. They give examples of hybrid images formed from the visibility amplitudes of one image and the visibility phases of another. The hybrid image closely resembles the original image with the same phase. This effect was also noticed by Baldwin & Warner (6) in one of their tests. A hybrid map made with the correct phase but an amplitude of unity bears much more resemblance to the true map than one made with the correct amplitude and zero phase. As in the amplitude-only case, it appears that phase-only mapping can produce a unique solution (apart from a scale factor), or equivalently that the magnitude can be recovered from the phase, in many practical situations in which the source is confined to a finite area.

The mathematical question of uniqueness in self-calibration is unsolved. In real observations, the situation is confused by the incomplete *uv*-coverage and the errors in the measurements. But in practice it appears that the self-calibration algorithms do converge to a unique solution in cases where the *uv*-coverage is adequate. This is supported by the results of "blind tests" of the self-calibration methods (85, 86). As in normal aperture synthesis, it is not clear what constitutes "adequate" sampling of the *uv*-plane, but heuristically at least it appears that self-calibration requires more complete sampling than normal aperture synthesis.

6.2   *Noise*

In aperture synthesis with good *uv*-plane coverage, the fundamental limitation to the quality of the map is statistical noise in the measurements. Noise is introduced by the receivers, the microwave background, galactic background emission, atmospheric emission, pickup from the ground, and other causes. To a good approximation it may be regarded as white, Gaussian noise. When a linear image restoration procedure is used, it is straightforward to compute the effect on the map of Gaussian noise in the visibility measurements: it appears as Gaussian noise on the map with rms (in order of magnitude) $\sigma/\sqrt{M}$ per beam area, where σ is the rms noise in an individual visibility measurement, and M is the number of independent visibility measurements. The noise deflections on the map are correlated on a scale of the synthesized beamwidth. A more complete discussion can be found in the general reviews of aperture synthesis referred to earlier; a particularly good treatment is that of Napier (72). The noise behavior of the nonlinear image restoration methods is less well understood, and self-calibration is inherently nonlinear, so a rigorous understanding of the effect of noise on self-calibration is difficult. Self-calibration methods increase the noise level on the map because of errors in the determination of the complex antenna gain factors, which are themselves determined from the noisy data. Cornwell (25, 27) has shown that the rms noise level on the map is increased by a factor of $\sqrt{(N-1)/(N-2)}$ if only phases are corrected, or $\sqrt{(N-1)/(N-3)}$ if both phases and amplitudes are corrected. These factors are small when the number of antennas (N) is large. This, however, is only true if each individual visibility measurement can be made with a good signal-to-noise ratio in a time less than the time scale of variation of the antenna gain factors. If the signal-to-noise ratio is too small, it is not possible to make a reliable estimate of the antenna gain factors before they change, and self-calibration is impossible.

6.3   *Convergence*

A robust self-calibration algorithm should converge uniformly toward a unique solution and should be insensitive to the choice of starting model. In general, an arbitrary starting model can be used, though in practice convergence is greatly improved by starting with a sensible model; for example, a point source is a good starting model for mapping a source that is known to be barely resolved. When mapping with amplitudes only, the starting phases can be chosen to be zero or can be entirely random; when mapping with phases only, the amplitudes can be chosen to be unity or, even better, to decrease with increasing baseline in the manner of realistic sources. In a partially phase-stable array, like the VLA, an adequate starting model can be obtained from a conventionally cleaned map; when the visibility phases are completely unknown, as in VLBI, a starting model must be obtained by least-squares model-fitting techniques. We have found that it is important to choose a model that fits the closure phases well; the fit to the amplitudes is less critical. This seems to be related to the observation mentioned earlier that phase carries more information about the source structure than does amplitude. It also helps to adjust only the phases in the first few iterations of self-calibration, and to only adjust the amplitudes when a good fit to the closure phases has been obtained. In a typical observation with the VLA, two iterations of phase adjustment, followed by two iterations of amplitude and phase adjustment, will be sufficient for the self-calibration process to converge. In a VLBI experiment, with larger initial errors, poorer uv-coverage, and a poorer starting model, 20 iterations may be needed for a complex source. Perhaps the most critical requirement for convergence is that the map-plane constraints be applied rigorously to exclude negative features, and features outside as small a window as possible, from the trial map (30). This is particularly important when the uv-coverage is poor. The window can be chosen in a variety of ways (for example, by inspection of maps made with other instruments or at other frequencies, or from the known field of view of the instrument). An objective method is to examine the autocorrelation function of the map, made by Fourier transformation of the squares of the visibility amplitudes (6, 38). Use of a small window in early iterations and a larger one in later iterations has sometimes been found to be helpful. These and other practical considerations have been discussed by Cornwell & Wilkinson (27, 30, 124).

6.4   Closure Errors

The self-calibration techniques discussed above are all based on the assumption that the gain and phase errors can be factorized by antenna. In practice there are instrumental errors that cannot be factorized. These "closure errors" may be regarded as systematic errors in the measurements of closure phase and closure amplitude. Unless the closure errors can somehow be removed, they are the limiting factor in the quality of the maps produced by self-calibration methods. A fractional error f in the estimated gains will introduce sidelobes on the map at a level of about $f/\sqrt{M}$, where M is the number of independent errors. For example, with constant 1% errors on 10 baselines, the dynamic range of the map would be limited to about 300:1. Thus, to achieve a dynamic range of 1000:1 or 10,000:1 (comparable to the best VLA maps), it is important to keep uncorrectable closure errors well below a level of 1% in gain or 0.5° in phase.

We should distinguish errors that remain constant in time from those that vary. Constant errors can in principle be measured by observations of a point source and then eliminated in the data processing. Alternatively, if the additional number of unknowns is not large, they can be solved for in the hybrid mapping in the same way as the station gains (29). But if the errors are variable on a short time scale, they can be neither measured nor corrected, and thus they can only be tackled by improving the quality of the equipment.

A number of causes of closure errors in existing interferometer arrays have been identified (22). Some errors may arise in the hardware of the correlators themselves and in the subsequent electronics and software; but if these are identical for all the correlators in the array, they may be ignored. Thompson & D'Addario (113) have shown that the factorization of gains by antenna is, in general, not possible if the antenna transfer functions are different functions of frequency, and they have investigated in detail the effects of mismatched transfer functions. Usually, identical rectangular bandpasses are used at each antenna; as an example of the sort of error that can occur, a shift of the center frequency by 0.7% of the bandwidth can introduce a gain closure error of 1%. It is clear that when self-calibration methods are to be used, the tolerances on bandpass distortions are very stringent. In present VLBI systems, designed before the advent of self-calibration, large bandpass mismatches are not uncommon and are probably a limiting factor in image quality. Another serious problem in present VLBI systems has been pointed out by Rogers (91): a phase closure error results if the antennas differ in their polarization purity; for example, a 0.5% error in polarization purity can cause a 1° phase closure error.

If the antenna gains are varying on a short time scale, one must be careful not to average the data for too long a period, as this will introduce closure errors (124). This can be a problem in VLBI, where the coherence time of the atmosphere and oscillators is short and where there is a temptation to average for as long as possible to improve the signal-to-noise ratio.

We have implicitly assumed that the complex antenna gains are the same for all parts of the source being observed. In VLBI, the field of view is small, and this is usually a good assumption, but it breaks down when mapping large fields. For example, the atmosphere may not be homogeneous across the source. It is usually assumed that the atmosphere is "isoplanatic" over regions ~ 1° at radio wavelengths, but this is not true in bad weather. Also, if the antenna pointing is not constant, gain errors are introduced that vary across the field of view, with the worst errors occurring where the slope of the antenna pattern is greatest. These are difficult to correct by self-calibration, though it may be possible to make some progress by self-calibrating different isoplanatic patches independently.

7.   THE FUTURE

The development of self-calibration methods for radio astronomy has already proved extremely valuable. By eliminating sidelobes due to unavoidable amplitude and phase errors, it has enabled arrays such as the VLA, the WSRT, and MERLIN to far surpass initial expectations of their imaging quality; and it has been the single most important factor in transforming VLBI into a true imaging technique.

In principle the techniques of self-calibration are not confined to radio astronomy, and we can expect to see them applied at other frequencies in the future. Michelson (68) demonstrated the feasibility of optical astronomical interferometers in 1920, but the effects of atmospheric turbulence impose severe limitations on this technique. Problems arise because both the time scale of atmospheric fluctuations is short (10 ms) and the transverse coherence length of the wavefront is small (10 cm). If we assume an aperture of diameter 10 cm, an integration time of 10 ms, and a bandwidth of 10 nm, and if we allow for the imperfect photon efficiency of even the best CCD detectors, the limiting magnitude of a ground-based Michelson interferometer is 8. There are a number of projects in progress to construct ground-based Michelson interferometers (e.g. 63, 108), but so far these efforts have met with limited success due to the stringent rigidity requirements and the difficulty of tracking fringes. It would clearly be of great value to be able to use self-calibration methods with such instruments. Although methods have been suggested for measuring closure phase with optical arrays (46, 88, 93), they have not yet been applied in astronomy. The application of self-calibration to heterodyne infrared interferometers (61, 115) will be more straightforward, since the techniques involved are similar to those of radio interferometry.

In order to observe faint astronomical objects, apertures larger than 10 cm are needed; transverse coherence then no longer obtains across the aperture, and the image is spread out over an angle λ/d, where d is the transverse coherence length (about 10 cm). A short exposure yields a "speckle pattern" image, which is caused by interference of rays from different portions of the wavefront. Labeyrie (62) first pointed out that these speckle patterns can be used to derive source structure down to the resolution limit of the aperture. That this must be so can be seen from the fact that the speckle pattern is formed in part by the interference of rays from the extreme edges of the aperture, and these rays clearly contain information about the transverse coherence due to the source over these distances. If the instantaneous telescope response to a point source, the so-called instantaneous "point-spread function," were known, it would be possible to obtain an image of the object by deconvolution. However, only the time-averaged point-spread function is known, which makes it difficult to extract phase information, although it is easy to construct the autocorrelation function of the image.

If two large apertures are operated as an interferometer, the speckles are modulated by interference fringes. Thus it might at first sight seem possible to use closure to extract phase information from these fringes. However, there is a random phase relationship between the fringes in different speckles. It might prove possible to remove this random component by the techniques of adaptive optics (e.g. 20, 51, 71). These techniques are closely related to the self-calibration methods used in radio astronomy (30): the adjustment of the phase of small elements of the telescope pupil is analogous to the adjustment of the complex antenna gains in a radio array. Unfortunately, the adjustment must be made in the 10-ms time scale of the atmospheric fluctuations, which limits the technique to bright objects. Radio astronomers have the advantage that they can record the visibility measurements and make the gain adjustments in a computer after all the measurements have been accumulated.

In the next decade we may expect to see new instruments developed to exploit the possibilities opened up by self-calibration in the radio and other wavebands. In the United States, the Astronomy Survey Committee of the National Research Council (73) has recommended the construction of a Very Long Baseline Array (VLBA) of radio telescopes designed to produce images with an angular resolution of 0.3 milliarcsec, and it is hoped that construction of this instrument will begin in 1984. Similar projects are under way in Canada (64) and Australia. The Astronomy Survey Committee has also recommended placing a VLBI antenna in space to increase the resolving power of ground-based arrays. We can also anticipate the development of large millimeter-wavelength arrays and infrared and optical interferometers on the ground or in space. All of these projects will rely on the self-calibration methods developed for radio astronomy to make high-quality images.

ACKNOWLEDGMENTS

We thank our colleagues who contributed to this review by discussion and comment, and by sending us preprints of their work. We are especially grateful to J. W. Dreher, D. H. Hough, T. W. B. Muxlow, R. T. Schilizzi, and R. S. Simon for providing the examples of self-calibration that appear in the figures. This work was supported in part by the US National Science Foundation (AST 8210259).

Literature Cited

1. Ables, J. G. 1974. *Astron. Astrophys. Suppl.* 15:383–93
2. Armstrong, J. W., Sramek, R. A. 1982. *Radio Sci.* 17:1579–86
3. Baars, J. W. M., Houghoudt, B. G. 1974. *Astron. Astrophys.* 31:323–31
4. Baker, P. L. 1981. *Astron. Astrophys.* 94:85–90
5. Baldwin, J. E., Warner, P. J. 1976. *MNRAS* 175:345–53
6. Baldwin, J. E., Warner, P. J. 1978. *MNRAS* 182:411–22
7. Baldwin, J. E., Warner, P. J. 1979. See Ref. 118, pp. 67–82
8. Bare, C., Clark, B. G., Kellermann, K. I., Cohen, M. H., Jauncey, D. L. 1967. *Science* 157:189–91
9. Bates, R. H. T. 1982. *Optik* 61:247–62
10. Boerner, W.-M., Jordan, A. K., Kay, I. W. 1981. *IEEE Trans. Antennas Propag.* AP-29:185–89
11. Born, M., Wolf, E. 1975. *Principles of Optics.* Oxford: Pergamon. 808 pp. 5th ed.
12. Bracewell, R. N. 1958. *Proc. IRE* 46:97–105
13. Bracewell, R. N. 1979. *Ann. Rev. Astron. Astrophys.* 17:113–34
14. Branham, R. L. 1982. *Astron. J.* 87:928–37
15. Bridle, A. H., Perley, R. A. 1984. *Ann. Rev. Astron. Astrophys.* 22:319–58
16. Broten, N. W., Legg, T. H., Locke, J. L., McLeish, C. W., Richards, R. S., et al. 1967. *Science* 156:1592–93; *Nature* 215:38
17. Brouw, W. N. 1975. In *Methods in Computational Physics*, ed. B. Alder, S. Fernbach, M. Rotenberg, 14:131–75. New York: Academic
18. Bruck, Yu. M., Sodin, L. G. 1979. *Opt. Commun.* 30:304–8
19. Bryan, R. K., Skilling, J. 1980. *MNRAS* 191:69–79
20. Buffington, A., Crawford, S., Pollaine, S. M., Orth, C. D., Muller, R. A. 1978. *Science* 200:489–94
21. Clark, B. G. 1980. *Astron. Astrophys.* 89:377–78
22. Clark, B. G. 1981. *VLA Sci. Memo. No. 137*, Natl. Radio Astron. Obs., Charlottesville, Va. 10 pp.
23. Cohen, M. H. 1969. *Ann. Rev. Astron. Astrophys.* 7:619–64
24. Cohen, M. H. 1973. *Proc. IEEE* 61:1192–97
25. Cornwell, T. J. 1981. *VLA Sci. Memo. No. 135*, Natl. Radio Astron. Obs., Charlottesville, Va. 4 pp.
26. Cornwell, T. J. 1982. See Ref. 114, Lect. No. 9
27. Cornwell, T. J. 1982. See Ref. 114, Lect. No. 13
28. Cornwell, T. J. 1983. *Astron. Astrophys.* 121:281–85
29. Cornwell, T. J., Wilkinson, P. N. 1981. *MNRAS* 196:1067–86
30. Cornwell, T. J., Wilkinson, P. N. 1984. *Proc. URSI/IAU Symp. Meas. Process. Indirect Meas., Sydney, 1983*, ed. J. A. Roberts. Cambridge: Cambridge Univ. Press. In press
31. Cotton, W. D. 1979. *Astron. J.* 84:1122–28
32. D'Addario, L. R. 1980. *Proc. Soc. Photo-Opt. Instrum. Eng.* 231:2–9
33. Davies, J. G., Anderson, B., Morison, I. 1980. *Nature* 288:64–66
34. Dreher, J. W., Feigelson, E. D. 1984. *Nature* 308:43–45
35. Elgaroy, O., Morris, D., Rowson, B. 1962. *MNRAS* 124:395–403
36. Fiddy, M. A., Brames, B. J., Dainty, J. C. 1983. *Opt. Lett.* 8:96–98
37. Fienup, J. R. 1978. *Opt. Lett.* 3:27–29
38. Fienup, J. R. 1982. *Appl. Opt.* 21:2758–69
39. Fomalont, E. B. 1973. *Proc. IEEE* 61:1211–18
40. Fomalont, E. B., Wright, M. C. H. 1974. In *Galactic and Extragalactic Radio Astronomy*, ed. G. L. Verschuur, K. I. Kellermann, pp. 256–90. New York: Springer-Verlag
41. Fort, D. N., Yee, H. K. C. 1976. *Astron. Astrophys.* 50:19–22
42. Frieden, B. R. 1972. *J. Opt. Soc. Am.* 62:511–18
43. Genzel, R., Reid, M. J., Moran, J. M., Downes, D. 1981. *Ap. J.* 244:884–902
44. Gerchberg, R. W. 1974. *Opt. Acta* 21:709–20
45. Gerchberg, R. W., Saxton, W. O. 1972. *Optik* 35:237–46
46. Goodman, J. W. 1970. In *Progress in Optics*, ed. E. Wolf, 8:1–50. Amsterdam: North-Holland
47. Gorenstein, M. V., Shapiro, I. I., Cohen, N. L., Corey, B. E., Falco, E. E., et al. 1983. *Science* 219:54–56
48. Gull, S. F., Daniell, G. J. 1978. *Nature* 272:686–90
49. Gull, S. F., Daniell, G. J. 1979. See Ref. 118, pp. 219–25
50. Hamaker, J. P. 1978. *Radio Sci.* 13:873–91
51. Hardy, J. W. 1978. *Proc. IEEE* 66:651–97
52. Heeschen, D. S. 1981. In *Telescopes for the 1980s*, ed. G. Burbidge, A. Hewitt, pp. 1–61. Palo Alto, Calif: Ann. Rev. Inc.

53. Hinder, R., Ryle, M. 1971. *MNRAS* 154:229–53
54. Hjellming, R. M., Bignell, R. C. 1982. *Science* 216:1279–85
55. Högbom, J. A. 1974. *Astron. Astrophys. Suppl.* 15:417–26
56. Högbom, J. A., Brouw, W. N. 1974. *Astron. Astrophys.* 33:289–301
57. Huiser, A. M. J., van Toorn, P. 1980. *Opt. Lett.* 5:499–501
58. Ishiguro, M. 1974. *Astron. Astrophys. Suppl.* 15:431–43
59. Jennison, R. C. 1958. *MNRAS* 118:276–84
60. Jennison, R. C. 1966. *Introduction to Radio Astronomy.* London: Newnes, New York: Philos. Libr., Inc. 160 pp.
61. Johnson, M. A., Betz, A. L., Townes, C. H. 1974. *Phys. Rev. Lett.* 33:1617–20
62. Labeyrie, A. 1970. *Astron. Astrophys.* 6:85–87
63. Labeyrie, A. 1978. *Ann. Rev. Astron. Astrophys.* 16:77–102
64. Legg, T. H. 1982. *Phys. Can. (La Physique au Canada)* 38:3–7
65. Lonsdale, C. J., Morison, I. 1983. *MNRAS* 203:833–51
66. Mathur, N. C., Grossi, M. D., Pearlman, M. R. 1970. *Radio Sci.* 5:1253–61
67. McReady, L. L., Pawsey, J. L., Payne-Scott, R. 1947. *Proc. R. Soc. London Ser. A* 190:357–75
68. Michelson, A. A. 1920. *Ap. J.* 51:257–62
69. Moran, J. M. 1976. In *Methods of Experimental Physics,* ed. M. L. Meeks, 12C:228–60. New York: Academic
70. Moran, J. M., Papadopoulos, G. D., Burke, B. F., Lo, K. Y., Schwartz, P. R., et al. 1973. *Ap. J.* 185:535–67
71. Muller, R. A., Buffington, A. 1974. *J. Opt. Soc. Am.* 64:1200–10
72. Napier, P. J. 1982. See Ref. 114, Lect. No. 3
73. National Research Council. 1982. *Astronomy and Astrophysics for the 1980s. Volume 1: Report of the Astronomy Survey Committee.* Washington DC: Natl. Acad. Press. 189 pp.
74. Nityananda, R., Narayan, R. 1982. *J. Astrophys. Astron.* 3:419–50
75. Noordam, J. E. 1981. *Proc. ESO Conf. Sci. Importance High Angular Resolut. Infrared Opt. Wavelengths, Garching,* ed. M. H. Ulrich, K. Kjär, pp. 257–61. Garching bei München, FRG: ESO
76. Noordam, J. E., de Bruyn, A. G. 1982. *Nature* 299:597–600
77. Norris, R. P. 1983. *Proc. Int. Conf. VLBI Tech., Toulouse, Fr., 1982,* pp. 341–44. Toulouse: Cent. Natl. Étud. Spat. 488 pp.
78. Oldenburg, D. W. 1976. *Geophys. J. R. Astron. Soc.* 44:413–31
79. Oppenheim, A. V., Lim, J. S. 1981. *Proc. IEEE* 69:529–41
80. Palmer, H. P., Rowson, B., Anderson, B., Donaldson, W., Miley, G. K., et al. 1967. *Nature* 213:789–90
81. Parker, R. L. 1977. *Ann. Rev. Earth Planet. Sci.* 5:35–64
82. Peckham, R. J. 1973. *MNRAS* 165:25–38
83. Readhead, A. C. S., Hough, D. H., Ewing, M. S., Walker, R. C., Romney, J. D. 1983. *Ap. J.* 265:107–31
84. Readhead, A. C. S., Napier, P. J., Bignell, R. C. 1980. *Ap. J. Lett.* 237:L55–60
85. Readhead, A. C. S., Walker, R. C., Pearson, T. J., Cohen, M. H. 1980. *Nature* 285:137–40
86. Readhead, A. C. S., Wilkinson, P. N. 1978. *Ap. J.* 223:25–36
87. Reid, M. J., Haschick, A. D., Burke, B. F., Moran, J. M., Johnston, K. J., Swenson, G. W. 1980. *Ap. J.* 239:89–111
88. Rhodes, W. T., Goodman, J. W. 1973. *J. Opt. Soc. Am.* 63:647–57
89. Rogers, A. E. E. 1976. In *Methods of Experimental Physics,* ed. M. L. Meeks, 12C:139–57. New York: Academic
90. Rogers, A. E. E. 1980. *Proc. Soc. Photo-Opt. Instrum. Eng.* 231:10–17
91. Rogers, A. E. E. 1983. *VLB Array Memo. No. 253,* Natl. Radio Astron. Obs., Charlottesville, Va. 2 pp.
92. Rogers, A. E. E., Hinteregger, H. F., Whitney, A. R., Counselman, C. C., Shapiro, I. I., et al. 1974. *Ap. J.* 193:293–301
93. Rogstad, D. H. 1968. *Appl. Opt.* 7:585–88
94. Rogstad, D. H., Shostak, G. S. 1971. *Astron. Astrophys.* 13:99–107
95. Ryle, M. 1952. *Proc. R. Soc. London Ser. A* 211:351–75
96. Ryle, M. 1975. Nobel Lecture delivered in Stockholm on 12 Dec. 1974. In *Les Prix Nobel en 1974.* Stockholm: P. A. Norstedt & Söner. Reprinted in *Science* 188:1071–79 and *Rev. Mod. Phys.* 47:557–66
97. Ryle, M., Hewish, A. 1960. *MNRAS* 120:220–30
98. Ryle, M., Vonberg, D. D. 1946. *Nature* 158:339–40
99. Sanroma, M., Estalella, R. 1983. *Proc. Int. Conf. VLBI Tech., Toulouse, Fr., 1982,* pp. 391–407. Toulouse: Cent. Natl. Étud. Spat. 488 pp.
100. Schafer, R. W., Mersereau, R. M., Richards, M. A. 1981. *Proc. IEEE* 69:432–50

101. Schilizzi, R. T., de Bruyn, A. G. 1983. *Nature* 303:26–31
102. Schwab, F. R. 1980. *Proc. Soc. Photo-Opt. Instrum. Eng.* 231:18–25
103. Schwab, F. R. 1981. *VLA Sci. Memo. No. 136*, Natl. Radio Astron. Obs., Charlottesville, Va. 20 pp. and Erratum (Feb. 1982)
104. Schwab, F. R., Cotton, W. D. 1983. *Astron. J.* 88:688–94
105. Schwarz, U. J. 1978. *Astron. Astrophys.* 65:345–56
106. Schwarz, U. J. 1979. See Ref. 118, pp. 261–75
107. Schwarz, U. J., Cole, D. J., Morris, D. 1973. *Aust. J. Phys.* 26:661–73
108. Shao, M., Staelin, D. H. 1980. *Appl. Opt.* 19:1519–22
109. Simon, R. S., Readhead, A. C. S., Wilkinson, P. N. 1984. *Proc. IAU Symp. 110, VLBI and Compact Radio Sources.* Dordrecht: Reidel. In press
110. Smith, F. G. 1952. *Proc. Phys. Soc. London Sect. B* 65:971–80
111. Swenson, G. W., Mathur, N. C. 1968. *Proc. IEEE* 56:2114–30
112. Thompson, A. R., Clark, B. G., Wade, C. M., Napier, P. J. 1980. *Ap. J. Suppl.* 44:151–67
113. Thompson, A. R., D'Addario, L. R. 1982. *Radio Sci.* 17:357–69
114. Thompson, A. R., D'Addario, L. R., eds. 1982. *Synthesis Mapping, Proc. NRAO–VLA Workshop, Socorro, N. Mex.*, Green Bank, W. Va: Natl. Radio Astron. Obs.
115. Townes, C. H., Sutton, E. C. 1981. *Proc. ESO Conf. Sci. Importance High Angular Resolut. Infrared Opt. Wavelengths, Garching*, ed. M. H. Ulrich, K. Kjär, pp. 199–223. Garching bei München, FRG: ESO
116. Twiss, R. W., Carter, A. W. L., Little, A. G. 1960. *Observatory* 80:153–59
117. Twiss, R. W., Carter, A. W. L., Little, A. G. 1962. *Aust. J. Phys.* 15:378–86
118. Van Schooneveld, C., ed. 1979. *Proc. IAU Colloq. 49, Image Formation from Coherence Functions in Astronomy, Groningen, The Neth., 1978.* Dordrecht: Reidel. 340 pp.
119. Walker, R. C. 1981. *Astron. J.* 86:1323–31
120. Walker, R. C., Burke, B. F., Haschick, A. D., Crane, P. C., Moran, J. M., et al. 1978. *Ap. J.* 226:95–114
121. Wells, D. C. 1980. *Proc. Soc. Photo-Opt. Instrum. Eng.* 264:148–56
122. Wernecke, S. J. 1977. *Radio Sci.* 12:831–44
123. Wernecke, S. J., D'Addario, L. R. 1977. *IEEE Trans. Comput.* C-26:351–64
124. Wilkinson, P. N. 1983. *Proc. Int. Conf. VLBI Tech., Toulouse, Fr., 1982*, pp. 375–89. Toulouse: Cent. Natl. Étud. Spat. 488 pp.
125. Wilkinson, P. N., Readhead, A. C. S. 1979. See Ref. 118, pp. 83–91
126. Wilkinson, P. N., Readhead, A. C. S., Purcell, G. H., Anderson, B. 1977. *Nature* 269:764–68

Ann. Rev. Astron. Astrophys. 1984. 22:131–55

SOLAR ROTATION

Robert Howard

Mount Wilson and Las Campanas Observatories,
Carnegie Institution of Washington, Pasadena, California 91101

1. INTRODUCTION

The accurate determination of the solar rotation rate and its variations with latitude and time are topics that have taken on increased importance in the last decade. This has resulted from the realization that it is the interaction of rotation and convection in an ionized medium that causes the dynamo action that is responsible for the solar activity cycle. The rotation is, therefore, an important element of solar activity. Our knowledge of the subsurface phenomena that constitute the basic cycle mechanism is very limited, so differential rotation data are a vital part of the information that we have available in this area.

Unfortunately, the modern data from various observatories are at times not in good agreement. In this review, these discrepancies are discussed, and generally the observational side of the subject is emphasized. The determination of the rotation rate of the Sun is not as simple a task as it may seem at first sight; each technique has its difficulties and uncertainties.

Several reviews have appeared in recent years in this field (Gilman 1974, Howard 1978, Noci 1978).

2. SURFACE ROTATION

The rotation of the surface layers is determined spectroscopically. All rotation tracers, such as sunspots or plages, are believed to be linked by magnetic field lines to deeper layers, and their rotation rates probably represent depths greater than the visible surface (Foukal 1972, Stix 1976).

Modern photoelectric determinations of spectroscopic line shifts started in the late 1960s. Before that time, photographic techniques were used. The earlier measurements suffered from influence by supergranular surface velocities because of the small number of measurements that were made for

131

0066–4146/84/0915–0131$02.00

each rotation determination. Now that velocity measurements cover at least a large part of the solar surface, the influence of supergranular motions is reduced but not eliminated. Supergranular motions contribute to the limit on the detection of giant convection cells [which is currently around 10 m s^{-1} (LaBonte et al. 1981)] and provide a few meters per second of the daily variations seen now in the spectroscopically determined equatorial rotation rate.

2.1   Early Measurements

The earliest spectroscopic measurements were as much verifications of the Doppler effect as determinations of solar rotation (Vogel 1872). Early in this century solar spectrographs were greatly improved, and fairly consistent results began to appear (e.g. Adams 1911, J. S. Plaskett 1915, H. H. Plaskett 1916, Storey 1932, St. John 1932). Nevertheless, systematic errors [notably scattered light (De Lury 1939)] plagued the early observers.

The technique used in the early measurements was to expose spectra near the east and west limbs at various latitudes and subtract the Doppler shifts of one or more spectrum lines at the two limbs to obtain the rotation rate. Although individual measurements could be made with great precision, this method was particularly sensitive, as mentioned above, to influence by supergranular motions, which have amplitudes of the order of 1 km s^{-1}. Also, because of the rapid variation of the limb redshift (Adam 1959) with radial distance from the disk center near the solar limb, small errors in position led to sizable errors in calculated angular velocity, especially near the west limb.

Because of these difficulties and the cumbersome task of exposing and processing many high-dispersion spectra, spectroscopic rotation rates are now almost all measured using photoelectric detectors of one sort or another.

2.2   Modern Measurements

Photoelectric measurements from full-disk scans promised to improve considerably the precision of such observations (Howard & Harvey 1970). The sensitivity of rotation results obtained in this way to influences by other surface velocity fields is less than that of the photographic method simply because of the larger number of velocity measurements that are obtained for each rotation determination. The various geometrical corrections that must be made to these measurements for effects such as the rotation and orbital motion of the Earth are described in detail by Howard & Harvey (1970). A vector formulation has recently been given by Kubičela & Karabin (1983). Indeed, the internal consistency of the new observations was excellent; however, variations of the measured rotation rate from day to day posed a problem in interpretation. Earlier observers, who also saw rapid variations in the rotation rate (e.g. Plaskett 1916, Evershed 1931), were convinced that these variations were solar in origin. Later, Svalgaard et al. (1978) showed that at least a large part of the short-term variation seen in the modern measurements was due to instrumental and atmospheric scattered light.

Other possible sources of systematic error in photoelectric measurements include interference fringes in the spectrograph and instrumental polarization (Howard et al. 1980). More recent observations have been made with such effects either eliminated or compensated. Recently, an error in dispersion due to small errors in published line wavelengths was discovered (Snodgrass et al. 1984). This has lowered previous rotation rates from Mount Wilson measurements by 0.55%.

The interpretation and elimination of large-scale background velocity fields is another serious problem in the analysis of photoelectric Doppler data. Such effects may also exist, of course, for some photographic measurements, although measurements made simply at either end of a parallel of latitude are in principle insensitive to velocity patterns symmetric about the central meridian. A time-varying velocity pattern seen at an amplitude of several tenths of a kilometer per second, symmetric about the central meridian, was named "ears" by the Mount Wilson observers (Howard et al. 1980). Beckers (1978) suggested that a meridional flow toward the poles could explain the effect. Such a flow had been detected earlier by Duvall (1977). This poleward flow was observed in the Mount Wilson data also (Howard 1979), although Balthasar & Wöhl (1980) find no such effect from sunspot motions. The combination of limb redshift and meridional motion leads to a large-scale background velocity field that must be accounted for in the analysis of differential rotation measurements. In general, when data covering the full disk are used, velocity patterns that are symmetric about the central meridian are not a serious problem for the analysis of the rotation, which is an antisymmetric pattern. However, at the level of a few meters per second, all patterns interact in the analysis, and since the origin of the weak patterns is not certain, the assumption that they are truly symmetric about the central meridian is not a very secure one. On the whole, the problem of the background velocity pattern on the Sun has not been satisfactorily resolved. In particular, the existence of meridional flows from equator to poles, although fairly well established, is still somewhat in doubt because of the uncertainties inherent in the limb shift. It is still not known if the limb shift—or the meridional flow, as measured spectroscopically (see also Tuominen 1961, 1976, Balthasar & Wöhl

1980)—varies on any time scale. In addition, the proper functional representation of the limb shift has not yet been determined. This whole area is one where a great deal of progress can be made in the next few years.

2.3   The Rotation Rate

In spite of the modern methods and the care taken at all observatories to correct for systematic errors, there is not good agreement between the various observers on even the equatorial rotation rate of the Sun. A photospheric rate that at times during the cycle is significantly different from the often-quoted Newton & Nunn (1951) sunspot rate was first suggested from modern observations by Howard & Harvey (1970). This has been confirmed by some other investigators (Duvall 1982, Livingston & Duvall 1979), although the agreement between these groups is by no means good. The discrepancies between the groups are due at least in part to their having observed in different epochs. The Stanford group (Svalgaard et al. 1978, Scherrer & Wilcox 1980, Scherrer et al. 1980) find a value for the equatorial rotation rate that is slightly higher than the others referenced above. Because the other observatories have found that the solar rotation rate varies slowly with time, this difference has varied over the years, but it is currently near 1%. The Stanford group finds an equatorial rate of 2.899 ± 0.019 μrad s^{-1} (Scherrer & Wilcox 1980). Their value shows essentially no variation with time. This is nearly identical to the sunspot group results of Newton & Nunn (1951), a rate that refers to 136 single-spot groups in the interval 1934–44. The Newton & Nunn result is $\omega = 2.905(\pm 0.002)$ $-0.598(\pm 0.018) \sin^2 B$ μrad s^{-1}. The Stanford rate is also in fairly good agreement with that of magnetic fields and magnetic-field-related features. The Stanford value should be decreased by 0.55% because of the dispersion error mentioned above (J. M. Wilcox, private communication), and this further supports a difference between the rates observed at Stanford and some of those derived elsewhere.

Although currently the equatorial rotation values from Stanford and other observatories are close, in past years the discrepancies amounted to several percent, even when scattered-light corrections were applied to the other values. These differences remain unexplained. The recently discovered correlation between sunspot and photospheric rotational velocities over intervals of some years lends credence to the measurements that show the slow variations (Gilman & Howard 1984a).

A further indication of the reality of a difference between photospheric and spot rotation is provided by the work of Foukal (1976, 1979), who measured the Doppler rotation rate of sunspots and the surrounding photosphere and found a difference comparable with the observed

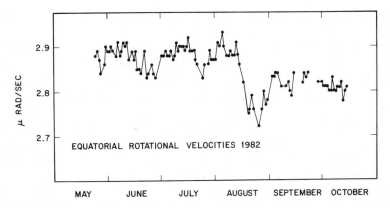

Figure 1   Daily equatorial rotational velocities of the Sun from Mount Wilson Doppler data for a six-month period in 1982.

difference from separate photospheric Doppler and sunspot motion observations.

The measured short-term temporal variations of the equatorial rotation rate that were so prominent in earlier analyses (e.g. Howard & Harvey 1970) have decreased in amplitude as scattered light has been taken into account and as instruments have improved (Howard et al. 1983b). Nevertheless, some variations in the rate remain. Figure 1 shows the equatorial rotation during an interval of several months from Mount Wilson magnetograph observations. The daily variations are of the order of 1% (20 m s^{-1}). Variations over periods of about a month are several times this amplitude. It is not presently known if the variations seen in this plot are solar in origin. There does seem to be a periodicity in the data that is close to the rotation rate. This problem is now under study. The recent addition of a reference wavelength has further improved the quality of the velocity signal at Mount Wilson, and variations of the sort seen in Figure 1 continue to be observed.

2.4   *Differential Rotation*

It has been known since the remarkable work of Carrington (1863) that the Sun rotates differentially : lower latitudes rotate faster than higher latitudes. In the first studies of this effect, using sunspots, only latitudes up to about 30° could be measured, because this is about as far poleward as spots are seen. Spectroscopic measures confirm, however, that this shear continues without interruption to at least within 10° or so of the pole (Howard & Harvey 1970).

The variation of rotation with latitude has traditionally been expressed as a series $\omega = a - b \sin^2 B - c \sin^4 B$. In general, for low-latitude features such as sunspots the last term is omitted.

Differential rotation can be generated in numerical models of a rotating spherical gas by the action of convection and rotation (e.g. Gilman 1979, Gilman & Miller 1981, Glatzmaier 1983). For the most part in these models, the rotation tends to be approximately constant on cylinders concentric with the rotation axis. This implies a rotation rate that decreases with depth. On the other hand, sunspots and other magnetic-field-related features show a faster rotation than the photosphere, and it is generally assumed that they are connected to faster-rotating subsurface layers. There is a discrepancy here that is not yet resolved.

In Figure 2, there are two examples of differential rotation prediction from the compressible global convection model of P. A. Gilman (unpublished). The profiles are compared to the grand average profile of Howard & Harvey (1970), plotted relative to the same rotating reference frame. In model 1 the eddy viscosity and thermal diffusivities are assumed independent of depth with a mixing-length-related value of $5 \times 10^{12} \text{ cm}^2 \text{ s}^{-1}$. In model 2 these parameters decrease with depth at a rate proportional to the square root of the fluid density. This simulates crudely the fact that the model resolves mixing-length-scale eddies at the bottom of the convection zone but not at the top. The convection transports one solar luminosity through the convection zone, and the zone is assumed to rotate at 2.6 μrad s^{-1}. The convection zone is deep ($0.6R$). The deeper zone gives better agreement with observational results but is in conflict with the current best estimates of this quantity. The model convection zone begins 10% below the surface in order to avoid small-scale convection near the surface, which the model could not resolve. Subsurface rotation is found to be nearly constant on cylinders concentric with the rotation axis. The increase of rotational velocity at high latitudes decreases with longer calculations. The calculations were cut off at 60° latitude in order to avoid computational instabilities near the pole. From this work it is clear that current compressible models are able to reproduce accurately the rotation profile of the Sun.

Figure 3 shows the rotation of the Sun as a function of latitude from various studies. The differences are evident, although it should be pointed out that the time intervals covered in each of these studies are different. As is discussed below, there is convincing evidence for real variations with time in the rotation rate of the Sun. Thus, comparing results from different epochs is not a proper procedure. Unfortunately, it is often necessary to make such comparisons because not all techniques are available from any one epoch. In Figure 3 the basic shape of the differential rotation is similar in the various curves, although generally the sunspot curves show slightly more differential rotation than do the Doppler curves, i.e. the variation with latitude is greater. Note that the Newton & Nunn (1951) results refer to single-spot groups, while the Howard et al. (1984) results refer to all individual sunspots.

In Figure 3 the north and south hemispheres of the Sun are combined. Actually, at times small differences are seen (using various techniques) between the rotation rates of the two hemispheres (Howard & Harvey 1970, Schröter et al. 1978, Godoli & Mazzucconi 1979, Arevalo et al. 1982,

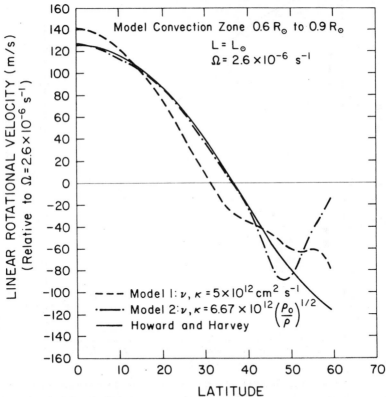

COMPRESSIBLE CONVECTION MODEL DIFFERENTIAL ROTATION COMPARED TO SOLAR OBSERVATIONS

Figure 2 Rotation results from two compressible models of the solar convection zone, compared with the Howard & Harvey (1970) observational results from Doppler data. In the first model, the eddy viscosity (ν) and thermal diffusivity (κ) are assumed independent of depth. In the second model, these quantities decrease with depth proportional to the square root of the density. Reproduced with permission from Dr. P. A. Gilman.

Howard et al. 1983a, Snodgrass 1983). In nearly all these cases the northern hemisphere has been found to rotate slightly slower than the southern hemisphere. It is interesting to note that during the last 15 years the northern hemisphere of the Sun has contained about 10% more magnetic flux than the southern (Howard 1974a, Tang et al. 1984), and that the active regions in the north have been about 10% larger than those in the south (Tang et al. 1984). It is tempting to speculate that these two asymmetries are related, i.e. the slower rotation of the northern hemisphere is due in some way to the excess magnetic flux there. However, generally the magnetic

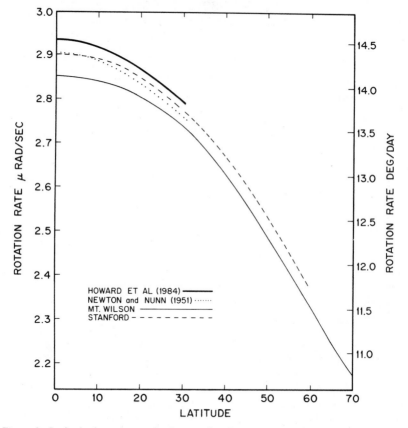

Figure 3 Latitude dependence of solar rotation from several sources and over several intervals. The Newton & Nunn (1951) results are for 136 single-spot recurrent groups in the interval 1934–44. The Stanford results are from Scherrer et al. (1980), with a correction applied for the dispersion error (Snodgrass et al. 1984). These results refer to the interval 1976–79. The Mount Wilson sunspot results (Howard et al. 1984) refer to the interval 1921–82. The spectroscopic results from Mount Wilson (Howard et al. 1983b) are for the interval 1967–82. The dispersion correction of Snodgrass et al. (1984) has been applied to these results.

fields are seen (at least at some observatories) to rotate slightly faster than the nonfield photosphere, so the situation is not clear.

The rotation curves shown in Figure 3—and in other figures in this review—are smoothed representations of the rotation using a simple series solution. Complicated latitude variations cannot be represented in these displays. In particular, a decrease in the rotation rate within a few degrees latitude of the equator is seen in latitude rotation data, but it is not visible in these figures (Howard & LaBonte 1980, Howard et al. 1980). This equatorial deceleration is within 2–3° of the equator and has an amplitude of the order of 5 m s^{-1}. Such an effect, with the same amplitude, has also been seen in the magnetic rotation data (Snodgrass 1983). Deubner & Vazquez (1975) found low-latitude fine structure in the rotation rate in data covering a relatively short time interval. This pattern is not seen in the Mount Wilson data and may be attributed to supergranular or other local velocity fields.

2.5 Torsional Oscillations

From Mount Wilson Doppler observations it has been determined that the Sun undergoes low-amplitude torsional oscillations about the rotation axis in several modes (Howard & LaBonte 1980, LaBonte & Howard 1982a,b). The first to be discovered has a wave number of 2 hemisphere^{-1}. This mode is illustrated in Figure 4. The amplitude is of the order of 5 m s^{-1}. This is a traveling wave that originates near the poles and moves equatorward, requiring 22 years to reach the solar equator. In the lower latitudes the torsional shear zone between the fast stream on the equator side and the slow stream on the pole side is the locus of solar activity. The well-known latitude drift of activity during the cycle, characterized by the "butterfly diagram," fits nicely the latitude drift of the torsional shear zone, i.e. the activity at any epoch occurs centered on the torsional shear latitude for that epoch. This may be seen in Figure 4 by a comparison of the magnetic flux data and the torsional streams. The total magnetic flux is a good indicator of solar activity, and its latitude variation follows closely that of active regions. At any time, the shear zone for the current cycle may be seen at low latitudes in Figure 4, and the torsional oscillations for the next cycle may be seen at high latitudes. The latitude drift speed of the shear zone is about 2 m s^{-1}.

These motions have also been detected in Doppler measurements at low latitudes by Scherrer & Wilcox (1980). In addition, it has been claimed that the torsional pattern is present in sunspot motions (Godoli & Mazzucconi 1982, Tuominen et al. 1984), although in both cases the result is marginally significant, judging from the errors.

It may be (and has been) argued that the torsional oscillation result is

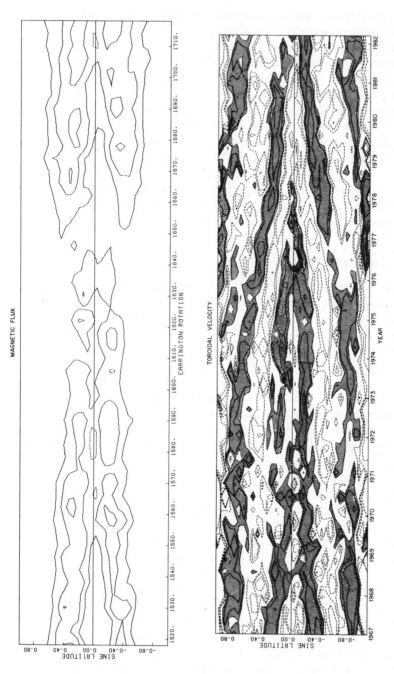

MAGNETIC FLUX

CARRINGTON ROTATION

TOROIDAL VELOCITY

YEAR

Figure 4 (*Lower plot*) Torsional oscillations of the Sun as seen in the velocity signal of the Mount Wilson magnetograph for the interval shown. The velocity contours are ±1.5, 3, 6 m s^{-1}. This represents, over 4-rotation averages, the excess or deficit of the rotational velocity at each latitude compared with a smoothed latitude curve. The solid contours and stippling indicate positive velocities, or westward flows. The dashed contours represent slower than normal rotation, or eastward flows. A constant rotation deficit near the equator has been removed. (*Upper plot*) For comparison, the total magnetic flux of the Sun, also in 4-rotation averages. The contour levels are 1.5, 3, 6 × 10^{21} Mx. Both plots represent data averaged over 34 equal intervals in sine latitude [from Howard & LaBonte (1983)]. Reproduced with permission from B. J. LaBonte and the International Astronomical Union.

spurious because the method used to obtain the torsional waves seen in Figure 4 can lead to a nonphysical result. In order to get such a low-amplitude signal, a difference technique is used. The series solution for each day's full-disk data is one representation of the rotation, but in a separate reduction using the same raw data, the rotation of the Sun is determined in a number of narrow latitude zones. The series rotation solution for each day is subtracted from the latitude zone solutions, and the result is the torsional pattern seen in Figure 4. The reason the pattern shows up is that the series cannot represent patterns with as many inflections in latitude as are seen in Figure 4. In order to examine the torsional oscillations in a way that does not depend upon the peculiarities of the $\sin^2 B$, $\sin^4 B$ series, LaBonte & Howard (1982a) used the latitude zone data alone to look for the oscillations. They subtracted the full-interval average rotation rate for each latitude zone from the time-varying values during a long interval to form a plot similar to Figure 4. In this plot (Figure 8 of LaBonte & Howard 1982a), the low-latitude deceleration described above was not seen because the average for each latitude zone had been subtracted, and the deceleration is a relatively constant phenomenon. Although this plot is noisy, the basic elements of the torsional oscillation can be seen, but these are superposed upon a larger torsional oscillation with about the same amplitude as the first one. This 1-hemisphere^{-1} oscillation is characterized by a faster rotation at high latitudes near solar activity maximum and a faster rotation at low latitudes near solar activity minimum. The high-latitude variation was discovered earlier by Livingston & Duvall (1979).

An additional criticism of the torsional oscillation reduction is that the high-latitude portion of the pattern may be generated by the reduction process from the low-latitude pattern, which itself could then be due to the velocity fields of active regions. LaBonte & Howard (1982b) have examined this possibility by generating a low-latitude pattern similar to the real torsional pattern, and then following the normal reduction procedures used for the solar data. The high-latitude pattern that is generated by this procedure is quite weak and bears no resemblance to that found from the solar data.

There is evidence for a 1/2-hemisphere^{-1} oscillation in a comparison of equatorial rotation values determined from north and south data separately (Howard et al. 1983a). This amplitude is comparable to those of the two modes discussed above.

The cyclic variation of the sunspots and Doppler rotation rate (discussed below) is essentially another torsional oscillation. The amplitude is about 200 m s^{-1}, which is much larger than the other modes.

It is evident that the Sun is very easily excited in torsional oscillation modes, i.e. a low-amplitude modulation of the rotation rate. So far there is no theoretical explanation for how such motions interact with activity, and even the restoring force for these modes is unknown, although it may be presumed to be subsurface magnetic fields.

2.6   Variation of Rotation Rate with Height in the Atmosphere

Starting with some of the earliest spectroscopic measurements, it has been noted that the rotation rates derived from various Fraunhofer lines may differ (Adams 1911). More recent studies of a number of lines have often explained these variations as being due to a rotational height gradient within the photosphere (Aslanov 1963, Solonsky 1972, Balthasar 1983). In all cases the observations are satisfied by a rotation that decreases with depth in the photosphere. Livingston (1969) and Gasanalizade (1980) have independently criticized these results—Livingston because he does not find a significant variation with optical depth using a very sensitive photoelectric method, and Gasanalizade because of alleged errors in the determinations of optical depths. In addition, Livingston & Milkey (1972) have suggested that blends with telluric lines have led to systematic errors in some of the studies of depth dependence of rotation.

The notion of a significant variation of the rotation rate with depth in the photosphere [Aslanov (1963) found 14%!] is difficult to reconcile with the model of a photosphere that is (at least in the upper layers) convectively stable and that in the lower layers may experience some convective overshoot from the hydrogen convection zone below. What would maintain the shear? In addition, the magnetic field lines that thread the photosphere are known to be very nearly vertical, but the small inclination that is seen is in the direction of trailing the rotation—just the opposite to what one would expect if the flux tubes were sheared by lower layers that rotated slower than higher layers (Howard 1974b).

A rotation rate difference between the photosphere and the Hα chromosphere (determined spectroscopically) has been seen for some years (Adams 1911, Livingston 1969). The magnitude of this effect is about 4% as seen by Livingston, and 2% as seen by Adams. The Hα measurements give a faster rotation rate. Both observers noted that the differential rotation is significantly decreased in Hα observations. At times the rate is nearly flat over the active latitudes. The variation of rotation across the sunspot zones is quite small in Hα.

A significant gradient from photosphere to chromosphere is nearly as hard to believe as a gradient within the photosphere, for the same reasons as listed above. On the other hand, the amplitude of the effect is so large that it is difficult to find an error in the observations or the analysis. Moreover, the chromospheric rotation from tracers (discussed below) confirms the faster chromospheric rate, so this is a confirmation of the height gradient if one can assume that the results from tracers and the spectroscopic technique measure the same quantity—which in the photosphere is not the case.

3. ROTATION FROM TRACERS

3.1 *Sunspots*

Sunspots gave the first indication that the Sun rotated at all. Numerous observers in the seventeenth and eighteenth centuries observed the rotation of the Sun from the motions of the sunspots (for recent analyses of older observational data, cf. Abarbanell & Wöhl 1981, Arevalo et al. 1982, Eddy et al. 1976, Herr 1978, Yallop et al. 1982), but it was not until the work of Carrington (1863) that a systematic study was made of the rotation derived from sunspots—and it was not until that work that differential rotation was noticed. His value for the orientation of the solar rotation axis is still used for ephemeris calculations. Carrington observed from 1853 to 1861. The Greenwich Observatory began synoptic observations of sunspots in 1874 and continued this homogeneous data set until 1976. A similar series was started in 1977 at the Heliophysical Observatory in Debrecen, Hungary. The Greenwich observers (e.g. Newton & Nunn 1951) analyzed the data for rotation of the Sun. More recently, numerous studies have used a digitized version of these data (e.g. Arevalo et al. 1982, Balthasar & Wöhl 1980, Godoli & Mazzucconi 1979, Ward 1966).

3.1.1. THE GREENWICH DATA   For over a century, the Greenwich observers, using white-light photographs from Greenwich, Kodaikanal, and elsewhere, maintained a thorough coverage of facular and sunspot group characteristics, including group positions from which rotation rates may be calculated. The earliest publications contained information on individual sunspots. This is the longest homogeneous telescopic solar record in existence. This invaluable data set has seen much use over the years. The Greenwich observers (e.g. Newton 1924, Newton & Nunn 1951) analyzed the data for rotation rates of sunspot groups, and Ward (1966) reanalyzed the Greenwich data, after digitizing it. A great deal of what we know about solar rotation comes from these data. Unfortunately, the Greenwich data from this century have contained information only on sunspot groups, not individual spots. Spot results from these data have relied on single-spot groups, which are not a representative sample of individual sunspots. Thus, many of the results on "sunspots" in the published literature contain biased data that do not represent the characteristics of individual sunspots. An exception is the work of Kearns (1979), who used individual-spot data over a short interval.

Tuominen & Kyröläinen (1982) have discussed the differences between the rotation-latitude curve at solar maximum and solar minimum, using the Greenwich data. The slope of the curve is found to be more steep at minimum.

3.1.2 THE MOUNT WILSON DATA　More recently, the Mount Wilson white-light data have been measured and analyzed (Howard et al. 1984). These data, although they do not represent as thorough a coverage as do those of Greenwich, were all obtained with the same instrument and were all measured by the same person in a relatively short interval. The reduction covers the years 1921 through 1982. A further advantage of these data is that they represent information on individual sunspots, not just sunspot groups.

The spot position and area data were obtained as follows. Positions and areas for individual spot umbras were recorded for each day. The spots for each day were sorted into groups by proximity. The groups were then matched with data from the next day to find the return of each group. There is very little ambiguity in such a process. The spots within the group on two successive days were matched, using a simple pattern recognition technique, to find returns of individual spots. This technique gave a match the next day of slightly less than 30% of the spots identified. This low rate of returns is due in part to the short lifetimes of spots and in part to the conservative nature of the matching technique. Another indication of the conservativeness of the technique is the fact that about 10% of the groups that returned the next day contained no spots that were found to match on the next day. In all, 96,283 sunspots were identified as next-day returns in the 62-yr interval, out of a total of 32,046 groups. All the rotation results discussed here from these data represent determinations from positions on two consecutive days; no full rotation return data were calculated.

As may be seen in Figure 5, the difference between spots and spot groups, when discussing rotation, is not merely semantic. There is a significant difference between the rotation rates of sunspots and sunspot groups. Figure 5 shows the rotation rates of sunspots in three size categories and of sunspot groups. The rotation curve for all spots is very close to that of the smallest spots, but it is slightly (0.2%) slower. From an inspection of Figure 5, one can determine why there is a difference between the rates of groups and all spots. The groups, whose positions are area-weighted, reflect very strongly the rotation characteristics of the largest spots. Spots of different sizes rotate at different speeds—the largest spots rotate slowest, as may be seen easily in Figure 5. Thus the groups, which are strongly influenced by the largest spots, rotate slower by about 0.8% than the average of the spots. This influence may be seen by the peculiar shape of the latitude curve at

0–5°N in both the large-spot data and the group data. The different rotation rates of spots of different sizes may result from different depths of the subsurface flux ropes of the two types of spots (Gilman & Foukal 1979). Alternatively, the viscous drag of the larger spots may be greater than that of the smaller spots as they drift through the slower-moving photosphere. The latter hypothesis seems unlikely because a study has shown that the difference between the rotation rates of the fast and slow spots for a year does not depend upon the annual average of the difference between the Doppler and spot rotation rates for that year (Howard et al. 1984). Thus, if there is a drag that differs with spot size, it does not vary as the difference between the photospheric and spot rates vary, which is contrary to what one would expect.

Annual averages of the sunspot rotation rate (for all spots) show variations that are correlated with phase in the solar activity cycle (Gilman

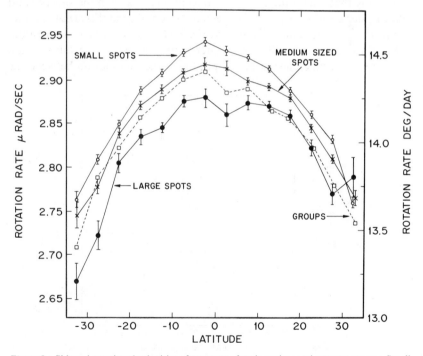

Figure 5 Sidereal rotational velocities of sunspots of various sizes and sunspot groups. Small spots are less than 5 millionths of the area of the hemisphere. Medium-sized spots are from 5 to 15 millionths. Large spots are greater than 15 millionths. The data are divided into 5° latitude zones, and each error bar represents one standard deviation from all the spots or groups within that zone in the 62-yr interval covered (1921–82). These results are all from the work of Howard et al. (1984).

& Howard 1984a). Figure 6 shows this effect. There is a clear peak in the rotation curve near both the maximum (at +4 and −7 yr) and minimum phases. Note that the rotation rate averaged over the sunspot latitudes is scaled for each latitude zone so that the shift of the mean latitude of the spots with phase in the cycle will not bias the results. The Mount Wilson Doppler results show a similar variation for data over the last 15 years. This is shown in Figure 7, where the agreement between the two curves may be seen. Note that the amplitude of the variation is greater in the photosphere than in the spots by about a factor of 2.

This is, in a sense, a torsional oscillation—a periodic speeding up and slowing down of the rotation rate. If we assume that the angular momentum variation at the surface results from an exchange with lower layers, we may speculate about which layers are involved. The depth involved in the exchange cannot be that where the sunspots are "anchored" (assuming that anchoring is a valid concept) because then the sunspot and photospheric variations would be seen out of phase, and this is not what we see in Figure 7. It is most likely that the exchange of angular momentum takes place with depths that are greater than those where the sunspots are anchored. The difference in the amplitude of the modulation between the

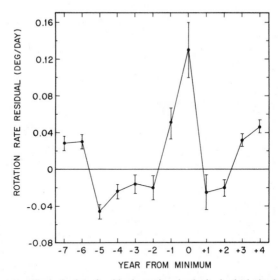

Figure 6   Superposed epoch plot of residual rotational velocity in the latitude interval −30° to +30°. The residual is calculated in each 5° latitude zone by subtracting the average rotational velocity over the full 62-yr period for that zone from each year's value. Then the zones are averaged. The zero phase refers to the minimum year of sunspot activity. The error bars represent one standard deviation, calculated from the annual rotation values at that cycle phase.

spots and the photosphere would be explained by the reaction of material of differing densities, and therefore inertias, to variations in the angular momentum. An alternative explanation is that there is no variation in momentum at the depths where the sunspots are anchored, and the observed modulation in the sunspot rotation rate is due to the drag of the faster-moving plasma on the spots at the photospheric level. When the plasma is moving faster, the sunspots are somewhat accelerated, but not to the same speed as the plasma. This would also explain the difference in the amplitude of the modulation between the plasma and the spots.

The fact that spots of different sizes rotate at different rates gives further credence to results showing that the photosphere rotates at a different rate than the sunpots. It is clear that the photosphere cannot rotate at the same rate shown by *all* the spots.

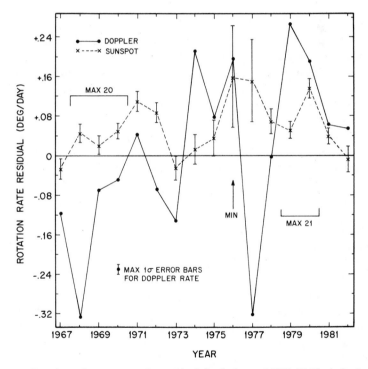

Figure 7 Doppler and sunspot rotation residuals for the interval 1967–82. The latitude zone covered is −30° to +30°. The residual is calculated in each latitude zone by subtracting the average rotational velocity over the full 62-yr period for that zone from each year's value. Then the zones are averaged. The maxima of cycles 20 and 21 are marked, as well as the minimum between them. The error bars represent one standard deviation, calculated from the contribution of the individual sunspots during the year.

The latitude drifts of the sunspots and sunspot groups were calculated from the Mount Wilson data. The correlation of the derived rotation rates and latitude drifts of spot groups is similar to the results of Ward (1965) computed from the Greenwich data. But for individual sunspots, the correlation is lower than that of Ward (or of the Mount Wilson group data) by a factor of 14. We conclude that the criticism by R. B. Leighton (unpublished manuscript) of the Ward result is correct. The correlation obtained when using sunspot groups is a spurious one because of the well-known properties of groups. Because of the tilt of the axes of spot groups, variations of the area-weighted group positions will be measured as variations in both rotation rate and latitude drift. The low correlation that is found implies that the eddy motions that were thought to transport angular momentum to the lower latitudes do not exist at depths where the motions affect the sunspot velocities (Gilman & Howard 1984b).

3.2   Rotation of Photospheric Magnetic Fields

The rotation of nonspot magnetic fields in the solar photosphere was first examined by Wilcox & Howard (1970) using autocorrelation functions in different latitude zones. Stenflo (1974, 1977) also used the Mount Wilson data for such an autocorrelation study. More recently, Snodgrass (1983) has looked at 15 years of Mount Wilson data for such an analysis. The earlier papers established that the solar magnetic fields rotate at a rate similar to that found by Newton & Nunn (1951) for large recurrent sunspots. At higher latitudes, the magnetic field rate is closer to that of the K corona and filaments, i.e. faster than sunspots. Stenflo (1977) found regions of large latitude shear and low angular velocity that drift equatorward at the spot zones.

Wilcox et al. (1970) showed that the autocorrelations of the magnetic field data, when carried over lags of several rotations or more, approached a rigid-body rotation curve. This effect could be due to the tendency of magnetic flux from more than one active region to coalesce to form large patterns that show rigid rotation (Bumba & Howard 1965). Coronal holes are an example of such magnetic patterns (Timothy et al. 1975). This tendency for magnetic patterns to rotate rigidly is not understood. In general, individual small magnetic features do rotate differentially, and it is only the large patterns that show rigid rotation (Bumba & Howard 1965, 1969).

Snodgrass (1983) used cross-correlations of successive days' data in narrow latitude zones to determine the rotation of magnetic fields at the solar surface. His errors varied from 0.1% at low latitudes to 1.1% at polar latitudes. He finds close agreement with the Newton & Nunn (1951) sunspot results near the equator and fair agreement with the Mount Wilson

Doppler results at high latitudes. His result is $\omega = 2.902 - 0.464 \sin^2 B$ $-0.328 \sin^4 B$ μrad s^{-1}. Figure 8 shows the Snodgrass results compared with the Mount Wilson Doppler results (Howard et al. 1983b), as well as the rotation rate for all sunspots and for all sunspot groups (Howard et al. 1984). In all cases the relationships shown in Figure 8 are smoothed curves from the series solutions. Note that the differential rotation slope for the magnetic data is intermediate between that for the spots and that for the Doppler data. Day-to-day variations of magnetic rotation are less than the errors. In particular, the cycle-related changes seen in the spot and Doppler data are not present in the magnetic rotation results within the error limits, although the error limits are sufficiently high for such an analysis that a variation close to the magnitude found for the spots could escape detection. An analysis such as that done by Snodgrass, which uses cross-correlations over one or a few days, tends to show the rotation rate of individual

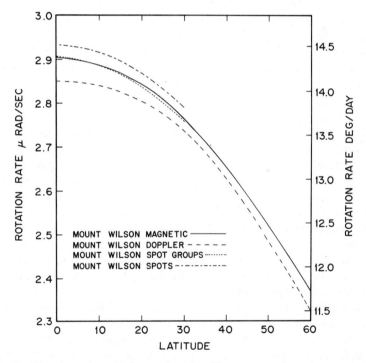

Figure 8 Sidereal rotation rates from various data. The sunspot and sunspot group results are from Howard et al. (1984) and represent the interval 1921–82. The magnetic results are from Snodgrass (1983) and represent the interval 1967–82. The Mount Wilson Doppler data are from the work of Howard et al. (1983b), covering the interval 1967–82. The dispersion correction of Snodgrass et al. (1984) has been applied.

magnetic features, not the large magnetic patterns, which show a rotation curve that is close to that of a solid body (Wilcox et al. 1970).

4. CHROMOSPHERE AND CORONA

The many types of chromospheric features seen in filtergrams and spectroheliograms are all linked closely to magnetic fields. Thus, one expects—and finds—that the rotation of chromospheric features closely resembles that of surface magnetic fields measured in the photosphere. In the corona the situation is similar. All coronal features are known to be related to local magnetic fields.

The rotation of plages and faculae was studied many years ago (Fox, 1921, Hale 1908, Maunder & Maunder 1905), and the results are generally similar to those for sunspots and surface magnetic fields. The quiet background network emission pattern in the K line of Ca II was analyzed by Schröter & Wöhl (1975) and Schröter et al. (1978), and their results also agree well with those for sunspots. For the most part, these results agree best with the rotation results from all spots rather than the slower rate of large spots (Howard et al. 1984). The later data of Schröter et al. (1978) gave a significantly slower rotation rate than did the earlier data. The authors attributed this to a real slowing of the rotation rate. Figure 9 shows a comparison of some of these rates.

The rotation rates of polar faculae are the slowest among the surface features (Müller 1954, Waldmeier 1955). These rates cluster around $10°$ day^{-1} sidereal.

In general, the rotation of features in the chromosphere agrees with the Hα spectroscopic results (Livingston 1969). The chromospheric rotation rate is faster by about 3% than the photosphere, and the differential rotation is less. This indicates that in the chromosphere, unlike the photosphere, the rotation of the magnetic fields dominates the plasma motions.

Belvedere et al. (1978) have reported that the rotation rate of chromo-spheric features is a function of their lifetimes (cf. El-Raey & Scherrer 1972, Antonucci & Svalgaard 1974, Belvedere et al. 1977). The smaller, younger, more compact faculae rotate faster than the larger, older, more diffuse faculae. One may speculate that the younger features are more rigidly connected to the faster-rotating subsurface magnetic flux tubes. As the features age, the subsurface connections become weaker, and the faculae reflect more closely the rotation rate of the surface plasma. Kearns (1979) finds that recurrent sunspots rotate slower than the average of all spots. (This is essentially equivalent to stating that large spots rotate slower than smaller spots, because recurrent spots tend to be larger than nonrecurrent spots.) An exception to these results is found in the work of Golub & Vaiana (1978), who find that the shortest-lived X-ray emission features rotate near the photospheric rate, while the larger, longer-lived features rotate faster by about 5%. This is a puzzling discrepancy, although it may be explainable as reflecting the difference between the photospheric and spot rate—the smallest features are influenced by the photospheric plasma, and the largest features are essentially sunspot groups.

Coronal rotation has been measured by several methods. Hansen et al. (1969) used digitized measures of the K corona in an autocorrelation analysis of coronal rotation at many latitudes. The features seen in the corona with such an instrument are poorly defined, and the resulting rotation rates are somewhat uncertain. Nevertheless, it is clear from this study that the rotation at the equator is in fair agreement with the sunspot results and shows less decrease with latitude at higher latitudes. This is similar to the magnetic pattern rotation seen in chromospheric features, as one might expect.

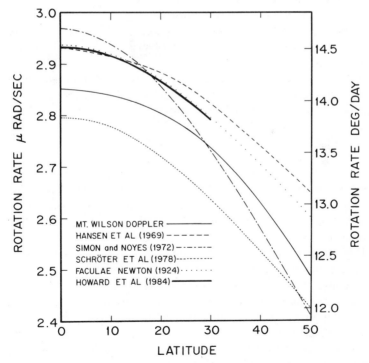

Figure 9   Sidereal rotation results from various sources. The Mount Wilson Doppler results are those of Howard et al. (1983b), corrected for dispersion after Snodgrass et al. (1984).

The coronal green line has been measured for a few individual coronal features at high latitudes (Waldmeier 1950, Cooper & Billings 1962). These results also indicate a faster rate than that of sunspots, suggesting less differential rotation. Green-line measurements by Trellis (1957) point to a somewhat faster rate at all latitudes than that of sunspots.

More recently, it has been possible to use satellite EUV and X-ray data to examine well-defined coronal features for rotation information. Simon & Noyes (1972) used bright emission points in the Lyman continuum and in the Mg X line $\lambda625$ from *Skylab* EUV spectroheliograms to measure daily motions. They found rates that were quite similar to those of the underlying plages in spite of the great difference in heights between the two sets of features (2000 and 11,000 km). However, the latitude dependence of these results is significantly greater than that of other features, as may be seen in Figure 9. Dupree & Henze (1972) used the same Lyman continuum data and a number of EUV lines from the same experiment to carry out an autocorrelation analysis for rotation. They found a significantly slower rotation ($\omega = 2.72 - 0.604 \sin^2 B \ \mu$rad s^{-1}), but also with a steep latitude dependence of the rotation. They also found no height gradient. A later analysis using better angular resolution confirmed these results (Henze & Dupree 1973). The explanation for the discrepancy in the two sets of results lies in the difference between the rotation of magnetic features and magnetic patterns (discussed above). The measurements of daily motions from the features seen on the spectroheliograms gives the true rotation of individual features. The autocorrelation technique tends to emphasize the magnetic patterns because longer time intervals (one or more rotations) are involved.

5.   ROTATION OF THE SOLAR INTERIOR

As discussed above, the best current convection models tend to lead to solutions where the rotation is roughly constant on cylinders concentric with the rotation axis. Thus, the rotation would be expected to decrease inward. This is in conflict with the kinematic dynamo model, where the magnetic flux tubes below the surface are believed to reflect a more rapid rotation rate in those layers. Yoshimura (1971) has suggested that the faster rotation of the magnetic features is due to an interaction of rotation with global convection that selectively accelerates the spots.

The interior rotation is, in principle, measurable from frequency shifts of solar global pressure-mode oscillations (Rhodes et al. 1979, Ulrich et al. 1979, Deubner et al. 1979). The first observational results in this area (Deubner et al. 1979; see also Deubner 1978) were ambiguous, although they did show a slight tendency for faster rotation than the photosphere, at least down to about 15 Mm depth. Below that, at least from these early observations, the situation becomes unclear. Claverie et al. (1981), using an optical resonance scattering technique, have detected splitting of lower pressure modes, which they interpret as the effects of a solar core rotating 2 to 9 times as fast as the surface.

Dicke (1976, 1977, 1982) has reanalyzed the Princeton solar distortion observations from the summer of 1966 and has found in the noise a recurrent signal with a sidereal period of 12.2 days. Secondary resonances also found in the data have provided further proof of the reality of this signal. Dicke interprets this periodic signal as due to a distortion in the gravitational potential induced by the rotation at this period of a core distorted by strong internal magnetic fields. Dicke (1982) further suggests that the magnetically distorted core must undergo torsional oscillations with a period of some years. This elaborate model rests on very little observational data. In an effort to remedy this, the Princeton group has undertaken further solar distortion observations with an upgraded version of the same telescope used in 1966. These new observations should shortly settle once and for all the reality of the rotating distortion. If the effect is real, this will be a very important diagnostic for the interior structure and dynamics of the Sun.

The study of solar oscillations of various sorts ("solar seismology") promises to provide a great deal of information about the interiors of the Sun and other stars within the next decade (Fossat 1981). We are on the threshold of a new era in solar and stellar research, with potential results that were undreamed of a decade ago.

Literature Cited

Abarbanell, C., Wöhl, H. 1981. *Sol. Phys.* 70: 197–203

Adam, M. G. 1959. *MNRAS* 119:460–74

Adams, W. S. 1911. *An Investigation of the Rotation Period of the Sun by Spectroscopic Methods, Carnegie Inst. Washington Publ. No. 138*, pp. 1–132

Antonucci, E., Svalgaard, L. 1974. *Sol. Phys.* 34:3–10

Arevalo, M. J., Gomez, R., Vazquez, M., Balthasar, M., Wöhl, H. 1982. *Astron. Astrophys.* 111:266–71

Aslanov, I. A. 1963. *Astron. Zh.* 40:1036–46. Transl., 1964, in *Sov. Astron. AJ* 1:794–800 (From Russian)

Balthasar, H. 1983. *Sol. Phys.* 84:371–76

Balthasar, H., Wöhl, H. 1980. *Astron. Astrophys.* 92:111–16

Beckers, J. 1978. *Proc. Workshop Sol. Rotation, Catania, Oss. Astrofis. Catania Publ. No. 162*, pp. 166–79

Belvedere, G., Godoli, G., Motta, S., Paternò, L., Zappalà, R. A. 1977. *Ap. J. Lett.* 214:L91–95

Belvedere, G., Zappalà, R. A., D'Arrigo, C., Motta, S., Pirronello, V., et al. 1978. *Proc. Workshop Sol. Rotation, Catania, Oss. Astrofis. Catania Publ. No. 162*, pp. 189–203

Bumba, V., Howard, R. 1965. *Ap. J.* 141:1502–12

Bumba, V., Howard, R. 1969. *Sol. Phys.* 7:28–38

Carrington, R. C. 1863. *Observations of the Spots on the Sun.* London: Williams & Norgate. 412 pp.

Claverie, A., Isaak, G. R., McCleod, C. P., van der Raay, H. B. 1981. *Nature* 293:443–45

Cooper, R. H., Billings, D. E. 1962. *Z. Ap.* 55:24–28

De Lury, R. E. 1939. *J. R. Astron. Soc. Can.* 33:345–78

Deubner, F.-L. 1978. *Proc. Workshop Sol. Rotation, Catania, Oss. Astrofis. Catania Publ. No. 162*, pp. 247–54

Deubner, F.-L., Vazquez, M. 1975. *Sol. Phys.* 43:87–90

Deubner, F.-L., Ulrich, R. K., Rhodes, E. J. Jr. 1979. *Astron. Astrophys.* 72: 177–85
Dicke, R. H. 1976. *Sol. Phys.* 47: 475–515
Dicke, R. H. 1977. *Ap. J.* 218: 547–51
Dicke, R. H. 1982. *Sol. Phys.* 78: 3–16
Dupree, A. K., Henze, W. Jr. 1972. *Sol. Phys.* 27: 271–79
Duvall, T. L. Jr. 1977. PhD thesis. Stanford Univ., Calif.
Duvall, T. L. Jr. 1982. *Sol. Phys.* 76: 137–43
Eddy, J. A., Gilman, P. A., Trotter, D. E. 1976. *Sol. Phys.* 46: 3–14
El-Raey, M., Scherrer, P. H. 1972. *Sol. Phys.* 26: 15–20
Evershed, J. 1931. *MNRAS* 92: 105–7
Fossat, E. 1981. In *Solar Phenomena in Stars and Stellar Systems,* ed. R. M. Bonnet, A. K. Dupree, pp. 75–98. Dordrecht: Reidel
Foukal, P. 1972. *Ap. J.* 173: 439–44
Foukal, P. 1976. *Ap. J. Lett.* 203: L145–48
Foukal, P. 1979. *Ap. J.* 234: 716–22
Fox, P. 1921. *Publ. Yerkes Obs.* 3: 67–203
Gasanalizade, A. G. 1980. *Izv. Pulkovo Obs. No.* 197, pp. 145–51
Gilman, P. A. 1974. *Ann. Rev. Astron. Astrophys.* 12: 47–70
Gilman, P. A. 1979. *Ap. J.* 231: 284–92
Gilman, P. A., Foukal, P. V. 1979. *Ap. J.* 229: 1179–85
Gilman, P. A., Howard, R. 1984a. *Ap. J.* In press
Gilman, P. A., Howard, R. 1984b. *Ap. J.* In press
Gilman, P. A., Miller, J. 1981. *Ap. J.* 46: 211–38
Glatzmaier, G. A. 1983. *Bull. Am. Astron. Soc.* 15(2): 716 (Abstr.)
Godoli, G., Mazzucconi, F. 1979. *Sol. Phys.* 64: 247–54
Godoli, G., Mazzucconi, F. 1982. *Astron. Astrophys.* 116: 188–89
Golub, L., Vaiana, G. S. 1978. *Ap. J. Lett.* 219: L55–57
Hale, G. E. 1908. *Ap. J.* 27: 219–29
Hansen, R. T., Hansen, S. F., Loomis, H. G. 1969. *Sol. Phys.* 10: 135–49
Henze, W. Jr., Dupree, A. K. 1973. *Sol. Phys.* 33: 425–29
Herr, R. B. 1978. *Science* 202: 1079–81
Howard, R. 1974a. *Sol. Phys.* 38: 59–67
Howard, R. 1974b. *Sol. Phys.* 39: 275–87
Howard, R. 1978. *Rev. Geophys. Space Phys.* 16: 721–32
Howard, R. 1979. *Ap. J. Lett.* 228: L45–50
Howard, R., Harvey, J. 1970. *Sol. Phys.* 12: 23–51
Howard, R., LaBonte, B. J. 1980. *Ap. J. Lett.* 239: L33–36
Howard, R., LaBonte, B. J. 1983. In *Solar and Stellar Magnetic Fields: Origins and Coronal Effects,* ed. J. O. Stenflo, pp. 101–11. Dordrecht: Reidel
Howard, R., Boyden, J., LaBonte, B. 1980.

Sol. Phys. 66: 167–85
Howard, R., Boyden, J. E., Bruning, D. H., Clark, M. K., Crist, H. W., LaBonte, B. J. 1983a. *Sol. Phys.* 87: 195–203
Howard, R., Adkins, J. M., Boyden, J. E., Cragg, T. A., Gregory, T. S., et al. 1983b. *Sol. Phys.* 83: 321–38
Howard, R., Gilman, P. A., Gilman, P. I. 1984. *Ap. J.* In press
Kearns, M. 1979. *Sol. Phys.* 62: 393–99
Kubičela, A., Karabin, M. 1983. *Sol. Phys.* 84: 389–93
LaBonte, B. J., Howard, R. 1982a. *Sol. Phys.* 75: 161–78
LaBonte, B. J., Howard, R. 1982b. *Sol. Phys.* 80: 373–78
LaBonte, B. J., Howard, R., Gilman, P. A. 1981. *Ap. J.* 250: 796–98
Livingston, W. C. 1969. *Sol. Phys.* 9: 448–51
Livingston, W. C., Duvall, T. L. Jr. 1979. *Sol. Phys.* 61: 219–31
Livingston, W. C., Milkey, R. 1972. *Sol. Phys.* 25: 267–73
Maunder, E. W., Maunder, A. S. D. 1905. *MNRAS* 65: 813–25
Müller, R. 1954. *Z. Ap.* 35: 61–66
Newton, H. W. 1924. *MNRAS* 84: 431–42
Newton, H. W., Nunn, M. L. 1951. *MNRAS* 111: 413–21
Noci, G. 1978. *Proc. Workshop Sol. Rotation, Catania, Oss. Astrofis Catania Publ. No.* 162, pp. 55–73
Plaskett, H. H. 1916. *Ap. J.* 43: 145–60
Plaskett, J. S. 1915. *Ap. J.* 42: 373–93
Rhodes, E. J. Jr., Deubner, F.-L., Ulrich, R. K. 1979. *Ap. J.* 227: 629–37
St. John, C. E. 1932. *Trans. IAU* 4: 42–44
Scherrer, P. H., Wilcox, J. M. 1980. *Ap. J. Lett.* 239: L89–90
Scherrer, P. H., Wilcox, J. M., Svalgaard, L. 1980. *Ap. J.* 241: 811–19
Schröter, E. H., Wöhl, H. 1975. *Sol. Phys.* 42: 3–16
Schröter, E. H., Soltau, D., Wöhl, H., Vazquez, M. 1978. *Proc. Workshop Sol. Rotation, Catania, Oss. Astrofis. Catania Publ. No.* 162, pp. 180–88
Simon, G. W., Noyes, R. W. 1972. *Sol. Phys.* 26: 8–14
Snodgrass, H. B. 1983. *Ap. J.* 270: 288–99
Snodgrass, H. B., Howard, R., Webster, L. 1984. *Sol. Phys.* In press
Solonsky, Y. A. 1972. *Sol. Phys.* 23: 3–12
Stenflo, J. O. 1974. *Sol. Phys.* 36: 495–515
Stenflo, J. O. 1977. *Astron. Astrophys.* 61: 797–804
Stix, M. 1976. *Astron. Astrophys.* 47: 243–54
Storey, J. 1932. *MNRAS* 92: 737–41
Svalgaard, L., Scherrer, P. H., Wilcox, J. M. 1978. *Proc. Workshop Sol. Rotation, Catania, Oss. Astrofis. Catania Publ. No.* 162, pp. 151–58

Tang, F., Howard, R., Adkins, J. M. 1984. *Sol. Phys.* In press
Timothy, A. F., Krieger, A. S., Vaiana, G. S. 1975. *Sol. Phys.* 42:135–56
Trellis, M. 1957. *Ann. Astrophys. Suppl.* 5:1–81
Tuominen, J. 1961. *Z. Ap.* 51:91–94
Tuominen, J. 1976. *Sol. Phys.* 47:541–50
Tuominen, J., Kyröläinen, J. 1982. *Sol. Phys.* 79:161–72
Tuominen, J., Tuominen, I., Kyröläinen, J. 1984. *Sol. Phys.* In press
Ulrich, R. K., Rhodes, E. J. Jr., Deubner, F.-L. 1979. *Ap. J.* 227:638–44

Vogel, H. C. 1872. *Astron. Nachr.* 78:248–50
Waldmeier, M. 1950. *Z. Ap.* 27:24–41
Waldmeier, M. 1955. *Z. Ap.* 38:37–54
Ward, F. 1965. *Ap. J.* 141:534–47
Ward, F. 1966. *Ap. J.* 145:416–25
Wilcox, J. M., Howard, R. 1970. *Sol. Phys.* 13:251–60
Wilcox, J. M., Schatten, K. H., Tanenbaum, A. S., Howard, R. 1970. *Sol. Phys.* 14:255–62
Yallop, B. D., Hohenkerk, C., Murdin, L., Clark, D. H. 1982. *Q. J. R. Astron. Soc.* 23:213–19
Yoshimura, H. 1971. *Sol. Phys.* 18:417–33

Ann. Rev. Astron. Astrophys. 1984. 22: 157–84

ALTERNATIVES TO THE BIG BANG

G. F. R. Ellis

Department of Applied Mathematics, University of Cape Town, Rondebosch 7700, Cape, South Africa,[1] and Department of Physics, University of Texas, Austin, Texas 78712

1. INTRODUCTION

At issue is one of the most significant questions in astronomy: What is the nature of the beginning of the Universe? While there is at present a widespread opinion that only one view (the "standard hot big bang," or SHBB) gives a viable picture of the physical origin of the Universe, there are alternatives that need consideration in any dispassionate review of models compatible with current evidence.

The concept of an expanding universe evolving from an initial singular state arose from the work (1922–30) of A. A. Friedmann, G. E. Lemaître, and A. S. Eddington, together with E. P. Hubble's (1929) determination of a linear velocity-distance relation for distant galaxies. With the work of H. P. Robertson and A. G. Walker (1933–35), the spatially homogeneous, isotropic Friedmann-Lemaître-Robertson-Walker (or FLRW) universes were established as the standard universe models. The steady-state universe (1952) of H. Bondi, T. Gold and F. Hoyle posited continuous creation in an expanding universe without a beginning. However, the discovery (1965) of the cosmic microwave background radiation (CMBR) firmly established the SHBB as the accepted theory, accounting not only for that radiation and the velocity-distance relation, but also for the observed abundances of elements.

In recent times, (*a*) alternative theories have been proposed for the redshifts observed for galaxies and QSOs; (*b*) new physical theories (such as the grand unified theories, or GUTs) have resulted in a reappraisal of the

---

[1] Permanent address.

0066–4146/84/0915–0157$02.00

effective equation of state of matter in the early Universe; (c) new theories of gravity have been proposed as alternatives to Einstein's general theory of relativity; and (d) consideration has been given to more complex geometries than the FLRW models that underlie the standard picture.

In each case, new possibilities arise in which the origin of the Universe is different from that in the standard picture. It is important to consider such alternatives, particularly as there are a number of problems that are not resolved by the SHBB; without such an examination, the argument for the standard view is clearly incomplete. (One cannot make a rational choice between alternative explanations on the basis of an examination of only one of them.)

2.   BASIC RESULTS

2.1   *The Nature of Redshift*

The standard theory assumes that electromagnetic radiation ("light" for short) travels on null geodesics $(x^a(\lambda): k^a_{;b}k^b = 0, \; k^a k_a = 0$, where $k^a \equiv dx^a/d\lambda)$ in the space-time of general relativity and is characterized by the relations $c = \nu\lambda, E = h\nu$, relating the wavelength λ, frequency ν, speed c, and energy E of a photon. Then the redshift $z \equiv \Delta\lambda/\lambda$ (where $\Delta\lambda$ is the change in wavelength measured from spectra for light emitted at wavelength λ) is a direct consequence of measurable time dilation between the source and the observer (25, 91, 111). There can be various sources for this time dilation.

In the cosmological context, there is assumed to be a well-defined average motion of matter at each point in the Universe, representing the "standard" motion of matter there (24, 25); galaxies moving in this standard way, characterized by a four-velocity u^a (where $u_a u^a = -1$), are called "fundamental particles," and an observer moving with four-velocity u^a is denoted as a "fundamental observer." The *cosmological redshift* z_c is the redshift observed when both the source and observer move with this velocity (at their respective space-time positions) and there are no local gravitational redshift effects. However, there will be *local Doppler redshifts* z_{Ds}, z_{Do} if the source or observer is not moving at the standard velocity, and there may be *local gravitational redshifts* z_{Gs}, z_{Go} generated at the source or the observer by local gravitational potential wells. (Notice that in each case the effect could be a blueshift rather than redshift; this corresponds to a negative value of z, and we shall assume this possibility without explicitly mentioning it.) These contributions to the total redshift z are not distinguishable from each other by any direct measurement of the received radiation; the measurable quantity is the total redshift z, given in terms of the other quantities by

$$(1 + z) = (1 + z_{Ds})(1 + z_{Gs})(1 + z_c)(1 + z_{Do})(1 + z_{Go}).$$ 1.

The cosmological redshift depends on the path of the light ray from the source to the observer. One can distinguish two contributions. The first arises from the increase in relative distance of fundamental observers due to integrated local relative motion of matter [characterized by the expansion tensor $\theta_{ab} = h_a^c h_b^d u_{(c;d)}$ (24, 25), where $h_{ab} \equiv g_{ab} + u_a u_b$ projects into the rest-space orthogonal to u^a]. The second arises from the noninertial motion of the fundamental observers, characterized by a nonzero acceleration vector $\dot{u}^a \equiv u^a_{;b} u^b$ (indicating that their world-lines are nongeodesic). The change of redshift when moving a distance dl, measured by a fundamental observer along a null geodesic in direction $e^a = h_b^a k^b/(u_c k^c)$, obeys the equation (24, 25)

$$dz = d\lambda/\lambda = (\theta_{ab} e^a e^b - \dot{u}_a e^a) \, dl.$$ 2a.

Thus one can write

$$(1 + z_c) = (1 + z_{Ec} + z_{Gc}),$$ 2b.

where z_{Ec} is the usual Hubble (expansion) term (obtained by integrating the first term in Equation 2a), and z_{Gc} is what might be termed a "cosmological gravitational redshift" (obtained by integrating the second term).

In the FLRW models, it is assumed that the other redshift contributions are very small compared with the basic cosmological expansion effect. Then

$$z = z_c = z_{Ec}, \qquad 1 + z = R(t_0)/R(t_e),$$ 3.

where $R(t)$ is the "radius function" of the Universe that represents the time behavior of the distance between any pair of fundamental particles as the Universe evolves (by the homogeneity and isotropy of these models, the distances between all pairs of fundamental particles scale in the same way); and t_e, t_0 are the times of emission and observation of the light, respectively. The observation of systematic redshifts indicates that the Universe is expanding at the present time t_0; so $H_0 \equiv (R^{-1}dR/dt)|_0 > 0$.

2.2 Basic Equations

The standard results rest on three pillars: the conservation equations, the Raychaudhuri equation, and the energy conditions. The form of these equations depends on the nature of the matter in the Universe. It is assumed here that the usual description of matter on a cosmological scale as a "perfect fluid" is adequate at most times. Then (24, 25) the stress tensor of each component of matter or energy present takes the form

$$T_{ab} = \mu u_a u_b + p h_{ab}, \qquad \text{4a.}$$

where μ is the (relativistic) energy density of the fluid and p its pressure. To complete the description of matter, we must give suitable equations of state for the matter variables; in the hot early Universe, the "standard" theory assumes (except during phases of pair annihilation) that these are

$$p = (1/3)\mu, \qquad \mu = aT^4, \qquad \text{4b.}$$

describing a "Fermi gas" with temperature T (107, 109).

THE CONSERVATION EQUATIONS   The stress tensor T_{ab} of matter that does not interact with other matter or radiation obeys the equations $T^{db}{}_{;b} = 0$. If this matter is a perfect fluid (Equation 4a), the energy conservation equation $u_d T^{db}{}_{;b} = 0$ takes the form

$$d\mu/dt + (\mu + p)(3/R)\, dR/dt = 0 \qquad \text{5.}$$

where t is proper time measured along the fluid flow lines, showing the effect of expansion or compression of the matter on its energy density. (The radius R is related to the volume V of a fluid element by the relation $V \propto R^3$.) The momentum conservation equation $h^a_d T^{db}{}_{;b} = 0$ is

$$(\mu + p)\dot{u}^a = -h^{ab}\partial p/\partial x^b, \qquad \text{6.}$$

which shows how spatial pressure gradients generate acceleration (and thus observable gravitational redshifts).

THE RAYCHAUDHURI EQUATION   Contracting the Ricci identity $u^a{}_{;bc} - u^a{}_{;cb} = R^a{}_{dbc}u^d$ for u^a and using Einstein's field equations

$$R_{ab} - (1/2)Rg_{ab} = \kappa T_{ab}, \qquad \text{7.}$$

it follows that the volume behavior of the fluid is controlled by the Raychaudhuri equation (82, 24, 25, 45)

$$3R^{-1}d^2R/dt^2 = 2(\omega^2 - \sigma^2) + \dot{u}^a{}_{;a} - (\kappa/2)(\mu + 3p), \qquad \text{8.}$$

where ω^2 is the squared magnitude of the fluid vorticity and σ^2 the squared magnitude of the shear (or rate of distortion). Because d^2R/dt^2 is the curvature of the curve $R(t)$, this equation shows how rotation tends to hold matter apart and distortion to make it collapse, with $(\mu + 3p)$ the "active gravitational mass density" that tends to cause collapse.

ENERGY CONDITIONS   Various "energy conditions" have been proposed as physical restrictions on the matter (45); for present purposes, the major such conditions are

$$\mu + p \geq 0,$$ 9a.

$$\mu + 3p \geq 0,$$ 9b.

with Equation 9a needed for physical stability of matter (as follows from Equations 5 and 6) and Equation 9b implying that the gravitational effect of matter is attractive (see Equation 8). It is usually assumed that "reasonable" matter obeys these conditions.

2.3   The Standard Models

The FLRW universe models (109, 45) are isotropic and spatially homogeneous, so their shear, vorticity, and acceleration are all zero. Thus the Raychaudhuri equation (8) reduces to

$$3R^{-1}d^2R/dt^2 = -(\kappa/2)(\mu + 3p).$$ 10.

By energy condition Equation 9b, this implies that the curve $R(t)$ of the radius function against time always bends down. As the Universe is expanding at the present time t_o, a singular origin must have occurred less than a Hubble time previously:

$$\exists \tau_o : |t_o - \tau_o| < 1/H_o, \qquad R \underset{t \to \tau_o}{\to} 0.$$ 11a.

Because of the definition of $R(t)$,

(*A*) at this time τ_o, the distance between all fundamental particles goes to zero like $R(t)$.

At early times, Equations 4b and 5 show that $T = T_o(R_o/R)$, where (from present observations of the CMBR) $T_o \approx 3$ K; thus,

$$t \to \tau_o \Rightarrow T \to \infty, \mu \to \infty, p \to \infty.$$ 11b.

This is the basic origin of the hot big-bang prediction in the FLRW universes:

(*B*) The Universe originates, and all physics breaks down, at the time $t \to +\tau_o$ where the temperature, energy density of matter, and space-time curvature diverge.

(For a more careful discussion, see 18.) Note that it is not merely the matter in the Universe that originates here: The space-time itself (and indeed the laws of physics) do not exist before this time. Now, Equations 4b, 5, and 10 show that

$$w(t) = \int_{\tau_0}^{t} R(t)^{-1} \, dt$$

converges; this implies that

(C) for each fundamental observer O, there exist particle horizons limiting possible communication with other fundamental observers in each direction at each time t; those further than the particle horizon will be unobservable by any radiation, and no causal influence from them can affect events in O's history up to the time t (45, 75, 85, 102).

The major properties characterizing the SHBB are (A)–(C) above and Equation 11, which follow from Equations 4, 5, 9, and 10. The standard results on the CMBR and element formation (i.e. the CMBR is relic radiation from the hot early phase when $T > 10^3$ K, and the light elements were synthesized when $T \approx 10^9$ K in the early Universe) can then be deduced from these properties (107, 109). If current ideas on the GUT are correct, at earlier times baryon creation will take place (42, 73, 116).

These models explain many features of the observed Universe in a satisfactory way, in particular the CMBR and primordial element abundances. However, they leave various other issues obscure, e.g. why the Universe is so nearly uniform, and yet not uniform [physical processes taking place after the creation of the Universe cannot be responsible because of the particle horizons (C); (38, 57, 64)], and why the density of matter is so close to the critical value separating recollapsing universes from ever-expanding universes (20, 38). Thus, one has the problem of explaining why the Universe started off from very special initial conditions. Additionally, the very concept of the creation of the Universe at such a singular beginning is philosophically objectionable to some scientists; they wish to find an alternative picture of its origin (23, 41, 102).

2.4   The Cosmological Constant

When the cosmological constant Λ is taken into account, the field equations (Equation 7) are altered to

$$R_{ab} - (1/2)Rg_{ab} + \Lambda g_{ab} = \kappa T_{ab}, \qquad\qquad 12.$$

which can be rewritten in the form

$$R_{ab} - (1/2)Rg_{ab} = \kappa T'_{ab}, \qquad\qquad 13a.$$

where

$$T'_{ab} \equiv T_{ab} - (\Lambda/\kappa)g_{ab}; \qquad\qquad 13b.$$

that is, the field equations can be regarded as the same as before but with an altered stress-energy tensor. Writing the effective stress-energy tensor T'_{ab} in the form of Equation 4a, one finds that

$$\mu' = \mu + \Lambda/\kappa, \qquad p' = p - \Lambda/\kappa, \qquad (\mu + 3p)' = (\mu + 3p) - 2\Lambda/\pi, \qquad \text{14a.}$$

and so Equation 10 becomes

$$3R^{-1}d^2R/dt^2 = -\frac{\kappa}{2}(\mu + 3p)' = -\frac{\kappa}{2}(\mu + 3p) + \Lambda. \qquad \text{14b.}$$

If the Λ-term is large enough, the argument by which Equation 10 leads to Equation 11 no longer applies, for the effective active gravitational density $(\mu + 3p)'$ can violate the energy condition Equation 9b. If it does so at all times, a singularity in the past can be avoided (cf. Equation 14b), and indeed it has long been known (86) that there are FLRW universes with a large positive cosmological constant that (a) start expanding from a state asymptotically like the Einstein static universe [Eddington (23) suggested that this was preferable to a "big bang" origin]; or (b) collapse from infinity to a minimum radius value $R_{min} > 0$ and then reexpand to infinity.

There are two problems with these models. Firstly, the deceleration parameter $q_o \equiv -(1/R)_o(d^2R/dt^2)_o/(H_o^2)$ must be negative at all times in these universes (for if it ever becomes positive, the universes must come from a singular origin), but observations suggest that the present value of q_o is positive (114). Secondly, an alternative explanation for the CMBR must be found in these cases, for it is not possible for this radiation to arise from a hot early stage in such a universe. The reason is that if such a hot early stage were to explain the CMBR, the asymptotic phase or turnaround caused by Λ would have to occur during the early plasma phase before decoupling when $T > 3000$ K; thus inequality Equation 9b would have to be violated by the effective matter (see Equation 14) for values of R less than $(R_o/1000)$. But at that time, $(\mu + 3p)$ would be at least 10^9 times larger than at present, while Λ would have the same value as at present (because it is constant); violation of Equation 9b at that time therefore implies a value of Λ so large that we would undoubtedly have detected it from q_o. It is difficult to find alternative explanations for the CMBR (54). In addition, these models require element abundances to be arbitrarily set by initial conditions in the Universe (for the argument above shows the turnaround would occur long before nucleosynthesis temperatures were achieved). Thus, these models leave so many questions open as to not be presently viable.

One should note here that it follows from Equations 5, 9, and 14b that a static FLRW universe model will necessarily be unstable. [Perturbing R from a static value R_c will lead either to collapse or expansion (23, 25).] This implies that even if we could find some other explanation for redshift (abandoning Equations 1 and 2) we would not expect a FLRW universe to be static unless we abandon either Einstein's field equations or the energy conditions.

2.5 Implications

The discussion above makes clear the nature of the available alternatives if one is to avoid the conclusion that the Universe originates in a SHBB. One can question

1. the nature of the observed redshifts (Equations 1–3) by adopting either a different theory of light propagation or a different astrophysical interpretation;
2. the conservation laws and/or gravitational field equations (Equations 5–7);
3. the nature of matter in the Universe, e.g. by assuming some effective contribution to the matter stress tensor that violates the energy conditions (Equation 9).

In each case one can avoid the existence of an initial singularity but must carefully consider if the resulting theory gives a satisfactory account of the microwave background radiation and element abundances. An additional possibility is to consider different geometric assumptions. One can question

4. the assumption of exact spatial homogeneity and isotropy. Then at least one of the shear, vorticity, and acceleration will be nonzero (so Equation 10 does not follow from Equation 8). Singularities will still occur in the past (45, 102), but they can be so different from the SHBB in their geometry and physics as to represent quite different initial situations.

3. ALTERNATIVE VIEWS OF REDSHIFT

3.1 Alternative Theories of Light

TIRED LIGHT   An alternative theory of light propagation is the "tired light" theory, in which light loses energy progressively while traveling across large distances of extragalactic space. Thus we abandon Equation 2; then the observed redshifts might occur in a static universe, where no SHBB occurs. When a detailed mechanism for such an effect is proposed, various problems occur (65); and attempts to check if it is true are negative (34).

ALTERNATIVE GEODESICS   Two-metric theories may involve light traveling on geodesics of another metric than that specifying length and time measurements; thus we effectively abandon Equation 2a. In general, these theories will still require different field equations or energy violation to avoid a singularity (see below). Furthermore, any such dynamic two-metric theory is likely to run into causality problems (77).

A particular case is Segal's theory (92), where in effect two metrics are laid down a priori from geometrical rather than physical principles (and are singularity free). The theory therefore proposes a cosmology independent of any gravitational equations; it demands a quadratic (magnitude, redshift) distance relation rather than the linear one predicted by the FLRW universes. The interpretation of the observations is the subject of a dispute centering on the nature of the selection effects and the statistics used (70, 93, 98); but the weight of the evidence seems to be against the quadratic effect. Additionally, the distribution of absorption lines in QSO spectra appears to rule these models out (81).

3.2 Local Redshift Theories

LOCAL CAUSES   The effective proposal is that for QSOs, the major source of redshift is either z_{Ds} or z_{Gs} (13, 90); thus, Equation 1 remains true but we abandon Equation 3. Then the redshift observed for these objects is no longer necessarily a result of the expansion of the Universe. However, the instability of the Einstein static universe still would lead us to expect an expanding Universe if we stay within the general relativity/FLRW framework; and as usual, it is accepted that galactic redshifts are evidence of cosmological expansion. Thus, this proposal results in local variation of the explanation, rather than an overall abandonment of the concepts of an expanding universe and the hot big bang.

NEW LOCAL REDSHIFT PHYSICS   In the second case, because of observed associations between galaxies and QSOs with differing redshifts (2, 33), the observed QSO redshifts are attributed to local effects of unknown nature: thus we abandon Equation 1 in the case of QSOs. This interpretation depends critically on the statistics of the image associations (117). Until the nature of this "new physics" is specified, this is not a theory in competition with the SHBB. As in the previous case, one again ends up with the plausibility of an expanding universe with a singular origin anyhow.

4. ALTERNATIVE PHYSICS

4.1 Alternative Matter

ENERGY VIOLATION   There are various proposals of alternative effective equations of state for matter, which result in violations of the energy condition (Equation 9b) through the existence of sufficiently large negative pressures. Thus, as well as transient solutions, one also obtains steady-state and periodic solutions (48); it is possible to avoid the initial singularity, e.g. by starting from a static initial situation that then becomes unstable or in a universe that goes from a collapse phase to a state of expansion (86). We discuss in turn several ways of generating negative pressures.

If the matter in the Universe has a bulk viscosity with the coefficient of bulk viscosity proportional to density, negative pressures arise such that energy violations sufficient to avoid an initial singularity occur (66, 113). However, it is not expected that this coefficient would have such a dependence on density in the case of ultrarelativistic fluids.

A classical conformal massive scalar field can generate violations of the energy conditions (45), and a massless conformal field coupled to pressureless matter can cause energy violations sufficient to cause a "bounce" instead of a singularity in a FLRW universe model (6). No such field is known to exist, but this situation may be regarded as analogous to the pion field, which mediates strong interactions in nuclear matter. The field would only be significant when very high densities had been attained, so the energy violation and bounce would occur for nuclear densities. Insofar as this is an adequate model of the fields dominant at these early times (which is unknown), the possibility arises of avoiding the singularity itself; indeed, if $k = +1$ the universe model could "bounce" at each of a series of singularities, as envisaged in Tolman's oscillating universe (103).

Energy violation may be expected to arise when quantum effects are significant, for such violations may be regarded as the driving force for the Hawking radiation emitted by black holes (14, 44). For example, if one uses as the source of the gravitational field the expectation value of a quantum field theoretical energy-momentum tensor for a quantized scalar field possessing mass, coherence effects can give rise to negative pressures sufficiently large to violate the energy conditions of the singularity theorems. Then a Friedmann-like collapse can be arrested at elementary particle dimensions and changed to an expansion (74).

A particular case where energy violation is expected to arise is in the inflationary universe scenario (38, 53) of the GUTs, where a supercooled metastable state of unbroken symmetry (a "false vacuum") gives rise to an effective cosmological constant Λ dominating other fields present. This in turn gives rise to an effective energy density $\mu' = \Lambda/\kappa$ and effective pressure $p' = -\Lambda/\kappa$; so $(p + \mu)' = 0$ (the limiting case of Equation 9a), which allows energy increase while conserving energy-momentum. (By Equation 5 the matter expands while its density stays constant, as in the steady-state universe, because of the negative pressures.) Now $(\mu + 3p)' = -2\Lambda/\kappa$, so if $\Lambda > 0$, then a violation of the energy condition (Equation 9b) *must* happen, causing an exponential expansion ("inflation"). This expansion solves the causal problems raised by horizons in the Universe because causally connected regions are now much larger than a Hubble radius (38, 53). While this energy violation cannot lead to avoidance of the initial singularity if thermal equilibrium is maintained (39), this condition is not necessarily fulfilled (79), and there is a possibility that the Universe could bounce and thereby avoid the initial singularity because of the energy violation (79). In the standard scenario, the GUTs phase is preceded by a radiation-dominated phase, with the usual singularity and horizons at early times; but one could conceive of the Universe as being created in the exponentially expanding state. Then its early properties will be those of the steady-state universe, without a curvature singularity or horizon.

Overall, it is clear that energy violation, in principle, could allow for avoidance of the initial singularity and permit a study of the Universe in its pre-expansion phase (101). When quantum effects dominate the equation of state, it is quite likely that such energy violation can occur.

OTHER EQUATIONS OF STATE   There are other proposals for alternative equations of state that do not violate the energy conditions but that disagree with the radiation form (Equation 4b); one still has an initial singularity, but it can be a cold or cool big bang, rather than a hot one (e.g. 40, 47, 84). The SHBB statements (A–C) and Equation 11 remain, except that the temperature T does not diverge; thus the basic view of the origin of the Universe remains, with only relatively small details being varied.

4.2   Alternative Gravity

CLASSICAL THEORIES   There are a large variety of classical (i.e. non-quantum) theories of gravity proposed as alternatives to the general theory of relativity (36, 112). In each case, one can work out the corresponding application of the theory to the Robertson-Walker metrics and so obtain a corresponding version of the origin of the Universe.

Many of these alternative theories can be written as close variants of general relativity, with field equations of the form

$$R_{ab} - (1/2)Rg_{ab} + \Lambda_{ab} = \zeta T_{ab} \qquad\qquad 15.$$

for some suitable choice of Λ_{ab} and ζ [which will be given in terms of other auxiliary fields, e.g. the scale function ϕ in scale-covariant theories such as the Brans-Dicke theory (12, 106, 109), the creation field C in the original Hoyle-Narlikar theory (46, 109), or the second metric γ_{ab} in Rosen's bimetric theory (87, 88); these auxiliary fields will satisfy their own field equations]. Now just as in the transition from Equation 12 to Equation 13, one can rewrite this equation in the form

$$R_{ab} - (1/2)Rg_{ab} = \kappa T'_{ab}, \qquad\qquad 16a.$$

where

$$T'_{ab} \equiv (\zeta T_{ab} - \Lambda_{ab})/\kappa; \qquad\qquad 16b.$$

that is, we can regard the new theory as described by Einstein's equations with a new form of matter. Although T_{ab} satisfies the energy condition (Equation 9b), T'_{ab} may not; if so, one may (as in the case of alternative matter) find cosmological solutions with a nonsingular origin. A further feature is now of importance. As is well known, the tensor $(R_{ab} - (1/2)Rg_{ab})$ has vanishing divergence; and indeed this fact is responsible for the standard form of the field equations (Equation 7). Equations 15 and 16 show that

$$\Lambda^{ab}_{;b} = (\zeta T^{ab})_{;b},$$ 17a.

$$T'^{ab}_{;b} = 0$$ 17b.

Now Λ_{ab} may have a nonzero divergence or $\zeta_{,b}T^{ab}$ may be nonzero. Thus, although T'_{ab} satisfies the conservation equations, in general T_{ab} does not, so the energy or momentum of the matter fields are not conserved in this theory; this is the origin of the possibility of matter creation in theories such as the Hoyle-Narlikar theory. However, one should note the comment by Lindblom & Hiscox (52): This procedure effectively constructs non-conserved quantities within a conservative theory. One can rewrite such theories in conservative form by putting the equations in the form of Equation 16 (cf. 61).

If an alternative theory of gravity leads to sufficiently large equivalent energy condition violations, a singularity can be avoided. This is possible, for example, in the Hoyle-Narlikar theory (46), in theories with torsion (83, 105), and in bimetric theories (80, 87). If the gravitational field equations are derived from a Lagrangian more complex than the general relativity Lagrangian, in general we may expect energy violations to be possible. Thus, if we introduce terms quadratic in the curvature invariants, a "bounce" may be possible (67); when nonlinear Lagrangians are considered that are arbitrary functions $f(R)$ of the scalar curvature R, the singularity can be avoided through a suitable choice of $f(R)$ (4, 50); and if the singularity is not avoided, one may nevertheless avoid the occurrence of horizons (4). Again, if the matter fields are nonminimally coupled to the curvature, singularity avoidance is, in general, possible (71).

CONFORMAL THEORIES   A class of alternative theories that have attracted particular attention are the variable-G or scale-covariant theories, such as Dirac's theory (21), the Brans-Dicke theory, the later Hoyle-Narlikar theories, and the theory of Canuto and coworkers. In general, different conformal gauges can be chosen in these theories that relate different metric representations through appropriate choice of a conformal factor (15, 89). This allows alternative explanations for observed redshifts; effectively one can switch terms in Equations 1 and 2, for when conformal transformations are allowed, one can no longer clearly distinguish between expansion and acceleration of the fundamental particles. [These depend on the conformal frame used (66).] In fact, conformal transformations can be used to transform the isotropic expansion of FLRW fundamental observers to zero; so one can remove the singularities in the matter flow in FLRW models by appropriate conformal transformation, i.e. by appropriate choice of a conformal frame. (In fact, it is well known that all the FLRW universe models are conformal both to flat space-time and to the Einstein static universe.) The effective matter tensor in the field equations (Equation 16) will therefore, in general, violate the energy conditions, for this is what makes it possible to transform the expansion to zero without violating the Raychaudhuri equation. This may explicitly be seen to be true in the revised Dicke theory for $(2\omega + 3) < 0$ (83). It is therefore not surprising that in some cases the massless scalar field in the Brans-Dicke theory can prevent the occurrence of initial singularities in the Universe (37). The remarks above lead us to expect that this will be a general feature of gauge-covariant theories.

QUANTUM COSMOLOGY   When investigating the behavior of the Universe at very early times, a quantum cosmological model may be used to take quantum gravitational effects into account; however, the construction of such models is considerably hampered by the lack of a complete and manageable quantum theory of gravity. Space limitations do not allow an adequate discussion of such models in this article; rather, the reader is referred to a recent excellent review article and the references therein (43).

Overall, it is easy to find alternative theories of gravitation where energy violation may occur. There is no compelling reason to support these theories against general relativity, except that it is clear that in the high-density and curvature limit a quantum theory of gravity is ultimately needed. Thus the quantum cosmology results (when agreement is reached on them) are clearly of considerable significance for any discussion of the origin of the Universe. Singularities might be avoided in the quantum case, but if not, it seems possible that they will be less severe than in the classical case; and at least horizons may not occur in these models.

5.   ALTERNATIVE GEOMETRIES

There are two basically different variations of the standard FLRW geometries: models that are spatially homogeneous but anisotropic, and models that are spatially inhomogeneous.

5.1   *Anisotropies*

The family of spatially homogeneous but anisotropic evolving universe models have been extensively studied (see 49, 55, 56 for recent reviews).

While these universes may at some stage in their history be very similar to the FLRW universes, they can have totally different singularity structures. The kinds of singularity that can occur depend on whether the universes are orthogonal or tilted.

ORTHOGONAL UNIVERSES   These are spatially homogeneous universes in which the matter flow vector is orthogonal (in the space-time sense) to the surfaces of homogeneity; thus, there is no motion of the matter relative to these homogeneous space sections. The fluid flow is geodesic and rotation free, and the surfaces of homogeneity are surfaces of simultaneity for all the fundamental observers. The models comprise Bianchi orthogonal models (the general case) and Kantowski-Sachs universes (a special "rotationally symmetric" case).

In all these cases, because $\omega = \dot{u}^a = 0$, the existence of a singularity follows directly from the energy conditions (Equation 9) and the Raychaudhuri equation (Equation 8); in essence, the Raychaudhuri equation shows that the normals to the surfaces of homogeneity must approach a point of intersection, but the fluid is moving along these normals and so runs into a singularity where the density of matter diverges.[2] The existence of the distortion term σ^2 in Equation 8 shows that the singularity is in general worse than in the FLRW case; indeed, usually the singularity is shear dominated, the effect of the matter term being negligible at early times. Limits on the amount of anisotropy come from observed limits on the anisotropy of the CMBR and from observed light-element abundances, compared with model predictions (55, 89, 100).

In these models, many different kinds of singularity structure can occur. In the simplest (Bianchi I) case, the solution resembles the vacuum Kasner solution; the singularity may be a "cigar" or "pancake" singularity (55, 100). If the cigar singularity occurs, horizons are broken in one spatial dimension (45, 64). Generically (in Bianchi VIII or IX) there will be oscillatory behavior, with a series of Kasner-like epochs taking place between "bounces" off a potential wall (7, 64). These possibilities have been discussed in a previous volume of this series (72). It seems agreed that the oscillatory behavior continues right back to the origin of the Universe, so that the initial behavior of the models can be characterized in a strict sense as "chaotic" (3); unfortunately, this does not suffice to break the horizons in all directions, as was once hoped (57).

While these singularities are geometrically quite different from the SHBB through their anisotropic expansion (and in many cases oscillatory behavior), which modifies (A), the essential features (B) remain; the most

---

[2] See (18) for a careful discussion of the Bianchi cases and (16) for the Kantowski-Sachs case.

significant difference is the possibility of horizon breaking in some directions [so that (C) no longer holds in all directions].

TILTED UNIVERSES   In this case, the surfaces of homogeneity are not orthogonal to the matter flow. Thus, although they are strictly spatially homogeneous in a rigorous sense, these universes appear inhomogeneous to the fundamental observers, who move relative to the homogeneous spatial sections (35). One can view this either as a result of their surfaces of instantaneity not coinciding with the surfaces of homogeneity (51) or because Doppler effects due to the peculiar velocity of the observer make a uniform distribution of matter appear anisotropic (30).

In tilted universes the matter can move with vorticity and will have nonzero acceleration unless the pressure is constant. Thus, the Raychaudhuri equation for the fluid no longer directly implies that a singularity occurs. However, the Raychaudhuri equation for the normals to the surfaces of homogeneity implies that these normals become singular within a finite proper distance; this result shows that the spatially homogeneous part of the Universe must end within a finite time. A careful analysis then demonstrates that the Universe must be singular at this time (45).

There are two rather different ways the spatially homogeneous region of the Universe can end in the past. (a) The first possibility is that an infinite density (spatially homogeneous) singularity can bound this region; then the situation is essentially the same as in the case of the orthogonal universes, except that the geometry can be more complex (because of the rotation and acceleration of the matter). The Universe originates at this infinite density singularity. (b) The second possibility (94) is that the density is finite at the boundary of the spatially homogeneous region of space-time (which represents late times in the Universe). The matter originates at early times in a spatially inhomogeneous, stationary region of space-time, separated from the spatially homogeneous region by a (Cauchy) horizon (18, 31).

It is easiest to understand this situation as follows (26, 27): consider the action of boosts of the Lorentz group through a hyperbolic angle β in two-dimensional Minkowski space-time (Figure 1a). If a single matter flow line through the origin P is spread over the space-time by repeating this action infinitely often, the (infinite) family of flow lines thus generated intersect at P and we have a model of Case (a): the spatially homogeneous region I comes to an end at a boundary where the matter density is infinite. This region comprises the entire Universe (Figure 1b). Suppose, however, a single flow line of matter that crosses from region I to region II is spread around by the same action of the Lorentz group. Then the surfaces of constant density change from being spacelike to being timelike as the fluid crosses the horizon $H^-(\mathscr{S})$ (Figure 1c), so while the late part I of the model is spatially homogeneous as before, the early part II is spatially inhomogeneous. This

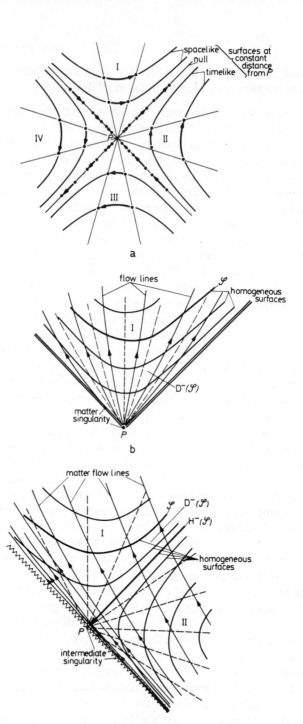

a

b

c

provides a model of Case (*b*). The limiting value of the density at the boundary of these regions of space-time, shown by a zigzag line in Figure 1*c*, is finite: it is the same as the density on the horizon $H^-(\mathscr{S})$ (which is a surface of homogeneity). However, the Universe cannot be extended beyond this boundary, for a singularity occurs here. One can see this by noting that in Figure 1*c* the light cone of any point is at 45°. The past null geodesic to the left from any point in the space-time runs into the boundary in a finite distance, but in that distance it crosses an infinite number of world lines. Consequently, the density of matter measured in a frame of vectors parallel propagated down this null geodesic will diverge. This parallel-propagated frame is related to a group-invariant frame by a Lorentz transformation that diverges at the boundary; and the density is perfectly finite at each point if measured in this group-invariant frame. Thus, this singularity is a nonscalar singularity (45) [colloquially known as a "Whimper" (31)]. All curvature and matter invariants are regular as one approaches the boundary; the divergence occurs when the Ricci and matter tensors are measured in a parallel-propagated frame.

This model gives a good representation of what happens in the full four-dimensional Case (*b*) tilted Bianchi cosmologies. The spatially homogeneous region is bounded by a Whimper singularity on the one hand, and a Cauchy horizon $H^-(\mathscr{S})$ on the other; the fluid crosses the Cauchy horizon from an earlier spatially inhomogeneous region, which is also partially bounded by the Whimper singularity. This singularity is causally null; correspondingly (45, 75), there will be no particle horizon in the direction of the Whimper singularity (but there will be one in the opposite direction; cf. the causal diagrams in 18). It is a "mild" singularity in that the temperature and density of the matter is finite there. The fluid itself originates at a second singularity (not shown in Figure 1*c*) that is timelike. Here, in general, the density will diverge, but this will not always be the case: situations are known (18) where the density is finite at this initial singularity (but where the conformal geometry is singular). Thus, in such exact solutions of Einstein's equations—which have reasonable equations of state and obey the energy conditions—two singularities occur, but there is a finite maximum density for matter in the Universe; hence, there is also a bound to the temperature that will occur.

The plausibility of these universe models rests on their ability to give

---

Figure 1 Two-dimensional Minkowski space-time as a model for singularities. (*a*) The action of "boosts" about *P* is shown. (*b*) A matter line through *P*, moved over the space by the boosts about *P*, leads to a flow diverging from *P*. A matter singularity occurs. (*c*) A matter line not through *P*, moved over the space by the boosts about *P*, leads to a flow that "piles up" toward *P* but that does not converge there. A nonscalar singularity occurs.

reasonable predictions for element production and compatibility with the microwave anisotropy. Initial investigation (31, 60) is encouraging; it seems that current observational evidence is quite compatible with such singularities. It should be mentioned that it is known that these models are of zero measure in the set of all spatially homogeneous models (95, 96), and this fact has been used as an argument against their occurrence. While this is an interesting line of argument, its status is unclear because of the uniqueness of the Universe. Additionally, it should be mentioned that if one pursues this line, the SHBB FLRW models are an even smaller subset of the spatially homogeneous models and thus, according to this argument, are even more unlikely to occur than the Whimpers.

OTHER EFFECTS   As in the previous case, one can consider spatially homogeneous, anisotropic universes with other equations of state or governed by other gravitational theories. Because of the shear in these cosmologies, in many cases (e.g. for orthogonal fluids) one may expect the singularities to be worse than in corresponding FLRW universes. Thus, for example, while torsion can prevent singularities in FLRW universes, it cannot do so in Bianchi I universes (99).

Where quantum field effects seem to make a significant difference is in the evolution forward in time from the singularity; for in anisotropic universes there will be significant pair creation of conformally invariant particles, and this will dissipate anisotropy and so drive the Universe toward a state of isotropy (43, 115). Similarly, the effect of a false-vacuum inflationary expansion in a Bianchi universe will be to make it evolve exponentially toward the de Sitter (isotropic) solution (108). However, if the anisotropy in the Universe is too large at the end of the quantum gravity era, the inflationary phase will never take place (5a). Thus, quantum effects can contribute to a limited extent toward Misner's (63) vision of a situation where any anisotropy or nonuniformity will be dissipated by physical processes in the early Universe, so giving an "explanation" of presently observed isotropy.

5.2   Inhomogeneities

As the fluid can now move with acceleration and rotation, the Raychaudhuri equation does not directly imply the existence of a singularity in the past. However, the powerful Hawking-Penrose singularity theorems (45, 102) show that as long as suitable energy and causality conditions are satisfied, singularities will indeed occur in a general inhomogeneous universe model [essentially because the matter that thermalized the microwave background radiation has sufficient energy to cause a refocusing of all the null geodesics that generate our past light cone (45)].

In many cases the singularity will be essentially of the same character as in the spatially homogeneous case, often showing the same kind of oscillatory behavior as the Bianchi IX cosmologies (7, 9; but see 5, 8, 57). The major difference is that one can now additionally get timelike singularities.

TIMELIKE INITIAL SINGULARITIES   The existence of inhomogeneities at early phases in a universe that is like a FLRW universe at late times implies that "the initial big bang is not necessarily simultaneous. One can formulate initial conditions in such a way that expansion of some parts of the universe are delayed" (116). It has been proposed that such "lagging cores" could be responsible for violent astrophysical phenomena such as QSOs (69, 71a). The situation could be the time-reverse of the collapse to a black hole: this case is called a "white hole" (71a). In this case, the singularity will be surrounded for most of its life by an absolute particle horizon, into which no particles may fall but from which particles may be ejected (76). Difficulties arise concerning the existence of "white holes," particularly in view of the particle creation process that would tend to make them rapidly evaporate away (76, 78).

However, there appears to be nothing preventing the existence of more general timelike singularities that do not occur instantaneously, but continue over a time (22), and that are not associated with such horizons (104); the timelike part of the singularity may have the nature of either a negative- or positive-mass singularity (62, 104). A new dynamics now occurs, for the singularity both emits matter and radiation to the Universe and receives them from it; thus, it both acts on and can be acted upon by the Universe. This relation of the singularity to the Universe is quite unlike that in the spacelike singularity case where the singularity emits matter and radiation to the Universe, and indeed determines all its initial conditions, but cannot be affected by it. One now has the possibility of investigating the dynamics of interaction of the singularity with the Universe.

These timelike singularities can occur as inhomogeneous segments of a singularity that is spacelike elsewhere (62, 104). Indeed, it is clear that unless one chooses very special conditions in the Universe (by setting to zero the "decaying modes" of density perturbations), inhomogeneities will dominate at early stages (97). Penrose argues on entropy grounds that the early Universe should obey such rather special conditions (77, 78); his argument depends on probability assumptions about the set of all possible universes—assumptions that may be queried in view of the uniqueness of the Universe.

The most intriguing possibility is if one or more isolated timelike singularities occur, with most of the matter passing between them rather than originating at them (45). This situation is closely related to the hope that many people have that the present expanding phase of the Universe should result from a previous collapse phase, with the formation of isolated singularities and subsequent reexpansion. Then the hot big bang phase is like an inhomogeneous SHBB but is preceded by a previous collapsing phase. The overall picture is like "cyclic" FLRW universes collapsing from infinity or from a previous expansion phase and then reexpanding (with most of the matter avoiding the singularities; these must form because the energy conditions are satisfied).

We do not have exact cosmological models where this situation occurs; but two examples indicate it may be possible, namely the Reissner-Nordstrom solution with test matter in it (45), and certain classes of Whimper singularities where two horizons may occur (31) (if these exist; no exact solutions of this kind are known at present). In both these cases the situation is probably unstable, but they show that it is not obviously prohibited by the field equations. If it does occur, the conditions of the present phase of expansion are determined by the two-way interaction between the Universe and the singularity formed from the previous collapse phase; no energy violations or quantum effects are needed to cause such a "bounce."

ESSENTIALLY INHOMOGENEOUS UNIVERSES   These are universe models containing no epochs where the space-time is approximately spatially homogeneous; that is, they are never like a FLRW universe model. This assumption, of course, goes against the usual "cosmological principle," which states a priori that the Universe is spatially homogeneous (109); but that is an unverified and, indeed, possibly unverifiable (28) philosophical statement that could be incorrect.[3]

One possibility is asymptotically flat models—an isolated dense universe surrounded by empty space. An example of this kind is the recent Narlikar & Burbidge (68) proposal, in which all detectable matter lies in an expanding super-supercluster. While this kind of situation is a possibility, many features remain to be explained, e.g. the element abundances and the microwave background radiation. The specific Narlikar-Burbidge version appears to be ruled out by QSO absorption-line measurements (81).

A related concept is that of a hierarchical universe, where clustering occurs over all possible scales, so that each observer is within an effective spherical inhomogeneity but the average density of matter goes to zero in larger and larger volumes. In this model the Universe is spatially homogeneous (19). However, this is difficult even to describe mathematically (as the description used depends completely on the scale of averaging); it is not clear that the models proposed so far (10, 110) in fact represent the hierarchical situation fully.

---

[3] It is false in the currently fashionable inflationary universe scenario, where major inhomogeneities occur at the bubble walls.

If we abandon statistical homogeneity, the only universe models that give isotropic observations[4] are inhomogeneous, spherically symmetric models, where we are near a center of symmetry (11). In order to explain the microwave background radiation in such models, an intriguing possibility is that there are *two* centers of symmetry in the Universe, with one in our neighborhood and the other at the location of a timelike singularity. This singularity would be surrounded by a massive fireball emitting blackbody radiation; thus the space sections of the Universe would look like Figure 2b, with the fireball—like a supermassive star—at the antipodes. A space-time diagram can be drawn as in Figure 2a. The timelike singularity at the center of the fireball would continually emit radiation and particles into the Universe as well as receive them from the Universe—it would be dynamically interacting with the Universe (as discussed above). A remarkable duality between such models and the FLRW universes is possible, with each stage of the FLRW universes that occurred in the past as a result of time evolution also occurring in these universes spatially as a result of spatial inhomogeneity (32).

Perhaps the most radical such proposal is the completely static two-centered universe possibility (17, 32), where the redshifts observed for distant galaxies are purely cosmological gravitational redshifts (cf. Equation 2b). High pressures are needed in these models to cause the acceleration that underlies the gravitational redshift; this is not impossible but is perhaps uncomfortable. Two problems arise: Firstly, because of the spherical symmetry, the (redshift, distance) law in such a universe must be quadratic at the center, so (as in Segal's theory discussed in Section 3.1) we can test this model by seeing if a quadratic relation is indeed observed. The usual data show a linear relation, contrary to expectation in such universes; however, these data are analyzed on the assumption the Universe is a FLRW universe, and selection effects play a crucial role in determining which is the correct law. (Essentially the same analysis will apply here as in the Segal theory.) The data still have not been examined using an analysis that takes surface brightness effects fully into account. Secondly, philosophical problems may arise because of the Earth being near the center in these models. However, these qualms can be offset by noting that physical conditions can favor the occurrence of life at this center in such universes and by considering the a priori improbability of the FLRW universe models (27a). On the positive side, it is intriguing to note that the background radiation, still our best evidence for isotropy of the Universe, would appear isotropic at *every* point in such universes (because its temperature will be determined by the ratio of gravitational potentials at the points of emission and reception of the radiation; and this is

---

[4] See, for example, (55) for a summary of the evidence that the Universe is isotropic about us.

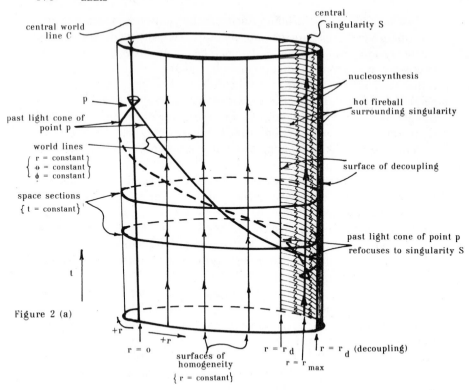

central world
line C

central
singularity S

nucleosynthesis

hot fireball
surrounding singularity

p

past light cone of
point p

world lines
$\left\{ \begin{array}{l} r = constant \\ \theta = constant \\ \phi = constant \end{array} \right\}$

surface of decoupling

space sections
$\{ t = constant \}$

past light cone of point p
refocuses to singularity S

t

Figure 2 (a)

+r

+r

$r = 0$

surfaces of
homogeneity
$\{ r = constant \}$

$r = r_d$ $r = r_d$ (decoupling)

$r = r_{max}$

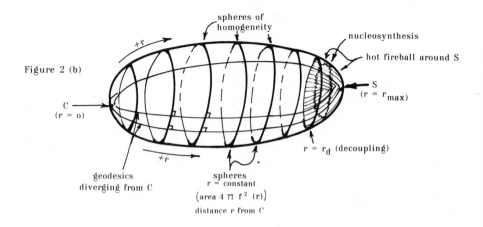

spheres of
homogeneity

nucleosynthesis

+r

hot fireball around S

Figure 2 (b)

C
$(r = 0)$

S
$(r = r_{max})$

$r = r_d$ (decoupling)

+r

geodesics
diverging from C

spheres
$r = constant$
$\left(area \ 4\pi \ f^2 \ (r) \right)$

distance r from C

independent of the path the light follows to the observer). Thus, observed isotropy of this radiation does not, in this case, imply that the Universe is isotropic about our position. On the other hand, isotropy of galactic redshifts would carry this implication.

The major interest of this model is the way in which it highlights the quite different role the singularity can play as a continuing influence in the evolution of the Universe. It is not a once-and-for-all event, but is instead a continuing source and sink of matter and information; in this case, it enables a steady state. The singularity has a different *meaning* than in the FLRW universe.

While it is difficult to make a convincing case for a static universe of this kind, it seems likely that expanding versions maintaining the essential features of Figure 2 could account for all present observations. If the expansion originated in a singularity, the model would then be essentially like the delayed core models mentioned above; but the expansion could, for example, conceivably be preceded by a stationary phase or by a contraction phase to some minimum radius with the singularity always localized in one region of the Universe.

FURTHER EFFECTS As in the previous cases, one can consider the nature of inhomogeneous cosmologies if energy violation, other gravitational theories, or quantum effects are taken into account.

Perhaps the most intriguing new possibility is that the energy conditions may be violated for essentially geometric reasons. The basic point here is that the Einstein field equations are tested, and believed to hold, on a particular scale (say, that of the solar system). In an inhomogeneous situation it is not clear that the geometrically averaged space-time metric representing the situation at other scales (say, that of clusters of galaxies) will obey the same field equations; indeed, one would expect correction terms (similar to the polarization terms in electromagnetic theory) to allow for the spatial averaging that is taking place (29). However, there is no guarantee that these terms will obey the energy conditions; and, indeed, in a turbulent situation they might be able to cause effective negative energy densities so large as to avoid a singularity (59). Thus the space-time metric averaged out to describe the Universe at the larger scale could be singularity free (but small-scale singularities could still occur, e.g. local black holes resulting from gravitational collapse of stars).

Figure 2 A two-centered universe, with our Galaxy near one of the centers. (*a*) Space-time diagram of the Universe (two spatial dimensions suppressed). It is a cylinder, with a timelike singularity at the other center. Our past light cone is shown. (*b*) Space section of the Universe (one spatial dimension suppressed). The singularity lies behind a fireball, which is like a supermassive star "over there."

6. CONCLUSION

If we assume the standard FLRW geometries, then different equations of state or field equations resulting in sufficiently large effective energy violation imply the possibility that there is no singularity at the beginning of the present phase of expansion of the Universe. This expansion could start from a stationary state (either a static phase or an exponentially expanding phase) or could be a bounce from a previous collapse. Any "bounce" that occurs must have taken place at very early times, if our present picture of the origin of the microwave background radiation and the elements is correct. It is plausible that quantum field effects could cause a bounce at extremely early times, but the foundations of quantum cosmology need clarification before such a statement is on firm footing; if this does not happen, the quantum field effects probably at least make the singularity less severe (e.g. by eliminating particle horizons).

A different geometry implies the possibility of quite different kinds of singularity, which entail different concepts of creation. Finite density singularities can occur at the beginning of the Universe in the case of tilted homogeneous cosmologies, lessening the severity of the initial singularity. (One must have some real singularity at any "beginning" of the Universe, for otherwise we can continue the Universe past this event to earlier times, thus showing that it was a "fictitious" singularity rather than a real one.) A major difference in the case of inhomogeneous universes is the possibility of timelike singularities, which continue to interact with the Universe over a period of time—perhaps for the entire history of the Universe—rather than the once-and-for-all interaction of the singularity in the standard theory, where creation takes place at one time and then the singularity is only an incident in the past.

There are two kinds of steady-state universe that seem to have the potential of giving a reasonable description of the astrophysical evidence: the original steady-state universe of Bondi, Gold, and Hoyle; and the two-centered static universe of Ellis, Maartens, and Nel. In both cases creation is a continuously proceeding process—in the first case diffused through space and in the second localized at a singular center of the Universe. However, both of these models run into difficulties when compared in detail with observational evidence.

The next major possibility is that the Universe started off in a steady-state situation and then changed to an evolutionary phase, as in the Eddington-Lemaître expansion from an Einstein static state or the Starobinski expansion from a de Sitter steady-state phase; there may also be suitable models expanding from an Ellis-Maartens-Nel static phase. A major problem with any such stationary-state origin is how the Universe chooses when to break this phase and start the present evolutionary phase: "Since the initial metric has a finite life-time, it could not exist as $t \to -\infty$, and therefore some other metric should exist before it" (53).

If a bounce takes place, it either comes from a state that has existed for indefinitely long in the past, perhaps collapsing from an infinite radius; or else it is a rebounce following a previous state of expansion from a state of high density, as in the "phoenix" universes that perpetually oscillate, and in many ways give an attractive understanding of the history of the Universe (20, 58). Each rebirth can take place either through effective energy violation or quantum effects (when no singularity occurs) or through the occurrence of isolated singularities that are sidestepped by most of the matter and that interact with the Universe for a brief period.

A singularity will exist unless effective energy violation takes place; it will provide an initial boundary to the Universe as we know it. The singularity could be "weak," so that quantum mechanically one can follow the Universe through to the era preceding the initial singularity; that is, one may be able to provide a theory of creation. Any such theory of creation of the Universe must by necessity postulate some preexistent structure that provides the basis for the equations used to describe creation (e.g. a previously existing oscillating universe, Minkowski space, a de Sitter universe, superspace, or pregeometry). These issues cannot be pursued further here; we refer the reader to the Appendix of Linde's (53) paper and the references therein.

There are then a variety of possibilities available, both within standard general relativity (where a variety of geometries are considered) and in variations of that theory, for understanding the nature of the origin of the Universe. The SHBB is one of this family but is by no means the only conceivable member; further possibilities are not completely excluded experimentally or theoretically.

Literature Cited

1. AAPT. 1979. *Cosmology*, ed. L. C. Shepley, A. A. Strassenburg. SUNY Stony Brook, N.Y.: Am. Assoc. Phys. Teach.[1]
2. Arp, H. C. 1980. *Ann. NY Acad. Sci.* 336:94–112
3. Barrow, J. D. 1982. *Phys. Rep.* 85:1–49
4. Barrow, J. D., Ottewill, A. C. 1983. *J. Phys. A* 16:2757–76

[1] Various key papers on cosmology are reprinted in this volume; they are indicated in this list of references by the designation "AAPT".

5. Barrow, J. D., Tipler, F. J. 1979. *Phys. Rep.* 56:371–402
5a. Barrow, J. D., Turner, M. S. 1981. *Nature* 292:35–38
6. Bekenstein, J. D. 1975. *Phys. Rev. D* 11:2072–75
7. Belinsky, B. A., Khalatnikov, I. M., Lifshitz, E. M. 1970. *Adv. Phys.* 19:525–73
8. Belinsky, B. A., Khalatnikov, I. M., Lifshitz, E. M. 1980. *Phys. Lett. A* 77:214–16; 83:321
9. Belinskii, V. A., Khalatnikov, I. M.,

182   ELLIS

Lifshitz, E. M. 1982. *Adv. Phys.* 31:629–67
10. Bonnor, W. B. 1972. *MNRAS* 159:261–68
11. Bonnor, W. B. 1974. *MNRAS* 167:55–61
12. Brans, C., Dicke, R. 1961. *Phys. Rev.* 124:925–31 (AAPT)
13. Burbidge, E. M. 1967. *Ann. Rev. Astron. Astrophys.* 5:410–11, 433–37
14. Candelas, P., Sciama, D. W. 1977. *Phys. Rev. Lett.* 38:1372–75
15. Canuto, V. M., Narlikar, J. V. 1980. *Ap. J.* 236:6–23
16. Collins, C. B. 1977. *J. Math. Phys.* 18:2116–24
17. Collins, C. B. 1983. *J. Math. Phys.* 24:215–19
18. Collins, C. B., Ellis, G. F. R. 1979. *Phys. Rep.* 56:65–105
19. de Vaucouleurs, G. 1970. *Science* 157:1203–13 (AAPT)
20. Dicke, R. W., Peebles, P. J. E. 1979. In *General Relativity: An Einstein Centenary Survey*, ed. S. W. Hawking, W. Israel, pp. 504–17. Cambridge Univ. Press
21. Dirac, P. A. M. 1938. *Proc. R. Soc. London Ser. A* 165:199–208
22. Eardley, D., Liang, E., Sachs, R. K. 1972. *J. Math. Phys.* 13:99–107
23. Eddington, A. S. 1930. *MNRAS* 90:668–78
24. Ehlers, J. 1961. *Abh. Akad. Wiss. Lit. Mainz, Math. Naturwiss. Kl.* 11:791–837
25. Ellis, G. F. R. 1971. In *General Relativity and Cosmology*, ed. R. K. Sachs, pp. 104–82. New York: Academic
26. Ellis, G. F. R. 1978. *Comments Astrophys. Space Phys.* 8:1–7
27. Ellis, G. F. R. 1975. *Ann. NY Acad. Sci.* 262:231–40
27a. Ellis, G. F. R. 1979. *Gen. Relativ. Grav.* 11:281–89
28. Ellis, G. F. R. 1980. *Ann. NY Acad. Sci.* 336:130–60
29. Ellis, G. F. R. 1984. *GR10 Conf. Proc.*, ed. A. Pascolini, et al. In press
30. Ellis, G. F. R., Baldwin, J. 1984. *MNRAS* 206:377–81
31. Ellis, G. F. R., King, A. R. 1974. *Commun. Math. Phys.* 38:119–56 (AAPT)
32. Ellis, G. F. R., Maartens, R., Nel, S. D. 1978. *MNRAS* 184:439–65
33. Field, G. B., Arp, H., Bahcall, J. N. 1973. *The Redshift Controversy.* Reading, Mass: Benjamin
34. Geller, M., Peebles, P. J. E. 1972. *Ap. J.* 174:1–5
35. Gödel, K. 1952. *Proc. Int. Congr. Math.*, ed. L. M. Graves, et al., pp. 175–81.

Providence, R.I.: Am. Math. Soc.
36. Goenner, H. 1984. *GR10 Conf. Proc.*, ed. A. Pascolini, et al. In press
37. Gurevich, L. E., Finkelstein, A. M., Ruban, V. A. 1973. *Astrophys. Space Sci.* 22:231–42
38. Guth, A. H. 1983. In *The Very Early Universe*, ed. G. W. Gibbons, S. W. Hawking, S. T. C. Siklos, pp. 171–204. Cambridge Univ. Press
39. Guth, A. H., Sher, M. 1983. *Nature* 302:505–6
40. Harrison, E. R. 1970. *Nature* 228:258–60
41. Harrison, E. R. 1981. *Cosmology.* Cambridge: Cambridge Univ. Press
42. Hartle, J. B. 1981. In *Quantum Gravity 2*, ed. C. J. Isham, R. Penrose, D. W. Sciama, pp. 313–28. Oxford Univ. Press
43. Hartle, J. B. 1983. In *The Very Early Universe*, ed. G. W. Gibbons, S. W. Hawking, S. T. C. Siklos, pp. 59–89. Cambridge Univ. Press
44. Hawking, S. W. 1975. In *Quantum Gravity: An Oxford Symposium*, ed. C. J. Isham, R. Penrose, D. W. Sciama, pp. 219–67. Oxford Univ. Press
45. Hawking, S. W., Ellis, G. F. R. 1973. *The Large-Scale Structure of Space-Time.* Cambridge Univ. Press
46. Hoyle, F., Narlikar, J. V. 1964. *Proc. R. Soc. London Ser. A* 278:465–78
47. Huang, K., Weinberg, S. W. 1970. *Phys. Rev. Lett.* 25:895–97
48. Jones, J. E. 1974. *Proc. R. Soc. London Ser. A* 340:263–86
49. Jantzen, R. T. 1983. *Adv. Ser. Astrophys.*, Vol. 1
50. Kerner, R. 1982. *Gen. Relativ. Grav.* 14:453–69
51. King, A. R., Ellis, G. F. R. 1973. *Commun. Math. Phys.* 31:209–42
52. Lindblom, L., Hiscox, W. A. 1982. *J. Phys. A* 15:1827–30
53. Linde, A. D. 1983. In *The Very Early Universe*, ed. G. W. Gibbons, S. W. Hawking, S. T. C. Siklos, pp. 205–49. Cambridge Univ. Press
54. Longair, M. S., Rees, M. J. 1973. In *Cargese Lectures in Physics*, ed. E. Schatzmann, 6:340–45. New York: Gordon & Breach
55. MacCallum, M. A. H. 1979. In *General Relativity: An Einstein Centenary Survey*, ed. S. W. Hawking, W. Israel, pp. 533–80. Cambridge Univ. Press
56. MacCallum, M. A. H. 1979. In *Physics of the Expanding Universe, Lecture Notes in Physics No. 109*, ed. M. Demianski, pp. 1–59. Berlin/New York: Springer

57. MacCallum, M. A. H. 1981. In *The Origin and Evolution of Galaxies*, ed. B. J. T. Jones, J. E. Jones, pp. 9–39. Dordrecht: Reidel
58. Markov, M. A. 1983. In *The Very Early Universe*, ed. G. W. Gibbons, S. W. Hawking, S. T. C. Siklos, pp. 353–71. Cambridge Univ. Press
59. Marochnik, L. S., Pelikov, N. V., Vereshkov, G. M. 1975. *Astrophys. Space Sci.* 34:249–95
60. Matravers, D. R., Vogel, D. L. 1983. Preprint, Dept. Appl. Math., Univ. Cape Town
61. McCrea, W. H. 1951. *Proc. R. Soc. London Ser. A* 206:562–75
62. Miller, B. D. 1979. *J. Math. Phys.* 20:1356–61
63. Misner, C. W. 1968. *Ap. J.* 151:431–57 (AAPT)
64. Misner, C. W. 1969. *Phys. Rev. Lett.* 22:1071–74 (AAPT)
65. Misner, C. W., Thorne, K. S., Wheeler, J. A. 1973. *Gravitation*. San Francisco: Freeman
66. Murphy, G. L. 1973. *Phys. Rev. D* 8:4231–33
67. Nariai, H., Tomita, K. 1971. *Prog. Theor. Phys.* 46:776–86
68. Narlikar, J. V., Burbidge, G. R. 1981. *Astrophys. Space Sci.* 74:111–29
69. Ne'eman, Y. 1965. *Ap. J.* 141:1303–5
70. Nicholl, J. F., Segal, I. E. 1982. *Proc. Natl. Acad. Sci. USA* 79:3913–17
71. Novello, M., Heintzmann, H. 1983. Preprint CBPF-NF-010/83, Cent. Bras. Pesqui. Fis., Rio de Janeiro
71a. Novikov, I. D. 1964. *Sov. Astron. AJ* 8:857, 1075
72. Novikov, I. D., Zel'dovich, Ya. B. 1973. *Ann. Rev. Astron. Astrophys.* 11:387–412
73. Parker, L. 1982. *Fundam. Cosmic Phys.* 7:201–39
74. Parker, L., Fulling, S. A. 1973. *Phys. Rev. D* 7:2357–74 (AAPT)
75. Penrose, R. 1964. In *Relativity, Groups, and Topology*, ed. C. de Witt, B. de Witt, pp. 565–84. New York: Gordon & Breach
76. Penrose, R. 1979. *General Relativity: An Einstein Centenary Survey*, ed. S. W. Hawking, W. Israel, pp. 600–11. Cambridge Univ. Press
77. Penrose, R. 1980. In *Essays in General Relativity*, ed. F. Tipler, pp. 1–12. New York: Academic
78. Penrose, R. 1981. *Quantum Gravity 2*, ed. C. J. Isham, R. Penrose, D. W. Sciama, pp. 244–72. Oxford Univ. Press
79. Petrosian, V. 1982. *Nature* 298:506–7, 805–8
80. Petry, W. 1981. *Gen. Relativ. Grav.* 13:1057–71
81. Phillipps, S., Ellis, R. S. 1983. *MNRAS* 204:493–506
82. Raychaudhuri, A. K. 1955. *Phys. Rev.* 98:1123–26
83. Raychaudhuri, A. K. 1980. *Relativistic Cosmology*. Cambridge Univ. Press
84. Rees, M. J. 1978. *Nature* 275:35–37, 343
85. Rindler, W. 1956. *MNRAS* 116:662–77
86. Robertson, H. P. 1933. *Rev. Mod. Phys.* 5:62–90
87. Rosen, N. 1980. *Gen Relativ. Grav.* 12:493–510
88. Rosen, N. 1983. *Found. Phys.* 13:363–72
89. Rothman, T., Matzner, R. 1982. *Ap. J.* 257:450–55
90. Schmidt, M. 1969. *Ann. Rev. Astron. Astrophys.* 7:540–44
91. Schrödinger, E. 1956. *Expanding Universes*. Cambridge Univ. Press
92. Segal, I. E. 1976. *Mathematical Cosmology and Extragalactic Astronomy*. New York: Academic
93. Segal, I. E. 1982. *Ap. J.* 252:37–38
94. Shepley, L. C. 1969. *Phys. Lett. A* 28:695–96
95. Siklos, S. T. C. 1977. *Commun. Math. Phys.* 58:255–72
96. Siklos, S. T. C. 1981. *Gen. Relativ. Grav.* 13:433–41
97. Silk, J. 1977. *Astron. Astrophys.* 59:53–58
98. Soneira, R. M. 1979. *Ap. J. Lett.* 230:L63–65
99. Stewart, J., Hajicek, P. 1973. *Nature Phys. Sci.* 244:96
100. Thorne, K. S. 1967. *Ap. J.* 148:51–68
101. Tipler, F. J. 1978. *Phys. Rev. D* 17:2521–28
102. Tipler, F. J., Clarke, C. J. S., Ellis, G. F. R. 1980. In *General Relativity and Gravitation*, ed. A. Held, 2:97–206. New York: Plenum
103. Tolman, R. C. 1934. *Relativity, Thermodynamics and Cosmology*. Oxford Univ. Press
104. Tomita, K. 1978. *Prog. Theor. Phys.* 59:1150–69
105. Trautman, A. 1973. *Nature Phys. Sci.* 242:7–8
106. van den Bergh, N. 1983. *Gen. Relativ. Grav.* 15:441–48
107. Wagoner, R. V. 1980. In *Physical Cosmology*, ed. R. Balian, J. Audouze, D. N. Schramm, pp. 394–442. Amsterdam: North-Holland
108. Wald, R. M. 1983. Preprint MPI-PAE/PTh 38/83, Max Planck Inst. Phys., Munich

184   ELLIS

109. Weinberg, S. E. 1972. *Gravitation and Cosmology*. New York: Wiley
110. Wesson, P. S. 1975. *Astrophys. Space Sci.* 32:273–284, 305–30
111. Weyl, H. 1923. *Phys. Z.* 24:230–32
112. Will, C. M. 1979. In *General Relativity: An Einstein Centenary Survey*, ed. S. W. Hawking, W. Israel, pp. 24–89. Cambridge Univ. Press
113. Woszczynn, A., Betkowski, W. 1982. *Astrophys. Space Sci.* 82:489–93
114. Yahil, A., Sandage, A., Tammann, G.

1980. In *Physical Cosmology*, ed. R. Balian, J. Audouze, D. N. Schramm, pp. 127–59. Amsterdam: North-Holland
115. Zeldovich, Ya. B. 1970. *JETP Lett.* 12:307
116. Zeldovich, Ya. B. 1979. In *Physics of the Expanding Universe, Lecture Notes in Physics No. 109*, ed. M. Demianski, pp. 60–80. Berlin/New York: Springer
117. Zuiderwijk, E. J., de Ruiter, H. R. 1983. *MNRAS* 204:675–89

Ann. Rev. Astron. Astrophys. 1984. 22:185–222

THE EVOLUTION OF GALAXIES IN CLUSTERS

A. Dressler

Mount Wilson and Las Campanas Observatories,
Carnegie Institution of Washington, Pasadena, California 91101-1292

1. INTRODUCTION

Even though they contain only a small fraction of the galaxies in space, rich clusters stand out distinctly against the fabric of galaxies that make up the visible Universe. Because of their high surface densities and large number of very luminous galaxies, clusters can be identified out to distances comparable to the present horizon of the Universe, making them important tools in the study of cosmology (Hubble 1936, Zwicky 1938). Because of this legacy, research on clusters of galaxies has traditionally centered on the measurements of standard candles (e.g. the luminosity of the brightest cluster member, or the "knee" in the cluster luminosity function) or standard metrics (radii of cluster galaxies or the cluster distribution as a whole). The review article by Bahcall (1977) reflects a goal of researchers at that time: to provide a morphological description of the sizes, shapes, and galaxy content of clusters, and whenever possible, the applicability of these parameters to cosmological investigations.

In the 1970s a new course of research on clusters of galaxies began to emerge with the resurgence of interest in galaxy evolution. It was recognized that certain clusters contain "supergalaxies" unlike any objects seen in the general field. How had these objects evolved in the unique environment of clusters? The general population of galaxies in clusters is highly skewed toward elliptical and S0 galaxies, a population quite unlike the spiral-dominated field where most galaxies are actively forming new stars. What could this obvious difference tell us about the influences of environment both in the formation and evolution of the different morphological types? It was discovered that many rich clusters contain a pervasive, hot intergalactic gas. Might interactions between this gas and galaxies be

185

0066–4146/84/0915–0185$02.00

strong enough to alter galaxy properties such as their own gas fraction and rates of star formation? Clusters, it was realized, are laboratories for the study of galaxy evolution and may become as useful as star clusters are in the study of stellar evolution.

This review is restricted to recent work on clusters and groups of galaxies that is relevant to the study of galaxy evolution. Such subjects as cluster morphology, structure (e.g. spatial distribution of galaxies, core radii), luminosity functions, mass-to-light ratios, catalogs, and dynamics are discussed only in connection with how galaxies may form and change in response to the cluster environment. Also beyond the scope of this review is the rapidly growing area of research that uses large-scale clustering properties to model the evolution of the early Universe (e.g. the distribution of clusters in space, the multiplicity function for clusters, the evolution of clusters and superclusters from primordial fluctuations). Two other areas have grown so rapidly since the Bahcall review that they warrant their own chapters: X-ray emission from clusters (Forman & Jones 1982) and the H I content of cluster galaxies (Haynes et al. 1984). These reviews are themselves highly relevant to the topics discussed here.

The following discussion is divided into four sections: the laboratory—a brief introduction to clusters of galaxies; mergers, tidal stripping, and accretion—the evolution of cD galaxies; the development of the morphological types; and the populations of clusters as a function of cosmological look-back time.

2. THE LABORATORY

A small fraction (about 5%) of the galaxies in the present-epoch Universe are collected into groups and clusters whose space density is larger than one galaxy per cubic megaparsec (about two orders of magnitude greater than the average). Some of these aggregations are both dense and populous. These are the rich clusters cataloged by Abell (1958) during the original Palomar Sky Survey. Typically they contain 100 galaxies with a luminosity spread of two orders of magnitude (and perhaps a larger number of fainter systems) in a volume several megaparsecs across. A roughly equal number of galaxies inhabit relatively poor groups with less than 30 galaxies, which also have relatively high densities (Bahcall 1980).

The Virgo cluster, the rich cluster nearest to our Galaxy, is a representative example, with the typical number and density of galaxies. Its structure is also quite ordinary, for although it has the beginnings of a well-developed core, a general lack of symmetry and significant subclustering indicates that dynamical relaxation on a large scale has just begun. Its population of galaxies is normal, with all morphological types [ellipticals, S0s, spirals (and irregulars)] more or less equally represented. Thus, in all these characteristics measured by optical techniques, the Virgo cluster is a very average cluster, as one might expect of our nearest neighbor.

Perhaps it has been a mixed blessing that another of the nearest clusters, the Coma cluster, is a very rare specimen. Its proximity has allowed detailed studies of a cluster that is unusually populous and rich in early-type galaxies, but it has also contributed to a widespread notion that Coma is a typical rich cluster. In fact, it is no more typical of rich clusters than M87 is typical of galaxies. The Coma cluster, with its population of several hundred bright galaxies, is richer than 95% of the clusters catalogued by Abell. Its galaxy distribution is highly concentrated and very symmetric, with little or no sign of subclustering; this probably indicates that it virialized many cluster crossing times ago [several $\times$ 10^9 yr (Gunn & Gott 1972)]. The population of its core is extreme, as it contains nearly equal numbers of elliptical and S0 galaxies and virtually no spirals or irregulars. The great advantage in having a cluster like Coma so nearby is that one can study an *extreme* environment where galaxy evolution may have been most influenced by the formation and evolution of the cluster. To the extent that the formation and evolution is a function of the *local* environment (e.g. the local density and characteristic velocity of the galaxies), the processes that have influenced galaxy development in the Coma cluster are a smooth extrapolation of what has happened in the less populous but more numerous groups of galaxies.

It is the long-term goal of the studies reviewed here to understand how galaxy-galaxy and galaxy-cluster interactions have influenced the development of different morphological types, the nuclear and star-forming activity of galaxies, and the structure and luminosity function of member galaxies. The processes that have been proposed are reviewed here and compared with observational data, chiefly from optical measurements. Most of these processes depend on one or more of the following parameters: (a) the local density of galaxies and/or gas, (b) the kinematics of the galaxy distribution, and (c) the types, structures and internal kinematics of the galaxies present. Given the wide range of environments, from the low-density field to the richest clusters, it should be possible to evaluate the domain of importance for each model and thus arrive at a better understanding of how galaxies form and evolve.

3.   MERGERS AND TIDAL STRIPPING: THE EVOLUTION OF cD GALAXIES

3.1   *Criteria for Identifying cD Galaxies*

The study of cD galaxies began with the work of W. W. Morgan in the 1960s. Matthews et al. (1964) called attention to the existence in some clusters of bright elliptical galaxies surrounded by extensive, amorphous stellar envelopes (Morgan & Lesh 1965). These they called D galaxies. The *largest* (and thus the most luminous) of these D galaxies, those with an extent 3–4 times larger than the largest lenticulars (S0s) in the cluster, were dubbed supergiant D: cD galaxies. The large sizes and extensive stellar envelopes (~ 100 kpc)[1] are the only *primary* attributes of cD galaxies, and they are the proper criteria to use when classifying clusters (Struble & Rood 1982).

These and later studies (see Bahcall's review for a comprehensive list) have called attention to other *secondary characteristics* that are neither sufficient nor necessary for a galaxy to be considered a cD. For example, cD galaxies usually reside in the cores of rich, regular clusters (Oemler 1974, 1976); however, both they and their more modest counterparts, the D galaxies, are sometimes found in more irregular clusters at *local density enhancements*. The cD galaxies often have very extensive (~ 1 Mpc) stellar envelopes of low surface brightness (Oemler 1976) and are frequently found to be at the kinematical centers of their clusters (Quintana & Lawrie 1982). They are often very flat galaxies and are aligned to a flattened distribution of cluster members (Sastry 1968, Dressler 1978c, 1981, Carter & Metcalfe 1980, Binggeli 1982); however, as with other giant ellipticals, their flattening is not primarily due to rotation (Faber et al. 1977, Dressler 1979). Some of the extreme examples appear to have low central and average surface brightness and, usually in concordance, large core radii of many kiloparsecs (e.g. A2029; Dressler 1979). A significant fraction (about 25–50%) have multiple nuclei (Rood & Leir 1979, Hoessel 1980, Schneider et al. 1983).

The evolution of cD galaxies is the best place to begin analyzing the influences of cluster environment on galaxy evolution because *cD galaxies do not occur in the low-density field* ($\rho \leq 1$ galaxy Mpc^{-3}). The fact that objects with the *primary* characteristics of cD galaxies have been found in very poor but high-density groups (Morgan et al. 1975, Albert et al. 1977) is a clear indication that *local conditions* such as density and velocity dispersion, rather than global parameters like cluster richness or size, are key to understanding the formation of these unique systems.

Unfortunately, much of the discussion about how cD galaxies might form is often muddled by a careless application of the original definition of Matthews et al. For example, two quite dissimilar objects like the cD in A2029 (Dressler 1978b, 1979) and NGC 1316, the brightest member of the Fornax cluster [labeled a cD by Schweizer (1980), but only a D galaxy in Matthews et al.], are compared with the models of cD formation. The cD in

---

[1] Linear scales have been determined assuming a value of $H_0 = 50$ km s^{-1} Mpc^{-1}. This value has been adopted for convenience throughout this review.

A2029 is about a factor of 5 more luminous than NGC 1316, and it is more than an order of magnitude more luminous when the extensive envelope, absent in NGC 1316, is included. In this early stage of comparison of observations with models of formation, it is preferable to restrict one's attention to the largest and most luminous examples of the class, where the predictions made by the models should be manifest in the extreme.

3.2 Simulations of Cluster Evolution and cD Models

The long-standing debate over whether the brightest members of galaxy clusters are simply the tail of the bright end of the luminosity function (Peebles 1968, Geller & Peebles 1976) or the products of special formation and/or evolution (Peach 1969, Sandage 1976, Ostriker & Tremaine 1975, Tremaine & Richstone 1977, Dressler 1978a) is not fully resolved (Godwin & Peach 1979), but it seems certain that the most luminous cD galaxies are indeed the special products of dense environments. This can be argued statistically, since the luminosity function for cluster galaxies does not differ significantly from the field (Schechter 1976, Tammann et al. 1979), and yet no luminous cD galaxies are found in the low-density environment of the field. Furthermore, such galaxies seem to occur almost exclusively at local density maxima, even inside the already dense environments of rich clusters (Beers & Geller 1983; cf. N4389, Oemler 1976). The argument for their uniqueness can also be made on a case-by-case basis in clusters like A2029 or A2218 (Dressler 1978b), where the cD is much too luminous to be a continuation of the cluster luminosity function. The clinching piece of evidence that cD galaxies are unique is the high incidence (about one half) of multiple nuclei, compared with the order-of-magnitude lower frequency for the second- and third-ranked ellipticals in the clusters (Schneider et al. 1983).

Variations of three basic models have been proposed to explain the properties of cD galaxies. (a) Mergers of the brightest cluster members and/or accretion of smaller galaxies are facilitated by dynamical friction between the galaxies and the stellar envelopes of their neighbors (Lecar 1975). This so-called cannibalism model can build a large, luminous, centrally located cD galaxy that will experience runaway growth until its supply of victims is depleted (Ostriker & Tremaine 1975, White 1976, Ostriker & Hausman 1977, Hausman & Ostriker 1978). (b) Many giant ellipticals form in the central parts of rich clusters, but only a centrally located one can retain its extensive stellar envelope against disruption by the mean tidal field of the cluster (Merritt 1984). Tidal debris (stellar envelopes stripped from cluster galaxies), moving in the potential well of the cluster, enhances the appearance of the central galaxy (Richstone 1976). (c) Vast amounts of cooling gas ($\sim 100\, M_\odot\, \mathrm{yr}^{-1}$) from the intracluster medium

(ICM) form a cD "in situ" at the cluster center (Mushotzky et al. 1981), perhaps precipitated by a large concentration of "dark" matter (Blandford & Smarr 1984).

The first two schemes can be studied by comparing the results of computer simulations with observed parameters for cD galaxies and clusters. Early attempts to model cluster evolution (Richstone 1976, Roos & Norman 1979, Aarseth & Fall 1980) demonstrated that tidal interactions could whittle down cluster galaxies, releasing mass and luminosity into the ICM, while mergers could build a significant number of more massive galaxies.

The work of Merritt (1983, 1984), Miller (1983), and Malumuth & Richstone (1984) has significantly expanded the predictive ability of computer simulations of clusters. The Monte Carlo simulation of Malumuth & Richstone (1984) includes the effects of tidal stripping of galaxies by their neighbors, dynamical friction of galaxies with the ICM, and merging. The model galaxies, which initially contain all of the cluster mass, are distributed with an isotropic velocity dispersion and follow a King model with $R_{core} = 500$ kpc. After a simulation time of 10^{10} yr, about 20% of these model clusters develop (by mergers and accretion) central massive galaxies with extensive stellar envelopes composed of the material stripped from cluster galaxies. Malumuth & Richstone use the Bautz-Morgan (1970) type to characterize the prominence of the brightest cluster member, and they find that the development of such cD galaxies is independent of cluster richness, in agreement with Leir & van den Bergh's (1977) classification of the Abell catalog. On the other hand, the luminosity of the *envelope* of tidal debris *is* a strong function of cluster richness, again in agreement with observations of poor and rich cD clusters by Thuan & Romanishin (1981), who find that cDs in poor clusters lack the extensive envelopes present in cDs in rich clusters. Another successful prediction of this simulation is that little or no luminosity segregation develops in the galaxy distribution (Oemler 1974, Chincarini & Rood 1977, Dressler 1978c, Sarazin 1980; cf. Quintana 1979, Capelato et al. 1980), even when the galaxies initially carry a significant amount of the cluster mass (cf. Rood 1969, White 1977).

In spite of these successes, however, the simulations by Malumuth & Richstone (1984) fail to produce a high enough *frequency* of cD galaxies. Although their value of 20% is close to the Leir & van den Bergh figure, most clusters in the Abell catalog have not been relaxed for 10^{10} yr. Probably only $\sim 10\%$ of present-epoch clusters are as evolved as the clusters modeled in the simulations, which suggests that the predicted cD frequency is actually an order of magnitude too low.

Merritt's (1984) simulations are even less efficient in producing cD galaxies from dynamical interactions. This appears to be the result of his adopted setup, in which a strong tidal field produced by the cluster mass distribution pares down the galaxies to a maximum size of only 30 kpc. With cross sections this small, mergers between galaxies become very rare, and thus they are neglected in Merritt's simulations. Merritt therefore concludes that if mergers build cD galaxies, they must occur *at cluster collapse* and only for galaxies on bound orbits (Aarseth & Fall 1980). Little evolution is expected after cluster collapse, as first suggested by Roos & Norman (1979) (cf. Wielen 1979, Cooper & Miller 1982).

The strong tidal limits imposed in Merritt's simulations may be too severe. Early work by Bahcall (1975) suggesting that the core radii of concentrated clusters are constant at about 250 kpc may have been unrepresentative, since subsequent studies by Dressler (1978c), Chincarini (1979), Sarazin (1980), Kent & Gunn (1982), Semeniuk (1982) [see also Baier (1978), and references therein] find larger values of the core radius, typically between 300 and 600 kpc. Dressler (unpublished) has used Sarazin's maximum likelihood technique to obtain similar large values for the most concentrated of his sample of 55 rich clusters (Dressler 1980a). An X-ray-emitting gas should give a good indication of the distribution of the binding mass in the cluster, which is an important agent of tidal limitation. Such observations (Forman & Jones 1982) usually imply core radii of ~ 500 kpc, except when massive central galaxies are present, and in these cases the potential of the central galaxy itself may be responsible for the small apparent size. Of course, the mean tidal fields will be weaker if core radii are larger. Similarly, Merritt's assumption that each galaxy receives the maximum tidal force (i.e. comes to perihelion near one core radius) for an essentially infinite time may also be too stringent. Miller (1983) also points out that if the outer stars have large tangential velocity components, they should be much more difficult to remove.

If one or more of these factors result in an easing of the tidal cutoff, then merging of cluster galaxies cannot be ignored, and significant evolution *after virialization* can be expected. If Merritt's parameters are correct, however, the effects of merging and accretion are sufficiently weakened that they will only be important in the context of special "initial" conditions (for example, bound orbits and low velocity dispersions). Consequently, Merritt suggests an alternative interpretation of a cD as the chance survivor of many large, massive galaxies that were ripped apart by the mean tidal field. If one of these giant ellipticals comes to rest in the center of the cluster, it will *not* be torn apart, because of its symmetrical placement within the gravitational field. Inside the core of the cluster, it will retain its envelope and, in addition, will be identified with the more extensive tidal debris belonging to the cluster as a whole (Richstone 1976).

Miller's (1983) simulations are unique in their use of "collision rules" taken from N-body simulations of galaxy encounters (see references in Miller 1983). He models a variety of initial conditions, including collapsing clusters from isothermal perturbations (galaxies form before clusters and have initially radial orbits) and pancakelike models (galaxies form late and have an isotropic velocity distribution). A number of runs covering a range of tidal-stripping efficiency and initial conditions show that as little as 10% to as much as 70% of the initial mass in galaxies is released into the cluster potential. Though mergers are important in Miller's simulations, particularly in those where tidal stripping is difficult, he agrees with Merritt that most of the action takes place during cluster collapse, when densities are high and velocity dispersions are still relatively low. Miller emphasizes the importance of measuring the intracluster light that is predicted by all three studies for both cD and non-cD clusters. The intracluster light has been measured in Coma (Thuan & Kormendy 1977, Mattila 1977, Melnick et al. 1977), in Abell 801 and 1132 (Baum 1973), and in Abell 2670 (Oemler 1973) at a typical level of 10–40%, but the measurements are difficult and the results obtained are very uncertain. These types of measurements are much needed and should be facilitated by the new generation of linear area detectors.

Reviewing these cluster simulations, it is apparent that various techniques agree quite well on the qualitative and even quantitative aspects of cluster evolution, but that the results are very sensitive to the orbits and velocities of the member galaxies and the degree to which they are tidally limited. All of the simulations discussed here are relatively *inefficient* in producing cD galaxies, even given a longer time than is available to most clusters. Unless tidal stripping is much less efficient than thought, the high frequency of cD occurrence must be telling us that the primary evolution of such systems occurs in subcondensations (small groups) *before virialization*, when tidal stripping is less effective and velocity dispersions are significantly lower. The idea that most of the evolution took place in this "small group" phase may also be more compatible with the morphology-density relation for clusters (see Section 4).

A completely different mechanism for forming cD galaxies has followed from the X-ray observations that show large amounts of gas flowing from the ICM into certain centrally located cluster galaxies (Fabian & Nulsen 1977, Canizares et al. 1979, Mushotzky et al. 1981, Fabian et al. 1981a,b). In the context of simple models, the cooling flows that are detected imply an accretion of tens or even hundreds of solar masses of material per year. Continuation of this process for anything approaching a Hubble time will obviously result in the accumulation of mass comparable to a cD galaxy. Since the insides of cD galaxies are much like normal ellipticals (Thuan &

Romanishin 1981), one could conclude that either (a) the accretion is taking place onto a preexisting galaxy and has not significantly altered its structure; or (b) the accretion is directed to the center of dark matter binding the cluster ($R_{core} \sim 500$ kpc), but dissipation in the gas causes it to form a much more condensed structure; or (c) a condensed structure in the dark matter, such as the "black pit" suggested by Blandford & Smarr (1984), forms a skeleton over which the cD is built from inflowing gas.

Blandford & Smarr's model attempts to explain the large number of fast-moving companion ellipticals found within ~ 10 kpc of many cD galaxies. They propose that the dark binding material of the cluster actually has a much smaller radius (~ 10 kpc) than the radius value $R_{core} \sim 500$ kpc of the galaxy distribution. Binding these companion galaxies with such high orbital velocities would make the time scale for accretion by the cD quite long and would therefore explain why such multiple systems are common. It is, of course, known from observations of the velocity dispersion as a function of radius (Dressler 1979, Carter et al. 1981) that some cD galaxies do *not* contain such black pits, since the velocity dispersion at 10 kpc would rise to ~ 700 km s^{-1}. Recent observations by Tonry (1984a), however, also seem to rule out black pits with core radii less than 12 kpc in A2199 (NGC 6166) and A2634 (NGC 7720), which are Blandford & Smarr's best cases. The velocity dispersions in these cDs do not rise as is predicted by the model. Tonry's numerical simulations (1984b) show that the orbits of galaxies remain highly elliptical as they are captured. Thus, these simulations are able to reproduce the proper frequency of such companions without an extremely concentrated distribution of dark matter.

Further evidence against the black pit model is that the X-ray core structures of A2199 and A2634 are quite different, as A2634 has a much larger core than is typical for the class (Forman & Jones 1982). It is tantalizing, however, that one prediction made by Blandford & Smarr— that there would be a strong X-ray source at the position of the binary system NGC 6041a,b in A2151 (Hercules)—is dramatically confirmed (Dressler et al. 1984b).

Let us return to the cases of accretion flows onto a preexisting galaxy or into a dark-matter distribution with a large core (~ 500 kpc). It is well established that the interiors of cD galaxies have properties similar to normal ellipticals (Thuan & Romanishin 1981), including velocity dispersions, mass-to-light ratios, stellar populations (optical spectra and colors), and densities (Faber et al. 1977, Dressler 1979, Malumuth & Kirshner 1981) Valentijn's (1983) report to the contrary of strong color gradients and extremely red nuclei in cDs is in strong disagreement with all previously published data on the color and color gradients of cD galaxies (e.g. Sandage 1972, Schild & Davis 1979, Gallagher et al. 1980, Wirth & Shaw 1983,

Lugger 1984) which imply that cD galaxies have the colors of normal ellipticals. These colors are compatible with a burst of star formation some 10^{10} yr old, as is usually assumed for an elliptical galaxy. The color of a galaxy with a constant rate of star formation over 10^{10} yr, such as a cD built steadily over a Hubble time, should be markedly different (Searle et al. 1973, Sarazin & O'Connell 1983) unless the initial stellar mass function is truncated at high mass. If the gas is instead forming low-mass, largely undetectable stars (Fabian et al. 1982, Sarazin & O'Connell 1983, Valentijn 1983), then the mass-to-light ratio of cD galaxies should be significantly higher than normal ellipticals. The evidence, however, is to the contrary, since measured mass-to-light ratios for cD galaxies do not differ from those of ellipticals (Malumuth & Kirshner 1981, Thuan & Romanishin 1981).

Because direct evidence for the cooling flow is observed in the form of temperature inversions and optical filaments (Heckman 1981, Fabian et al. 1981b), it is reasonable to conclude that gas accretion rates of order $10 \, M_\odot \, \mathrm{yr}^{-1}$ are occurring in the centers of some clusters. However, there are no optical observations that support the cases for accretion of hundreds of solar masses per year, a rate that should be capable of building a giant galaxy in a Hubble time. This suggests either that the models of such huge accretion flows are incorrect or that accretion on such a scale has not persisted for such long periods of time. It is worth pointing out, however, that if cD galaxies *were* built primarily from inflow of this type, then it is probable that *all* galaxies formed in a similar way, since the properties of cDs and other ellipticals are so similar.

3.3   *Optical Observations Relevant to Merging, Accretion, and Tidal Stripping*

ARE cD GALAXIES BUILT FROM HOMOLOGOUS MERGERS? Ostriker & Hausman (1977) and Hausman & Ostriker (1978; hereinafter HO) have proposed that mergers and accretion build giant galaxies with radial profiles that scale simply as $R \propto M$. This assumption allows them to predict the surface brightnesses and metric magnitudes of cD galaxies as a function of their total luminosities. This approach has been sharply criticized by Schweizer (1981), who claims that both N-body simulations (White 1978, Duncan et al. 1983, Farouki et al. 1983) and observations show that mergers are *not* homologous, but that rather they build cores with higher central concentrations than the progenitors. Because the observational data are extensive and varied, only a subset can be examined here. It is sufficient to show, however, that although the specific question of whether giant ellipticals and cDs have small core radii is unresolved, as Schweizer (1979) might like to put it, the available observations are roughly consistent with homologous growth for cD galaxies.

Schweizer is correct in pointing out that if all giant ellipticals have R_{core} ~ 100 pc, like NGC 1316, then few galaxies for which it has been claimed that core radii have been measured have *actually* been measured. Several overenthusiastic observers, including Dressler, have misinterpreted data like Oemler's (1976) as indicating large core radii for cD galaxies, when the true core radii may well have been hidden by the atmospheric seeing disk (as Schweizer notes). He is probably also correct in stating that Hoessel's (1980) analysis of 108 first-ranked cluster galaxies includes a large number, perhaps a majority, of cases whose Hubble core radii are only artifacts of the seeing. In some cases, however, the cores have radii of several kiloparsecs and are well resolved.

All of this is probably irrelevant, however, since it is the *large-scale structure* that is indicative of whether the cannibalism model adequately describes the evolution of cD galaxies. Hoessel (1980) and Schneider et al. (1983) were really after the logarithmic intensity gradient $\alpha = d$ $(\ln L)/d$ $(\ln r)$ at $r = 16$ kpc. The latter study included a better (double-Gaussian) seeing model and fit the galaxies to both a modified Hubble law and a de Vaucouleurs profile that has *no core*. These new data show that it is possible to measure α quite accurately for galaxies with $z \leq 0.2$, since the radius at which α is being measured is an order of magnitude larger than the seeing. Schneider (private communication) has repeated observations of 41 galaxies in a variety of seeing conditions, and his data show that α changes by an order of 10% over a range of seeing of a factor of two, again indicating that these measurements are basically unaffected by seeing. The core radii that are derived when a Hubble law is fit may be artifacts of the technique, but only *Space Telescope* observations will tell if the core radii are actually much smaller.

In the samples of Hoessel and Schneider et al., α increases with absolute magnitude in a way that is roughly consistent with the HO models of homologous growth. Morbey & Morris (1983) reach the same conclusion independently for another sample of cD galaxies. Thuan & Romanishin (1981) and Morbey & Morris (1983) both find that giant cDs have much lower surface brightnesses within their *effective radii*, again in agreement with the HO model.

Since it is clear that cDs do seem to mimic the model of homologous growth on the large scale, it is worthwhile to reconsider Schweizer's comments about core radii and *central* surface brightnesses. Although the N-body models consistently predict higher central surface brightness as evolution proceeds, these simulations have a small number of particles (~ 1000), and it is possible that two-body relaxation effects dominate in the central regions and render these models inappropriate. On the observational side, it is relatively easy to find examples of giant galaxies whose central surface brightness is low. For example, despite Schweizer's early report to the contrary, four of the five cases studied by Kron & Albert (1982) have *lower central surface brightness* than other bright ellipticals in the clusters. A2029, A2218, and A1413 are additional excellent examples of cDs whose central surface brightness is very low. The lower surface brightnesses of these archetype cDs are striking, and comparisons with puny galaxies like NGC 1316 are, therefore, very misleading. Though there may be good counterexamples (A2634, A2670), the general rule seems clear: the most luminous cD galaxies have low central surface brightnesses. This is again consistent with the homologous merger picture. The exceptions might be due to the remnants of a captured galaxy.

Perhaps the best piece of evidence that the merger process is responsible for building the insides of cD galaxies is the observation that a large fraction (25–50%) of such objects have multiple nuclei (Matthews et al. 1964, Hoessel 1980, Schneider et al. 1983). This high frequency cannot be dominated by projection effects (Schneider et al.), so the conclusion that at least some galaxies are being consumed by the cD seems unavoidable. Schneider et al. point out that that the frequency of multiple nuclei for the second- and third-ranked cluster galaxies is *an order of magnitude less*. This datum seems to confirm the special place of the cD in the bottom of the cluster potential well and its unique evolution compared with other bright galaxies in the cluster.

If none of this evidence is compelling for the case that cD galaxies grow by "cannibalism," it is sufficient to say that most optical observations now available are consistent with the general features of this model.

ARE THE TIME SCALES FOR DYNAMICAL FRICTION CORRECT?   Rood & Leir (1979) have pointed out that the brightest galaxies in Bautz-Morgan (1970; hereinafter BM) type I and II clusters (one or two dominant galaxies, respectively) are multiple systems about 25% of the time, and that this is much higher than the frequency in BM III clusters (no dominant galaxy). They and Tremaine (1981) agree that the relevant time scale for coalescence of these systems (as indicated by N-body simulations) is 1 or 2 orbital periods. Assuming these binary galaxies have circular orbits with typical separations of 40 kpc and relative velocities of order 500 km s^{-1}, this time scale is several times 10^8 yr. Tremaine uses this time scale, a frequency of occurrence of 25%, and a cluster age of order 5×10^9 yr to conclude that there are *several* such short-lived binaries in the lifetime of *each* cluster. Rood & Leir (1979) reject this idea on the grounds that it would result in very dissimilar brightnesses for the two components, but one could imagine that *each* galaxy is the result of a previous coalescence. However, unless orbital angular momentum can be dumped into the outer envelope, mergers from bound circular orbits would result in rapidly rotating, bright ellipticals (White 1979) and none are observed (Davies et al. 1983). The time scale for orbital decay could also be several times longer if (a) the pairs seen are actually on highly eccentric orbits whose semimajor axes are larger than the projected separation implies (this also mitigates the rotation problem), and/or (b) the mass of each component is much less than the 10^{13} $M_\odot$ value used by Rood & Leir (i.e. the visible galaxies are moving in a common envelope formed from their dark halos).

This possible discrepancy is reminiscent of similar discussions about cD galaxies and the frequency of their multiple nuclei, which are typically 10 kpc from the cD center. Hoessel (1980) and Schneider et al. (1983) have argued that the frequency of multiple nuclei is consistent with the expected time scale for dynamical friction of 10^9 yr. (The victims are low mass.) On the other hand, Blandford & Smarr (1984) claim that a much larger central mass than a cD galaxy (a black pit) is needed to hold these victims on their spiraling orbits. Furthermore, a smaller core radius and a larger orbital velocity will result in a longer infall time, consistent with the high frequency of occurrence of multiple nuclei. Tonry's (1984b) simulations imply, however, that black pits may not be necessary if the galaxies are infalling on highly elliptical orbits.

Such simulations and observations suggest that there is much to be learned about the processes that go under the label "dynamical friction," and, therefore, discrepancies between "predicted" and "observed" time scales of less than an order of magnitude should be viewed with interest, but not despair.

OTHER OBSERVATIONS RELEVANT TO CANNIBALISM   Many other optical observations are useful for testing the cannibalism model. Quintana & Lawrie (1982) review the evidence that cDs are at the kinematical centers of their clusters, and Beers & Geller's (1983) study indicates that D and cD galaxies are usually found at local density maxima.

Related to this topic are studies by Dressler (1979) of the cD in A2029 and by Carter et al. (1981) of the dumbbell galaxy IC 2082, which show that the velocity dispersions of these galaxies *rise* with increasing radius, in contrast with the steady or declining velocity dispersions in normal ellipticals (Illingworth 1983). Dressler (1979) interprets this rise in velocity dispersion as the gravitational effect of the cluster's dark matter on the stars in the cD's envelope. Thus the rise in velocity dispersion is, in itself, *neither evidence for nor against the cannibalism model*, save that it requires the galaxies in question to lie at the true centers of their clusters. Dressler's model for the cD in A2029 includes a normal bright elliptical centered in the cluster potential, but it also requires a component of intermediate M/L and velocity dispersion to supply the additional light at $R \sim 100$ kpc that distinguishes this cD. This extra component *is* suggestive of the cannibalism model, since it can be attributed to the remains of accreted galaxies.

McGlynn & Ostriker (1980) show that the relaxation time is a function of Bautz-Morgan type (BM I clusters have shorter times), as is expected with the cannibalism model, but the dependence is weak.

Van den Bergh (1983) has argued that the number of globular clusters per unit luminosity is very high in M87 (Virgo) and NGC 3311 (Hydra), so that it is unlikely that these central galaxies are "star piles" made from more normal galaxies. Unfortunately, no data are available on the globular cluster frequency in the truly outstanding cD galaxies, and it is for these examples that the cannibalism model is most appealing. Van den Bergh also cites the fact that cDs are often quite elongated and aligned with flattened clusters (see references above) as evidence that the formation of the cD and its cluster were simultaneous (see also Carter & Metcalfe 1980, Adams et al. 1980). On the other hand, Binney (1977) has pointed out that a strong anisotropy in the velocity field of the cluster will result in an anisotropic stellar envelope in the cD as the victims are cannibalized.

Lugger (1984) has presented the most extensive data on colors of the nuclei of bright ellipticals and cDs, from which she concludes that there is a monotonic reddening with increasing luminosity (cf. Thuan & Romanishin 1981). This is in contradiction with the HO prediction that the cDs should break from this relation and be about $0^{m}.1$ bluer in $U - B$ than an extrapolation from the lower luminosity ellipticals.

Reliable measurements of *color gradients* are only available for a small sample of elliptical and cD galaxies (Strom & Strom 1978a, Gallagher et al. 1980, Wirth & Shaw 1983). At present, there are insufficient data to make a meaningful comparison, as well as uncertainty in the models as to what extent the color gradients will be erased in the merging process (e.g. HO, White 1980).

EVOLUTION OF THE LUMINOSITY FUNCTION   The luminosity function (LF) specifies the number distribution as a function of absolute magnitude for a volume-limited sample of galaxies. Abell (1962) showed that there is remarkably little variation of the shape and characteristic luminosity of the LF from cluster to cluster, consistent with a "universal" function that he characterized as two power laws. Subsequent studies by Bautz & Abell (1973) and Oemler (1974) supported this claim of universality, and luminosity functions for the field (Felten 1977, Kirshner et al. 1979) and small groups (Turner & Gott 1976) showed that variations in luminosity functions are small even in very different environments. These LFs all show a characteristic luminosity L^{*} that varies by less than factors of 2–3 (Austin et al. 1975, Schechter 1976). Brighter than L^*, the numbers of galaxies fall rapidly with increasing magnitude; fainter than L^*, the counts level off.

This result is quite remarkable. It means that the processes that determined the luminosities of galaxies were either very insensitive to local conditions such as density, temperature, and turbulence (angular momentum), or that these conditions varied little from protocluster to protofield regions of space. Furthermore, it implies that the *evolution* of the distribution, in dense regions for example, is not substantial.

Nevertheless, the processes of merging, tidal stripping, and accretion discussed above are expected to produce some evolution of the LF, and those early-Universe models in which galaxies form *after* clusters (within adiabatic perturbations) might be expected to produce some *initial* variations as a function of environment. Therefore, it is worthwhile to look beyond "first-order universality" for second-order differences. A large number of LFs for rich clusters, produced mainly by photographic surface photometry, are now available for such a comparison.

First indications that there are systematic differences from cluster to cluster came from Oemler (1974), who showed what he considered to be weak evidence that the composite LF for "spiral-rich" clusters is flatter than those for "spiral-poor" or "cD" clusters. The latter had a more sudden rise at the bright end and then leveled off fainter than some characteristic magnitude.

Dressler (1978b) applied statistical tests to his sample of 12 very rich clusters to see if such variations were significant. He fit the cluster LFs to a functional form suggested by Schechter (1976), and used Monte Carlo models to test for significant variations of the characteristic magnitude M^* and the faint-end (power-law) slope α. Dressler found three types of variation from the universal form. Statistically significant deviations of M^* were found in at least two clusters. Several clusters were found to have an unusually flat slope at the faint end, and there was once again marginal evidence that the form itself varies, much like Oemler's comparison of "spiral-rich," "spiral poor," and "cD" clusters.

Unfortunately, most of the LFs produced subsequently have not been subjected to these types of statistical tests. Only Schneider (1982) has checked on the universality of M^*. For his large sample he finds variations in M^* similar to those found by Dressler, but because the small field size limited his study to $\sim 50-100$ galaxies per cluster, these variations are not statistically significant.

On the other hand, there is some supporting evidence of variations in faint-end slopes and steepness of the bright end. Among the 8 clusters (plus Virgo) studied by Bucknell et al. (1979), there are clear examples of clusters in which the LF bright end is very steep (A2065, A2670, A2199, and A426), and others where it is much shallower and a distinct "break" at M^* is not obvious (A2151, A2147, and Virgo). Similar steep bright ends have been noted for A1146 (Carter & Godwin 1979), A1413 (Austin & Peach 1974), and A1930 (Austin et al. 1975), while A1367 (Godwin & Peach 1982) has been found to have a flat bright end. The clusters A1656 (Godwin & Peach 1977) and CAO 340 − 538 (Quintana & Havlen 1979) appear intermediate. These differences in shape may reflect different mixes of morphological types, since the clusters with the steep bright ends are rich in E and S0 galaxies, and the shallower LFs with less obvious breaks are often found in spiral-rich clusters. Thompson & Gregory's (1980) study of the Coma cluster shows that a composite E + S0 LF differs from a spiral LF in just this way. They predict that if the Coma populations are representative, an ensemble of clusters from spiral rich to spiral poor should have $\sigma(M^*)$ ~ $0\overset{m}{.}5$, which is consistent with both Dressler's and Schneider's results. New luminosity functions for clusters with morphological types like Dressler's (1980a) sample are necessary to test this putative dependence of the LF on cluster population.

The LF bright end is usually very steep in clusters that have a luminous cD galaxy (see, for example, Austin & Peach 1974, Oemler 1974, Dressler 1978b)—that is, there are few other bright galaxies. This could be interpreted as the result of dynamical evolution, where the brightest galaxies have all merged to form the cD. Because of the small-number statistics in each case (only a few bright galaxies are "missing"), it is difficult to test this suggestion without a much larger sample.

Some large variations have been reported in the faint-end slope of the luminosity function, as described by the power-law slope α (Schechter 1976). The typical situation in rich clusters is for the LF to continue to rise fainter than M^*, with α ~ 1.25. This result has been verified to M_v ~ -14 in Virgo by Sandage & Binggeli (1984) and in Coma by Beckman (1982), although the faint-end LF in Coma is rather complex. Certain other rich clusters, however, particularly A2670 (Oemler 1973, Dressler 1978b), A401 (Dressler 1978b), and A1146 (Carter & Godwin 1979), have flat LFs ($\alpha = 1.0$) several magnitudes below M^*. Such measurements are plagued, unfortunately, by the uncertainties in field and incompleteness corrections, and a thorough study of cluster membership (i.e. redshifts) may be necessary to confirm this effect. Nevertheless, the data are suggestive, particularly in light of simulations by Miller (1983), who shows that tidal stripping can have a severe effect on faint galaxies, resulting in the flattening of the faint-end LF. (These simulations are usually of the *mass function*, however, so conclusions about the *luminosity function* include the additional un-certainty of the distribution of light in a galaxy relative to its mass distribution.) It is therefore worth noting that the rich clusters with flat faint-end LFs are dense and have a high degree of symmetry, properties thought to be indicative of clusters whose dynamical evolution is very advanced.

Recently, evidence of even more dramatic variations in the faint-end LF have been reported (Heiligman & Turner 1980, White & Valdes 1980), but these variations are likely to be due to selection effects.

These indications of variation in the "nearly universal luminosity function" are suggestive, but not yet compelling. The rapid growth in the technology of digitizing photographic plates and the advent of linear area detectors should result in many new LFs in this decade; these should settle the issue.

EVIDENCE FOR TIDAL STRIPPING   Encounters and collisions between galaxies are expected to remove the loosely bound stars in their outer regions (Gallagher & Ostriker 1972, Richstone 1976, Dekel et al. 1980). Unfortunately, the ease with which a galaxy is tidally stripped is not well understood because of the uncertainties in the distribution of dark matter relative to the luminous stars and the orbits of those stars (Miller 1983). A poor knowledge of the velocity dispersions, ages, and mean tidal fields of groups and clusters makes it even more difficult to estimate the amount of tidal damage.

A few observations are available, however, and they tentatively support the standard models quoted above that predict that galaxies in the densest regions of space may have lost up to 50% of their luminosity (and most of their mass). A monumental data set for 400 elliptical galaxies was constructed by Strom & Strom for A1656 (1978a), A1367 and A426 (1978b), and A1228, A2151, and A2199 (1978c) using photographic photometry from Mayall 4-m plates. The primary data compiled are the R_{26} radii and the M_v magnitudes within R_{26}. Strom & Strom (1978d) conclude that similar $\log(R_{26})$ vs. M_v relations exist for all clusters, but that elliptical galaxies in the denser, spiral-poor clusters are, on average, 10% smaller at a given M_v than those in spiral-rich clusters. There is also a gradient in size *within* the spiral-poor clusters, such that ellipticals within 0.5 Mpc of the cluster center are about 5% smaller than those farther out. A similar conclusion concerning the size of S0 galaxies in Coma as a function of radial distance is also reported by Strom & Strom (1978e). They also find that at a given central surface brightness (within a 2″ aperture), the smaller galaxies are about 0^m5 fainter. Both of these results concerning size and brightness are consistent with the effects of tidal stripping predicted in the studies cited above.

There is, of course, considerable observational and cosmic scatter in these relationships, and the shortcomings of photographic surface photometry add uncertainty to these results. It is disturbing, for example, that two apparently dense (and evolved) clusters, A401 and A2670 (Strom & Strom 1979), have a similar radial dependence of galaxy size, but that the absolute sizes of the ellipticals in these clusters are as large at a given M_v as those in the spiral-rich clusters studied. More data, preferably done with linear area detectors, are necessary before any firm conclusions can be drawn on the sizes of E and S0 galaxies as a function of environment.

Also consistent with the tidal truncation expected from galaxy encounters is the correlation found by Hickson et al. (1977) between the intergalactic spacing in dense groups and the size of the largest galaxy. These data emphasize the importance of reliable measurements of intergalactic light in these systems for use as an *unambiguous* measure of tidal damage. For example, Rose (1979) used the *absence* of intergalactic light to argue that the compact appearance of certain groups of galaxies must be temporary, since there is little evidence of tidal debris.

3.4 *Summary*

Searching for compelling evidence of mergers, tidal stripping, and accretion has not proven easy. A case can perhaps be built, but it is likely to rest on a large body of circumstantial evidence. More surface photometry for cluster luminosity functions, galaxy sizes and colors, and intergalactic light measurements are crucial to future progress in this area.

4. THE EVOLUTION OF DIFFERENT MORPHOLOGICAL TYPES

The previous section reviewed the major processes that can alter the mass functions and spatial distributions of galaxies in regions of high galaxy density. In this section, the level of complexity is raised to include the *forms* of the luminous matter, described by the morphological type of the galaxy. This requires additional knowledge of how the processes described above, combined with initial conditions, will vary the density, angular momentum, and gas content in a way that produces the observed range and distribution of morphological types. Again, clusters are ideal laboratories in which to study these effects, since they contain a different mix of galaxy types than is seen in the low-density field.

4.1 *A Basic Description of Galaxies*

The following is a list of the primary attributes of galaxies and their distributions that a comprehensive theory of formation and evolution should explain.

1. Different morphological types exist. There are spheroidal galaxies, as

well as disk galaxies that include a very flat distribution of stars in addition to a spheroid (Morgan 1958, Sandage 1961, de Vaucouleurs et al. 1976). There is a continuous range of disk-to-bulge luminosity ratios (D/B) among disk galaxies, probably melding smoothly into the elliptical galaxies, which have negligible disks. When gas is present, a luminous spiral pattern, related to the sites of star formation in the disk, is usually seen. Most gas-free disk systems (S0s and some early-type spirals) and ellipticals have normally had little star formation within the last several billion years (Searle et al. 1973, Larson & Tinsley 1978, Caldwell 1983).

2. All morphological types are found in both low- and high-density environments. Hubble & Humason (1931), Morgan (1961), and Abell (1965) described the transition to earlier-type galaxies (Es and S0s) in rich clusters, and Oemler (1974) quantified the relationship by identifying characteristic global mixes of E, S0, and spiral galaxies. Dressler (1980b) showed that such global descriptions derive from a tight relation of galaxy morphology to local galaxy density, and this behavior has been shown to extend all the way to the low-density field (Bhavsar 1981, de Souza et al. 1982, Postman & Geller 1984), about five orders of magnitude in space density. The fraction of spiral galaxies decreases as the fraction of S0s and Es increases with local galaxy density, almost independently of global cluster characteristics. This "morphology-density" relation is monotonic but extremely slow (roughly logarithmic), so that the low-density field is dominated by 80–90% spirals and the highest-density regions are composed of 80–90% elliptical and S0 galaxies, but all types are represented in all environments.

3. Elliptical galaxies appear to be the simplest systems. To first order they form a one-parameter family in luminosity, presumably proportional to mass. Their sizes, colors (metal abundances?), and characteristic internal velocities (central velocity dispersions) scale monotonically with luminosity (see, for example, Faber & Jackson 1976, Tonry & Davis 1981). There is some evidence of an additional parameter, which could be galaxy ellipticity, surface brightness, or mass-to-light ratio (Terlevich et al. 1981, de Vaucouleurs & Olson 1982, Efstathiou & Fall 1983). The average surface brightness (and true mass density?) of ellipticals rises slowly with increasing brightness, levels off between $-19 > M_B > -21$, and then begins to fall again (Binggeli et al. 1984). Most elliptical galaxies can be represented by models of oblate isotropic rotators (flattening due to rotation); however, with rising luminosity $(M_B < -21)$, an increasing fraction owe their flatness to anisotropic velocity dispersions (Binney 1981, Illingworth 1983, Davies et al. 1984).

4. S0 galaxies almost always have small D/B (~ 1) and appear to have a "thick disk" component in addition to the thin disks found in both S0s and spirals.

5. Spirals are classified by their D/B and by the detailed form of their spiral pattern (e.g. openness of the arms, arm thickness, contrast of the arms to the underlying disk; Sandage 1961). Their forms do not correlate well with luminosity, but the spiral pattern and gas content seem related to D/B (Strom 1980). At least a two-parameter family is indicated (Brosche 1973, Whitmore 1984).

6. More-detailed morphological and kinematical structure is reviewed by Kormendy (1982). These numerous constraints are beyond the scope of this review and are more detailed than the general models that are discussed here.

7. Along with these characteristics, it is important to remember that the dominant structure in some or all types may be a massive, unseen halo, and interactions among halos may have a critical effect on the luminous matter within them.

4.2   Three Classes of Models

The models that are reviewed here fall roughly into three classes. In the first type, similar initial conditions are assumed for all galaxies, regardless of their future destiny as cluster or field galaxies. This is the case, for example, in the hierarchical clustering model (Peebles 1974a,b, 1980) with a perturbation spectrum that is nearly white and randomly phased, because in this model the fluctuations that become galaxies reached the nonlinear growth phase long before the cluster-size perturbations. In this sense, galaxies could not "know" their future environments. The aim is to reproduce all the morphological variation with fairly late evolution (after clusters became important); for example, S0s might be produced from stripping and ellipticals from merging an initial population of spiral galaxies.

In the second class, later evolution is retained as the primary modifier of galaxy type, in particular through the truncation of disk development, but initial conditions or very early evolution are added to account for the prominence of bulge-dominated galaxies in regions of high galaxy density. This might be accomplished, for example, by dropping the condition of random phasing, so that galaxies with higher central concentration were destined to inhabit regions of high galaxy density (Dressler 1980b), or by including mergers in the early evolution of clusters (the small group phase) to build up a population of more massive spheroids.

If these types of models are unable to explain the morphological data described above (along with the more detailed constraints that will undoubtedly be forthcoming), there is still the possibility that *initial conditions* were primarily responsible for galaxy morphology. The challenge with this third class of models is to understand how galaxies "knew" at the time of their formation about their eventual environments. Recent cosmological models may provide a solution to this problem.

CLASS 1 : LATE EVOLUTION   Models that explain the existence of S0 galaxies as "normal spirals" that have lost their gas as a result of environmental influences attempt to explain variation in morphological types by recent environment alone. Spitzer & Baade (1951) originally proposed that spiral galaxies in dense clusters might strip gas from each other in direct collisions, but today ram-pressure stripping (Gunn & Gott 1972) and gas evaporation (Cowie & Songaila 1977) by a hot ($T \sim 10^{7-8}$ K) intracluster gas are considered to be more important. Strong evidence that this is not the primary mechanism for the production of S0 galaxies is presented by Dressler (1980b), who points out that most ($\sim 80\%$) S0s are found in low-density environments (the outskirts of clusters and in the field), where these processes are ineffective. It seems extravagant to propose a different production mechanism for the remaining fraction, which are indistinguishable from their field counterparts in properties like D/B, color, luminosity function, kinematics, and surface density. Detailed comparisons of spirals and S0s by Burstein (1979), Dressler (1980b), and Gisler (1980) confirm that field and cluster S0s have $D/B \sim 1$, much lower than the value ($D/B \sim 3$–10) typical of most present-day spirals, as noted by Faber & Gallagher (1976) and Sandage & Visvanathan (1978). In addition, Burstein (1979) has demonstrated that S0s have a "thick disk" component not found in spirals. This component has a larger scale height and *decreases more slowly with radius than the thin disk.* In this latter respect, thick disks are more like the bars and lenses described by Kormendy (1982). These structural differences are further evidence that most S0s have not descended from late-type spiral galaxies.

This is not to say that gas ablation is unimportant in the evolution of spiral galaxies. Although there is still disagreement over the ease with which gas can be removed from a spiral (cf. Gisler 1979, 1980, Kent 1980, Nulsen 1982), there is now observational evidence that such processes could be responsible for the some of the "anemic" spirals (van den Bergh 1976) in clusters (Strom & Strom 1979, Wilkerson 1980, Sullivan et al. 1981). These observations show a tendency for spirals in the intermediate-density environments (there are very few spirals in the densest environments) to be gas poor by factors of 2–3 relative to their field counterparts at the same Hubble type. These data have been compiled from H I measurements (Bothun et al. 1982, Giovanelli et al. 1981, and references therein) and from optical measurements of integrated Hβ flux (Kennicutt 1983). Gas depletion, whether by ablation or exhaustion by star formation (accelerated by interactions with neighboring galaxies, for example), looks to be a rather slow process, since the relationships among galaxy gas fraction, color, and star formation rate are maintained as the gas is removed (Bothun 1982, Kennicutt 1983).

Gas "deficiencies" of spirals are *modest* (factors of 2–3) in intermediate-density environments, where S0s are quite common. Since S0s are deficient in gas by factors of 100 or more compared with spirals, this is further evidence that this type of environmental influence does not explain the existence of most S0 galaxies. Only in the Coma cluster has a deficiency of order 10 been found, and Bothun (1981) claims that in this one case there is a population of small-bulge S0s in the very core, possibly the true remains of stripped spirals. It is doubtful that a spiral could survive in such an environment; thus the rarity of small-bulge, gasless systems is probably a sign that (*a*) few large D/B spirals have plunged into these conditions up to the present-epoch or (*b*) disk formation is slowest for these systems, so that their development will be inhibited by the cluster virialization (see below). On the other hand, ablation may be very important for gas removal from relatively gas-poor systems like S0s and ellipticals (Gisler 1978, Fabian et al. 1980, Nulsen 1982).

Toomre & Toomre's (1972) suggestion that ellipticals form by mergers of disk galaxies is another attempt at a class 1 model. This idea has been more fully developed by Roos (1981), who attempts to build both ellipticals and the bulges of disk galaxies from mergers of what are initially purely disk systems. The attractions and problems of such merger models are briefly discussed in the next section.

CLASS 2: LATE EVOLUTION PLUS INITIAL CONDITIONS   Kent (1981) has provided a prescription for altering disk luminosity as a function of density that can reproduce the fractional increase in elliptical and S0 galaxies in dense environments as found by Dressler (1980b). Dressler pointed out that a "fading" of disks of 1.5–2.0 magnitudes (this could be dimming due to changes in the stellar population or to prevented formation) could account for the difference between the distributions of bulge luminosity for spirals and S0s. Kent develops this idea and shows that if this fading is a particular function of the local galaxy density, the morphology-density relation is reproduced as well. This is a class 2 model, since Kent requires an ad hoc "initial condition" that large-bulge systems preferentially "become" S0s, while small-bulge systems remain spirals. Kent's model is very successful in reproducing the morphology-density relation, and the identification of the critical parameter as the virialization time at a given density is particularly attractive. It fails only for the reason that Dressler rejected the 2-magnitude fading: the LF (luminosity function) of the "faded" galaxies should be quite different from the LF of the galaxies that have finished building their disks (the low-density clusters or field). Kent notes this problem, but he points out correctly that Dressler's data suffer from incompleteness just fainter than M^* and thus cannot be used to rule out the *single* power-law LF he adopts. (A single power law has no characteristic magnitude, of course, so "before" and "after" LFs are indistinguishable.) However, comparisons of spiral and S0 LFs that do not suffer the incompleteness problem (Tammann et al. 1979, Thompson & Gregory 1980) also show little or no difference in M^*. This issue is discussed in more detail below.

Larson et al. (1980; hereinafter LTC) concentrate their efforts on a different aspect of the same scheme by suggesting a mechanism for the "fading." Their calculations indicate that gas exhaustion times for spirals are short, typically a few billion years; therefore, they suggest that spirals must be refueled by infall from tenuous gas envelopes. These envelopes would be easily "stripped" by tidal encounters, so a morphology-density relation is expected. This mechanism should still be effective in lower-density regions (while those that remove gas directly from the disks are not). Furthermore, star formation would continue for a few billion years in these "stripped spirals," and this might explain the "survival" of blue galaxies observed by Butcher & Oemler in dense, high-redshift clusters (see Section 5).

Although gas envelopes of the type required have yet to be observed (Sancisi 1981, 1983, and references therein), the LTC model, like Kent's, has attractive aspects, and it is the first to make detailed predictions that can be checked by observations. Like Kent's model, however, it cannot easily explain the similarity of spiral and S0 LFs. LTC brushed this problem aside by noting that spheroids in dense environments appear to be more luminous to begin with (for example, there are all those ellipticals), so it seems reasonable that the S0s will come preferentially from *large-bulge spirals*.

Comparing luminosity functions to test disk truncation and initial conditions   The evidence that brighter spheroids, in addition to fainter disks, are necessary to explain the morphology-density relation rests on a comparison of LFs in high- and low-density regions. This comparison is a decisive test of whether initial conditions or very early evolution must be invoked in addition to (or in place of) later evolutionary processes like disk truncation. At present, this LF comparison is suggestive, though perhaps not compelling. In Dressler's (1980a) data, for example, Hercules (Abell 2151) has a 55% spiral population and a $\langle D/B \rangle = 3.70$ (weighted by luminosity and averaged over all types), comparable to a value of $\langle D/B \rangle$ ~ 5 for a volume-limited sample of the *Revised Shapley-Ames Catalog* (Sandage & Tammann 1981). Coma (A1656) has a 12% spiral population in

Dressler's sample and $\langle D/B \rangle = 1.18$. A "fading" of the disks by a factor of 3.14 (1.25 magnitudes) is therefore required to bring Hercules to the same $\langle D/B \rangle$ as Coma. In this sample, D/B is not a strong function of luminosity near M^*, so this 1.25-magnitude disk fading will result in a change $\Delta M^* = 0.85$.[2]

Schechter's (1976) analysis of Oemler's (1974) data gives M^*(Coma) $- M^*$(Hercules) $= 0.26$, considerably smaller than the value predicted in the disk-fading model. Indeed, for the average of Oemler's four "spiral-rich" clusters compared with the five "cD" clusters (lowest proportion of spirals), $\Delta M^* = -0.21$, which goes in the *wrong sense for the disk-fading model*. An even larger value of $\Delta M^* \sim 1.2$ magnitudes is to be expected in comparing a "faded field" population ($\langle D/B \rangle \sim 5$) with the population of the Coma core ($\langle D/B \rangle \sim 0.8$), but, in fact, there is no significant difference in M^* for any region of Coma compared with the field.

If these data are corroborated by better-quality photometric observations, then the inference will be clear: *a brightening of spheroids by factors of 2–3 in dense regions is needed* to compensate for the lower luminosities of disks and thus leave the LF basically unchanged. It is perhaps encouraging that early attempts to model cluster evolution (Miller 1983, Malumuth & Richstone 1984) do show an approximate balance between the opposing effects of mergers and tidal stripping.

In summary, it is difficult to account for the data on morphological types, D/B, and LFs as a function of environment without appealing to both disk truncation and larger spheroidal components in dense regions. Since a disk luminosity function for the Coma cluster would include few luminous systems like those found in the low-density field, it seems reasonable to conclude that a disk truncation mechanism is at work. Furthermore, as pointed out by Ostriker (1977), the mass of intracluster gas in Coma is roughly what would remain if disks of the size found in field spirals were prevented from forming. (This implies, moreover, that the fraction of mass in "left over" gas will be found to be a function of spiral fraction, and that *little residual gas should be found in the field.*) By implication, the spheroidal LF, like the disk LF, must also be different in clusters than in the field (since the spheroidal + bulge = total LFs are so similar).

Building up spheroids by mergers   Is there a way to account for the larger bulges in dense environments, short of appealing to initial conditions? Mergers may provide a method of producing ellipticals from spiral galaxies (Toomre & Toomre 1972, Toomre 1977, White 1978, 1979, Fall 1979), and perhaps bulges of disk galaxies are "built up" by some kind of merger as well

---

[2] The author has checked this by fading each galaxy in the Hercules sample individually and finds the expected shift of $\Delta M^* \sim 1.0$.

(Roos 1981). It is too early to tell if mergers can account for the observed parameters of ellipticals, but some of the attractive aspects, as well as some of the difficulties, are mentioned here. A comprehensive review can be found in White (1982). Fall & Efstathiou (1980) have shown that the dissipation of gas in large halos can reasonably account for the properties of disk galaxies, but Fall (1983) argues that it is difficult to produce the more slowly rotating ellipticals in a similar manner. Ellipticals have low values of the dimensionless spin parameter λ that range from ~ 0.07 for the most luminous systems to 0.2–0.3 for the flattened, low-luminosity galaxies (Davies et al. 1983). Simulations by Efstathiou & Jones (1979) and Aarseth & Fall (1980) show that low values of λ result from mergers of galaxies bound on highly eccentric orbits. (These simulations refer to the nonluminous halos; it is unknown if the more concentrated luminous matter would also acquire the low-λ values typical of ellipticals.) It is particularly attractive to identify the brightest systems with multiple mergers, since the average surface brightness of elliptical galaxies falls with increasing luminosity $M < -21$ (Binggeli et al. 1984) and the frequency of ellipticals flattened by anisotropic velocity fields is rising (Davies et al. 1983).

Some of the early objections to making ellipticals from basically stellar mergers (see, for example, Ostriker 1980) appear to have weakened. Numerical simulations (e.g. White 1978, Farouki et al. 1983) indicate that central densities can grow as mergers proceed (nonhomology), and central velocity dispersions may also rise. Both are necessary if ellipticals are to be produced. A study by Dressler & Lake (1984) of about 15 disturbed systems from the Arp (1966) and Arp & Madore (1984) catalogs of peculiar galaxies confirms that the central velocity dispersions are in reasonable accord with the Faber-Jackson relation (1976) for ellipticals, i.e. these probable mergers already match ellipticals in the L-σ diagram. Schweizer (1982, 1983) has presented empirical evidence that mergers of field galaxies are common and may result in elliptical galaxies. For example, he claims that the luminosity profile of the highly interactive NGC 7252 system will eventually match the $R^{1/4}$ form of typical of ellipticals. The shells and plumes around many ellipticals (Arp 1966, Malin & Carter 1980) may be further evidence that "field ellipticals" (those in low-density regions) are in environments that favor mergers, since such features may result from the infall of disk galaxies (Quinn 1983).

The objection that ellipticals are common in rich clusters, where the high relative velocities suppress mergers, is probably negated by the likelihood that the merging occurred when the cluster was a collection of small groups (see, for example, Geller & Beers 1982) in which the velocity dispersions were much lower. This could also explain the existence of field ellipticals as the remnants of mergers of compact groups of galaxies (Carnevali et al.

1981). (In general, it is likely that the evolution of the different morphological types must have begun at a time when the *density contrast* from the field to the clusters was *much smaller than it is today*, since morphology is a very slow function of local density.)

Nevertheless, serious problems remain for the theory. Harris (1981) claims that a difference in globular cluster frequency per unit luminosity for spirals compared with ellipticals may be evidence against a merger origin for ellipticals. Sandage (1983) points to a continuity of properties from dwarf ellipticals to luminous systems (a factor of 10^4 in luminosity) as an argument against mergers, since it is unlikely that the dwarf ellipticals were also made by mergers. The observational data on color (abundance) gradients are still sparse, and the extent to which such gradients would be erased by merging is unclear (White 1980, Villumsen 1982); however, the general trend of redder colors and higher metallicities for brighter systems (Faber 1973, Sandage & Visvanathan 1978) must certainly be explained.

Are mergers responsible for building the bulges of disk galaxies as well? Bulges and low-luminosity ellipticals are isotropic rotators; therefore, they could have developed along the lines of the classical collapse models (Eggen et al. 1962, Larson 1976, Gott & Thuan 1976). In this simple picture, mergers are perhaps responsible only for the "giants," the most massive ellipticals, which are slowly rotating and have low surface brightnesses. But if mergers are to help account for all large spheroids in dense regions, then they must produce some larger-bulge disk galaxies as well. This raises a number of additional problems, such as explaining why disk systems held on to some of their gas while the ellipticals did not. If the gas that formed the disk fell in well *after* the merger that produced the bulge, can the angular momenta of disk and bulge be expected to align as well as they do (see Gerhard 1981)? Why are even the most luminous bulges flattened by rotation (Kormendy & Illingworth 1982, Dressler & Sandage 1983), while the most luminous ellipticals are not? In the case of our own Galaxy, it has been proposed that gas lost from a spheroid that formed *prior to the disk* explains otherwise puzzling features of the chemical enrichment history (Ostriker & Thuan 1975). Are these data consistent with a model where disks formed *first*, and bulges followed afterward from mergers or accretion? A merger model that can explain all of these observations, together with a limitation on disk building as a function of density, can probably account for all basic properties of the morphological types, and the morphology-density relation as well.

CLASS 3: INITIAL CONDITIONS   The models described above rely on later evolution to produce the range in morphological types and their dependence on present environment. It is also possible that these later influences are *minor* compared with the role played by initial conditions. The first attempts to identify initial conditions that would lead to different galaxy types identified angular momentum (Sandage et al. 1970) and density (Gott & Thuan 1976) as the variables responsible for the differentiation. The models have grown in complexity in recent years (see, for example, Larson 1976, Rees & Ostriker 1977, Silk 1978, White & Rees 1978, Tinsley & Larson 1979, Ostriker & Cowie 1981, Silk & Norman 1981, Struck-Marcell 1981, Kashlinsky 1982, Faber 1982, and Gunn 1982 if you doubt it), but a dependence of galaxy concentration, and possibly angular momentum, on the amplitude of the initial density perturbation is still generally expected. Observations of present-day galaxies show that more-concentrated systems are dominated by spheroidal components with lower specific angular momenta than their disks. If density perturbations of high amplitude consistently lead to the conversion of most of the available gas into stars before large-scale dissipation forms a global disk, then the basic structures of present-epoch galaxies could be established at very early epochs. Judging from the small fraction of gas in field S0s and early type spirals, it seems possible that the presence of a large spheroidal component *alone* can induce increased rates of star formation that lead to gas exhaustion, *independent of external environment* (Roberts et al. 1975). Thus the LTC-type model of truncating the formation of disks may be superfluous, and the fraction of gas "left over" in the Coma cluster (roughly twice the luminous matter) may be typical of *all* environments.

The deficiency in these models that employ only initial conditions is that they provide no obvious explanation for the observed dependence of morphology on later environment. To achieve this, the perturbation that becomes the protogalaxy must be affected by the larger scale perturbation that grows into a group or cluster. In a crude sense, they must form at the same time. For a white noise spectrum, this can be accomplished by abandoning random phasing so that the largest amplitude galaxy perturbations are always found within the largest amplitude cluster perturbations. For the isothermal perturbation model (hierarchical clustering; Peebles 1974a,b, 1980), the same results can be achieved if the initial spectrum contained more power at large scales (pink noise). The adiabatic perturbation models (Doroshkevich et al. 1980, and references therein) have a built-in preferred scale of large mass. These conditions result in earlier growth of the groups and clusters relative to the galaxies; therefore, the final amplitude of a protogalaxy fluctuation will depend on whether it is in a protocluster or protofield region. The recent emphasis on cosmologies with exotic particles that dominate the mass of the Universe has produced several models where these preferred mass scales for galaxies and clusters are expected [see Primack & Blumenthal (1983) for a review], and within these types of models such coupling of galaxies to clusters is likely, if not unavoidable. High-energy physics, then, could supply the extra ingredient that enables the construction of galaxy formation models that rely primarily on initial conditions.

Alternatively, it has been suggested that differences might arise in the angular momentum gained through tidal torques (Peebles 1969) in protocluster and protofield regions (DiFazio & Vagnetti 1979, Shaya & Tully 1984). Large-scale variations in the distribution of angular momentum would provide a correlation with environment for the model of galaxy differentiation first suggested by Sandage et al. (1970).

At present, it might be best to consider these types of models as "last resorts." Our ignorance of initial conditions is greater than our ignorance of later evolutionary effects; therefore, such models are rather ad hoc and offer few predictions or tests. Until they are able to do so, it seems advisable to embrace this alternative only if the models that stress later external influences fail to explain the data on morphological types and their environments.

5.   LOOKING FORWARD TO LOOKING BACK

Astronomers have an advantage over other historians: they can observe history *directly*—if not their own, at least someone else's. Substantial gains in detector technology have recently enabled photometric and spectroscopic measurements of average luminosity galaxies at redshifts approaching unity. This means that we can now view galaxies half as old as local galaxies.

This review article has concentrated on traditional approaches to the study of galaxy evolution, which involve looking at the structure, dynamics, and the stellar/gas content of *present-epoch* galaxies in order to test models of formation. Including the dimension of time will greatly enhance our ability to choose among models—for example, to discover if S0 galaxies have been dormant in star formation for more than the last few billion years. (Consider how much more difficult it would be to study stellar evolution if only old galactic globular clusters could be observed.)

It would be presumptuous to consider this last section a review. The subject is too young and too confused to allow anything more than an introduction to what should become the most powerful tool in the study of galaxy evolution. The discussion is divided into two sections, photometry and spectroscopy, which, as usual, reflect the choice between obtaining a small amount of information for a large number of objects or vice versa.

5.1   Galaxy Color as a Function of Look-Back Time

Butcher & Oemler (1978a) pioneered the study of the color evolution of average luminosity (L^*) galaxies by obtaining V and r images with an ISIT vidicon of two distant clusters: Cl0024 + 1654 ($z = 0.39$; Humason & Sandage 1957) and the cluster containing the luminous radio galaxy 3C 295 ($z = 0.46$; Minkowski 1960). These clusters are rich and centrally concentrated like the Coma cluster. However, Butcher & Oemler found that unlike Coma, whose population is ~95% *red* E and S0 galaxies, these distant clusters contained many galaxies with $V - r$ colors (a color between $U - B$ and $B - V$ in the rest frame at $z = 0.4$) bluer than S0s or ellipticals. Specifically, in a square box about 1 Mpc on a side, only a few percent of the galaxies in the Coma cluster are significantly bluer in $B - V$ than the average elliptical or S0 galaxy, but in a comparable area of the distant clusters, the fraction ranges from 30% for the 3C 295 cluster to 50% for Cl0024 + 1654.

From this comparison of the distribution of galaxy colors in two distant clusters with that of the nearby Coma cluster, Butcher & Oemler concluded that a very striking evolution in the galaxy content of rich clusters had occurred in only the last one third of a Hubble time. This "excess of blue galaxies" in distant clusters compared with nearby clusters is commonly referred to as "the Butcher-Oemler effect."

Butcher & Oemler, along with others in the field, were understandably surprised that galaxy evolution appeared to be so dramatic over such a small fraction of the age of the Universe. Rapid evolution of normal galaxies had been considered unlikely, perhaps because colors and luminosities of early-type galaxies for $z \lesssim 1$ are consistent with an initial, *single* burst of star formation lasting ~10^9 yr (see, for example, Bruzual 1980, and references therein). In fact, there is no convincing evidence for galaxy evolution from spectrophotometry of either field or cluster galaxies [see Kron (1982) and Ellis (1983) for reviews]. On the other hand, finding a large population of blue galaxies in clusters at $z \sim 0.5$ would imply that many of the early-type galaxies seen in present-epoch clusters, especially S0s, were actively forming stars until quite recently, but that over a relatively short period of time this activity has ceased.

It is not surprising, then, that the Butcher-Oemler effect has been greeted with a degree of skepticism and caution. DeGioia-Eastwood & Grasdalen (1980) suggested that the range of ultraviolet colors of early-type galaxies, the difference in K-corrections for red and blue galaxies, and the rather large photometric errors associated with these very faint objects all might conspire to produce a spread in $V - r$ color at $z \gtrsim 0.4$ as large as that seen by Butcher & Oemler. Koo (1981) published photometry of Cl0016 + 16, an even more distant cluster ($z = 0.54$) that appeared to have *no excess of blue galaxies*. Mathieu & Spinrad (1981) obtained new photometry from photographic plates of the 3C 295 cluster and concluded that the contamination of the cluster counts by field galaxies was underestimated by Butcher & Oemler. This was later verified by Dressler & Gunn (1982), who obtained spectra for a sample of the blue galaxies in the Butcher-Oemler field and found 11 out of 20 to be either foreground or background galaxies.

On the other hand, subsequent studies have confirmed the original observations of Butcher & Oemler for larger samples. Couch & Newell (1984a,b; for an initial discussion, see Couch 1981) obtained photographic photometry in J and F for 14 rich clusters in the range $0.18 < z < 0.39$ and concluded that the clusters with $z > 0.26$ have 2–5 times as many blue galaxies as do nearby clusters of similar type. A recent compilation by Butcher & Oemler (1984a), which includes data for 12 nearby clusters ($z < 0.1$), 16 intermediate-distance clusters ($0.1 < z < 0.3$; data primarily from Butcher et al. 1983), and 5 distant clusters, shows the blue fraction rising continuously for $z > 0.1$, although there is considerable scatter at any epoch. In this latest study, Butcher & Oemler define the blue fraction f_B as those galaxies that are (a) of magnitude $M_v < -20$, (b) within the radius that contains 30% of the cluster galaxies, and (c) bluer by more than $0^m.2$ in (rest-frame) $B - V$ than an elliptical galaxy of the same absolute magnitude. According to this definition, the nearby clusters in their sample with high central concentrations have $f_B < 0.05$, and clusters in the intermediate distance range have $0.00 < f_B < 0.20$, with a median of about 0.10. The four distant clusters have f_B values of 0.02 (Cl0016 + 16), 0.16, 0.21, and 0.22, although one distant irregular cluster has a value near 0.37. (These percentages of blue galaxies are lower by a factor of 1.5–2.0 than the numbers quoted in earlier studies, reflecting a change in the accounting system discussed in the next section.) The scatter from any mean line $f_B \propto z$ is so large that it is just as consistent at this time to interpret the relationship as an envelope diagram where the *upper limit* to the blue galaxy fraction rises with redshift, although Butcher & Oemler argue that most of the scatter from a mean line is accounted for by sampling statistics.

It seems, then, that samples from several sources confirm the Butcher-Oemler effect for $z < 0.5$. (Koo's study of Cl0016 + 16 at $z = 0.54$ provides the most deviant point.) Is it safe, therefore, to conclude *from photometry alone* that significant evolution has occurred since $z = 0.5$ in the populations of rich clusters? Unfortunately, the answer is probably no. The difficulty lies in selection effects that may have been present in the identification of clusters at high redshift.

First and foremost, the problem of subtracting the superposed field galaxies, most of which are blue, is a serious one for these types of studies.

The contrast of distant clusters over the field is not large, which means that an underestimation of the field contribution by a factor of ~ 2 could account for the entire Butcher-Oemler effect. The photometry referenced above is of high quality, so that errors this large in the field determination should not arise from improper treatment of the data. Furthermore, if the fields that contained the clusters were *randomly selected*, fluctuations this large in the field counts would be rare. The problem is that these fields were *not* randomly selected, but chosen because of their contrast against the field. This means that if, by chance, the field counts were relatively low around a distant cluster, or relatively high in the cluster region, the probability that the cluster would be noticed and included in a catalog might be greatly increased. Therefore, such a sample of distant clusters might include a disproportionate number of cases with large fluctuations in the field counts. From Shectman & Dressler's (1984) redshift study of 14 of the 55 clusters chosen by Dressler (1980a), it is clear that even in a sample of nearby clusters, where contrast with the field is high, $\sim 30\%$ of the clusters contain a significant contamination (of order one third) by a superposed cluster or group. Since these alignments should be less common than this, it follows that the increase in richness caused by the contamination has biased Dressler to include them in his catalog. This type of problem will be much more severe in distant samples, and although it is possible to estimate the expected fluctuations of the field counts and the probability of super-position, it is nearly impossible to determine how much the biases discussed above increase the chances of inclusion in a catalog of distant clusters.

The problem is also complicated by the fact that rich clusters are apparently embedded in superclusters, which increases the chance of cluster superposition and may influence the field counts. For example, a prolate supercluster with one or more embedded clusters might have a very high probability of inclusion in a catalog if it were viewed pole-on. This could result in the superposition of many blue galaxies with nearly the cluster redshift that are nevertheless quite distant from the primary cluster.

These are examples of how photometry alone provides suggestive, but not yet compelling, evidence that greater percentages of blue galaxies inhabited the central regions of rich, concentrated clusters only one third of a Hubble time ago. The definitive evidence must come from knowledge of the redshifts of the galaxies in question.

5.2   Verifying and Interpreting the Butcher-Oemler Effect by Spectroscopy

Low-resolution (10–20 Å FWHM) spectroscopy of L^* galaxies at $z \sim 0.3$–0.5 has recently become possible with the high throughput CCD spectro-graphs available on the Palomar 5-m, the KPNO 4-m, and the AAT 4-m telescopes. Depending on the strengths of emission and absorption features in the spectra, a redshift of a typical cluster galaxy can be obtained in 1–4 hr with these systems. Multiaperture masks, which allow the simultaneous observation of ~ 10 galaxies over a field of a few arcminutes, have made it feasible, though by no means easy, to obtain redshifts for the brightest 30–50 galaxies in such a cluster.

At the present time, redshifts of 20 or more galaxies have been determined for only 3 fields. These are the 3C 295 field (Dressler & Gunn 1982, 1983), Cl0024 + 1654 (Dressler & Gunn 1982, Dressler et al. 1984a), and Cl1446 + 2619 (Butcher & Oemler 1984b). In the 3C 295 field, Dressler & Gunn obtained redshifts for 6 red and 17 blue galaxies. All 6 of the red galaxies were cluster members at $z \sim 0.46$, but only 6 of the 17 blue galaxies were actually at the cluster redshift. This indicates a blue cluster population of about 20%, or about half of Butcher & Oemler's estimate (1978a; see Table 3 therein), in agreement with Mathieu & Spinrad's claim (1981) of a larger field correction. Dressler et al. (1984a) have found 11 out of 12 red galaxies in Cl0024 + 1654 to be cluster members at $z \sim 0.39$, and 13 out of 21 blue galaxies to be at the cluster redshift. This implies a blue galaxy population of $\sim 45\%$, much larger than in the 3C 295 cluster, though somewhat smaller than Butcher & Oemler's field-corrected estimate of 56%. Butcher & Oemler (1984b) have a similar result for the irregular cluster Cl1446 + 2619, where they find 4 out of 4 red galaxies to be cluster members and at least 7 out of 12 blue galaxies to be at the cluster redshift of $z \sim 0.37$. In these last two cases, then, there does appear to be a significant blue galaxy population at the cluster redshift, but an assessment of how unusual these populations are compared with present-epoch clusters is more difficult.

In their first paper, Butcher & Oemler (1978a) assumed that the blue galaxies in the distant clusters were normal spirals of types Sb and later, although there was no evidence for this other than the single broadband color. (Of course, it will require the spatial resolution of the *Space Telescope* to obtain direct images of these galaxies suitable for morphological classification.) Since there were more data on morphological types than colors of the galaxies in nearby clusters, they attempted to use the mean relation of colors and morphological types for nearby field galaxies to predict a "spiral fraction" from the blue galaxy fraction of the distant clusters. This approach is more fully developed in Butcher & Oemler's second paper (1978b), where the evolution of the spiral fraction as a function of look-back time is presented. However, this tack of normalizing to the spiral fraction is not a safe procedure. Dressler & Gunn (1982) were able to show from a relation between the [O II] strength and galaxy color for nearby spirals that most of the blue galaxies in the 3C 295 field could *not* be spirals at the cluster redshift, a result later verified through extensive spectroscopy (Dressler & Gunn 1983). Secondly, Wirth & Gallagher (1980) showed that the spiral fraction of nearby clusters is sensitive to the resolution and depth of the images used for the classification.

A better procedure has now been adopted by Butcher & Oemler (1984a). They compare the fraction of blue galaxies in *both* nearby and distant clusters. The new definition of f_B, as described earlier, requires that the galaxy be quite blue, in the central part of the cluster, and brighter than $M_v = -20.0$. With this more stringent definition, the fraction of blue galaxies in the distant clusters has dropped from the $\sim 50\%$ level of the early papers (which implies an even higher spiral fraction due to a correction for red spirals) to values $\sim 20\%$. One is tempted to conclude that the Butcher-Oemler effect has virtually disappeared, since many nearby clusters, even concentrated ones, have *spiral fractions* of this order. But Butcher & Oemler now argue (1984a) that most of the spirals in the central regions of nearby clusters are *red*. From their definition and new data sample, they claim that 3 out of 4 *field* spirals are blue, but only 1 out of 4 of the nearby *cluster spirals* is blue. This result, which might be called the "Oemler-Butcher effect," is perhaps as surprising as the Butcher-Oemler effect, and although it seems to be in qualitative agreement with the H I deficiency data (e.g. Giovanelli & Haynes 1983), it is in marked disagreement with the color-H I-morphology relations of Bothun et al. (1982). These relations indicate that spirals in clusters are normally indistinguishable from field spirals. Cast in this new way, the Oemler-Butcher effect might be described as follows: Distant, concentrated clusters have spirals with the colors of today's field galaxies, but these spirals have become red by the present epoch.

If Butcher & Oemler's new data are correct, it might be that the ancestors of S0s and early-type spirals in today's clusters were actively forming stars at $z \sim 0.5$, and that they have only begun to "run down" in recent times as a result of cluster influences and/or simply the exhaustion of gas by star formation. Observations of many more distant clusters with a range of environments, as well as more color and morphological data for nearby clusters, are essential to test even these simplest notions.

But even the few data available at this time seem to defy any attempt at a consistent picture of the evolution of ordinary spirals. From the study of the 3C 295 cluster, Dressler & Gunn (1982) concluded that the blue population was small and made up of only Seyfert and starburst galaxies, with no normal spirals detected. These observations imply that Seyfert and starburst galaxies occurred with a frequency about an order of magnitude higher in $z \sim 0.5$ clusters than in nearby clusters. It is tempting, therefore, to regard this evolution of "active galaxies" as a possible explanation of the Butcher-Oemler effect (see also Henry et al. 1983). But Dressler et al.'s (1984a) study of Cl0024 + 1654 picked up only one definite Seyfert and no starburst galaxies, and, in contrast to the 3C 295 cluster, the blue cluster members *do* have the spectra of spirals (proper emission-line strengths as a function of integrated color). Butcher & Oemler (1984b) also find typical spiral spectra in Cl1446+2619. Thus, the strong activity in the 3C 295 cluster may be connected to a global event in this cluster instead of a more general evolution with look-back time; without this activity, the cluster population might be very red like the Coma cluster. Similarly, Koo's photometry of Cl0016+16 also identifies a distant cluster with a very small percentage of blue galaxies, although confirming spectroscopy is not yet available.

Clearly, then, no consistent interpretation of the data is possible at this time. Even the data for clusters at the same epoch contain some puzzling inconsistencies. Much more photometry is needed to decide if both red and blue clusters inhabited the $z \sim 0.5$ epoch and to resolve the question of whether nearby cluster spirals have different colors than their field counterparts. The approach by Couch et al. (1983) of narrow-band photometry in 7 bandpasses may be an effective compromise between spectroscopy and broadband photometry, since it provides more reliable information on the spectral energy distribution than either of these methods and therefore is best suited to look for evidence of spectral evolution. Unfortunately, the more time-consuming spectroscopy is necessary to provide the redshifts that separate field galaxies from cluster members.

We can expect the study of galaxy populations as a function of look-back time to make a substantial impact on the field of galaxy evolution in the near future. This is an area of research where new instruments and larger telescopes, both on the Earth and in space, will make an enormous difference on the type, quality, and quantity of the data needed to advance the subject. It is a time of great promise.

ACKNOWLEDGMENTS

It is a pleasure to thank James Binney, Dave Burstein, Roger Davies, Alfonso Cavaliere, George Efstathiou, Sandra Faber, Mike Fall, Gus Oemler, Paul Schechter, Don Schneider, Steve Shectman, and Roberto Terlevich for valuable discussions and comments about the material in this review.

Literature Cited

Aarseth, S. J., Fall, S. M. 1980. *Ap. J.* 236:43
Abell, G. O. 1958. *Ap. J. Suppl.* 3:211
Abell, G. O. 1962. In *Problems of Extragalactic Research*, ed. G. C. McVittie, p. 213. New York: MacMillan
Abell, G. O. 1965. *Ann. Rev. Astron.*
Astrophys. 3:1
Adams, M. T., Strom, K. M., Strom, S. E. 1980. *Ap. J.* 238:445
Albert, C. E., White, R. A., Morgan, W. W. 1977. *Ap. J.* 211:309
Arp, H. C. 1966. *Ap. J. Suppl.* 14:1

Arp, H. C., Madore, B. F. 1984. *A Catalogue of Southern Peculiar Galaxies and Associations.* Toronto: Clark-Irwin. In press

Austin, T. B., Godwin, J. G., Peach, J. V. 1975. *MNRAS* 171:135

Austin, T. B., Peach, J. V. 1974. *MNRAS* 168:591

Bahcall, N. A. 1975. *Ap. J.* 198:249

Bahcall, N. A. 1977. *Ann. Rev. Astron. Astrophys.* 15:505

Bahcall, N. A. 1980. *Ap. J. Lett.* 238:L117

Baier, F. W. 1978. *Astron. Nachr.* 299:311

Baum, W. A. 1973. *Publ. Astron. Soc. Pac.* 85:530

Bautz, L. P., Abell, G. O. 1973. *Ap. J.* 184:709

Bautz, L. P., Morgan, W. W. 1970. *Ap. J. Lett.* 162:L149

Beckman, B. C. 1982. PhD thesis. Princeton Univ., N.J.

Beers, T. C., Geller, M. J. 1983. *Ap. J.* 274:491

Bhavsar, S. P. 1981. *Ap. J. Lett.* 246:L5

Binggeli, B. 1982. *Astron. Astrophys.* 107:338

Binggeli, B., Sandage, A., Tarenghi, M. 1984. *Astron. J.* 89:64

Binney, J. 1977. *MNRAS* 181:735

Binney, J. 1981. In *The Structure and Evolution of Normal Galaxies,* ed. S. M. Fall, D. Lynden-Bell, p. 55. Cambridge: Cambridge Univ. Press

Blandford, R. D., Smarr, L. 1984. Preprint

Bothun, G. D. 1981. PhD thesis. Univ. Wash., Seattle

Bothun, G. D. 1982. *Publ. Astron. Soc. Pac.* 94:774

Bothun, G. D., Schommer, R. A., Sullivan, W. T. III 1982. *Astron. J.* 87:731

Brosche, P. 1973. *Astron. Astrophys.* 23:259

Bruzual, G. 1980. PhD thesis. Univ. Calif., Berkeley

Bucknell, M. J., Godwin, J. G., Peach, J. V. 1979. *MNRAS* 188:579

Burstein, D. 1979. *Ap. J.* 234:435

Butcher, H., Oemler, A. Jr. 1978a. *Ap. J.* 219:18

Butcher, H., Oemler, A. Jr. 1978b. *Ap. J.* 226:559

Butcher, H., Oemler, A. Jr. 1984a. *Ap. J.* Submitted for publication

Butcher, H., Oemler, A. Jr. 1984b. In preparation

Butcher, H., Oemler, A. Jr., Wells, D. C. 1983. *Ap. J. Suppl.* 52:183

Caldwell, N. 1983. *Ap. J.* 268:90

Canizares, C. R., Clark, G. W., Markert, T. H., Berg, C., Smedira, M., et al. 1979. *Ap. J. Lett.* 234:L33

Capelato, H. V., Gerbal, D., Mathez, G., Mazure, A., Salvador-Sole, E., Sol, H. 1980. *Ap. J.* 241:521

Carnevali, P., Cavaliere, A., Santangelo, P. 1981. *Ap. J.* 249:449

Carter, D., Efstathiou, G., Ellis, R. S., Inglis, I., Godwin, J. G. 1981. *MNRAS* 195:15p

Carter, D., Godwin, J. G. 1979. *MNRAS* 187:711

Carter, D., Metcalfe, N. 1980. *MNRAS* 191:325

Chincarini, G. 1979. *Cluster parameters.* Presented at Int. Sch. Astrophys., Erice, Italy, July

Chincarini, G., Rood, H. J. 1977. *Ap. J.* 214:351

Cooper, R. G., Miller, R. H. 1982. *Ap. J.* 254:16

Couch, W. J. 1981. PhD thesis. Aust. Natl. Univ., Canberra

Couch, W. J., Ellis, R. S., Godwin, J., Carter, D. 1983. *MNRAS* 205:1287

Couch, W. J., Newell, E. B. 1984a. *Ap. J. Suppl.* Submitted for publication

Couch, W. J., Newell, E. B. 1984b. In preparation

Cowie, L. L., Songaila, A. 1977. *Nature* 266:501

Davies, R. L., Efstathiou, G., Fall, S. M., Illingworth, G., Schechter, P. L. 1983. *Ap. J.* 266:41

DeGioia-Eastwood, K., Grasdalen, G. L. 1980. *Ap. J. Lett.* 239:L1

Dekel, A., Lecar, M., Shaham, J. 1980. *Ap. J.* 241:946

de Souza, R. E., Capelato, H. V., Arakaki, L., Logullo, C. 1982. *Ap. J.* 263:557

de Vaucouleurs, G., de Vaucouleurs, A., Corwin, H. G. 1976. *Second Reference Catalog of Bright Galaxies.* Austin: Univ. Tex. Press

de Vaucouleurs, G., Olson, D. W. 1982. *Ap. J.* 256:346

DiFazio, A., Vagnetti, F. 1979. *Astrophys. Space Sci.* 64:57

Doroshkevich, A. G., Khlopov, M. Yu., Sunyaev, R. A., Szalay, A. S., Zeldovich, Ya. B. 1980. *Texas Symp. Relativ. Astrophys., 10th,* ed. R. Ramaty, F. C. Jones, p. 32. New York: N.Y. Acad. Sci.

Dressler, A. 1978a. *Ap. J.* 222:23

Dressler, A. 1978b. *Ap. J.* 223:765

Dressler, A. 1978c. *Ap. J.* 226:55

Dressler, A. 1979. *Ap. J.* 231:659

Dressler, A. 1980a. *Ap. J. Suppl.* 42:565

Dressler, A. 1980b. *Ap. J.* 236:351

Dressler, A. 1981. *Ap. J.* 243:26

Dressler, A., Faber, S. M., Jones, C., Forman, W. 1984b. In preparation

Dressler, A., Gunn, J. E. 1982. *Ap. J.* 263:533

Dressler, A., Gunn, J. E. 1983. *Ap. J.* 270:7

Dressler, A., Gunn, J. E., Schneider, D. P. 1984a. In preparation

Dressler, A., Lake, G. 1984. In preparation

Dressler, A., Sandage, A. 1983. *Ap. J.* 265:664

Duncan, M. J., Farouki, R. T., Shapiro, S. L. 1983. *Ap. J.* 271:22

Efstathiou, G., Fall, S. M. 1983. Preprint

Efstathiou, G., Jones, B. J. T. 1979. *MNRAS* 186:133

Eggen, O. J., Lynden-Bell, D., Sandage, A. 1962. *Ap. J.* 136:748

Ellis, R. S. 1983. In *The Origin and Evolution of Galaxies*, ed. B. J. T. Jones, J. E. Jones, p. 255. Dordrecht: Reidel

Faber, S. M. 1973. *Ap. J.* 179:731

Faber, S. M. 1982. In *Astrophysical Cosmology*, ed. H. A. Bruck, G. V. Coyne, M. S. Longair, pp. 191, 219. Vatican: Pontif. Acad.

Faber, S. M., Burstein, D., Dressler, A. 1977. *Astron. J.* 82:941

Faber, S. M., Gallagher, J. S. 1976. *Ap. J.* 204:365

Faber, S. M., Jackson, R. E. 1976. *Ap. J.* 204:668

Fabian, A. C., Hu, E. M., Cowie, L. L., Grindlay, J. 1981a. *Ap. J.* 248:47

Fabian, A. C., Ku, W. H.-M., Mahlin, D. F., Mushotzky, R. F., Nulsen, P. E. J., Stewart, G. C. 1981b. *MNRAS* 196:35p

Fabian, A. C., Nulsen, P. E. J. 1977. *MNRAS* 180:479

Fabian, A. C., Nulsen, P. E. J., Canizares, C. R. 1982. *MNRAS* 201:933

Fabian, A. C., Schwarz, J., Forman, W. 1980. *MNRAS* 192:135

Fall, S. M. 1979. *Nature* 281:200

Fall, S. M. 1983. In *Internal Kinematics and Dynamics of Galaxies*, ed. E. Athanassoula, p. 391. Dordrecht: Reidel

Fall, S. M., Efstathiou, G. 1980. *MNRAS* 193:189

Farouki, R. T., Shapiro, S. L., Duncan, M. J. 1983. *Ap. J.* 265:597

Felten, J. 1977. *Astron. J.* 82:861

Forman, W., Jones, C. 1982. *Ann. Rev. Astron. Astrophys.* 20:547

Gallagher, J. S., Faber, S. M., Burstein, D. 1980. *Ap. J.* 235:743

Gallagher, J. S., Ostriker, J. P. 1972. *Astron. J.* 77:288

Geller, M. J., Beers, T. C. 1982. *Publ. Astron. Soc. Pac.* 94:421

Geller, M. J., Peebles, P. J. E. 1976. *Ap. J.* 206:939

Gerhard, O. E. 1981. *MNRAS* 197:179

Giovanelli, R., Chincarini, G. L., Haynes, M. P. 1981. *Ap. J.* 247:383

Giovanelli, R., Haynes, M. P. 1983. *Astron. J.* 88:881

Gisler, G. R. 1978. *MNRAS* 183:633

Gisler, G. R. 1979. *Ap. J.* 228:385

Gisler, G. R. 1980. *Astron. J.* 85:623

Godwin, J. G., Peach, J. V. 1977. *MNRAS* 181:323

Godwin, J. G., Peach, J. V. 1979. *Nature* 277:364

Godwin, J. G., Peach, J. V. 1982. *MNRAS* 200:733

Gott, J. R., Thuan, T. X. 1976. *Ap. J.* 204:649

Gunn, J. E. 1982. In *Astrophysical Cosmology*, ed. H. A. Bruck, G. V. Coyne, M. S.

Longair, p. 233. Vatican: Pontif. Acad.

Gunn, J. E., Gott, J. R. 1972. *Ap. J.* 176:1

Harris, W. E. 1981. *Ap. J.* 251:497

Hausman, M., Ostriker, J. P. 1978. *Ap. J.* 224:320

Haynes, M. P., Giovanelli, R., Chincarini, G. L. 1984. *Ann. Rev. Astron. Astrophys.* 22:445

Heckman, T. M. 1981. *Ap. J. Lett.* 250:L59

Heiligman, G. M., Turner, E. L. 1980. *Ap. J.* 236:745

Henry, J. P., Clarke, J. T., Bowyer, S., Lavery, R. J. 1983. *Ap. J.* 272:434

Hickson, P., Richstone, D. O., Turner, E. L. 1977. *Ap. J.* 213:323

Hoessel, J. G. 1980. *Ap. J.* 241:493

Hubble, E. 1936. *The Realm of the Nebula.* New Haven: Yale Univ. Press

Hubble, E., Humason, M. L. 1931. *Ap. J.* 74:43

Humason, M. L., Sandage, A. R. 1957. *Carnegie Inst. Washington Yearb.* 56:61

Illingworth, G. 1983. In *Internal Kinematics and Dynamics of Galaxies*, ed. E. Athanassoula, p. 257. Dordrecht: Reidel

Kashlinsky, A. 1982. *MNRAS* 200:585

Kennicutt, R. C. Jr. 1983. *Astron. J.* 88:483

Kent, S. M. 1980. PhD thesis. Calif. Inst. Technol., Pasadena

Kent, S. M. 1981. *Ap. J.* 245:805

Kent, S. M., Gunn, J. E. 1982. *Astron. J.* 87:945

Kirshner, R. P., Oemler, A., Schechter, P. 1979. *Astron. J.* 84:951

Koo, D. C. 1981. *Ap. J. Lett.* 251:L75

Kormendy, J. 1982. In *Morphology and Dynamics of Galaxies*, ed. L. Martinet, M. Mayor, pp. 130–37. Sauverny, Switz.: Geneva Obs.

Kormendy, J., Illingworth, G. 1982. *Ap. J.* 256:460

Kron, R. G. 1982. *Vistas Astron.* 26:37

Kron, R. G., Albert, C. E. 1982. *Publ. Astron. Soc. Pac.* 94:887

Larson, R. B. 1976. *MNRAS* 176:31

Larson, R. B., Tinsley, B. M. 1978. *Ap. J.* 219:46

Larson, R. B., Tinsley, B. M., Caldwell, C. N. 1980. *Ap. J.* 237:692

Lecar, M. 1975. In *Dynamics of Stellar Systems*, ed. A. Hayli, p. 161. Dordrecht: Reidel

Leir, A. A., van den Bergh, S. 1977. *Ap. J. Suppl.* 34:381

Lugger, P. M. 1984. *Ap. J.* In press

Malin, D. F., Carter, D. 1980. *Nature* 285:643

Malumuth, E. M., Kirshner, R. P. 1981. *Ap. J.* 251:508

Malumuth, E. M., Richstone, D. O. 1984. *Ap. J.* 276:413

Mathieu, R. D., Spinrad, H. 1981. *Ap. J.* 251:485

Matthews, T. A., Morgan, W. W., Schmidt, M. 1964. *Ap. J.* 140:35

Mattila, K. 1977. *Astron. Astrophys.* 60:425

McGlynn, T. A., Ostriker, J. P. 1980. *Ap. J.* 241:915

Melnick, J., White, S. D. M., Hoessel, J. G. 1977. *MNRAS* 180:207

Merritt, D. 1983. *Ap. J.* 264:24

Merritt, D. 1984. *Ap. J.* 276:26

Miller, G. E. 1983. *Ap. J.* 268:495

Minkowski, R. 1960. *Ap. J.* 132:908

Morbey, C., Morris, S. 1983. *Ap. J.* 274:502

Morgan, W. W. 1958. *Publ. Astron. Soc. Pac.* 70:364

Morgan, W. W. 1961. *Proc. Natl. Acad. Sci. USA* 47:905

Morgan, W. W., Kayser, S., White, R. A. 1975. *Ap. J.* 199:545

Morgan, W. W., Lesh, J. R. 1965. *Ap. J.* 142:1364

Mushotzky, R. F., Holt, S. S., Smith, B. W., Boldt, E. A., Serlemitsos, P. J. 1981. *Ap. J. Lett.* 244:L47

Nulsen, P. E. J. 1982. *MNRAS* 198:1007

Oemler, A. Jr. 1973. *Ap. J.* 180:11

Oemler, A. Jr. 1974. *Ap. J.* 194:1

Oemler, A. Jr. 1976. *Ap. J.* 209:693

Ostriker, J. P. 1980. *Comments Astrophys.* 8:177

Ostriker, J. P., Cowie, L. L. 1981. *Ap. J.* 247:908

Ostriker, J. P. 1977. In *The Evolution of Galaxies and Stellar Populations*, ed. B. M. Tinsley, R. B. Larson, p. 369. New Haven: Yale Univ. Press

Ostriker, J. P., Hausman, M. A. 1977. *Ap. J. Lett.* 217:L125

Ostriker, J. P., Thuan, T. X. 1975. *Ap. J.* 202:353

Ostriker, J. P., Tremaine, S. D. 1975. *Ap. J. Lett.* 202:L113

Peach, J. V. 1969. *Nature* 223:1140

Peebles, P. J. E. 1968. *Ap. J.* 153:13

Peebles, P. J. E. 1969. *Ap. J.* 155:393

Peebles, P. J. E. 1974a. *Ap. J. Lett.* 189:L151

Peebles, P. J. E. 1974b. *Astron. Astrophys.* 32:197

Peebles, P. J. E. 1980. *The Large-Scale Structure of the Universe.* Princeton, N.J.: Princeton Univ. Press

Postman, M., Geller, M. J. 1984. *Ap. J.* In press

Primack, J. R., Blumenthal, G. R. 1983. In *Workshop on Grand Unification, 4th*, ed. H. A. Weldon, P. Langacker, P. J. Steinhardt, p. 256. Boston: Birkhaüser

Quinn, P. 1983. In *Internal Kinematics and Dynamics of Galaxies*, ed. E. Athanassoula, p. 347. Dordrecht: Reidel

Quintana, H. 1979. *Astron. J.* 84:15

Quintana, H., Havlen, R. J. 1979. *Astron. Astrophys.* 79:70

Quintana, H., Lawrie, D. G. 1982. *Astron. J.* 87:1

Rees, M. J., Ostriker, J. P. 1977. *MNRAS* 179:541

Richstone, D. O. 1976. *Ap. J.* 204:642

Roberts, W. W., Roberts, M. S., Shu, F. H. 1975. *Ap. J.* 196:381

Rood, H. J. 1969. *Ap. J.* 158:657

Rood, H. J., Leir, A. A. 1979. *Ap. J. Lett.* 231:L3

Roos, N. 1981. *Astron. Astrophys.* 95:349

Roos, N., Norman, C. A. 1979. *Astron. Astrophys.* 76:75

Rose, J. A. 1979. *Ap. J.* 231:10

Sancisi, R. 1981. In *The Structure and Evolution of Normal Galaxies*, ed. S. M. Fall, D. Lynden-Bell, p. 149. Cambridge: Cambridge Univ. Press

Sancisi, R. 1983. In *Internal Kinematics and Dynamics of Galaxies*, ed. E. Athanassoula, p. 55. Dordrecht: Reidel

Sandage, A. 1961. *The Hubble Atlas of Galaxies.* Washington DC: Carnegie Inst. Washington

Sandage, A. 1972. *Ap. J.* 178:1

Sandage, A. 1976. *Ap. J.* 205:6

Sandage, A. 1983. In *Internal Kinematics and Dynamics of Galaxies*, ed. E. Athanassoula, p. 327. Dordrecht: Reidel

Sandage, A., Binggeli, B. 1984. In preparation

Sandage, A., Freeman, K. C., Stokes, N. R. 1970. *Ap. J.* 160:831

Sandage, A., Tammann, G. A. 1981. *A Revised Shapley-Ames Catalog of Bright Galaxies.* Washington: Carnegie Inst. Washington

Sandage, A., Visvanathan, N. 1978. *Ap. J.* 225:742

Sarazin, C. L. 1980. *Ap. J.* 236:75

Sarazin, C. L., O'Connell, R. W. 1983. *Ap. J.* 268:552

Sastry, G. N. 1968. *Publ. Astron. Soc. Pac.* 80:252

Schechter, P. L. 1976. *Ap. J.* 203:297

Schild, R., Davis, M. 1979. *Astron. J.* 84:311

Schneider, D. P. 1982. PhD thesis. Calif. Inst. Technol., Pasadena

Schneider, D. P., Gunn, J. E., Hoessel, J. G. 1983. *Ap. J.* 268:476

Schweizer, F. 1979. *Ap. J.* 233:23

Schweizer, F. 1980. *Ap. J.* 237:303

Schweizer, F. 1981. *Ap. J.* 246:722

Schweizer, F. 1982. *Ap. J.* 252:455

Schweizer, F. 1983. In *Internal Kinematics and Dynamics of Galaxies*, ed. E. Athanassoula, p. 319. Dordrecht: Reidel

Searle, L., Sargent, W. L. W., Bagnuolo, W. G. 1973. *Ap. J.* 179:427

Semeniuk, I. 1982. *Acta Astron.* 32:337

Shaya, E., Tully, R. B. 1984. *Ap. J.* In press

Shectman, S. A., Dressler, A. 1984. In preparation

Silk, J. 1978. *Ap. J.* 220:390

Silk, J., Norman, C. A. 1981. *Ap. J.* 247:59

Spitzer, L., Baade, W. 1951. *Ap. J.* 113:413

Strom, S. E. 1980. *Ap. J.* 237:686

Strom, K. M., Strom, S. E. 1978a. *Astron. J.* 83:73

Strom, S. E., Strom, K. M. 1978b. *Astron. J.* 83:732

Strom, K. M., Strom, S. E. 1978c. *Astron. J.* 83:1293

Strom, S. E., Strom, K. M. 1978d. *Ap. J. Lett.* 225:L93

Strom, K. M., Strom, S. E. 1978e. In *Structure and Properties of Nearby Galaxies*, ed. E. M. Berkhuijsen, R. Wielebinski, p. 69. Dordrecht: Reidel

Strom, S. E., Strom, K. M. 1979. In *Photometry, Kinematics, and Dynamics of Galaxies*, ed. D. Evans, p. 37. Austin: Univ. Tex. Press

Struble, M. F., Rood, H. J. 1982. *Astron. J.* 87:7

Struck-Marcell, C. 1981. *MNRAS* 197:487

Sullivan, W. T. III, Bothun, G. D., Bates, B., Schommer, R. A. 1981. *Astron. J.* 86:919

Tammann, G. A., Yahil, A., Sandage, A. 1979. *Ap. J.* 234:775

Terlevich, R., Davies, R. L., Faber, S. M., Burstein, D. 1981. *MNRAS* 196:381

Thompson, L. A., Gregory, S. A. 1980. *Ap. J.* 242:1

Thuan, T. X., Kormendy, J. 1977. *Publ. Astron. Soc. Pac.* 89:466

Thuan, T. X., Romanishin, W. 1981. *Ap. J.* 248:439

Tinsley, B. M., Larson, R. B. 1979. *MNRAS* 196:503

Toomre, A., Toomre, J. 1972. *Ap. J.* 178:623

Tonry, J. 1984a. *Ap. J.* In press

Tonry, J. 1984b. In preparation

Tonry, J., Davis, M. 1981. *Ap. J.* 246:680

Toomre, A. 1977. In *The Evolution of Galaxies and Stellar Populations*, ed. B. M. Tinsley, R. B. Larson, p. 401. New Haven: Yale Univ. Press

Tremaine, S. 1981. In *The Structure and Evolution of Normal Galaxies*, ed. S. M. Fall, D. Lynden-Bell, p. 67. Cambridge: Cambridge Univ. Press

Tremaine, S. D., Richstone, D. O. 1977. *Ap. J.* 212:311

Turner, E. L., Gott, J. R. 1976. *Ap. J.* 209:6

Valentijn, E. 1983. *Astron. Astrophys.* 118:123

van den Bergh, S. 1976. *Ap. J.* 206:883

van den Bergh, S. 1983. *Publ. Astron. Soc. Pac.* 95:275

Villumsen, J. V. 1982. *MNRAS* 199:493

White, S. D. M. 1976. *MNRAS* 174:19

White, S. D. M. 1977. *MNRAS* 179:33

White, S. D. M. 1978. *MNRAS* 184:185

White, S. D. M. 1979. *MNRAS* 189:831

White, S. D. M. 1980. *MNRAS* 200:733

White, S. D. M. 1982. In *Morphology and Dynamics of Galaxies*, ed. L. Martinet, M. Mayor, p. 289. Sauverny, Switz.: Geneva Obs.

White, S. D. M., Rees, M. J. 1978. *MNRAS* 183:341

White, S. D. M., Valdes, F. 1980. *MNRAS* 190:55

Whitmore, B. C. 1984. *Ap. J.* In press

Wielen, R. 1979. *Mitt. Astron. Ges.* 45:16

Wilkerson, M. S. 1980. *Ap. J. Lett.* 240:L115

Wirth, A., Gallagher, J. S. 1980. *Ap. J.* 242:469

Wirth, A., Shaw, R. 1983. *Astron. J.* 88:171

Zwicky, F. 1938. *Publ. Astron. Soc. Pac.* 50:218

Ann. Rev. Astron. Astrophys. 1984. 22: 223–65

SAGITTARIUS A AND ITS ENVIRONMENT

Robert L. Brown and Harvey S. Liszt

National Radio Astronomy Observatory,[1] Charlottesville, Virginia 22901

1. INTRODUCTION

One need not look further than the pages of this review series for evidence of the important role that galactic nuclei play in the energetics, structure, and evolution of galaxies (normal and otherwise). When we assign culpability for radio structures many hundreds of kiloparsecs in extent to "nuclear activity" and then ascribe that activity to a massive nuclear black hole, we appear to be basing our conclusions in large measure on informed, or perhaps inspired, speculation. We may be correct, but we also may be simply engaged in clever legerdemain. Fortunately, we have one convenient place to assess our ideas—in the nucleus of our own Galaxy. Our proximity to this specific galactic nucleus provides us with an unparalled perspective.

Although it was only seven years ago that Oort (1977) reviewed the galactic center in this series, the observations in the intervening years have revolutionized many of our views. The galactic center is a region exhibiting bizarre energetic activity (Yusef-Zadeh et al. 1984) and harboring objects, such as the time-variable source of positron annihilation radiation (Riegler et al. 1981), that are unique in the Galaxy. On the other hand, the gas motion at the center, while disturbed, presently shares few characteristics with the emission-line gas in Seyfert, starburst, or other active nuclei (Lacy et al. 1979, 1980). At present, we find ourselves in the unsatisfying position of having remarkable new observational insight into the nature of the galactic center but lacking a sturdy interpretative framework. Thus, it is unavoidable that this review will become dated as fast as the ink dries on the page. But this situation also means that great progress is being made.

[1] The National Radio Astronomy Observatory is operated by Associated Universities, Inc., under contract with the National Science Foundation.

0066–4146/84/0915–0223$02.00

Here, we concentrate on the inner ~ 50 pc of the Galaxy. Specifically, we focus our attention on Sgr A and material extending to the edge of the radio continuum arc [see Oort (1977) and Burton & Liszt (1983b) for a discussion of the larger-scale structure]. We use "Sgr A" to refer to the $\sim 2'$ diameter association of the nonthermal shell source Sgr A East and the elongated thermal region Sgr A West (see Ekers et al. 1983); "Sgr A*" refers specifically to the compact nonthermal radio point source within Sgr A West.

2. THE ENVIRONMENT OF SGR A

The source Sagittarius A is embedded in a region showing a great wealth of structure in both ionized and neutral gas. Although we are still very far from constructing explanations of the observed phenomena, or from a coherent interpretation of the region as a whole, the observations themselves have increased in quantity to the point where many new associations and relationships are apparent.

2.1 Large-Scale Structure of the Ionized Gas

Radio continuum maps of the inner few degrees of longitude (Downes & Maxwell 1966) show that the Sgr A complex at longitudes $-0.10 \leqslant l \leqslant 0.3°$ is immersed in a diffuse envelope extending about $1°$ on either side; the radio continuum is remarkably well centered a few arcminutes below $b = 0°$ (i.e. at about the latitude of Sgr A itself) and is strongly weighted to positive longitudes. Within this context, the morphology of radio continuum emission inside the Sgr A complex is remarkable because of its prominent misalignment with the galactic plane. As mapped at the highest resolution attainable with single-dish radio telescopes (Pauls et al. 1976), the stronger extended emission lies in a bent arc, a large segment of which is perpendicular to the galactic equator. This structure, which we refer to as the spur, is joined to Sgr A by a bifurcated bridge of emission that is itself oriented north-south in celestial (not galactic) coordinates.

The internal structure of the radio arc has been mapped in remarkable detail by Yusef-Zadeh et al. (1984), as shown in Figure 1. Previous interpretations of the arc, which was sometimes referred to as an essentially continuous belt of H II regions (see Figure 32 of Oort 1977), did not hint at its highly unusual filamentary nature. The spur is composed of a few (~ 4) distinct strands that have the appearance of being encircled several times by fainter loops. The bridge is seen to consist partly of diffuse gas (perhaps an extension of the material seen around Sgr A itself) and more prominently of several curved filaments. Also visible in the radiograph are the nonthermal shell of material in Sgr A (see Sections 2.4 and 5) and the compact source Sgr A* (Section 6).

Clearly, the radio arc is an exceedingly complex structure, and there are several reasons to believe that rather different physical conditions prevail over its various portions. Recombination lines—one means of ascertaining the nature of radio emission mechanisms—are present over most of the bridge but cannot be found over much of the length of the spur (Figure 10 of Güsten & Downes 1980). Figure 2 shows a stylized representation of the single-dish radio continuum structure (taken from Downes et al. 1978) on which are superposed contours representing the 55-μm and 125-μm infrared emission mapped by Dent et al. (1982) (see also Hildebrand et al. 1978). Again, the spur is largely absent, and it is the regions within the IR contours over which the recombination lines are detectable: the propor-

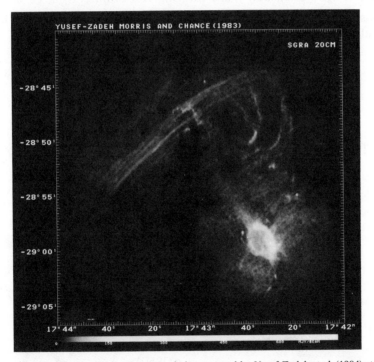

Figure 1   The 20-cm radio continuum emission mapped by Yusef-Zadeh et al. (1984) at the VLA with 5′ × 9″ (α × δ) resolution; map coordinates are 1950 celestial. Sgr A itself is at the lower right. Note the elongated shell of nonthermal radiation against whose western edge the compact source Sgr A* appears as a bright dot. The remarkable filamentary structures north and northeast of Sgr A comprise the arc familiar from earlier single-dish data.

tions of thermally and nonthermally emitting material vary greatly between the spur and the bridge. The higher-resolution, longer-wavelength IR map shows several instances of detailed correspondence with the radio data. Near Sgr A, a secondary IR peak overlays the positions of several newly discovered H II regions (Ekers et al. 1983), and the bifurcation of the radio bridge is also suggested. Alternatively, a minimum in the radio continuum distribution just north of Sgr A is seen as a peak at either IR wavelength.

The indication, then, is that a spur of nonthermally emitting gas is joined to Sgr A by a bridge of (more nearly) thermal material. Despite the difference in radiation mechanisms, the general connectedness of the arc feature suggests that the two regions have a common origin. One factor linking them can be found in the neutral gas reservoir mapped in ^{12}CO (Section 2.3).

2.2 X-Ray Emission from the Sgr A Complex

Watson et al. (1981) have mapped the Sgr A complex in the 0.5–4.5 keV band at 1.6 resolution; their most important result is, of course, the

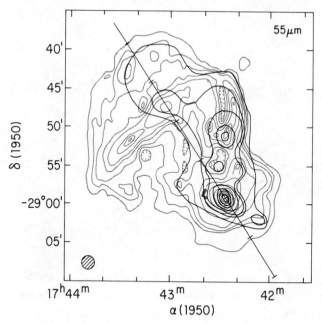

Figure 2a   The distribution of 55-μm infrared radiation (bold contours) mapped by Dent et al. (1982) at 2′ resolution, superposed on the 2.8-cm radio continuum distribution of Downes et al. (1978). Those portions of the shaded radio map not overlain by infrared emission are more nearly nonthermal in origin: radio recombination lines are also not detected there. The solid diagonal line shows the direction of the galactic equator.

detection of a source coincident with Sgr A West. The X-ray distribution shows both an extended low-level component, whose integrated emission actually dominates the total X-ray luminosity with $L_x = 2.2 \times 10^{36}\,\mathrm{erg\,s^{-1}}$ in the observed band, and about a dozen discrete sources with total luminosity $L_x = 6.9 \times 10^{35}$ erg s^{-1}. The Sgr A source has luminosity $L_x = 1.5 \times 10^{35}$ erg s^{-1}.

The X-ray distribution resembles that of the radio continuum (and many other tracers) to the extent that (a) its diffuse component is of roughly the same size and (b) the X-ray emission appears mostly at positive longitudes (Figure 3). More detailed correspondence is, at best, subtle. It is curious that one of the discrete sources falls in the bifurcated region of the radio bridge structure, but this is most likely only a coincidence.

Except for their luminosities, lack of strong variability, and reasonably hard spectra, we have little insight into the nature of the discrete sources; Watson et al. (1981) prefer an explanation in terms of luminous early stars because the discrete sources roughly follow the total gas distribution in their longitude asymmetry. The lack of obvious variability argues against their being binary or cluster sources, and the lack of obvious radio

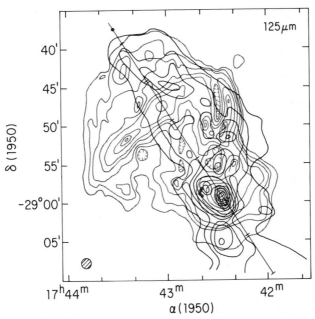

Figure 2b   As in Figure 2a, but with a higher-resolution (1′) 125-μm infrared map. The shaded dots represent compact H II regions (Ekers et al. 1983) just outside the Sgr A shell. These sources, also clearly visible in Figure 1, are associated with a secondary peak in the infrared.

counterparts argues against their being supernova remnants; the latter line of reasoning might also be applied to OB stars, as radio H II regions might then be expected. Possible explanations of the diffuse flux involve a large number (10^4) of T Tauri stars, synchrotron emission, and inverse Compton scattering of IR photons. The latter two processes require reacceleration of electrons on time scales of 500–3000 yr.

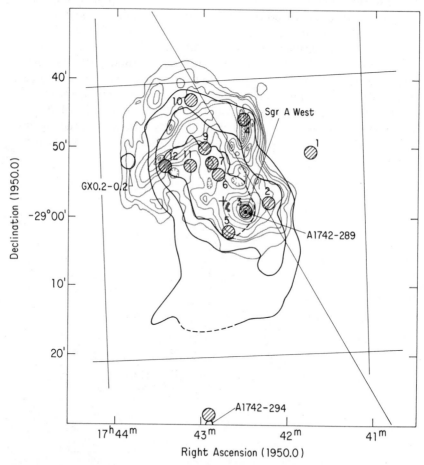

Figure 3   The location of discrete sources in the 0.5–4.5 keV X-ray band, as represented schematically by Watson et al. (1981) and shown superposed on the radio continuum as in Figure 2. Only the shaded circles are presumably associated with the galactic center. The approximate centroid of diffuse X-ray emission (bold contours) is marked by a cross.

2.3 The Relationship Between Neutral and Ionized Gas in the Sgr A Complex

The association between massive neutral gas clouds and star formation (or dense, thermally emitting ionized gas) is well established in the disk of the Milky Way, and the galactic center region is the source of copious emission from neutral atomic (Burton & Liszt 1983b) and molecular gas clouds (Bania 1977, Liszt & Burton 1978). Oort (1983), in particular, has stressed the importance of the fact that most of the molecular material seen near the galactic center occurs at positive longitudes, and this is a common characteristic of all the tracers previously discussed in this section. It is natural then to compare the radio, X-ray, and IR work with observations of the neutral gas reservoir; such a comparison is unusually rewarding in this case.

Although the distribution of H I cannot be adequately traced over the regions represented by the radio continuum, emission studies of many molecular transitions have been undertaken, and very thorough discussions and results of absorption mapping are presented by Bieging et al. (1980), Güsten & Downes (1980), and Whiteoak & Gardner (1979) for H_2CO, and by Cohen & Few (1976) for OH (see also the earlier work of McGee et al. 1970). Of the molecules studied, ^{12}CO emission shows the fullest range of behavior and can be traced most easily over the largest region. Other emission species, which are less easily excited, more clearly delineate the highest-density regions; such observations are discussed later in this section. Absorption results, which are used to infer the inner Galaxy geometry, are discussed briefly at the end of this section and in more detail in Section 2.4.1.

One component of the molecular gas distribution of particular interest over the Sgr A complex, and more immediately around Sgr A itself, occurs at 40–60 km s^{-1} in a body we refer to as the "50 km s^{-1} cloud." The structure of this cloud, whose existence as a coherent entity is a point of some possible controversy (see the Introduction to Schwarz et al. 1977), is shown in Figure 4; contours of integrated ^{12}CO intensity are superposed on the representation of the radio continuum taken from Downes et al. (1978) (see Liszt & Burton 1984). This diagram provides an immediate rationale for the contrary orientation (perpendicular to the galactic equator) of the radio spur. The spur occurs projected against a fairly abrupt positive-longitude boundary in the molecular gas, a boundary that is not evident in maps of other neutral gas tracers [as in Fukui et al. 1977 (reproduced by Oort 1977) and in Figures 5 and 6 here].

The manner in which the molecular cloud is traced by radio continuum emission is unusually detailed, perhaps unique in the Galaxy. Clearly, any explanation of the morphology of the ionized gas must invoke the neutral gas structure as well. The ^{12}CO peak nearest Sgr A overlays the H II region cluster of Ekers et al. (1983) and a peak at $\lambda 125\ \mu m$; it is similar in character to behavior seen near many dense H II regions in the galactic disk. The positive-longitude ^{12}CO peak, however, occurs outside the IR distribution, near portions of the radio continuum that are probably not thermal, and is exceptional. Sgr A itself occurs just at the periphery of the 50 km s^{-1} cloud, but the relative positioning of neutral and ionized gas on arcminute scales near that source has been a continuing subject of controversy, as is discussed at length in Section 2.4.

Several molecular transitions have recently been mapped in emission in the Sgr A complex, i.e. $\lambda 2$-mm H$_2$CO (Sandqvist 1982, and private communication), $\lambda 3$-mm HC$_3$N and $\lambda 2$-mm H$_2$CO (Fukui et al. 1982), $\lambda 2.8$-cm

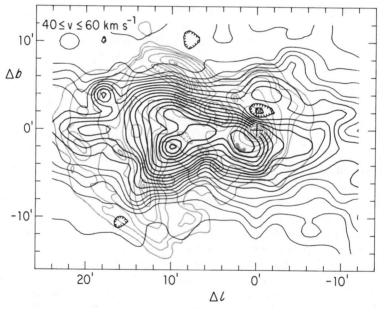

Figure 4   Contours of ^{12}CO intensity integrated over the range $40 \leqslant v \leqslant 60$ km s^{-1}, showing the extent of the 50 km s^{-1} molecular cloud: these are again superposed on a shaded representation of the single-dish radio continuum distribution. The ^{12}CO data, from the survey of Liszt & Burton (1984), were taken with 2′ spacing of a 1′ telescope beam. Coordinates are galactic, measured from Sgr A*. Essentially all of the radio continuum appears projected against the borders of the molecular gas. The abrupt positive-longitude boundary of the molecular cloud provides a rationale for the peculiar orientation, perpendicular to the galactic equator, of much of the radio continuum.

NH_3 (Güsten et al. 1981 ; see also Güsten & Henkel 1983), and λ3-mm HCN and HCO^+ (Fukui et al. 1977, 1980). The densest and hottest portions of the molecular gas are outlined in Figure 5 from Güsten (1982). The peak $M - 0.13 - 0.08$ is a feature at 20 km s^{-1} that coincides with a pronounced minimum in the λ2-μm light distribution (see Güsten et al. 1981) and so is presumed to lie somewhat closer than the bulk of the stars providing the near-IR radiation. Other peaks in this figure are associated with the 50 km s^{-1} cloud and can be seen in the ^{12}CO map. Velocity-position diagrams show that 20 km s^{-1} and 50 km s^{-1} gas components merge around Sgr A (Fukui et al. 1982, Sandqvist 1982), giving the impression of a continuous belt. This situation is outlined in Figure 6. Note, however, that the apparently continuous velocity gradient in neutral gas across Sgr A may be an artifact, because the 50 km s^{-1} cloud certainly has a well-defined velocity and because 20 km s^{-1} gas appears on both sides of Sgr A. There is also a separate, very widely distributed component of the molecular gas that has a strong velocity gradient across Sgr A and is at most indirectly related to the behavior under discussion here (see Section 3).

The mass of the 50 km s^{-1} cloud found from ^{12}CO is about 10^6 $M_\odot$, but higher values ($\simeq 2 \times 10^7$ $M_\odot$ for $M - 0.13 - 0.08$ and 2×10^6 $M_\odot$ for each of several peaks within the 50 km s^{-1} cloud) were derived by Güsten & Henkel (1983). The mass of neutral material seen in the vicinity of Sgr A is uncertain but is clearly very large. Do these clouds actually exist in proximity to the galactic center? Güsten & Downes (1980) argued that the inner regions of the Galaxy would be swept free of neutral material because of strong galactic tides and intense ionizing particle and photon fluxes, but Güsten & Henkel (1983) conclude that densities in the molecular material are high enough to permit them to exist at even very small galactocentric distances. The coincidences between neutral and ionized gas seen in many observing bands require that either all or none of the phenomena observed around Sgr A be associated with the central region. Although there appear to be no objections in principle to the presence of a large neutral gas reservoir in the innermost regions of the Galaxy, the geometry of the 50 km s^{-1} cloud is somewhat controversial.

In many cases, absorption spectra alone suffice to constrain the geometry quite substantially; the radio spur, for instance, can be said to lie within $\simeq 100$ pc of the galactic nucleus because of the observed apparent optical depths of features at -135 km s^{-1} and $+165$ km s^{-1} (Bieging et al. 1980). These are believed to be displaced rather symmetrically about the galactic center (Section 3). In more complicated cases where the column density of the absorbing gas might be expected to change substantially over the region of interest, a comparison of absorption patterns with emission-line intensities may suffice to constrain the geometry because the latter can

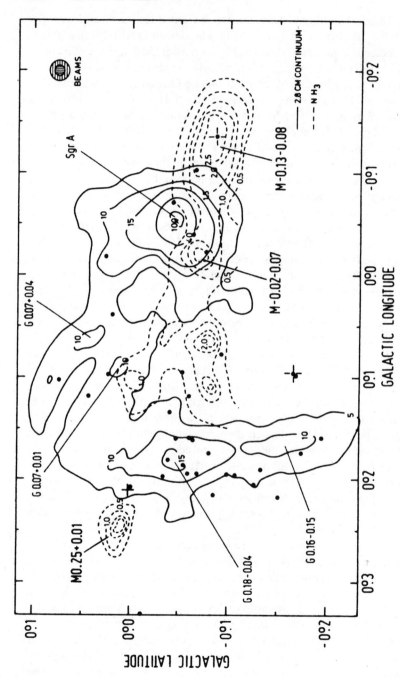

Figure 5 A diagram [from Güsten (1982)] showing the distribution of ammonia emission intensity in the (1, 1) transition (dashed contours) and the 2.8-cm radio continuum. Dots represent compact radio sources (Downes et al. 1978) and crosses are H$_2$O masers. The source M$-0.13-0.08$ occurs at 20 km s^{-1}. Other peaks in the map trace the densest and warmest portions of the 50 km s^{-1} cloud.

provide information on the total gas column density, independent of geometry. In the most general situation, where the column density varies and the continuum background may not be contiguous, even emission-absorption comparisons may lead to controversy; such is the case around Sgr A.

The morphologies of species observed in emission and absorption around Sgr A mimic each other quite closely; Figure 6 [from Güsten et al. (1981) and Güsten & Henkel (1983)] illustrates this observation by using NH_3 emission and $\lambda 2$-cm H_2CO absorption. Nonetheless, detailed consideration of the differences between the two patterns led these authors to conclude that the $\simeq 50$ km s^{-1} gas lies behind Sgr A West and is sandwiched between this source and the rest of the radio continuum (i.e. Sgr A East). An exactly opposite conclusion was reached by Sandqvist (1982), who compared emission and absorption from H_2CO. A third geometry, in which all of Sgr A lies behind the gas, was inferred by Liszt et al. (1983) after comparison of H I absorption and CO emission. This controversy has its roots in the small-scale behavior of neutral gas close to Sgr A, which we now discuss.

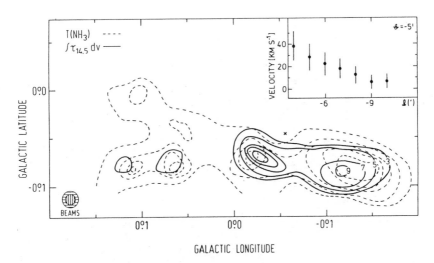

Figure 6   This diagram [from Güsten & Henkel (1983)] compares the ammonia emission (Figure 5) with the integrated optical depth in 2-cm H_2CO. Such emission-absorption comparisons are required in order to reconstruct the geometry of the inner Galaxy material, but they lead to different pictures depending on whether similarities or differences in the maps are stressed. The insert shows the mean gas velocity as a function of longitude, which varies very smoothly (see also Figure 12).

2.4   Associations Between Neutral and Ionized Gas Near Sgr A

2.4.1 ABSORPTION MEASUREMENTS   The relationship between Sgr A and the 50 km s^{-1} cloud is shown in more detail in Figure 7, where the 6-cm maps of Ekers et al. (1983) are superposed on a higher-resolution ^{12}CO map. The continuum, which is discussed in more detail in Sections 5 and 6, consists here of four parts. Several small sources, probably compact *H II regions*, overlay the molecular cloud peak. These sources also occur just outside the edge of a *shell* of nonthermal emission whose portions nearest

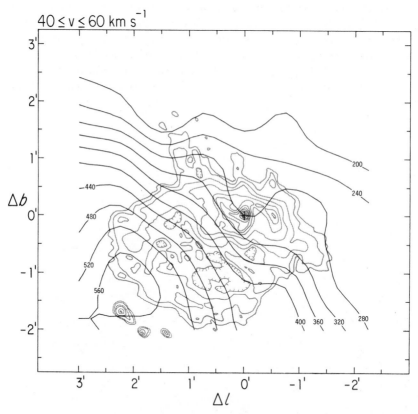

Figure 7   A small-scale version of Figure 4. In this figure, the shaded contours are from the 6-cm VLA radio continuum maps of Ekers et al. (1983) with 5″ × 8″ resolution (α × δ), and the ^{12}CO data from Liszt et al. (1983) were taken with 45″ spacing of a 1′ beam. Sgr A* at the origin of the galactic coordinate system is surrounded by the bar-spiral thermal emission pattern comprising Sgr A West (Section 5). Sgr A East is the portion of the nonthermal shell structure nearest the molecular cloud peak.

the molecular cloud were previously known as Sgr A East; the other edge of the shell merges smoothly (in projection) with the *thermal emission region* Sgr A West. The compact source *Sgr A** at the map origin is surrounded by a pattern of thermal gas that is described as having a spiral or bar and spiral pattern. The actual positioning of the various components of the continuum with respect to each other and to the neutral gas can be explored by detailed study of the neutral gas characteristics in emission and absorption.

The behavior of neutral gas observed in absorption near Sgr A was discussed in detail by Oort (1977); here, we briefly summarize the situation to that time. When the only absorption spectra available were those against all of Sgr A, the presence of positive-velocity absorption could be taken as evidence of infall toward the nucleus. After the aperture synthesis of H_2CO absorption by Whiteoak et al. (1974) showed that only Sgr A East was strongly absorbed, it appeared that this gas had to be between the various components (the nonthermal source Sgr A East was required to lie behind Sgr A West), and Oort (1974, but see 1977) suggested that the positive velocity had its origin in ejection of material from the nucleus. The geometrical interpretation of these absorption measurements was challenged by Liszt et al. (1975) because rapid variations in the ^{12}CO emission intensity, as in Figures 4 and 7, would by themselves cause a strong variation in optical depth in the observed sense, independent of the source geometry.

Now that the shell morphology of Sgr A East has been demonstrated by Ekers et al. (1983), its placement significantly beyond the galactic nucleus may be somewhat less attractive; if it is a supernova remnant, the inferred rapid decline in stellar density renders the most likely site of such an event closer to the nucleus. More complete mapping of the 50 km s^{-1} cloud does not support the idea that this material has been ejected from the nucleus, although (as we shall see) there may actually be more than one such component to contend with.

Synthesis of H I absorption at Westerbork with resolution $\simeq 0.4' \times 2.0'$ by Schwarz et al. (1977, 1983) confirmed the rapid east-west decline in opacity at $\simeq 50$ km s^{-1} (and first detected a portion of the positive-velocity absorption, which we associate below with the presence of a rotating ring around Sgr A); in their work, Schwarz et al. found no absorption from this gas specifically toward Sgr A West. Whiteoak et al. (1983), using the VLA to achieve 3″5 resolution in the λ6-cm absorption line at H_2CO, detected absorption at 40–50 km s^{-1} (and near 0 velocity) toward portions of the thermal spiral around Sgr A*; however, they found no occultation of the compact radio source itself. There are several possible interpretations of this result. (*a*) The compact source is in front of both the molecular gas and the thermal spiral. (*b*) The molecular gas is very near the nucleus and is dissociated around Sgr A*. (c) The molecular gas is clumped, and the compact source is not occulted. The first interpretation may become more palatable if it continues to be impossible to prove that Sgr A* is coincident with the nucleus as represented by IRS 16. The last, which is somewhat ad hoc, cannot be applicable to the gas observed slightly to the east (which has high opacity), but clearly there are strong variations in gas properties across Sgr A.

Absorption across Sgr A has also been mapped by Liszt et al. (1983) in H I at the VLA at 12″, yielding somewhat different results. These longer-wavelength measurements are generally more sensitive to nonthermal emission and to the more extended structure in which the thermal emission pattern is embedded. The atomic gas can be traced over both Sgr A East and Sgr A West and has features over the entire range $-200 \leqslant v \leqslant 140$ km s^{-1}; one feature that is conspicuously absent is that at 165 km s^{-1} discovered by Sanders & Wrixon (1974). The absence of this widely distributed gas implies that Sgr A* does not lie very far beyond the nucleus. Actually, the H I and H$_2$CO absorption measurements taken together imply that if Sgr A* is not at the nucleus, it is a foreground object. However, the "expanding molecular ring" feature at -135 km s^{-1} [see Scoville (1972) and Liszt & Burton (1978) for two interpretations] does appear in its spectrum, which indicates that it also cannot be displaced very far in the near direction either.

The large apparent optical depths observed in H I and H$_2$CO toward the eastern portions of the nonthermal shell source prove conclusively that the 50 km s^{-1} cloud occults this material; no background radiation can provide the observed behavior. Because this shell can be traced continuously around to Sgr A West, 50 km s^{-1} absorption would be expected to appear ubiquitously over and around the thermal emission region if the shell were uniformly and strongly absorbed. Such is not the case, of course, and placement of the nonthermal shell behind Sgr A West is not sufficient by itself to explain the absence of the 50 km s^{-1} gas.

Some of the VLA H I results are shown in Figure 8 (Liszt & Burton 1984). In this right ascension–velocity diagram through the position of Sgr A*, the apparent optical depth at 40–60 km s^{-1} does not vanish toward the compact source. These data and an independent absorption spectrum toward Sgr A* at higher resolution led Liszt et al. (1983) to conclude that all of Sgr A lay behind the 50 km s^{-1} cloud. A somewhat different inter-pretation arises, however, when the gas kinematics are sampled in the orthog-onal direction (Figure 9). With varying declination, the 40–60 km s^{-1} gas absorbing Sgr A* is seen to have a substantial velocity gradient, more than 100 km s^{-1} (arcmin)$^{-1}$, which is not a property of the 50 km s^{-1} cloud and does not appear on larger angular scales around the nucleus in that

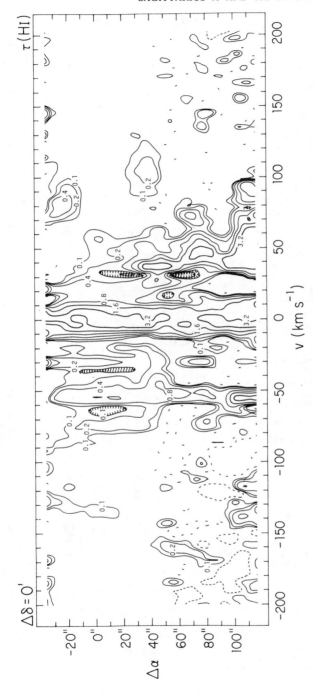

Figure 8 A position-velocity map of 21-cm H I opacity at 12″ resolution, taken along a line of constant declination through Sgr A* (at $\Delta\alpha = 0$″). The spatial track of this map moves diagonally from lower left to upper right in the previous figure, crossing the nonthermal shell region Sgr A East some 100″ from the origin. Along this line, the optical depth of 40–60 km s^{-1} gas diminishes steadily but does not vanish at Sgr A*. Data are those of Liszt et al. (1983).

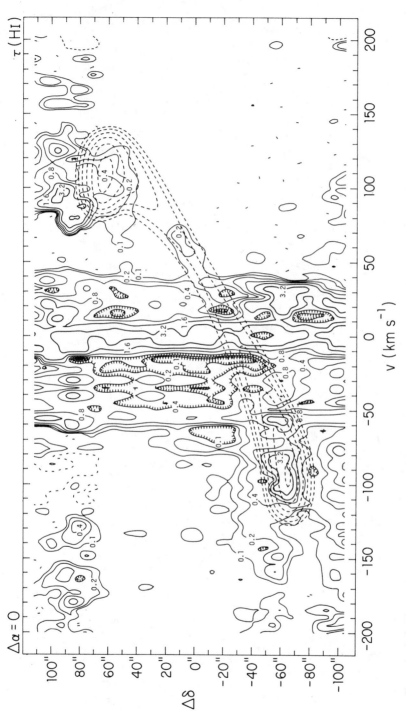

Figure 9 As in Figure 8, but now along a line of constant right ascension through Sgr A*. As viewed in this way, the H I absorption at 40–60 km s^{-1} is isolated and possessed of a large velocity change in position. The dashed contours represent the kinematic behavior of the front side of a model ring having mean radius 3.2 pc (65″) and (conventional) inclination 70°. The ring rotates at 110 km s^{-1} in the normal sense, but it is infalling toward Sgr A* at 50 km s^{-1}. It is more likely that the 40–60 km s^{-1} gas seen toward Sgr A West arises from this structure than from the 50 km s^{-1} cloud, the velocity of which is constant with position. Figure from Liszt & Burton (1984).

object. Its presence suggests that the 40–60 km s^{-1} gas absorbing Sgr A*
and surrounding regions is unrelated, or at most only indirectly related, to
the 50 km s^{-1} cloud. If this is actually the case, the inner Galaxy geometry
probably cannot be reconstructed from absorption measurements in the
manner previously supposed (not that such studies were obviously bound
to converge on a single interpretation in any case).

2.4.2   COMPARISON OF EMISSION AND ABSORPTION MEASUREMENTS   One of
the salient features of the galactic center gas distribution is the similarity
between emission and absorption characteristics in many species. This
similarity is particularly revealing for the gas at $|v| \geqslant 70$–80 km s^{-1}, which
in H I is rather symmetrically disposed about Sgr A*; a ^{12}CO map is shown
in Figure 10. The molecular emission exhibits a double-lobed structure that

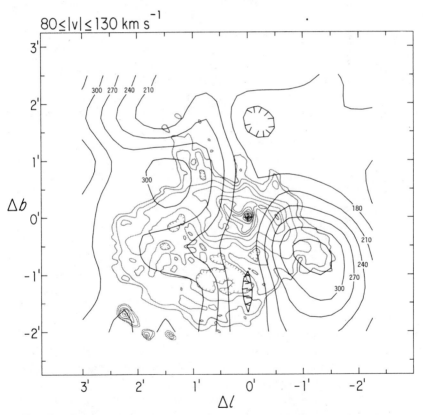

Figure 10   Similar to Figure 6, but with the range of integration taken as $80 \leqslant |v| \leqslant 130$
km s^{-1}. Even at the coarse resolution of the CO observations, the ends of the H I ring structure in the previous figure are quite prominent. Figure from Liszt & Burton (1984).

suggests that a ring of material is being observed nearly edge-on (else the double structure would not appear so clearly). The two-sided high-$|v|$ H I absorption would result from increases in projected column density and indicate a projected major axis for the ring that lies nearly north-south. The dashed lines superposed on the observed optical depth contours of Figure 9 arise from the front side of a model circular ring of radius 65″ (3.2 pc), rotating in the normal sense at 110 km s^{-1} and falling toward the galactic nucleus at 50 km s^{-1}. Clearly, the apparent motion of much of the H I is explained by such kinematics, although the observed optical depths are much patchier than can be reproduced by a uniform model. The ring model used to produce the dashed contours had (conventional) inclination 70°, but this can be varied somewhat by altering the ring thickness parallel to its symmetry axis; the orientation should nonetheless be fairly close to edge-on, and high-resolution emission measurements may show a more nearly filled structure, one more akin to a disk, if both its front and back halves are detected along the same line of sight. It is almost certainly the case that the kinematics of a circular model involving superposed circular and radial motion can be reproduced by circulation along elliptical paths without actual infall of material [compare the discussions of Burton & Liszt (1978) and Liszt & Burton (1980) for an illustration of this point].

The mass interior to the ring is $9 \times 10^6 \, M_\odot$ if the rotation speed is the equilibrium circular value, while estimates of the mass interior to 1 pc yield 2–$5 \times 10^6 \, M_\odot$ (as discussed in Section 4.2). The mass of the ring itself is uncertain for a variety of reasons (Liszt & Burton 1984), but it is at least several hundred solar masses, much greater than that of the radio-emitting ionized gas within it. The ring detected in H I and CO is roughly twice as large as a similar double-lobed structure seen at longer IR wavelengths (Section 4.2), and its inferred speed of rotation is slightly larger than is seen either in the O I kinematics of that gas or at the ends of the thermal radio spiral (Section 5). Both double-lobed structures have the unusual property that the line joining the lobes is tilted out of the galactic plane, to positive latitudes at more positive longitude. The kinematic major axis of the H I occurs nearly along a north-south line, and so is parallel to the radio continuum bridge and to the IR shown and discussed in Section 2. This orientation, which makes a very large angle with the tilt supposed to describe the neutral gas on very large (kiloparsec) scales (Burton & Liszt 1978), can, however, be seen in the more widely distributed molecular emission in the inner 100 pc of the Galaxy (Section 3.2).

As we have seen, such a ring highly complicates the interpretation of the absorption measurements from which one would like to infer the placement of various continuum components. Alternatively, the recognition of its presence as an additional feature of Sgr A will have a bearing on models of the smaller-scale structure.

2.4.3 GAS AT -190 KM S^{-1} Güsten & Downes (1981) detected and discussed a feature at this velocity that appears at and around Sgr A* in H I and H$_2$CO absorption at 3–9′ spatial resolution; they suggested that it had been ejected relatively recently from the galactic center. This feature is present in the VLA H I work also, which proves that it occurs in front of Sgr A, and it was partially mapped in ^{12}CO by Liszt et al. (1983). Much, but not all of the gas occurs in a narrow ridge $\leqslant 1' \times \geqslant 3'$ at a longitude just below Sgr A* (45″). An intimate association with the nucleus is certainly suggested, but much more complete mapping is required.

3.   AN OVERVIEW OF NEUTRAL GAS PATTERNS OBSERVED IN THE INNER GALAXY

3.1   *The Geometry and Kinematics of the Inner Few Kiloparsecs*

A significant fraction of Oort's (1977) earlier review was dedicated to discussion of the kinematics observed in neutral atomic and molecular gas within some 10° of Sgr A. Since that summary, some new trends in interpretation have arisen, and a variety of new observations have become available; a very comprehensive tabulation of inner Galaxy survey observations is given by Burton & Liszt (1983a). Because the neutral gas behavior is so clearly related to the morphology of other features near Sgr A, and because the larger-scale gas distribution represents an enormous reservoir ($\geqslant 10^9 \ M_\odot$) of material whose processing may eventually be manifested as stronger nuclear activity, we briefly summarize recent work in this area.

In the disk of the Milky Way, purely rotational motion in a circular, thin, plane-stratified geometry provides a strong interpretative framework. In the inner regions of the Galaxy, however, strong departures from the galactic equator are relatively common in the neutral gas. Many features having prominent noncircular motions have been observed and ascribed to ejection from the galactic nucleus.

A coherent tilt in the gas layer was demonstrated by Kerr (1968) for the extreme permitted velocities occurring 5–10° from the nucleus, and by van der Kruit (1970) for portions of the forbidden ones. The former make an angle of 10° or less with the galactic equator (in the sense of increasingly negative latitude in the first quadrant), the latter somewhat more than this at lower longitudes. Cohen & Davies (1979; see also their earlier references) found similar values, and a feature of ejective explanations is collimation of the ejecta along a line making a substantial angle with both the galactic plane and the galactic rotation axis.

Burton & Liszt (1978) and Sinha (1979) showed that the kinematic symmetry axis of the inner Galaxy gas layer is also coherently tilted. The first group found that a very substantial fraction of the overall behavior could be explained if most of the gas were contained in a disk of radius 1.5 kpc making a $\simeq 30°$ angle with the galactic rotation axis, with the major axis of this body appearing along a line $b = -l \tan 22°$. This description, which invoked superposed circular and noncircular motions of roughly equal magnitude ($\simeq 170$ km s^{-1}), could account for many features simply as projection effects of the (smoothly varying, large-scale) velocity field. Such a model is unattractive because of the very large mass and energy flow across its outer borders, but we can circumvent these difficulties if the gas kinematics is described as circulation along paths of varying eccentricity (Liszt & Burton 1980), as (perhaps) in an inclined bar.

In such models, the rotating nuclear disk and other features having large noncircular motions arise from the same body, and it is one of the main results of more recent observations that the nuclear disk appears *not* to be purely of rotational origin. The considerations that lead to this conclusion are given by Sinha (1979) and Liszt & Burton (1980). This being the case, the inner Galaxy mass distribution cannot be directly derived from the observed maximum line-of-sight H I velocities, and the inner Galaxy potential must deviate strongly from the previous picture. There seems little doubt at present concerning the importance of the tilt and the general contribution of noncircular motions in describing the bulk of the inner Galaxy phenomena. However, there is uncertainty concerning the tilt angle, which appears to be smaller in gas observed farther from the nucleus, and there is no unique model of the large-scale velocity field or geometry. Some care must be taken in inferring the global geometry from observations over a limited portion of the gas distribution. Kinematical effects of projection can be highly misleading in such cases.

A general field of noncircular motions is expected of a barlike figure, and its presence in the inner Galaxy is not surprising. The tilt is a less expected phenomenon. Extrapolation from the observable behavior to a description of the underlying gravitational potential represents the ultimate challenge of studies of the neutral gas. Orbits in triaxial potentials (as discussed by Lake & Norman 1982) and bending wave theory (Blitz et al. 1981) provide possible explanations.

3.2   *A Framework for the 50 km s^{-1} Cloud*

As discussed in Section 2, it is the 50 km s^{-1} molecular cloud to which the behavior of Sgr A is most directly related; thus, placing this object within a larger context would be highly desirable. Unfortunately, the innermost few degrees of longitude have not been fully mapped and are more easily described than explained. For specifics, we refer the reader to Bania (1977, 1980), Liszt et al. (1977), Liszt & Burton (1978), Bieging et al. (1980), and

Güsten & Downes (1980) [see Oort (1977) for a review of earlier work]. Our purpose here is only to point out a few salient characteristics that may aid in understanding the 50 km s^{-1} cloud.

Most mapping of the inner $\pm 2°$ has been done along or very near the galactic plane, as in the ^{12}CO longitude-velocity diagram in Figure 11. Most of the extreme-velocity behavior in that figure occurs in a feature crossing zero longitude at -135 and $+165$ km s^{-1} that does show the tilt characteristics of large-scale descriptions (see Liszt & Burton 1978) and so can be reproduced by models with or without a net outward flow. This feature, which contains little of the observed intensity in the map (although it may actually represent a very large gaseous mass), is the only portion of the observed structure to which the tilt is demonstrably relevant.

Most of the strong emission occurs in a feature having a pronounced but probably misleading rotation signature over much ($-2 < l < 0.5°$, $-200 < v < 80$ km s^{-1}) but not all of its length. This gas is the molecular analogue of the well-known H I rotating nuclear disk, but the lack of a substantial velocity gradient at $l > 0.5°$, $v > 80$ km s^{-1} implies a crescent shape if the motions within it are assumed to consist only of "normal" galactic rotation (Güsten & Downes 1980). Another complication hindering a simple explanation is the fact that this feature crosses zero-velocity not toward Sgr A but at a longitude some 5–10′ more negative; the gas has a positive velocity of > 20–30 km s^{-1} toward the nominal rotation center of the Galaxy. Of course, the 50 km s^{-1} cloud also exhibits this peculiarity, and even the apparently "expanding" gas at -135, $+165$ km s^{-1} is offset in similar fashion.

The structure of the stronger molecular emission at positive longitudes can also be interpreted as outflow, as in the fan model of Fukui (1980). However, in a manner very similar to that of H I on large scales, the kinematics seen near the galactic equator are deceptive because of an unexpected geometry. When ^{12}CO is mapped above the plane at positive longitudes, there is a strong velocity gradient in latitude, and the higher positive velocities $v \geqslant 80$–160 km s^{-1} (which are absent in Figure 11) can be found at $b < 0.4°$ (Burton & Liszt 1983b). The small velocity gradient in the galactic plane at $l > 0.5°$ probably arises geometrically, in the sense that the circulation axis of the gas is substantially oblique to the rotation axis of the Galaxy at large. The sense of this tilt is opposite to that mentioned in Section 3.1, but we have encountered it previously in the radio continuum bridge, in the more extended IR flux around Sgr A, and very close to Sgr A* in the neutral gas ring at 3 pc.

The velocity structure of strong CO emission is shown in Figure 12, a map of intensity-weighted mean velocity. Note that the zero-velocity contour occurs (at low latitudes) at somewhat negative longitudes relative

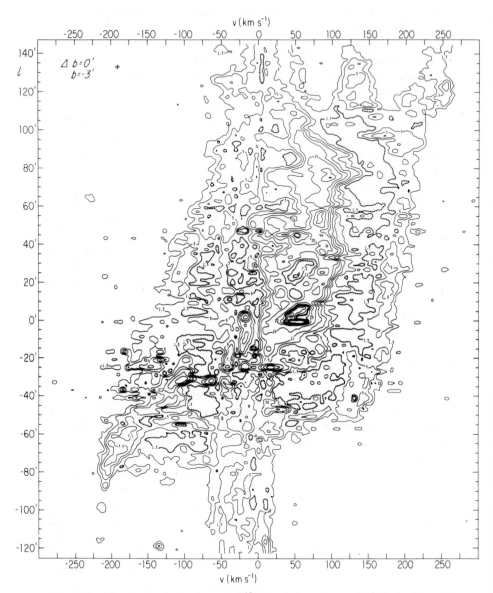

Figure 11   A longitude-velocity diagram of ^{12}CO emission, taken at the latitude of Sgr A*
with 2′ beam spacings. Figure from Liszt & Burton (1978).

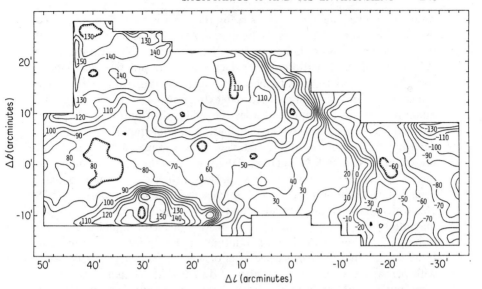

Figure 12 A map of the intensity-weighted mean velocity of ^{12}CO emission, made with 2′ beam spacings (Liszt & Burton 1984). Coordinates are measured relative to Sgr A*. Note the upward slant of the velocity gradient in the vicinity of the map origin, as well as the substantially positive mean velocities that occur there.

to Sgr A*. The velocity gradient seen around Sgr A is sharply inclined. The 50 km s^{-1} cloud, however, fits well into the overall kinematic organization. The contours are not disrupted over the limited extent of this feature.

What then of the 50 km s^{-1} cloud? If its velocity were construed as largely random, it might be plunging across the galactic center region, eventually to cross the nucleus itself and fuel one of a series of episodic outbursts of nuclear activity (Sanders 1981). Alternatively, because the cloud does not really appear anomalous when considered along with the more general run of molecular gas behavior, it might be better viewed as only a small part of a geometrical and kinematical structure whose nature remains to be explored. The 50 km s^{-1} cloud certainly deserves special attention because of its unique and intimate relationship with Sgr A, but the crucial problem is to fit both of these sources into a satisfying picture of the inner Galaxy structure.

4.   THE CENTRAL 2 PARSECS OF THE GALAXY

4.1   *The Stellar Distribution*

A discussion of the innermost region of the Galaxy begins with a definition of what we mean by *the* galactic center. For the purpose of this discussion, we interpret the galactic center as being the centroid of the galactic mass distribution as described by galactic dynamics. Since the preponderance of the galactic mass is contained in stars, we use stellar observations for a definition and identification of the galactic center. In practice this is not as straightforward as it appears, because the stellar radiation at $\lambda < 1$ μm suffers large and spatially structured extinction ($A_v = 15$–30 mag). To minimize the extinction, one studies the stellar distribution in the near-infrared ($\lambda\lambda 1$–2 μm), and so the distribution of cool stars is preferentially determined. The first suitably detailed map at these wavelengths was presented by Becklin & Neugebauer (1975). Most of the radiation shown on their map comes from discrete sources, stars, rather than from a spatially extended background. Of particular interest in this regard is source 16 (IRS 16), which has a finite spatial extent of $\sim 2''$ and an absolute 2.2-μm magnitude near -8.7. As we expect K or M giants to dominate the 2.2-μm radiation, the finite extent and magnitude of IRS 16 interpreted in these terms suggest that it consists of a cluster of at least 100 such giants, together with perhaps a much larger number of dwarfs that dominate the cluster mass but contribute insignificantly to the luminosity. Interpreted as a stellar cluster, IRS 16 marks the highest concentration of stars in the nuclear region, and hence it has been presumed to be *the* galactic center. However, unlike a cluster of late-type stars, IRS 16 shows no 2.3-μm CO bandheads, and its colors are much bluer than those of any other 2-μm continuum source in the inner Galaxy. We return to this point in Section 6.

Using IRS 16 as the center, the stellar distribution in the inner Galaxy can be inferred from the radial distribution of near-infrared surface brightness. Bailey (1980) used Becklin & Neugebauer's (1968) 2.2-μm map to estimate that the surface brightness decreases away from IRS 16 as $r^{-0.8}$ from 2 to 1000 arcsec (0.1 to 50 pc). More recently, Allen et al. (1983) used a high-resolution map of the inner 45'' of the Galaxy to confirm Bailey's result and to show that the radial variation, which is more nearly $\propto r^{-0.68}$, continues down to less than $1''.25$. Thus, to the angular limit of the observations, no nuclear core, no flattening of the radial surface brightness distribution, is evident. The important point here is that the stellar mass distribution inferred from the surface brightness monotonically increases as $\rho(r) \propto r^{-1.68}$ inward toward IRS 16.

4.2   The $R \sim 2$ pc Ring of Dust and Gas

Investigations of the gaseous component of the galactic center, which are again restricted to long wavelengths owing to the large visual extinction, indicate that the gas distribution is apparently not at all as simple as the monotonically increasing stellar distribution. Here the far-infrared observations ($\lambda > 30$ μm) are particularly instructive. The unfortunately large extinction that prohibits visual observations has the virtue of ensuring that all the ultraviolet or visual luminosity emitted by the galactic center region is absorbed by interstellar grains and reradiated in the observable far-infrared. Since the volume emissivity of heated dust is proportional to both the number density of grains and to the heating flux of optical and UV radiation incident on the grains, the far-infrared maps can be used to determine both the column density of dust and the distribution of sources that heat the dust. Figure 13 illustrates such maps as made at 30, 50, and 100 μm by Becklin et al. (1982); in each of these maps, the position of IRS 16 is marked with a cross.

There are two points to be made here. First, the 30-μm map is made at the peak of the 2–1000 μm energy distribution from the Sgr A region, and as

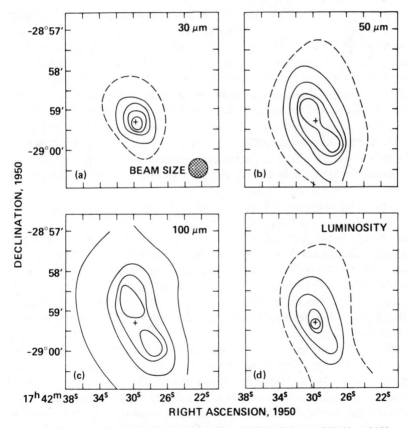

Figure 13   The infrared surface brightness of the inner 4' of the Galaxy at λ30, 50, and 100 μm, together with the luminosity as deduced from these maps. From Becklin et al. (1982).

such it traces the regions of peak IR luminosity; these are centrally condensed near IRS 16. Second, the 50- and 100-μm maps show no such central concentration. Rather, they have a double-lobed structure symmetrically placed about IRS 16 along the galactic plane. The surface brightness, and hence the column density, of the cool dust mapped at 50 and 100 μm peaks at two complementary positions well displaced from IRS 16. Since a dust grain radiates more power at every wavelength as its temperature is increased, the only way to reconcile these 3 maps is to conclude that the dust is distributed in an annular (not spherical) region about IRS 16 that lies very nearly in the galactic plane. Thus, the IR dust observations indicate the presence of the same annular structure that was inferred from the CO and H I observations (Section 2.4.2). Interior to this annulus the dust density *decreases*, while the distribution of heat sources *increases* inward toward IRS 16.

In order for a central concentration of luminous sources to provide the optical and UV photons that heat the dust in an annulus of inner radius ~ 20–$40''$ (1–2 pc), it follows that the dust density, and hence the visual extinction interior to the annulus, is very low ($A_v \ll 1$). The thinly distributed dust grains in the interior are heated efficiently by the high-energy density of trapped Lyα radiation as well as directly from the central source of optical and UV luminosity. These hot grains dominate the 10–30 μm luminosity. The observation that the hot grains do not dominate the 50–100 μm maps argues compellingly for the $R \sim 2$ pc ring of high dust density noted earlier. The grains in the ring are cooler and denser than those in the interior, but they are heated by the same central source (or sources) that heats the tenuous interior dust.

The total observed IR luminosity, obtained by integrating over a projected area 2 pc $\times$ 4 pc in extent (Figure 13), is $5 \times 10^6 \, L_\odot$. Since this luminosity is the heat deposited in dust grains by the central luminous source, the optical/UV luminosity of that source is at least of this order. But we must also correct for the fact that the dust annulus subtends a finite solid angle as seen from the luminous centroid. Much of the optical/UV radiation will not be intercepted by the ring. Making this correction, Becklin et al. (1982) estimate that the total luminosity of the central source is $(1$–$3) \times 10^7 \, L_\odot$.

We can further delineate the properties of the dust ring by studying those neutral atomic and molecular species that are commonly found in cool galactic dust clouds. The observations of both the 63-μm $^3P_1$–$^3P_2$ [O I] fine-structure line (Lester et al. 1981, Genzel et al. 1984) and the $v = 1$–$0 \, S(1)$ line of molecular hydrogen (Gatley et al. 1984) show these neutral species to be distributed within the dust ring. The H$_2$ distribution is particularly striking because it appears to exist only in a very thin region on the inner edges of the more extensive 50–100 μm structure. Additionally, the relative strengths of the H_2 $v = 1$–0 and $v = 2$–1 lines indicate that the molecular gas is shock excited. We return to this point below.

The overriding importance of the spectroscopic observations of the dust ring is that they provide a measure of the ring kinematics and therefore an estimate of the mass interior to 2 pc. Genzel et al. (1984) find that the [O I] lines observed with a 44″ beam have an average width (FWHM) of about 300 km s^{-1}. Parallel to the galactic equator, the line centroid changes from a value $v_{lsr} = -70$ km s^{-1} at points 40–80″ toward negative longitude from IRS 16 to $v_{lsr} = +70$ km s^{-1} at points equal distance from IRS 16 toward positive longitude. The change in sign of the velocity centroid occurs very near the longitude of IRS 16. Thus, it appears from the rapid change in sign around the central position and the approximately constant value away from the center that the [O I] emission indeed arises in the dust ring with $R \sim 2$ pc and that the neutral oxygen is rotating about the center in the sense of galactic rotation. The somewhat scantier and lower spectral resolution H_2 data complement this same picture (Gatley et al. 1984).

Genzel et al. (1984) estimate the mass interior to the dust ring (i.e. interior to 2 pc) two ways: by interpreting the [O I] rotation curve as indicative of circular rotation, and by using the observed line widths and assuming that the virial theorem applies. These two estimates give consistent results, and from them Genzel et al. infer that the total mass interior to 1 pc is $M(R < 1) = (2$–$5) \times 10^6$ $M_{\odot}$.

At this point, we conclude that the distribution of cool stars that dominate the 2.2-μm surface brightness in the inner parsec (as noted above) cannot account for the $(1$–$3) \times 10^7$ $L_{\odot}$ total luminosity from the central parsec. The total mass of such stars is $(2$–$5) \times 10^6$ $M_{\odot}$; this, together with a characteristic mass-to-luminosity ratio $(M/L) \sim 3$, falls at least an order of magnitude short of supplying the needed luminosity. Some other source is needed.

A second intriguing property of the 63-μm [O I] line emission is its surprisingly high luminosity ($\sim 10^5$ $L_{\odot}$; Genzel et al. 1984), which is on the order of 1% of the total IR luminosity; this implies that the excitation mechanism must be collisional. The most likely collisions are those with hydrogen atoms in the neutral or weakly ionized ($n_e/n < 0.1$) gas in which the atomic oxygen is found (Lester et al. 1981). Adopting this description and assuming that the O I and warm annular dust are coextensive and optically thin, Genzel et al. use the observed ratio of [O I] intensity to 63-μm continuum intensity to infer that the pressure in the [O I] region is

$$n_H T = 10^{7 \pm 0.2} \text{ cm}^{-3} \text{ K}.$$

To separately disentangle the elements of this product requires a model of the O I heating mechanism: If either UV photodissociation of H_2 or dust photoelectric heating is dominant, then $n \geqslant 10^5$ and $T = 100-300$ K; but if the O I is heated via shock dissipation of noncircular motions in the dust ring, then lower densities and warmer temperatures can be accommodated.

5.   THE REGION INTERIOR TO THE 2-PC RING

5.1   *Ionized Gas Associated with the Ring*

As noted above, the interior of the ring is a region of increasing stellar density, and of decreasing dust density but increasing dust temperature; it is also a region apparently devoid of molecular H_2 or atomic [O I] emission. The gas in the inner parsec is ionized.

The distribution of ionized gas within obscured galactic H II regions is traced either by its radio continuum (bremsstrahlung) emission or by the 10–20 μm thermal emission from nebular dust that is heated by trapped Lyα radiation within the nebula. In general, the radio and 10-μm continuum maps appear very similar because density enhancements ("clumps") in both appear bright: The local radio bremsstrahlung emissivity and the local Lyα intensity that heats the dust (and which depends directly on the recombination rate) are proportional to n_e^2. This similarity is readily evident in the ionized gas interior to the dust ring of Sgr A West.

In Figure 14 we show both an infrared 10.6-μm continuum map (Rieke et al. 1978) and a high-resolution 14.9-GHz VLA continuum map of the inner parsec. Excluding the bright nonthermal point radio source Sgr A* (seen at $\alpha = 17^h42^m29^s335$, $\delta = -28°59'18''6$) and the stellar IR sources numbered 3 and 7, the two maps are in striking agreement. This correspondence admits the conclusions that the hot dust is mixed with ionized gas and that the various 10.6-μm peaks are sites of local density enhancement. Note, however, that the ratio of radio to 10.6-μm IR brightness is not constant from one peak to another. For example, in the IR map, IRS 1 is noticeably brighter than IRS 2, which is itself brighter than IRS 6; but in the radio map, IRS 2 is by far the brightest source, with IRS 1 and IRS 6 being of approximately equal brightness. In the absence of a significant extinction variation across the IR map, this suggests that the gas-to-dust ratio differs markedly from one peak to another.

The morphology of the ionized gas bears an intriguing relationship to the larger-scale morphology of the 2-pc dust ring. Figure 15a is a superposition of a 4.9-GHz VLA map, constructed so as to emphasize low surface brightness features (Brown & Johnston 1983), and the 100-μm map (Becklin et al. 1982). Note that the bright radio emission falls within the dust ring in a manner such that the NE and SW radio "spiral arm" features terminate at the projected inner boundary of the dust ring. Moreover, the collisionally

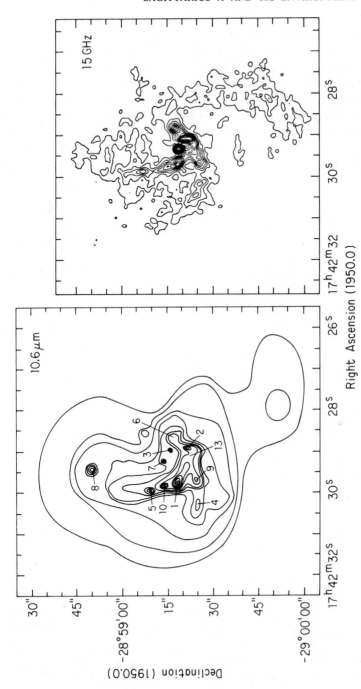

Figure 14   A comparison of the 10.6-μm continuum map of the inner ~ 1.5 of the Galaxy (Rieke et al. 1978) with a similar map of the 15-GHz radio continuum emission.

excited H_2 emission reported by Gatley et al. (1984) is seen at the positions where the radio arms and the dust ring appear to coincide along the line of sight. This suggests that at least some of the radio emission arises in an ionized region at the edge of the dust ring.

We pursue this association further by using the high-resolution 14.9-GHz VLA map to outline a possible sketch of the inner boundary of the dust ring. The double-lobed 100-μm brightness distribution suggests that any such ring is being viewed at a large inclination ($\sim 70°$); the ring sketched from the radio map is consistent with this conclusion.

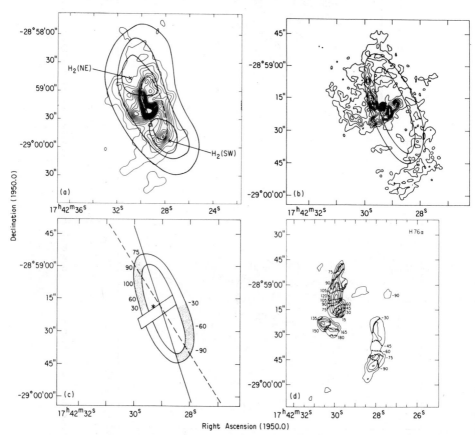

Figure 15   An overlay of the 100-μm map (which indicates the presence of the 2-pc dust ring) on the 5-GHz radio map is shown in panel (a). The positions at which shocked H_2 have been detected are noted (Gatley et al. 1984). Using (a) and the high-resolution 15-GHz map, we sketch an outline of the ring in (b). The resultant model is drawn in (c), while (d) is the van Gorkom et al. (1984) map of radio recombination line emission. Note that the radio lines are seen only in the ionized gas associated with the ring—this is noted by the shaded areas in (c).

Furthermore, the radio "spiral arm" features are wholly accounted for in such a picture as being the ionized inner boundary of a much larger structure, viz. the dust ring. Questions involving the long-term maintenance and stability of the filamentary radio structures are therefore cast in quite a different light (cf. Ekers et al. 1983). But if this description bears any validity, the H II regions on the inner edge of the dust ring should share in the ring kinematics, as deduced from the [O I] observations. They appear to do so.

Figure 15d is a map of the integrated H76α (14.7 GHz) radio recombination line emission from the inner parsec made by van Gorkom et al. (1984). *Line emission is seen from all those H II regions in the ring (as sketched in Figure 15c) but from none of the regions interior to the ring.* The kinematics of the H II regions in the ring are consistent with the [O I] inference that there is a coherent rotation of material about an axis passing near IRS 16. The motion, of order 100 km s^{-1}, is in the sense of galactic rotation and is accompanied by significant noncircular velocities. This result lends credence to the suggestion that the line-emitting H II regions are indeed the ionized inner edge of the 2-pc dust ring. The lack of circular symmetry in the radio map implies that either the density or the ionization in the ring is structured in azimuth.

Referring again to the radio map in Figure 15b, we see that nearly all of the radio emission that is not part of the ring is found in a barlike structure extending from SE to NW nearly through the center of the map. No radio recombination line emission is evident from the bar, as noted in Figures 15c and 15d. But there is unequivocally ionized hydrogen in the bar: the Bγ line has been detected and mapped within and around the region of the bar by Neugebauer et al. (1976), Nadeau et al. (1981), Wollman et al. (1982), Hall et al. (1982), and Storey & Allen (1983); the Bα line has also been mapped along the bar (Bally et al. 1979, Geballe et al. 1982). We return to this point below.

The global excitation of the interior ionized gas has been investigated by Lacy et al. (1980) using a compendium of IR emission lines from various ionic species of H, He, S, Ne, and Ar. They find that in the Sgr A ionized gas, sensitive ionization diagnostics such as the ratios [Ar^{+2}/Ar] $\sim$ 0.1 and [S^{+3}/S] $\sim$ 0.02 are markedly lower than in typical galactic H II regions. This suggests to them that the gas is photoionized and that the blackbody temperature of the ionization source is $T \leqslant 35{,}000$ K if multiple sources of excitation are distributed interior to the dust ring, or $T \leqslant 31{,}000$ K if a single central source is involved. A similar result can be obtained simply by deriving the total Lyman continuum flux needed to ionize the gas in the central parsec, which would be $F(\text{Lyc}) = 2 \times 10^{50}$ s^{-1} from the 20.7 Jy of thermal emission at 4.9 GHz noted by Brown & Johnston (1983). This implies a luminosity of Lyc $= 1.3 \times 10^6$ $L_\odot$, while the total bolometric luminosity is $L_{Bol} = (1-3) \times 10^7 \, L_\odot$. The ratio $Lyc/L_{Bol} \sim 0.03-0.10$ is characteristic of a $T = 33,000$ K 09.5V star.

5.2 Ionized Gas in the Bar

The most noteworthy property of the gas in the bar is its peculiar kinematical structure, as revealed most clearly in the remarkable [Ne II] 12.8-μm spectroscopic observations presented by Lacy et al. (1979, 1980). Their map is reproduced as Figure 16, with two radio contours overlaid so as to delineate the spatial extent of the bar. In this figure note the following:

1. The systematic velocity bifurcation, in the sense that large positive velocities are found to the SE whereas equally large negative velocities are seen in the NW.
2. The demarcation between positive and negative velocity emission (the line of zero velocity) along position angle $\sim 45°$ through the map, i.e. nearly parallel to the galactic plane and perpendicular to the bar.

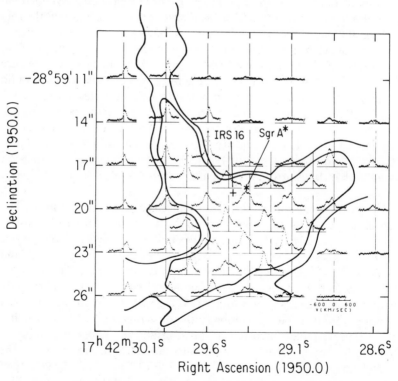

Figure 16   The [Ne II] 12.8-μm spectral profiles in the inner parsec (Lacy et al. 1980), projected against an outline of the bright 15-GHz radio emission from this region.

3. The uncommonly large velocity width (Δv = 200–400 km s^{-1}).
4. The increase of the line width toward the center of the bar.

The velocity separation across the galactic equator suggested to Lacy et al. (1979, 1980, 1982) that the ionized gas was clumped in short-lived ($\sim 10^3$ yr) clouds that were orbiting about the galactic center in a disklike structure. But rotation implies that as one observes closer to the rotation axis, the velocity dispersion should decrease; this is contrary to the sense of the [Ne II] kinematics (Figure 16). Thus, if rotation is present, it must be accompanied by a significant radial velocity gradient. But any radial motion will skew the line of zero velocity away from the apparent rotation axis unless the disk is viewed precisely edge-on. Finally, it is clear that the centroid of the bar coincides with neither IRS 16, the putative dynamical center of the Galaxy, nor with Sgr A*, the compact radio source at the galactic center; both of these are evident at the northern boundary of the bar. The kinematics of the bar are a puzzle.

The temperature of the ionized gas in the bar is also peculiar for a galactic H II region. Here we can estimate the temperature by using the observed radio brightness temperature, together with a determination of the optical depth obtained from a comparison of radio maps made at two frequencies with the same angular resolution. Figure 17 shows radio maps of the inner parsec of the Galaxy made at 4.9 and 14.9 GHz, with various peaks on the

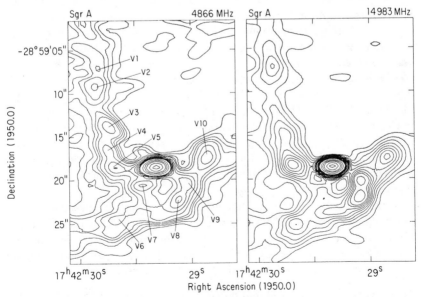

Figure 17   5- and 15-GHz radio continuum emission from the inner parsec.

4.9-GHz map labeled. The flux densities at the positions of the peaks are obtained at both frequencies by integrating over the identical region in the two maps. Two of the catalogued peaks, V2 and V7, are nonthermal, as indicated by their steep spectral index. All the other peaks have the flat spectra characteristic of thermal bremsstrahlung emission (Brown 1984). Assuming that the radio emission is indeed thermal and that over the ~ 0.1 pc diameter typical of these peaks the gas is homogeneous and isothermal, we can estimate the 4.9-GHz optical depth $\tau_{4.9}$ from the two-point spectral index α. This technique is applicable, of course, only to those regions in which the spectral index reflects the effect of partial opacity, i.e. where $\alpha > -0.08$ or so. Finally, from the observed 4.9-GHz brightness temperature we can use $\tau_{4.9}$ to solve for the electron temperature as $T_B = T_e(1 - e^{-\tau_{4.9}})$.

The electron temperatures so derived fall in two classes. The regions V4 and V5—the latter identifiable with IRS 1—have a temperature of ~ 5300 K; they are among the ionized structures included in Figure 15b as being the inner edge of the $R = 2$ pc dust ring, and radio recombination line emission at $v \sim 0$ km s^{-1} is detectable at their positions (Figure 15d). In contrast, the regions V6, V8, V9, and V10 have considerably higher temperatures (12,000–20,000 K). These are the H II regions in the bar. They have the very large [Ne II] velocity dispersions (Figure 17), and from them no radio recombination line emission is detectable (van Gorkom et al. 1984). The latter point is in fact a verification of their higher temperatures, since it reflects the strong temperature dependence of the radio recombination line optical depth, $\propto T^{-5/2}$ (e.g. Brown et al. 1978).

Thus, we are led to distinguish between the ionized gas in the 2-pc dust ring and the ionized gas in the bar. We can make this distinction three ways: (a) *thermodynamically*—the temperature of the 2-pc gas in the ring is ~ 5000 K (as noted above; see also Rodriguez & Chaisson 1979, Pauls 1980), whereas the gas in the bar is $> 12,000$ K; (b) *kinematically*—the exceptionally wide $\Delta v = 200$–400 km s^{-1} [Ne II] lines are found only in the bar, while in IRS 1 and along the ring $\Delta v \sim 100$ km s^{-1} (Figure 17); (c) *dynamically*—at the positions where the ends of the bar appear to merge with the dust ring, the velocity is highly discontinuous [to the NE (SW) the Ne II velocity in the bar is -160 km s^{-1} ($+160$ km s^{-1}), but the velocity of the dust ring is 0 km s^{-1}].

As the ionized gas in the bar is the hottest, most clearly disturbed material in the galactic center region, it apparently is most affected by the object (or objects) in the center and as such provides the best diagnostic of that region. Models of the central gas dynamics in terms of either mass infall or mass outflow can be constructed with the following constraints:

Mass infall   Ekers et al. (1983) noted that the overall "three-arm spiral" pattern of the radio brightness of Sgr A (Figure 15b) could be interpreted as the ionization and subsequent disruption of the 2-pc dust ring followed by infall to the center. A specific form of this model has been outlined by Lo & Claussen (1983). As Ekers et al. (1983) noted, the appeal of an infall model is that it can accommodate the observed spatial displacement of the ionized gas in the bar (the infalling gas in this model) from the central object onto which the gas is falling—presumably IRS 16 or the compact nonthermal radio source Sgr A*. Material stripped off from the ring has a finite angular momentum that prevents it from falling directly to the center, and spatial separation corresponding to a degenerating orbit is a reasonable consequence.

The central difficulty with the infall description lies in its account of the gas dynamics. The velocity at the ends of the bar, where it appears to merge with the ring, is discontinuous with that of the ring. Kinematically, how does the ring material enter the bar? Furthermore, if material at the two ends of the bar, both $\sim 7'' = 0.35$ pc from IRS 16, is falling inward and we observe a velocity of ± 160 km s^{-1} at these points, then the bar cannot be in the plane of the sky; rather, it must be inclined at some angle i. If so, then the mass of the object toward which this gas falls is

$$M = 2.1 \times 10^6 \ M_\odot/(\sin^2 i \cos i) > 5.5 \times 10^6 \ M_\odot,$$

which exceeds the total mass interior to the 2-pc dust ring as derived from the [O I] kinematics (Genzel et al. 1984).

Mass outflow   An alternative possibility is that the ionized gas in the bar is collimated outflow. Here the picture is that mass is expelled preferentially from the poles of a central object at a rate $\sim 10^{-3} \ M$ yr^{-1}. The outflow is directed along position angle $+130°$ ($-50°$), that is, perpendicular to the galactic plane. In this description, the bar is wholly distinct from the dust ring—distinct dynamically, kinematically, and thermodynamically.

The most persuasive evidence for mass loss from the galactic center is the observation toward IRS 16 of an enormously broad ($\Delta v = 1500 \pm 300$ km s^{-1}) line identified by Hall et al. (1982) as the $2p$-$2s$ 2.06-μm line of He I. This line, localized within a few arcseconds of IRS 16, is much broader than any of the [Ne II] lines, and indeed it is very similar to the He I 2.06-μm lines commonly found in WC mass-loss stars (Williams et al. 1980). Recent observations by Geballe et al. (1984) confirm the broad He I line and, in addition, report the detection of a similarly broad Bα line from the same vicinity; the spectra of these two lines are shown in Figure 18. The mass loss inferred from these observations is $10^{-4} < M < 3 \times 10^{-2}$ $M_\odot$ yr^{-1} at ~ 700 km s^{-1}. Additionally, mass loss at this rate is implied by, and hence is consistent with, that needed to shock-excite the H_2 vibrational-line emission observed by Gatley et al. (1984).

The principal difficulty with a mass-loss interpretation for the central bar

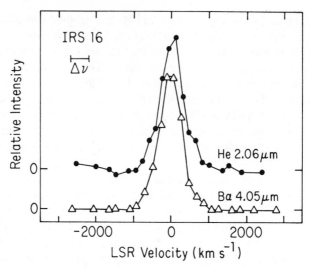

Figure 18   Broad He I and Bα lines toward IRS 16, as observed with a 4″ telescope beam by Geballe et al. (1984).

is that the two obvious sources of that mass loss, IRS 16 and Sgr A*, are both found on the northern edge of the bar (Figure 16). In contrast, the centroid of the bar is approximately 2″ south and 1″ west of Sgr A*. If mass outflow is to provide a viable explanation for the central bar, then either the centroid of the broad He I and Bα emission must be localized SW of IRS 16 and attributable to an unseen object, or else an account must be given for the misalignment of the "central" sources.

6.   IRS 16 AND SGR A*

The identification proposed by Becklin & Neugebauer (1975) of the 2.2-μm continuum source IRS 16 as the galactic center resulted from (*a*) the observation that IRS 16 is spatially extended at a resolution of 2″.5, in contrast to the other 2.2-μm sources in the inner parsec, and (*b*) the apparent association of IRS 16 with the compact nonthermal source Sgr A* (Balick & Brown 1974). More recent observations confirm the extended structure of IRS 16, but they also reveal a significant angular displacement between IRS 16 and Sgr A*.

Allen et al. (1983) show that at a resolution of 1″ the inner parsec breaks up into a distribution of point sources, some of which (including IRS 16) are partially blended. These sources appear to be stellar, late-type giants. The stellar density increases rapidly toward IRS 16, but it does not favor the galactic plane; and, importantly, there are a significant number of dark regions, indicating locally elevated 2.2-μm obscuration. IRS 16 itself is resolved into 3 components aligned roughly parallel to the galactic equator. The two outlying components, IRS 16-NE and IRS 16-SW, have detectable Bγ emission (Storey & Allen 1983), which supports their interpretation as locally excited H II regions. Additionally, the central and SW components show CO absorption; this is consistent with the presence of the K–M giants presumed on the basis of IR colors and surface brightnesses to constitute IRS 16.

With the exception of the likely association of IRS 16 with the broad He I and Bα mass-outflow region, there is no evidence for energetic activity on a scale of $\sim 10^7 \ L_\odot$ associated with IRS 16. On the other hand, there is an energetic source at the galactic center—the compact radio source Sgr A*. This object, which has a rising spectrum from 1 to 20 GHz indicative of a very compact, optically thick source (Brown et al. 1978, Backer 1982), exhibits a variation of angular size with wavelength [$\theta = 0\!''\!.05 \ (\lambda/6 \ \text{cm})^2$], consistent with the effect of angular broadening in the interstellar medium (Lo et al. 1981). At frequencies sufficiently high that interstellar scattering should be negligible, the source has an extent of $\sim 0\!''\!.015 \times 0\!''\!.004$ oriented along a position angle of $\sim 107°$; the corresponding brightness temperature is 4×10^8 K (Lo 1984). The radio source is variable on a time scale of one day, and its spectrum has been slowly evolving for several years (Brown & Lo 1982). The dimension of Sgr A* (5×10^{14} cm), together with the variability time scale, implies that changes occur at a substantial fraction of the speed of light. Models of such a source involving relativistic winds have been constructed by Reynolds & McKee (1980); in these models, the luminosity implied (in mass motion and relativistic particles) is large (10^5–$10^6 \ L_\odot$).

It is appealing to associate IRS 16, the apparent stellar centroid of the Galaxy, with the energetic object Sgr A* and to conclude on this basis that the two thereby define *the* galactic center. This may well be the case. From such an association follows the conclusion that Sgr A* likely is a massive object—a black hole—at the galactic center (Rees 1982). But recent evidence points to a small angular displacement between IRS 16 and Sgr A*. At 2.2 μm, Storey & Allen (1983) find Sgr A* to be $1\!''\!.5$ NW of IRS 16 (center), a result confirmed by E. E. Becklin et al. (private communication, 1983) and from CCD observations at 0.98 μm by Henry et al. (1984); these results are illustrated in Figure 19. It may be that the separation is real and yet IRS 16 and Sgr A* are associated. Here the work of Gurzadyan (1984) is particularly interesting. He uses the dynamics of Bahcall & Wolfe (1976) to demonstrate that if Sgr A* is bound to the star cluster IRS 16, then the small angular separation between the two is reasonable as long as the mass of Sgr A* [$M(A^*)$] satisfies $10^2 \ M_\odot < M(A^*) \ll M(\text{IRS 16})$, where $M(\text{IRS 16})$

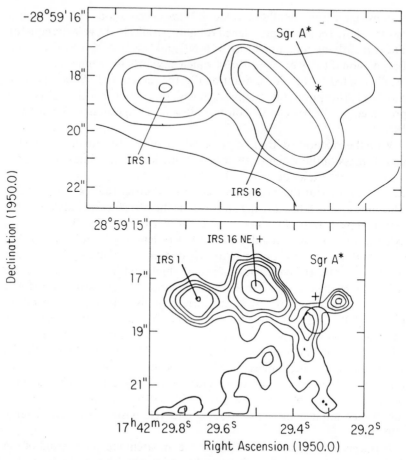

Figure 19   Maps of the IRS 16 region at 2.2 μm [E. E. Becklin et al., private communication, 1983 (*top*)] and 1.0 μm [Henry et al. 1984 (*bottom*)].

$\sim 10^5$–10^6 $M_\odot$. Thus, Sgr A* is a low-mass ($\sim 10^2$–10^4 $M_\odot$) nonstellar object. The relation of Sgr A* to IRS 16 remains a central question.

7.   EVIDENCE FOR AN ENERGETIC OBJECT

The most persuasive evidence for the existence of a compact, very energetic object at the galactic center comes from observations of the positron annihilation line at 511 keV. This line, observed for over a decade beginning in 1970 from the direction of the galactic center (e.g. MacCallum & Leventhal 1983), was surprisingly luminous ($\sim 6 \times 10^{37}$ erg s^{-1}).

Furthermore, the line observed with high-energy resolution by the Jet Propulsion Laboratory group (Riegler et al. 1981) has a width even smaller than 2.5-keV instrumental resolution. Such a narrow feature corresponds to a temperature in the annihilation region of $<5 \times 10^4$ K and requires that the positrons thermalize prior to forming the positronium from which annihilation subsequently ensues. Positrons are efficiently thermalized in a medium with a large fractional ionization, $n_e/n > 0.1$ (Bussard et al. 1979). Thus, the annihilation occurs in ionized gas.

The strongest constraints on the annihilation source were provided by observations beginning in 1980 in which the annihilation line was absent. In less than 6 months, the source had turned off (Riegler et al. 1981). Thus, the dimension of the region emitting the annihilation line is less than $\sim 10^{18}$ cm, and we are dealing with a restricted emission region. To stop and thermalize positrons in less than a half-year constrains the region to a density of $> 10^5$ cm^{-3}. This is reminiscent of the conditions in the ionized gas in the inner parsec. Furthermore, the ratio of annihilation-line luminosity to $E > 511$ keV continuum luminosity ($\sim 30\%$) suggests an efficiency of positron production that can be provided only by photon-photon pair production (Lingenfelter & Ramaty 1982). This in turn implies the existence of an extremely luminous ($> 10^{38}$ erg s^{-1}) but very compact ($<2 \times 10^8$ cm) central source, viz. a black hole. But it does not imply a particularly massive black hole: for an isotropic source the mass is $M \sim 10^2\ M_\odot$ (Lingenfelter & Ramaty 1983), whereas for a beamed source the mass could be $\sim 10^6\ M_\odot$ (Burns 1983).

The angular resolution of the gamma-ray detectors was sufficient only to place the source of the annihilation line within $4°$ of the galatic center. The association of this source with the galactic center is only circumstantial. Nevertheless, as this is a unique object—the brightest gamma-ray source in the Galaxy and the only source of annihilation-line radiation—lying in a unique direction, it cannot be ignored; and, indeed, it may prove to be the key to the galactic center.

8. SUMMARY

On the largest scales considered here (tens of parsecs), the inner Galaxy is characterized by a strong sense of asymmetry. The thermal and nonthermal radio continuum emission, as well as the IR emission from heated dust and the molecular and atomic spectral line emission from many species, are all preferentially disposed to positive longitudes with respect to Sgr A. Furthermore, the peculiar morphology of the distribution of several constituents unequivocally spatially associated with Sgr A—especially the radio continuum arc—appears to be intimately related to the presence of one molecular cloud, the $+50\,\mathrm{km\,s^{-1}}$ cloud. The manner in which the radio continuum emission traces the periphery of this 2×10^6–$2 \times 10^7\,M_\odot$ cloud is unique in the Galaxy. Sgr A itself is but part of the much larger radio structure seen outlined against one edge of the $+50\,\mathrm{km\,s^{-1}}$ cloud.

Closer to Sgr A, within 3–5 pc, H I and molecular CO observations at $|v| \geqslant 70\,\mathrm{km\,s^{-1}}$ exhibit a double-lobed structure, which suggests that a ring of material is being seen nearly edge-on. The ring appears to be rotating in the sense of galactic rotation at $\sim 110\,\mathrm{km\,s^{-1}}$ at a projected radius of 3.2 pc. This double-lobed structure is also manifested in $\lambda 50$–100 μm continuum radiation and in [O I] and H_2 IR spectral observations, although the apparent radius of the IR ring is ~ 2 pc. It is likely that the IR emphasizes the heated inner edge of the larger molecular ring.

Interior to 2 pc is a region of increasing stellar density, decreasing dust density, and clumpy ionized gas. A summary of the properties of this region is given in Table 1. Some of these properties appear as complementary parts of a single picture; others are disparate. For example, the gas ring must be a transient structure unless a mechanism exists for its maintenance and stability. The mass outflow (stellar wind) of $\sim 10^{-3}\,M_\odot\,\mathrm{yr^{-1}}$ associated with the region of broad He I and Bα emission will provide an adequate stabilizing pressure, but at the expense of causing the gas at the interface

Table 1   The inner 2 parsecs

Parameter	Value	Observation	Reference[a]
Mass			
Stellar	$(2$–$5) \times 10^6\,M_\odot$	[O I]	1
Ionized gas	$70\,M_\odot$	15-GHz map	2
Luminosity			
Bolometric	$(1$–$3) \times 10^7\,L_\odot$	IR	3
$S(\mathrm{Lyc})$: Flux	$2 \times 10^{50}\,\mathrm{s^{-1}}$	15-GHz map	2
Lyc	$\geqslant 1.3 \times 10^6\,L_\odot$	15-GHz map	2
Pressure in the ring			
$n_H T$	$10^{7 \pm 0.2}\,\mathrm{cm^{-3}\,K}$	[O I]	1
$n_H T$	$10^{7.3}\,\mathrm{cm^{-3}\,K}$	[H II]	2
Pressure on the ring			
Mass loss "$n_H T$"	$10^{7.8}\,\mathrm{cm^{-3}\,K}$	[H_2]	4

[a] References:
  1. Genzel et al. (1984)
  2. From Figure 14 via $M = (m_H/\alpha n_e)\,S(\mathrm{Lyc})$
  3. Becklin et al. (1982)
  4. Gatley et al. (1984) via $(n_H T) = \dot{M}v/4\pi R^2 k$.

region to be Rayleigh-Taylor unstable; this could, of course, be the source of the observed clumpiness along the ring. On the other hand, the observation that the ionization along the ring is incomplete in azimuth means that this simple picture is not sufficient (cf. Brown 1984).

The remarkable observations of Sgr A in recent years have provided answers to a wealth of questions but have not yet, unfortunately, given us the answer to the one crucial question: What is the object at the galactic center? Part of the difficulty lies in the ambiguity of *the* galactic center. Here it is important to emphasize that the kinematical center of the Galaxy, as measured by *stellar* kinematics, has not been established. So it is not entirely clear what object we should look at. But whatever the object at the center—possibilities include, but are not limited to, IRS 16 and Sgr A*—it must have a luminosity of $\sim 10^7 \, L_\odot$, a mass-to-luminosity ratio of (M/L) ~ 0.05–0.5 (Table 1), and a temperature not in excess of 35,000 K.

These properties suggest to van Buren (1978), Rieke & Lebofsky (1982), Oort (1983), and others that the nucleus of the Milky Way is a "starburst" nucleus in which the burst of star formation occurred sufficiently long ago that all stars formed in the burst with spectral types earlier than O8 (and hence temperatures higher than 35,000 K) have completed their evolution. The ~ 100 O8 stars remaining provide the ionizing luminosity of Sgr A but are undetectably faint at 2.2 μm. We have no information as to their present distribution.

Alternatively, we could be dealing with a single energetic object, a black hole (Rees 1982), or perhaps a cooler and somewhat less luminous superstar similar to R136, the central source of 30 Doradus (Cassinelli et al. 1981). If so, then the energetic, time-variable source of positron annihilation radiation, which necessarily demands a black hole for its production (Lingenfelter & Ramaty 1983), is naturally included in the picture. And most importantly, our Galaxy is then one of the many (if not all) spiral galaxies that according to Keel (1983) exhibit the nuclear activity common to Seyferts and active radio galaxies, but it does so at a much lower luminosity. A determination of the extent to which this correspondence applies to Sgr A is the challenge and the opportunity of the future.

ACKNOWLEDGMENT

We gratefully acknowledge the generosity of those colleagues who have communicated their current or unpublished results, in particular Drs. Becklin, Burns, Gatley, Geballe, Genzel, Henry, Lo, Sandqvist, van Gorkom, and Werner; special thanks are due to Mark Morris and Farhad Yusef-Zadeh for permission to reproduce their wonderful map as Figure 1. We also thank R. D. Ekers, R. H. Sanders, and Professor J. H. Oort for discussions of the inner Galaxy material. The collaborative effort of HSL with W. B. Burton, source of many of the ideas in Sections 2 and 3, is supported by Grant No. 008.82 of the North Atlantic Treaty Organization.

Literature Cited

Allen, D. A., Hyland, A. R., Jones, T. J. 1983. *MNRAS* 204:1145
Backer, D. C. 1982. In *Extragalactic Radio Sources*, ed. D. S. Heeschen, C. M. Wade, p. 389. Dordrecht: Reidel. 490 pp.
Bahcall, J. N., Wolfe, R. A. 1976. *Ap. J.* 209:214
Bailey, M. E. 1980. *MNRAS* 190:217
Balick, B., Brown, R. L. 1974. *Ap. J.* 194:265
Bally, J., Joyce, R. R., Scoville, N. Z. 1979. *Ap. J.* 229:917
Bania, T. M. 1977. *Ap. J.* 216:381
Bania, T. M. 1980. *Ap. J.* 242:95
Becklin, E. E., Gatley, I., Werner, M. W. 1982. *Ap. J.* 258:134
Becklin, E. E., Neugebauer, G. 1968. *Ap. J.* 151:145
Becklin, E. E., Neugebauer, G. 1975. *Ap. J. Lett.* 200:L71
Bieging, J., Downes, D., Wilson, T. L., Martin, A. H. M., Güsten, R. 1980. *Astron. Astrophys. Suppl.* 42:163
Blitz, L., Mark, J. W.-K., Sinha, R. P. 1981. *Nature* 290:120
Brown, R. L. 1984. In preparation
Brown, R. L., Johnston, K. J. 1983. *Ap. J. Lett.* 268:L85
Brown, R. L., Lo, K. Y. 1982. *Ap. J.* 253:108
Brown, R. L., Lo, K. Y., Johnston, K. J. 1978. *Astron. J.* 83:1594
Burns, M. L. In *Positron-Electron Pairs in Astrophysics*, ed. M. L. Burns, A. K. Harding, R. Ramaty, p. 267. New York: Am. Inst. Phys. 447 pp.
Burton, W. B., Liszt, H. S. 1978. *Ap. J.* 225:815
Burton, W. B., Liszt, H. S. 1983a. *Astron. Astrophys. Suppl.* 52:63
Burton, W. B., Liszt, H. S. 1983b. In *Surveys of the Southern Galaxy*, ed. W. B. Burton, F. P. Israel, pp. 149–58. Dordrecht: Reidel. 309 pp.
Bussard, R. W., Ramaty, R., Drachman, R. J. 1979. *Ap. J.* 228:928
Cassinelli, J. P., Mathis, J. S., Savage, B. D. 1981. *Science* 212:1497
Cohen, R. D., Davies, R. 1979. *MNRAS* 186:453
Cohen, R. D., Few, R. W. 1976. *MNRAS* 176:495
Dent, W. A., Werner, M. W., Gatley, I., Becklin, E. E., Hildebrand, R. H., et al. 1982. In *The Galactic Center*, ed. G. R.

Riegler, R. D. Blandford, pp. 33–41. New York: Am. Inst. Phys. 216 pp.
Downes, D., Goss, W. M., Schwarz, U. J., Wouterloot, J. G. A. 1978. *Astron. Astrophys. Suppl.* 35:1
Downes, D., Maxwell, A. 1966. *Ap. J.* 146:653
Ekers, R. D., van Gorkom, J. H., Schwarz, U. J., Goss, W. M. 1983. *Astron. Astrophys.* 122:143
Fukui, Y. 1980. In *Interstellar Molecules*, ed. B. H. Andrew, pp. 209–13. Dordrecht: Reidel, 704 pp.
Fukui, Y., Iguchi, T., Kaifu, N., Chikada, Y., Morimoto, M., et al. 1977. *Publ. Astron. Soc. Jpn.* 29:643
Fukui, Y., Kaifu, N., Morimoto, M., Miyaji, T. 1980. *Ap. J.* 241:147
Fukui, Y., Ozawa, H., Deguchi, S., Suzuki, H. 1982. In *The Galactic Center*, ed. G. R. Riegler, R. D. Blandford, pp. 18–24. New York: Am. Inst. Phys. 216 pp.
Gatley, I., Hyland, A. R., Jones, T. J., Beattie, D. H., Lee, T. J. 1984. In preparation
Geballe, T. R., Persson, S. E., Lacy, J. H., Neugebauer, G., Beck, S. C. 1982. In *The Galactic Center*, ed. G. R. Riegler, R. D. Blandford, pp. 60–66. New York: Am. Inst. Phys. 216 pp.
Geballe, T. R., Krisciunas, K. L., Lee, T. J., Gatley, I., Wade, R., et al. 1984. In preparation
Genzel, R., Watson, D. M., Townes, C. H., Dinerstein, H. L., Hollenbach, D., et al. 1984. *Ap. J.* 276:551
Gurzadyan, V. G. 1984. Preprint
Güsten R. 1982. In *The Galactic Center*, ed. G. R. Riegler, R. D. Blandford, pp. 9–11. New York: Am. Inst. Phys. 216 pp.
Güsten, R., Downes, D. 1980. *Astron. Astrophys.* 87:6
Güsten, R., Downes, D. 1981. *Astron. Astrophys.* 99:27
Güsten, R., Henkel, C. 1983. *Astron. Astrophys.* 126:136
Güsten, R., Walmsley, C. M., Pauls, T. 1981. *Astron. Astrophys.* 103:197
Hall, D. N. G., Kleinmann, S. G., Scoville, N. Z. 1982. *Ap. J. Lett.* 260:L53
Henry, J. P., DePoy, D. L., Becklin, E. E. 1984. In preparation
Hildebrand, R. H., Whitcomb, S. E., Winston, R., Stiening, R. F., Harper, D. A.,

Moseley, S. H. 1978. *Ap. J. Lett.* 219:L101

Keel, W. C. 1983. *Ap. J.* 269:466

Kerr, F. J. 1968. In *Radio Astronomy and the Galactic System, IAU Symp. No. 31*, ed. H. van Woerden, pp. 239–52. Dordrecht: Reidel. 501 pp.

Lacy, J. H., Baas, F., Townes, C. H., Geballe, T. R. 1979. *Ap. J. Lett.* 227:L17

Lacy, J. H., Townes, C. H., Geballe, T. R., Hollenbach, D. J. 1980. *Ap. J.* 241:132

Lacy, J. H., Townes, C. H., Hollenbach, D. J. 1982. *Ap. J.* 262:120

Lake, G. R., Norman, C. 1982. In *The Galactic Center*, ed. G. R. Riegler, R. D. Blandford, pp. 189–93. New York: Am. Inst. Phys. 216 pp.

Lester, D. F., Werner, M. W., Storey, J. W. V., Watson, D. M., Townes, C. H. 1981. *Ap. J. Lett.* 248:L109

Lingenfelter, R. E., Ramaty, R. 1982. In *The Galactic Center*, ed. G. R. Riegler, R. D. Blandford, p. 148. New York: Am. Inst. Phys. 216 pp.

Lingenfelter, R. E., Ramaty, R. 1983. In *Positron-Electron Pairs in Astrophysics*, ed. M. L. Burns, A. K. Harding, R. Ramaty, p. 267. New York: Am. Inst. Phys. 447 pp.

Liszt, H. S., Burton, W. B. 1978. *Ap. J.* 226:790

Liszt, H. S., Burton, W. B. 1980. *Ap. J.* 236:779

Liszt, H. S., Burton, W. B. 1984. *Ap. J.* Submitted for publication

Liszt, H. S., Burton, W. B., Sanders, R. H., Scoville, N. Z. 1977. *Ap. J.* 213:38

Liszt, H. S., Burton, W. B., van der Hulst, J. M., Ondrechen, M. 1983. *Astron. Astrophys.* 126:341

Liszt, H. S., Sanders, R. H., Burton, W. B. 1975. *Ap. J.* 198:537

Lo, K. Y. 1984. Preprint

Lo, K. Y., Claussen, M. J. 1983. *Nature* 306:647

Lo, K. Y., Cohen, M. H., Readhead, A. C. S., Backer, D. C. 1981. *Ap. J.* 249:504

MacCallum, C. J., Leventhal, M. 1983. In *Positron-Electron Pairs in Astrophysics*, ed. M. L. Burns, A. K. Harding, R. Ramaty, p. 211. New York: Am. Inst. Phys. 447 pp.

McGee, R. X., Brooks, J. W., Sinclair, M. W., Batchelor, R. A. 1970. *Aust. J. Phys.* 23:777

Nadeau, D., Neugebauer, G., Matthews, K., Geballe, T. R. 1981. *Astron. J.* 86:561

Neugebauer, G., Becklin, E. E., Matthews, K., Wynn-Williams, C. G. 1976. *Ap. J. Lett.* 205:L139

Oort, J. H. 1974. In *The Formation and Dynamics of Galaxies, IAU Symp. No. 58*, ed. J. R. Shakeshaft, pp. 378–82. Dordrecht: Reidel. 441 pp.

Oort, J. H. 1977. *Ann. Rev. Astron. Astrophys.* 15:295

Oort, J. H. 1983. In *The Milky Way Galaxy, IAU Symp. No. 106*. In press

Pauls, T. A. 1980. In *Radio Recombination Lines*, ed. P. A. Shaver, p. 159. Dordrecht: Reidel. 284 pp.

Pauls, T., Downes, D., Mezger, P. G. 1976. *Astron. Astrophys.* 46:407

Rees, M. J. 1982. In *The Galactic Center*, ed. G. R. Riegler, R. D. Blandford, p. 166. New York: Am. Inst. Phys. 216 pp.

Reynolds, S. P., McKee, C. F. 1980. *Ap. J.* 239:893

Riegler, G. R., Ling, J. C., Mahoney, W. A., Wheaton, W. A., Willett, J. B., et al. 1981. *Ap. J. Lett.* 248:L13

Rieke, G. H., Lebofsky, M. J. 1982. In *The Galactic Center*, ed. G. R. Riegler, R. D. Blandford, p. 194. New York: Am. Inst. Phys. 216 pp.

Rieke, G. H., Telesco, C. M., Harper, D. A. 1978. *Ap. J.* 220:556

Rodriguez, L. F., Chaisson, E. J. 1979. *Ap. J.* 228:734

Sanders, R. H. 1981. *Nature* 294:427

Sanders, R. H., Wrixon, G. T. 1974. *Astron. Astrophys.* 33:9

Sandqvist, Aa. 1982. In *The Galactic Center*, ed. G. R. Riegler, R. D. Blandford, pp. 12–17. New York: Am. Inst. Phys. 216 pp.

Schwarz, U. J., Ekers, R. D., Goss, W. M. 1983. *Astron. Astrophys.* 110:100

Schwarz, U. J., Shaver, P. A., Ekers, R. D. 1977. *Astron. Astrophys.* 54:863

Scoville, N. Z. 1972. *Ap. J. Lett.* 175:L127

Sinha, R. P. 1979. In *The Large-Scale Characteristics of the Galaxy, IAU Symp. No. 84*, ed. W. B. Burton, pp. 341–42. Dordrecht: Reidel. 611 pp.

Storey, J. W. V., Allen, D. A. 1983. *MNRAS* 204:1153

van Buren, H. G. 1978. *Astron. Astrophys.* 70:707

van der Kruit, P. C. 1970. *Astron. Astrophys.* 4:462

van Gorkom, J. H., Schwarz, U. J., Bregman, J. D. 1984. In preparation

Watson, M. G., Willingale, R., Grindlay, J. E., Hertz, P. 1981. *Ap. J.* 250:142

Whiteoak, J. B., Gardner, F. F. 1979. *MNRAS* 188:445

Whiteoak, J. B., Gardner, F. F., Pankonin, V. 1983. *MNRAS* 202:11p

Whiteoak, J. B., Rogstad, D. H., Lockhart, I. A. 1974. *Astron. Astrophys.* 36:245

Williams, P. M., Adams, D. J., Arakaki, S., Beattie, D. H., Born, J., et al. 1980. *MNRAS* 192:25p

Wollman, E. R., Smith, H. A., Larson, H. P. 1982. *Ap. J.* 258:506

Yusef-Zadeh, F., Morris, M., Chance, D. 1984. *Ap. J. Lett.* Submitted for publication

Ann. Rev. Astron. Astrophys. 1984. 22: 267–89

CORONAL MASS EJECTIONS

William J. Wagner

High Altitude Observatory, National Center for Atmospheric Research, Boulder, Colorado 80307

1. Introduction

Sudden expulsions of dense clouds of plasma from the outer atmosphere of the Sun, termed "coronal mass ejections" (CMEs), are the focus of intense observational and theoretical efforts. CMEs are a type of coronal transient, the general name given the disruption of coronal structure. The mass ejection phenomenon, known for barely a dozen years, is rapidly generating interest within the areas of solar, stellar, and solar-terrestrial physics.

Flares, eruptive prominences, and nonequilibrium magnetic field configurations are among the postulated origins of CMEs. Not surprisingly, large-scale magnetic fields are also related to mass ejections. The slowly evolving coronal density structure is now recognized as marking the location of fields and, ultimately, the instantaneous state of the solar dynamo. A new class of CMEs has been found that perhaps shows the continuous renewal of surface fields by interior circulation. Understanding CMEs, therefore, may lead to insights about flares and other solar phenomena.

The CME may be described as an ejection of magnetized plasma out of the gravitational potential well of a central star (56). Thus, phenomena analogous to CMEs should also be occurring in other stars whose magnetic fields and gravitational potentials are Sun-like. Those techniques at radio wavelengths that allow ground-based CME observations at large elongation angles may have applicability to the detection of stellar mass ejections.

Finally, the importance of understanding CME effects on the interplanetary medium has long been appreciated. Up to 2×10^{16} g may be injected into the solar wind by a CME, and the total energy of the event (kinetic, enthalpic, and magnetic) may exceed 10^{32} erg. Long-standing interest, together with the recent availability of space probes, means that the

267

0066–4146/84/0915–0267$02.00

most immediate impact of CME research is certain to be in the area of solar-terrestrial physics.

Progress toward an understanding of CMEs has been steady since their discovery (32, 60, 99). CME results from the *Skylab*-era coronagraphs near the minimum of the solar activity cycle were interpreted by MacQueen (58). Dryer (12) reviewed data and models of mass ejections, particularly as related to shock waves in the solar wind. Additional useful compendia exist on the CME physical properties of radio emission (15, 91, 101) and mass ejection from the Sun (33, 80), and on the CME as a magnetohydrodynamic phenomenon (3).

Information on the behavior of CMEs during the maximum of the solar cycle is now available. Observations are being provided by three new telescopes that detect the light scattered from free electrons: the Naval Research Laboratories Solwind on satellite *P78-1* [with field of view from 3 to 10 radii from Sun center $(R_\odot)$], the High Altitude Observatory (HAO) Coronagraph/Polarimeter on the *Solar Maximum Mission* spacecraft (1.5–6 $R_\odot$), and the HAO Mark III K-coronameter (1.2–2.2 $R_\odot$) at Mauna Loa Observatory. Recent work from these experiment systems constitutes the principal input for this review.

The origin of CMEs is not clearly understood. In the next section, some models are briefly described. Section 3 reports information derived from CME observations, and Section 4 reviews the implications for the physics of CMEs of other solar phenomena. Section 5 offers suggestions for solar-terrestial physics based on the preceding discussion. Finally, in Section 6, future directions of research are suggested from an observational viewpoint, and comments are offered on the most promising model for CME initiation.

2.   Models of the CME Origin

Three general approaches to a theoretical description of the origin of CMEs may be found in the literature (56). The first set of models assumes that the CME is the result of an energetic input pressure pulse that consists of increased temperature, magnetic flux, or mass, such as might accompany a flare or eruptive prominence. It solves the standard magnetohydrodynamic equations (55, Equations 1–5) numerically to predict the reaction of the corona. Calculations under this first interpretation of the CME produce a nonlinear wavelike coronal response that some authors have identified with the CME (13, 96, 113). This identification has been questioned by others on the basis of the density predictions (90) and because of the observed separation of the flare blast wave from the CME (102).

A second approach uses analytic approximations to model possibly relevant magnetic structures. Envisaged are magnetic flux loops with twist

(65) or with axial current (2), buoyant magnetic loops immersed in the ambient coronal total pressure field (115, 116) or in the expanding solar wind (59), and closed magnetic regions driven by pressure gradients resulting from field-line reconnection in arcades (73) or from diamagnetic plasmoids (74).

Finally, Low (56) has found exact analytic solutions for the magnetohydrodynamic equations and has identified the CME with a large-scale, fully developed magnetohydrodynamic outflow. The mass ejection is initiated when magnetic tension and gravity no longer balance the natural tendency for the magnetized plasma to expand. The CME is the result of evolution into magnetostatic nonequilibrium in this interpretation.

3.   Observations of the CME

In this section, empirical information about the CME is presented without consideration of its relation to other phenomena. An attempt is made to draw a picture of CME characteristics and to establish a prototype using data for several hundred events from six different coronagraphs.

3.1   PHENOMENOLOGICAL DESCRIPTION   The CME is best visualized as a framework that transports mass. The mass is distributed on the frame in various ways from event to event and at different times in the progress of an individual event. The forms of *Skylab* CME events near the minimum of the solar activity cycle were categorized as follows: one third were called loops; one fourth, clouds or amorphous plasmoids; and the remainder were not given a special name (58). In these 1973 data, the CMEs were superimposed on the quiescent coronal structure at low latitudes, giving relatively poor CME-to-background contrast. With a wider latitude distribution of events, almost one half of the Solwind 1979–81 CMEs (87) seen higher in the corona (between 3–10 $R_\odot$) have the appearance of loops or of spikes (loops that have been overwhelmed by the radiance of the dust corona). Inspection of *Solar Maximum* data by this reviewer shows that lower in the corona (1.8–3.0 $R_\odot$) fully 80% of the Coronagraph/Polarimeter CMEs are of the type that is led by a bubble or loop shape. About 14% are clouds and 6% seem to resemble radial tongues. The most common type, the loop/bubble CME (similar to that shown in Figure 1), is addressed in this review.

The strength of the CME density enhancement relative to the electron corona appears to evolve. Seen earliest in its evolution by K-coronameters (20, 21), the CME may first appear at 1.2 $R_\odot$ as a fully formed bright loop (generally characteristic of flare events) or, for events associated with only an eruptive prominence, as a dark bubble-shaped depletion protruding into a preexisting streamer (55). In the lower corona (1.2–2.2 $R_\odot$), lateral and radial speeds are comparable, as are width and height (57, 59), but lateral

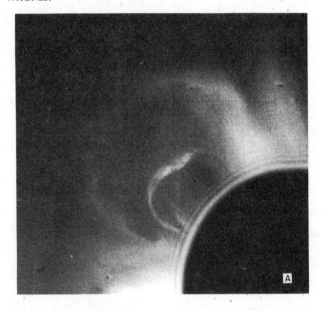

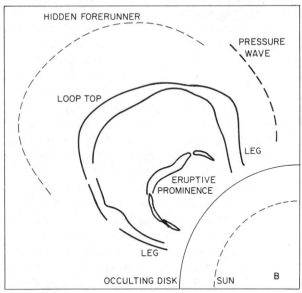

Figure 1   A coronal mass ejection is described by Low (56) as an expulsion of magnetized plasma out of the gravitational potential well of a central star. This CME, with underlying eruptive prominence (*A*), was recorded by the Coronagraph/Polarimeter on the *Solar Maximum Mission* spacecraft on 14 April 1980. The dark occulting disk has a radius of 1.51 $R_\odot$. Long-lived coronal structure that was visible prior to this loop CME extends vertically above the disk and horizontally to the left. (*B*) A sketch of salient CME features.

expansion soon ceases. As the eruptive-prominence-associated CME accelerates and increases its altitude, bright excess mass legs develop from the bottom upward at the cavity edges until a complete loop has been formed at about 1.8 $R_\odot$. In the higher corona, after achieving a constant speed, the loop top tends to fade again, leaving two dense outer legs or spikes. Such two-legged CMEs are especially common above 3 $R_\odot$ in the Solwind field of view. Indeed, for eruptive prominence events, measurements show that the fractional electron density enhancement of the moving loop top above 2 $R_\odot$ decreases to considerably less than that of the two stationary legs (90).

Within the underdense cavity enclosed by the loop, there usually appear clouds of plasma. Concentric loop events ("filled bottles") are moderately common; this reviewer generally finds the filled bottle to be an "unassociated" CME (see below), insofar as it does not appear to occur with any significant near-surface event. For CMEs associated with eruptive prominences, a few of the entrained plasmoids are cool Hα material (37) or remnant clouds of rapidly ionizing prominence matter, as exemplified by the brightest arch in Figure 1a.

The material from which the CME loop forms seems to have been present in the preexisting coronal structure (34). Dense coronal X-ray loops that have erupted have been shown on at least one occasion (79) to contain mass equal to that in the accompanying CME loop. Similarly, a symmetrical three-dimensional volume of corona seen up to 2.34 $R_\odot$ overlying a flare-CME site could have provided about one half of the mass of another CME, with the remainder coming from the unobserved region below the 1.2-$R_\odot$ occulting disk (19).

Because the loop preexists as coronal material, it is certainly at coronal, rather than chromospheric or prominence, temperatures (34, 37, 75, 84). More precise values of the electron temperature and its spatial distribution do not appear to be needed for any of the current models of CMEs. In contrast to temperature, the CME electron density distribution (discussed next) is controversial and poorly understood.

3.2 EMPIRICAL DENSITY DISTRIBUTIONS A key to understanding mass ejections lies in their shape. Discussion continues on whether the observed CMEs are planar or bubble shaped. With a given CME geometry, we can calculate an empirical electron density distribution and compare it with models. A two-dimensional planar loop seems to agree only with the model of an expanding magnetic flux tube (2, 65). However, a spherical shock (resembling a bubble with significant line-of-sight depth) can be initiated either by a pressure pulse (13) or by rapid passage through the corona of a magnetic tube (or an entire arcade of flux; 28, 77). Early images of CMEs evoked the concept of a "loop" transient (30, 35), a term that may have begged the issue and prevented serious consideration of the bubble geometry (33, 58). Because the orientation of preevent fields was apparently nonrandom (11), it was argued (100) that a distribution of viewing aspects had not been seen. For this reason, no examples were available of edge-on planar loop CMEs.

The most direct measure of CME shape would be provided by careful studies of the percentage of white-light polarization and its dependence on the distance of the scattering center from the plane of the sky; such measurements have rarely been tried. The only attempt (9) to carry out such an analysis for a moving mass ejection (deduced to be bubble shaped) is somewhat compromised by approximations. For a source extended in depth, careful computer modeling of the changing polarization along the line of sight is required for comparison with precise polarization data.

Arguments for the CME being three dimensional follow from the construction of empirical models based on comparisons of emission measures in white light versus radio (23, 27) and on deconvolutions of white-light voids (19, 20, 37). If "magnetic bottles," clouds (6) detected in situ by multiple spacecraft between 1–2 AU, are the interplanetary signature of mass ejections, their shape in the ecliptic, spanning 30° longitude by 0.5 AU radial extent, must be three dimensional. Noncompressive density enhancements in the solar wind (29) may also offer such evidence.

The Solwind coronagraph observed an Earth-directed mass ejection, associated with a small flare and a large eruptive prominence at Sun center, as a bright halo around the occulting disk. This CME was interpreted as a three-dimensional conical shape whose interior was filled with material (38). An alternative model was that of a hollow conical shell with a flattened and dimmed front; this would appear as in Figure 1a, when seen in profile off the limb of the Sun.

In summary, the crucial test for determining the distribution of material in a CME—polarization analysis—has not been done. However, there is little evidence that CMEs are planar loops. The search for clues that would differentiate between three-dimensional bubble models must be continued.

3.3   KINEMATICS   Previous analyses have provided insufficient consistency and precision in CME velocity histories to allow a clear definition of the driving forces or their duration. Above 2 $R_\odot$, flare CMEs have higher speeds (measuring 300–1200 km s^{-1}) than do CMEs associated with eruptive prominences (less than about 600 km s^{-1}) (31, 33). The class of CMEs that are unassociated with other activity average 100 km s^{-1} slower than eruptive events in the high corona. With K-coronameter observations of the inner corona, however, an important pattern appears in CME kinematics (59). Unlike constant-speed flare CMEs, ejections with prominences show substantial observable acceleration in the lower corona. This acceleration (sometimes > 50 m s^{-2}) decreases with height until approximately constant speed is reached, usually above 2–3 $R_\odot$. Only rarely are events seen with appreciable acceleration at 10 $R_\odot$ (87).

3.4   CME FREQUENCY   The rate of occurrence of CMEs is important for two reasons. Firstly, the flux of mass, energy, angular momentum, and magnetic field from the Sun is partly determined by the frequency of occurrence of coronal mass ejections. Secondly, CME frequency is an extremely interesting characteristic by itself. Changes of frequency through the solar activity cycle might be expected to follow those of the proposed causal mechanisms of mass ejections. Also, CME rates, especially in the ecliptic plane (Section 4.2), should match those of hypothesized effects of coronal mass ejections at the Earth.

Establishing a rate for CMEs is a difficult task that must consider the relative sensitivity of individual coronagraphs, together with their duty cycles. Today, the most thoroughly documented estimate (40) gives rates of CMEs occurring from the limb of the Sun at 0.74 day^{-1} for the *Skylab* period in 1973, and 0.90 day^{-1} for the 1980 Coronagraph/Polarimeter epoch. This is a surprisingly small 20% frequency change over most of a sunspot cycle for the supposedly activity-related CMEs, but its magnitude may be explained in two ways. First, while the flare rate changed by factors of 6 (and surge/spray by factors of 4) for 1973 versus 1980, the eruptive prominence frequency was constant. A count by association over the activity cycle is shown in Figure 2 [compiled from data of (40) and (83)]. Noted in Figure 2 is the addition of flare- and eruptive-prominence-associated CMEs to a rather large and constant population of CMEs unassociated with near-surface events. The variable-rate flare events, a small component of the entire CME population, have little overall effect on the net CME frequency change from 1973 to 1980.

The second explanation for the nearly constant CME frequency over the cycle relates to the question of why flares are not the most common Hα event associated with CMEs, or why the flare-associated CME rate did not also vary by a factor of 6 as did the flare rate. Flares cluster in time and space, while eruptive prominences and CMEs do not. An absence was noted (41, 105) of subsequent mass ejections when additional flares followed a prior flare-CME. Steinolfson (97) suggests, in the context of the shock model, that later impulses may occur, but he shows that the ensuing density enhancement and, hence, the visibility of later CMEs will be decreased as a result of the second shock propagating into a medium of disturbed magnetic field following the first transient. Indeed, transient coronal holes are observed to last about 10 hr astride the sites of X-ray long-decay events (78). These holes presumably mark high-speed stream flows from field regions opened for approximately 10 hr by CMEs (see also 29). Such a 10-hr suppression period has been confirmed statistically for *Skylab* and *Solar Maximum* CMEs (103). Evidence is offered (though sparse) that the delay results not from outflow conditions, but from the slow reformation of prominence or magnetic field systems. The extrapolation of CME rates from solar minimum to maximum (36), based on a 500% increase in Zürich sunspot number, was vitiated by the then-unknown recovery period, necessary prior to a second CME.

Post-CME outflow may inhibit the visibility of repetitive flare disturbances in the corona (based on the CME-as-impulse interpretation). Another explanation is that CME production requires a properly restored corona on the margin of magnetic equilibrium. Observations cannot yet discriminate between these explanations. But if suppression is caused by outflow, shock disturbances without observed CMEs and perhaps without type II radio bursts should frequently emanate from flares at the maximum of the activity cycle, thereby complicating correlations with terrestrial and other effects.

3.5 FRINGES OF THE CME   Important information may be gleaned from investigations of the leading spatial fringes of coronal mass ejections. A

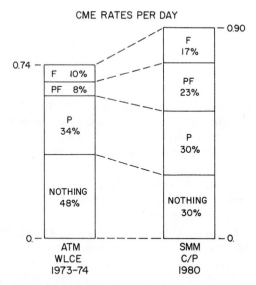

Figure 2   The frequency of coronal mass ejections increases by about 20% as the Sun evolves from the minimum to the maximum of the sunspot cycle, chiefly as a result of enhanced flare activity. Shown above is a breakdown of CMEs by association with flares (F), eruptive prominences (P), prominences and flares (PF), or no near-surface event (NOTHING). Data are developed from those of Sawyer et al. (83) and Hundhausen et al. (40). Rates refer to CMEs visible at the limb; whole-Sun rates are about twice as high. Estimated behind-the-limb activity has also been included.

CME traveling faster than ~400 km s^{-1} should be supersonic in the expanding solar wind. The dense mass ejection might thus be led by a driven shock wave. Some models of CMEs interpret the mass ejection itself as a shock process. Hence, evidence should be sought of shock phenomena at the leading edge of the CME. Finally, an understanding of this fringe region will help in interplanetary studies also. Shocks leaving the Sun (such as a flare blast wave and a degenerate driven shock from a CME) may be more easily distinguished from effects that have been generated in situ by virtue of interactions at fractions of 1 AU.

3.5.1 *Mass forerunners*   The greatest progress to date on the fringe phenomena that lead a CME was provided by the discovery of forerunners in coronagraph data during a *Skylab* workshop (44). Representing a region of density enhanced slightly over the pre-CME corona, the forerunner was defined (44) to be the excess mass wedge from the CME leading edge to the distance where the added brightness of the excess mass blended into the background corona. The average ratio of the excess mass in forerunners to total event mass was 0.15. The outer boundary of the forerunners, sketched in Figure 1b, lay 0.7–3.0 $R_\odot$ (average 1.3 $R_\odot$) ahead of the highest outer edge of the CMEs, with somewhat less offset at the sides of the mass ejection. A forerunner was found with each CME for which reasonable data existed (44). In a selection of control periods that were examined, no forerunner-type event was seen without an accompanying CME. Forerunners had also been noted in two previous events (16, over part of the CME; 35). In *Solar Maximum Mission* images, one forerunner was found (24); a second reported Coronagraph/Polarimeter forerunner (95) did not resemble the accepted description.

Forerunner behavior has thus far not been shown to resemble that of a shock. The forerunner offset from event to event is clearly not a function of the CME speed or mass. Furthermore, the 21 *Skylab* CMEs with fore-runners showed speeds that ranged from 50 to greater than 1000 km s^{-1}. For the seven events that were measurable, offset distances were constant in time (or slightly decreased).

3.5.2 *Outriding pressure pulses*   In addition to the forerunner, a magneto-hydrodynamic or pressure signal that distorts streamer structure at the fringe of the CME is also observed above or outside the typical CME (30, 35), most noticeably above 2 $R_\odot$. Numerous examples of this signal are known to exist in Solwind, *Skylab*, and *Solar Maximum* data. It is not clear whether this outriding bow wave (Figure 1b) is a different phenomenon from the forerunner, or whether it corresponds to the toe of the excess mass wedge (whose location must be defined arbitrarily by the detection threshold at the 2σ level). If an arc visible to Gary et al. (24) is indeed the pressure pulse, it is probably not intimately related to the forerunner in that event, as judged by the dissimilar position angles and propagation speeds. Few studies have been made comparing the speed of distortions due to the pressure pulse to the speed of the forerunner, nor are reports available detailing the restoration characteristics of the structures. With such information, it might be possible to distinguish single outriding impulses from steady forerunner material flows. Crucial low-corona data from K-coronameters on pressure pulses, forerunners, and precursors (Section 3.6) are thus far missing.

3.6 CME PRECURSORS   Whereas the outriding pressure wave and the forerunner are frequently seen to lead the CME in space, other less well-established effects are reported to precede the CME in time. With such precursors, the effects of processes that are physically unrelated may appear sequentially, rather than superposed as in the fully developed CME. This underexplored problem area may provide a physical understanding not only of CMEs, but of associated phenomena that include flares or eruptive prominences as well.

An increase (by a factor of 2.5) is reported (47) in the production rate of type III radio bursts during the 10 to 5 hr before the first observations of CMEs (with less lead time for faster events; 45). A commencement of soft X-ray and radio noise storm signals 30–10 min before a filament disruption/CME start (54), and a cessation of type I emission (23) 10 min before another CME, are offered as other harbingers of mass ejections.

In further work on a subset of 16 of the Jackson & Hildner (44) forerunner events, Jackson (43) suggested that for 12 cases density enhancements were visible *before* any accompanying Hα prominence eruption or flare. In 7 events out of the 16 with offsets measurable in time, motion in the corona above $2 R_\odot$ (not necessarily material motion, but rather that of the 2σ point) was detected hours earlier than the associated near-surface event. Jackson described the acceleration of these forerunners as slower than that of the eventual eruptive prominence. Eleven of the 12 early events involved only eruptive prominences, while one was flare associated, with material injected into the corona as a surge. The overall CME process was subsequently described (43) as beginning in the outer corona with preexisting structure and ending with the near-surface manifestation.

On early CME manifestations, it is suggested that work be done that separates classical forerunner effects (44), particular perhaps to the CME motion itself, from the preevent processes in existing coronal structure. This reviewer has inspected the *Skylab* raw data and can confirm that forerunners as CME excess mass halos are not usually visible in the enhanced images. What *is* apparent in the raw images are cases of sudden structural evolution, nonimpulsive streamer-related changes, which precede many CMEs. This is not to deny the presence of forerunners rimming CMEs, but rather to raise a caveat concerning the difficulty of establishing their preevent existence and motion convincingly in view of other larger, more significant rearrangements of mass. Gross redistributions of mass are evident before most of the Jackson & Hildner (44) CMEs, all of the Jackson (43) samples, and one event from Hansen et al. (32). Such fast change (or evolution) appears to be more the effect of a streamer filling and swelling (35, 42) than an appendage (the forerunner) of the later CME itself. In the opinion of this reviewer, Jackson (43) did not establish that eruptive prominence CMEs are preceded in time by classical forerunners, but he did discover that the structure of the corona changes before a CME.

3.7 OBSERVING THE FINISH OF A CME   No general study for many cases has been made of the aftermath of mass ejections. Post-CME structure is commonly believed to take the form of radial striations, often characterized by denser legs at each edge of the CME volume. The leg separation quickly reaches a constant distance early in the event (90). With careful photometry, the legs could be detected for as long as 2–3 days (4), but such long lifetimes are questioned for the solar maximum period (20, 64). The density striations marking the interior of a departing CME usually smooth out within 8 hr.

Until recently, no white-light, X-ray, or forbidden emission-line loops had ever been reported returning to the Sun after the CME. Such a contraction, or restoration, might have been expected if the field lines had been dragged outward until they reconnected perhaps to form a closed interplanetary "magnetic bottle" (7). Thus, the recent report (42) of a possible magnetic field reconnection event very late in an extremely slow (≤ 50 km s^{-1}) CME is an important observation. An inverted arch (a large, slightly enhanced density front, convex when seen from the Sun) was identified with the disconnection of magnetic loops that had been seen rising 15 hr earlier. The remnant of the CME that was left at the Sun after the event was seen to pinch down into a narrow ray or streamer.

The passage of an inverted front plus streamer contraction (42) has not been commonly observed in the corona. Additional examples of this possible observation of reconnected magnetic structure should be expected from Solwind, with its larger field of view. Even if the rapidly moving front is too difficult to be seen normally, the contraction of the post-CME structure to form a remnant streamer should be a common phenomenon, especially visible on synoptic electron corona brightness maps. Also inconsistent is the report (81) that temporary *enlargements* of active region X-ray plasmas (rather than contractions) are typically seen for 3–25 hr after eruptives. Thus, attempts should be made to confirm the disconnection observation (42). Failing this, it may still be believed that post-CME reconnection occurs—high in the corona or interplanetary medium above a coronagraph field of view. The detected secondary brightening of the legs 12–24 hr after the CME start (4) may mark the only evidence usually available in white light of the final closure of reconnecting field lines.

Certain general conclusions on the characteristics of mass equations may be made, despite wide variations in CME appearance. Most CMEs are three dimensional, although details of their density distributions are lacking. Outriding features, such as mass forerunners and pressure pulses, could be manifestations of a driven shock wave structure. However, no empirical evidence of this nature has been offered. The observation of a CME seems to depend on the prior history of the coronal region, either because rising impulses are unable to accumulate excess mass in a previously disturbed corona or because a finite time is required to reestablish the proper magnetic field configuration prior to the initiation of a second CME. Signs of rapid evolution apparently exist in preevent coronal structure before the start of a mass ejection, as do hints of possible field-line reconnection at the "bottom" of the expelled CME plasma.

4.   CMEs Related to Other Solar Phenomena

The previous section documented a pattern of form, behavior, and appurtenances for coronal mass ejections. However, in order to model mass ejections, information beyond what can be learned from the CME itself is essential. Next follows a study of CME relations with both active and quiet Sun processes.

4.1   THE ACTIVE SUN AND CMEs   An understanding of mass-ejection association with other near-surface active Sun phenomena will be of particular utility in identifying the origin of the CME, especially if it is uncertain as to whether the CME is a cause or effect of other activity. A catalog of the associations will contain evidence of what physical processes must be present for the triggering of a mass ejection.

4.1.1   *Flares and eruptive prominences*   An extensive study of the 1973 *Skylab* data (67) found 34 of 77 mass ejections to be definitely or probably associated with flares or, more frequently, with eruptive prominences (see also 76). Using probabilities to eliminate random associations, together with logic to account for behind-the-limb events, studies were performed (83) of both *Skylab* and *Solar Maximum* coronagraph CME data. This work (83) verified the abundance of eruptive prominence associations (67), but it also found that nonassociated ejections appeared to constitute a real population equal in number to those accompanied by prominences (Figure 2). Correlations between CMEs and flare phenomena result in part from observational selection. For example, observers may use flares as a guide in deciding which coronal patrol data to retain and search for CMEs, or they may choose to observe on days when an active region condensation is especially bright in emission lines.

Flares indisputably *can* be linked to CMEs. Especially-large flares are often seen as the event accompanying the first CME to recur after the suppression period following an earlier CME. However, the ejection of chromospheric Hα material (as a spray or surge) seems to mark the flares that are more effective in producing CMEs (67), and it must be emphasized that CMEs in general are more commonly associated with eruptive prominences. Theories of CMEs that invoke flares thus seem needlessly complicated, if not completely inappropriate. In fact, the easily observed "unassociated" CMEs without near-surface activity may represent the unembellished CME in its slowest-moving, most basic form.

4.1.2 *High-temperature emission phenomena*   Disturbances seen below 1.2 $R_\odot$ in forbidden emission lines should be equivalent to Thomson-scattering CMEs, in view of their analogous associations with Hα and radio events (10, 67). Nevertheless, attempts to find film records of simultaneous CMEs and λ5303 Fe XIV transients have not been productive. A strangely shaped, fleeting λ6374 Fe X transient observed by R. N. Smartt at Sacramento Peak Observatory was, however, seen by both Solwind and the Coronagraph/Polarimeter (106).

A noteworthy distinction between Thomson-scattering CMEs and transients seen in high-temperature emissions is that 50% of the λ5303 transients (10) are whips (17, 18) that show a rapidly expanding loop breaking open, often near one foot-point. Such action, also reported in X-ray emission with a transient (79), is virtually unknown in continuum CMEs. If the whip demonstrates field-line reconnection, as has been suggested (104), it is seen much earlier in the Fe XIV emission-line event and considerably lower in the corona (at heights < 1.02 $R_\odot$) than the very different reconnection phenomena reported late in a white-light CME (42).

In spite of this discrepancy between white-light and forbidden emission-line observations, CMEs are associated with certain types of X-ray events. From soft X-ray data on 43 limb flares near solar minimum, Pallavicini et al. (71) defined two distinct types of flares: low-lying, high-energy density compact flares (Class I); and six other events (Class II) in larger, more diffusive loop arcade systems. The latter showed very long (> 3 hr) X-ray durations (49, 88, 110), with strong CME and eruptive prominence association (together with Hα flaring in 5 of the 6 cases). Such a clear distinction in X-ray events was difficult to find during solar maximum observations, perhaps owing to a lack of true Class II flare events during this period (1). A second search for Class I/II differences (86) apparently admitted among its data X-ray bursts from nonactive region prominence eruptions without classical flare effects. The Pallavicini et al. flare distinction probably still obtains, and should be especially evident (78) for the CMEs associated with large proton shower-producing flares in the rise to and fall from solar maximum.

4.1.3 Implications from radio physics

Metric type II and type IV bursts (111) near the solar limb are usually accompanied by CMEs (67). The reverse correlation is much weaker, however, with only 15–25% of the CMEs seen to appear with type IIs or IVs (83). Few mass ejections traveling slower than 400 km s^{-1} are accompanied by a type II (30, 31); conversely, occasional CMEs at 700 km s^{-1} also fail to show type IIs, perhaps as a result of source occultation behind the limb (16) or extremely high Alfvén speeds in the ambient corona (105). For a CME to be accompanied by a type IV event, the sudden insertion of prominence or chromospheric (spray) material into the corona seems to be required.

The tidy *Skylab*-era concept of shock-associated type II sources traveling in front of or at the CME loop (15, 58, 61, 91) has proven untenable. While type IIs that lead the CME may be found in future analyses, enough contrary cases now exist, in events for which simultaneous white-light images and radioheliograms are available, to allow a physical dissociation of the type II exciter from the loop (102, a variation of the ideas of 62). Recent work from radio and optical telescopes has been reviewed (101). The results show discrepancies in the CME–type II source timing (23, 28, 82, 95, 114) and positioning (23, 28, 95), as well as the relative velocities (25, 61) and directions (26, 28, 92).

One cannot intimately relate the type II to the CME except by invoking overly complicated shock models (101). In such constructs, accelerated particles produce radio emission in suitable coronal regions distant from the instantaneous position of the shock front. A simpler interpretation is that the CME is not physically connected to the type-II-producing flare blast shock. A CME-driven stand-off shock could also exist, but it is not the shock observed to excite the type II source.

Stationary type IV bursts, beyond their possible role (Section 5.1) in the acceleration of energetic particles (50), hold a unique importance for CMEs (58), namely, as the only diagnostic currently available for magnetic field strength in the mass ejection itself. Three 1980 stationary type IV events, at last with true overlap of radio and visible data (23, 95, 105), are added to those reviewed by MacQueen (58). Only two events to date (105; and perhaps 16) give information on the CME itself. These authors find that magnetic energy densities exceed thermal energy densities by factors of 17 and 140 in the two cases. The only remaining cases (27, 53, 94) show stationary type IV emission at locations not on the CME.

The moving type IV cannot be dismissed as peripheral to a study of mass ejections (58). The recent radio and visible data show apparent spatial coincidence of moving type IV bursts with the CME. Two events with moving type IV bursts and dense plasmoids were observed simultaneously in 1980 in radio and white light. One moving type IV (95) was found in a radially elongated plasmoid that apparently intersected the expanding CME. An arcade of loops was described (28), with a dense plasmoid at its top emitting a moving type IV. These recent data show clearly that moving type IV sources are hot plasmoids (as opposed to 10^4 K neutral hydrogen knots), high in the CME (usually *on* its apparent surface). The origin of the moving type IV material may be either a spray or the highest and fastest plasmoid in an eruptive prominence that has been rapidly ionized (53, 93, 94). Although no classical fast-moving, high brightness temperature type IV was observed in this solar cycle maximum, even these smaller events near the CME top revealed the existence of impressive amounts of convected free magnetic energy high in the corona being released as mechanical energy of expansion.

In summary, radio observations no longer permit the CME to be physically associated with the flare blast wave marked by its type II burst. Evidence from stationary type IV emission suggests that magnetic rather than hydrodynamic forces control not only the legs, but also the top of the CME. Type IV emission also indicates that magnetic energy may be the dominant energy in some events. The high-energy plasmoids of moving type IV emission are now seen to be at coronal temperatures, and they are apparently cospatial with the top of the CME.

A caveat may be appropriate here. The emission mechanism for the type IV has usually been ascribed to gyrosynchrotron radiation (15). Now, white-light observations of high densities cospatial with type IV bursts (16, 28, 95, 105) seem to suggest that emission is due to the same mechanism (plasma radiation) common to most other burst types. This interpretation may reduce the promise of the type IV as a diagnostic for magnetic field strengths.

4.2 CMEs AND LARGE-SCALE MAGNETIC FIELDS The latitude distribution of the total CME population widens from solar minimum to solar maximum (40). More particularly, in terms of their associations, CMEs with flares and most of those with eruptive prominences were in the sunspot belts in 1980, as shown in Figure 3. However, there is now postulated a class of unassociated CMEs, events that are not correlated with near-surface Hα, X-ray, or metric radio activity (83). Figure 3 shows that the unassociated events were found at low latitude in 1973 but much nearer to the poles at the 1980 solar maximum. Replacing the unassociated CMEs at low latitudes were an added population of activity belt flare-associated CMEs. Significantly, the frequency of all visible mass ejections *in the ecliptic* plane appears to be nearly constant through the activity cycle (40), with perhaps an important change in their character (a greater fraction of flare and prominence CMEs at maximum).

The unassociated *Skylab* CMEs observed just before solar minimum occurred mainly in a very narrow zone within $\pm 15°$ of the equator (Figure 3). This region lay between the mean sunspot belts and apparently was at the confluence of drift of the preceding polarity fields from each hemisphere. In 1973, these fields were erasing the remnant equatorial flux of the earlier cycle. Conversely, shortly after sunspot maximum, Coronagraph/Polarimeter 1980 data showed that the unassociated CMEs lay entirely above the northern sunspot belt in the northern hemisphere, and were evenly distributed up to 80°S in the southern. Thus, this unassociated component of the CME population seems related to the neutralization of large-scale magnetic fields [stage 4 of Babcock (4a)]. The rapid evolution of intense, complex fields above active regions was suggested (33) as playing a

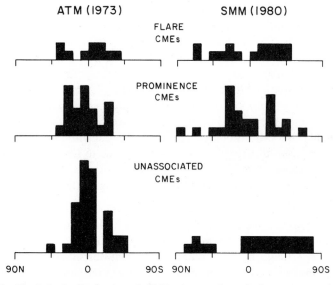

Figure 3 The latitude distribution of CMEs changes through the sunspot cycle chiefly because the locations of unassociated CMEs, perhaps marking regions of neutralization of the large-scale solar magnetic field (4a), change from low to high latitudes. These data, showing number of CMEs versus latitude, are generated from Sawyer et al. (83). Adjusting these known associations statistically for behind-the-limb flares and prominences would decrease the unassociated distribution by only 2–3 CMEs; a considerable number of CMEs would be shifted from the prominence-without-a-flare category to the flare class.

role in CME production. Mass ejections may not be related simply to intense active region fields over neutral lines, but rather to the occurrence of field disappearance anywhere, whether through emergence of active region flux with reconnection or through poorly understood processes of neutralization at the interface of old and new large-scale background fields.

In conclusion, CME associations have been seen to involve not only classical active-Sun phenomena, but also possibly steady-state solar magnetic processes. In this latter respect, unassociated CMEs may prove to be a feature of any star showing intrinsic magnetic cycles.

5. Association with Interplanetary Phenomena

So far we have discussed observational evidence on the nature of CME events. We now turn to the implications of coronal mass ejection research for other disciplines. Several interplanetary phenomena discussed in this section can apparently be related to mass ejections by virtue of CME characteristics measured in the corona. The section closes with a brief mention of other effects of CMEs that are detectable far from the Sun in the solar wind and that may also have stellar analogues.

5.1 PROMPT PROTON EVENTS   The CME has been associated with second-stage particle acceleration in flares (98). It has been suggested (51) and reaffirmed (52, 70) that CMEs or their disk equivalent, X-ray long-decay events (49, 88, 110), may be a necessary condition for proton showers. Arguments are provided (8, 50) that it is a driven shock standing out from the CME bow that accelerates particles, and that the evanescent flare blast shock, which is suggested (61) as producing the more localized type II bursts, is not effective in proton acceleration. However, the CME cannot be a sufficient cause, at least as a driven shock that accelerates particles, because far fewer shocks than CMEs are seen in the interplanetary medium (78). It must be stressed that no evidence is found from *Helios* space probes of driven shocks with 13 flare-associated CMEs seen by Solwind at the Sun; instead, only impulsive shocks (blast-type, without continuous input of energy) are detected (85). In addition, while driven shocks leaving the Sun could be degenerating into blast-type shocks before reaching spacecraft, it is argued above that no credible cases exist of radio type II emission from driven shocks ahead of the CME in the inner corona either.

From statistical data, Kahler (50) concludes that other requirements for a proton flare, besides the common type II radio association, are long duration (≥ 10 min) microwave emission, Hα importance ≥ 2, and a metric stationary type IV burst. These evidences of trapped nonthermal electrons imply the growth of postflare loop prominence systems (see also 109). *Solar Maximum Mission* gamma-ray observations indicate that high energies are reached immediately at the flare site itself (22), although gamma-ray flares per se correlate poorly with interplanetary showers of prompt > 10 MeV protons (72). This reviewer believes that the CME does not necessarily drive the particles with a leading shock, but rather that it opens the coronal fields, thereby providing access to the interplanetary field for previously energized flare protons (see also 66).

5.2   GALACTIC COSMIC-RAY MODULATION   Coronal mass ejections that are accompanied by flare/eruptive prominence/X-ray/type II–IV radio events remain at relatively low latitude throughout the activity cycle, as indicated by Figure 3. Figure 2 implies that these energetic events then vary in frequency in the ecliptic plane by a factor of 3.5 through the cycle, unlike the almost constant overall CME count. This variation of 3.5 in the particularly energetic CMEs and the low limb frequencies themselves (0.21 day^{-1} for the Coronagraph/Polarimeter and 0.06 day^{-1} for *Skylab*) may begin to match those of the Forbush decreases of galactic cosmic rays. However, for the long-term modulation of cosmic rays by magnetic inclusions in the solar wind, models (68) seem to require a higher as well as more cycle-dependent frequency (varying by factors of 3–10) for the nearly constant total CME population than is actually observed.

5.3   NONCOMPRESSIVE DENSITY ENHANCEMENTS   Noncompressive density enhancements (NCDEs) seen in the solar wind have been suggested (29) to be the slower CMEs detected at 1 AU. NCDEs are regions of higher density in which proton and electron temperatures are low and inversely proportional to density. About one third of the NCDEs showed some evidence of a slow reversal in the magnetic field direction, which was of normal strength during their passage. Comparing NCDE characteristics with those of flare- and eruptive-prominence-associated CMEs, Gosling et al. (29) noted that the low speed of the latter better matched NCDEs. Major objections with this identification were raised because of the closed, nonenhanced fields of NCDEs (58) and the lack of frequency dependence of NCDEs on the solar cycle (68) With the recent discovery (40) that the frequency of all CMEs in the ecliptic plane is nearly constant, the second objection is removed. Of more concern is the report (85) that far *fewer* NCDEs were observed in 1979–81 than in 1971–74. It is suggested that the NCDE events should be identified with the unassociated CMEs that originated near the solar equator in 1973 but away from the low solar latitudes (and the ecliptic plane) by 1980 (Figure 3). Unresolved at this time is whether the unassociated CME/NCDE is also the "magnetic bottle" (7) mentioned in Sections 3.2 and 3.7.

5.4   MASS AND ENERGY COMPARISONS TO INTERPLANETARY SHOCKS   The quantity of mass lost with a CME and the free energy available from a mass ejection in the interplanetary medium remain largely the same at solar maximum as near solar minimum. Both flare and eruptive prominence CMEs seen from the *Skylab* typically averaged about 3.5×10^{15} g; one extraordinary event (with an origin far onto the solar disk, giving poor observing circumstances) skews the overall flare average above this number (79). Excess mass has now been determined for about 35 CMEs as 2×10^{14} to 2×10^{16} g (19, 24, 35, 44, 95, 105, 108). This seems considerably less than the 3.5×10^{16} g averaged for 22 shock waves with assumed geometry (39) that were measured near Earth, perhaps after sweeping up ambient mass.

Of the energy available for work from a CME, the thermal exceeds kinetic for all mass ejections slower than 750 km s^{-1} (assuming 2×10^6 K coronal material with mean particle mass of 1.2 times that of a proton). CME free energies in the interplanetary medium range from about 2×10^{29} to 5×10^{31} erg. However, with certain interpretations of radio emission mechanisms (63, 107), both the kinetic and the thermal energy of some CMEs are dwarfed by the convected magnetic energy. These estimates have been made for only a few cases, but they are said to obtain for both the plasmoids emitting radio type IV bursts (10^{32} erg; 14, 16) and the apparent loop itself (105). This magnetic energy, from fields that are transported from the lower corona, is presumably released by later field-line reconnection at the loop "bottom," and it is thus available for conversion to mechanical energy. In comparison to CMEs, the 22 shocks in 1965–67 are reported (39) to show a (geometry-dependent) average of 7×10^{31} erg at 1 AU.

In sum, if CMEs correspond to the shocks observed at Earth, then either mass is swept up or the inner corona CME mass estimates must be judged to be overly conservative. Instead, those measured from 6–10 $R_\odot$, which tend to be higher (38, 76, 93, 94), must be adopted. Also, the mass ejection free energy can be reconciled with the shock energies only if the major energy in a CME does reside in a convected magnetic field.

5.5   CMEs FROM THE NEAREST STAR   CMEs have been detected at large angles from the Sun through the scintillation of pulsar radio signals caused by density or bulk velocity perturbations in the solar wind (89). Mass ejections are observed to contribute velocity dispersion proportional to total electron content to signals transmitted by spacecraft occulted by the corona (112). Faraday rotation is also provided by the CME magnetic plasma to such linearly polarized radio signals (5). Signatures similar to those described above may also be available for investigations of stellar CMEs, provided that strong nonspatially resolved radio sources lie behind candidate stars.

Finally, of particular interest for heliospheric physics are the recent detections of coronal mass ejections with zodiacal light photometers (46, 77a). Comet-tail anomalies (48) thus far appear difficult to associate with the momentum flux changes in a CME (6). However, Niedner & Brandt (69) report that tail disconnection events result from the impingement of an oppositely directed field, such as that found at interplanetary magnetic sector boundaries. Some analogous effect might be expected if indeed the CME convects its own insurgent closed magnetic fields, as is commonly believed.

6. *Perspectives*

One decade and six coronagraphs after the discovery of coronal mass ejections, the nature of a CME is still not fully understood. Perhaps the most valuable interpretive contribution over these years has been an extension of our thinking into three dimensions. Building on the analyses of data from 1971–74, observations taken near the 1980 solar cycle maximum have suggested that CMEs are probably bubble shaped. Empirical density models based on careful polarimetric analyses of CMEs are needed to confirm such an interpretation. Observed distributions of material could then be compared with those predicted from the very disparate theoretical descriptions of mass ejection origins.

Observational progress on the coronal mass ejection phenomena is apparent today, but further efforts are required to resolve specific uncertainties. Preevent changes in helmet streamers (as yet incompletely documented) seem to presage prominence eruptions with associated CMEs. Some eruptive prominence CMEs are now known to begin in the low corona as dark voids showing strong acceleration. Such CMEs should be investigated to verify that they culminate in diffuse field-line reconnection fronts leaving the Sun.

Little change in CME frequency accompanied the growth of flare activity at the solar maximum. This was found to be the result of an apparent 10-hr period of local suppression of CMEs following any earlier event. CMEs accompanied by flares now appear to be divorced from the flare blast wave (and CME wave model) in view of the relative timing of the Hα event with the inferred CME launch time, as well as the location and movement of type II shock emission.

To the classes of CMEs associated with flares and with eruptive prominences, a third group is added, the equally abundant category of CMEs unassociated with near-surface activity. The significance of unassociated CMEs lies in their promise of extending the argument (33) that the neutralization of magnetic fields is accompanied by CMEs. Earlier, it was suspected that flare and eruptive prominence CMEs signaled field reconnection. Unassociated CMEs suggest that old large-scale remnant fields may be cancelled by new insurgent flux (4a) during the phase of the solar cycle in which the "march to the poles (and equator)" occurs.

The magnetic field nonequilibrium model (56) holds promise, not only for its versatility, but also for its ability to explain possible precursors and the frequently observed unassociated CMEs. Conversely, the unassociated CMEs, unaccompanied by sudden energy releases, pose a challenge for the impulse wave model (13, 96, 113). The familiar cases of CMEs associated with the sudden perturbations that accompany flares and eruptive prominences are still admitted in this interpretation (56), provided that the preevent coronal systems have approached their limit of stability.

Data on coronal mass ejections from the maximum of the solar cycle are just now beginning to be analyzed. The data flow from these same coronagraphs may be expected to continue toward the 1985 solar minimum. We can anticipate an exciting period of analysis and theory testing lying immediately ahead.

ACKNOWLEDGMENTS

This review (or more properly, progress report) of CME research during the last decade was provided considerable guidance and encouragement by the Coronagraph/Polarimeter group and E. Hildner. I am grateful to R. Fisher, A. Hundhausen, R. MacQueen, and G. Newkirk, Jr. for comments on the manuscript. This work was supported in part by the Contracts NASA S55989 and S55989A. The National Center for Atmospheric Research is sponsored by the National Science Foundation.

Literature Cited

1. Acton, L. W. 1982. *Observatory* 102: 123–24
2. Anzer, U. 1978. *Sol. Phys.* 57:111–18
3. Anzer, U. 1980. In *Solar and Interplanetary Dynamics*, ed. M. Dryer, E. Tandberg-Hanssen, pp. 263–77. Dordrecht: Reidel. 558 pp.
4. Anzer, U., Poland, A. I. 1979. *Sol. Phys.* 61:95–113
4a. Babcock, H. W. 1961. *Ap. J.* 133:572–87
5. Bird, M. K. 1982. *Space Sci. Rev.* 33: 99–126
6. Burlaga, L. F., Rahe, J., Donn, B., Neugebauer, M. 1973. *Sol. Phys.* 30: 211–22
7. Burlaga, L., Sittler, E., Mariani, F., Schwenn, R. 1981. *J. Geophys. Res.* 86:6673–84
8. Cliver, E. W., Kahler, S. W., McIntosh, P. S. 1983. *Ap. J.* 264:699–707
9. Crifo, F., Picat, J. P., Cailloux, M. 1983. *Sol. Phys.* 83:143–52
10. DeMastus, H. L., Wagner, W. J., Robinson, R. D. 1973. *Sol. Phys.* 31: 449–59
11. Dodson, H. W., Hedeman, E. R., Roelof, E. C. 1982. *Eos, Trans. Am.*

Geophys. Union 63:156 (Abstr.)
12. Dryer, M. 1982. *Space Sci. Rev.* 33:233–75
13. Dryer, M., Wu, S. T., Steinolfson, R. S., Wilson, R. M. 1979. *Ap. J.* 227:1059–71
14. Dulk, G. A. 1973. *Sol. Phys.* 32:491–503
15. Dulk, G. A. 1980. In *Radio Physics of the Sun*, ed. M. Kundu, T. Gergely, pp. 419–33. Dordrecht: Reidel. 475 pp.
16. Dulk, G. A., Smerd, S. F., MacQueen, R. M., Gosling, J. T., Magun, A., et al. 1976. *Sol. Phys.* 49:369–94
17. Dunn, R. B. 1971. In *Physics of the Solar Corona*, ed. C. Macris, pp. 114–29. Dordrecht: Reidel. 345 pp.
18. Evans, J. W. 1957. *Publ. Astron. Soc. Pac.* 69:421–26
19. Fisher, R. R., Munro, R. H. 1984. *Ap. J.* In press
20. Fisher, R. R., Poland, A. I. 1981. *Ap. J.* 246:1004–9
21. Fisher, R. R., Garcia, C. J., Seagraves, P. 1981. *Ap. J. Lett.* 246:L161–64
22. Forrest, D. J., Chupp, E. L. 1983. *Nature* 305:291–92
23. Gary, D. E. 1982. *Radio emission from solar and stellar coronae.* PhD thesis. Univ. Colo., Boulder. 241 pp.

24. Gary, D. E., Dulk, G. A., House, L. L., Illing, R., Sawyer, C., et al. 1984. *Astron. Astrophys.* In press
25. Gergely, T. E. 1984. *Proc. Maynooth Conf., 1982.* In press
26. Gergely, T. E., Kundu, M. R., Hildner, E. 1983. *Ap. J.* 268:403–11
27. Gergely, T. E., Kundu, M. R., Munro, R. H., Poland, A. I. 1979. *Ap. J.* 230:575–80
28. Gergely, T. E., Kundu, M. R., Erskine, F. T., Sawyer, C., Wagner, W. J., et al. 1984. *Sol. Phys.* In press
29. Gosling, J. T., Hildner, E., Asbridge, J. R., Bame, S. J., Feldman, W. C. 1977. *J. Geophys. Res.* 82:5005–10
30. Gosling, J. T., Hildner, E., MacQueen, R. M., Munro, R. H., Poland, A. I., Ross, C. L. 1974. *J. Geophys. Res.* 79:4581–87
31. Gosling, J. T., Hildner, E., MacQueen, R. M., Munro, R. H., Poland, A. I., Ross, C. L. 1976. *Sol. Phys.* 48:389–97
32. Hansen, R. T., Garcia, C. J., Grognard, R. J.-M., Sheridan, K. V. 1971. *Proc. Astron. Soc. Aust.* 2:57–60
33. Hildner, E. 1977. In *Study of Traveling Interplanetary Phenomena*, ed. M. Shea, D. Smart, S. Wu, pp. 3–21. Dordrecht: Reidel. 439 pp.
34. Hildner, E., Gosling, J. T., Hansen, R. T., Bohlin, J. D. 1975. *Sol. Phys.* 45:363–76
35. Hildner, E., Gosling, J. T., MacQueen, R. M., Munro, R. H., Poland, A. I., Ross, C. L. 1975. *Sol. Phys.* 42:163–77
36. Hildner, E., Gosling, J. T., MacQueen, R. M., Munro, R. H., Poland, A. I., Ross, C. L. 1976. *Sol. Phys.* 48:127–35
37. House, L. L., Wagner, W. J., Hildner, E., Sawyer, C., Schmidt, H. U. 1981. *Ap. J. Lett.* 244:L117–21
38. Howard, R. A., Michels, D. J., Sheeley, N. R. Jr., Koomen, M. J. 1982. *Ap. J. Lett.* 263:L101–4
39. Hundhausen, A. J. 1972. *Solar Wind and Coronal Expansion.* New York: Springer-Verlag. 238 pp.
40. Hundhausen, A. J., Sawyer, C., House, L., Illing, R. M. E., Wagner, W. J. 1984. *J. Geophys. Res.* In press
41. Illing, R. M. E. 1982. *Observatory* 102:122–23
42. Illing, R. M. E., Hundhausen, A. J. 1983. *J. Geophys. Res.* 88:10210–14
43. Jackson, B. V. 1981. *Sol. Phys.* 73:133–44
44. Jackson, B. V., Hildner, E. 1978. *Sol Phys.* 60:155–70
45. Jackson, B. V., Dulk, G. A., Sheridan, K. V. 1980. In *Solar and Interplanetary Dynamics*, ed. M. Dryer, E. Tandberg-Hanssen, pp. 379–80. Dordrecht: Reidel. 558 pp.
46. Jackson, B. V., Howard, R. A., Koomen, M. J., Michels, D. J., Sheeley, N. R. 1983. *Bull. Am. Astron. Soc.* 15:705
47. Jackson, B. V., Sheridan, K. V., Dulk, G. A., McLean, D. J. 1978. *Proc. Astron. Soc. Aust.* 3:241–42
48. Jockers, K., Lust, Rh. 1973. *Astron. Astrophys.* 26:113–21
49. Kahler, S. 1977. *Ap. J.* 214:891–97
50. Kahler, S. W. 1982. *Ap. J.* 261:710–19
51. Kahler, S. W., Hildner, E., van Hollebeke, M. A. I. 1978. *Sol. Phys.* 57:429–43
52. Kahler, S. W., McGuire, R. E., Reames, D. V., von Rosenvinge, T. T., Sheeley, N. R. Jr., et al. 1983. *Proc. Int. Cosmic Ray Conf., 18th.* In press
53. Kosugi, T. 1976. *Sol. Phys.* 48:339–56
54. Lantos, P., Kerdraon, A., Rapley, G. G., Bentley, R. D. 1981. *Astron. Astrophys.* 101:33–38
55. Low, B. C. 1982. *Ap. J.* 254:796–805
56. Low, B. C. 1984. *Ap. J.* In press
57. Low, B. C., Munro, R. H., Fisher, R. R. 1982. *Ap. J.* 254:335–42
58. MacQueen, R. M. 1980. *Philos. Trans. R. Soc. London Ser. A* 297:605–20
59. MacQueen, R. M., Fisher, R. R. 1983. *Sol. Phys.* 89:89–102
60. MacQueen, R. M., Eddy, J. A., Gosling, J. T., Hildner, E., Munro, R. H., et al. 1974. *Ap. J. Lett.* 187:L85–88
61. Maxwell, A., Dryer, M. 1982. *Space Sci. Rev.* 32:11–25
62. McLean, D. J. 1967. *Proc. Astron. Soc. Aust.* 1:47–49
63. McLean, D. J., Dulk, G. A. 1978. *Proc. Astron. Soc. Aust.* 3:251–52
64. Michels, D. J., Howard, R. A., Koomen, M. J., Sheeley, N. R. Jr., Rompolt, B. 1980. In *Solar and Interplanetary Dynamics*, ed. M. Dryer, E. Tandberg-Hanssen, pp. 387–91. Dordrecht: Reidel. 558 pp.
65. Mouschovias, T. C., Poland, A. I. 1978. *Ap. J.* 220:675–82
66. Mullan, D. J. 1983. *Ap. J.* 269:765–78
67. Munro, R. H., Gosling, J. T., Hildner, E., MacQueen, R. M., Poland, A. I., Ross, C. L. 1979. *Sol. Phys.* 61:201–15
68. Newkirk, G. Jr., Hundhausen, A. J., Pizzo, V. 1981. *J. Geophys. Res.* 86:5387–96
69. Niedner, M. B. Jr., Brandt, J. C. 1978. *Ap. J.* 223:655–70
70. Nonnast, J. H., Armstrong, T. P., Kohl, J. W. 1982. *J. Geophys. Res.* 87:4327–37
71. Pallavicini, R., Serio, S., Vaiana, G. S. 1977. *Ap. J.* 216:108–22
72. Pesses, M. E., Klecker, B., Gloeckler, G., Hovestadt, D. 1981. *Proc. Int. Cosmic Ray Conf., 17th*, 3:36–39

73. Pneuman, G. W. 1980. *Sol. Phys.* 65: 369–85
74. Pneuman, G. W. 1983. *Adv. Space Res.* 2: 233–36
75. Poland, A. I., Munro, R. H. 1976. *Ap. J.* 209: 927–34
76. Poland, A. I., Howard, R. A., Koomen, M. J., Michels, D. J., Sheeley, N. R. Jr. 1981. *Sol. Phys.* 69: 169–75
77. Priest, E. R., Milne, A. M. 1980. *Sol. Phys.* 65: 315–46
77a. Richter, I., Leinert, C., Planck, B. 1982. *Astron. Astrophys.* 110: 115–20
78. Rust, D. M. 1983. *Space Sci. Rev.* 34: 21–36
79. Rust, D. M., Hildner, E. 1976. *Sol. Phys.* 48: 381–87
80. Rust, D. M., Hildner, E. 1980. In *Solar Flares*, ed. P. Sturrock, pp. 273–339. Boulder: Colo. Assoc. Univ. Press. 513 pp.
81. Rust, D. M., Webb, D. F. 1977. *Sol. Phys.* 54: 403–17
82. Sawyer, C. 1983. *Adv. Space Res.* 2: 265–70
83. Sawyer, C., Wagner, W. J., Hundhausen, A. J., House, L. L., Illing, R. M. E. 1984. *J. Geophys. Res.* Submitted for publication
84. Schmahl, E., Hildner, E. 1977. *Sol. Phys.* 55: 473–90
85. Schwenn, R. 1983. *Space Sci. Rev.* 34: 85–99
86. Sheeley, N. R. Jr., Howard, R. A., Koomen, M. J., Michels, D. J. 1983. *Ap. J.* 272: 349–54
87. Sheeley, N. R. Jr., Howard, R. A., Koomen, M. J., Michels, D. J., Harvey, K. L., Harvey, J. W. 1982. *Space Sci. Rev.* 33: 219–31
88. Sheeley, N. R. Jr., Bohlin, J. D., Brueckner, G. E., Purcell, J. D., Scherrer, V. E., et al. 1975. *Sol. Phys.* 45: 377–92
89. Sime, D. G. 1983. *Proc. Am. Geophys. Union Chapman Conf., Solar Wind, 1982*, 5: 453–67
90. Sime, D. G., MacQueen, R. M., Hundhausen, A. J. 1984. *J. Geophys. Res.* In press
91. Stewart, R. T. 1980. In *Solar and Interplanetary Dynamics*, ed. M. Dryer, E. Tandberg-Hanssen, pp. 333–55. Dordrecht: Reidel. 558 pp.
92. Stewart, R. T., Magun, A. 1980. *Proc.*

Astron. Soc. Aust. 4: 53–55
93. Stewart, R. T., Howard, R. A., Hansen, F., Gergely, T., Kundu, M. 1974. *Sol. Phys.* 36: 219–31
94. Stewart, R. T., McCabe, M. K., Koomen, M. J., Hansen, R. T., Dulk, G. A. 1974. *Sol. Phys.* 36: 203–17
95. Stewart, R. T., Dulk, G. A., Sheridan, K. V., House, L. L., Wagner, W. J., et al. 1982. *Astron. Astrophys.* 116: 217–23
96. Steinolfson, R. S. 1982. *Astron. Astrophys.* 115: 39–49
97. Steinolfson, R. S. 1982. *Astron. Astrophys.* 115: 50–53
98. Svestka, Z. 1976. *Solar Flares.* Dordrecht: Reidel. 399 pp.
99. Tousey, R. 1973. *Space Res.* 13: 713–30
100. Trottet, G., MacQueen, R. M. 1980. *Sol. Phys.* 68: 177–86
101. Wagner, W. J. 1983. *Adv. Space Res.* 2: 203–19
102. Wagner, W. J., MacQueen, R. M. 1983. *Astron. Astrophys.* 120: 136–38
103. Wagner, W. J., Wagner, J. J. 1984. *Astron. Astrophys.* In press
104. Wagner, W. J., Hansen, R. T., Hansen, S. F. 1974. *Sol. Phys.* 34: 453–59
105. Wagner, W. J., Hildner, E., House, L. L., Sawyer, C., Sheridan, K. V., Dulk, G. A. 1981. *Ap. J. Lett.* 244: L123–26
106. Wagner, W. J., Illing, R. M. E., Sawyer, C., House, L. L., Sheeley, N. R. Jr., et al. 1983. *Sol. Phys.* 83: 153–66
107. Webb, D. F. 1980. In *Solar Flares*, ed. P. Sturrock, pp. 471–99. Boulder: Colo. Assoc. Univ. Press. 513 pp.
108. Webb, D. F., Jackson, B. V. 1981. *Sol. Phys.* 73: 341–61
109. Webb, D. F., Kundu, M. R. 1978. *Sol. Phys.* 57: 155–73
110. Webb, D. F., Krieger, A. S., Rust, D. M. 1976. *Sol. Phys.* 48: 159–86
111. Wild, J. P., Smerd, S. F. 1972. *Ann. Rev. Astron. Astrophys.* 10: 159–96
112. Woo, R., Armstrong, J. W. 1981. *Nature* 292: 608–10
113. Wu, S. T., Dryer, M., Nakagawa, Y., Han, S. M. 1978. *Ap. J.* 219: 324–35
114. Wu, S. T., Wang, S., Dryer, M., Poland, A. I., Sime, D. G., et al. 1983. *Sol. Phys.* 85: 351–73
115. Yeh, T. 1982. *Sol. Phys.* 78: 287–316
116. Yeh, T., Dryer, M. 1981. *Ap. J.* 245: 704–10

Ann. Rev. Astron. Astrophys. 1984. 22: 291–317

ASTRONOMICAL FOURIER TRANSFORM SPECTROSCOPY REVISITED

S. T. Ridgway and J. W. Brault

Kitt Peak National Observatory,[1] Tucson, Arizona 85726

1. INTRODUCTION

The Fourier transform spectrometer (FTS) has joined the ranks of a small number of general purpose, high-performance spectroscopic instruments now available to astronomers. As with any technique, the FTS has advantages and disadvantages in comparison with other current or potential experimental methods. The principal objective of this review is to introduce the FTS and to discuss both its demonstrated capabilities and its limitations for astronomical measurements. A very brief introduction to the optical characteristics of the FTS is followed by a point-by-point consideration of the distinctive features of the FTS that make it attractive or unattractive for particular astronomical applications. To further clarify the possible uses of the technique, the final section of the review describes in some detail numerous successful applications of the FTS.

It is more than a decade since Pierre Connes (17) reviewed Fourier spectroscopy in this annual. At that time he remarked, "Thus the time for reviewing the subject seems ripe; hopefully a few years from now it may be impossible to cover entirely." This has indeed proved to be the case, and it is not feasible to include within this review all of the important developments of the past 14 years. Thus we frequently refer to other recent reviews that cover some topics in more depth.

[1] Operated by the Association of Universities for Research in Astronomy, Inc., under contract with the National Science Foundation.

291

0066–4146/84/0915–0291$02.00

2. THE INSTRUMENT

2.1   *The Optical and Mechanical Implementation*

An FTS is simply a Michelson interferometer with variable path difference between the two arms. Nevertheless, the FTS has long had the reputation of being both esoteric and phenomenally expensive. The former assessment is no doubt a consequence of the hesitation of many astronomers to think in Fourier transform space a decade or more ago. This faintness of heart is less widespread today. The cost of the technique derives from the necessity to provide physical motion with interferometric tolerances (which can still be expensive), and the need for a computer to extract a spectrum from the data (which is no longer so costly). But fundamentally, when the FTS was introduced, astronomers simply were not used to the cost of the high technology that is now taken for granted in image intensifiers, array detectors, and customized electronics. By modern standards the FTS can no longer be considered highly esoteric, exceptionally difficult to fabricate, or remarkably expensive.

In the actual fabrication of an FTS, the driving factors in cost and complexity are the design wavelength and the resolution. The former determines the required tolerances, the latter the required scale. The principal design problems are the support of the beam splitter and the mechanisms for varying the optical path difference between the two arms. Some examples of several operating FTS systems demonstrate the range of possibilities:

1. A simple and inexpensive FTS for modest resolution in the IR has been developed and operated at Steward Observatory by Thompson (88). Plane mirrors (as in the classical Michelson) are transported on a commercial mechanical slide with a screw under torque motor control.
2. A high-resolution FTS with cat's-eye retroreflectors transported on air bearings has been developed and operated at the University of Arizona Lunar and Planetary Laboratory (18); the instrument was built for airborne and ground-based use. This group has also operated a moderate-resolution FTS at many telescopes over a period of more than a decade (56).
3. Hall and Ridgway have developed a high-resolution FTS for coudé operation at the KPNO Mayall telescope. It employs cat's-eye retroreflectors translated on a wheeled carriage and flexible parallelogram, respectively (38).
4. Brault has developed a high-resolution FTS designed for any wavelength from the UV to the IR. Cat's-eye retroreflectors are translated on

oil bearings, with the whole installed in a vacuum tank. It operates in a laboratory environment at the KPNO McMath Solar Telescope (9).

Other designs have been executed for astronomical use, both on the ground and in a host of other environments, including balloon, satellite, and spacecraft operation (e.g. 4, 43).

2.2   Characteristics of a Modern FTS

A modern FTS has a number of instrumental characteristics that make it a powerful tool for astronomical observations. None of these characteristics is unique to the FTS method: some have been realized in other types of spectroscopy; some that have not been realized elsewhere could be. For a more extended discussion of the comparison between the FTS and alternative methods of high-resolution spectroscopy, see Brault (11).

2.2.1   THROUGHPUT   A classic feature of the FTS, shared with the Fabry-Perot interferometric spectrometer, is a large throughput or étendue (the product of solid angle and aperture of the spectrometer). In practical terms, the FTS has a circular entrance aperture rather than an entrance or exit slit. In any large FTS, the field of view for even very high resolution is substantially greater than the typical seeing disk. In fact, the tolerable solid angle is in practice usually limited not by the desired resolution but by other considerations. For extended sources, the scientific objective generally compels a restricted aperture (near the seeing limit) because, whether for the Sun, planets, or nebulae, spatial resolution is useful in the observation. (One notable exception is the study of the Sun as a star, where no image at all is formed.) For point sources, the photon noise of the sky background discourages the use of a large field of view. In laboratory spectroscopy, on the other hand, where we are dealing with homogeneous extended sources, the extra throughput of a large field of view can be extremely valuable, as the FTS can often accept 10–1000 times more source flux than the grating spectrograph.

Another factor contributing to high throughput is the inherent high efficiency of the FTS technique. A carefully selected beam splitter can easily have an efficiency exceeding 90%. Overall telescope plus FTS transmission of 25% of the incident flux may be reasonably expected.

2.2.2   SIGNAL-TO-NOISE   A number of characteristics tend to protect FTS spectra from degradation due to external disturbances. Rapid scanning or modulation of the FTS places the data frequencies in selected ranges of tens to thousands of hertz. Since frequencies associated with atmospheric scintillation are typically lower than this and have a power spectrum generally proportional to $1/f$, it is convenient to place the data in a frequency range where the atmospheric disturbances (not to mention telescope and building vibrations) are negligible.

Another related factor is the simultaneity of the observation. Although the FTS is a scanning instrument, it is scanning in *Fourier space*, and hence all spectral elements are observed simultaneously. This greatly reduces the effect of guiding errors in astronomical observations, or of source variations in laboratory measurements. In some cases the simultaneity is an essential feature of the experimental objective.

Internal normalization of the data further reduces the sensitivity of a measurement to variations of any kind in source flux (e.g. guiding errors): since the two detectors always detect all the flux in the passband, a ratio of the interferogram to the current total flux level provides a complete correction for source intensity variations that are common to all wavelengths.

Another method of reducing instrument sensitivity to irregularities in the atmosphere is the use of the common dual input aperture scheme, which provides continuous internal sky subtraction within the optical system prior to detection of the signal. This technique typically reduces the sky contamination by a factor of 50, and standard aperture switching techniques readily accommodate the residual second-order sky correction, resulting in a total cancellation factor of 10^3–10^4.

A very important consideration is the dynamic range of the measurement. The discrete diode detectors commonly used in an FTS have an extremely large dynamic range (typically 10^6). Furthermore, the FTS is inherently insensitive to nonlinearities in its detector, since their main effect is to produce sidebands that are typically outside the spectral passband.

All of the factors just cited, and some of those discussed below, contribute to the capability of the FTS to obtain spectra of very high signal-to-noise ratio. Of course, the FTS cannot exceed the fundamental limit due to a finite number of photons per spectral element. Some of the factors listed above help to maximize the use of the photons that enter the telescope. Other spectrometric techniques used in astronomy are inherently limited in the obtainable signal-to-noise, even in the case of arbitrarily large source flux. The fact that the FTS records data that are self-normalized, simultaneous throughout the passband, and rapidly scanned in Fourier space contributes to the very high values of signal-to-noise. As seen in the examples below, signal-to-noise ratios of hundreds, thousands, even tens of thousands to one are achieved.

2.2.3 QUALITY   A spectrograph must be judged in part by the quality of the spectra it produces, measured by how closely they approximate the result achieved by an ideal instrument. Several features of the FTS technique contribute to make it the standard against which other techniques must be judged.

The instrumental line profile is explicitly determined by easily controlled characteristics of the instrument and the observation, and in general it is possible to obtain a measurement in which the form of the profile is known analytically.

Scattered light, the bane of all grating spectrographs, does not exist in an FTS. Its closest analogue is a low-resolution feature due to nonlinearity in the detector system, which produces second harmonic and intermodulation products. If the region of interest covers a wavelength range of no more than $3:2$, then all of these unwanted sidebands may be ignored.

The FTS offers a unique capability for frequency calibration. The simultaneous measurement in Fourier space of all frequencies ensures a uniform frequency calibration throughout the passband. Frequency precision of a few parts in 10^7 is readily obtained; this corresponds directly to a velocity of $\sim 100 \text{ m s}^{-1}$. The accuracy of relative frequency determinations is usually signal-to-noise limited, and, as indicated above, the signal-to-noise can be very high; for the Sun, frequency accuracy of $1:10^8$ is possible.

Another feature of the FTS, and one of the most important, is the ease with which high resolution may be obtained. In fact, existing FTS instruments readily achieve resolution sufficient to completely resolve the most narrow spectral structures known in stellar, planetary, and nebular spectra. In such cases, the instrumental profile becomes irrelevant, since no distortion of the spectrum is introduced by the spectrometer.

The combination of these characteristics permits the FTS to obtain spectra of unparalleled quality. Combined with the earlier characteristics of high throughput and high signal-to-noise, these properties enable the FTS to produce the nearest thing available to definitive spectra of astronomical sources.

2.2.4 PHOTOMETRIC ACCURACY   A number of characteristics of the FTS make it a suitable instrument for recording photometric information. Rapid scanning and simultaneity of coverage reduce photometric errors. The typical use of a field of view substantially greater than the seeing disk minimizes photometric errors due to guiding. The internal normalization of the interferogram improves the accuracy of relative photometry within the bandpass (though contributing nothing to the calibration of the integrated bandpass). In addition, the very wide bandwidth typically achieved in FTS spectroscopy permits relative photometry over a substantial frequency range during a single measurement. It is also significant that the resolution of an FTS measurement is continuously variable from very low to very high in a simple and natural way. This means that low-resolution measurements of even faint photometric standards can be obtained in a fraction of the time spent on high-resolution measurements of program objects.

The large achievable bandwidth of high-resolution spectra also contributes unique capabilities. Photometric measurement of fluxes can be obtained for the narrow continuum points detected only with the highest resolution. Accurate interpolation of a continuum from distant portions of the spectrum is possible.

2.2.5 INSTRUMENTAL FLEXIBILITY   The FTS possesses many features that make it very flexible and adaptable for a variety of spectroscopic measurements, including the photometric character of the data, the variable resolution, and the availability of large field of view. Polarimetry may be added to the measurement in an elegant way to provide high-resolution polarimetric information (9, 87). Furthermore, the standard FTS configuration produces a stigmatic image of the input field at the output, permitting a variety of multiple-detector arrangements.

2.2.6 MULTIPLEXING CHARACTER OF THE FTS   As traditionally implemented, the FTS puts the full spectral band on a single detector in each of two output beams. Since each spectral element is encoded in the data at a distinct electrical frequency (multiplexed), the spectrum can be extracted. If the sensitivity is limited by photon noise (from noise or background), the speed of the FTS observation will be comparable to the speed of a single-detector scanning spectrometer. Therefore, in the visible the FTS may be slow compared with multidetector instruments. Use of the FTS at short wavelengths is thus motivated by other instrumental features (e.g. throughput, resolution, wavelength accuracy).

For the range 1.1–2.5 μm, spectroscopic measurements of faint sources are often limited by detector noise. In this case the FTS has a "multiplex advantage," owing to the fact that it does record a large range of spectrum simultaneously with one detector. The gain in speed compared with a scanning spectrometer can be about 10^4 with current detectors. This speed advantage will diminish as improved IR detectors become available.

3.   EXAMPLES OF FTS LABORATORY AND ASTRONOMICAL OBSERVATIONS

In this section we provide a range of examples that will allow the reader to extrapolate to different observational or laboratory projects. First, the proportionalities that govern the determination of signal-to-noise in the FTS method are indicated, followed by a presentation of some illustrated examples for a diversity of measurements. Details of instrumental and observational parameters are provided in the figure captions. In organizing the observational results, the emphasis is on the astronomical objective in preference to instrumental considerations.

For the purposes of scaling the performance of the FTS, the following expression will be useful:

$$\frac{S}{N} \approx \delta_\sigma t^{1/2} \begin{cases} F_\sigma \Delta_\sigma^{-1/2}, & \text{1.} \\ F_\sigma, & \text{2.} \\ F_\sigma^{1/2} \Delta_\sigma^{-1/2}, & \text{3.} \end{cases}$$

where δ_σ is the resolution element width, t is the integration time, F_σ is the flux per spectral element, and Δ_σ is the full spectral bandpass on a single detector. The three equations refer to observations where the dominant noise source is (1) thermal background, (2) detector noise, and (3) source photon noise. To assist in determining which is the case, Figure 1 shows the boundaries between these three situations for a variety of possible parameters.

3.1 Solar and Stellar Studies

Nowhere is the close relation between solar, stellar, and laboratory science more evident than in the area of high-resolution spectroscopy. It is now possible to obtain spectral resolution and signal-to-noise on bright stars that were obtainable only for the Sun just a few years ago. And for both the Sun and stars, spectra of laboratory quality are often achieved.

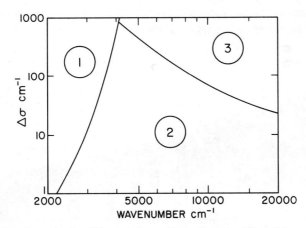

Figure 1 Signal-to-noise should be scaled according to Equation 1, 2, or 3 depending on the region of this figure in which the observation lies. The boundary between regions 2 and 3 is for magnitude $M = 7$. For other magnitudes, shift the boundary vertically according to $\Delta_\sigma \approx F^{-1/2}$. The boundary between regions 1 and 2 is for telescope/instrument temperature of 15°C. Other parameters: NEP $= 2 \times 10^{-16}$W Hz$^{-1/2}$, QE $= 0.80$.

Figure 2 is a low-dispersion plot of a high-resolution solar spectrum. This type of data is readily obtained for the Sun in any spectral region, or for bright stars from 1 to 5 μm. The relative flux calibration was obtained from a laboratory lamp, but it could equally well have been determined from a standard star. This capability of producing a very-high-resolution spectrum with broadband photometric integrity is a new and valuable tool for astronomers. The continuum can be determined by interpolation (perhaps with the aid of a synthetic model spectrum) from distant frequencies.

Such spectra have been used in a variety of abundance analyses for the Sun and stars. Results of abundance studies by FTS methods have been reviewed recently (66, 74). Examples include studies of C, N, O and isotopic abundances and of the H/H_2 ratio (48, 106) in cool stars. Among the few solar abundance studies was the determination of the solar nickel isotope ratio in a combined laboratory-solar investigation (10).

The Sun has been used to empirically calibrate the strengths of certain molecular transitions that are important in stellar spectroscopy. Recent examples include OH (36), C_2 (12), and CN (86). FTS laboratory spectroscopy continues to provide line positions and strengths for such molecules as OH (65), H_2O (29), and many, many others (for examples, see recent issues of the *Journal of Molecular Spectroscopy*). At present, most atomic lines in infrared stellar spectra are still unidentified (e.g. 75), but FTS laboratory measurements are producing large numbers of new identifications and are also radically improving our knowledge of line positions and atomic energy levels.

Sophisticated model atmospheres are indispensable for interpretation of solar and stellar spectra. Equally important are diagnostics for the choice

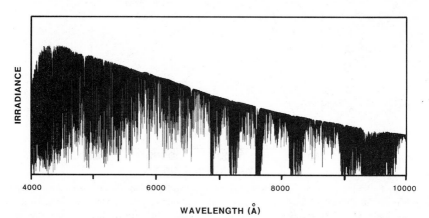

Figure 2 A compressed, high-resolution plot of the spectrum of the Sun in integrated light. Parameters: $\delta\sigma = 0.04 \text{ cm}^{-1}$, $t = 1$ h, tel = 1.5 m.

and verification of the models. Use of resonance-line profiles in the blue is complemented by measurement of the strong CO bands in the infrared. Heasley et al. (46) studied the 5-μm CO fundamental in Arcturus, and Tsuji (105) and Ayres & Testerman (3) observed these bands in the Sun. All three groups inferred the need for multicomponent models in the upper solar and stellar atmospheres.

Our empirical knowledge of solar and stellar convection, stellar rotation, and magnetic fields comes almost exclusively from line profiles. The FTS is well suited to record fully resolved profiles of low noise and high wavelength precision for the Sun and brighter stars. The subtle asymmetries in a line profile are often indiscernible in a normal spectral plot, even at high dispersion. A common method for emphasizing the deviations from symmetry is the bisector technique. The bisectors of chords across the profile at various depths may be plotted at greatly enhanced dispersion, as in Figure 3. These bisector curves by Livingston (61) show the characteristic C-shape observed in the Sun. The upward convecting regions of the photosphere, which contribute somewhat more than half of the emergent intensity, produce a blueshift in the central regions of the line profiles, while the stationary layer above the convection zone and the slowly convecting deep layers yield core and wing frequencies near the center-of-mass velocity (20). The velocity noise in a one-day sequence of such full-disk spectra is only 5 m s^{-1}.

A study of cool giants (72) found much more extreme effects, and an S-shaped bisector, suggestive of mass loss above the convective region. Hinkle et al. (47) have reported the time evolution of complex, multicomponent line profiles in Mira variables. They are able to distinguish numerous atmospheric regimes, including a rising pulsation wave, falling material, stationary overlying material, and several shells of ejected gas.

The IR spectrum is favorable for magnetic field studies, as the ratio of Zeeman splitting to Doppler width increases with wavelength. The unexpected discovery of atomic emission lines near 12 μm in the infrared solar spectrum (13, 68) provides an excellent example of line profiles useful as a probe of magnetic fields. The strongest of these lines have recently been shown by Chang & Noyes (15) to be due to high angular momentum states in neutral Mg and Al, with a Landé g-value of 1. Even with this relatively low value of g, they have the highest Zeeman sensitivity of any known lines. Figure 4 shows the Zeeman-split profiles for a number of magnetic regions on the solar surface. The emission is presumed to be chromospheric within a broader photospheric line; thus information about the field variation with depth is encoded in the profiles. In a stellar observation, Giampapa et al. (35) have measured the splitting of an Fe I line at 6388.65 cm^{-1} in the RS CVn star λ And; they detected a large field (1300 G).

In a unique series of observations, Stenflo et al. (87) have modified an FTS to function as a spectropolarimeter, and they have used this to survey the solar limb spectrum from 4200 to 9950 Å at high resolution. They report polarization effects in many lines, and discuss the results with reference to resonant and fluorescent scattering and quantum mechanical interference effects.

Low-resolution IR spectroscopy offers considerable scope for the FTS in stellar studies, but this area is not well developed (66, 74).

3.2   *Planetary Astronomy*

In 1969, Connes et al. (16) published their remarkable infrared atlas of the spectra of Venus, Mars, and Jupiter, and in so doing demonstrated the

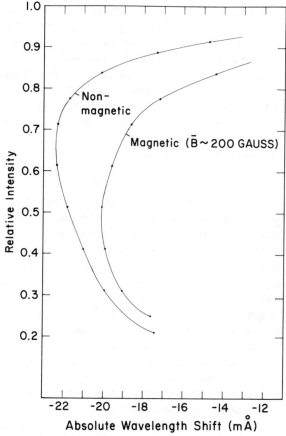

Figure 3   Typical FTS line bisectors for magnetic and nonmagnetic regions (61). The line is Fe 5250.6 Å. Parameters: aperture = $1' \times 2'$, $\delta\lambda = 0.008$ Å, $t = 6$ min.

information-gathering capability of an FTS. The further opening of the infrared in subsequent years has produced many pioneering investigations and contributed much to the revival of planetary spectroscopy.

Results obtained by Fourier transform spectroscopy include a large increase in the number of known chemical constituents of the solar system. Measured elemental abundances in the outer planets may determine the "cosmic" values, or they may instead reveal something about the planetary formation process; it is not yet clear which will be the case (33). The best estimate of the deuterium abundance should eventually come from the outer planets, although not without some effort (21, 23, 26, 53). Other work on planetary atmospheres has revealed stratospheric emission due to airglow on Mars and photochemistry on Jupiter, Saturn, and Titan (52, 64). Infrared spectra led to the discovery of the temperature inversion in the atmospheres of the giant planets, and they have since been employed to map the thermal structures of these atmospheres (70). Abundances of PH_3 and H_2O reveal the presence of nonequilibrium chemistry deep in the atmospheres of Jupiter and Saturn (53, 59). FTS spectra of Saturn have been used to study NH_3 in both gaseous and solid form (28) in an attempt to understand its low abundance there. Remote mineralogical analysis of solid planetary, ring, satellite, and asteroid materials is a rich field, with detections of various ices in the outer solar system and of a number of

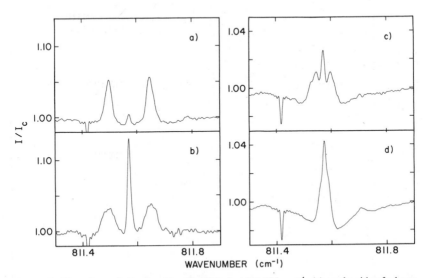

Figure 4   The solar emission line $7i \to 6h$ of Mg I at 811.575 cm^{-1}: (*a*) on the side of a large sunspot penumbra toward disk center; (*b*) on the side of the penumbra toward the limb; (*c*) and (*d*) in two locations in a plage near disk center (13). Parameters: $\delta\sigma = 0.005$ cm^{-1}, $S/N \sim 10^3$, aperture ≤ 60 arcsec^2, $t = 10$–60 min, tel = 1.5 m.

mineralogical groups that correlate with known meteoritic materials (24, 25, 30, 31, 57, 60). The only available IR spectrum of a comet was also obtained with an FTS (49).

Excellent recent reviews of planetary spectroscopy, with emphasis on FTS techniques, have been given by Fink & Larson (27), Larson (58), and Encrenaz & Combes (22). The results of the last few years are too numerous to list, but they are mostly concerned with the composition of the atmospheres of the outer planets. Lutz et al. (63) recorded a spectrum of Titan in the infrared H band. A portion of this spectrum (shown in Figure 5) shows the evidence for the detection of CO. This molecule was first sought as a possible vestigial remnant of the primordial solar nebula, although contemporary production by photochemical processes appears possible. In the course of a related laboratory study of methane, also employing Fourier spectroscopy, Lutz et al. (62) discovered a previously unknown band of CH_3D near 6427 cm^{-1}. It is present in their Titan spectrum and should yield an improved estimate of the D/H ratio.

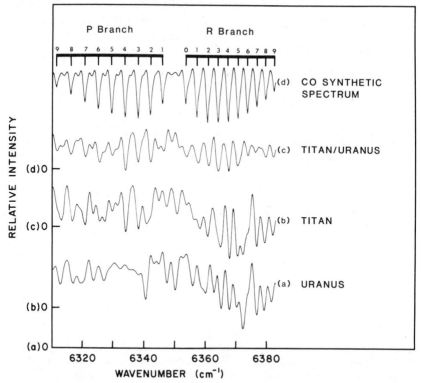

Figure 5   The detection of CO in the atmosphere of Titan (63). Parameters: $\delta\sigma = 1.2$ cm^{-1}, H filter, $m_H \sim 8.4$, $t = 7.5$ h, tel = 4 m.

Each Voyager spacecraft carried a Fourier spectrometer and obtained superb spectra of Jupiter, Saturn, and their satellites, with good spatial resolution and low noise, though only modest spectral resolution [Hanel et al. (41, 42, 44, 45)]. The spectrum of the North Equatorial Belt of Jupiter is shown in Figure 6 [Kunde et al. (53)]. This spectrum covers a wide spectral range with high photometric accuracy. It has been used to study the thermal balance of the atmosphere (70), the He/H ratio, and the possible existence of trace constituents. Ground-based spectroscopy, limited to atmospheric windows, can produce complementary information. In a co-ordinated observation from the ground, Tokunaga et al. (102) observed three regions on Jupiter with higher spectral resolution in a small bandpass. Their spectrum of the North Tropical Zone is shown in Figure 7. Although noisier, this spectrum reveals details of the molecular band structure and shows trace gases such as $^{15}NH_3$ and C_2H_6.

Continuing activity shows that much remains to be done in planetary spectroscopy, and substantial improvements in sensitivity are badly needed. Many detected molecular bands have not yet been analyzed, and there is a persistent need for supporting laboratory spectroscopy. Since wide spectral coverage, high resolution, and accurate line profiles are important in this work, it is likely that Fourier spectroscopy will continue to serve this field.

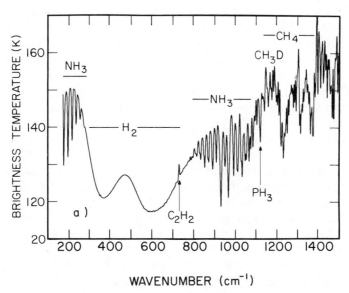

Figure 6 The mean spectrum of Jupiter's North Equatorial Belt, from *Voyager 1* (53). Parameters: $\delta\sigma = 4.3$ cm^{-1}.

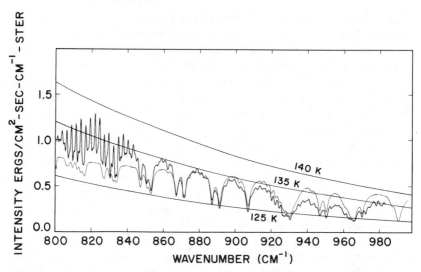

Figure 7 The ground-based mean spectrum of Jupiter's North Tropical Zone (dark curve), compared with a synthetic spectrum (light curve) (102). Parameters: $\delta\sigma = 1.1$ cm^{-1}, aperture = 3.5″, $t = 3$ h, tel = 4 m.

3.3   *Young Stars, Molecular Clouds, and Gaseous Nebulae*

The high extinction of molecular clouds precludes most visible observations. However, both the infrared and the radio regimes have been fruitfully employed to penetrate the obscuring dust. FTS spectroscopy has been particularly successful in revealing new phenomena associated with regions of star formation, especially including hydrogen Bγ emission from compact H II regions and stellar winds, and H$_2$ emission from shock fronts.

The Steward Observatory FTS (88) has been used in a survey program to reveal the major emission features from obscured sources in the 4000–6500 cm^{-1} region (54, 90–93, 99, 101). An example is shown in Figure 8. The resolution is ~ 130 km s^{-1}, which maximizes the sensitivity to lines of this width. Evaluation of the extinction by Thompson (94, 95) revealed excess line emission relative to the spectral type estimated from the total luminosity. These results have stimulated lively discussion of the deviations from recombination theory in a stellar wind or an ultradense H II region optically thick in the hydrogen lines, and of possible accretion disks (51, 84, 95, 96). In evaluating the line-formation process, line profiles are again of great importance. Hall et al. (37), Simon et al. (83, 84), and Scoville et al. (81) have examined profiles of Bα and Bγ for several of these sources and found wide wings > 100 km s^{-1}, suggestive of a stellar wind.

Only one obscured young star has been the subject of extensive study.

This is the BN source in the Orion molecular cloud OMC-1. Note that this cloud is behind the famous Orion H II nebula. A series of high-resolution FTS spectra has been obtained in the accessible portions of the 2–5 μm region (37, 78, 79, 81). These spectra reveal complex absorption and emission profiles in the CO transitions.

Figure 9, from Scoville et al. (81), shows the CO overtone vibration-rotation band. The emission from excited vibrational states indicates the remarkable excitation temperature $T_{vib} > 19,000$ K. This suggests that the molecular material is a useful probe to a region quite close to the exciting star. The absorption in Figure 9 shows a low excitation temperature, appropriate for the foreground molecular cloud.

Figure 10 [Scoville et al. (81)] shows a selection of line profiles from the CO vibration-rotation fundamental band near 5 μm. Two prominent absorption features are associated with the undisturbed foreground molecular cloud and with the expanding shell that is pushed out by some unknown energetic activity within the cloud. A strong emission feature,

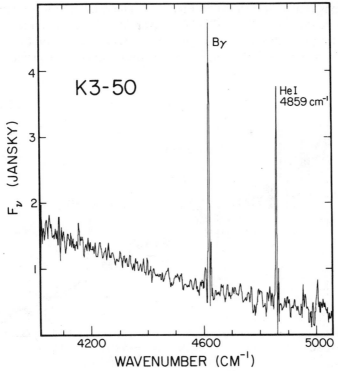

Figure 8 The obscured IR object K3-50 [Thompson & Tokunaga (93)]. Variation in continuum indicates typical noise. Parameters: $\delta\sigma = 2.0$ cm^{-1}, K band, $t = 1$ h, tel = 2.3 m.

with low excitation temperature, is quite distinct from the hot emission seen in Figure 9. Yet another possible absorption feature has the dynamical signature appropriate to infalling material. The disentanglement of this spectral detail appears to require multiple sources or a nonspherical geometry. Possibilities include a remnant disk and a bow shock (81). A particular puzzle from the high-resolution data is that the sources studied thus far (BN, CRL 490, and M17 IRS1) all have substantial velocities relative to their surrounding molecular cloud, suggestive of rapid escape from the presumed place of birth.

The discovery of molecular hydrogen emission from the Orion molecular cloud (32) provided a new and unexpected probe of the cloud interior. It is generally believed that these emission lines are a cooling mechanism for shock-heated material, although this picture does not provide a fully satisfactory understanding of the phenomenon. Active observational work by the FTS, the Fabry-Perot, and other means has extended the known sites of H_2 emission to include T Tauri stars, Herbig-Haro objects, planetary nebulae, and supernova remnants (6). An airborne observation by Davis et al. (19) is shown in Figure 11. The resolution is ~ 90 km s^{-1}, similar to the known line widths. The advantage of working at high altitude lies in the possibility of observing lines that are partially or totally obscured from the ground, as is the case for some of the O- and Q-branch lines. In this

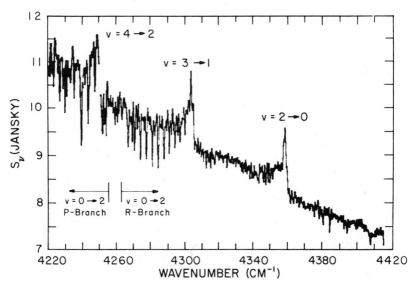

Figure 9   CO $v = 2 \rightarrow 0$ absorption and emission from the BN source (81). Parameters: $\delta\sigma = 0.3$ cm^{-1}, K band, $m_K \sim 5.0$, $t = 4.8$ h, tel = 4 m.

spectrum the H_2 emission is from deep within the cloud, while the H I emission is from the foreground H II region.

Considerable attention has been given to the question of the extinction in front of the H_2 emission regions. This is critical in estimating the energetics and the geometry. High-resolution measurements of the S- and Q-branch lines have been obtained in order to estimate reddening. A selection of resolved profiles is shown in Figure 12 from Scoville et al. (80). The authors consider the possibility that the blue and red wings arise in different regions of the cloud and have different extinction. They find that the most highly blueshifted emission is more reddened than the low-velocity emission.

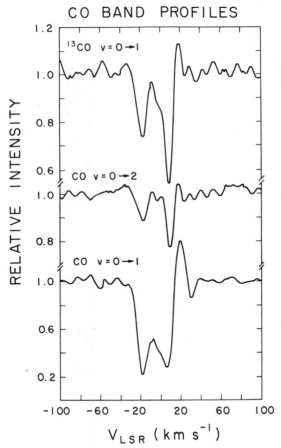

Figure 10   CO $v = 1 \rightarrow 0$ and $2 \rightarrow 0$ line profiles in the BN spectrum (81). Parameters: res = 7 km s^{-1}, $m_K \sim 5.0$, $m_L \sim 1.7$, tel = 4 m.

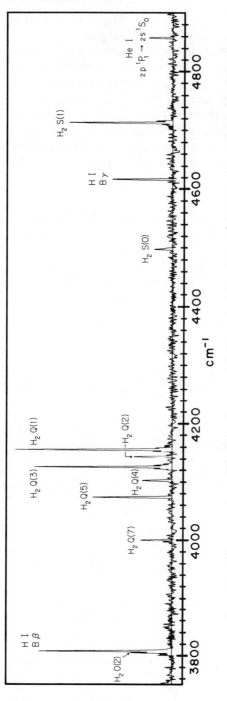

Figure 11 Airborne spectrum of the Orion H₂ source (19). Parameters: $\delta\sigma = 1.2$ cm^{-1}, 2.1–2.7 μm, $t = 150$ min, tel $= 0.9$ m.

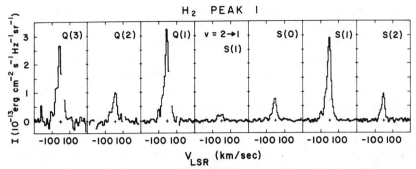

Figure 12 High-resolution spectrum of the H_2 band in emission peak 1 in Orion (80). Parameters: $\delta\sigma = 0.3$ cm^{-1}, K band, aperture = 3.75″, $t = 70$ min, tel = 4 m.

FTS studies in the IR have also been extended to galactic nebulae. These studies have included detection and study of H_2 emission (85, 104), measurement of H I and He I emission-line widths (100), and unsuccessful attempts to resolve line structure in the mysterious 3.3-μm emission feature (103). Airborne, high-resolution spectra in the far-IR have been obtained to study the abundances and cooling processes in H II regions [for example, in M8, M17, and NGC 7538 (5, 67)].

Substantial observational effort in the radio and IR is aimed at identifying protoplanetary nebulae. FTS studies near 2 μm have produced evidence that one candidate, HM Sag, almost certainly is not one, as it shows evidence of a cool-star spectrum (98). CRL 618, however, which has both H I and H_2 emission from a complicated structure, might be (97).

3.4 *Mass Loss from Evolved Stars*

The study of mass loss is an ideal area for the application of the FTS technique. Line profiles at the highest resolution are required, and many transitions are needed to explore a range of excitation temperatures. This subject is still in a primitive state of development because of the difficulties of coupling observation and theory. On the theoretical side, it is not clear how mass loss occurs. In cool, luminous stars, mass loss is observed to be accompanied by a dusty circumstellar envelope. Grains can be expelled by radiation pressure under some conditions, but static stellar models do not produce grains in sufficient abundance to account for observed mass loss rates. Interpretation of the observational material is at an impasse because of the inability to assign a radius to the inner boundary of the circumstellar shell. Infrared studies hold the promise of deducing the spatial extent of the shell from the excitation temperature of the spectrum. Particularly attractive is the availability of many tens of transitions of the CO molecule from the ground vibrational state. Owing to its high dissociation energy,

CO is expected to be stable from the photosphere of the cool star out through the shell.

The most extensive case studies of cool-star mass loss based on IR FTS data are for χ Cyg (47), α Ori (7), and IRC + 10216 (50, 73). Figure 13 shows a line profile extracted from a spectrum of the CO fundamental band in the M supergiant α Ori. This profile reveals several components. The telluric CO line, which has not been removed in the reduction, shows the position of the unshifted line. The very broad stellar photospheric line reaches a depth of only about 70%, typical of strong photospheric lines in this spectral region. Narrow circumstellar absorption appears with two distinct components, S1 and S2, which clearly differ in Doppler shift. From inspection of many lines of differing lower-state excitation, it is possible to estimate the rotational temperature of the CO. The S1 component has $T_{rot} \sim 200$ K, and S2 has $T_{rot} \sim 70$ K. This reveals a common feature of late-type stellar shells: The material is apparently clumped in discrete velocity regimes, with the rotational temperature decreasing with increasing expansion velocity (8, 71).

A long observational study of χ Cyg, an S-type long-period variable, has provided a possible clue to the cool-star mass loss process by discovering a warm, stationary shell of gas above the pulsating photosphere [Hinkle et al. (47)]. The authors speculate that dynamic activity may support a grain-forming region at several stellar radii, from which outward acceleration may begin. The detection of discrete shells suggests that either the outflow is modulated and episodic, or that the acceleration is accomplished in some sense impulsively, rather than continuously (71).

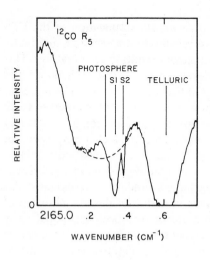

Figure 13 CO vibration-rotation $1 \rightarrow 0$ $R(5)$ line in α Ori. [Bernat et al. (7)]. Parameters: $\delta\alpha = 0.006$ cm^{-1}, 2110–2200 cm^{-1}, $S/N = 95$, $t = 180$ min, tel $= 4$ m.

An extensive series of observations of IRC + 10216, as well as a detailed interpretation by numerical modeling, has been used to estimate parameters of mass loss in this extreme case (50). Figure 14 shows an interesting comparison between line profiles of CO rotational transitions, measured in the millimeter radio spectrum (69), and profiles of CO vibration-rotation transitions from the infrared. The millimeter profiles are of purely emission lines, taken with an antenna that did not resolve the source structure (typical in such measurements). Emission is detected equally from the front and back sides of the shell. The two CO species show profiles characteristic of optically thick and optically thin emission in an expanding spherical shell. The IR profiles show velocity structure in the foreground shell, and P Cygni emission from near the source.

The infrared profiles contain much more, though different, information. First, the absorption component is formed against the continuum source, and thus the profiles are not blurred seriously by projection effects. Consequently, it is possible to distinguish considerable structure in the shell dynamics, as discussed above. In this case, IRC + 10216 has 3 distinct shells, with $T_{rot} \sim 700, 300,$ and 120 K, respectively, and with increasing expansion velocity. Second, the emission component of the 1-0 lines is formed close to the star, and it shows the effects of obscuration of the back side of the

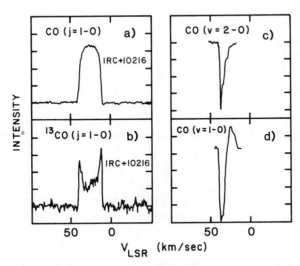

Figure 14   CO emission and absorption in the IRC + 10216 circumstellar shell [Keady (50)]. Millimeter emission from (*a*) the optically thick $^{12}CO\ J = 1 \rightarrow 0$ line and (*b*) the optically thin $^{13}CO\ J = 1 \rightarrow 0$ line. Infrared absorption-emission profiles from (*c*) the $v = 2 \rightarrow 0$ band and (*d*) the $v = 1 \rightarrow 0$ band.

emission shell. Finally, a broad, shallow absorption, attributed to the photospheric line, may show effects of scattering from the expanding dust shell. In sum, the IR line shows one emission and four absorption components, whereas the millimeter spectrum showed one in emission. However, for measuring the total shell mass, or for detecting minor constituents, the millimeter measurement is far more sensitive to the large volume of material at large radius from the source. Keady (50) estimates a mass loss of $10^{-4} M_\odot \, yr^{-1}$, and he finds that several regimes of acceleration can account for the peculiar velocity structure of the lines.

A number of F supergiants have been determined to have high mass loss rates. It seems unlikely that dust plays a role in the ejection process. The infrared region, especially the CO molecule, is nevertheless a good diagnostic for the shells of these stars. Thompson & Boroson (89) studied infrared Na I and Mg I emission in the wind or shell of IRC + 10420. They attributed the emission to UV pumping of the upper levels and inferred a high density of material above the photosphere. Lambert et al. (55) used IR spectra of CO to investigate the photospheric and circumstellar mass motions around ρ Cas and HR 8752. In 1975 the latter star apparently ejected a discrete shell, which has decelerated and is now falling back. Pulsations in the underlying photosphere have also been monitored.

3.5 Galactic Nuclei

The nuclear regions of our own and other galaxies are within reach of an FTS, although at very modest spectral resolution. The center of our Galaxy is sufficiently close that we may distinguish several discrete components, both extended and pointlike. In spite of an estimated 30 magnitudes of visual extinction to the galactic center, infrared spectroscopy of these sources is possible. Several of the point sources are found to be late-type supergiants or star clusters (2, 40, 107). Doppler shifts of these may be used to probe the virial mass of the galactic center and eventually, perhaps, some of the stellar parameters.

In Figure 15, from Hall et al. (40), the spectrum of IRS 7 shows the CO bands characteristic of a cool, high-luminosity star. Figure 15 also displays the spectrum of IRS 16, the slightly extended source at the center of the far-infrared emission region. Weak stellar CO indicates a contribution from a spectral type earlier than K0, or perhaps spectral dilution by continuum emission from a nonstellar source. Emission from ionized gas, also evident in the spectrum of IRS 16, provides another indication of activity in the galactic core. The interest in the gaseous emission hinges on the line profiles. Wollman et al. (107) found a complex Bγ profile, with narrow and broad components, while Hall et al. (40) determined a remarkable velocity width of 2000 km s^{-1} for the He II line at 2.06 μm. These flows are probably determined by an unknown combination of gravitational, radiative, and other forces, and the interpretation may not be straightforward.

A spectrum of the central region of the spiral galaxy M81 was obtained by Aaronson (1). The CO bands at 4300 cm^{-1} were detected, confirming the assumption of earlier filter photometry that a depression there is due to CO in cool stellar photospheres. A number of active galactic nuclei have been studied. Rieke et al. (76) measured CO absorption and Bγ emission in spectra of the IR luminous galaxies M82 and NGC 253. They found that star-burst models with a dearth of low-mass stars could account for many of the observed characteristics of these nuclei.

Infrared spectra of the central regions of the bright Seyfert galaxies NGC 4151 and NGC 1068 have been recorded. Rieke & Lebofsky (77) combined a variety of spectrally and spatially resolved data to interpret the continuum and line emission detected from NGC 4151. They gave particular emphasis to the role of dust, which evidently is present in both the broad and narrow emission-line regions of the nucleus. An interesting comparison is found in the NGC 1068 observations of Hall et al. (39). Their

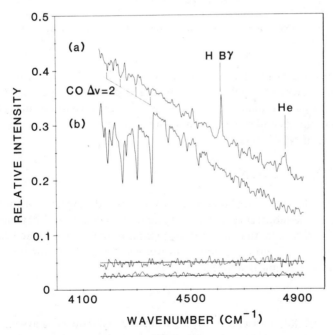

Figure 15   Galactic center sources: (*a*) IRS 16; (*b*) IRS 7 (40). Parameters: $\delta\sigma = 3.0$ cm^{-1}, K band, $m_K \sim 6.7, 8.4, t = 70, 170$ min, tel = 4 m.

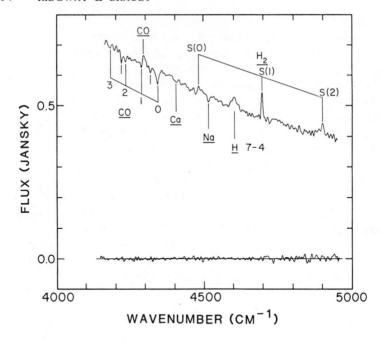

Figure 16   NGC 1068 (39). Parameters: $\delta\sigma = 5.0\,\mathrm{cm}^{-1}$, K band, $m_K \sim 7.8$, $t = 8$ h, tel = 4 m.

spectrum appears in Figure 16. It shows cool-star absorption, broad Bγ emission, narrow H_2 emission, hot dust, and possibly CO emission. The authors discussed analogies with sites of star formation in galactic molecular clouds and suggested that similar activity takes place on a very large scale in NGC 1068.

4. CONCLUSION

FTS methods have contributed in depth to a broad range of astrophysical research in the laboratory, the solar system, the Galaxy, and beyond. The FTS has surely more than fulfilled the dreams of the pioneers who brought it to fruition; it is now a unique and well-established tool for the observation of astronomical sources.

ACKNOWLEDGMENTS

We are grateful to Richard Joyce, John Keady, Harold Larson, John Leibacher, Michael Merrill, and Rodger Thompson for reading and commenting on a draft of this paper.

Literature Cited

1. Aaronson, M. 1981. See Ref. 108, pp. 297–316
2. Augason, G. C., Smith, H. A., Wollman, E. R., Larson, H. P., Johnson, H. R. 1982. In *The Galactic Center*, ed. G. R. Riegler, R. D. Blandford, pp. 82–84. New York: Am. Inst. Phys. 216 pp.
3. Ayres, T. R., Testerman, L. 1981. *Ap. J.* 245:1124–40
4. Baluteau, J. P., Anderegg, M., Moorwood, A. F. M., Coron, N., Beckman, J. E., et al. 1977. *Appl. Opt.* 16:1834–40
5. Baluteau, J. P., Moorwood, A. F. M., Biraud, Y., Coron, N., Anderegg, M., Fitton, B. 1981. *Ap. J.* 244:66–75
6. Beckwith, S. 1981. See Ref. 108, pp. 167–78
7. Bernat, A. P., Hall, D. N. B., Hinkle, K. H., Ridgway, S. T. 1979. *Ap. J. Lett.* 233:L135–39
8. Bernat, A. P. 1981. *Ap. J.* 246:184–92
9. Brault, J. W. 1979. *Ossni. Mem. Oss. Astrofis. Arcetri* 106:33–50
10. Brault, J. W., Holweger, H. 1981. *Ap. J. Lett.* 249:L43–46
11. Brault, J. W. 1982. *Philos. Trans. R. Soc. London Ser. A* 307:503–11
12. Brault, J. W., Delbouille, L., Grevesse, N., Roland, G., Sauval, A. J., Testerman, L. 1982. *Astron. Astrophys.* 108:201–5
13. Brault, J. W., Noyes, R. W. 1983. *Ap. J. Lett.* 269:L61–66
14. Carrington, A., Ramsay, D. A., eds. 1982. *Molecules in Interstellar Space.* London: R. Soc. 167 pp.
15. Chang, E. C., Noyes, R. W. 1983. *Ap. J. Lett.* 275:L11–13
16. Connes, J., Connes, P., Maillard, J.-P. 1969. *Atlas of Near Infrared Spectra of Venus, Mars, Jupiter and Saturn.* Paris: CNRS. 475 pp.
17. Connes, P. 1970. *Ann. Rev. Astron. Astrophys.* 8:209–30
18. Davis, D. S., Larson, H. P., Williams, M., Michel, G., Connes, P. 1980. *Appl. Opt.* 19:4138–55
19. Davis, D. S., Larson, H. P., Smith, H. A. 1982. *Ap. J.* 259:166–79
20. Dravins, D., Lindegren, L., Nordlund, Å. 1981. *Astron. Astrophys.* 96:345–64
21. Drossart, P., Encrenaz, T., Kunde, V., Hanel, R., Combes, M. 1982. *Icarus* 49:416–26
22. Encrenaz, T., Combes, M. 1981. See Ref. 108, pp. 1–34
23. Encrenaz, T., Combes, M. 1982. *Icarus* 52:54–61
24. Feierberg, M. A., Lebofsky, L. A., Larson, H. P. 1981. *Geochim. Cosmo-chim. Acta* 45:971–81
25. Feierberg, M. A., Larson, H. P., Chapman, C. R. 1982. *Ap. J.* 257:361–72
26. Fink, U., Larson, H. P. 1978. *Science* 201:343–45
27. Fink, U., Larson, H. P. 1979. In *Fourier Transform Infrared Spectroscopy: Applications to Chemical Systems*, eds. J. R. Ferraro, L. J. Basile, 2:243–315. New York: Academic. 336 pp.
28. Fink, U., Larson, H. P., Bjoraker, G. L., Johnson, J. R. 1983. *Ap. J.* 268:880–88
29. Flaud, J.-M., Camy-Peyret, C., Maillard, J.-P., Guelachvili, G. 1977. *J. Mol. Spectrosc.* 65:219–28
30. Fraundorf, P., Patel, R. I., Freeman, J. J. 1981. *Icarus* 47:368–80
31. Gaffey, M. J., McCord, T. B. 1979. See Ref. 34, pp. 688–723
32. Gautier, T. N., Fink, U., Treffers, R., Larson, H. P. 1976. *Ap. J. Lett.* 207:L129–33
33. Gautier, D., Bezard, B., Marten, A. Baluteau, J. P., Scott, N., et al. 1982. *Ap. J.* 257:901–12
34. Gehrels, T., ed. 1979. *Asteroids.* Tucson: Univ. Ariz. Press. 1181 pp.
35. Giampapa, M. S., Golub, L., Worden, S. P. 1983. *Ap. J. Lett.* 268:L121–25
36. Goldman, A., Murcray, F. J., Gillis, J. R., Murcray, D. G. 1981. *Ap. J. Lett.* 248:L133–35
37. Hall, D. N. B., Kleinmann, S. G., Ridgway, S. T., Gillett, F. C. 1978. *Ap. J. Lett.* 223:L47–50
38. Hall, D. N. B., Ridgway, S. T., Bell, E. A., Yarborough, J. M. 1979. *Proc. Soc. Photo-Optical Instrum. Eng.* 172:121–29
39. Hall, D. N. B., Kleinmann, S. G., Scoville, N. Z., Ridgway, S. T. 1981. *Ap. J.* 248:898–905
40. Hall, D. N. B., Kleinmann, S. G., Scoville, N. Z. 1982. *Ap. J. Lett.* 260:L53–57
41. Hanel, R., Conrath, B., Flasar, M., Herath, L., Kunde, V., et al. 1979. *Science* 206:952–56
42. Hanel, R., Conrath, B., Flasar, M., Kunde, V., Lowman, P., et al. 1979. *Science* 204:972–76
43. Hanel, R., Crosby, D., Herath, L., Vanous, D., Collins, D., et al. 1980. *Appl. Opt.* 19:1391–1400
44. Hanel, R., Conrath, B., Flasar, F. M., Kunde, V., Maguire, W., et al. 1981. *Science* 212:192–200
45. Hanel, R., Conrath, B., Flasar, F. M., Kunde, V., Maguire, W., et al. 1982. *Science* 215:544–48

316   RIDGWAY & BRAULT

46. Heasley, J. N., Ridgway, S. T., Carbon, D. F., Milkey, R. W., Hall, D. N. B. 1978. *Ap. J.* 219:970–78
47. Hinkle, K. H., Hall, D. N. B., Ridgway, S. T. 1982. *Ap. J.* 252:697–714
48. Johnson, H. R., Goebel, J. H., Goorvitch, D., Ridgway, S. T. 1983. *Ap. J. Lett.* 270:L63–67
49. Johnson, J. R., Fink, U., Larson, H. P. 1983. *Ap. J.* 270:769–77
50. Keady, J. J. 1982. *The circumstellar envelope of IRC + 10216.* PhD thesis. N. Mex. State Univ., Las Cruces
51. Krolik, J. H., Smith, H. A. 1981. *Ap. J.* 249:628–36
52. Kunde, V. G., Aikin, A. C., Hanel, R. A., Jennings, D. E., Maguire, W. C., Samuelson, R. E. 1981. *Nature* 292:686–88
53. Kunde, V., Hanel, R., Maguire, W., Gautier, D., Baluteau, J. P., et al. 1982. *Ap. J.* 263:443–67
54. Lada, C. J., Gautier, T. N. 1982. *Ap. J.* 261:161–69
55. Lambert, D. L., Hinkle, K. H., Hall, D. N. B. 1981. *Ap. J.* 248:638–50
56. Larson, H. P., Fink, U. 1975. *Appl. Opt.* 14:2085–95
57. Larson, H. P., Veeder, G. J. 1979. See Ref. 34, pp. 724–44
58. Larson, H. P. 1980. *Ann. Rev. Astron. Astrophys.* 18:43–75
59. Larson, H. P., Fink, U., Smith, H. A., Davis, D. S. 1980. *Ap. J.* 240:327–37
60. Lebofsky, L. A., Feierberg, M. A., Tokunaga, A. T., Larson, H. P., Johnson, J. R. 1981. *Icarus* 48:453–59
61. Livingston, W. C. 1982. *Nature* 297:208–9
62. Lutz, B. L., de Bergh, C., Maillard, J.-P., Owen, T., Brault, J. 1981. *Ap. J. Lett.* 248:L141–45
63. Lutz, B. L., de Bergh, C., Owen, T. 1983. *Science* 220:1374–75
64. Maguire, W. C., Hanel, R. A., Jennings, D. E., Kunde, V. G., Samuelson, R. E. 1981. *Nature* 292:683–86
65. Maillard, J.-P., Chauville, J., Mantz, A. 1976. *J. Mol. Spectrosc.* 63:120–41
66. Merrill, K. M., Ridgway, S. T. 1979. *Ann. Rev. Astron. Astrophys.* 17:9–41
67. Moorwood, A. F. M., Baluteau, J.-P., Anderegg, M., Coron, N., Biraud, Y., Fitton, B. 1980. *Ap. J.* 238:565–76
68. Murcray, F. J., Goldman, A., Murcray, F. H., Bradford, C. M., Murcray, D. G., et al. 1981. *Ap. J. Lett.* 247:L97–99
69. Olofsson, H., Johansson, L. E. B., Hjalmarson, Å., Rieu, N.-G. 1982. *Astron. Astrophys.* 107:128–44
70. Orton, G. S. 1981. See Ref. 108, pp. 35–56
71. Ridgway, S. T. 1981. In *Physical Processes in Red Giants*, ed. I. Iben, A. Renzini, pp. 305–9. Dordrecht: Reidel. 488 pp.
72. Ridgway, S. T., Friel, E. D. 1981. In *Effects of Mass Loss on Stellar Evolution*, ed. C. Chiosi, R. Stalio, pp. 119–24. Dordrecht: Reidel. 566 pp.
73. Ridgway, S. T., Keady, J. J. 1982. See Ref. 14, pp. 33–38
74. Ridgway, S. T. 1984. In *Galactic and Extragalactic Infrared Spectroscopy*, ed. M. F. Kessuer, J. P. Phillips, pp. 309–30. Dordrecht: Reidel
75. Ridgway, S. T., Carbon, D. F., Hall, D. N. B., Jewell, J. 1984. *Ap. J. Suppl.* 54:177–210
76. Rieke, G. H., Lebofsky, M. J., Thompson, R. I., Low, F. J., Tokunaga, A. T. 1980. *Ap. J.* 238:24–40
77. Rieke, G. H., Lebofsky, M. J. 1981. *Ap. J.* 250:87–97
78. Scoville, N. Z. 1981. See Ref. 14, pp. 23–32
79. Scoville, N. Z. 1981. See Ref. 108, pp. 187–205
80. Scoville, N. Z., Hall, D. N. B., Kleinmann, S. G., Ridgway, S. T. 1982. *Ap. J.* 253:136–48
81. Scoville, N. Z., Kleinmann, S. G., Hall, D. N. B., Ridgway, S. T. 1983. *Ap. J.* 275:201–24
82. Deleted in proof
83. Simon, M., Righini-Cohen, G., Fischer, J., Cassar, L. 1981. *Ap. J.* 251:552–56
84. Simon, M., Felli, M., Cassar, L., Fischer, J., Massi, M. 1983. *Ap. J.* 266:623–45
85. Smith, H. A., Larson, H. P., Fink, U. 1981. *Ap. J.* 244:835–43
86. Sneden, C., Lambert, D. L. 1982. *Ap. J.* 259:381–91
87. Stenflo, J. O., Twerenbold, D., Harvey, J. W., Brault, J. 1983. *Astron. Astrophys. Suppl.* 54:505–14
88. Thompson, R. I., Reed, M. A. 1975. *Publ. Astron. Soc. Pac.* 87:929–32
89. Thompson, R. I., Boroson, T. A. 1977. *Ap. J. Lett.* 216:L75–77
90. Thompson, R. I., Tokunaga, A. T. 1978. *Ap. J.* 226:119–23
91. Thompson, R. I., Tokunaga, A. T. 1979. *Ap. J.* 229:153–57
92. Thompson, R. I., Tokunaga, A. T. 1979. *Ap. J.* 231:736–41
93. Thompson, R. I., Tokunaga, A. T. 1980. *Ap. J.* 235:889–93
94. Thompson, R. I. 1981. See Ref. 108, pp. 153–66
95. Thompson, R. I. 1982. *Ap. J.* 257:171–78
96. Thompson, R. I., Thronson, H. A., Campbell, B. 1983. *Ap. J.* 266:614–22
97. Thronson, H. A. 1981. *Ap. J.* 248:984–91

98. Thronson, H. A., Harvey, P. M. 1981. *Ap. J.* 248:584–90
99. Thronson, H. A., Thompson, R. I. 1982. *Ap. J.* 254:543–49
100. Thronson, H. A. 1983. *Ap. J.* 264:599–604
101. Tokunaga, A. T., Thompson, R. I. 1979. *Ap. J.* 229:583–86
102. Tokunaga, A. T., Knacke, R. F., Ridgway, S. T. 1980. *Icarus* 44:93–101
103. Tokunaga, A. T., Young, E. T. 1980. *Ap. J. Lett.* 237:L93–96
104. Treffers, R. R., Fink, U., Larson, H. P., Gautier, T. N. 1976. *Ap. J.* 209:793–99
105. Tsuji, T. 1977. *Publ. Astron. Soc. Jpn.* 29:497–510
106. Tsuji, T. 1983. *Astron. Astrophys.* 122:314–21
107. Wollman, E. R., Smith, H. A., Larson, H. P. 1982. *Ap. J.* 258:506–14
108. Wynn-Williams, C. G., Cruikshank, D. P., eds. 1981. *Infrared Astronomy.* Dordrecht: Reidel. 376 pp.

Ann. Rev. Astron. Astrophys. 1984. 22:319–58

EXTRAGALACTIC RADIO JETS

Alan H. Bridle

National Radio Astronomy Observatory,[1] Charlottesville, Virginia 22901

Richard A. Perley

National Radio Astronomy Observatory, Socorro, New Mexico 87801

1. INTRODUCTION

Powerful extended extragalactic radio sources pose two vexing astrophysical problems [reviewed in (147) and (157)]. First, from what energy reservoir do they draw their large radio luminosities (as much as 10^{38} W between 10 MHz and 100 GHz)? Second, how does the active center in the parent galaxy or QSO supply as much as 10^{54} J in relativistic particles and fields to radio "lobes" up to several hundred kiloparsecs outside the optical object? New aperture synthesis arrays (68, 250) and new image-processing algorithms (66, 101, 202, 231) have recently allowed radio imaging at subarcsecond resolution with high sensitivity and high dynamic range; as a result, the complexity of the brighter sources has been revealed clearly for the first time. Many contain *radio jets*, i.e. narrow radio features between compact central "cores" and more extended "lobe" emission. This review examines the systematic properties of such jets and the clues they give to the physics of energy transfer in extragalactic sources. We do not directly consider the jet production mechanism, which is intimately related to the first problem noted above—for reviews, see (207) and (251).

1.1 Why "Jets"?

Baade & Minkowski (3) first used the term *jet* in an extragalactic context, describing the train of optical knots extending $\sim 20''$ from the nucleus of M87; the knots resemble a fluid jet breaking into droplets. They suggested

---

[1] The National Radio Astronomy Observatory is operated by Associated Universities, Inc. under contract with the National Science Foundation.

0066–4146/84/0915–0319$02.00

that "the jet was formed by ejection from the nucleus" (even though its continuous spectrum gave no clue to its velocity) and that an [OII] $\lambda3727$ emission line in the nucleus whose centroid is blueshifted by several hundred kilometers per second from the systemic velocity is "emitted by a part of the material which forms the jet and is still very close to, if not still inside, the nucleus." The "optical wisp" (227) near the QSR 3C 273, which resembles the M87 knots, was also called a *jet* without direct evidence for outflow. Radio detection of these optical "jets" (114, 160) prompted description of narrow features in other 3C sources (168, 253) as "radio jets." In 1973, refined versions of the continuous outflow, or "beam," models for extragalactic sources proposed earlier by Morrison (162) and Rees (204) were developed (25, 147, 221). The new models (a) obviated adiabatic losses, which led "explosive" models to require that the compact precursors of extended sources be far more luminous than any actually observed; and (b) they explained how the synchrotron lifetimes of electrons in bright lobe "hot spots" can be less than the light travel time to the hot spots from the parent galaxy or QSR (e.g. 113). Continuous flow models and jet data have kept close company ever since.

1.1.1   CAVEAT   The term *jet* connotes continuous outflow of fluid from a collimator, but there is no direct evidence for flow in any continuous extragalactic "jet." VLBI studies of proper motions of knots in some compact radio sources—reviewed in (57)—suggest outflow of jetlike features from stationary "cores," but only in 3C 345 (7) has this been tested in an external reference frame. Such proper motions cannot be monitored in truly continuous emission. Narrow kiloparsec-scale features are therefore called *jets* mainly because they occur where "beam" models required collimated outflow from active nuclei.

1.2   What Makes a "Narrow Feature" a "Jet"?

Terminology that so prejudges source physics should be used sparingly, so we require (as in 27) that to be termed a *jet*, a feature must be

1. at least four times as long as it is wide,
2. separable at high resolution from other extended structures (if any), either by brightness contrast or spatially (e.g. it should be a narrow ridge running through more diffuse emission, or a narrow feature in the inner part of a source entering more extended emission in the outer part),
3. aligned with the compact radio core where it is closest to it.

1.2.1   EXAMPLES   Figures 1 to 5 show examples of jets of various powers and sizes. They also illustrate some ambiguities—with less sensitivity, the NGC 6251 jet (Figure 2) breaks into discrete knots, not all of which are

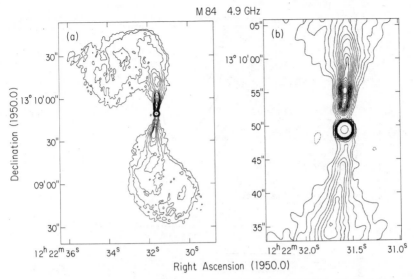

M 84   4.9 GHz

Figure 1   VLA maps of the jets in the weak radio galaxy M84 at 4.9 GHz, with the right panel showing detail of the central region. The peak on these maps is the radio core. Note the one-sided bright base of the northern jet and the faint cocoon of emission flaring from both jets beyond 5″ from the core (data of R. A. Laing and A. H. Bridle, in preparation).

elongated along it. We call a train of knots a *jet*, however, only if it has more than two knots or if some knots are elongated along it (e.g. Figures 3 and 5). (We prefer to exclude some blobby jets temporarily than to apply the prejudicial term *jet* too liberally.) The elongated outer lobes of some edge-darkened sources, e.g. 3C 31 (245), may equally plausibly be termed broad jets (87), so dividing them into "jet" and "lobe" segments by morphology alone may be subjective. We ask that they contain a "spine" of bright emission meeting criterion (2.) before we call them *jets*.

2. THE INCIDENCE OF EXTRAGALACTIC RADIO JETS

2.1 *A List of Known Jets*

Table 1 lists data on 125 radio sources known to us (in mid-August 1983) to have *jets* by our criteria. (We use $H_0 = 100$ km s^{-1} Mpc^{-1} and $q_0 = 0.5$.) Column 2 gives the identification—galaxy (G) or quasar (Q)—and its redshift. Columns 3 and 4 measure the observer's frame core and total monochromatic powers—$\log_{10} P_{core}^5$ at 5 GHz and $\log_{10} P_{tot}^{1.4}$ at 1.4 GHz. (The flux densities S_{core}^5 and $S_{tot}^{1.4}$ are those most often available.) We use "typical" values for variable cores. Some S_{core}^5 and $S_{tot}^{1.4}$ values are estimated

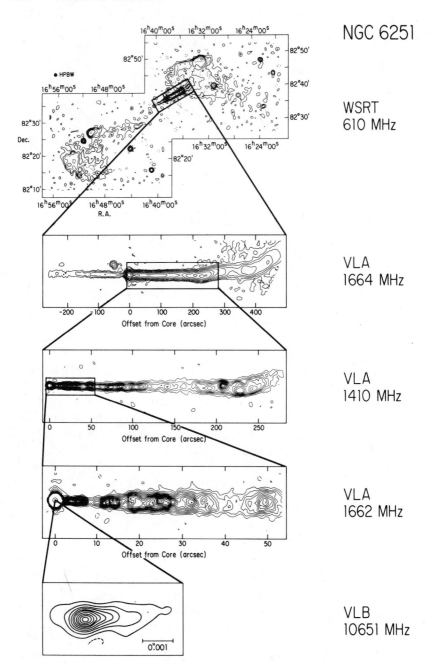

NGC 6251

WSRT
610 MHz

VLA
1664 MHz

VLA
1410 MHz

VLA
1662 MHz

VLB
10651 MHz

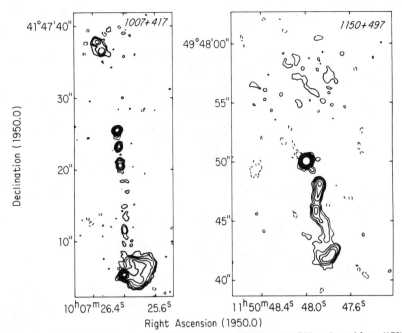

Figure 3   VLA maps of two one-sided jets in extended QSRs at 4.9 GHz, adapted from (173). Note the knotty structure of the jet in the left panel and the "gap" and the "wiggle" in the jet in the right panel. The peak on each map is the radio core.

from neighboring frequencies, assuming v^0 spectra for the cores and $v^{-0.7}$ for the total emission. Column 5 gives the projected length d_j of the brighter jet in kiloparsecs. Column 6 categorizes the jet sidedness (see Section 3.1). Columns 7 and 8 codify the origin of the data (see the footnotes below the table). We refer to VLBI data on jetlike extensions of the cores in sources with larger-scale jets whether or not the extensions separately meet criterion (1.) of Section 1.2. Table 2 lists 73 sources with features meeting only some of our criteria; modest increases in data quality could promote them to Table 1.

Jets occur in extragalactic sources of all luminosities, sizes, and structure types, always accompanied by detectable emission in the inner kiloparsec of the parent object. It is therefore reasonable to associate jets (*a*) with a

Figure 2   The structure of the jet and counterjet in NGC 6251 over a wide range of angular scales. Note the knotty substructure of the jet, and the large-scale "wiggle" (*middle panel*). The nuclear "core-jet" (*bottom panel*) and the mean position angle of the larger-scale jet are misaligned by $4.5 \pm 1°$ [data from (279)—top panel; R. A. Perley and A. H. Bridle (in preparation)—second and third panels; (183)—fourth panel; and (56)—bottom panel].

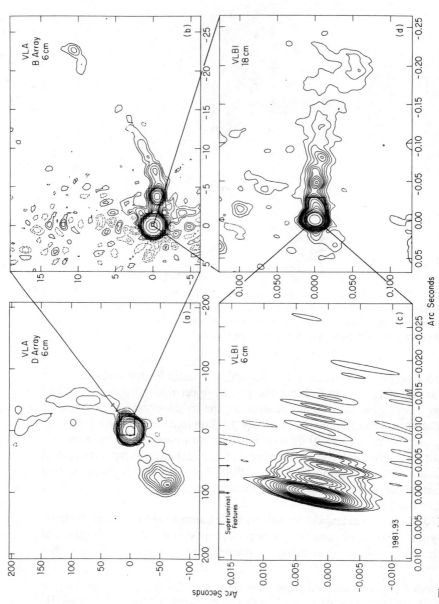

Figure 4 The structure of the core and jet in 3C 120 over a wide range of angular scales. The three brightest features in the lower-left panel exhibit superluminal expansion (268). Note the continuity between the different scales and the large misalignment between the smallest and largest structures (*left panels*). Montage provided by Drs. R. C. Walker and J. M. Benson.

process common to all extragalactic radio sources and (*b*) with continuing activity in their nuclei. This supports the view that jets result from inefficiencies in the energy transport from the cores to the lobes of extragalactic sources, whether their emission originates in the primary flow itself or in a dissipative sheath around it.

2.2   Occurrence Rates of Jets

The rates of occurrence of detectable radio jets (by our criteria) can be determined in several samples representing different extragalactic source types.

Twelve (55%) of the 22 radio galaxies in the re-revised 3C sample (140)—

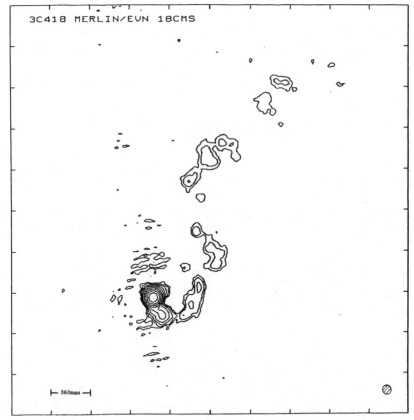

Figure 5   MERLIN/EVN map of the one-sided jet in the strongly core-dominated QSR 3C 418, provided by Dr. T. B. Muxlow. This is the most powerful core known to be associated with a jet. Note the sharp curvature of the jet near the core and its knotty structure. The tic marks are 0.36″ apart.

Table 1 125 extragalactic radio jets

Source name	ID, z	P^5_{core}	$P^{1.4}_{tot}$	d_j	SI	Data[a]	References[b]
0017+15 = 3C 9	Q, 2.012	25.45	28.16	30.	1	A	248, K1
0033+18 = 3C 14	Q,					A	L4
0055+30 = NGC 315	G, 0.0167	23.24	24.08	240	B	ABW	27, 29, 30, 99, 142, 278
0055+26 = NGC 326	G, 0.0472	22.29	24.61	25.	2	AW	97
0104+32 = 3C 31 (NGC 383)	G, 0.0169	22.45	24.21	14.	2	ACOW	32, 42, 52, 99, 245
0106+72 = 3C 33.1	G, 0.181	23.76	26.11	140	1	AW	258, R1
0123−01 = 3C 40	G, 0.018	22.35	24.38	38.	2	A	O1
0130+24 = 4C 24.02	Q, 0.457	25.11	26.21	92.	?	A	W1
0149+35 = NGC 708	G, 0.0160	21.31	22.62	4.5	2	A	D1, L3
0206+35 = UGC 1651	G, 0.0373	23.15	24.52	18.	2	A	D1, L3
0220+42 = 3C 66B	G, 0.0215	22.59	24.69	45.	2	ACOW	52, 168, 258
0238+08 = NGC 1044	G, 0.0214	22.54	23.80	43.	2	AN	100, C1
0240−00 = 3C 71 (NGC 1068)	G, 0.004	20.99	22.94	0.3	2	ABM	62, 180, 264, 282, 283
0255+05 = 3C 75A,B	G, 0.0241	22.40	24.61	30.	2	A	O2
0256+13 = 4C 13.17B	G, 0.0748	22.30	24.07	15.	2	A	O1
0305+03 = 3C 78 (NGC 1218)	G, 0.0289	23.77	24.83	0.6	1	ABM	126, P2, U1
0314+41 = 3C 83.1B (NGC 1265)	G, 0.0255	22.15	24.76	18.	2	ACW	161, 171, 272
0316+41 = 3C 84 (NGC 1275)	G, 0.0177	24.87	24.66	5.0	1	BM	176, 179, 200
0320−37 = For A (NGC 1316)	G, 0.0063	21.23	24.69	2.7	B	A	96
0326+39 = VV 7.08.14	G, 0.0243	22.70	24.06	41.	2	AW	27, 174
0336−35 = PK (NGC 1399)	G, 0.0049	20.41	22.71	8.1	2	A	E1
0415+37 = 3C 111	G, 0.0485	24.47	25.59	78.	1	AB	97, 142, 145
0430+05 = 3C 120	G, 0.0334	24.93	24.76	83.	1	ABMW	2, 5, 6, 38, 126, 224, 268, 269
0445+44 = 3C 129	G, 0.0208	22.19	24.58	8.8	B	ACW	44, 214, 258, 262
0449−17 = PK	G, 0.0313	22.03	23.93	10.	2	A	E1
0459+25 = 3C 133	G, 0.2775	25.33	26.72	14.	1	AM	138, 212, L1
0514−16 = PK	Q, 1.278	27.32	27.25	33.	1	A	P2
0538+49 = 3C 147	Q, 0.545	26.78	27.95	0.8	1	BM	193, 194, 201, 203, 233, 234, 275
0546−32 = PK	G, 0.147	23.90	25.47	200	?	A	E1

Source	Class, z				N	Type	Notes
0658+33 = B2	G, 0.127	23.98	24.82	55.	2	A	O1
0704+35A = 4C 35.16A	G, 0.078	21.83	24.28	17.	2	A	O1
0712+53 = 4C 53.16	G, 0.064	22.96	24.83	13.	?	A	49
0723+67 = 3C 179	Q, 0.846	26.62	27.39	18.	1	ABM	38, 173
0742+31 = 4C 31.30	Q, 0.462	26.24	26.55	210	1	A	163
0812+36 = B2	Q, 1.025	27.10	27.23	30.	1	A	185, P2
0812+02 = 4C 02.23	Q, 0.402	25.60	26.68	32.	?	A	102, 285, H1
0824+29 = 3C 200	G, 0.458	24.98	26.73	41.	1	A	B2
0833+65 = 3C 204	Q, 1.112	25.71	27.39	52.	1	A	L1
0833+58	Q, 2.101	27.64	27.52	45.	?	A	P2
0838+13 = 3C 207	Q, 0.684	26.45	27.23	25.	1	A	W1
0844+31 = 4C 31.32	G, 0.0675	23.35	24.88	61.	?	AW	258, 262
0850+14 = 3C 208	Q, 1.11	26.44	27.64	22.	?	A	L1
0855+14 = 3C 212	Q, 1.049	25.38	27.61	22.	1	A	L1
0908+37 = B2	G, 0.1047	23.50	24.89	23.	2	A	153, D1
0917+45 = 3C 219	G, 0.1744	24.18	26.45	36.	1	ACNW	184, 253
0938+39 = 4C 39.27	Q, 0.618	25.00	26.99	96.	1	A	190, W1
0957+00 = 4C 00.34	Q, 0.907	26.02	27.03	76.	?	A	H1
0957+56 = Double QSO	Q, 1.405	26.28	27.15	22.	1	AB	108, 109, 189
1001+22 = 4C 22.26	Q, 0.974	25.73	26.94	31.	?	A	W1
1003+35 = 3C 236	G, 0.0989	24.64	25.78	0.4	1	A(?)B	98, 225
1004+14 = NGC 3121	Q, 0.031	22.97	24.07	77.	2	AC	125, C1
1004+13 = 4C 13.41	Q, 0.240	23.87	25.92	60.	1	A	96
1007+41 = 4C 41.21	Q, 0.613	25.86	26.91	77.	1	A	173
1029+57 = HB 13	G, 0.034	22.50	23.70	280	2	CW	156, S2
1033+00 = PK	G,					A	216
1100+77 = 3C 249.1	Q, 0.311	25.00	26.41	21.	1	A	149, B2, L1
1122+39 = NGC 3665	G, 0.0067	20.46	21.76	3.9	2	A	D1, H2
1131+49 = IC 708	G, 0.0321	22.74	24.13	35.	2	A	257
1137+18 = NGC 3801	G, 0.0105	20.59	23.06	2.1	2	AC	125, L1
1150+49 = 4C 49.22	Q, 0.334	25.88	26.43	23.	1	A	173, 181
1209+74 = 4C T.74.17.1	G, 0.107	23.26	24.99	120	1	AW	263, P2
1216+06 = 3C 270 (NGC 4261)	G, 0.0073	22.25	24.01	31.	2	A	126, K1

Table 1 (continued)

Source name	ID, z	P^5_{core}	$P^{1.4}_{tot}$	d_j	SI	Data[a]	References[b]
1217+02 = PK	Q, 0.240	25.33	25.68	120	?	A	163
1222+13 = 3C 272.1 (M84)	G, 0.0031	21.72	23.24	3.3	2	ABCW	126, 127, L2
1226+02 = 3C 273	Q, 0.158	26.92	27.12	39.	1	ABMOX	2, 38, 58, 64, 175, 178, 181, 199, 226, P2
1228+12 = Vir A (M87)	G, 0.0043	22.92	24.78	1.8	1	ABCMOX	1, 2, 18, 54, 67, 71, 135, 141, 150, 164, 165, 172, 209, 226, 230, 237, 244, 247, 249
1241+16 = 3C 275.1	Q, 0.557	25.74	27.08	36.	?	A	243
1250−10 = NGC 4760	G, 0.0138	22.14	23.27	2.9	2	A	L1
1251+273 = NGC 4789	G, 0.027	21.16	23.55	6.7	2	A	L3
1251+278 = 3C 277.3 (Com A)	G, 0.0857	22.98	25.37	11.	1	AO	31, 158, 159, V3
1251−12 = 3C 278 (NGC 4783)	G, 0.0138	22.13	24.23	14.	2	A	C1
1253−05 = 3C 279	Q, 0.536	27.56	27.53	9.9	1	AB	38, 70, 175
1256+28 = NGC 4869	G, 0.0235	21.08	22.89	2.6	?	A	O1
1258+40 = 3C 280.1	Q, 1.659	26.21	27.84	42.	1	A	248
1258−32 = PK	G,					A	P2
1315+34 = B2	Q, 1.050	26.76	26.98	13.	1	A	181
1316+29 = 4C 29.47	G, 0.728	23.25	24.85	110	2	A	63
1317+52 = 4C 52.27	Q, 1.060	26.79	27.37	60.	?	A	173, H1
1321+31 = NGC 5127	G, 0.0161	21.77	23.85	55.	2	ACW	83, 88, 125
1322−42 = Cen A (NGC 5128)	G, 0.0012	22.20	24.62	5.2	1	ABOX	20, 34, 48, 77, 91, 106, 107, 170, 187, 188, 192, 228, 229, 235
1328+30 = 3C 286	Q, 0.849	27.88	28.18	0.2	?	AB	177, 234, P1
1333−33 = IC 4296	G, 0.0129	22.46	24.05	128	2	A	103, E1
1407+17 = NGC 5490	G, 0.0163	22.00	23.68	5.5	2	C	125
1414+11 = 3C 296 (NGC 5532)	G, 0.0237	22.67	24.43	50.	2	AC	19, L1
1441+52 = 3C 303	G, 0.141	24.53	25.75	26.	?	A	148, K1
1448+63 = 3C 305	G, 0.041	22.57	24.73	0.9	2	A	116
1450+28 = B2	G, 0.1265	22.56	24.56	37.	2	A	D1
1451−37 = PK	Q, 0.314	26.24	26.36	17.	1	A	181
1458+71 = 3C 309.1	Q, 0.904	27.62	28.01	3.8	?	BM	274, S1

1615+42	G, 0.131	23.20	24.20	14.	2	A	O1
1618+17 = 3C 334	Q, 0.555	25.78	26.97	63.	?	A	271, H1
1626+27 = 3C 341	G, 0.448	23.49	26.80	112	B	A	This paper, Figure 6
1637+82 = NGC 6251	G, 0.0230	23.66	24.14	161	1	ABCW	33, 56, 183, 198, 220, 267, 280
1638+53 = 4C 53.37	G, 0.1098	23.21	24.93	40.	2	A	45, 50
1641+39 = 3C 345	Q, 0.594	27.62	27.52	9.5	1	AB	38, 58-60, 177, 199, 200, 224, 241, 255, P2
1642+69 = 4C 69.21	Q,					M	39
1648+05 = 3C 348 (Her A)	G, 0.154	23.61	27.10	118	1	A	D2
1752+32 = B2 1752+32B	G, 0.0449	22.67	23.46	30.	2	A	D1
1759+21 = PK	G,					A	216
1807+27 = 4C 27.41	Q, 1.76	27.29	27.66	13.	?	A	P2
1807+69 = 3C 371	G, 0.050	24.60	24.84	2.0	1	AB	39, 176, P1
1842+45 = 3C 388	G, 0.0908	23.76	25.73	18.	1	A	46, 47
1857+56 = 4C 56.28	Q, 1.595	26.34	27.57	62.	1	A	173, 216, 217
1919+47 = 4C 47.51	G, 0.103	23.19	24.86	265	1	A	43, 211
1924+50 = 4C 50.47	Q,					A	173
1939+60 = 3C 401	G, 0.201	24.14	26.37	24.	1	A	B2, L1
1940+50 = 3C 402N	G, 0.0247	22.08	24.31	6.2	2	C	210
1957+40 = 3C 405 (Cyg A)	G, 0.057	24.12	27.73	47.	?	AB	132, 142, P2
2037+51 = 3C 418	Q, 1.686	28.24	28.41	9.3	1	ABM	274, M1
2116+26 = NGC 7052	G, 0.0164	22.12	22.72	26.	2	A	125, D1, L3
2121+24 = 3C 433	G, 0.1016	22.76	26.15	30.	?	A	259
2153+37 = 3C 438	G, 0.292	23.99	26.86	27.	2	A	L1
2221-02 = 3C 445	G, 0.057	23.51	25.30	210	1	A	V1
2229+39 = 3C 449	G, 0.0171	22.07	24.03	19.	2	ACW	19, 65, 186, P2
2236+35 = B2	G, 0.0277	21.88	23.40	7.7	2	A	D1
2251+15 = 3C 454.3	Q, 0.859	28.02	28.10	21.	1	AM	38, 177, 181, 274
2300-18 = PK	Q, 0.129	24.90	25.45	68.	1	A	122
2316+18 = OZ 127	G, 0.0395	22.41	23.70	16.	2	A	O1
2318+07 = NGC 7626	G, 0.0112	21.31	23.17	6.4	2	A	L1
2325+29 = 4C 29.68	Q, 1.015	26.37	27.34	85.	?	A	W1
2335+26 = 3C 465 (NGC 7720)	G, 0.0293	23.37	24.85	24.	1	ACW	44, 81, 258
2337+26 = NGC 7728	G, 0.0314	23.15	23.49	39.	2	ACW	125, 126, 127, L1, V2

Table 1 (*continued*)

Source name	ID, z	P^5_{core}	$P^{1.4}_{tot}$	d_j	SI	Data[a]	References[b]
2338+04 = 4C 04.81	Q, 2.594	27.15	27.98	4.6	?	M	B1
2349+32 = 4C 32.69	Q, 0.671	25.15	26.57	99.	1	A	75, 190, 191, 271
2354+47 = 4C 47.63	G, 0.046	22.49	24.63	37.	1	A	49

[a] Data codes: A—VLA; B—VLB; C—Cambridge; M—MERLIN; N—NRAO; O—optical; X—X-ray; W—WSRT.
[b] Unpublished references:

B1: Barthel, P. D., Lonsdale, C. J. 1983. Preprint
B2: Burns, J. O., Basart, J. P., De Young, D. S., unpublished data
C1: Cornwell, T. J., unpublished data
D1: de Ruiter, H., Parma, P., Fanti, C., Fanti, R., unpublished data
D2: Dreher, J. W., Feigelson, E. D. 1983. Preprint
E1: Ekers, R. D., unpublished data
H1: Hintzen, P., Ulvestad, J., Owen, F. N. 1983. Preprint
H2: Hummel, E., Kotanyi, C., unpublished data
K1: Kronberg, P. P., unpublished data
L1: Laing, R. A., unpublished data
L2: Laing, R. A., Bridle, A. H., unpublished data
L3: Laing, R. A., Kotanyi, C., Hummel, E., unpublished data
L4: Laing, R. A., Owen, F. N., Puschell, J., unpublished data
M1: Muxlow, T. W. B., Jullian, M., Linfield, R., unpublished data
O1: O'Dea, C. P., unpublished data
O2: Owen, F. N., unpublished data
P1: Pearson, T. J., Perley, R. A., Readhead, A. C. S., unpublished data
P2: Perley, R. A., unpublished data
R1: Rudnick, L., Edgar, B. K., unpublished data
S1: Simon, R. S., unpublished data
S2: Strom, R. G., unpublished data
U1: Unger, S. V., Booler, R. V., Pedlar, A. 1983. Preprint
V1: van Breugel, W. J. M., unpublished data
V2: van Breugel, W. J. M., Fomalont, E. B., Bridle, A. H., unpublished data
V3: van Breugel, W. J. M., Miley, G. K., Heckman, T., Butcher, H. R., Bridle, A. H., unpublished data
W1: Wardle, J. F. C., Potash, R. I., Roberts, D. H., unpublished data.

henceforth $3CR^2$—with $\delta \geq 10°$, $b \geq 10°$, and $z \leq 0.05$ have definite jets, and two more (9%) have possible jets. We exclude from the sample the weak source 3C 231 = M82, whose emission comes mainly from a galactic disk. A fifteenth galaxy (3C 338) has structure resembling a jet except that it does not align with the radio core (51). Jets are thus detected in at least 55%, and perhaps 65%, of this sample, whose median $P_{tot}^{1.4}$ is $10^{24.43}$ W Hz^{-1}. Jets were also found in 82% (9 of 11) of well-resolved sources in a complete sample of B2 radio galaxies with $m_{pg} < 15.7$ (83). R. A. Laing (private communication) finds definite jets in 13 sources (55%), and possible jets in 5 more (20%), in an unbiased sample of 24 E and S0 galaxies with $0° < \delta < 37°$, $m_{pg} \leq 14.0$, and $S_{tot}^{2.3} > 0.035$ Jy.

Forty-two extended $3CR^2$ galaxies or probable galaxies with $z > 0.4$ or $V > 20$ have been mapped at the VLA with good dynamic range (Table 1, ref. L4). Only two (5%) of these powerful radio galaxies (median $P_{tot}^{1.4}$ = $10^{27.36}$ W Hz^{-1}) have continuous jets; one other has an elongated knot between its core and one lobe.

Twenty-two extended $3CR^2$ QSRs have been mapped at the VLA with

Table 2   73 possible extragalactic radio jets

0120 + 33 = NGC 507	0913 + 38 = B2	1508 − 05 = 4C − 05.64
0131 − 36 = NGC 612	0915 − 11 = 3C 218 (Hyd A)	1510 − 08 = PK
0134 + 32 = 3C 48	0926 + 79 = 3C 220.1	1529 + 24 = 3C 321
0137 + 01 = 4C 01.04	0947 + 14 = 3C 228	1548 + 114 = 4C 11.50
0212 + 17 = MC 3	0956 − 26 = NGC 3078	1615 + 35 = NGC 6109
0300 + 16 = 3C 76.1	1015 + 49	1626 + 39 = 3C 338 (NGC 6166)
0327 + 24 = B2	1103 − 00 = 4C − 00.43	1636 + 47
0333 + 32 = NRAO 140	1104 + 16 = 4C 16.30	1704 + 60 = 3C 351
0448 + 51 = 3C 130	1108 + 27 = B2	1712 + 63
0457 + 05	1113 + 29 = 4C 29.41	1850 + 70
0518 + 16 = 3C 138	1144 + 35 = B2	1828 + 48 = 3C 380
0531 − 36	1208 + 39 = NGC 4151	1830 + 28 = 4C 28.45
0549 − 07 = NGC 2110	1218 + 33 = 3C 270.1	1833 + 32 = 3C 382
0609 + 71 = Markarian 3	1222 + 21 = 4C 21.35	1845 + 79 = 3C 390.3
0634 − 20 = MSH 06 − 210	1254 + 27 = NGC 4839	1928 + 73
0703 + 42 = 4C 42.23	1317 + 25 = 4C 25.42	2019 + 09 = 3C 411
0716 + 71	1319 + 64 = 4C 64.18	2040 − 26 = PK
0722 + 30 = B2	1336 + 39 = 3C 288	2201 + 31 = 4C 31.63
0754 + 12	1346 + 26 = 4C 26.42	2203 + 29 = 3C 441
0755 + 37 = NGC 2484	1347 + 60 = NGC 5322	2216 − 03 = 4C − 03.79
0802 + 10 = 3C 191	1350 + 31 = 3C 293	2223 − 05 = 3C 446
0814 + 54 = 4C 54.16	1351 + 36 = NGC 5352	2247 + 11 = NGC 7385
0837 + 29 = 4C 29.30	1404 + 34 = 3C 294	2305 + 18 = 4C 18.08
0903 + 16 = 3C 215	1415 + 25 = NGC 5548	
0905 − 09 = 26W20	1419 + 41 = 3C 299	

good dynamic range (Table 1, refs. L4 and W1). Ten (45%) of this group (whose median $P_{tot}^{1.4}$ is $10^{27.43}$ W Hz^{-1}, similar to that of the distant 3CR^2 galaxies) have definite jets, and five more (23%) have structure resembling the brighter parts of jets. Ten (40%) of 24 extended QSRs in a 966-MHz survey have jets or structure resembling them (173), while five of the eight most extended sources in a complete sample of 4C QSRs have jets (271).

Jets are thus detected in 65 to 80% of weak radio galaxies, and in 40 to 70% of extended QSRs, but (with similar instrumental parameters) in only < 10% of distant galaxies similar to the QSRs in radio power (49, 139, 173). Among the extended 3CR^2 QSRs, the jet detection rate increases with the relative prominence $f_C = S_{core}^5/S_{tot}^{1.4}$ of the radio core, apparently regardless of redshift: all six QSRs with $f_C > 0.03$ (but only two of the six with $f_C < 0.005$) have jets or features resembling the brightest parts of jets. The lack of detectable jets in distant 3CR^2 galaxies may be related to their lower values of f_C—the median f_C in the distant 3CR^2 galaxy sample is only 0.0005. The two with detected jets are 3C 200, with $f_C = 0.018$ and weak emission lines, and 3C 341 (Figure 6), with $f_C = 0.0005$ and strong emission lines. The relations between jet, core, and emission-line fluxes of 3CR^2 galaxies need clarifying, but the multivariate ($P_{tot}, P_{core}, P_{optical}$) luminosity function for radio jets may contain clues to the physics of energy transport in these sources. We urge observers to publish integrated flux densities of jets and lobes separately, to allow study of this function.

2.3 Jets in Weak Sources?

2.3.1 SPIRAL GALAXIES   Several Seyfert galaxies (for a review and references, see 282) with $10^{21.5} < P_{tot}^{1.4} < 10^{23}$ W Hz^{-1} have S-shaped kiloparsec-scale radio structures that may be low-thrust jets being bent and disrupted by the ram pressure of a rotating gas disk. If this is correct, the radio sources in Seyfert spirals may differ from those in ellipticals and QSRs mainly by (a) the smaller power output of the central "engine" and (b) the

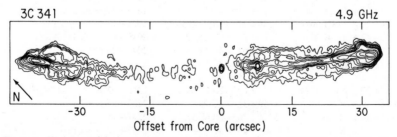

Figure 6   VLA map of the jets and cocoon in the distant radio galaxy 3C 341 at 4.9 GHz. Note the narrowing of the cocoon to the right of the figure, and the irregular brightness distribution of the jets (data of A. H. Bridle and E. B. Fomalont, in preparation).

inability of their jets to escape from the dense, rotating interstellar medium (ISM) of a spiral disk. Interpreting the Seyfert sources as continuous jets (282) is not yet obligatory, however (264). Several edge-on spirals with unusually bright compact cores (121, and references therein) also have radio features extending from their nuclei apparently near the galactic minor axes. The features are as yet too poorly resolved to meet our criteria for jethood (Section 1.2), but if they do result from nuclear activity, they could be weak analogs of jets in elliptical galaxies.

2.3.2 GALACTIC JETS   The S-shaped structure of Sgr A West (36, 84) and its relation to the velocities of the [Ne II] lines in the region can be interpreted as the result of collimated outflow from the galactic center at ~ 300 km s^{-1} (35, 84). The extent to which Sgr A West is a weak parsec-scale analog of more active galactic nuclei is unclear, however; the S-structure has a thermal spectrum (unlike extragalactic jets; Section 4.2), and models involving tidal distortion of infalling clouds also fit the data (84). The binary star SS 433 has the only astrophysical jet whose velocity (0.26c) and precessional geometry relative to us are unambiguous (for reviews and references, see 120, 154). Both the ~ 100 light-day scale of the known radio structure and its typical radio luminosity ($P_{tot}^{1.4} = 10^{15.8}$ W Hz^{-1}) are much less than those of extragalactic jets, but SS 433 allows us to observe the evolution of synchrotron-emitting plasma in precessing supersonic jets directly. There is also evidence for bipolar, and perhaps well-collimated, outflows from star-forming regions in dense molecular clouds—see (55) and (232) for reviews and references. The ability to measure velocities, densities, and temperatures in and around these nearby flows may help to test models of extragalactic jet production and propagation (133, 207).

3. TRENDS WITH LUMINOSITY

3.1 *Sidedness*

Most extended extragalactic sources have lobes of similar powers and sizes on each side of the parent object, but not all jets have detectable counterjets. We denote by $\mathscr{S}$ the ratio of intensities between the brighter and fainter jet measured at the same distance from the parent object at low transverse resolution (to minimize the influence on $\mathscr{S}$ of differences in their expansion rates). We classify as *one sided* those for which $\mathscr{S} > 4$ wherever the dynamic range of the map allows this to be determined, and as *two sided* those for which $\mathscr{S} < 4$ everywhere. We use $\mathscr{S} = 4$ as the break point solely because it separates two equally numerous groups of jets on available maps, not because it has special physical meaning.

Most jets are one-sided close to their parent object. Jets in weak radio galaxies (e.g. Figure 1) become two sided after a few kiloparsecs; the one-sided bases are typically $< 10\%$ of their length, and the jet with the one-sided base is generally somewhat brighter on the large scale. Table 1 (Column 6) classifies the sidedness of the outer 90% of the jets—"1" if $\mathscr{S} > 4$, "2" if $\mathscr{S} < 4$, and "B" if $\mathscr{S} > 4$ at some distances from the core but $\mathscr{S} < 4$ at others. Most jets in sources with $P_{core}^5 \leq 10^{23}$ W Hz^{-1} or $P_{tot}^{1.4} \leq 10^{24.5}$ W Hz^{-1} are mainly two sided, while most jets in more powerful sources, whether radio galaxies or QSRs, are entirely one sided. The value $P_{tot}^{1.4} \approx 10^{25}$ W Hz^{-1} marks the transition between the structural classes I and II of Fanaroff & Riley (86; henceforth FR): the jets in FR I sources (weak, edge darkened, lacking hot spots) are mainly two sided, while those in FR II sources (powerful, edge brightened, strong hot spots) are one sided. All jets in QSRs, whether core dominated or lobe domi-nated, are entirely one sided.

3.2 Magnetic Configuration

Degrees of linear polarization up to 40% are common in radio jets at centimeter wavelengths (e.g. 19, 99, 186, 191, 220, 248, 263), and local values $> 50\%$ are not unusual (183, 184, 256, 278). Jets with low ($< 5\%$) polarization at centimeter wavelengths (31, 47) are exceptional. High degrees of polarization imply ordering of the component of the jet magnetic field B_j perpendicular to the line of sight, but this order need not be three dimensional (137). Field ordering is seen directly on well-resolved maps of the "apparent" (synchrotron emissivity weighted) field B_a derived from multifrequency polarimetry. Three magnetic configurations are common in the 40 jets for which adequate polarimetry exists:

1. $B_\parallel$, i.e. B_a is predominantly parallel to the jet axis all across the jet.
2. $B_\perp$, i.e. B_a is predominantly perpendicular to the jet axis all across it.
3. $B_{\perp\text{-}\parallel}$, i.e. B_a is predominantly perpendicular to the jet axis at the center of the jet, but becomes parallel to the axis near one or both of its edges.

Most two-sided regions of straight jets have either the $B_\perp$ or the $B_{\perp\text{-}\parallel}$ field configuration, while most one-sided regions of jets have the $B_\parallel$ configura-tion (27, 277). In most straight jets in sources with $P_{core}^5 < 10^{23}$ W Hz^{-1} (or $P_{tot}^{1.4} < 10^{24.5}$ W Hz^{-1}), B_a turns from $B_\parallel$ to $B_\perp$ or $B_{\perp\text{-}\parallel}$ in the first 10% of the jet, but most jets in more powerful sources are $B_\parallel$-dominated for their entire length (27, and A. H. Bridle, in preparation). This observation, in combination with those of Section 3.1, identifies two main classes of (straight) radio jet—two-sided $B_\perp$- or $B_{\perp\text{-}\parallel}$-dominated jets in weak FR I sources (edge darkened, no hot spots) and one-sided $B_\parallel$-dominated jets in powerful FR II sources (edge brightened, stronger hot spots).

Two departures from this trend may arise when jet flows are perturbed:

1. Two-sided jets often have the $B_{\perp\text{-}\parallel}$ configuration where they bend. The $B_\parallel$ edge is often deeper (and more strongly polarized) on the outside of the bend [e.g. 3C 31 (42, 245), NGC 6251 (183), M84 (Table 1, ref. L2)], as if $B_\parallel$ is amplified there by stretching and shearing (4, 208). The bent jets in the C-shaped head-tail source NGC 1265 are $B_\parallel$-dominated even though they are two sided (Table 1, ref. O1). Such fields may be an extreme example of the $B_{\perp\text{-}\parallel}$ configuration resulting from viscous interaction with the ambient medium.

2. Some knots in one-sided jets have $B_\perp$-fields although fainter emission near them is $B_\parallel$-dominated [e.g. knot A in M87 (172), a knot 50″ from the core of NGC 6251 (183), and knot A2 in Cen A (48)]. These "magnetic anomalies" at bright knots may be due to oblique shocks that accelerate relativistic particles and amplify the component of $\mathbf{B_j}$ parallel to the shock. The fields also become dominated by $B_\perp$ components where one-sided $B_\parallel$ jets end at bright hot spots, and the physics there may be similar.

The degree of polarization is generally highest at the edges of $B_\parallel$-regions, but near the centers of $B_\perp$-regions. Two three-dimensional field configurations can fit these polarization distributions: (a) tangled field loops confined to a plane perpendicular to the jet axis near the center of the jet (136, 137) but stretched along the axis toward its edges (21, 183), or (b) "flux ropes" with helical fields of variable pitch (53) and some random component (183).

3.3 Size, Curvature, and Misalignments

Jets in weak galaxies (Figure 1) and in powerful core-dominated sources (Figure 5) are generally short ($< 10\%$ of all jets in sources with P^5_{core} $< 10^{22.5}$ W Hz^{-1}, and only 13% of those in sources with $P^5_{\text{core}} > 10^{26.5}$ W Hz^{-1}, have $d_j > 40$ kpc, but $\sim 50\%$ of those in sources of intermediate powers exceed this length). The jets in core-dominated sources may be shortened by projection effects if the cores are Doppler boosted (Section 6.1.7), but those in weak galaxies are two sided (e.g. Figure 1) and so are probably short intrinsically. Strong jet curvature is also common at the two extremes: C-shaped jets in weak "head-tail" cluster galaxies, and jets in powerful core-dominated sources (Figure 5). The curvature in head-tail sources is probably due to bending by ram pressure (10, 128); that in powerful sources may be due to confinement or to wandering of the central collimator (see Section 6.2.2).

The misalignments between parsec- and kiloparsec-scale jets, summarized in (200), increase with core prominence f_C. Several lobe-dominated radio galaxies with kiloparsec-scale jets have cores with one-sided parsec-scale jetlike extensions on the same side as the large jets (Section 6.2.3) and aligning with them to $\leq 10°$ (56, 67, 132, 142, 197–199, 209; Figure 2). In core-dominated sources, however, parsec- and kiloparsec-scale structures are often misaligned by $> 20°$ (57, 69; Figure 4), and in 3C 345 and 3C 418 (Figure 5) by $> 90°$. The jets in core-dominated sources bend continuously, with the sharpest curvature occurring closest to the core (58–60, 200, 274; Figure 5). These data are consistent with the short jets in core-dominated sources being close to the line of sight—small bends in jets near the line of sight may project as large apparent misalignments (e.g. 201).

4.   STRUCTURAL DETAILS

4.1   *Collimation, Freedom, and Confinement*

The jets in over a dozen radio galaxies (but in few QSRs) have been resolved transversely well enough to show their lateral expansions directly. They are generally center brightened, supporting the view that jets radiate by dissipation in the energy transport region itself, not in a static cocoon around it. The variation of (deconvolved) synchrotron FWHM Φ with angle Θ from the radio core may then track the variation of flow radius R_j with distance z from the nucleus. A steady free jet (whose pressure $p_j \gg p_e$, the sum of all external pressures) would expand with a constant lateral velocity v_r equal to its internal sound speed c_S where it first became free. It would widen at a constant rate $dR_j/dz = v_r/v_j$, unless the flow velocity v_j is slowed by gravity. If $d\Phi/d\Theta = 2(dR_j/dz) \sec (i)$, where i is the angle of the jet to the plane of the sky, nonlinearities in $\Phi(\Theta)$ reflect changes in the balance between p_j and p_e with distance z. The $\Phi(\Theta)$ data for well-resolved jets show that few are free at all z. Their expansion rates are not set once and for all on parsec scales, even though VLBI jets are first collimated on such scales.

4.1.1   WEAK RADIO GALAXIES   The first kiloparsec or so of well-resolved jets in weak radio galaxies typically expand with $d\Phi/d\Theta \leq 0.1$ (e.g. 27, 48, 198). Between 1 and 10 kpc, these jets "flare", with $d\Phi/d\Theta$ reaching values of 0.25 to ~ 0.6 (e.g. 27, 32, 48, 183). On still larger scales they may recollimate (27, 32, 33, 88, 183, 214, 220, 278). In NGC 315 (278) and NGC 6251 (33, 183), $d\Phi/d\Theta$ oscillates where the jets recollimate; these jets re-expand > 100 kpc from their cores.

The jet pressure is given by $p_j = p_{jt} + p_{jr} + p_{jm}$, where p_{jt} and p_{jr} are the pressures of the jet's thermal and relativistic particles and p_{jm} is the pressure of its magnetic field B_j. The external pressure is $p_e = p_{et} + B_\phi^2/8\pi$, where p_{et} is the thermal pressure and $B_\phi^2/8\pi$ represents confinement by $\mathbf{J} \times \mathbf{B}$ forces of toroidal magnetic fields B_ϕ on any current carried by the jet (11, 14, 16, 28, 53, 61, 207, 208). Recollimation requires $p_j \approx p_e$ over many kiloparsecs, but it is unclear which component of p_e dominates. Both halves of two-sided jets tend to recollimate at similar distances from their cores (32, 88, 186, 278).

Those in 2354 + 47 decollimate as they descend intensity gradients in its soft X-ray halo (49). The synchrotron properties of weak radio galaxy jets set lower limits to p_j ranging from $\sim 10^{-10}$ dyne cm^{-2} in the inner few kiloparsecs to $\sim 10^{-13}$ dyne cm^{-2} ~ 100 kpc from the galactic nuclei. These data suggest, but do not confirm, that weak radio galaxy jets can be collimated solely by p_{et} in hot galactic haloes. Confinement by gas at ~ 1–3 $\times 10^7$ K [cf. the M87 halo (85)] is (just) compatible with the *Einstein* IPC detections of, or upper limits to, extended soft X-ray sources around several jets, e.g. NGC 315 (278), 3C 66B (152, 168), Cen A (48), and NGC 6251 (183). The contribution of compact nuclear X-ray sources to the IPC data is unclear in some cases, however. *Einstein* and VLA data for M87 (18, 85) show that the minimum p_j in the knots (in this case a few times 10^{-9} dyne cm^{-2}) exceeds p_{et} at their projected distances in the X-ray halo by at least a factor of 10; only the first few hundred parsecs of this jet can be thermally confined by the X-ray halo, unless the jet is relativistic with $\gamma_j \geq 50$ (18). Nevertheless, its first kiloparsec expands at a constant rate, but the expansion slows beyond knot A; the B_ϕ term has been invoked (18) to explain this behavior.

If the longer rapidly expanding segments of these jets are free, the observed $d\Phi/d\Theta \ll 1$ implies that they are supersonic. The data suggest the jets are collimated initially (and become transonic) < 1 kpc from the nuclei, and that they then escape into regions where p_e drops rapidly. If p_e falls faster than $\sim z^{-2}$, continued confinement of a supersonic jet eventually requires that $v_r > c_s$ (236), so the jet becomes free by "detaching" from p_e at an oblique shock (219). If p_e again falls slower than z^{-2} farther out, as in the X-ray halo of M87 (85, 230), the free jet may be reconfined. Conical shocks would propagate into it from its surface, where it first "feels" the declining gradient of p_e, reheating it and possibly (re)accelerating relativistic particles in it (76, and references therein). The shock structure downstream from the reconfinement may be quasi-periodic, leading (*a*) to oscillations in the jet's expansion rate and (*b*) to regularly spaced knots along it (219). These phenomena may have been observed in NGC 315 (278) and particularly in NGC 6251 (33, 183; Figure 2), whose jet is limb brightened near its first reconfinement, consistent with particle acceleration in the conical shocks.

4.1.2 POWERFUL RADIO GALAXIES AND QUASARS   The jets in more powerful sources expand more slowly than those in weaker radio galaxies—Table 3 gives the average, minimum, and maximum expansion rates $d\Phi/d\Theta$ for 25 transverse-resolved jets. Several in powerful sources show little systematic expansion, e.g. 3C 33.1 (Table 1, ref. R1), 3C 111 (145), and 3C 219 (184). The small median angle ($< 1°$) subtended at the radio cores by "hot spots" in powerful doubles (e.g. 238) supports the trend, if the sizes of the hot spots indicate (roughly) the diameters of Mach disks where jets terminate (166, 167). The narrower collimation of the jets in stronger sources, coupled with their greater distances, means that their $\Phi(\Theta)$ forms are only crudely known. The data are adequate to show, however, that jets in powerful sources must be either (*a*) free with Mach numbers ≥ 50, (*b*) confined by much larger external pressures than those in nearby radio galaxies, or (*c*) the approaching sides of relativistic twin jets, whose minimum p_j is overestimated by the conventional calculation due to Doppler boosting (Section 6.1.7); they are all one sided (Section 3.1), so this interpretation is permitted.

Thermal confinement of the parsec-scale jets in several powerful radio galaxies (but not in Cyg A) is compatible with the X-ray data (144), but for several large-scale QSR jets (271) the *Einstein* data rule out pure thermal confinement at ~ 1–3×10^7 K unless the jets are Doppler boosted. Wardle & Potash (271) argue that freedom is inconsistent with energy and thrust balance (Section 6.1). Eichler (78, 79) discusses balancing p_{et} against the

Table 3   Expansion data for radio jets

Jet name	$\log_{10} P_{core}^5$	$\langle d\Phi/d\Theta \rangle$	$[d\Phi/d\Theta]_{min}$	$[d\Phi/d\Theta]_{max}$
1321+31 SE	21.77	0.30	≈ 0	0.4
1321+31 NW	21.77	0.25	< 0.07	0.27
3C 449 N	22.07	0.20	0.1	0.80
3C 449 S	22.07	0.20	0.1	0.45
3C 129 E	22.19	0.13	0.1	0.35
Cen A	22.20	0.19	0.05	0.20
3C 31 N	22.45	0.30	0.08	0.38
3C 31 S	22.45	0.28	0.18	0.36
3C 296 (mean)	22.67	0.16	—	—
0326+39 E	22.70	0.22	0.10	0.34
0326+39 W	22.70	0.25	0.10	0.26
M87	22.92	0.07	—	—
NGC 315 SE	23.24	0.11	0.06	0.6
NGC 315 NW	23.24	0.11	≈ 0	0.58
4C T74.17	23.26	0.12	—	—
Her A W	23.61	< 0.1	—	—
NGC 6251 NW	23.66	0.08	≈ 0	0.17
3C 33.1	23.76	0.06	≈ 0	0.09
Cyg A	24.12	0.03	—	—
3C 219	24.18	0.07	≈ 0	0.15
3C 111	24.47	0.04	0.01	0.06
4C 32.69	25.15	0.06	—	—
3C 280.1	26.21	0.05	—	—
3C 273	26.92	0.013	≈ 0	0.018
3C 279	27.56	< 0.02	—	—

inertia of low-entropy jets to collimate them. Magnetic confinement is also frequently invoked (11, 14, 16, 53, 191, 207, 208). It requires jet currents of $\sim 10^{17-18}$ A if the fields are near equipartition; the return currents are assumed to lie outside the observed radio emission regions. The QSR jets are $B_{\parallel}$-dominated (Section 3.2), so the toroidal B_{ϕ} must also be supposed to lie (frustratingly unobserved) outside the main synchrotron-emitting regions.

4.1.3   COCOONS   The study of jet collimation is complicated by sources such as M84 (Figure 1), 3C 341 (Figure 6), 1321 + 31 (88), 4C 32.69 (75), and 2354 + 47 (49) with faint emission "cocoons" around brighter jets. The collimation properties of cocoons may differ radically from those of their jets, e.g. that in M84 (Figure 1) expands much faster than the jets at $\Theta > 5''$. At what level of brightness (if any) in such sources does the synchrotron expansion rate $d\Phi/d\Theta$ indicate streamline shapes in an underlying flow? The minimum cocoon pressures are only $\sim 0.1\ p_{\mathrm{j}}$ (if the jets are unbeamed), so thermal confinement of the jets should crush the cocoons (75). The relationship of cocoons to the brighter structure—whether they are faint "outer jets," static sheaths, or backflows such as those in simulations of thermal matter flows in jets (166, 167)—is unclear. Polarimetry of the cocoons may test whether they contain the B_{ϕ} required for magnetic confinement of the jets, by detecting radial changes in $\mathbf{B_a}$ or transverse rotation measure gradients (183).

4.2   Radio Spectra

About 40% of jets have spectra between $v^{-0.6}$ and $v^{-0.7}$ near 1.4 GHz, and $\geq 90\%$ have spectra between $v^{-0.5}$ and $v^{-0.9}$. Spectral gradients along most jets are small, but where they have been detected the spectra steepen away from the cores (48, 54, 70, 75, 245), consistent with synchrotron depletion of the higher-energy electrons in the outer jets (277).

4.3   Intensity Evolution

Both the magnetic field strengths and the relativistic particle energies will decrease along an expanding laminar jet, with no magnetic flux amplification or particle reacceleration. If (a) magnetic flux is conserved and (b) the radiating particles do work, as a jet with the typical $v^{-0.65}$ spectrum (Section 4.2) both expands laterally and responds to variations in its flow velocity v_{j}, then the jet's central brightness I_v will vary as $R_{\mathrm{j}}^{-5.2}v_{\mathrm{j}}^{-1.4}$ in $B_{\parallel}$-dominated regions, or $R_{\mathrm{j}}^{-3.5}v_{\mathrm{j}}^{-3.1}$ in $B_{\perp}$-dominated regions (88, 183). Note that neither $B_{\parallel}$ varying as R_{j}^{-2} nor $B_{\perp}$ as $R_{\mathrm{j}}^{-1}v_{\mathrm{j}}^{-1}$ to conserve magnetic flux are compatible with equipartition of energy between radiating particles and B_{j} in a confined jet if the particles do work and are not reaccelerated;

equipartition requires B_j to vary as $R_j^{-4/3}v_j^{-0.3}$ and I_v to decline as $R_j^{-4.1}v_j^{-1.9}$ for a $v^{-0.65}$ spectrum. Actual variations of I_v with jet FWHM Φ (assumed proportional to R_j) are often much slower than these "adiabats" over large regions of the jets (27, 118, 183, 278). Near the core, I_v often increases with Φ—the jets "turn on" following regions of diminished emission, or "gaps" (19, 29, 186, 277; Figures 1, 3, 6). The "turn-on" is often followed by regimes many kiloparsecs long in which I_v declines as $\sim \Phi^{-x}$, with $x = 1.2$–1.6; the value of x reaches ~ 4 in the outer regions of some jets (33, 117, 183, 278), but in NGC 6251 the "adiabatic" decline ≥ 100 kpc (200″) from the core is repeatedly interrupted by the "turning on" of bright knots (see Figure 2 and 183).

It is likely that some of the bulk kinetic energy of the jets (which is not lost by adiabatic expansion) is converted to magnetic flux and relativistic particles through dissipative interactions with the surrounding ISM. Indeed, if B_j is near equipartition on kiloparsec scales, $B_{\parallel}$ must be amplified locally (instead of falling as R_j^{-2}) or else long $B_{\parallel}$-dominated jets would have unreasonably high fields on parsec scales. Models for "reheating" of jets include shock formation (24, 66, 167, 219) and various mechanisms following the development of large-scale vortical turbulence (8, 13, 15, 17, 73, 82, 94, 112, 118) from the growth of instabilities at the jet surface. Some models based on large-scale turbulence link the synchrotron emissivity directly to the turbulent power, and hence to the jet spreading rate as $(d\Phi/d\Theta)^n$ with $1.5 \leq n \leq 3$ (17, 82, 118). They can thereby explain why the rapidly expanding jets in weak sources (Table 3) are so conspicuous, and why a jet's most rapidly expanding segments are often those of its most "subadiabatic" intensity evolution (118, 183). Initially laminar jets may also propagate far from their sources before becoming turbulent; rapid fading in the laminar ("adiabatic") regime (as in parsec-scale VLBI jets; Figure 4) can be followed by "turning on" of a large-scale jet in the same direction once turbulence becomes well developed. This may explain the "gap" phenomenon (8, 13, 17, 118, 129). Velocity variations may also keep jets bright in two distinct ways. Fluctuations in v_j at the core can produce strong shocks that locally enhance the synchrotron emissivity (206). Entrainment of surrounding material will decrease v_j along a jet—the resulting axial compression may partly compensate the effects of lateral expansion, particularly where $B_\perp$ dominates (88, 183).

Detailed understanding of what keeps large-scale jets lit up requires self-consistent modeling of their collimation, intensity evolution, and magnetic field configurations. Abrupt changes in $\mathbf{B_a}$ from $B_{\parallel}$ to $B_\perp$ at bright knots (Section 3.2) may indicate particle acceleration at oblique shocks, particularly if the knots have their sharpest brightness gradients on their coreward sides, as in M87 (18, 54) and NGC 6251 (183). The degrees of linear polarization in, and the depths of, $B_{\parallel}$ edges on $B_{\perp}$-dominated jets may indicate the extent of viscous interactions with the surrounding ISM. The observations provide copious constraints for the models: jet expansion rates, "turn-on" heights, transverse intensity profiles, field orderliness and orientation, as well as the $I_\nu(\Phi)$ evolution. Models of jet propagation are not yet sufficiently versatile to confront the data at all of these points self-consistently, however.

5. RADIO JETS AT OTHER WAVELENGTHS

5.1 Optical and Infrared Wavelengths

5.1 CONTINUUM   Optical continuum emission coincides with bright knots in the radio jets of 3C 31 and 3C 66B (52), M87 (249, and references therein), 3C 273 (2, 141, and references therein), 3C 277.3 (159), and possibly Cen A (34). The spectral index between 4500 Å and 5 GHz is generally within 0.1 of 0.7 (52). The M87 knots all have essentially the same connected optical-infrared-radio spectra, with slopes in the radio, infrared, and optical of ~ 0.6, 0.8, and 1.7, respectively; the latter spectral break occurs near 6000 Å in every bright knot (150, 165, 237, 244). In 3C 273 the radio jet brightens toward its tip (64, 181) but the optical jet is more uniformly bright, except for knots at each of its bends (141), so the spectrum above 5 GHz steepens with distance from the QSO (cf. Section 4.2).

The optical continuum is up to 20% linearly polarized in M87 (226, and references therein) and $\sim 14\%$ linearly polarized in 3C 277.3 (158, 159). This polarization, the positional coincidence with the radio knots, and the connected optical-radio spectrum in M87 are evidence that the optical continuum is synchrotron radiation from the same regions as the radio jets. The overall linear polarization of the optical jet in 3C 273 is only $3.7 \pm 4.1\%$ (226), whereas the radio jet is $\sim 12\%$ polarized at both 1.4 and 5 GHz at 2″ resolution (Table 1, ref. P2). The radio E vectors in the "head" and "tail" of this jet are nearly perpendicular, however, so high-resolution optical polarimetry is needed to test whether the optical emission is synchrotron radiation there also. If B_j is near equipartition in these jets, the electrons radiating at optical wavelengths are several synchrotron lifetimes from the radio cores. The distribution of optical emission marks the sites of relativistic electron reacceleration, or possibly of pitch angle scattering (239, 240), better than the radio data because of the much longer synchrotron lifetimes of the radio electrons. Studies of optical jets with the *Space Telescope* will show how discrete, or continuous, the reacceleration regimes are.

5.1.2   EMISSION LINES   Data on emission lines from the vicinity of radio jets are reviewed in (158) and (260). The line-emitting regions generally lie beside or beyond the jets, particularly on the outer edges of bends, and are often brightest near, but not at, knots or hot spots. The spectra are not those of normal H II regions, but instead resemble those of Seyfert 2 nuclei. Typical densities are $\sim 10^2$ to 10^3 cm^{-3} and temperatures $\sim 20{,}000$ K, leading to pressures near the lower limits to p_j in adjacent radio features. Typical bulk velocities are a few hundred kilometers per second, and line widths are 300–500 km s^{-1}. In Cen A, some emission-line filaments beyond the jet have internal differential velocities up to 800 km s^{-1} (106); they cannot stay intact long enough to have been convected out from the nucleus. The line strengths suggest photoionization by a power-law continuum, and possibly also shock heating, with different mixtures of excitation mechanisms in different sources. The occurrence of this extranuclear line emission at the edges of radio features, and the increase of line widths toward them (116), suggest interaction between jets and the ambient ISM. The line-emitting gas may be clouds in normal galactic rotation that have encountered jets, becoming heated, ionized, and accelerated by them (26, 73, 74, 116, 158, 159, 282). The uncertain dynamics of the emission-line gas preclude using its bulk velocities to infer jet velocities directly, as suggested in (159).

A continuum and emission-line feature in DA 240 has been reported as an "optical jet" blueshifted relative to the galactic nucleus by 3400 km s^{-1} (40, 41). This has been invoked (41, 246) as direct evidence that $v_j = 3400$ km s^{-1} in DA 240, but the feature is not a radio jet and may be a confusing spiral galaxy (261).

5.2   X-Ray Wavelengths

Three radio jets are known to be X-ray sources—M87 (230), Cen A (228), and 3C 273 (276). [See (90) for a review and further references.] The region near the M87 jet has a luminosity of $\sim 10^{41}$ erg s^{-1} in the *Einstein* HRI band. Individual knots are not resolved, but this integrated X-ray luminosity is consistent with extrapolating the steep spectrum of the knots above 6000 Å to the X-ray regime. If the synchrotron interpretation favored in (230) is correct, electrons with $\gamma \approx 10^{7.3}$ are required to produce the observed X-rays in the equipartition magnetic field of the knots; this provides a severe test for particle acceleration models. The radiative lifetimes of such electrons would be ≤ 200 yr, comparable to the light crossing time in the knots, but much less than the light travel time to the knots from the nucleus of M87. The X-ray and radio structures of the jet in Cen A are also very similar (48), suggesting that this is also synchrotron emission, though the case is not as strong as for M87 (90). The detection of the X-ray jet in 3C 273 depends heavily on deconvolution of the point-source response from the data, so it has not yet been analyzed in detail.

6. JET LAG-MISSING PHYSICS

6.1 *Velocity Estimates*

Without direct velocity indicators such as emission lines from material in the jet flows, estimates of the average flow velocities v_j are indirect and sensitive to the initial assumptions. Most methods assume jet properties to be stationary in time and estimate v_j from observables using one or more of the following arguments.

6.1.1 ENERGY FLUX   If synchrotron losses from a lobe of luminosity L_{lobe} are continuously replenished by the energy influx from a jet at an efficiency ε, the energy flux supplied to the lobe must be $L_j = L_{lobe}/\varepsilon$. The flux L_j is related to other jet parameters by

$$L_j = A_j \rho_j v_j^3 \gamma_j^2 \left[\frac{\gamma_j}{\gamma_j + 1} + \frac{h_j}{\rho_j v_j^2} \right],$$

where A_j, v_j, ρ_j, γ_j, and h_j are the cross-sectional area, flow velocity, density, Lorentz factor, and enthalpy per unit rest mass of the jet at some point along it, respectively. (The transverse structure of the jet is usually ignored when estimating v_j.)

6.1.2 MOMENTUM FLUX   For jets which terminate at hot spots, the thrust $T_{j/h}$ in the rest frame of the hot spot must balance $p_h A_h$, where p_h and A_h are the minimum pressure and cross-sectional area of the hot spot estimated from its synchrotron parameters. The thrust $T_j = A_j \rho_j v_j^2 \gamma_j^2$ in the galaxy or QSR frame can be calculated from this by assuming the dynamics of the interaction between the hot spot and the ambient density ρ_e to relate v_j to the velocity v_h of advance of the hot spot. Estimates for T_j can also be made for C-shaped jets in "head-tail" sources if these jets are bent by the ram pressure of an intergalactic density ρ_e through which their parent galaxy moves at velocity v_g (10, 128). If the radius of curvature of the C-structure is R_c and the scale over which the ram pressure is transmitted to the jet is H, momentum balance requires that $T_j = \rho_e v_g^2 A_j R_c/H$. The parameter H is the radius R_j of the jet in the "naked" jet-bending model of (10), but it is a scale associated with the ISM of the parent galaxy in the "shielded" model of (128).

6.1.3 MASS FLUX   The mass flux $dm/dt = A_j \rho_j v_j \gamma_j$ down the jet must meet either reasonable constraints on the rate of ejection from the "central engine" or constraints from depolarization data on the total mass injected into the lobes over the lifetime of the source. (The latter constraints are generally less stringent, but they could be tightened by high-resolution polarimetry of lobes at frequencies below 1 GHz.)

6.1.4   JET EXPANSION   The Mach number M_j of a free jet where it detached from its confinement (Section 4.1) can be estimated from its synchrotron expansion rate via $d\Phi/d\Theta = (2/M_j)$ sec (i), where i is the (assumed) angle between the jet axis and the plane of the sky. Then $v_j = M_j\sqrt{\Gamma p_j/\rho_j}$, where Γ is the ratio of principal specific heats in the jet, and a lower limit to p_j is obtained from the synchrotron parameters. If the jet is actually confined, M_j is overestimated. If the weak radio galaxy jets are alternately free and confined (Section 4.1.1), M_j is best indicated by the rapid expansions at $z \approx 1$–10 kpc, which imply that $M_j \approx 10$ there.

6.1.5   ELIMINATING THE JET DENSITY   Usage of the techniques of Sections 6.1.1–6.1.4 alone requires estimates of ρ_j, generally from centimeter-wavelength Faraday depolarization data that are hard to obtain and to interpret (87). High signal-to-noise is needed to reduce Ricean bias in the polarized signal (183, 270). The configuration of $\mathbf{B}_j$ is unclear (Section 3.2), particularly the scale distribution of its reversals, which may "hide" thermal gas. Some jets are surrounded by emission-line filaments (Section 5.1.2) and magnetoionic media with clumping scales of ~ 1–5 kpc (33, 183; Table 1, ref. L2). Differential Faraday rotation across the radio beam by such media may decouple the observed depolarization from ρ_j, so even setting limits to ρ_j from low-resolution depolarization data without mapping the rotation measure gradients is hazardous, especially at low frequencies. The methods in Section 6.1.1–6.1.4 all permit $v_j \rightarrow c$ if $\rho_j \rightarrow 0$, but Table 4 shows how v_j can be constrained using combinations of these methods to eliminate ρ_j for "cold" $(h_j \ll \rho_j v_j^2)$ jets. As $h_j/\rho_j v_j^2 = \Gamma/M_j^2(\Gamma-1)$ from the gas laws and Section 6.1.4, this is a good assumption for jets with $M_j \geq 4$. These combinations also eliminate A_j, bypassing the (uncertain) relationship between the jet's synchrotron width Φ and its flow radius R_j. They normally yield velocities in the range $1000 < v_j < 30{,}000$ km s^{-1} unless low efficiencies ε or high mass fluxes dm/dt are assumed.

Table 4   Parameter combinations for cold jets

Combination	$v_j \ll c$	$\gamma_j \gg 1$
$\left[\dfrac{L_j}{cT_j}\right] = \sqrt{\dfrac{\gamma_j-1}{\gamma_j+1}}$	$\dfrac{\beta_j}{2}$	1
$\left[\dfrac{L_j}{c^2(dm/dt)}\right] = \gamma_j - 1$	$\dfrac{\beta_j^2}{2}$	γ_j
$\left[\dfrac{T_j}{c(dm/dt)}\right] = \sqrt{\gamma_j^2-1}$	β_j	γ_j

6.1.6   SUPERLUMINAL MOTION   A simple model for the observed proper motions of knots in compact sources (57) is that apparent "superluminal" motion at $\beta_{app} = v_{app}/c > 1$ arises for features in the approaching side of a high-γ_j jet at a large angle i to the plane of the sky, whereupon $v_j = v_{app}/[\beta_{app} \sin (i) + \cos (i)]$.

6.1.7   DOPPLER BOOSTING   With the above notation, v_j is related to the ratio of intensities $\mathscr{S}_{app}$ of the approaching and receding sides of an intrinsically symmetric ($\mathscr{S} = 1$) jet as $v_j \sin (i) = c[\mathscr{S}_{app}^{\delta} - 1]/[\mathscr{S}_{app}^{\delta} + 1]$, where $\delta = 1/(2 + \alpha)$ for a continuous jet with a $v^{-\alpha}$ spectrum (23). Assuming $\mathscr{S}_{app}$ to be due entirely to Doppler boosting therefore constrains $\beta_j \sin (i)$. Note that with the typical value $\alpha = 0.65$ (Section 4.2), $\mathscr{S}_{app}$ varies as $\gamma_j^{5.3}$ if the line of sight is $< 1/\gamma_j$ radians from the jet axis; also, note that a jet can be "one sided" as in Section 3.1 ($\mathscr{S}_{app} > 4:1$) if $\beta_j \sin (i) > 0.26$, which at $i = 30°$ (the median value for randomly oriented sources) requires only that $\beta_j > 0.52$.

6.1.8   JET WIGGLING   Many jets wiggle around their mean direction (e.g. 29, 172, 183, 214, 220, 262, 263). Mechanisms for periodic lateral deflections Δ as a function of angle Θ from the core [reviewed in (252)] include (a) orbital motion of the primary collimator around a companion mass in the parent nucleus (9, 151) or a nearby member of the same group or cluster (22, 263); (b) precession of the primary collimator or of a larger-scale recollimating atmosphere due to interaction with another body (9, 104, 123, 143, 205, 284); and (c) growth of helical Kelvin-Helmholtz instabilities at the boundary of a confined jet (Section 6.4). Pure orbital motion leads to C-symmetry between the two sides of a jet, with fixed wiggle amplitude Δ and a period τ_o. Pure precession of the source of a free nonrelativistic jet leads to S-symmetry, linear growth of Δ with Θ, and a period τ_p. The analogue for relativistic jets is more complicated as the S-symmetry is broken by light travel time effects, which might themselves indicate v_j if other distortions were absent (63, 104, 143). Helical surface instabilities on a confined expanding jet make wiggles whose amplitudes Δ and wavelengths λ_i both grow with Θ; linear theory has been used to estimate the most rapidly growing wavelength λ_i as a function of jet radius R_j, Mach number M_j, and density contrast ρ_j/ρ_e (Section 6.4).

Attempts to constrain v_j from jet-wiggling data "match" an observed pattern $\Delta(\Theta)$ to one of these pure forms to find a characteristic wavelength λ_0 and a self-consistent estimate of the characteristic period τ_o or τ_p. Then, v_j or M_j is derived from one of the following: $v_j = \lambda_0/\tau_o$, $v_j = \lambda_0/\tau_p$, or $\lambda_0 \approx \lambda_i = R_j F_1(M_j) F_2(\rho_j/\rho_e)$, where the functions F_1 and F_2 are provided by (linear) instability theory. These methods are fraught with uncertainties. Well-studied jets rarely match simple orbital or ballistic precessional shapes convincingly (22, 104, 105, 257), so additional poorly constrained parameters (e.g. multiple or eccentric orbits, variation of precession cone angle with time) are invoked. Even goodness of fit to a simple C- or S-shape does not guarantee uniqueness of the model (63, 122). Bending and buoyancy effects (e.g. 119, 235, 283) may also be present and—unless the jet is denser than the ambient medium—lateral motions may excite surface instabilities, whose growth also alters the shape of the jet (12). Linear instability theory may be inadequate to describe any mode that grows sufficiently to become detectable on radio maps.

6.2   The Velocity Dilemma

The above methods give velocities ranging from $v_j \approx 1000$ km s^{-1} in C-shaped jets in head-tail sources (using Section 6.1.2) to $\approx c$ (using Section 6.1.6 to interpret one sidedness or Section 6.1.7 to interpret superluminal motion). This uncertainty in v_j seriously obstructs progress in elucidating the physics of radio jets.

6.2.1   ARGUMENTS FOR $v_j \approx c$ ON PARSEC SCALES   Five arguments favor $v_j \approx c$ on parsec scales in some sources:

1. The superluminal separations of knots in VLBI "core-jet" structures can be explained if $\gamma_j \approx 2.5$–10 (for $H_0 = 100$) and these structures are within $\sim 1/\gamma_j$ radians of the line of sight (Section 6.1.6).
2. The same parameters entail Doppler boosting (Section 6.1.7), which accounts for the one sidedness of these VLBI core-jets.
3. Similar assumptions (but with higher values of γ_j) may explain the high brightness temperatures implied by the rapid low-frequency variability of some compact radio sources (89, and references therein).
4. The same assumptions explain the low Compton X-ray fluxes from compact radio sources (e.g. 155).
5. The small angles to the line of sight ($i \approx 90°$) required by this interpretation of the properties of strong compact cores are consistent with the large apparent bending of the jets in core-dominated sources (Section 3.3).

There is little evidence against $v_j \approx c$ on parsec scales: 3C 147 has a complex, two-sided parsec-scale structure (194), but two sidedness may be ascribed to bending a one-sided jet across the line of sight, in a suitably small number of cases, without endangering the relativistic-jet picture of compact sources.

6.2.2   ARGUMENTS AGAINST $v_j \approx c$ ON KILOPARSEC SCALES   The sensitivity of Doppler boosting (Section 6.1.7) to $v_j \sin(i)$ argues against $v_j \approx c$ in the C-shaped jets in "narrow head-tail" sources. If these are indeed swept back by ram pressure of the intergalactic medium (Section 6.1.2), $\mathbf{v}_j$ changes direction along them by as much as 90° (e.g. 171, 214, 257). If $v_j \approx c$, they would (a) have large side-to-side asymmetries and (b) brighten or fade dramatically as they bend, in conflict with observation (265). The orientations of dust lanes in some weak radio galaxies also suggest that brightness asymmetry at the bases of their two-sided jets (Section 3.1) is unlikely to be due to Doppler favoritism. The jets are generally $> 70°$ from the dust lanes in projection (134 and Table 1, ref. L1), so they should generally be nearly perpendicular to them in three dimensions. The orientation of the dust lane (266) in M84 thereby suggests that the northern jet, which is the brighter and has the one-sided base (Figure 1), is either receding or very close to the plane of the sky (if it is an outflow). Both this constraint and the fact that it becomes two sided without bending argue that its greater brightness is due to greater power output or greater dissipation on its side of the nucleus, rather than to Doppler boosting. R. A. Laing (personal communication) finds similar results in NGC 3665 and for the possible jet in NGC 612, although in Cen A the peculiar velocities of the optical filaments (Section 5.1.2) argue that the bright radio jet is approaching.

This evidence against $v_j \approx c$ in sources with $P_{tot}^{1.4} < 10^{25}$ W Hz^{-1} (FR I structures) leaves open the possibility that v_j increases with P_{tot}, so that the long one-sided jets in powerful sources might be Doppler-boosted flows with $\gamma_j \gg 1$. Some bent one-sided jets have smooth brightness variations [e.g. 1150+497 (Figure 3), 4C 32.69 (191)], which are inconsistent with changing Doppler boosts in high-γ_j jets if they bend because they are confined. Such jets could be ballistic, however, with their shapes arising from wobble (precession?) of the primary collimator; v_j would then not follow the bends, but the wiggle pattern would move radially as a whole. Changes in $v_j \sin (i)$ and in the Doppler boosting (Section 6.1.7) may then be small. We must know whether or not such jets are confined (Section 4.1), and if so where, to decide whether their brightness distributions argue against $v_j \approx c$. Doppler boosting models for long one-sided jets also require large angles i, so boosted one-sided jets would be significantly longer in three dimensions than they appear in projection. It is unclear whether this seriously conflicts with $v_j \approx c$ in these jets, as the existing statistics of QSR source sizes come from samples containing significant numbers of one-sided jets (Section 2.2). Maps with greater dynamic range are needed to assess the degree of one sidedness of these jets (we do not know by how much $\mathscr{S} > 4$ in most cases), as the average deprojection increases with the average asymmetry.

6.2.3   ARGUMENTS FOR $v_j \approx c$ ON KILOPARSEC SCALES   Table 1 lists 22 sources with VLBI jets or elongations. Of these, five exhibit superluminal expansion—3C 120, 3C 179, 3C 273, 3C 279, 3C 345 (57). In all five, the kiloparsec- and parsec-scale jets start on the same side of the core, as in Figure 4 (see references in Table 1). Sixteen others have VLBI elongations and kiloparsec-scale jets, but their proper motions on parsec scales are unknown. In 11 of these (NGC 315, 3C 78, 3C 84, 0957 + 56, 3C 111, M87, Cen A, NGC 6251, 3C 371, 3C 405, and 3C 418), the larger-scale jet starts on the same side as the small, e.g. Figures 2 and 5. Of the remaining five, two (3C 147 and 3C 236) have two-sided small-scale structure, two (M84 and 3C 454.3) do not have closure-phase VLBI maps, and 3C 309.1 has complex structure. The correlation between small- and large-scale sidedness argues that one sidedness has the same cause on both scales. It supports the idea that v_j can be high enough on kiloparsec scales for Doppler favoritism to be important, if one is convinced by the case for $\gamma_j \gg 1$ on parsec scales (Section 6.2.1). This case is strongest for the five superluminal sources, but it is not yet impregnable. Since there are no known coreless large-scale jets, either both the cores and the jets are Doppler boosted or the luminosities of intrinsically one-sided jets are coupled to those of the cores; the reason for such coupling over such a wide range of linear scales is unclear if the sidedness is due to asymmetric dissipation. If the kiloparsec-scale sidedness is intrinsic (124, 213, 218, 273, 280), these data require a switching time-scale $\tau > d_j/v_j$ and an alternative model for superluminal expansions. The constraint $\tau > d_j/v_j$ is often hard to reconcile with τ being less than the synchrotron lifetimes in the hot spots (e.g. 113, 265). On balance, the correlation between parsec and kiloparsec sidedness favors $\gamma_{core} \approx \gamma_j \geq 1$ in powerful sources.

Other (weaker) arguments for $v_j \approx c$ on kpc scales are the following:

1. Relativistic jets need less confinement, since Doppler boosting and projection effects at large angles to the plane of the sky mean that the standard synchrotron calculation may overestimate p_j.
2. It is difficult to brake a jet by entrainment between the parsec and kiloparsec scales without converting much of its energy into heat, unless M_j is low (222). Losing $\sim c/v_j$ times the total power of a strong extended radio source by dissipation as "waste heat" near its nucleus is a daunting prospect for source models if $v_j \ll c$.

6.3 Jet/Hot-Spot Symmetries and the Sidedness Dilemma

The symmetries of the regions where powerful jets end may also offer clues to the reasons for their one sidedness (215). If it is always due to Doppler favoritism, the jetted and unjetted lobes should look similar—unless high-γ_j jets push the hot spots out at $v_h \approx c$, in which case the brighter jet should appear to feed the brighter and more distant hot spot if the two sides of the source have the same history (146). (In the extreme case of a "young" high-γ_j source, radiation from the receding side may also not yet have reached us.)

In 34 of the 46 FR II sources in Table 1 with one-sided jets, one lobe has a significantly brighter hot spot than the other on the highest resolution map available. Seventeen of the 34 have $f_C = S_{core}^5/S_{tot}^{1.4} > 0.05$; the brighter jet points to the brighter hot spot in 16 of these. Unless the jets are "young", either the brighter jet has a higher thrust or the jets and the hot spots in these sources are both Doppler boosted. In the 17 cases with $f_C < 0.05$, the brighter jet points to the brighter hot spot in ten and to the weaker in seven. This is consistent with one sidedness due either to differential dissipation or to Doppler boosting. There is no trend in either group for the jetted hot spot to be more distant, so if boosting is important the hot-spot separations must not reflect travel time differences from simultaneous ejecta. They might instead be determined by the history of the source, e.g. by a wandering or intermittent jet illuminating different parts of a lobe at different times. These trends imply that either (a) jet one sidedness has different causes in FR II sources with different f_C, or (b) the jets, but not the hot spots, are boosted in sources with $f_C < 0.05$, while both are boosted if the core is strong. The relative brightnesses of hot spots are sensitive to linear resolution, however, so the trends must be checked with more uniform data.

6.4   Stability

Jets are surprisingly stable. They can extend for hundreds of kiloparsecs or bend through $\geq 90°$ (in C-shaped "head-tails") without disruption. Early analyses of the stability of confined cylindrical jets to helical, fluting, and pinching perturbations analogous to the Kelvin-Helmholtz instabilities of a vortex sheet (15, 92, 93, 95, 110, 195) suggested that jets are generally unstable to modes with wavelengths of a few jet radii. The growth rates are less for $M_j > 1$ and for $v_j \approx c$, but the stability of observed jets forces re-examination of simplifying assumptions made in these analyses. The stabilizing influence of a surface shear layer on modes with wavelengths less than its scale depth was examined in (92) and (196), and that of jet expansion on long-wavelength modes in (111). Within a thermally confined jet, $B_\parallel$ may stabilize long-wavelength pinching modes (12, 92, 195). The firehose instability can be inhibited by sizable B_j and by linking the inertia of a plasma cocoon around the jet to B_j (12). The stability of magnetically confined jets has yet to be studied thoroughly, although first steps have been taken (12, 16, 61). Progress here is hampered by ignorance of basic MHD parameters in jets: we know little about ion or electron temperatures, field strengths, particle densities, and sound or Alfvén speeds, independent of the assumption of equipartition. Currently favored models of jet production from rotating disks or tori near supermassive objects (e.g. 251) may produce flows with net helicity. The influence of such helicity on jet stability merits attention, as helicity can lead to efficient generation of large-scale B_j by turbulent amplification of small seed fields (72).

Instabilities in real jets may grow algebraically, rather than exponentially. Exponential growth can be stopped in many ways—shock formation when the perturbation velocities become supersonic, shifting of the modes to longer wavelengths as their amplitudes grow, or saturation of the instabilities by in situ particle acceleration (15, 93, 94). Nondestructive instabilities might dominate the observed shapes and relative brightnesses of radio jets and lobes (15, 112, 286): algebraic growth of short-wavelength instabilities may help to keep jets bright (Section 4.3), long-wavelength helical modes to explain jet wiggling (Section 6.1.8), and pinching modes to form knots (Sections 4.1.4 and 4.3). Instability growth may also determine overall source sizes. MHD stability analyses including jet expansion, velocity and density gradients, and realistic B_j configurations and velocity profiles are needed; the analytical difficulties are great, and numerical simulations that do not legislate axisymmetry may be required.

6.5   Unified Models

"Unified" models seek to relate differences between sources with weak and powerful radio cores solely to differences in viewing angle. If the arguments in Section 6.2.1 indeed support $\gamma_j \gg 1$ and $i \approx 90°$ in the jets of core-dominated sources, a randomly oriented sample should contain $\sim 2\gamma_j^2$ unboosted sources for every boosted one if the jets have narrow cone angles. For $\langle \gamma_j \rangle \approx 5$ (169), core-dominated sources would number only a few percent of their parent population in the plane of the sky, which may therefore be a well-known class of object. Proposed parent populations for the core-dominated QSRs are radio-quiet QSOs (223) and QSRs with lobe-dominated extended radio sources (23, 169). The latter proposal is not encouraged by the fact that the lobe-dominated sources have weaker [FeII] emission and broader lines than core-dominated sources (242, 281). It is hard to see how such differences in the line strengths could be produced by the small aspect differences ($\Delta i \approx 1/\gamma_j$ radians) over which the Doppler-boosting factor varies markedly (115). Although the flux density distribution of strong radio sources in optically selected QSO samples conflicts with the unified models (131, and references therein), VLA studies of the Schmidt-Green QSO sample to a limiting flux density of 250 μJy at 6 cm are consistent with them over most of the flux density range. The "excess" of strong sources may be due to a separate population of extended sources, most of whose emission is presumably unbeamed (K. Kellermann et al., in preparation). As the emission lines cannot be beamed, models that beam the optical continuum luminosities of core-dominated QSRs predict the existence of emission-line QSOs without nonthermal continua; these have not been detected.

About half of all core-dominated sources have detectable kiloparsec-scale secondary structure, which is generally one sided, as in Figure 5 (182, 185, 274). If the parent population is to be "radio quiet" (223) or to have normal two-sided lobes (169), both "unified models" must assert that most of this one-sided secondary structure is also boosted. The parent population of most core-dominated sources must then be a class of numerous weak extended radio sources with at least mildly relativistic kiloparsec-scale jets and relatively strong Fe lines.

6.6   A Broader Unified Model

As about half of all radio galaxies and nearly all QSRs have detectable radio cores (131, and references therein), there must be a mixture of boosted and unboosted contributions to the core emission. Furthermore, some kiloparsec-scale jets emerging from weak cores must have nonrelativistic velocities and intrinsic emission asymmetries (Section 6.2.1), while some jets emerging from powerful cores may be relativistic (Section 6.2.3). It may be that $\gamma_{core} \geq \gamma_j > \gamma_h$ in all sources, while all three tend to increase with the actual source power (measured by the luminosity of the most extended radio features). The correlations between P_{core} and f_C, the occurrence rates of jets (Section 2.3), their sidedness (Section 3.1), their magnetic field configuration (Section 3.2), and the large-scale source structure (FR class) might be assimilated in a broader unified model as follows.

The kiloparsec-scale jets in most weak sources have $v_j \ll c$, and so appear two sided, with minor asymmetries that are either intrinsic or the result of asymmetric internal dissipation of flow kinetic energy to synchrotron radiation. They expand rapidly, so $B_\perp$ dominates over $B_\parallel$ except at their bases. They have low thrusts and so are readily bent, sometimes maintaining $B_\parallel$ layers at their edges by shearing or stretching as they bend. Low Mach numbers allow them to become turbulent, to entrain material and thus to decelerate [all effects that keep them well lit up (Section 4.3)], and to terminate gently without forming hot spots. These characteristics lead to FR I morphology (Figure 1). Weak sources with $\gamma_j = \gamma_h \approx 1$ but $\gamma_{core} \gg 1$ would be strongly core dominated if oriented near the line of sight; their extended low-brightness FR I structure would be detected only on maps with high dynamic range. Such sources could be BL Lac objects with very weak large-scale structure (23, 37, 254). There cannot, however, be large numbers of sources with $\gamma_{core} \gg \gamma_j$, or else we would see many "coreless jets."

The kiloparsec-scale jets in more powerful sources may have higher v_j. They may also have higher Mach numbers, leading to narrower cone angles where they are free and to prominent hot spots where they end. They may be more stable, less turbulent, and thus dimmer relative to their lobes, leading to FR II morphology. Higher v_j may lead, however, to deeper boundary layers with the intergalactic medium, in which $B_{\parallel}$ is maintained by shearing (129, 208). The combination of such shearing and good collimation could make the jets that do stay lit up appear $B_{\parallel}$-dominated (Section 3.2) at low transverse resolution. If $v_j \rightarrow c$ in the more powerful sources, Doppler boosting may contribute to correlations between jet detectability, f_C, and jet/hot-spot symmetries (Section 6.3). The jets and some core emission in powerful sources near the plane of the sky would be beamed away from us, producing "jetless" FR II sources with weak cores, as in the distant $3CR^2$ radio galaxies (Section 2.2). Similar sources turned toward us would have strong cores and one-sided jets, as in the extended $3CR^2$ QSRs. The ~ 40 to 50% detection rate of jets in $3CR^2$ QSRs requires, however, that only mild boosting ($\gamma_j \leq 2$) is usually involved, and the lack of "coreless jets" again implies that $\gamma_{core} \approx \gamma_j$ in general. Intrinsic asymmetries may therefore still be significant in the powerful sources. There are weak relationships between f_C and projected linear size, misalignments, and lobe separations among extended QSRs (130); these relationships are consistent with some core boosting in these sources.

Such "unified models" of extragalactic radio sources may ultimately be judged by whether or not the optical and X-ray differences between different source types can be correlated with intrinsic source power and with indicators of the viewing angle.

7. SUMMARY AND SOME KEY EXPERIMENTS

Jets occur often, in a wide range of extragalactic sources, and with properties well correlated with those of the compact radio cores; thus, it is reasonable to relate them to the fundamental process of energy transport from the cores to the lobes. Their presence supports continuous flow source models and shows that collimation, particle acceleration, and magnetic field amplification probably all occur on both parsec and kiloparsec scales in extragalactic sources. Beyond this, knowledge of jet physics is fragmentary, mainly because we lack credible estimates of jet densities and have only loose, model-dependent constraints on their velocities. Some important questions may be answerable, however, by observations with present or planned instruments:

1. How well does sidedness on parsec and kiloparsec scales correlate with core superluminal motion? Does superluminal motion occur in the cores of sources that should be oriented toward the plane of the sky (e.g. very large lobe-dominated sources)? (Both require sensitive, high dynamic range phase-closure VLBI mapping of cores that are not selected for high flux density alone.)

2. How asymmetric are one-sided kiloparsec-scale jets? Mild brightness

asymmetries $\mathscr{S}$, compatible with weak Doppler boosting (Section 6.1.7) or small differences in the ratio of radiative losses to bulk energy flux, are much easier to explain in large samples than $\mathscr{S} \geq 100:1$ (Sections 6.1.7 and 6.5). (This requires high dynamic range maps of large jets whose cores are not too dominant.)

3. Are jets brighter relative to the lobes when the core is also brighter? Does the answer vary with FR class or optical identification? (This requires unbiased statistics of core, jet, and lobe powers for identified sources.)

4. Can studies of the lobes distinguish the Doppler boosting, asymmetric dissipation, or "flip-flop" models of jet sidedness (Sections 6.2 and 6.3)? (This requires studies of the shapes, spectra, and degrees of polarization of hot spots in jetted and unjetted lobes.)

5. Are jets confined thermally or magnetically (Section 4.1)? Thermal confinement can be tested by high-resolution X-ray imaging and temperature determinations of the environs of recollimating jets, and magnetic confinement may be checked by radio polarimetry of jet cocoons (Section 4.1.3).

6. Can sharp brightness gradients in knots in kiloparsec-scale jets be used to constrain models for jet one sidedness (80)? (This requires proper-motion studies of knots in nearby kiloparsec-scale jets.)

7. Do any jets unambiguously show depolarization that cannot be attributed to foreground Faraday screens, and that might therefore be used to indicate jet densities (Section 6.1)?

Finally, radio, optical, or X-ray spectroscopic evidence for outflow in jets will be welcome now that jets are being interpreted as tracers of the paths of energy transfer in all extragalactic sources.

ACKNOWLEDGMENTS

We are indebted to many colleagues who sent us unpublished data on jets, and who are credited individually in the footnotes to Table 1. We also thank Robert Laing, Bob Sanders, and Dick Henriksen for many invigorating discussions, and Peter Scheuer for valuable criticism of an early draft of this review.

Literature Cited

1. Arp, H. C. 1967. *Astrophys. Lett.* 1:1
2. Arp, H. C. 1981. *Proc. ESO/ESA Workshop, 2nd, Optical Jets in Galaxies,* ed. B. Battrick, J. Mort, p. 53. *ESA SP-162*
3. Baade, W., Minkowski, R. 1954. *Ap. J.* 119:215
4. Baan, W. A. 1980. *Ap. J.* 239:433
5. Baldwin, J. A., Carswell, R. F., Wampler, E. J., Smith, H. E., Burbidge, E. M., Boksenberg, A. 1980. *Ap. J.* 236:388
6. Balick, B., Heckman, T. M., Crane, P. C. 1982. *Ap. J.* 254:483

7. Bartel, N. 1984. *Proc. IAU Symp. 110, VLBI and Compact Radio Sources*, ed. G. Setti, K. I. Kellermann. Dordrecht: Reidel. In press
8. Begelman, M. C. 1982. See Ref. 117, p. 223
9. Begelman, M. C., Blandford, R. D., Rees, M. J. 1980. *Nature* 287:307
10. Begelman, M. C., Rees, M. J., Blandford, R. D. 1979. *Nature* 279:770
11. Benford, G. 1979. *MNRAS* 183:29
12. Benford, G. 1981. *Ap. J.* 247:792
13. Benford, G. 1982. See Ref. 117, p. 231
14. Benford, G. 1983. See Ref. 92a, p. 271
15. Benford, G., Ferrari, A., Trussoni, E. 1980. *Ap. J.* 241:98
16. Bicknell, G. V., Henriksen, R. N. 1980. *Astrophys. Lett.* 21:29
17. Bicknell, G. V., Melrose, D. 1983. *Ap. J.* 262:511
18. Biretta, J. A., Owen, F. N., Hardee, P. E. 1983. *Ap. J. Lett.* 274:L27
19. Birkinshaw, M., Laing, R. A., Peacock, J. A. 1981. *MNRAS* 197:253
20. Blanco, V. M., Graham, J. A., Lasker, B. M., Osmer, P. S. 1975. *Ap. J. Lett.* 198:L63
21. Blandford, R. D. 1983. *Astron. J.* 88:245
22. Blandford, R. D., Icke, V. 1978. *MNRAS* 185:527
23. Blandford, R. D., Königl, A. 1979. *Ap. J.* 232:34
24. Blandford, R. D., Königl, A. 1982. *Astrophys. Lett.* 20:15
25. Blandford, R. D., Rees, M. J. 1974. *MNRAS* 169:395
26. Booler, R. V., Pedlar, A., Davies, R. D. 1982. *MNRAS* 199:229
27. Bridle, A. H. 1982. See Ref. 117, p. 121
28. Bridle, A. H., Chan, K. L., Henriksen, R. N. 1981. *J. R. Astron. Soc. Can.* 75:69
29. Bridle, A. H., Davis, M. M., Fomalont, E. B., Willis, A. G., Strom, R. G. 1979. *Ap. J. Lett.* 228:L9
30. Bridle, A. H., Davis, M. M., Meloy, D. A., Fomalont, E. B., Strom, R. G., Willis, A. G. 1976. *Nature* 262:179
31. Bridle, A. H., Fomalont, E. B., Palimaka, J. J., Willis, A. G. 1981. *Ap. J.* 248:499
32. Bridle, A. H., Henriksen, R. N., Chan, K. L., Fomalont, E. B., Willis, A. G., Perley, R. A. 1980. *Ap. J. Lett.* 241:L145
33. Bridle, A. H., Perley, R. A. 1983. See Ref. 92a, p. 57
34. Brodie, J., Königl, A., Bowyer, S. 1983. See Ref. 92a, p. 145
35. Brown, R. L. 1982. *Ap. J.* 262:110
36. Brown, R. L., Johnston, K. J., Lo, K. Y. 1981. *Ap. J.* 250:155
37. Browne, I. W. A. 1983. *MNRAS* 204:23p
38. Browne, I. W. A., Clark, R. R., Moore, P. K., Muxlow, T. W. B., Wilkinson, P. N., et al. 1982. *Nature* 299:788
39. Browne, I. W. A., Orr, M. J. L. 1981. *Proc. ESO/ESA Workshop, 2nd, Optical Jets in Galaxies*, ed. B. Battrick, J. Mort, p. 87. *ESA SP-162*
40. Burbidge, E. M., Smith, H. E., Burbidge, G. R. 1975. *Ap. J. Lett.* 199:L137
41. Burbidge, E. M., Smith, H. E., Burbidge, G. R. 1978. *Ap. J.* 219:400
42. Burch, S. F. 1979. *MNRAS* 187:187
43. Burns, J. O. 1981. *MNRAS* 195:523
44. Burns, J. O. 1983. See Ref. 92a, p. 67
45. Burns, J. O., Balonek, T. 1982. *Ap. J.* 263:546
46. Burns, J. O., Christiansen, W. A. 1980. *Nature* 287:208
47. Burns, J. O., Christiansen, W. A., Hough, D. H. 1982. *Ap. J.* 257:538
48. Burns, J. O., Feigelson, E. D., Schreier, E. J. 1983. *Ap. J.* 273:128
49. Burns, J. O., Gregory, S. A. 1982. *Astron. J.* 87:1245
50. Burns, J. O., Owen, F. N. 1980. *Astron. J.* 85:204
51. Burns, J. O., Schwendeman, E., White, R. A. 1983. *Ap. J.* 271:575
52. Butcher, H. R., van Breugel, W. J. M., Miley, G. K. 1980. *Ap. J.* 235:749
53. Chan, K. L., Henriksen, R. N. 1980. *Ap. J.* 241:534
54. Charlesworth, M., Spencer, R. E. 1982. *MNRAS* 200:933
55. Cohen, M. 1982. *Publ. Astron. Soc. Pac.* 94:266
56. Cohen, M. H., Readhead, A. C. S. 1979. *Ap. J. Lett.* 233:L101
57. Cohen, M. H., Unwin, S. C. 1982. See Ref. 117, p. 345
58. Cohen, M. H., Unwin, S. C., Lind, K. R., Moffet, A. T., Simon, R. S., et al. 1983. *Ap. J.* 272:383
59. Cohen, M. H., Unwin, S. C., Pearson, T. J., Seielstad, G. A., Simon, R. S., et al. 1983. *Ap. J. Lett.* 269:L1
60. Cohen, M. H., Unwin, S. C., Simon, R. S., Seielstad, G. A., Pearson, T. J., et al. 1981. *Ap. J.* 247:774
61. Cohn, H. 1983. *Ap. J.* 269:500
62. Condon, J. J., Condon, M. A., Gisler, G., Puschell, J. J. 1982. *Ap. J.* 252:102
63. Condon, J. J., Mitchell, K. J. 1984. *Ap. J.* 276:472
64. Conway, R. G. 1982. See Ref. 117, p. 167
65. Cornwell, T. J., Perley, R. A. 1982. See Ref. 117, p. 139
66. Cornwell, T. J., Wilkinson, P. N. 1981. *MNRAS* 196:1067
67. Cotton, W. D., Shapiro, I. I., Wittels, J. J. 1981. *Ap. J. Lett.* 244:L57

68. Davies, J. G., Anderson, B., Morison, I. 1980. *Nature* 288:64
69. Davis, R. J., Stannard, D., Conway, R. G. 1978. *MNRAS* 185:435
70. De Pater, I., Perley, R. A. 1983. *Ap. J.* 273:64
71. de Vaucouleurs, G., Nieto, J.-L. 1979. *Ap. J.* 231:364
72. De Young, D. S. 1980. *Ap. J.* 241:81
73. De Young, D. S. 1981. *Nature* 293:43
74. De Young, D. S. 1982. See Ref. 117, p. 69
75. Dreher, J. W. 1982. See Ref. 117, p. 135
76. Drury, L. O. 1983. *Rep. Prog. Phys.* 46:973
77. Dufour, R. J., van den Bergh, S. 1978. *Ap. J. Lett.* 226:L73
78. Eichler, D. 1982. *Ap. J.* 263:571
79. Eichler, D. 1983. *Ap. J.* 272:48
80. Eichler, D., Smith, M. 1983. *Nature* 303:779
81. Eilek, J. A., Burns, J. O., O'Dea, C. P., Owen, F. N. 1984. *Ap. J.* 278:37
82. Eilek, J. A., Henriksen, R. N. 1982. See Ref. 117, p. 233
83. Ekers, R. D., Fanti, R., Lari, C., Parma, P. 1981. *Astron. Astrophys.* 101:194
84. Ekers, R. D., van Gorkom, J. H., Schwarz, U. J., Goss, W. M. 1983. *Astron. Astrophys.* 122:143
85. Fabricant, D., Lecar, M., Gorenstein, P. 1980. *Ap. J.* 241:552
86. Fanaroff, B. L., Riley, J. M. 1974. *MNRAS* 167:31p
87. Fanti, R. 1983. See Ref. 92a, p. 253
88. Fanti, R., Lari, C., Parma, P., Bridle, A. H., Ekers, R. D., Fomalont, E. B. 1982. *Astron. Astrophys.* 110:169
89. Fanti, R., Padrielli, L., Salvati, M. 1982. See Ref. 117, p. 317
90. Feigelson, E. D. 1983. See Ref. 92a, p. 165
91. Feigelson, E. D., Schreier, E. J., Delvaille, J. P., Giacconi, R., Grindlay, J. E., Lightman, A. P. 1981. *Ap. J.* 251:31
92. Ferrari, A., Massaglia, S., Trussoni, E. 1982. *MNRAS* 198:1065
92a. Ferrari, A., Pacholczyk, A. G., eds. 1983. *Proc. Int. Workshop Astrophys. Jets.* Dordrecht: Reidel. 327 pp.
93. Ferrari, A., Trussoni, E., Zaninetti, L. 1978. *Astron. Astrophys.* 64:43
94. Ferrari, A., Trussoni, E., Zaninetti, L. 1979. *Astron. Astrophys.* 79:190
95. Ferrari, A., Trussoni, E., Zaninetti, L. 1981. *MNRAS* 196:1051
96. Fomalont, E. B. 1981. *Proc. IAU Symp. 94, Origin of Cosmic Rays*, ed. G. Setti, G. Spada, A. W. Wolfendale, p. 111. Dordrecht: Reidel
97. Fomalont, E. B. See Ref. 92a, p. 37
98. Fomalont, E. B., Bridle, A. H., Miley, G. K. 1982. See Ref. 117, p. 173

99. Fomalont, E. B., Bridle, A. H., Willis, A. G., Perley, R. A. 1980. *Ap. J.* 237:418
100. Fomalont, E. B. Palimaka, J. J., Bridle, A. H. 1980. *Astron. J.* 85:981
101. Fort, D. N., Yee, H. K. C. 1976. *Astron. Astrophys.* 50:19
102. Ghigo, F. D., Rudnick, L., Johnston, K. J., Wehinger, P. A., Wyckoff, S. 1982. See Ref. 117, p. 43
103. Goss, W. M., Wellington, K. J., Christiansen, W. N., Lockhart, I. A., Watkinson, A., et al. 1977. *MNRAS* 178:525
104. Gower, A. C., Gregory, P. C., Hutchings, J. B., Unruh, W. G. 1982. *Ap. J.* 262:478
105. Gower, A. C., Hutchings, J. B. 1982. *Ap. J. Lett.* 258:L63
106. Graham, J. A. 1983. *Ap. J.* 269:440
107. Graham, J. A., Price, R. M. 1981. *Ap. J.* 247:813
108. Greenfield, P. E., Burke, B. F., Roberts, D. H. 1980. *Nature* 286:865
109. Greenfield, P. E., Roberts, D. H., Burke, B. F. 1980. *Science* 208:495
110. Hardee, P. E. 1979. *Ap. J.* 234:47
111. Hardee, P. E. 1982. *Ap. J.* 257:509
112. Hardee, P. E. 1983. *Ap. J.* 269:94
113. Hargrave, P. J., Ryle, M. 1974. *MNRAS* 166:305
114. Hazard, C., Mackey, M. B., Shimmins, A. J. 1963. *Nature* 197:1037
115. Heckman, T. M. 1983. *Ap. J. Lett.* 271:L5
116. Heckman, T. M., Miley, G. K., Balick, B., van Breugel, W. J. M., Butcher, H. R. 1982. *Ap. J.* 262:529
117. Heeschen, D. S., Wade, C. M., eds. 1982. *Proc. IAU Symp. 97, Extragalactic Radio Sources.* Dordrecht: Reidel. 490 pp.
118. Henriksen, R. N., Bridle, A. H., Chan, K. L. 1982. *Ap. J.* 257:63
119. Henriksen, R. N., Vallée, J. P., Bridle, A. H. 1981. *Ap. J.* 249:40
120. Hjellming, R. M., Johnston, K. J. 1982. See Ref. 117, p. 197
121. Hummel, E., Kotanyi, C. G., van Gorkom, J. H. 1983. *Ap. J. Lett.* 267:L5
122. Hunstead, R. W., Murdoch, H. S., Condon, J. J., Phillips, M. M. 1984. *MNRAS* 207:55
123. Icke, V. 1981. *Ap. J. Lett.* 246:L65
124. Icke, V. 1983. *Ap. J.* 265:648
125. Jenkins, C. R. 1982. *MNRAS* 200:705
126. Jones, D. L., Sramek, R. A., Terzian, Y. 1981. *Ap. J.* 246:28
127. Jones, D. L., Sramek, R. A., Terzian, Y. 1981. *Ap. J. Lett.* 247:L57
128. Jones, T., Owen, F. N. 1979. *Ap. J.* 234:818
129. Kahn, F. D. 1983. *MNRAS* 202:553
130. Kapahi, V. K., Saikia, D. J. 1982. *J. Astrophys. Astron.* 3:465

356   BRIDLE & PERLEY

131. Kellermann, K. I. 1983. In *Highlights of Astronomy*, ed. R. M. West, 6:481. Dordrecht: Reidel
132. Kellermann, K. I., Downes, A. J. B., Pauliny-Toth, I. I. K., Preuss, E., Shaffer, D. B., Witzel, A. 1981. *Astron. Astrophys.* 97:L1
133. Königl, A. 1982. *Ap. J.* 261:115
134. Kotanyi, C. G., Ekers, R. D. 1979. *Astron. Astrophys.* 73:L1
135. Laing, R. A. 1980. *MNRAS* 193:427
136. Laing, R. A. 1980. *MNRAS* 193:439
137. Laing, R. A. 1981. *Ap. J.* 248:87
138. Laing, R. A. 1982. See Ref. 117, p. 161
139. Laing, R. A. 1983. In *Highlights of Astronomy*, ed. R. M. West, 6:731. Dordrecht: Reidel
140. Laing, R. A., Riley, J. M., Longair, M. S. 1983. *MNRAS* 204:151
141. Lelièvre, G., Nieto, J.-L., Horville, D., Renard, L., Servan, B. 1984. *Astron. Astrophys.* In press
142. Linfield, R. 1981. *Ap. J.* 244:436
143. Linfield, R. 1981. *Ap. J.* 250:464
144. Linfield, R. 1982. *Ap. J.* 254:465
145. Linfield, R., Perley, R. A. 1984. *Ap. J.* In press
146. Longair, M. S., Riley, J. M. 1979. *MNRAS* 186:625
147. Longair, M. S., Ryle, M., Scheuer, P. A. G. 1973. *MNRAS* 164:243
148. Lonsdale, C. J., Hartley-Davies, R., Morison, I. 1983. *MNRAS* 202:1p
149. Lonsdale, C. J., Morison, I. 1983. *MNRAS* 203:833
150. Lorre, J. J., Nieto, J.-L. 1984. *Astron. Astrophys.* 130:167
151. Lupton, R. H., Gott, J. R. 1982. *Ap. J.* 255:408
152. Maccagni, D., Tarenghi, M. 1981. *Ap. J.* 243:42
153. Machalski, J., Condon, J. J. 1983. *Astron. J.* 88:143
154. Margon, B. 1981. *Ann. NY Acad. Sci.* 375:403
155. Marscher, A. P., Broderick, J. J. 1981. *Ap. J. Lett.* 247:L49
156. Masson, C. R. 1979. *MNRAS* 187:253
157. Miley, G. K. 1980. *Ann. Rev. Astron. Astrophys.* 18:165
158. Miley, G. K. 1983. See Ref. 92a, p. 99
159. Miley, G. K., Heckman, T. M., Butcher, H. R., van Breugel, W. J. M. 1981. *Ap. J. Lett.* 247:L5
160. Miley, G. K., Hogg, D. E., Basart, J. 1970. *Ap. J. Lett.* 159:L19
161. Miley, G. K., Wellington, K. J., van der Laan, H. 1975. *Astron. Astrophys.* 38:381
162. Morrison, P. 1969. *Ap. J.* 157:L73
163. Neff, S. G. 1982. See Ref. 117, p. 137
164. Nieto, J.-L. 1983. See Ref. 92a, p. 113
165. Nieto, J.-L., Lelièvre, G. 1982. *Astron. Astrophys.* 109:95
166. Norman, M. L., Smarr, L., Winkler, K.-H. A., Smith, M. D. 1982. *Astron. Astrophys.* 113:285
167. Norman, M. L., Winkler, K.-H. A., Smarr, L. 1983. See Ref. 92a, p. 227
168. Northover, K. J. E. 1973. *MNRAS* 165:369
169. Orr, M. J. L., Browne, I. W. A. 1982. *MNRAS* 200:1067
170. Osmer, P. S. 1978. *Ap. J. Lett.* 226:L79
171. Owen, F. N., Burns, J. O., Rudnick, L. 1978. *Ap. J. Lett.* 226:L119
172. Owen, F. N., Hardee, P. E., Bignell, R. C. 1980. *Ap. J. Lett.* 239:L11
173. Owen, F. N., Puschell, J. J. 1984. *Astron. J.* In press
174. Parma, P. 1982. See Ref. 117, p. 193
175. Pauliny-Toth, I. I. K., Preuss, E., Witzel, A., Graham, D., Kellermann, K. I., Rönnäng, B. 1981. *Astron. J.* 86:371
176. Pearson, T. J., Readhead, A. C. S. 1981. *Ap. J.* 248:61
177. Pearson, T. J., Readhead, A. C. S., Wilkinson, P. N. 1980. *Ap. J.* 236:714
178. Pearson, T. J., Unwin, S., Cohen, M. H., Linfield, R., Readhead, A. C. S., et al. 1981. *Nature* 290:365
179. Pedlar, A., Booler, R. V., Davies, R. D. 1983. *MNRAS* 203:667
180. Pedlar, A., Booler, R. V., Spencer, R. E., Stewart, O. J. 1983. *MNRAS* 202:647
181. Perley, R. A. 1981. *Proc. ESO/ESA Workshop, 2nd, Optical Jets in Galaxies*, ed. B. Battrick, J. Mort, p. 77. *ESA SP-162*
182. Perley, R. A. 1982. See Ref. 117, p. 175
183. Perley, R. A., Bridle, A. H., Willis, A. G. 1984. *Ap. J. Suppl.* 54:291
184. Perley, R. A., Bridle, A. H., Willis, A. G., Fomalont, E. B. 1980. *Astron. J.* 85:499
185. Perley, R. A., Fomalont, E. B., Johnston, K. J. 1982. *Ap. J. Lett.* 255:L93
186. Perley, R. A., Willis, A. G., Scott, J. S. 1979. *Nature* 281:437
187. Peterson, B. A., Dickens, R. J., Cannon, R. D. 1975. *Proc. Astron. Soc. Aust.* 2:366
188. Phillips, M. M. 1981. *MNRAS* 197:659
189. Porcas, R. W., Booth, R. S., Browne, I. W. A., Walsh, D., Wilkinson, P. N. 1981. *Nature* 289:758
190. Potash, R. I., Wardle, J. F. C. 1979. *Astron. J.* 84:707
191. Potash, R. I., Wardle, J. F. C. 1980. *Ap. J.* 239:42
192. Preston, R. A., Wehrle, A. E., Morabito, D. D., Jauncey, D. L., Batty, M. J., et al. 1983. *Ap. J. Lett.* 266:L93
193. Preuss, E., Alef, A., Pauliny-Toth, I. I. K., Kellermann, K. I. 1982. See Ref. 117, p. 289

194. Preuss, E., Alef, W., Whyborn, A., Wilkinson, P. N., Kellermann, K. I. 1984. *Proc. IAU Symp. 110, VLBI and Compact Radio Sources*, ed. G. Setti, K. I. Kellermann. Dordrecht: Reidel. In press

195. Ray, T. P. 1981. *MNRAS* 196:195

196. Ray, T. P. 1982. *MNRAS* 198:617

197. Readhead, A. C. S. 1980. *Proc. IAU Symp. 92, Objects of High Redshift*, ed. G. O. Abell, P. J. E. Peebles, p. 165. Dordrecht: Reidel

198. Readhead, A. C. S., Cohen, M. H., Blandford, R. D. 1978. *Nature* 272:131

199. Readhead, A. C. S., Cohen, M. H., Pearson, T. J., Wilkinson, P. N. 1978. *Nature* 276:768

200. Readhead, A. C. S., Hough, D. H., Ewing, M. S., Walker, R. C., Romney, J. 1983. *Ap. J.* 265:107

201. Readhead, A. C. S., Napier, P. J., Bignell, R. C. 1980. *Ap. J. Lett.* 237:L55

202. Readhead, A. C. S., Wilkinson, P. N. 1978. *Ap. J.* 223:25

203. Readhead, A. C. S., Wilkinson, P. N. 1980. *Ap. J.* 235:11

204. Rees, M. J. 1971. *Nature* 229:312

205. Rees, M. J. 1978. *Nature* 275:516

206. Rees, M. J. 1978. *MNRAS* 184:61p

207. Rees, M. J. 1982. See Ref. 117, p. 211

208. Rees, M. J., Begelman, M. C., Blandford, R. D. 1981. *Proc. Texas Symp., 10th, Ann. NY Acad. Sci.* 315:254

209. Reid, M. J., Schmitt, J. H. M. M., Owen, F. N., Booth, R. S., Wilkinson, P. N., et al. 1982. *Ap. J.* 263:615

210. Riley, J. M., Pooley, G. G. 1975. *Mem. R. Astron. Soc.* 80:105

211. Robertson, J. G. 1980. *Nature* 286:579

212. Robson, D. W. 1981. *Nature* 294:57

213. Rudnick, L. 1982. See Ref. 117, p. 47

214. Rudnick, L., Burns, J. O. 1981. *Ap. J. Lett.* 246:L69

215. Saikia, D. J. 1981. *MNRAS* 197:11p

216. Saikia, D. J., Cornwell, T. J. 1983. See Ref. 92a, p. 53

217. Saikia, D. J., Shastri, P., Cornwell, T. J., Banhatti, D. G. 1983. *MNRAS* 203:53p

218. Saikia, D. J., Wiita, P. J. 1982. *MNRAS* 200:83

219. Sanders, R. H. 1983. *Ap. J.* 266:73

220. Saunders, R., Baldwin, J. E., Pooley, G. G., Warner, P. J. 1982. *MNRAS* 197:287

221. Scheuer, P. A. G. 1974. *MNRAS* 166:513

222. Scheuer, P. A. G. 1983. In *Highlights of Astronomy*, ed. R. M. West, 6:735. Dordrecht: Reidel

223. Scheuer, P. A. G., Readhead, A. C. S. 1979. *Nature* 277:182

224. Schilizzi, R. T., de Bruyn, A. G. 1983. *Nature* 303:26

225. Schilizzi, R. T., Miley, G. K., Janssen, F. L. J., Wilkinson, P. N., Cornwell, T. J., Fomalont, E. B. 1981. *Proc. ESO/ESA Workshop, 2nd, Optical Jets in Galaxies*, ed. B. Battrick, J. Mort, p. 97. *ESA SP-162*

226. Schmidt, G. D., Peterson, B. A., Beaver, E. A. 1978. *Ap. J. Lett.* 220:L31

227. Schmidt, M. 1963. *Nature* 197:1040

228. Schreier, E. J., Burns, J. O., Feigelson, E. D. 1981. *Ap. J.* 251:523

229. Schreier, E. J., Feigelson, E. D., Delvaille, J., Giacconi, R., Grindlay, J., et al. 1979. *Ap. J. Lett.* 234:L39

230. Schreier, E. J., Gorenstein, P., Feigelson, E. D. 1982. *Ap. J.* 261:42

231. Schwab, F. R. 1980. *Int. Opt. Comput. Conf., SPIE Vol. 231*, p. 18

232. Schwartz, R. D. 1983. *Ann. Rev. Astron. Astrophys.* 21:209

233. Simon, R. S., Readhead, A. C. S., Moffet, A. T., Wilkinson, P. N., Allen, B., Burke, B. F. 1983. *Nature* 302:487

234. Simon, R. S., Readhead, A. C. S., Moffet, A. T., Wilkinson, P. N., Anderson, B. 1980. *Ap. J.* 236:707

235. Slee, O. B., Sheridan, K. V., Dulk, G. A., Little, A. G. 1983. *Publ. Astron. Soc. Aust.* 5:247

236. Smith, M. D., Norman, C. A. 1981. *MNRAS* 194:771

237. Smith, R. M., Bicknell, G. V., Hyland, A. R., Jones, T. J. 1983. *Ap. J.* 266:69

238. Spangler, S. R. 1979. *Astron. J.* 84:1470

239. Spangler, S. R. 1979. *Ap. J. Lett.* 232:L7

240. Spangler, S. R., Basart, J. P. 1981. *Ap. J.* 243:1103

241. Spencer, J. H., Johnston, K. J., Pauliny-Toth, I. I. K., Witzel, A. 1981. *Ap. J. Lett.* 251:L61

242. Steiner, J. E. 1981. *Ap. J.* 250:469

243. Stocke, J. T., Christiansen, W. A., Burns, J. O. 1982. See Ref. 117, p. 39

244. Stocke, J. T., Rieke, G. H., Lebofsky, M. J. 1981. *Nature* 294:319

245. Strom, R. G., Fanti, R., Parma, P., Ekers, R. D. 1983. *Astron. Astrophys.* 122:305

246. Strom, R. G., Willis, A. G. 1981. *Proc. ESO/ESA Workshop, 2nd, Optical Jets in Galaxies*, ed. B. Battrick, J. Mort, p. 83. *ESA SP-162*

247. Sulentic, J. W., Arp, H. C., Lorre, J. J. 1979. *Ap. J.* 233:44

248. Swarup, G., Sinha, R. P., Saikia, D. J. 1982. *MNRAS* 201:393

249. Tarenghi, M. 1981. *Proc. ESO/ESA Workshop, 2nd, Optical Jets in Galaxies*, ed. B. Battrick, J. Mort, p. 145. *ESA SP-162*

250. Thompson, A. R., Clark, B. G., Wade,

C. M., Napier, P. J. 1980. *Ap. J. Suppl.* 44:151
251. Thorne, K. S., Blandford, R. D. 1982. See Ref. 117, p. 255
252. Trussoni, E., Ferrari, A., Zaninetti, L. 1983. See Ref. 92a, p. 281
253. Turland, B. D. 1975. *MNRAS* 172:181
254. Ulvestad, J. S., Johnston, K. J., Weiler, K. W. 1983. *Ap. J.* 266:18
255. Unwin, S. C., Cohen, M. H., Pearson, T. J., Seielstad, G. A., Simon, R. S., et al. 1983. *Ap. J.* 271:536
256. Vallée, J. P. 1982. *Astron. J.* 87:486
257. Vallée, J. P., Bridle, A. H., Wilson, A. S. 1981. *Ap. J.* 250:66
258. van Breugel, W. J. M. 1980. *Structure in radio galaxies.* PhD thesis. Univ. Leiden. 219 pp.
259. van Breugel, W. J. M., Balick, B., Heckman, T., Miley, G. K., Helfand, D. 1983. *Astron. J.* 88:40
260. van Breugel, W. J. M., Heckman, T. M. 1982. See Ref. 117, p. 61
261. van Breugel, W. J. M., Heckman, T. M., Bridle, A. H., Butcher, H. R., Strom, R. G., Balick, B. 1983. *Ap. J.* 275:61
262. van Breugel, W. J. M., Miley, G. K. 1977. *Nature* 265:315
263. van Breugel, W. J. M., Willis, A. G. 1981. *Astron. Astrophys.* 96:332
264. van der Hulst, J. M., Hummel, E., Dickey, J. M. 1982. *Ap. J. Lett.* 261: L59
265. van Groningen, E., Miley, G. K., Norman, C. 1980. *Astron. Astrophys.* 90:L7
266. Wade, C. M. 1960. *Observatory* 80:235
267. Waggett, P. C., Warner, P. J., Baldwin, J. E. 1977. *MNRAS* 181:465
268. Walker, R. C., Benson, J. M., Seielstad, G. A., Unwin, S. C. 1984. *Proc. IAU Symp. 110, VLBI and Compact Radio Sources,* ed. G. Setti, K. I. Kellermann.

Dordrecht: Reidel. In press
269. Walker, R. C., Seielstad, G. A., Simon, R. S., Unwin, S. C., Cohen, M. H., et al. 1981. *Ap. J.* 257:56
270. Wardle, J. F. C., Kronberg, P. P. 1974. *Ap. J.* 194:249
271. Wardle, J. F. C., Potash, R. I. 1982. See Ref. 117, p. 129
272. Wellington, K. J., Miley, G. K., van der Laan, H. 1973. *Nature* 244:502
273. Wiita, P. J., Siah, M. J. 1981. *Ap. J.* 243:710
274. Wilkinson, P. N. 1982. See Ref. 117, p. 149
275. Wilkinson, P. N., Readhead, A. C. S., Purcell, G. H., Anderson, B. 1977. *Nature* 269:764
276. Willingale, R. 1981. *MNRAS* 194:359
277. Willis, A. G. 1981. *Proc. ESO/ESA Workshop, 2nd, Optical Jets in Galaxies,* ed. B. Battrick, J. Mort, p. 71. *ESA SP-162*
278. Willis, A. G., Strom, R. G., Bridle, A. H., Fomalont, E. B. 1981. *Astron. Astrophys.* 95:250
279. Willis, A. G., Strom, R. G., Perley, R. A., Bridle, A. H. 1982. See Ref. 117, p. 141
280. Willis, A. G., Wilson, A. S., Strom, R. G. 1978. *Astron. Astrophys.* 66:L1
281. Wills, B. J. 1982. See Ref. 117, p. 373
282. Wilson, A. S. 1983. In *Highlights of Astronomy,* ed. R. M. West, 6:467. Dordrecht: Reidel
283. Wilson, A. S., Ulvestad, J. S. 1982. *Ap. J.* 263:576
284. Wirth, A., Smarr, L., Gallagher, J. S. 1982. *Astron. J.* 87:602
285. Wyckoff, S., Johnston, K., Ghigo, F., Rudnick, L., Wehinger, P., Boksenberg, A. 1983. *Ap. J.* 265:43
286. Zaninetti, L., Trussoni, E. 1983. See Ref. 92a, p. 309

Ann. Rev. Astron. Astrophys. 1984. 22:359–87

HIGH-ENERGY NEUTRAL RADIATIONS FROM THE SUN

E. L. Chupp

Department of Physics, University of New Hampshire, Durham, New Hampshire 03824

𝔄 hypothesis or theory is clear, decisive and positive but it is believed by no one but the man who created it. Experimental findings, on the other hand, are messy, inexact things which are believed by everyone except the man who did the work.

Harlow Shapley

1. INTRODUCTION

The solar flare phenomenon has for many years posed challenging and puzzling questions. New insights were gained with the observations of transient increases in the cosmic-ray flux at the Earth, lasting from a few minutes to hours, following major solar flares. In some outstanding events, the proton energies can be as high as 20 GeV (78). These observations suggest that the energetic charged particles (ions and electrons) could be accelerated at the time of the optical flare or result from processes initiated after the flare. It was realized quite early that if in fact the particles are accelerated in the solar atmosphere, they will interact with ambient material and generate uncharged secondaries (5, 79) that will be a direct probe of the charged-particle distributions near the flare site. Neutral solar radiations, such as gamma rays and neutrons, travel unimpeded to the Earth carrying the signature of their production history, while charged solar energetic particles (SEPs)[1] have their signature distorted as they

[1] We use solar energetic particles (SEPs) instead of the more common term solar cosmic rays.

0066–4146/84/0915–0359$02.00

traverse coronal and interplanetary magnetic fields. In addition, it is of particular significance that observations of gamma rays of nuclear origin and high-energy neutrons give direct information on the energetic ion component. Early searches, with the exception of a few dramatic instances (referenced below), did not unambiguously identify solar flare gamma rays or neutrons. Only recently, with the launch of the *Solar Maximum Mission* (*SMM*) and *Hinotori* satellites, has it been possible to routinely detect these neutral radiations, to identify them with confidence, and to study their properties. The new observations require revisions in earlier concepts, particularly those concerning particle acceleration.

This review describes *SMM* and other relevant observations, and discusses some of the exciting implications and questions that these data raise. Reviews of earlier work are in (11, 95).

2. OBSERVATIONAL STATUS (SOLAR GAMMA RAYS AND NEUTRONS)

Even though transient solar high-energy photon emissions $> 270 \, \text{keV}$ were detected as early as 1958 (83), for the next 15 years only a few observations were reported (41, 45, 46). Not until the detection of the first solar flare gamma-ray *lines* in 1972 (14, 15, 76), and reports of several subsequent observations referenced below, did the field of solar gamma-ray astronomy develop. Recently, the data base has been greatly augmented by *SMM* and *Hinotori* (118) satellite experiments, which continue to provide unprecedented information on solar flares. The Gamma Ray Spectrometer (GRS) (36) on *SMM* alone has observed over 150 solar flares with photon energies $E > 270 \, \text{keV}$ in the first four years of operation. This wealth of new data makes it possible to address fundamental questions on the acceleration of energetic particles in solar flares and their interactions with the solar atmosphere.

Figure 1 shows the GRS $> 270 \, \text{keV}$ event frequency compared with the smoothed sunspot number for Solar Cycle 21. Before the beginning of GRS observations (1980 February 17), there was no continuous coverage of solar activity with a gamma-ray spectrometer of sensitivity comparable to GRS; therefore, only a few sporadic observations are plotted in the figure (9, 48, 85, 101, 117). We now review the major characteristics of the gamma-ray and neutron flares observed primarily by the GRS.

2.1 *Identification and Analysis of Gamma-Ray Events*

Because gamma-ray spectrometers used for solar flare observations are nondirectional, transient neutral solar events are identified by time coincidence with an optical ($H\alpha$) or radio flare. The spectrometers currently in use record events resulting from both gamma-ray and neutron interactions. The measured count distribution during an event, in theory, makes it possible to determine which primary particles (electrons or ions) produced the source radiation (i.e. bremsstrahlung, nuclear gamma ray, or neutron). For example, the clear detection of a nuclear gamma-ray line indicates the occurrence of nuclear interactions of ions at the Sun.

To illustrate, we show in Figure 2 the net count spectrum >270 keV for the impulsive phase of a GRS flare event in which gamma-ray lines are clearly present (16). This spectrum results from a combination of primary electron bremsstrahlung and many gamma-ray lines due to ion interactions. For these data, the two components were separated by obtaining a

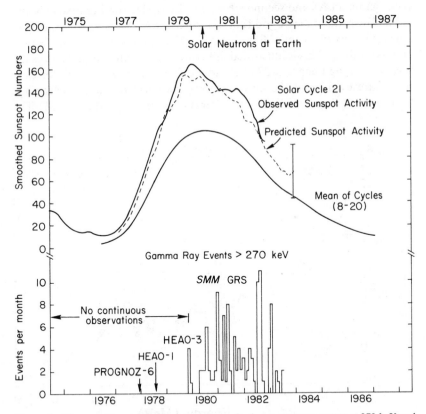

Figure 1   The monthly number of observations of solar gamma-ray events >270 keV and smoothed sunspot numbers are shown for Solar Cycle 21, observed between 1980 February and 1983 September by the GRS with ~50% duty cycle. The two GRS events for which high-energy neutrons were also observed are indicated. Only 5 of 22 *HEAO-3* events (117) are shown. The frequency scale applies only to *SMM* GRS observations.

first estimate of the bremsstrahlung contribution, assuming an input power-law photon spectrum folded through the known instrument response and fitted to the count spectrum below 1-MeV energy loss. The excess above this fit, especially in the MeV region, is a good estimate of the narrow and broad nuclear-line contribution (50, 93), modified by the GRS response. The cutoff above ~ 7.5 MeV, characteristic of a spectrum dominated by nuclear lines, supports this argument. For many GRS events there is an MeV excess consistent with the above description. The assumption of a power law with a single slope to above 1 MeV, as used to obtain Figure 2, should be considered tentative, since realistic calculations of the relativistic electron bremsstrahlung spectrum from a flare region have not been applied. Above 10 MeV, one must also consider contributions from π^0 decay gamma rays and secondary $\pi^{\pm}$-$\mu^{\pm}$ decay electron bremsstrahlung.

The GRS data indicate good correlation of the MeV excess with the time-integrated flux (fluence) of flare X-rays > 270 keV resulting from electron bremsstrahlung. This is illustrated in Figure 3, where the ordinate gives the bremsstrahlung fluence > 270 keV, found from the power-law fit for each solar flare event, and the abscissa gives the corresponding nuclear-line fluence above the power law for the energy-loss range 4–8 MeV (34, 37). The

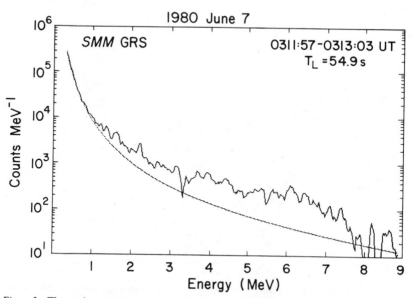

Figure 2   The total gamma-ray count spectrum above background is shown for the impulsive flare on 1980 June 7. The smooth curve is the counts spectrum produced by folding the best-fit power law through the instrument response. Excess counts above this curve are due to narrow and broad nuclear gamma-ray lines and termed MeV excess.

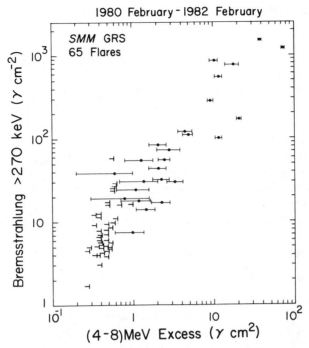

Figure 3   The fluence (total time-integrated flux) of gamma-rays > 270 keV is compared with the 4–8 MeV excess fluence (see Figure 2) for GRS events observed between 1980 February and 1982 February [from Forrest (34)].

statistical limit for the 4–8 MeV fluence is ~ 1 photon cm^{-2}, and Figure 3 indicates that at this level the bremsstrahlung fluence is still detectable. These data suggest that nuclear-line contributions could be present in all events with fluence > 270 keV, and that they would be detected if the GRS were more sensitive. Further confirmation of a nuclear origin for the abscissae in Figure 3 is given by the good correlation of the 2.223-MeV line fluence [from the reaction $^1H(n, \gamma)^2H$ in the solar photosphere] with the 4–8 MeV excess fluence (103).

2.2   *Comparison of GRS Gamma Rays with Other Flare Emissions*

The first 2.5 years of GRS data were correlated with a few other flare observations, taking into account observation duty cycle, with the following results: (*a*) A gamma-ray event > 270 keV may be associated with any Hα class; (*b*) 20% (10/50) of flares of class ≥ 2B have associated 4–8 MeV excess (19); (*c*) 75% of all *SMM* GRS events > 270 keV are from

Hα of B (brilliant) class; (d) 50% (13/26) of *GOES* 1–8 Å X-ray events with peak intensity $\geq X2$ have significant 4–8 MeV excess (19); (e) GRS events > 270 keV are always associated with a solar microwave burst (≥ 1 GHz); and (f) 53% (19/36) of 9-GHz bursts with peak flux density ≥ 1200 FU have significant 4–8 MeV excess (19).

There is no correlation between the peak SEP flux measured near the Earth and the GRS MeV excess fluence (20). This lack of correlation is quite unexpected, since peak flux is a classic indicator of proton fluence injected into space (109, 110); thus, it may indicate that the SEPs come from a different site, or that they are produced by a different mechanism than the ions required to produce the gamma-ray burst, or that storage and propagation effects mask the correlation. Finally, we note that SEP flares with an enhanced flux ratio of electrons (6–11 MeV) to protons (24–44 MeV) generally have associated 4–8 MeV excess (30, 31). The significance of the lack of correlation between SEP and gamma-ray events has recently been discussed in some detail (74).

2.3 Temporal Characteristics of Gamma-Ray Events

A comparison of flare emission time histories at different energies could reveal salient properties of the particle acceleration mechanism(s). For example, differing time histories of X-ray (29–41 keV) and gamma-ray (> 300 keV) emissions in the 1972 August 4 flare (14, 15) were interpreted as evidence for a two-stage acceleration process in flares (4).

The GRS gamma-ray observations show that the range of event durations runs from ~ 10 s to over 20 min. Most events include several emission pulses, which can be as short as 10 s in an event of ~ 1-min duration or as long as 2 min in an event of ~ 20-min duration (34; see Figure 4). The shortest pulses are similar to the reported "elementary flare bursts" (EFBs) in hard X-rays < 100 keV (27), with widths varying between 4 ± 1 and 24 ± 5 s for different flares. At lower photon energies (< 270 keV), structure is observed down to ~ 0.1 s within longer (~ 2 s) bursts (49, 57). In a preliminary study of several GRS events (40), it was found that the time of the maximum count rate in an individual burst in the 4.1–6.4 MeV energy band occurred between 2 and 45 seconds *later* than the corresponding maximum for the hard X-rays > 270 keV, with the retardation observed to be proportional to the pulse rise time.

Simultaneous peaking of low-energy (< 100 keV) and high-energy (> 10 MeV) photons is also observed, limited only by the GRS time resolution. Figure 5 shows the count rates for a flare with several pulses of photons extending from 40 keV to 25 MeV, all occurring within a time interval of 2 min. At ~ 1249 UT, photons having energies < 300 keV were strongly attenuated by the atmosphere, since *SMM* was just

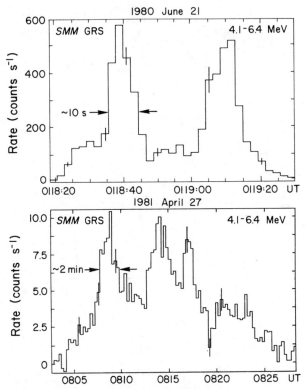

Figure 4   A comparison of the extremes of duration, the rise time, and the burst width for two impulsive flares in gamma-ray emission at 4.1–6.4 MeV [from Forrest (34)].

entering sunlight. The peak intensities for the third impulse at ~1250 UT occur simultaneously at all photon energies to within ±1 s. This means that either the particles producing the energetic photon emission up to >25 MeV were injected into and interacted with the target medium *at the same time* (≤1 s), or else they were produced (accelerated) in the target medium *at the same time*.

The earliest time at which photon emission is present at different energies in a flare has been determined in two intense bursts (35) by comparing the starting times in several energy bands from ~40 keV to 6.4 MeV. Event initiation in each band was simultaneous to within ±0.8 s in one flare and ±2.2 s for the other. Again, this close timing over such a wide energy range requires that all energetic particles be interacting in the target at *the same time*! In the same events, the time of the peak intensity at the higher photon energies (>4 MeV) was characteristically later (≥2 s) (35, 81). A similar

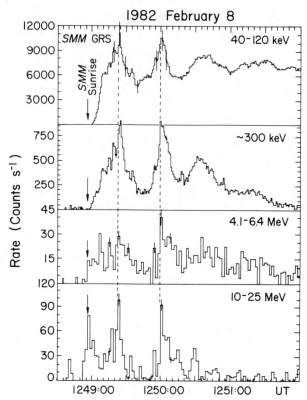

Figure 5 The time history of photon emissions from 40 keV to ∼25 MeV is shown for the flare on 1982 February 8, which occurred before 1249 UT. Early in the flare, low-energy photons ≲300 keV are occulted by the Earth's atmosphere. The times of the peak intensity for the pulse at ∼1250 UT are the same to within 2 s for all photon energies.

temporal pattern was observed in over 40 flares with emission >300 keV by the *Hinotori* satellite (119).

2.4   *Prompt Gamma-Ray Spectra*

Prompt gamma-ray line spectra give information on the temporal behavior, composition, and energy spectrum of accelerated ions, as well as on the composition of the ambient solar atmosphere. In about 30 GRS events, lines are strong enough to be individually detected. A few intense flares show the rich gamma-ray line spectrum expected for accelerated ions incident on elements extending from hydrogen to above iron. In Figure 6, the count spectrum above background is shown (solid curve) for a limb flare, after subtracting an estimated (see Figure 2) electron bremsstrahlung contribution (39). The figure also shows the count spectrum produced by an incident photon spectrum folded through the instrument response (dotted line) and normalized arbitrarily, such that measured and predicted counts at the 4.439 MeV peak are equal. The reader is referred to Section 4.4 for a discussion of this discrepancy. Several of the identified lines in this spectrum have been tested for evidence of Doppler broadening and shifts. There is no evidence for centroid shifts ≥ 10 keV; thus there is no evidence for detectable particle beaming in the gamma-ray line spectrum obtained for this limb flare. The data are, however, consistent with kinematic Doppler broadening of ~ 60 keV for the ^{16}O line at 6.129 MeV, a value in agreement with theoretical calculations (89, 94). On the other hand, the 1.634-MeV line from ^{20}Ne appears wider than expected. Several prominent lines and their parent nuclei are identified in the figure. The sharp features in this spectrum are superimposed on a greatly broadened nuclear-line component.

2.5   Delayed Gamma-Ray Lines (511 keV and 2.223 MeV)

The gamma-ray lines at 511 keV and 2.223 MeV are due to e^{+}-e^{-} annihilation and the reaction ^{1}H$(n, \gamma)^{2}$H, respectively. Emission of these *delayed* lines extends over a time period longer than the production time of the progenitor radioactive positron emitters or π^{+} mesons and neutrons.

The strongest individual line from disk flares is at 2.223 MeV; however, because of its photospheric origin and the flare location at N16W90, it is strongly attenuated and does not stand out in Figure 6. These characteristics of the 2.223-MeV line have been established in earlier observations (14, 15, 48, 85), as well as from recent GRS (12, 16, 86, 103) and *Hinotori* observations (120), and are in agreement with predictions (114). The fluence measured for this line ranges from ~ 1 to > 300 photons cm^{-2} for disk flares (85, 103). The decay time for the intensity of the 2.223-MeV line is $\tau_{loss} \simeq 100$ s (16, 86). Its energy has been measured from solar flares at 2.2248 ± 0.0010 MeV (86) and 2.224 ± 0.002 MeV (13). This clearly verifies that the line results from neutron capture in hydrogen, as discussed in Section 3.3. A measurement of the line profile with high spectral resolution gives an upper limit to the natural line width of 5 keV FWHM.

The e^{+}-e^{-} annihilation line at 511 keV is difficult to analyze because of a strong, highly variable background line at the identical energy. However, in two intense flares observed by GRS, the 511-keV line was studied in detail (104). Two important conclusions of this study are that (*a*) the upper limit line width is < 20 keV, indicating that the annihilation region temperature is $< 3 \times 10^{6}$ K (21, 104); and (*b*) the time decay of intensity for this line (104) is consistent with what is expected if the positrons are predominately from radioactive decay (97). More information will soon be forthcoming with the further analysis of GRS data.

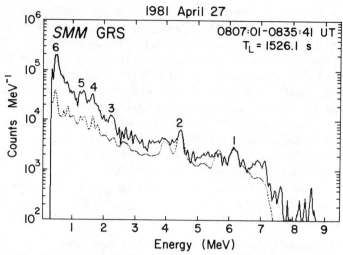

Figure 6 The time-integrated excess gamma-ray counts spectrum above background (solid curve), after removal of a bremsstrahlung continuum, is shown for the 1981 April 27 flare (39). Significant gamma-ray line features and their origin are (1) 6.129 MeV (^{16}O); (2) 4.439 MeV (^{12}C); (3) 2.313 MeV (^{14}N) and 2.223 MeV ($n\gamma$); (4) 1.634 MeV (^{20}Ne); (5) 1.369 MeV (^{24}Mg) and 1.238 MeV (^{56}Fe); and (6) 511 keV (β^+), 478 keV (^{7}Li), and 431 keV (^{7}Be). The 2.223-MeV line, normally the strongest for disk flares, is strongly suppressed in this limb flare. The dotted curve is based on a predicted incident photon spectrum folded through the instrument response. The incident spectrum was generated assuming an accelerated ion spectrum of the form $\alpha T = 0.02$, incident on an ambient thick target with "standard" photospheric abundances (97) (see Section 4.3).

2.6   The Species and Energies of Accelerated Solar Particles

The particle species produced in solar flares and the energies they attain are determined indirectly from measured, impulsive gamma-ray and high-energy neutron spectra, or by direct observation of SEPs. It should be noted that it is possible that the ions and electrons producing the GRS events that make up the SEPs are only causally related and are produced by different mechanisms (see Sections 2.2 and 5).

The production of prompt and delayed gamma-ray lines requires ions of energy ≥ 10 MeV nucleon^{-1} (see Section 3.2). The highest energies attained impulsively by ions have been determined by direct observation of neutrons at the Earth (17, 18, 25). The GRS has recorded a characteristic signature (70) for neutrons produced at the Sun in a time interval short compared with the neutrons' transit time to the Earth. This signature is shown in Figure 7. In the inset of the figure, the time behavior of photon interactions

> 10 MeV is shown; this chart reflects the production time history of gamma rays and neutrons at the Sun. The delayed signal, which peaks at ~ 500 s, is most likely the result of neutrons (17) that have energies between 50 MeV and ~ 500 MeV. The highest energy neutrons could be produced either by protons of significantly higher energy (> 700 MeV nucleon^{-1}) or by α-particles with energies of > 300 MeV nucleon^{-1}. A second event with neutrons of energy ~ 1 GeV has also been recorded by GRS (18) and by

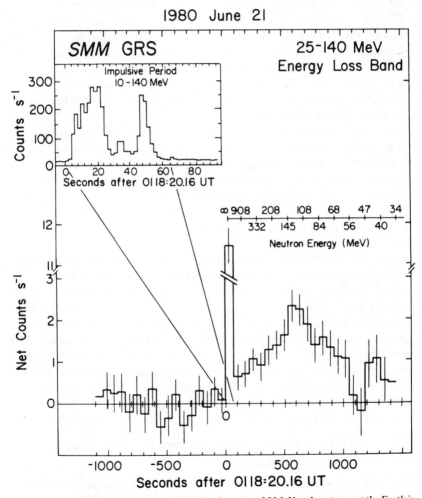

Figure 7   The arrival time history of energetic photons > 25 MeV and neutrons at the Earth is shown for the intense flare on 1980 June 21, as recorded by the GRS (17).

ground-level neutron monitors (25, 59). Neutron decay protons near the Earth, in the same event, have also been observed (32).

The GRS observed several flares with impulsive emission > 10 MeV (100), often accompanied by direct evidence for ion acceleration (i.e. nuclear-line emission < 10 MeV). This high-energy emission could be due to (a) highly relativistic electron bremsstrahlung ($\gamma > 20$), or (b) decay of π^0 mesons and bremsstrahlung from secondary $\pi^\pm$-$\mu^\pm$ decay electrons.

The SEPs, on the other hand, extend in energy to several hundred MeV nucleon^{-1} (112), and on rare occasions to 20 GeV (78). SEP observations (77) indicate the presence of nuclei with $Z > 56$ and with energies up to 100 MeV nucleon^{-1}. The high-Z elements have large abundance variations from flare to flare that may be related to large variations in the abundance ratio of ^{3}He/^{4}He (74). There are no confirmed observations of ^{2}H or ^{3}H nuclei associated with solar flares (74), but such nuclei would be expected if the ^{3}He were produced by nuclear reactions at the Sun (91). Relativistic electrons (> 1 MeV) associated with flares are commonly observed in space (23, 31, 69).

Since GRS observations have clearly established that impulsive photon emission from solar flares can extend to > 50 MeV and that neutrons can be produced with energies as high as 1 GeV, we must conclude that either an impulsive phase particle accelerator is operative or that preaccelerated particles are released impulsively into a target.

3. THEORETICAL STATUS

Theoretical models that predict gamma-ray and neutron yields and their production history are essential for a definitive interpretation of experimental observations. The required accelerated particle energy spectrum is difficult to calculate for any astrophysical situation (51) and has not been adequately treated for the solar flare case. Therefore, approximations are usually made in available calculations by assuming that an accelerated particle spectrum is produced isotropically at the flare site and undergoes interactions in one of two extreme models: (a) a *thin target approximation*, in which the accelerated particles lose a negligible amount of energy while in the interaction region, or (b) a *thick target approximation*, in which the accelerated particles lose all of their energy in the interaction region.

3.1 Basic Yield Calculations (Secondary Production, Ion and Electron Interactions)

ION INTERACTIONS   The basic equation for calculating the production rate s^{-1} of secondaries (type s) with energy between E_s and $E_s + dE_s$, generated in the interaction of incident particles (type j) with target species (type i) at time

t, is given by

$$q_s(E_s, t) = \sum_i \sum_j \int dE N_j(E, t) \{B_i n_H\} \sigma_{ijs}(E) f_s(E, E_s) c \beta_j, \qquad 1.$$

where B_i is the relative abundance of the target nucleus to hydrogen with number density n_H, $N_j(E, t)dE$ is the *instantaneous number* of accelerated particles with energy nucleon^{-1} between E and $E + dE$, $c\beta_j(E)$ is the relative velocity between incident and target particles, $\sigma_{ijs}(E)$ is the total reaction cross section (cm^2) for production of the secondary particle s, and $f_s(E, E_s)$ is the normalized production energy spectrum of the secondary with energy E_s to $E_s + dE_s$ by the primary particle j of energy E. The cross section and production spectra for a number of important processes have been compiled (70, 94).

The total yield of secondaries of a given type is determined by integrating Equation 1 over time (88). For the thin target, one obtains the time-averaged secondary yield Q'_s (thin) from Equation 1 by using $\overline{N_j(E)}dE$, the average instantaneous number of accelerated particles in the thin target for constant target number density. For the thick target case, the time-averaged secondary yield Q'_s (thick) is expressed in terms of $\overline{N^*_j(E)}dE$, the average *total number* of particles released into the thick target, using a modified Equation 1 to account for ionization loss (88).

The *normalized* yields Q_s (thin) and Q_s (thick) have been calculated using a number of different forms for accelerated particle spectra: (a) for a power-law form in kinetic energy with exponent s,

$$\bar{N}^*_j(E) \quad \text{and} \quad \bar{N}_j(E) = \begin{cases} B_j E^{-s}, & E > E_c, \\ B_j E_c^{-s}, & E \leq E_c, \end{cases} \qquad 2.$$

where E_c is an assumed cutoff energy below which the spectrum is flat (see 92–94); (b) for an exponential rigidity form,

$$\bar{N}^*_j(E) \quad \text{and} \quad \bar{N}_j(E) = B_j \exp(-R_j/R_{0j}) \, dR_j/dE, \qquad 3.$$

where R_j and R_{0j} are the rigidity and its characteristic value for the jth primary particle (70, 88); and (c) for a Bessel function form,

$$\bar{N}^*_j(E) \quad \text{and} \quad \bar{N}_j(E) = B_j(p_0, \alpha T)K_2[2(3p/mc\alpha T)]. \qquad 4.$$

Here, K_2 is the modified Bessel function of order 2, p and m are the momentum and mass of the ion, respectively, and p_0 is the initial momentum for ions continuously injected into a stochastic acceleration medium with an acceleration efficiency α and escape time T (33, 87, 88). In all cases, B_j is proportional to the relative abundance of an accelerated ion.

For the three spectral forms given by Equations 2, 3, and 4, the spectral shape parameters are determined by s, R_{0j}, and αT, respectively.

Equations 3 and 4 have been shown to be compatible with the observed SEP spectra (75), but we should keep in mind that the SEPs may not have the same origin as the particles producing the gamma rays. The Bessel function has been shown to be the nonrelativistic solution of a Fokker-Planck equation (87, 88) for stochastic Fermi acceleration in the case of continuous injection into the acceleration trap.

ELECTRON INTERACTIONS   When electrons have energies >270 keV, an exact relativistic treatment, using electron-ion and electron-electron cross sections, is necessary for bremsstrahlung yield calculations; however, this has been performed frequently for the thin target case (2, 4, 6, 10, 43, 84, 107). The necessary cross sections have been available for some time (42, 58), but the relativistic thick target case has only recently been considered for flare conditions (91a). It should be noted that inverse Compton scattering (60, 61) may be important for photons >1 MeV if the electron spectrum is extremely flat and the ambient density low ($n_H < 10^{10}$ cm^{-3}) (2) where bremsstrahlung would be suppressed. For electron energies <270 keV, nonrelativistic approximations for the bremsstrahlung cross section have been used (8, 24, 29, 55, 68, 84).

As with the determination of the ion source spectrum, it may be difficult to infer an exact solar electron source spectrum from the GRS bremsstrahlung observations, since no detailed information is available on conditions at the flare site.

3.2   Gamma-Ray Lines and Continua (Origin and Yield Estimates)

PROMPT LINES   Prompt gamma-ray lines arise from deexcitation of nuclear levels with radiative lifetimes that are short ($<10^{-11}$ s) compared with the time over which secondary production rates change. Therefore, from the time-dependent intensity of prompt lines, one obtains the time-dependent energetic particle interaction rate with matter. The strongest prompt lines of interest are below ~ 8 MeV, but weak lines above 10 MeV may be significant as well (22). These nuclear levels are populated through a number of processes: (a) inelastic scattering, as in $^{12}C(p, p')^{12}C^*(2^+)$; (b) spallation reactions, e.g. $^{16}O(p, x)^{12}C^*(2^+)$ both giving the 4.439 MeV line; or (c) radioactive decay, e.g. $^{16}O(p, p2n)^{14}O(\beta^+)T_{1/2}/(71$ s) $^{14}N(O^+)$ giving a line at 2.313 MeV. The important energy range for incident ions is 10–100 MeV nucleon^{-1}.

The observed gamma-ray line fluence can be related to the number of accelerated ions (see Section 2.4) using the following:

$$\phi_l(\text{narrow}) = 1/4\pi R^2 \begin{Bmatrix} N_p^*(>30 \text{ MeV}) \, Q_l(\text{thick}) \, \bar{f}_l \\ n_H \tau_r \bar{N}_p(>30 \text{ MeV}) \, Q_l(\text{thin}) \, \bar{f}_l \end{Bmatrix} \text{photons cm}^{-2}, \quad 5.$$

where $N_p^*(>30 \text{ MeV})$ is the *total* number of protons >30 MeV injected into a thick target, $\bar{N}_p(>30 \text{ MeV})$ is the time-averaged *instantaneous* number of protons in a thin target interaction region for a reaction time τ_r, $\bar{f}_l$ is the transfer efficiency from nuclear deexcitations to escaping photons, and R is the Earth-Sun distance. (Other terms were defined previously.) For isotropic conditions and no attenuation, $\bar{f}_l = 1$; but if interactions occur where $n_H > 10^{15} \text{ cm}^{-3}$, then $\bar{f}_l < 1$, with the value varying with heliocentric longitude. For ions beamed downward at central meridian, $\bar{f}_l \leq 0.8$ (88). Normalized values $Q(\text{thick})$ and $Q(\text{thin})$ are found in (70, 88, 92, 96).

The reaction kinematics determine the width and apparent energy of a nuclear line. Both can be used to establish the interaction geometry, i.e. whether the particles are beamed into the target or whether they are isotropic. For the isotropic case with all ambient constituents accelerated, narrow lines of width $\Delta E < 5 \text{ keV}$ result from excited ambient heavy nuclei ($Z > 8$), and weaker broad lines ($\Delta E > 200 \text{ keV}$) result from accelerated energetic heavy nuclei ($Z \geq 6$). Lines from ambient $^{12}C^*$ (4.439 MeV) and $^{16}O^*$ (6.129 MeV) have $\Delta E \simeq 100 \text{ keV}$ (89, 94). A shift can be expected in apparent energy (for instance, $\sim 40 \text{ keV}$ for the 6.129-MeV line from $^{16}O^*$) for strongly beamed energetic ions with a large velocity component along the line of sight (89). For a beam of α-particles moving transverse to the direction of observation, α-α reactions produce well-resolved prompt lines at 431 keV from $^7Be^*$, and at 478 keV from $^7Li^*$ (62, 88).

DELAYED GAMMA RAYS Delayed gamma-ray lines are emitted over a time interval long compared with the production time of secondaries. Three important examples of such lines are (a) annihilation radiation from positrons arising from radioactive nuclides, such as ^{11}C and ^{11}B or the π^+-μ^+ decay chain (105); (b) capture gamma-ray lines from neutrons thermalized in the photosphere, such as the 2.223-MeV line from $^1H(n, \gamma)^2H$ (52, 53, 115); and (c) deexcitation lines and bremsstrahlung continuum from radioactive secondaries ($T_{1/2} \lesssim 20 \text{ min}$) whose $\beta^{\pm}$ decay leads to excited states in daughter nuclides (94). Photospheric Compton scattering of these gamma rays could be an additional source of delayed gamma rays.

511-kev ANNIHILATION LINE AND CONTINUUM Positron annihilation takes place through free or bound state interactions with ambient electrons. In the first instance, a discrete gamma-ray line at 511 keV is emitted (105).

Bound singlet state annihilation results in the 511-keV line, while triplet state annihilation gives a three-photon continuum (21, 67, 105).

The annihilation-line fluence ϕ_{511} is found by substituting Q_+(thick) or Q_+(thin) (88, 97) and $\bar{f}_{511}$ for Q_l and $\bar{f}_l$ in Equation 5. If there is no attenuation, the number of escaping 511-keV photons per positron produced, $\bar{f}_{511}$, may vary from 0.5 to 2 depending on the fraction of positronium formed. If the density is low (thus making annihilation highly improbable) or if the positrons are produced in or transported to the photosphere (where $n_H > 10^{15}$ cm^{-3}), then $\bar{f}_{511} < 0.5$ because of absorption. The width of the 511-keV line depends on the temperature of the annihilation region according to ΔE (keV) $= 15.6\sqrt{T_6}$ (FWHM), where T_6 is the temperature in 10^6 K.

2.223-MeV CAPTURE LINE   The fluence $\phi_{2.223}$ is determined by using in Equation 5 a normalized neutron yield Q_n and $\bar{f}_{2.223}(\theta)$, the conversion efficiency of neutrons to escaping line photons; the latter parameter depends on the neutron capture depth, flare longitude (θ), and photospheric ^{3}He/^{1}H abundance. For $\theta = 0$ (central meridian), $\bar{f}_{2.223}(0) \leq 0.3$ (52, 88, 114).

GAMMA-RAY AND HARD X-RAY CONTINUUM (>270 keV)   The continuum radiation (>270 keV) is mainly prompt emission, and it can come from (a) lines that are severely Doppler broadened, (b) bremsstrahlung from primary electrons, and (c) meson decay. The > 10 MeV (prompt) continua result from π^0 decay $\pi^\pm$-$\mu^\pm$ decay electrons and relativistic primary electrons, but each continuum has a different spectral shape (22, 70, 73).

For gamma rays > 10 MeV, Equation 5 (with appropriate substitutions) is used. This gives $\phi_{>10}$ with $\bar{f}_l = 1$, or $\phi_{\pi^0\gamma}$ with $\bar{f}_l = 2$, using normalized Q (> 10 MeV) or $Q(\pi^0)$ values, respectively (97). The normalized yields have been calculated relative to Q (4–7 MeV), the normalized yield of all prompt (narrow and broad) gamma-ray lines between 4–7 MeV. The ^{12}C line at 4.439 MeV is one of the strongest prompt lines from solar flares (predicted and observed), and the yield ratio $[Q(4$–7 MeV$)/Q(4.439$ MeV$)]$ has been determined (93, 97). This value is then used to find $[Q(4$–7 MeV$)/Q_n]$. This gives us $Q(>10$ MeV$) = [Q(>10$ MeV$)/Q(4$–7 MeV$)] \times [Q(4$–7 MeV$)/Q_n] \times Q_n$ or $Q(\pi^0) = [Q(\pi^0)/Q(4$–7 MeV$)] \times [Q(4$–7 MeV$)/Q_n] \times Q_n$. The determination of Q_n is discussed below.

3.3   Solar Neutrons (Yield, Photosphere Interaction, Flux at the Earth)

Solar neutrons are unique compared with the charged particles produced in flares. They are not deflected by solar magnetic fields, and they travel directly to the photosphere or the Earth if they survive decay ($\tau \simeq 925$ s).

NEUTRON YIELD   To determine the solar neutron production spectrum, the differential production cross section $\sigma_n(E)f_n(E_p, E_n)$ is used in Equation 1. The total neutron production cross section $\sigma_n(E)$ is assumed to be isotropic. For each E_p, normalized differential neutron energy distribution functions $f_n(E_p, E_n)$, are estimated for the cases of proton-proton, proton-^{4}He, or proton-CNONe interactions (52, 53, 70, 71, 92). For "standard" solar abundances, the total neutron yield for all E_n per incident proton is greatest for p-α and α-p reactions for all incident energies up to ~ 1 GeV nucleon^{-1}; thereafter, p-p interactions become the main source of total neutrons from the reaction $p(p, \pi^+ n)p$. General results for the total neutron yield Q'_n may be expressed as follows: $Q'_n(>1$ MeV$) = \bar{N}^*_p(>30$ MeV$) Q_n(\text{thick})$; or Q'_n $(>1$ MeV$) = n_H \tau_r \bar{N}_p(>30$ MeV$) Q_n(\text{thin})$, where $Q_n(\text{thick})$ and $Q_n(\text{thin})$ are the normalized yields given in (70, 88, 92). For example, the Bessel function form for a thick target with $\alpha T = 0.02$ gives $Q'_n(>1$ MeV$)/\bar{N}^*_p$ $(>30$ MeV$) = Q_n(\text{thick}) \simeq 3.4 \times 10^{-3}$.

NEUTRON INTERACTION WITH THE PHOTOSPHERE   The bulk of secondary neutrons produced have energies of ~ 10 MeV; however, a few are produced with an energy comparable to that of the highest energy proton accelerated. For the typical neutron emitted in the direction of the photosphere, elastic scattering on protons rapidly ($\sim$ seconds) reduces its energy to thermal levels (~ 1 eV). This occurs within a column density of <10 g cm^{-2}, corresponding to a photospheric depth of ~ 100 km (99, 114), unless the neutrons are scattered out of the Sun. The thermalized neutrons can undergo nonradiative capture via ^{3}He$(n, p)^3$H or ^{1}H$(n, \gamma)^2$H, emitting the 2.223-MeV line. Both reactions are equally probable if the photospheric abundance ratio ^{3}He/H is $\simeq 5 \times 10^{-5}$ (52, 53, 114). For a flare at central meridian ($\theta = 0$), 25% of the 2.223-MeV photons moving toward the Earth are scattered in the photosphere, producing a low-energy Compton continuum (52, 53); for a limb flare, on the other hand, nearly all the earthward-bound photons will undergo scattering, reducing the line intensity by a large factor (>100) (113). For the solar flare case, neutron capture produces the strongest line for disk flares. The width of the line is expected to be intrinsically narrow (<1 eV), reflecting the photospheric temperature.

The neutrons that scatter out of the photosphere and have energies <2 keV remain gravitationally trapped and can reenter the photosphere. About 50% of all neutrons produced decay according to $n \rightarrow p + \beta^- + \bar{\nu}_e$ (53).

The time history of the intensity of the 2.223-MeV line is dependent on all of the mechanisms described above. The rise time of intensity is determined mainly by the neutron production time history, while the decay of intensity, which is essentially exponential with a time-constant τ_{loss}, is dependent on all loss processes, i.e. $^1H(n, \gamma)^2H$, $^3He(n, p)^3H$, and decay. In principle, then, a measurement of τ_{loss} can determine the $^3He/H$ ratio if the local density can be independently determined. For example, calculations (52, 114) show that for capture at a typical density $n_H = 10^{17}$ cm^{-3}, τ_{loss} ranges from 71 to 122 s for abundance ratios of 5×10^{-5} and 0, respectively. The observed value for τ_{loss} is $\simeq 100$ s [see Section 2.5 and (16, 86)].

SOLAR NEUTRONS AT THE EARTH   The main factors determining the neutron flux at the Earth are the energy-dependent survival probability $P(E_n)$ and the time dispersion factor dE_n/dt_e, a consequence of the finite mass of the neutron giving the energy-dependent transit time t_e to the Earth. If neutrons are produced in a time short compared with t_e (δ-function approximation), with a differential emissivity spectrum I_n(MeV^{-1} sr^{-1}) emitted into a unit solid angle toward the Earth, then the time-dependent neutron flux at the Earth is F_n(cm^{-2} s^{-1}) $= I_n$(MeV^{-1} sr^{-1})R^{-2} $(dE_n/dt_e)P_s(E_n)$, where R is the Sun-Earth distance. This model gives a unique relation between E_n and t_e. Predictions of the expected solar neutron flux for this case have been given in (70, 71). The emissivity function for a finite production interval can be determined by integration over a weighted set of δ-functions; however, in this general case, the unique energy–transit time relation does not hold.

4.   TOWARD AN INTERPRETATION

4.1   A Thick Target Model

Only an incomplete interpretation of observations is possible at this time, since detailed flare particle acceleration models for comparison with data are not available. Therefore, we describe here how the theoretical estimates can be used to interpret gamma-ray and neutron data for two impulsive flares: (a) a "modest" event on 1980 June 7 (Flare 1), and (b) an "intense" event on 1982 June 3 (Flare 2). We assume that an accelerator injects ions and electrons into a thick target in the lower chromosphere. This allows us to explain the data in a phenomenological sense and to point out some uncertainties. The following interrelated elements are considered: (a) the accelerator characteristics, (b) the absolute particle spectra, and (c) the interaction geometry, including abundances of the beam and target.

4.2   Accelerator Characteristics

The solar flare impulsive phase particle accelerator for electrons and ions could be described, if its location and basic parameters were known, by the energy and type of particle, the shape of the output pulse, and the duration

Table 1 Likely characteristics of a primary particle accelerator

Parameter	Characteristic
Location	Low corona–Chromosphere
Minimum risetime of pulse	5 s
Electron energy	> 10 MeV
Ion energy	Typical: 10 MeV nucleon^{-1}
	Occasional: > 1 GeV nucleon^{-1}
Interaction region	10^{12} cm^{-3} < n_H < 10^{16} cm^{-3}

of acceleration. Are we dealing with an accelerator that operates over a long period (e.g. 28) in the *low-density corona* and then impulsively injects electrons and ions, or instead with an accelerator that operates in a transient manner? Either model must give energetic particle interactions in the time pattern required by observations. A preimpulsive phase accelerator cannot be dismissed by current observations. Energetic (MeV) electrons produced before the relatively intense impulsive phase could reveal themselves through precursor gyro-synchrotron emission and bremsstrahlung. In this regard, the following observations must be considered : (a) there is evidence for preflare microwave emission as much as an hour before impulsive flares (63, 64, 66); (b) there is also evidence for precursor gamma-ray (> 3.5 keV) emission of a weak, impulsive character (108) many hours before a flare; and (c) in some intense flares seen by the GRS, there is an early low-level impulsive emission at MeV energies.

The predominant emissions observed by the GRS seem to naturally suggest the existence of a transient, rather than long-term, accelerator. The basic characteristics of a transient primary accelerator for electrons and ions, suggested by the gamma-ray observations, are shown in Table 1. The site of operation of this "primary accelerator" is tacitly taken to be the region (possibly extended) of impulsive energy release in the low corona or chromosphere where $n < 10^{15}$ cm^{-3}. The location would vary with each flare, as controlled by the magnetic topology of the active region (see the reports of *SMM* (111) and solar terrestrial physics (102) workshops). This location is consistent with the site of the impulsive microwave emission, a correspondence that in some flares suggests repeated acceleration of electrons within the same magnetic trap (65).

4.3   *Absolute Energetic Particle Spectra*

The normalized yields from ion interactions, e.g. Q(thick), are dependent on the assumed energy spectrum. It has been shown (92) that by comparison of predicted and observed fluence ratios of a prompt gamma-ray line and the

2.223-MeV line it is possible to estimate a spectral shape parameter, if a spectral form (Section 3.1) and an interaction model (thick or thin) are initially assumed. This result is then used to give the absolute accelerated ion energy spectrum by normalizing to the observed fluence for any secondary radiation (e.g. Equation 5). A similar procedure could be used to obtain a relativistic electron spectrum from the estimated bremsstrahlung spectrum, but as mentioned above, the relevant calculations are only available for the thin target model. For our thick target example, we can therefore estimate only the accelerated ion spectrum. Table 2 shows the GRS observations for the two flares denoted as Flare 1 and Flare 2, each occurring at $\sim 73°$ longitude from central meridian.

To obtain the accelerated ion spectral shape parameter at low energies < 400 MeV nucleon^{-1}, we determine the observed ratio of gamma-ray line fluence in the 4–7 MeV band to that at 2.223 MeV, corrected to central meridian by using $f(0-\theta) = 0.6$ (53, 88). This ratio gives an observed estimate of $[Q(4-7 \text{ MeV})/Q_n]$ after multiplying by $\bar{f}_{2.223}(0) = 0.14-0.30$ (88; see Section 3.2). The value of $[Q(4-7 \text{ MeV})/Q_n]$ depends strongly on the spectral shape parameter in the energy region 10–100 MeV nucleon^{-1} and on the target abundances used in the calculations. Comparison of the observed and calculated values of $[Q(4-7 \text{ MeV})/Q_n]$ gives a measure of the incident particle spectrum. The actual spectral form is unknown, so for convenience, we use the Bessel function predictions of (88) and obtain (for spectral shape parameters) $\alpha T = 0.016$ and 0.025 for Flare 1 and Flare 2, respectively. Then, using either the measured 4–7 MeV or 2.223-MeV fluence, we obtain values for the total number of accelerated protons $\bar{N}_p^*(> 30 \text{ MeV})$ (see Equation 5) of 8×10^{31} and 2×10^{33} for Flares 1 and 2, respectively. The total neutron yields for Flare 1 and Flare 2 then directly follow, giving 3×10^{29} and 1×10^{31} neutrons, respectively. (These are shown as *Inferred parameters* in Table 3).

β^+ YIELD   From the *Inferred parameters* in Table 3 we can predict the total positron yield from radioactive nuclei, as discussed previously. For Flares 1 and 2, the results are 3×10^{28} and 5×10^{29} β^+, respectively, and are shown under *Predictions* in Table 3. The conversion efficiency $\bar{f}_{511}$ from β^+ to 511-keV photons is difficult to estimate for reasons discussed in Section 3.2. However, if we assume $\bar{f}_{511}$ has the maximum value of 2, then the fluences at the Earth should be ~ 19 and ~ 400 (511-keV photons cm^{-2}) for Flares 1 and 2, respectively (see Table 3). Comparison of these values with the measured fluences in Table 2 shows more than 10 times as many 511-keV photons cm^{-2} predicted as observed for Flare 1. For Flare 2, the values differ by a factor of four. If 100% positronium were produced, then the discrepancy could be lowered by a factor of four in each case. This implies

Table 2 The observed *nominal* gamma-ray fluence at Earth (photons cm^{-2}) in several energy bands and the total neutron emission > 50 MeV at the Sun[a]

Event	Class GOES/Hα	Flare coordinates	Impulsive phase ~(UT)	Neutrons at Sun >50 MeV	Gamma-ray fluence (MeV^{-1} cm^{-2})[a] >270 keV	Gamma-ray fluence (cm^{-2})[a]			
						511 keV	2.223 MeV	4–7 MeV	>10 MeV
Flare 1 1980 June 7	M7.3/SN	N12W74	(0312–0313)	<10^{29}	~55 $E^{-2.6}$	<2	~8	≤16	<1
Flare 2[b] 1982 June 3	X8.0/2B	S09E72	(1143–1144)	*~4 × 10^{30}*	*1800 E^{-2}*	*100*	*314*	*<300*	*50*

[a] Errors that are model dependent are not shown on fluence values. Statistical count errors are <10%.
[b] Italicized fluence values may be significantly revised due to dead time and gain-shift corrections.

Table 3 The parameters inferred from data in Table 2 using the thick target interaction model[a]

Event	Inferred parameters		Predictions[a]		
	Low-energy ions	High-energy ions	Total neutrons >1 MeV	Yields and line fluence 511 keV	Yields and fluence >10 MeV
Flare 1	$\alpha T \simeq 0.016$ $\bar{N}_p^*(>30\ \text{MeV}) = 8 \times 10^{31}$	—	$Q_n = 3 \times 10^{29}$	$RA(\beta^+) = 3 \times 10^{28}\ (19)$[b] $\pi^+(\beta^+) = 2 \times 10^{24}(10^{-3})$[b]	π^0-yield $= 2 \times 10^{24}(2 \times 10^{-3})$[c] $(>10\ \text{MeV}) = 5 \times 10^{24}(2 \times 10^{-3})$[d]
Flare 2	$\alpha T \simeq 0.025$ $\bar{N}_p^*(>30\ \text{MeV}) = 2.3 \times 10^{33}$	$s = -5.5$ $\bar{N}_p^*(>100\ \text{MeV}) = 4 \times 10^{32}$	$Q_n = 1 \times 10^{31}$	$RA(\beta^+) = 5 \times 10^{29}\ (400)$[b] $\pi^+(\beta^+) = 6 \times 10^{27}(5)$[b]	π^0-yield $= 9 \times 10^{26}\ (\leq 1)$[c] $(>10\ \text{MeV}) = 4 \times 10^{27}(\sim 1)$[d]

[a] Predicted yields and fluences based on the inferred spectral shape parameters. Comparison should be made between predicted and observed fluences.
[b] The numbers in parentheses give the predicted 511-keV fluence at Earth for $\bar{J}_{511} = 2$ and the β^+ yield at Sun.
[c] The numbers in parentheses give the predicted π^0 gamma-ray fluence at Earth for $\bar{J}_1 = 2$ and the π^+ yield at Sun.
[d] The numbers in parentheses give the predicted >10 MeV gamma-ray fluence at Earth due to all meson contributions.

that positronium is in a region where $n_H < 10^{15}$ cm^{-3}; otherwise, it would suffer collisional breakup (21). Another possibility is that the annihilation emission region could be at a depth of > 8 g cm^{-2}, sufficient to suppress the 511-keV fluence. A third possibility is that the positrons are at a low density ($n_H \lesssim 10^{10}$ cm^{-3}), where the slowing-down time of β^+ is much greater than the typical observation time (> 600 s) available.

The other source of β^+, the π^+-μ^+ decay chain, provides 2×10^{24} and 6×10^{27} β^+ for Flares 1 and 2, respectively, using αT values obtained from the prompt gamma-ray/neutron fluence ratios. These results are shown under *Predictions* in Table 3. Therefore, the Bessel function spectral shape parameter needed to explain the prompt gamma-ray line emission for these flares predicts a negligible β^+ yield from π^+ meson decay. On the other hand, a power-law ion spectrum with spectral shape parameter $s \leq 2$ would produce more β^+ from π^+-μ^+ decay than from radioactive emitters (92). In Flare 1, there seem to be too few 511-keV photons if positron emitters are the source, but such emitters could be the primary source for this line in Flare 2.

PHOTONS > 10 MeV   Recent calculations (97) using the Bessel function spectral form give the normalized yield of photons $Q(> 10$ MeV) from both π^0 decay and secondary electron bremsstrahlung, the latter from the $\pi^\pm$-$\mu^\pm$ decay chain. For Flare 1 the total photon yield $Q'(> 10$ MeV) at the Sun is $\sim 5 \times 10^{24}$ photons, and for Flare 2 it is $\sim 4 \times 10^{27}$ photons. If accelerated ions follow a Bessel function form, then for the αT values given in Table 3 the predicted high-energy (> 10 MeV) photon fluences (Table 3) from meson decay, derived from these yields, are considerably below those observed (Table 2). This result (97, 100) has led to the suggestion that the impulsive photon emission > 10 MeV is due to bremsstrahlung from primary electrons ($\gamma > 20$). Before such a conclusion can be reached, however, the measured photon energy spectrum below and above 10 MeV must be tested against an input model that includes both secondary pion and primary electron contributions. This analysis is now in progress using GRS data for all events with measured emission > 10 MeV.

HIGH-ENERGY NEUTRONS   Observations of neutrons > 50 MeV in Flare 2 provide the experimental means to investigate the spectrum above ~ 100 MeV nucleon^{-1}. In the neutron energy region (~ 50–500 MeV) the observations indicate a spectrum with an approximate power-law form and with an exponent about -3.5. An accelerated proton power-law spectrum with exponent between -4 and -6 in the energy band ~ 100–1000 MeV is needed to produce the observed neutron spectrum (53). The observed yield of neutrons > 50 MeV (for Flare 2 only) is $\sim 4 \times 10^{30}$ (18) and requires for their production $\sim 4 \times 10^{32}$ protons > 100 MeV (see Tables 2 and 3). The

correlated with the flare gamma-ray emissions. The type II radio bursts that follow many impulsive flares by several minutes and that are sometimes well correlated with the appearance of SEPs indicate an area in which second "phase" concepts naturally fit. In this connection, it has been pointed out (116) that the type II bursts naturally suggest a shock-wave explanation ($V_{sk} \simeq 1000$ km s^{-1}). Since energetic particles and shocks are observed together in space following flares, a second "phase" acceleration process may exist in some flares. There could also be a role for stochastic mechanisms (for example, post impulsive phase acceleration) to explain greatly delayed radio emissions (64).

6. OUTSTANDING QUESTIONS

The observations and theoretical work reviewed here show the significant advances made in high-energy solar physics since 1972. Because of the necessary brevity of this paper, it is not possible to explore all the potential directions future work could take in order to resolve many of the questions implied in this review. Nevertheless, it is necessary to emphasize the direct observational evidence: The onset of impulsive high-energy photon emission from flares can be simultaneous to <1 s over a wide range of energies (e.g. 40 keV to ~ 50 MeV), sometimes with associated gamma-ray line and > 500-MeV neutron emission. This requires rapid acceleration of ions to 50 MeV (and likely to 1 GeV), and of electrons to > 10 MeV, all to within < 1 s. This circumstance raises a pressing question: Is it possible to find a single, primary, repetitive acceleration mechanism operative in a flare, or will only a fast, cyclic, multistage process meet observational constraints for both electrons and ions? In addition, a few related questions are the following: (a) How do we explain the apparent variability in the electron spectral shape from flare to flare while the ion spectral shape (mostly apparent in the less intense GRS flares) seems to be relatively constant? (b) Are flares composed of elementary flare bursts? (c) What is the relationship of low-level "precursor" emission ($E_\gamma \gtrsim 1$ MeV) to the intense impulsive emission following within 1 minute? (d) What is the relationship between the solar energetic particles observed in space and the gamma-ray producing ions and electrons? (e) Do the GRS observations imply significant deviations from normal solar abundances in the interaction region?

At the beginning of this review, I stated that the study of the neutral solar radiation will provide us with new insights regarding particle acceleration and energetic particle interaction with matter at the Sun during solar flares. The GRS findings presented here have provided some answers and have raised even more questions. I am confident that the answers to some of these questions will come from continued analysis of the GRS data.

ACKNOWLEDGMENTS

The author appreciates the stimulating discussions and the support he received, in preparing this review, from the many members of the *Solar Maximum Mission* Gamma Ray Spectrometer Team at the Max Planck Institute for Astrophysics in Germany; at the Naval Research Laboratory in Washington, DC; and at the University of New Hampshire. I wish to thank in particular W. T. Vestrand for critical comments, advice, and assistance; D. J. Forrest for helpful discussions; and J. Narayanaswamy for assistance. Thanks also goes to several Guest Investigators associated with the GRS data analysis effort. The critical editorial work by Mary M. Chupp was essential in meeting all exacting requirements of this work; her tireless efforts and those of Robert Hoffman and Jan Soller are especially acknowledged. Financial support for this work was partially provided by NASA and the US Air Force under Contract NAS5-23761, by NASA under Grant NGL 30-002-021, and by the University of New Hampshire.

Literature Cited

1. Arons, J., Max, C., McKee, C., eds. 1979. *AIP Conf. Proc. 56, Particle Acceleration Mechanisms in Astrophysics.* New York: Am. Inst. Phys. 433 pp.
2. Bai, T. 1977. *Studies on solar hard X-rays and gamma rays: Compton backscatter anisotropy, polarization and evidence for two phases of acceleration.* PhD thesis. Univ. Md., College Park. 147 pp.
3. Bai, T. 1982. See Ref. 72, pp. 409–17
4. Bai, T., Ramaty, R. 1976. *Sol. Phys.* 49:343–58
5. Biermann, L., Haxel, O., Schulter, A. 1951. *Z. Naturforsch.* 6a:47–48
6. Blumenthal, G. R., Gould, R. J. 1970. *Rev. Mod. Phys.* 42:237–70
7. Burns, M. L., Harding, A. K., Ramaty, R., eds. 1983. *AIP Conf. Proc.* 101, *Positron-Electron Pairs in Astrophysics.* New York: Am. Inst. Phys. 447 pp.
8. Brown, J. C. 1975. See Ref. 54, pp. 245–82
9. Chambon, G., Hurley, K., Niel, M., Talon, R., Vedrenne, G., et al. 1981. *Sol. Phys.* 69:147–59
10. Cheng, C. 1972. *Space Sci. Rev.* 13:3–123
11. Chupp, E. L. 1976. *Gamma Ray Astronomy.* Dordrecht: Reidel. 317 pp.
12. Chupp, E. L. 1982. See Ref. 72, pp. 363–81
13. Chupp, E. L. 1983. *Sol. Phys.* 86:383–93
14. Chupp, E. L., Forrest, D. J., Higbie, P. R., Suri, A. N., Tsai, C., et al. 1973. *Nature* 241:333–35
15. Chupp, E. L., Forrest, D. J., Suri, A. N. 1975. See Ref. 54, pp. 341–59
16. Chupp, E. L., Forrest, D. J., Ryan, J. M., Cherry, M. L., Reppin, C., et al. 1981. *Ap. J. Lett.* 244:L171–74
17. Chupp, E. L., Forrest, D. J., Heslin, J., Kanbach, G., Pinkau, K., et al. 1982. *Ap. J. Lett.* 263:L95–99
18. Chupp, E. L., Forrest, D. J., Share, G. H., Kanbach, G., Debrunner, H., et al. 1984. *Proc. Int. Cosmic Ray Conf., 18th, Bangalore.* In press
19. Cliver, E., Share, G., Chupp, E. L., Matz, S., Howard, R. 1983. *Bull. AAS Meet., 161st, Boston* 14:874 (Abstr.)
20. Cliver, E. W., Forrest, D. J., McGuire, R. E., Von Rosenvinge, T. T. 1984. *Proc. Int. Cosmic Ray Conf., 18th, Bangalore.* In press

21. Crannell, C. J., Joyce, G., Ramaty, R., Werntz, C. 1976. *Ap. J.* 210:582–92
22. Crannell, C. J., Crannell, H., Ramaty, R. 1979. *Ap. J.* 229:762–71
23. Datlowe, D. W. 1971. *Sol. Phys.* 17:436–58
24. Datlowe, D. W. 1975. See Ref. 54, pp. 191–208
25. Debrunner, H., Fluckiger, E., Chupp, E. L., Forrest, D. J. 1984. *Proc. Int. Cosmic Ray Conf., 18th, Bangalore.* In press
26. De Jager, C. 1969. In *Solar Flares and Space Research, Proc. Symp. Plenary Meet. Comm. Space Res., 11th*, pp. 1–15. Amsterdam: North-Holland. 419 pp.
27. De Jager, C., De Jonge, G. 1978. *Sol. Phys.* 58:127–37
28. Elliot, H. L. 1973. See Ref. 90, pp. 12–18
29. Emslie, A. G. 1983. *Ap. J.* 271:367–75
30. Evenson, P., Meyer, P., Yanagita, S. 1981. *Proc. Int. Cosmic Ray Conf., 17th, Paris* 3:32–35
31. Evenson, P., Meyer, P., Yanagita, S., Forrest, D. J. 1984. *Ap. J.* 282: In press
32. Evenson, P., Meyer, P., Pyle, K. R. 1983. *Ap. J.* 274:875–82
33. Forman, M. A., Ramaty, R., Zweibel, E. G. 1984. See Ref. 106. In press
34. Forrest, D. J. 1983. See Ref. 7, pp. 3–14
35. Forrest, D. J., Chupp, E. L. 1983. *Nature* 305:291–92
36. Forrest, D. J., Chupp, E. L., Ryan, J. M., Cherry, M. L., Gleske, I. U., et al. 1980. *Sol. Phys.* 65:15–23
37. Forrest, D. J., Chupp, E. L., Reppin, C., Rieger, E., Ryan, J. M., et al. 1981. *Proc. Int. Cosmic Ray Conf., 17th, Paris* 10:5–8
38. Forrest, D. J., Gardner, B. M., Matz, S. M., Chupp, E. L., Reppin, C., et al. 1981. *Bull. AAS Meet., 159th, Boulder* 13:903 (Abstr.)
39. Forrest, D. J., Matz, S. M., Chupp, E. L., Share, G., Rieger, E. 1983. *Bull. APS Meet., Baltimore* 28:730 (Abstr.)
40. Gardner, B. M., Forrest, D. J., Zolcinski, M. C., Chupp, E. L., Rieger, E., et al. 1981. *Bull. Am. Astron. Soc.* 13(4):903 (Abstr.)
41. Gruber, D. E., Peterson, L. E., Vette, J. I. 1973. See Ref. 90, pp. 147–61
42. Haug, E. 1975. *Z. Naturforsch.* 30a:1099–1113
43. Haug, E. 1975. *Sol. Phys.* 45:453–58
44. Heyvaerts, J. 1981. In *Solar Flare Magnetohydrodynamics*, ed. E. R. Priest, 1:429–555. New York/London/Paris: Gordon & Breach. 563 pp.
45. Hirasima, Y., Okudaira, K., Yamagami, T. 1969. *Acta Phys. Hung.* 29:683–88 (Suppl. 2)
46. Holt, S. S., Cline, T. L. 1968. *Ap. J.* 154:1027–38

47. Hoyng, P., Duijveman, A., Machado, M. E., Rust, D. M., Svestka, Z., et al. 1981. *Ap. J. Lett.* 246:L155–59
48. Hudson, H. S., Bai, T., Gruber, E., Matteson, J. L., Nolan, P. L., et al. 1980. *Ap. J. Lett.* 236:L91–L95
49. Hurley, K., Niel, M., Talon, R. 1983. *Sol. Phys.* 86:367–73
50. Ibrigamov, I. A., Kocharov, G. E. 1977. *Sov. Astron. Lett.* 3:211–22
51. Jokipii, J. R. 1979. See Ref. 1, pp. 1–9
52. Kanbach, G., Reppin, C., Forrest, D. J., Chupp, E. L. 1975. *Proc. Int. Cosmic Ray Conf., 14th, Munich* 5:1644–49
53. Kanbach, G., Chupp, E. L., Cooper, J., Forrest, D. J., Reppin, C., et al. 1984. In preparation
54. Kane, S. R., ed. 1975. *Solar Gamma, X-, and EUV-Radiation, IAU Symp. No. 68.* Dordrecht: Reidel. 439 pp.
55. Kane, S. R., Anderson, K. A. 1970. *Ap. J.* 162:1003–18
56. Kane, S. R., Fenimore, E. E., Klebesadel, R. W., Laros, J. G. 1982. *Ap. J. Lett.* 254:L53–57
57. Kiplinger, A. L., Dennis, B. R., Emslie, A. G., Frost, K. J., Orwig, L. E. 1983. *Sol. Phys.* 86:239–40 (Abstr.)
58. Koch, H. W., Motz, J. W. 1959. *Rev. Mod. Phys.* 31:920–55
59. Kocharov, G. E. 1983. *Invited Talks Eur. Cosmic Ray Symp., 8th, Bologna*, pp. 51–61
60. Korchak, A. A. 1967. *Sov. Astron. AJ* 11(2):258–63
61. Korchak, A. A. 1971. *Sol. Phys.* 18:284–304
62. Kozlovsky, B., Ramaty, R. 1977. *Astrophys. Lett.* 19:19–24
63. Kundu, M. R. 1982. *Rep. Prog. Phys.* 45:1435–1541
64. Kundu, M. R., Vlahos, L. 1982. *Space Sci. Rev.* 32:405–62
65. Kundu, M. R., Bobrowsky, M., Velusamy, T. 1981. *Ap. J.* 251:342–51
66. Kundu, M. R., Schmahl, E. J., Velusamy, T., Vlahos, L. 1982. *Astron. Astrophys.* 108:188–94
67. Leventhal, M. 1973. *Ap. J. Lett.* 183:L147–50
68. Lin, R. P., Hudson, H. S. 1976. *Sol. Phys.* 50:153–78
69. Lin, R., Mewaldt, R. A., Hollebeke, M. A. I. 1982. *Ap. J.* 253:949–62
70. Lingenfelter, R. E., Ramaty, R. 1967. In *High Energy Nuclear Reactions in Astrophysics*, ed. B. S. P. Shen, pp. 99–158. New York: Benjamin. 281 pp.
71. Lingenfelter, R. E., Canfield, E. H., Flamm, E. J., Kellman, S. 1965. *J. Geophys. Res.* 70:4077–95
72. Lingenfelter, R. E., Hudson, H. S., Worrall, D. M., eds. 1982. *AIP Conf.*

Proc. 77, Gamma Ray Transients and Related Astrophysical Phenomena. New York: Am. Inst. Phys. 500 pp.
73. Marscher, A. P., Brown, R. L. 1978. *Ap. J.* 221: 588–97
74. McGuire, R. E. 1983. *Rev. Geophys. Space Phys.* 21(2): 305–18
75. McGuire, R. E., Von Rosenvinge, T. T., McDonald, F. B. 1981. *Proc. Int. Cosmic Ray Conf., 17th, Paris* 3: 65–68
76. Meliorenski, A. S., Pissarenko, N. F., Shamolin, W. M., Likin, O. B., Gourdon, B., et al. 1975. *Astron. Astrophys.* 41: 379–84
77. Mewaldt, R. A. 1980. *Proc. Conf. Ancient Sun,* ed. J. A. Eddy, R. Pepin, R. B. Merrill, pp. 81–101. New York: Pergamon
78. Meyer, P., Parker, E. N., Simpson, J. A. 1956. *Phys. Rev.* 104(3): 768–83
79. Morrison, P. 1958. *Nuovo Cimento* 7: 858–65
80. Mullan, D. J. 1983. *Ap. J.* 269: 765–78
81. Nakajima, H., Kosugi, T., Kai, K., Enome, S. 1983. *Nature* 305: 292–94
82. Orwig, L. E., Frost, K. J., Dennis, B. R. 1980. *Sol. Phys.* 65: 25–37
83. Peterson, L. E., Winckler, J. R. 1959. *J. Geophys. Res.* 64: 697–707
84. Petrosian, V. 1973. *Ap. J.* 186: 291–304
85. Prince, T. A., Ling, J. C., Mahoney, W. A., Riegler, G. R., Jacobson, A. S. 1982. *Ap. J. Lett.* 255: L81–84
86. Prince, T. A., Forrest, D. J., Chupp, E. L., Kanbach, G., Share, G. H. 1984. *Proc. Int. Cosmic Ray Conf., 18th, Bangalore.* In press
87. Ramaty, R. 1979. See Ref. 1, pp. 135–54
88. Ramaty, R. 1982. *NASA Tech. Memo. 83904.* Also see Ref. 106. In press
89. Ramaty, R., Crannell, C. J. 1976. *Ap. J.* 203: 766–68
90. Ramaty, R., Stone, R. G., eds. 1973. *High Energy Phenomena on the Sun.* Washington DC: NASA. 641 pp.
91. Ramaty, R., Kozlovsky, B. 1974. *Ap. J.* 193: 729–40
91a. Ramaty, R., Murphy, R. J. 1984. *Proc. Summer Workshop Astron. Astrophys.,* Santa Cruz, Calif. In press
92. Ramaty, R., Kozlovsky, B., Lingenfelter, R. E. 1975. *Space Sci. Rev.* 18: 341–88
93. Ramaty, R., Kozlovsky, B., Suri, A. N. 1977. *Ap. J.* 214: 617–31
94. Ramaty, R., Kozlovsky, B., Lingenfelter, R. E. 1979. *Ap. J. Suppl.* 40: 487–526
95. Ramaty, R., Colgate, S. A., Dulk, G. A., Hoyng, P., Knight, J. W. 1980. In *Solar Flares,* pp. 117–85. Boulder: Colo. Assoc. Univ. Press. 513 pp.

96. Ramaty, R., Lingenfelter, R. E., Kozlovsky, B. 1982. See Ref. 72, pp. 211–29
97. Ramaty, R., Murphy, R. J., Kozlovsky, B., Lingenfelter, R. E. 1983. *Sol. Phys.* 86: 395–408
98. Ramaty, R., Murphy, R. J., Kozlovsky, B., Lingenfelter, R. E. 1983. *Ap. J. Lett.* 273: L41–45
99. Reppin, C., Chupp, E. L., Forrest, D. J., Suri, A. N. 1973. *Proc. Int. Cosmic Ray Conf., 13th, Denver* 2: 1577–82
100. Rieger, E., Reppin, C., Kanbach, G., Forrest, D. J., Chupp, E. L., et al. 1984. *Proc. Int. Cosmic Ray Conf., 18th, Bangalore.* In press
101. Riegler, G. R., Ling, J. C., Mahoney, W. A., Prince, T. A., Wheaton, W. A., et al. 1982. *Ap. J.* 259: 392–96
102. Rosner, R., Chupp, E. L., Gloeckler, G., Krimigis, Y., et al. 1983. *Particle acceleration.* Presented at Coolfont Sol.-Terr. Phys. Workshop, Berkeley Springs, W. Va
103. Share, G. H., Forrest, D. J., Chupp, E. L., Rieger, E. 1983. *Bull. APS Meet., Baltimore* 28: 730 (Abstr.)
104. Share, G. H., Chupp, E. L., Forrest, D. J., Rieger, E. 1983. See Ref. 7, pp. 15–20
105. Stecker, F. W. 1971. *Cosmic Gamma Rays.* Baltimore: Mono Book Corp. 246 pp.
106. Sturrock, P. A., Holzer, T. E., Mihalas, D., Ulrich, R. K., eds. 1984. *Physics of the Sun.* Dordrecht: Reidel. In press
107. Suzuki, I., Kawabata, K. 1983. *Sol. Phys.* 86: 253–57
108. Svestka, Z., Schadee, A. 1983. *Sol. Phys.* 86: 267–77
109. Van Hollebeke, M. A. I. 1979. *Rev. Geophys. Space Phys.* 17: 545–51
110. Van Hollebeke, M. A. I., Ma Sung, L. S., McDonald, F. B. 1975. *Sol. Phys.* 41: 189–223
111. Vlahos, L., Alissandrakis, C., Brunel, F., Bai, T., Batchelor, D., et al. 1984. *SMM Workshop.* In preparation
112. Von Rosenvinge, T. T., Ramaty, R., Reames, D. V. 1981. *Proc. Int. Cosmic Ray Conf., 17th, Paris* 3: 28–31
113. Wang, H. T. 1975. *2.2 MeV and 0.51 MeV gamma-ray line emissions from solar flares.* PhD thesis. Univ. Md., College Park. 94 pp.
114. Wang, H. T., Ramaty, R. 1974. *Sol. Phys.* 36: 129–37
115. Wang, H. T., Ramaty, R. 1975. *Ap. J.* 202: 532–42
116. Wild, J. P., Smerd, S. F., Weiss, A. A. 1963. *Ann. Rev. Astron. Astrophys.* 1: 291–366

117. Willett, J. B., Ling, J. C., Mahoney, W. A., Riegler, G. R., Jacobson, A. S. 1982. See Ref. 72, pp. 401–8
118. Yoshimori, M., Okudaira, K., Hirasima, Y. 1983. *Nucl. Instrum. Methods* 215: 255–59
119. Yoshimori, M., Hirasima, Y., Okudaira, K. 1983. *Sol. Phys.* 86: 375–82
120. Yoshimori, M., Okudaira, K., Hirasima, Y., Kondo, I. 1984. *Proc. Int. Cosmic Ray Conf., 18th, Bangalore.* In press

Ann. Rev. Astron. Astrophys. 1984. 22 : 389–424

ORIGIN AND HISTORY OF THE OUTER PLANETS:
Theoretical Models and Observational Constraints[†]

James B. Pollack

Space Science Division, NASA-Ames Research Center,
Moffett Field, California 94035

1. INTRODUCTION

The planets of our solar system can be subdivided into two major classes: the inner or terrestrial planets, and the outer or Jovian planets.[1] The four inner planets—Mercury, Venus, Earth, and Mars—are made almost entirely of "rock," a combination of iron- and magnesium-bearing silicates and metallic iron. Volatiles, such as water, carbon, and nitrogen-containing compounds, constitute only a tiny fraction ($\sim 10^{-4}$ for Earth) of these planets' mass, with the volatiles being partitioned among their interiors (e.g. carbonate rocks), surfaces (e.g. water oceans and ice caps), and atmospheres. Carbon dioxide and nitrogen are major components of the atmospheres of the three terrestrial planets with significant atmospheres (Venus, Earth, and Mars); these gases are derived most likely from a partial degassing of their planet's interior (57). The decay of long-lived radionuclides (U^{238}, U^{235}, K^{40}, Th^{232}) is a major internal heat source for the interior of the inner planets at the current epoch. In all cases, the interior heat flux reaching the surface is several orders of magnitude smaller than the flux of incident sunlight. Finally, the terrestrial planets possess, in toto, only three natural satellites: Earth's moon plus the two tiny (~ 10 km) Martian moons, Phobos and Deimos. These moons, like their parent planets, are made of "rock."

[†] The US Government has the right to retain a nonexclusive royalty-free license in and to any copyright covering this paper.

[1] We exclude from this classification the outermost planet Pluto, which resembles a large satellite of the outer planets and may in fact be an escaped moon.

The Jovian planets—Jupiter, Saturn, Uranus, and Neptune—differ in fundamental ways from the terrestrial planets. Hydrogen and helium, in approximately solar elemental proportions, constitute the major components of their atmospheres. Furthermore, these atmospheres extend over a significant fraction of their planet's radius and represent a large fraction of their planet's mass (ranging from 10–20% for Uranus and Neptune to 80–95% for Jupiter and Saturn). Their inner portions, or cores, are made of a mixture of "rock" and "ice," with the latter term referring to a combination of volatiles, such as water, methane, and ammonia, in either a solid or liquid phase. The Jovian planets are one to two and a half orders of magnitude more massive than Earth, the most massive terrestrial planet. All the outer planets, except Uranus, radiate to space several times more energy than they absorb from the Sun. The implied excess or internal energy is believed to be derived from a combination of a cooling of their hot interiors and the gravitational separation of major constituents (30). Finally, satellites and rings are common features of the outer solar system. Altogether, the four Jovian planets possess 41 known moons, with their dimensions ranging from that of a small asteroid (~10 km) to greater than that of the planet Mercury. "Ices" as well as "rock" constitute major components of many of these moons. All the outer planets except Neptune are now known to have rings within their classical Roche tidal limits.

Any theory of the origin of the solar system must explain the origin of all the components of the solar system: the Sun, the two classes of planets, their satellite systems, and the independent small bodies—asteroids and comets. While a number of attempts have been made to do just this, none has been universally accepted so far. In view of the complexity of this task, it is useful to focus on one of the major components, particularly one ripe with clues on its origin. In this paper, I review our current understanding of the origin and subsequent development of the Jovian planets and their associated satellite and ring systems within the context of plausible models of the early solar system and modern astrophysical concepts. Stress is given not only to delineating alternative hypotheses but also to comparing their predictions against relevant observations.

2. OBSERVATIONAL CONSTRAINTS

In this section, I summarize relevant properties of the solar system that theories of the origin and development of the Jovian planets need to satisfy and hopefully predict. These properties include the structure and composition of these planets, their current internal heat fluxes, their magnetic fields, the nature of their satellites and rings, and properties of other parts of the solar system that may have been effected during their formation.

Information about the bulk composition of the outer planets can be derived by constraining static models of their interiors to agree with their observed mass, radius (and hence mean density), rotation rate, and gravitational moments J_2 and J_4. Such models are based on the equation of hydrostatic equilibrium, a very good approximation because of the very high pressure of the deep interiors (0.1–100 Mbar), their being in a fluid phase, and their having very low convective velocities. Owing to the large internal heat fluxes (except possibly Uranus), the internal temperature gradient is typically taken to be adiabatic: Conductive and radiative heat transfer cannot carry these fluxes with subadiabatic temperature gradients. A key to the ability to construct accurate interior models has been the development of good equations of state for the materials of interest at the high pressures and temperatures of these planetary interiors (69). However, further work is still needed in this area (e.g. the equation of state of molecular hydrogen at pressures of 0.1–2 Mbar).

In accord with modern cosmogonic concepts that are discussed later, three basic materials have been used to construct interior models of the giant planets: a solar mixture of elements dominated by hydrogen and helium ("gas"); water, methane, and ammonia with O, C, and N in solar proportions ("ice")[2]; and magnesium- and iron-containing silicates and metallic iron, with Mg, Fe, and Si in cosmic proportions ("rock"). By matching observed properties of the Jovian planets, such models provide estimates of the relative proportions of these three building blocks and their locations.

Figure 1 summarizes the interior structures of the outer planets derived in the above manner (69). Several points should be noted. First, in the cases of Jupiter and Saturn (but not Uranus and Neptune), pressures become high enough (>few megabars) in the outer "gas" envelope for molecular hydrogen to be dissociated and ionized and thus converted into metallic hydrogen. Second, for none of the giant planets is it possible to construct acceptable models that contain only "gas." Rather, in order to produce the observed mean density, "rock" and/or "ice" must be added, with the relative proportion of high Z material ("rock" and "ice") steadily increasing from Jupiter to Saturn to Uranus and Neptune. Note that simply segregating high-Z material from the "gas" envelope into an inner core would not produce an acceptable model for any of the outer planets. Third, in all cases, the J_2 gravitational moment implies an at least partial segregation of "high-Z" material from the "gas," with the former residing, in part, in an inner core.

Table 1 presents a quantitative summary of our current knowledge

---

[2] The word "ice" is used here as a description of composition, not phase. The ice component of the outer planets is probably in a fluid phase at the high temperatures of their interiors.

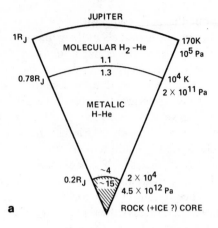

JUPITER

1R$_J$

MOLECULAR H$_2$-He

1.1

0.78R$_J$

1.3

METALIC
H-He

~4

0.2R$_J$ ~15

170K
10^5 Pa

10^4 K
2 × 10^{11} Pa

2 × 10^4
4.5 × 10^{12} Pa

a   ROCK (+ICE ?) CORE

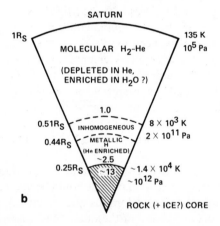

SATURN

1R$_S$

MOLECULAR H$_2$-He

(DEPLETED IN He,
ENRICHED IN H$_2$O ?)

1.0

0.51R$_S$ INHOMOGENEOUS

0.44R$_S$ METALLIC
H
(He ENRICHED)

~2.5

0.25R$_S$ ~13

135 K
10^5 Pa

8 × 10^3 K
2 × 10^{11} Pa

~1.4 × 10^4 K
~10^{12} Pa

b   ROCK (+ ICE?) CORE

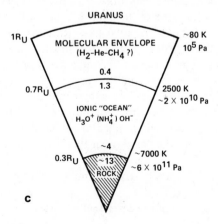

URANUS

1R$_U$

MOLECULAR ENVELOPE
(H$_2$-He-CH$_4$?)

0.4

0.7R$_U$

1.3

IONIC "OCEAN"
H$_3$O$^+$ (NH$_4^+$) OH$^-$

~4

0.3R$_U$ ~13
ROCK

~80 K
10^5 Pa

2500 K
~2 × 10^{10} Pa

~7000 K
~6 × 10^{11} Pa

c

concerning the interiors of the Jovian planets. A very fundamental property emerges from these data: The masses of the "gas" envelopes vary over about two orders of magnitude, while the masses of the "high-Z" material are the same within a factor of several. It has been a more difficult task to ascertain the relative proportion of "ice" and "rock" in the high-Z material and the degree to which the high-Z material is partially contained in the "gas" envelope. In the cases of Jupiter and Saturn, it has not been possible to separately evaluate the masses of the "ice" and "rock" components. But for Uranus and Neptune, where the "high-Z" material is the major component and is located (in part) closer to the surface, the models of Podolak & Reynolds (49, 50) imply that the mass ratio of "ice" to "rock" lies between 1 and 3.6, with the upper bound corresponding to the value expected from solar elemental abundances. The ability to determine this ratio is limited at the present time by the poorly known values of the rotation periods of Uranus and Neptune, a situation that is expected to be remedied in the near future by *Voyager* spacecraft observations and improved ground-based techniques. Improved periods will also permit meaningful estimates to be made of the partitioning of the "ice" component between the "gas" envelope and a separated inner zone.

Complementary information on giant planet composition (to that derived from interior models) comes from spacecraft and ground-based determinations of the composition of the uppermost part of the "gas" envelope, the observable atmosphere. These data are summarized in Table 1. According to this table, there may be several significant deviations in the compositions of these atmospheres from those expected for solar elemental abundances. First, while the He/H ratio for Jupiter is approximately solar, its value for Saturn appears to be smaller than both the value for Jupiter and the Sun. Since it is very difficult to conceive ways of fractionating H and He in the solar nebula, the source region for the giant planets, this result, if confirmed by further analyses of *Voyager* infrared spectra, implies a nonuniform mixing ratio of He with depth in Saturn's interior. As is discussed later, such a compositional gradient could be generated by the partial immiscibility of He in metallic H.

A second deviation from solar elemental abundances is manifested by the elevated C/H ratio for all four giant planets. Note that the factor of two enhancement in this ratio for Jupiter and Saturn is relative to Lambert's (35)

Figure 1   Schematic representation of typical interior models of Jupiter (*a*), Saturn (*b*), and Uranus (*c*). The numbers on the left side show distance from the center in units of the planet's radius, the numbers on the right side show temperature and pressure, and the numbers in the center show densities in g cm^{-3}. The corresponding interior model for Neptune is very similar to that of Uranus.

Table 1 Properties of the outer planets

Property	Jupiter	Saturn	Uranus	Neptune
Radius (km)	6.98×10^4	5.83×10^4	2.55×10^4	2.45×10^4
Mean density (g cm^{-3})	1.334	0.69	1.26	1.67
Total mass $(M_\oplus)$[a]	318.1	95.1	14.6	17.2
[b] Envelope mass $(M_\oplus)$[a]	288.1–303.1	72.1–79.1	1.3–3.6	0.7–3.2
[b] Core mass $(M_\oplus)$[a,c]	15–30	16–23	11–13.3	14–16.5
[d] Atmospheric composition:				
He/H	0.117 ± 0.025	0.062 ± 0.011	—	—
C/H[e]	$(1.09 \pm 0.08) \times 10^{-3}$	$(7.5\text{–}17.4) \times 10^{-4}$	$(1.9\text{–}9.4) \times 10^{-3}$	$(9.4\text{–}118) \times 10^{-4}$
N/H[f]	$(1\text{–}2) \times 10^{-4}$	$\sim 2 \times 10^{-4}$	$<1 \times 10^{-4}$	$<1 \times 10^{-4}$
D/H	$(2.0\text{–}4.1) \times 10^{-5}$	$(1.1\text{–}4.8) \times 10^{-5}$	$(3.3\text{–}6.3) \times 10^{-5}$	—
[g] Excess luminosity:				
Luminosity (W)	$(3.35 \pm 0.26) \times 10^{17}$	$(8.63 \pm 0.6) \times 10^{16}$	$<1.5 \times 10^{15}$	$(2 \pm 1) \times 10^{15}$
Luminosity/mass (W g^{-1})	$(1.76 \pm 0.14) \times 10^{-13}$	$(1.52 \pm 0.11) \times 10^{-13}$	$<2 \times 10^{-14}$	$(2 \pm 1) \times 10^{-14}$

[a] In units of the Earth's mass = 5.98×10^{27} g.
[b] Based on the models of Hubbard & MacFarlane (31) and Podolak & Reynolds (49).
[c] Includes all high-Z material in excess of solar elemental values, including high-Z material in the gas envelope.
[d] Based on data summarized by Gautier & Owen (19) and Hanel et al. (25). In all cases, ratios refer to mole fractions.
[e] Lambert's (35) value for the Sun is 4.7×10^{-4}.
[f] Lambert's (35) value for the Sun is 9.8×10^{-5}. The possible depletion of NH_3 in the atmospheres of Uranus and Neptune based on radio thermal emission has been called into question by a sizable variation in this emission for Uranus with orientation of the axis of rotation (24).
[g] Based on data summarized by Hanel et al. (26) and Hubbard (30).

value for the Sun, which is believed to be more accurate than a factor of 2. Also, the C/H ratio for Uranus and Neptune is poorly determined, in part, because temperatures in portions of their observable atmospheres fall below the condensation point of methane. The strong indication of an enhanced C/H ratio in all four atmospheres, with perhaps a greater enhancement for the atmospheres of Uranus and Neptune, implies a partial mixing of the C "ice" component from the core into the "gas" envelope. In the case of Uranus and Neptune at least, such a mixing may reflect the poor solubility of CH_4 in a hypothesized water ionic ocean [cf. Figure 1c (169)].

Finally, Table 1 implies that there is not a large difference in the D/H ratio for Uranus' atmosphere on the one hand and Jupiter's and Saturn's atmospheres on the other. Since it seems likely that convective motions in these planetary interiors and the high temperatures there would result in a homogenization of the D/H ratio between the cores and envelopes over the age of the solar system, the relative uniformity of the D/H ratio in these atmospheres may imply that there was no large fractionation of D between the source materials for the gas and ice components of these planets. (Recall that ice represents a significant fraction of Uranus' mass, but is at most a minor component of Jupiter and Saturn's mass.)

Both Jupiter and Saturn possess strong magnetic fields, while it is not yet known whether Uranus and Neptune have any. Jupiter's field is characterized by a dipole moment that is about a factor of 20 times larger than that of Saturn's; Jupiter's field has significant multipole moments, while Saturn's is much more dominated by its dipole component; and Jupiter's dipole is tilted by about 10° with respect to the planet's axis of rotation, while Saturn's dipole is aligned within 1° (65, 66). The first two differences might be attributed to the planets' magnetic fields being generated by a dynamo process within their metallic H regions: This region is deeper and encompasses a smaller fractional mass for Saturn than for Jupiter. As is discussed later, the final difference may provide additional support for a compositional gradient of He within Saturn's interior.

Table 1 also summarizes the internal heat flux that characterizes the four outer planets (30). In the case of Jupiter, Saturn, and Neptune, this heat flux at their surface is comparable to the amount of sunlight they are absorbing. The upper bound on Uranus' internal heat flux is equivalent to several tens of percent of the sunlight it absorbs. These fluxes place important boundary conditions on models of these planets' evolution. As such, they can help to set lower bounds on interior temperatures soon after the planets formed (the fluxes reflect a gradual cooling of the planets' interiors) and help to verify models that provide estimates of the planets' luminosity at times when their satellite systems were forming.

There are 16, 18, 5, and 2 known satellites in orbit about Jupiter, Saturn,

Table 2   Properties of the satellites of the outer planets[a]

	Jupiter	Saturn	Uranus	Neptune
Regular satellites:				
Number	8	17	5	1[c]
Distance[b]	1.8–26.6	2.3–59.0	5.1–23	14.6
Radius (km)	20–2630	15–2575	150–400	1600
Irregular satellites:				
Number	8	1	0	1[c]
Distance[b]	156–333	216	—	227
Radius (km)	5–85	110	—	150

[a] Based on data given by Stone & Miner (71), Morrison (46), and Morrison et al. (47).
[b] In units of planetary radii.
[c] Triton is assumed to be a regular satellite in view of its large mass and proximity to Neptune and despite its high orbital inclination and retrograde motion.

Uranus, and Neptune, respectively. Table 2 summarizes several properties of these satellite systems. Many of the satellites have orbits of low eccentricity and low inclination with respect to their planet's rotational equator, and they travel in prograde orbits about their planet. These "regular" satellites are therefore considered to have been generated in nebulae or accretion disks about their planets that may be analogous to the solar nebula within which the planets themselves formed. However, the outer 8 satellites of Jupiter and the outermost satellite of Saturn (Phoebe) and Neptune (Nereid) have highly eccentric and inclined orbits, and some of these "irregular" moons have retrograde orbits.[3] These orbital characteristics are suggestive of a capture origin.

The four large ($\sim$ several thousand kilometers diameter) Galilean satellites of Jupiter show a monotonic decrease in their mean density with distance from Jupiter, with the innermost Galilean moon, Io, having a mean density of 3.5 g cm^{-3} and the outermost one, Callisto, having a mean density of 1.8 g cm^{-3}. These densities imply that the inner two large moons are made almost entirely of "rock," while the outer two have amounts of water ice and rock similar to those expected from solar elemental abundances. This compositional gradient within the Jovian satellite system implies that temperatures in the regions of the Jovian nebula where the inner large satellites formed were sufficiently high throughout almost all of the satellite formation period to inhibit the condensation of any ices, including water ice (56).

[3] Strictly speaking, the massive inner satellite of Neptune, Triton, should also be considered an irregular satellite, since it has a retrograde orbit.

All the regular satellites of Saturn contain a significant ice component (e.g. 52). This conclusion is based on the low values of the mean densities of the larger Saturnian moons and the high reflectivity of the smaller ones (64). Also, near-infrared absorption bands of water ice are observed in the reflectivity spectra of their surfaces (12, 18). Furthermore, ices in addition to water ice (e.g. ammonia monohydrate, nitrogen, and methane clathrate) appear to have been present in the material that formed some of the satellites of Saturn. Direct evidence of these more volatile ices is provided by the nitrogen and methane components of Titan's dense atmosphere, and indirect evidence is supplied by the longevity of geological activity (e.g. tectonism, resurfacing events) on several of the other large moons and by the low albedo of a portion of Iapetus' surface (52). The above compositional inferences imply that temperatures in Saturn's primordial accretion disk were colder than those in Jupiter's during the period of satellite formation.

All of the moons of Uranus and Neptune whose infrared spectra have been measured have significant deposits of water ice on their surfaces (4). There is also evidence for surface and/or atmospheric methane and nitrogen on Triton, the larger of the two moons of Neptune (13). Thus, ices may constitute a significant fraction of these satellites.

The low visible albedo and visible and near-infrared spectral characteristics of the irregular satellites of Jupiter and Saturn imply that they are made of carbonaceous chondritic material, a type of meteoritic "rock" that is rich in carbon compounds and water of hydration (15). Thus, in addition to differences in orbital characteristics, the irregular satellites are apparently compositionally distinct from the ice-rich regular satellites.

Jupiter, Saturn, and Uranus have rings that lie interior to their satellite systems. However, searches for rings about Neptune have so far been unsuccessful (16). These rings are made of a myriad of tiny particles in independent orbit about their parent planet. By far, the most prominent and massive of these rings are those of Saturn. The particles of its rings are made almost entirely of water ice, i.e. the ice-to-rock ratio seems to be far in excess of that expected from solar elemental abundances (14). The rings of Saturn have a mass of about 6.4×10^{-8} Saturn masses, or about the mass of the intermediate-sized (~ 200 km radius) Saturnian moon Mimas (29).

In contrast to the broad main rings of Saturn—the A, B, and C rings—the nine known rings of Uranus are very narrow (width $\lesssim 100$ km), although they too have significant optical depths ($\sim$ unity; 17). Jupiter's ring system is very optically thin ($\sim 10^{-5}$) and consists of a thin, flat ring and a halo component that has a large thickness ($\sim 10^4$ km; 32). The composition of the particles in Jupiter's and Uranus' rings is not known.

Finally, certain characteristics of the inner solar system are perhaps pertinent to the outer solar system. The low mass of Mars, as compared with that of Earth and Venus, and the occurrence of many small-sized bodies in the asteroid belt, rather than a single large object, may reflect the gravitational influence of Jupiter during the early history of the solar system. Jupiter may have perturbed the orbits of planetesimals in the outer part of the inner solar system, and thus caused them to have high velocities with respect to one another that hindered their aggregation. If so, most of Jupiter's mass had to be accreted before the asteroids could accrete into a single object and before the formation of Mars was completed.

3. SOLAR NEBULA

In this section, I provide background information on the physical and chemical properties of the gas cloud within which the outer planets, as well as other parts of the solar system, formed and on the processes that controlled the evolution of this cloud. The latter subject may have relevance for the evolution of the nebulae of the outer planets, within which their regular satellites and perhaps their rings formed. The gas cloud of current interest is the solar nebula, the end product of a series of gravitational instabilities and contractions that isolated the primordial solar system from the giant molecular cloud of its parentage. The solar nebula (a) had dimensions comparable to that of the present solar system (i.e. tens to hundreds of astronomical units); (b) was a flattened disk, as implied by the angular momentum of the current planets; and (c) had a central concentration of mass, which would eventually become the Sun.

Although (for reasons to be shortly discussed) the properties of the solar nebula, including its mass and thermodynamic state variables, evolved with time, it is useful to bound their time-averaged values. A lower bound on the mass of this nebula, excluding that of the central object, can be obtained from the mass of the present-day planets. One augments each of their masses by an amount of hydrogen and helium sufficient to establish solar elemental abundances—the solar nebula presumably had the same composition as the Sun. In this way, a minimum mass on the order of $10^{-2} M_\odot$ is obtained, where $M_\odot$ is the Sun's present mass. An upper bound can be obtained by assuming that the mass of the central condensation was much less than 1 $M_\odot$, in which case an upper bound of 1 $M_\odot$ is derived. Much larger masses seem implausible (and perhaps energetically impossible!), since a means must be found of eliminating the extra mass from the solar system. Furthermore, more massive nebulae would probably be violently unstable (10).

Recent models of the structure and evolution of the solar nebula draw heavily on the physics of accretion disks about stars (e.g. 6, 39, 41). In such models, a prime role is assigned to turbulent viscosity. Because of this viscosity, there is a continual flux of mass inward in the inner part of the nebula, which is accreted onto the central object, and a flux of angular momentum outward, which expands the dimensions of the nebula. A characteristic lifetime for the nebula may be defined by the ratio of its mass to its mass flux. For models of the solar nebula in which turbulent viscosity follows an α-law[4] with $\alpha \sim 1$, this lifetime is on the order of 10^5 yr. Conceivably, however, the nebula's lifetime may also be influenced by continued accretion of mass from exterior portions of the subcondensation from which it formed and by the development of a "T Tauri" wind emanating from the primordial Sun.

Turbulent viscous dissipation may also act as the chief heating source for the nebula, in which case its midplane temperature T may have the following approximate dependence on radial distance r: $T \sim r^{-3/2}$ (39). The implied radial temperature gradient exceeds the adiabatic one ($T \sim r^{-1}$), but it is probably maintained by a rotational inhibition of radial convection. In such models, the luminosity of the central object is assumed to play a secondary role in directly (radiation) or indirectly (convection) heating the nebula. It is also assumed that heating due to mass accreted onto the accretion disk is of secondary importance.

Given the central role that turbulent viscosity plays in some theories of viscous accretion disks, it is most unfortunate that in many models it is essentially a free parameter. This state of affairs stems from two problems. First, the dominant source of the turbulence is not always known. However, Lin & Papaloizou (41) have proposed that vertical thermal convection is the chief source of turbulence in the solar nebula. But such a source of turbulence is dependent on there being a large optical depth in the vertical direction due to dust grains. The relatively rapid accretion of these grains to much larger objects could greatly reduce the particulate opacity and thus suppress this source of turbulence. Also, accretional heating of the nebula at its boundaries due to the continued infall of gas could inhibit vertical convection and generate its own brand of turbulence. A second problem area connected with turbulent viscosity has to do with calculating it, given a particular source of turbulence. Using a mixing length theory for thermal convection, Lin & Papaloizou (41) have made a crude estimate of this viscosity for their solar nebula model. Further progress has been achieved by Canuto et al. (8), who have developed an approximate method for determining turbulent viscosity based on the linear stability properties of turbulent flows.

The high-Z material of the cores of the outer planets was presumably

---

[4] $v = (1/3)\alpha V_s H$, where v, V_s, and H are viscosity, speed of sound, and nebula scale height, respectively.

derived from the condensed particulate matter in the solar nebula. Thus, the composition of this material is of considerable interest. Given a knowledge of the abundance of molecular species in the solar nebula and its temperature and pressure structure, the composition and abundance of condensed matter can be readily calculated by comparing the saturation vapor pressure of various particulate species with the partial pressure of their associated gas species. Three major compositional domains result in such calculations (36, 37): At temperatures between about 500 and 1500 K, high-temperature "rock" occurs, consisting of a mixture of metallic iron, troilite (FeS), and high-temperature silicates (e.g. enstatite). At temperatures between about 200 K and 500 K, lower-temperature "rock" appears, with this material being similar to ordinary and carbonaceous chondrites (i.e. this "rock" contains carbon compounds, water of hydration, and iron-bearing silicates). Finally, at temperatures below about 200 K, with the exact temperature depending on the nebular pressure, ices can condense. As the temperature decreases, first water ice forms, and at lower temperatures some of the condensed water combines with other nebular gases to form hydrates (e.g. $NH_3 \cdot H_2O$) and clathrates (e.g. $CH_4 \cdot 6H_2O$, $N_2 \cdot 6H_2O$, $CO \cdot 6H_2O$). It is not clear whether temperatures became cold enough in the solar nebula to reach the condensation points of *pure* gases other than H_2O.

For many years, it was thought that NH_3 and CH_4 were the dominant nitrogen- and carbon-containing gas species in the outer portion of the solar nebula where the outer planets formed. In such regions, where the temperature was less than ~ 500 K, local thermodynamic equilibrium implies a dominant role for NH_3 and CH_4. But Lewis & Prinn (38) have shown that the time scale for conversion of the high-temperature forms of N and C—N_2 and CO—into their low-temperature forms—NH_3 and CH_4— may exceed the lifetime of the solar nebula. This "kinetic inhibition" of low-temperature thermodynamic equilibrium is due in part to the low pressure of the solar nebula. According to Lewis & Prinn (38), a combination of the circulation of material between the inner and outer zones of the solar nebula and kinetic inhibition may result in a dominance of CO and N_2 in the colder, outer regions as well as the warmer, inner regions.

The chemistry of the C and N gases in the outer portion of the solar nebula could significantly affect the relative proportions of condensed "rock" and "ice" there, and thus the composition and mass of the outer planets' cores. First, the utilization of some O for CO means that the H_2O ice component has a smaller fractional abundance. Second, the occurrence of N_2 as the chief N species implies a much lower condensation temperature (~ 75 K vs. ~ 150 K) for the nitrogen-containing ice. Thus, water ice might have been the only major ice species throughout much of the outer solar nebula if kinetic inhibition resulted in N_2 and CO being the chief nitrogen-

and carbon-containing gases there. In this case, an ice-to-rock ratio of $\sim 1/2$ rather than ~ 3 might have characterized this region of the solar nebula (50). But, as discussed in Section 2, Neptune and Uranus have ice-to-rock ratios of 1–3.6. Therefore, either temperatures were very cold in the region of the solar nebula where they formed, so that N_2- and CO-containing ices could condense, or the kinetic time scale to convert N_2 and CO to NH_3 and CH_4 was much smaller in the solar nebula than has been estimated by Lewis & Prinn (38). Another possibility is that there was little exchange of material between hot and cold regions of the nebula, with NH_3 and CH_4 in the cold regions being derived from the protosolar cloud.

4. ORIGIN OF THE OUTER PLANETS

Gaseous and solid aggregates of matter are thought to form by different physical processes. Gaseous objects, such as stars, are the end products of one or several successive gravitational instabilities that occur in dense molecular clouds. In particular, a density perturbation (e.g. due to a shock wave) may cause the gravitational binding energy of a certain region of a cloud to exceed its thermal energy, in which case the region begins to contract (the "Jeans instability" criterion). Rotation, magnetic fields, and turbulent motions may also play a role in this initial instability, as well as in subsequent ones that may fragment the contracting object. By way of contrast, small solid objects, such as asteroids and comets, may form in part as the result of binary accretion events, in which the successive, gentle collisions of pairs of small "planetesimals" result in the development of a large solid body. However, even in this case, gravitational instabilities may be responsible for producing the smallest planetesimals, which begin the binary accretion process: Dust grains in a nebular disk may settle by sedimentation to the central plane of the nebula, so that the Jeans instability criterion is satisfied in the central plane. In this event, kilometer-sized planetesimals may result (21).

A very fundamental question concerning the formation of objects in the solar system is the mass at which they are no longer produced by the binary accretion of solid planetesimals, but rather by gas instabilities. The mass at which this transition occurs could lie between that of asteroids and terrestrial planets; or between that of terrestrial and Jovian planets; or between that of the Jovian planets and the Sun. Because the outer planets contain both a massive "high-Z" core ($\sim 15\ M_\oplus$) and an extensive "gas" envelope (~ 1–$300\ M_\oplus$), they could have originated by either process. If they formed as a result of a gas instability in the solar nebula, mechanisms need to be found for the gaseous protoplanet to acquire a high-Z core. Similarly, if they formed by binary accretion, the resultant massive core had to acquire a massive gaseous envelope. Below, we discuss the nature of these two types of models and compare their predictions with the observations of Section 2.

The occurrence of global gravitational instabilities within the solar nebula is dependent on the ratio of its mass, exclusive of that of the central condensation, to that of the central condensation. Cameron (6) estimates that for ratios in excess of about 0.1, the solar nebula is unstable to global azimuthal perturbations for radial scales comparable to the spacing between the planets. These global instabilities may result in the formation of rings of enhanced density. According to Cameron, as the gas density in these rings increases, they become susceptible to local Jeans-type instabilities that produce giant gas balls. Coalescence of giant gas balls in a given ring by binary accretion (or the elimination of some from the region of the ring by binary gravitational scattering) leads to the formation of a massive gaseous protoplanet. It should be noted that there is some uncertainty concerning the nature of the fastest-growing gravitational instability in the above hypothesized solar nebula. Spiral density waves might develop rather than rings (P. Cassen, private communication).

There are at least three ways in which a giant gaseous protoplanet might acquire a massive core of high-Z material. First, it might effectively capture by gas drag small planetesimals from the surrounding solar nebula (51). As is discussed in Section 6.1, such a mechanism may have been responsible for the capture of the irregular satellites. If so, a much larger amount of mass would have spiraled into the planet. Second, pressure and temperature conditions in the deep interiors of the evolving protoplanets may have liquified some of the solid grains in these regions, with the liquid droplets growing rapidly by coagulation and coalescence and becoming large enough to sediment rapidly to the protoplanet's center (62). In order for this process to lead to a substantial excess of high-Z material in the planet (in contrast to a mere redistribution in their location, which is inconsistent with interior models of the outer planets), there had to be an efficient resupply of grains from the solar nebula to the deep interiors of the protoplanets. Third, after such a sedimentation process had segregated most of the high-Z material into a central core, much of the "grain"-depleted envelope could have been lost by tidal stripping due to the developing Sun (6) or by thermal evaporation as the solar nebula heated up (7). While this latter mechanism has been evoked to explain the origin of the terrestrial planets (6), in which case almost all of the envelope is removed, it does not seem to be consistent with the trend of the core mass to envelope mass ratio increasing with distance from the Sun for the Jovian planets.

The alternative theory for the origin of the outer planets—the "core instability" hypothesis—postulates that large, solid objects formed by binary accretion in both the inner and outer regions of the solar nebula, resulting in the formation of the terrestrial planets in the inner zone and the high-Z cores of the Jovian planets in the outer zone. Because the solid objects in the outer solar system were over an order of magnitude more massive than their counterparts in the inner solar system, the former were able to permanently capture massive gas envelopes from the surrounding solar nebula, whereas the latter objects did not. The reasons for this difference in the mass of the large solid objects in the two zones is not clear. For example, ice as well as rock may have existed as a condensed phase in the colder, outer regions of the solar system. But this additional material would have augmented the core mass by only a factor between 1.5 and 4, depending on the form of the C and N gases in the outer solar system.

According to the core instability hypothesis, the ability of the cores to permanently capture a very large amount of gas from the solar nebula stemmed from the occurrence of a very rapid contraction of the gas gravitationally concentrated about them. Such a contraction would have two important effects: First, it would lead to a highly compacted envelope, which would therefore be far less susceptible than an extended envelope to loss by tidal stripping or changes in the pressure/temperature conditions in the solar nebula. Second, a massive amount of gas would rapidly enter the protoplanet's sphere of influence from the surrounding solar nebula, enabling the gas envelopes to reach masses of $\sim 1\text{--}300\ M_{\oplus}$ before the solar nebula dissipated.

Much of the work on the core instability model has centered about the determination of the critical or smallest core mass at which the gas envelope undergoes a rapid contraction. The first such calculation involved the construction of static core/envelope models that were in hydrostatic equilibrium. These models were assumed to have completely convective envelopes and were fitted to the pressure/temperature conditions of the surrounding solar nebula at the model's tidal radius, as determined by the mass of a fully formed Sun (48). Perri & Cameron (48) found that it was not possible to construct such models for a core mass in excess of a certain critical value and interpreted this result as implying that the envelopes associated with larger core masses would hydrodynamically collapse. They obtained critical core masses on the order of $100\ M_{\oplus}$, with these masses depending sensitively on the boundary conditions with the solar nebula. Both of these aspects of the critical core mass are incompatible with the values inferred for the giant planets—a nearly constant core mass of $15\ M_{\oplus}$.

Much more satisfactory predictions have been obtained in more recent calculations, in which the temperature structure of the gas envelope is calculated rather than assumed. The occurrence of a large zone in radiative equilibrium in the outermost portion of the envelope results in critical core masses on the order of 10 $M_\oplus$, with its exact value not depending sensitively on either the outer boundary conditions or the amount of grain opacity, the chief source of opacity in the cool outer envelope (45).

Unfortunately, there are two problems associated with the above calculations of the critical core mass. First and most important, the failure to construct a static model for core masses in excess of a critical value does not necessarily imply that the envelope undergoes a hydrodynamical collapse once the critical core mass is exceeded. Indeed, as the core mass grows by accretion and the envelope radiates to the surrounding solar nebula, its dimensions should change. Second, the outer boundary of the envelope need not necessarily be determined by tidal forces acting on it. Rather, it might be determined by the location at which the thermal energy of the gas has the same absolute magnitude as its gravitational binding energy ("accretionary" radius). At smaller values of protoplanet mass, the accretionary radius is less than the tidal radius and is therefore the appropriate one to use to define its outer boundary.

Very recently, Bodenheimer & Pollack (3) have performed calculations that address both of the above concerns. They studied the *evolutionary* growth of core/envelope configurations, in which the core mass grew with time, the envelope radiated to the solar nebula, and its mass increased in accord with boundary conditions determined by either the accretionary or the tidal radius, whichever was smaller. The envelope mass grew both because the outer boundary of the protoplanet increased as the core mass increased and because its radiative losses implied a steady contraction of constant mass surfaces within the envelope. They found that the envelope experienced a very rapid contraction but not a hydrodynamical collapse when the core mass reached a value of about 15 $M_\oplus$. This contraction occurred when the mass of the envelope became a significant fraction of that of the core, so that the rate of gravitational energy released by the contraction of constant mass surfaces in the envelope exceeded that associated with core accretion. The critical core mass was found to have a somewhat greater sensitivity to boundary conditions and other parameters than that obtained by Mizuno (45), due, in part, to the allowance made by Bodenheimer & Pollack (3) for the accretionary radius.

Once the very rapid contraction of the envelope occurs, its mass increases much more quickly than does the core mass: Gas from the surrounding solar nebula continually fills the protoplanet's sphere of influence and quickly contracts onto a now much more compact protoplanet. Safronov & Ruskol (59) have attempted to define the factors limiting subsequent growth of the protoplanet. These include the need to resupply gas from the more distant parts of the nebula once the nearby gas is depleted and the eventual dissipation of the solar nebula. However, perhaps an even more important constraint on the ultimate size of the outer planets, especially Jupiter and Saturn, stems from their gravitational interaction with the surrounding solar nebula. A protoplanet exerts a gravitational torque on the nebula that pushes the nebula gas away from it. When a protoplanet's mass becomes sufficiently large, such torques may create gaps on either side of it, thus essentially shutting off further accretion (e.g. 40, 73). Studies of this critical and interesting problem are just beginning.

We next compare the predictions of the gas instability and the core instability models with relevant properties of the solar system. The strongest point in favor of the "core instability" hypothesis is its prediction of core and envelope masses for the giant planets. As discussed above, a value of about 15 $M_\oplus$ has been found for the critical core mass at which its associated envelope first experiences a very rapid contraction. Furthermore, the critical core mass does not depend very sensitively on the outer boundary conditions of the envelope (i.e. its location in the solar nebula) or on the core accretion rate, whereas the ultimate envelope masses may vary by large factors, depending in part on the longevity of subsequent gas accretion and the density of the surrounding solar nebula. Thus, the core instability hypothesis predicts very well the magnitude of the core mass of the giant planets, as well as (in at least a qualitative fashion) the degree of variance of the core and envelope masses among the four outer planets. By way of contrast, the gas instability model does not as yet account for these properties, especially the core masses, in an obvious fashion.

The strongest point in favor of the gas instability model has to do with the relative times of formation of the outer and inner planets (68). If the outer planets formed by gas instabilities in the solar nebula, while Mars and the asteroids formed by binary accretion, it seems likely that at least the massive envelope of Jupiter would have formed before the accretion of these inner solar system objects was completed; therefore, through gravitational perturbations, Jupiter could have interfered with the growth of both types of objects. If so, an explanation would be provided for Mars' comparatively small mass (relative to Earth and Venus) and the multiplicity of asteroids. However, if the inner planets, the asteroids, and the cores of the outer planets formed by binary accretion, it is difficult to see how Jupiter could have formed first: The mass of Jupiter's core is very much greater than the mass of Mars or of the asteroids, and the volume density of planetesimals, like the gas density, was probably smaller in the outer solar system than in the inner solar system. Consequently, it should have taken longer to form Jupiter's core. If so, then during the formation of the objects in the inner solar system, proto-Jupiter's mass was much less than its current value, and thus it was far less effective in gravitationally perturbing its neighbors.

In principle, the compositional anomalies in the atmospheres of the giant planets (cf. Table 2) may provide additional tests of the two models of their formation. In practice, however, it has proven to be difficult to do this. For example, the elevated C/H ratios in the atmospheres of all four giant planets might be explained by either model. The core instability model might produce excess CH_4 in these atmospheres by a partial evaporation of carbon-containing ices of the planet's cores due to the high temperatures that occurred during the rapid contraction phase and/or the planet's subsequent evolution. The gas instability model might generate excess atmospheric CH_4 through the ablation of C ices and/or carbonaceous matter contained in captured planetesimals as they passed through the planets' envelopes on their way to the planets' developing cores.

5. EVOLUTION OF THE OUTER PLANETS

Numerical codes, patterned after those used to study stellar evolution, have helped to define the time histories of Jupiter and Saturn subsequent to their formation; much more simplified codes, analogous to those used to determine the cooling of white dwarf stars, have helped to evaluate past temperature conditions within Uranus and Neptune. In both cases, the present excess luminosity of these giant planets has served as an important constraint on these calculations.

Three major phases in the evolution of Jupiter and Saturn (and presumably also Uranus and Neptune) can be distinguished: an early quasi-hydrostatic phase in which the planet's radius varies slowly with time (phase 1); a collapse phase in which the planet rapidly contracts, eventually resuming a configuration that is nearly in hydrostatic equilibrium (phase 2); and a late, quasi-hydrostatic stage in which the planet slowly contracts to its present radius (phase 3). These three phases characterize both theories of the origin of the outer planets. But, the first two phases have rather different characteristics for the two origin theories. In the case of the "core instability" model, the protoplanet expands during phase 1 due to the increase in its mass and hence its outer boundary, while in the case of the "gas instability" model, the protoplanet most likely has an almost constant mass during this phase—at least for Jupiter and Saturn—and hence contracts. This contraction is driven by the protoplanet's luminosity and is analogous to the pre-main-sequence contraction of very young stars. Gravitational energy released from the contraction both steadily heats the interior and provides the radiation emitted by the protoplanet's photosphere to space.

During phase 1, the protoplanets have dimensions from ~ 2 to ~ 3 orders of magnitude larger than their current values, with the precise value depending on the mass of the early Sun at this time (it may not yet have been fully formed), the time elapsed since the start of this phase, and the mode of formation (2, 3). With regard to the first factor, the mass of the Sun helps to determine the tidal boundary of the protoplanet. For both modes of formation of the outer planets, interior temperatures steadily rise in their gas envelopes. In the case of the core instability model this increase is due chiefly to accretional energy during almost all of phase 1, while in the case of the gas instability model it is due chiefly to gravitational energy of contraction. A point is eventually reached for the gas instability models at which the temperatures in the deeper portions of the envelope exceed about 2500 K and molecular hydrogen dissociates. This dissociation initiates a hydrodynamical collapse, i.e. the start of phase 2. However, the envelopes of core instability models begin to experience a very rapid contraction, but not a hydrodynamical collapse at the end of phase 1 (see Section 4).

In the case of the gas instability model, the duration of phase 1 is set by the rate at which the protoplanet radiates to space part of the gravitational energy of contraction, i.e. it is a Kelvin-Helmholtz time scale. This time scale is about 4×10^5 yr for Jupiter and about 5×10^6 yr for Saturn. For this model, the radii of Jupiter and Saturn at the end of phase 1 are about $200\,R_J$ and $60\,R_s$, respectively, where R_J and R_s are the present radii of Jupiter and Saturn (2). In the case of the "core instability" model, the duration of phase 1 is set by the rate of accretion of core material. For this model, Jupiter and Saturn have radii that are several hundred to several thousand times their current dimensions at the end of phase 1, with this value being set by their tidal boundary and thus depending on the Sun's mass at this time (3).

For the gas instability model, phase 2, the hydrodynamical collapse stage, lasts for only ~ 0.25 and ~ 0.07 yr for Jupiter and Saturn, respectively, with this duration being comparable to a free-fall time scale (2). Collapse first ceases at the bottom of the protoplanet's envelope when the gas density there becomes sufficiently high (~ 1 g cm^{-3}) for it to become almost incompressible. Adjacent layers of the envelope fall on the static central region at progressively higher velocities, soon achieving supersonic values and thereby creating a shock wave near the outer boundary of the static region. At the end of phase 2, the protoplanet regains a quasi-hydrostatic configuration, with its radius equal to about $3\,R_J$ and $3.4\,R_s$ for Jupiter and Saturn, respectively (2, 53). In the case of the core instability theory, a much longer time may characterize phase 2, one determined by the continuing accretion of material (especially gas) from the solar nebula.

Figures 2a and b summarize the temporal behavior of models of Jupiter and Saturn, respectively, during phase 1 (left side of figure) and phase 2 (right side), according to the calculations of Bodenheimer et al. (2). These results pertain to the gas instability model. A rather different evolution characterizes the core instability model during these phases, as noted

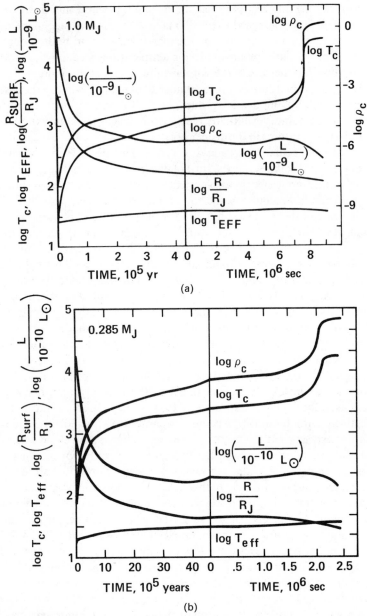

Figure 2 Properties of models of Jupiter (*a*) and Saturn (*b*) during phases 1 and 2 for origins involving gas instabilities. The quantities T_c, ρ_c, L, R, and T_{eff} refer to the central temperature, central density, internal luminosity, radius, and effective temperature, respectively, in units of K, g cm^{-3}, $10^{-9} L_\odot$, present radius, and K, respectively.

above. However, both models follow the same evolutionary path during phase 3.

During almost all of phase 3, the now fully formed outer planets have envelopes that are too incompressible for the gravitational energy released by continued contraction to compensate for the excess energy they radiate to space, i.e. the amount in excess of the amount of absorbed solar energy (22, 23, 55). Consequently, their excess or "internal" luminosity is balanced in large measure by the loss of internal energy built up during phases 1 and 2 and perhaps the earliest part of phase 3. Phase 3 has lasted almost the entire 4.6×10^9 yr age of the solar system and is continuing into the indefinite future. During phase 3, the radius of Jupiter and Saturn decreased by factors of several, with this contraction occurring much more rapidly during the early part of the phase.

Figure 3 illustrates the temporal behavior of Jupiter's and Saturn's internal luminosity during phase 3 (55). Although these results pertain to models having only envelopes, almost identical results were obtained for models containing cores as well (23). According to this figure, both planets had luminosities during the early portions of phase 3 that were several orders of magnitude higher than their current values. The potential importance of this early high luminosity for the composition of their regular satellites is discussed in the next section.

The square and circle in Figure 3 indicate the observed values of the

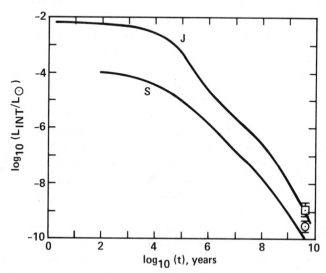

Figure 3   Intrinsic luminosity of Jupiter (J) and Saturn (S), in units of the Sun's luminosity, as a function of time during phase 3.

internal luminosities of Jupiter and Saturn, respectively, at the current epoch. The degree of agreement between theory and observation implied by this figure remains valid when core/envelope models are studied (23) and when recent spacecraft measurements of the planets' internal luminosities are used (25, 26): To within the error of observation (several tens of percent), the observed and predicted luminosities agree for Jupiter. Thus, the loss of internal energy from Jupiter's interior accounts for its current excess energy. However, the same model predicts a factor of 3 less internal luminosity for Saturn than is observed.

The success of the homogeneous contraction/cooling model for Jupiter and its lack of success for Saturn imply that Saturn alone has some additional internal energy source. However, only an energy source that is connected with the release of gravitational energy is quantitatively capable of maintaining Saturn's excess luminosity over a significant fraction of the age of the solar system. In particular, energy derived from the decay of long-lived radioactive nuclides in Saturn's core or from the accretion of interplanetary debris fails by orders of magnitude to produce the needed rate of energy release (30). Also, temperatures are too low in the interiors of any of the giant planets for nuclear fusion to be important (22). In the case of the homogeneous models discussed above, the present internal luminosity is ultimately derived from the rapid contraction of the planets in their early histories, which built up the internal energies that they are now dissipating. The only other way in which gravitational energy can be released in significant quantities is through a chemical differentiation of the planet, in which denser components sink toward its center and lighter ones rise toward its surface. In order to be able to generate a sufficient amount of energy, this chemical differentiation must involve the major constituents of Saturn. In particular, it must involve the segregation of He from H.

At sufficiently cool temperatures in the envelope of a giant planet, He may become partially immiscible in H, with helium-rich fluid elements forming and sinking toward the planet's center and hydrogen-rich fluid elements rising to take their place (60, 67). As illustrated in Figure 4, He immiscibility in Saturn's interior is expected to occur first in the outer region of the metallic H zone (60, 69, 70). When this occurs, helium-rich droplets nucleate and rapidly grow to a large enough size (~ 1 cm) to rapidly sink toward the planet's center, ultimately redissolving in deeper, unsaturated regions of the metallic H zone. Frictional dissipation of the droplets' fall velocity acts as the mechanism for converting gravitational energy into thermal energy. This process behaves like a servomechanism in that the interior cools only by an amount sufficient to cause the amount of He segregation needed to balance the excess energy radiated to space.

Initially, differentiation of He and H was proposed as the source of

Jupiter's excess luminosity (e.g. 60, 67). But it was later realized that this mechanism was more likely to be applicable to Saturn (55, 68). There are two reasons for this latter conclusion. First, the homogeneous cooling model worked for Jupiter, but not for Saturn. Second, the calculations of Stevenson & Salpeter (70) of the temperatures at which He first becomes immiscible in metallic H implied that temperatures in Jupiter's metallic H zone are too warm at the present epoch for phase separation to occur, but are at about the right temperatures in the outer part of Saturn's metallic H zone for such a separation to be occurring.

The above explanation of Saturn's current excess luminosity is consistent with several of the observations presented in Section 2. First, in order for such a segregation to supply an adequate amount of energy over the last several billion years, He needs to be freely exchanged between the more massive molecular portion of the envelope and the less massive outer fraction of the metallic H zone. Thus, the observable atmosphere should have about the same He mixing ratio as the upper part of the metallic H zone and should therefore have a He mixing ratio that is several tens of percent less than the solar value (55, 68). In fact, the *Voyager* measurements of the He/H ratio yield such a He deficiency in Saturn's observable

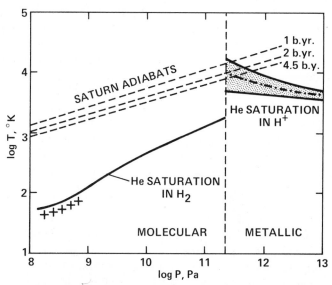

Figure 4   Saturation of helium in a cosmic mixture of elements as a function of pressure. At temperatures below the saturation temperature, helium partially separates from hydrogen. The dashed curves labeled "Saturn adiabats" show the run of temperatures through the planet's interior at various times from the beginning of phase 3.

atmosphere (cf. Table 2), although only at the $\sim 2\sigma$ level. A second piece of evidence in favor of He segregation may be given by the very small tilt of Saturn's dipole magnetic field with respect to the planet's axis of rotation. The gradient in the mean molecular weight of the gas in the outer part of the metallic H zone may strongly inhibit large-scale convection and the dynamo generation of the planet's magnetic field there. Consequently, differential rotation is possible there, and it might act to align the field (68).

However, the theoretical foundation for He immiscibility in Saturn's metallic H zone has been called into question by the recent thermodynamic calculations of MacFarlane & Hubbard (43). They find that He immiscibility first begins at temperatures that are about a factor of 10 smaller than the ones obtained by Stevenson & Salpeter (70), in which case Saturn's interior is too warm for He segregation to be occurring. The very large disagreement between these two calculations is a reflection of the very high accuracy with which the Gibbs free energy of H-He mixtures needs to be calculated ($\sim 0.1\%$) to achieve definitive predictions of the phase separation boundary (see 52).

Calculations of the evolutionary history of Uranus and Neptune are at a far less advanced level than those described above for Jupiter and Saturn. They have focused on the source of Neptune's excess luminosity and the lack of a comparable source for Uranus (cf. Table 1). Neptune's excess luminosity is much too large to be explained by the decay of long-lived radionuclides in the rock component of its core, with this component having chondritic elemental abundances of U, K, and Th relative to Si, Fe, and Mg (30). Trafton (72) suggested that this excess luminosity was produced by the tidal friction associated with the decay of the retrograde orbit of Neptune's massive satellite, Triton. But this seems unlikely, since the time for Triton's orbit to reach Neptune's surface in this case would be much less than the age of the solar system. As is the case for Jupiter, the most likely source of Neptune's excess energy is the loss of the planet's internal energy, which was derived from the release of gravitational energy in its early history (30). However, unlike Jupiter, most of Neptune's internal energy resides in its core rather than its envelope, simply because its core contains most of its mass.

The first law of thermodynamics can be used to calculate Neptune's and Uranus' excess luminosity as a function of time for various assumed initial temperatures (30). At least in the case of Neptune, the present and past internal luminosities are too high for molecular thermal conduction in its envelope and core to transport the implied fluxes for a subadiabatic temperature gradient; hence, adiabatic temperature profiles can be used (30). For such a situation, the rate of cooling is directly proportional to $T_e^4 - T_0^4$ and inversely proportional to the planet's mass and average heat capacity, where T_e and T_0 are the effective temperature of the planet's photosphere and the value of this temperature needed to radiate to space the amount of energy absorbed from the Sun. As T_e approaches T_0, the rate of internal cooling diminishes and, in the asymptotic limit of $T_e = T_0$, vanishes.

The difference between Uranus' and Neptune's present internal luminosities may be due to Uranus' closer distance to the Sun. This implies a higher value of T_0 and hence a shorter cooling time for Uranus beyond which solar heating partially shuts off interior cooling (30). Figure 5 illustrates quantitative cooling histories based on this concept (30). In accord with the large amount of gravitational energy released during the formation of Uranus and Neptune, initial temperatures within their interiors were assumed to be much higher than their current values. After cooling for a time equal to the age of the solar system, the two planets achieve effective temperatures comparable to their observed values (according to Figure 5). However, it is possible that these planets cooled somewhat faster than was the case for the time-invariant-composition models of this figure as a result of a partial dissolution of core material into their envelopes (69). Such a dissolution would move heavy material outward and therefore cause the gravitational energy of the planet to become *smaller* in absolute value. Hence, some of the energy of cooling might be used to compensate for this change in gravitational energy. Physically, such an energy transformation might take place by convective motions being used to carry core material aloft.

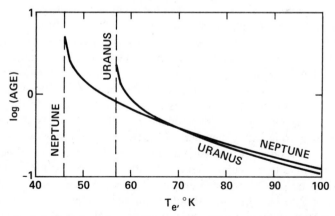

Figure 5   Cooling curves for Uranus and Neptune [log (Age) = 0 and −1 refer to the present epoch and the start of phase 3, respectively]. The vertical dashed lines show the planet's effective temperature T_e due solely to the absorption of sunlight.

6. ORIGIN OF THE SATELLITES OF THE OUTER SOLAR SYSTEM

In this section, we relate the origin of the irregular and regular satellites, as well as the rings, of the outer planets to models of these planets' origin and evolution, as discussed in Sections 4 and 5.

6.1 *Origin of the Irregular Satellites*

As discussed in Section 2, the orbital and compositional characteristics of the eight outermost satellites of Jupiter and the outermost satellite of Saturn and Neptune imply a capture origin: i.e. these moons, or their parent bodies, formed elsewhere in the solar system—presumably within the neighboring solar nebula—and were subsequently captured by their current planetary system. There are three major classes of capture theories: Lagrange point capture, collision between a stray body and a natural satellite, and gas drag capture.

According to the first of these theories—Lagrange point capture—a body in orbit about the Sun can be captured by a planet by passing through the interior Lagrange point at a low velocity relative to that of the planet (27). However, because the laws of motion are symmetric with respect to time, such capture is generally only *temporary*: The captured body will escape back out the Lagrange point in a time of the order of 100 orbital periods or less (28). Such capture could be made permanent if the Sun's mass decreased by several tens of percent or the planet's mass grew by a similar amount during the capture period. But such large changes in mass over such a short interval of time seem unlikely.

The collision model for the origin of the irregular satellites was developed for the irregular satellites of Jupiter, whose orbital semimajor axes and inclinations cluster into two groups: an inner prograde group and an outer retrograde group. According to this hypothesis, these two families formed as a result of the collision of a large stray body with an outer regular satellite of Jupiter (11). The large fragments of the regular satellite formed the prograde group, and the large fragments of the stray body formed the retrograde group. While this model successfully predicts the velocity dispersion among family members of the two groups, it has two problems. First, it is not obvious how such a model can be applied to Phoebe, the single, retrograde, irregular satellite of Saturn. Second, water ice does not appear to be present on the surfaces of either the irregular satellites of Jupiter or Phoebe, whereas it should be present on the surfaces of some of these bodies if their parent bodies were regular satellites: For both planetary systems, water ice is a major component of the surfaces and interiors of the outer, regular satellites.

The final theory—gas drag capture—postulates that the irregular satellites formed as independent bodies in the solar nebula and were captured by gas drag when they passed through their planet's "outer atmosphere" (33, 51, 61). In the modern version of this theory (51), such a capture is postulated to occur very close (within ~ 10 yr) to the end of phase 1, so that hydrodynamical collapse or very rapid contraction removes almost all the gas from the captured body's orbit soon after capture. Consequently, its orbit does not continue to decay until it is incorporated into the central body. Conversely, many more objects than the ones that end up as irregular satellites should have been incorporated into the giant planets, since the "window" for capturing an irregular satellite is short compared with the duration of phase 1 and since only the larger objects captured during this "window" would not be swept along with the gas during the collapse epoch.

The gas drag model correctly (but crudely) predicts several properties of the irregular satellites. First, the size of an irregular satellite can very roughly be estimated from the joint requirements that its velocity is diminished sufficiently in passing through the outer region of the protoplanet to permit capture (it must intercept a gas mass equal to at least 10% of its own mass) and that it not be dragged along with the gas during collapse. Based on Bodenheimer's (1) model of proto-Jupiter during phase 1, Pollack et al. (51) estimated that the irregular satellites so captured should have radii lying between about 0.1 and several hundred kilometers. The observed radii of the known irregular satellites lie between 5 and 150 km, consistent with this estimate. Second, the orbital distances of the irregular satellites should be comparable to the radius of the protoplanet at the inception of phase 2. For the gas instability model of the planets' origin, Jupiter and Saturn had radii of about 200 R_J and 60 R_s, respectively, at the start of hydrodynamical collapse. Somewhat larger sizes (factors of 2–5) characterize the size of Jupiter and Saturn at the end of phase 1 for the core instability model, with the exact value depending on the mass of the Sun at this time. (The protoplanet's outer boundary is set by tidal considerations.) The predictions of the latter model are in somewhat better agreement with the observed distances of the irregular satellites of Jupiter and Saturn (cf. Table 2). However, an in-depth analysis of the gas densities for the core instability model at the start of and during phase 2 is needed to determine how drag-free the environment of a captured object was at these times.

If for the moment we accept the gas drag theory as the correct one, then it has several interesting implications for the early solar system. First, the composition of the irregular satellites provides a guide to the temperature conditions in the solar nebula near the end of phase 1. In particular, temperatures at this time near Jupiter and Saturn may not have been cool enough (< 200 K) to permit the condensation of ice, but they may have been cool enough (< 500 K) to allow the formation of carbonaceous chondritic material. Second, as discussed above, a much greater mass of captured planetesimals was added to the planets than ended up as irregular satellites. The former may have helped enrich the planets' envelopes in high-Z materials during their passage to the planets' cores and, possibly, may have made a nontrivial contribution to the masses of the cores. Finally, the occurrence of a substantial volume density of planetesimals in the solar nebula near the end of phase 1—as implied by a probability of ∼ 1 for each outer planet capturing ∼ 1 object during a 10-yr interval—lends credence to the "core" instability model, but, of course, does not definitively prove the validity of this model.

6.2 Origin of the Regular Satellites and Ring Material

The regular satellites and the material that constitute their rings were formed by condensation and accretion processes in gaseous nebulae surrounding their parent planets. In a number of ways, these nebulae were similar to the solar nebula, with the outer planets playing the role that the early Sun played for the solar nebula. However, as discussed below, some important quantitative and even qualitative differences may have existed between these two classes of nebulae.

The present dimensions of the systems of regular satellites and rings (∼ several tens of planetary radii; cf. Table 2) imply that their nebulae of birth came into existence during phase 2 in the evolution of their parent planets, when the size of their planets changed from ∼ 100 R_p to ∼ several R_p (where R_p is the present size of their planets). These gaseous disks formed when the angular velocity in the outermost parts of the protoplanets' envelopes attained high enough values for centrifugal force to balance the planets' gravitational force.

Some indication of the radial distribution of angular momentum within the protoplanets at the time when their nebulae formed can be obtained from the distribution of angular momentum among the outer planets and their satellites at the current epoch. In contrast to the solar system, for which the orbits of the planets (especially those of Jupiter and Saturn) contain much more angular momentum than is present in the Sun's rotation, the reverse is true for the orbital motions of the satellites and the rotation of the outer planets. However, the specific angular momentum of the larger satellites is 1 to 2 orders of magnitude greater than that of their planets. This unequal distribution of specific angular momentum reflects the outward transfer of angular momentum that took place by viscous dissipation during phases 1 and 3. No such *transfer* was likely during the short-duration phase 2, but the increase in angular velocity that accompanied contraction during this phase led to the creation of the nebula disk.

Although, as discussed below, the nebula disks of the outer planets evolved with time, it is useful to consider initially static, "representative" models of them. "Ballpark" estimates of their density and pressure can be obtained by combining estimates of their mass and dimensions. A minimum mass may be found by adding a sufficient mass of H_2 and He to that of the satellites to obtain solar elemental ratios. In this way minimum masses of $\sim 10^{-2}\ M_p$ are found,[5] where M_p is the planet's mass (52). The mass of the parent planet represents an extreme upper limit to the mass of its nebula. The above bounds on nebula mass result in typical nebula densities of $\sim 10^{-5}$ to 10^{-3} g cm^{-3} and pressures of ~ 0.1 to 10 bars when the nebula mass is uniformly distributed throughout the current radial dimension of the satellite system, a temperature of ~ 100 K is assumed, and a vertical dimension that is 10^{-1} of the radial dimension (39) is used. These densities and pressures are several orders of magnitude larger than their counterparts for the solar nebula.

The above estimates of pressure in the nebulae of the outer planets have several interesting implications. First, kinetic inhibitions may have been far less severe in these nebulae than for the solar nebula for converting high-temperature, thermodynamic equilibrium species (e.g. CO and N_2) into their low-temperature counterparts (e.g. CH_4 and NH_3). In particular, the calculations of Prinn & Fegley (58) suggest that CH_4 and NH_3 were the dominant carbon- and nitrogen-containing species in the nebulae of the outer planets, although these nebulae may have also contained nontrivial, but minor, amounts of CO and N_2.

The high pressures of the nebulae and the domination of local thermodynamic equilibrium species constrain the composition and condensation temperatures of rocks and ices that helped to form the satellites and ring material of the outer planets. Figure 6 is such a phase diagram, in which the two dashed lines show possible adiabats in minimum and maximum mass nebulae of Saturn (52). At all pressures, water is included in the "ices" that condense at the highest temperatures. Note, however, that as the pressure approaches that of the maximum mass nebula, a liquid solution of water and ammonia rather than pure water ice becomes the first condensate. As the temperature decreases, ammonia monohydrate $(NH_3 \cdot H_2O)$ forms, and at still lower temperatures methane clathrate $(CH_4 \cdot 6H_2O)$ and other clathrates (e.g. $N_2 \cdot 6H_2O$, $Ar \cdot 6H_2O$, $CO \cdot 6H_2O$) form. Thus, most kinds of ices are produced by gas phase species combining with H_2O rather than forming pure ices of their own: Very low temperatures—probably lower than ones likely to be found in these nebulae—are required to produce pure ices of most species, such as CH_4. Thus, if the nebulae gradually cooled with time, so that water condensed

---

[5] However, the minimum mass for Neptune's nebula is $\sim 10^{-1}\ M_p$ based on Triton's mass.

first, the ability to incorporate other species into ice condensates may have been severely limited by the amount of small-sized water particles and the previous incorporation of other species into the water exposed on the surfaces of these particles.

The high pressures of the nebulae also imply that they were quite opaque to thermal radiation in both the radial and vertical directions because of grain or even just gas (e.g. pressure-induced H_2) opacity (42, 54). Consequently, thermal convection can be expected in at least the vertical direction. The resultant turbulent motions may have acted as the chief source of viscosity in these nebulae, by analogy to models of the solar nebula (41).

A key uncertainty about the nature of these nebulae involves the source of their heating and the resulting radial temperature gradient. On the one hand, the early high luminosity of the parent planets (see Figure 3) implies that their luminosity, in concert with grain and gas opacity, controlled the

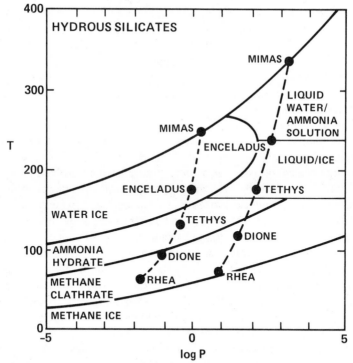

Figure 6   Phase boundaries for various rock and ice species as a function of temperature T (in K) and pressure P (in bars). The left and right dashed lines show hypothetical adiabats for "minimum" and "maximum" mass nebulae of Saturn, respectively.

temperature structure of these nebulae. If so, an adiabatic temperature structure ($T \sim r^{-1}$) can be expected in the inner part of the nebula, with the temperature eventually becoming subadiabatic in the outer part and joining smoothly onto the planet's photospheric temperature and that of the neighboring solar nebula at its inner and outer boundaries, respectively (54). On the other hand, as is thought to be the case for viscous accretion disks and the solar nebula, viscous dissipation may be the key source of heating, in which case an even stronger temperature gradient ($T \sim r^{-3/2}$) may be established in the inner region of the nebula (39). In this case, the temperatures in the innermost part of the nebula need not smoothly join onto that of the planet's photosphere.

Several qualitative trends in nebula temperature, and hence in the composition of satellite-forming condensates, characterize both sources of nebula heating. Temperatures are expected to monotonically decrease with increasing distance from the planet and, at a fixed distance, to monotonically decline with time (39, 54). Also, higher temperatures are expected within the more massive nebulae of the more massive outer planets, either because of the higher luminosity of the planet ($L \sim M_p^2$) or the higher viscous dissipation rates (39, 54). Thus, the increase in the bulk water ice to rock ratio from the inner to the outer large moons of Jupiter can be attributed to warmer temperatures and hence to an inhibition of water vapor condensation in the region of the Jovian nebula where the inner satellites formed. The presence of significant quantities of water ice in the innermost satellites of Saturn and in its rings, in contrast to the situation for Jupiter, implies cooler temperatures in the inner part of Saturn's nebula than in Jupiter's at the time of solid body formation. This deduction is consistent with differences in nebula temperature expected from differences in the mass of the two planets. The occurrence of ices other than water ice in some of the satellites of Saturn (e.g. Titan) is also consistent with Saturn having a cooler nebula. The composition of the regular satellites of Uranus and Neptune is also consistent—to the degree that their composition is known—with the expectation that their nebulae were even colder than Saturn's: As discussed in Section 2, water ice is an abundant material on the surfaces of the satellites of Uranus, and ices other than water ice appear to be present on Triton, Neptune's largest satellite.

In principle, the above compositional information can be used to set bounds on the duration of satellite formation and hence presumably the lifetime of the nebulae of the outer planets. However, in order to do this, it is necessary to choose a particular heating source for the nebulae. Figures 7a and b illustrate the time history of temperatures at the location of some regular satellites of Jupiter and Saturn, respectively, for optically thick nebulae heated by the luminosity of the central planet (54). The ice-to-rock ratio among the Galilean satellites implies that satellite formation in the Jupiter system occurred during the first $\sim 10^5$ yr of phase 3, while the presence of substantial quantities of water ice in the rings of Saturn and the likely corporation of methane clathrate into Titan imply minimum lifetimes for Saturn's nebula of $\sim 10^7$ and $\sim 10^6$ yr, respectively (52, 54).

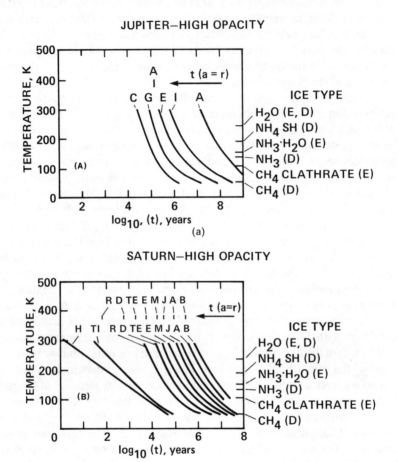

Figure 7   Temperature of a condensing ice grain as a function of time from the start of phase 3 for nebulae of Jupiter (*a*) and Saturn (*b*). Each curve refers to a fixed distance from the center of the planet and is labeled by the first letter of the name of the satellite or ring segment that is currently at that distance. The right-hand vertical scale denotes the temperatures at which various ice species condense under equilibrium (E) or disequilibrium (D) conditions between the gas and solid phases. Abbreviations for Jupiter: A—Amalthea, I—Io, E—Europa, G—Ganymede, C—Callisto. Abbreviations for Saturn: A—A ring, B—B ring, J—Janus, M—Mimas, E—Enceladus, Te—Tethys, D—Dione, R—Rhea, Ti—Titan, H—Hyperion.

If the material that makes up the rings of Saturn formed at the present distance of the main rings, a lower bound on the lifetime of Saturn's nebula can be derived that is independent of the identity of the chief heating source for the nebula. The planet's radius had to decrease sufficiently so that the planet was no longer located in the region of the rings: At the start of stage 3, the planet extended beyond the current region of the main rings. If we use the criterion that the planet needed to be situated entirely inside the inner edge of the B ring, a minimum nebula lifetime of $\sim 5 \times 10^5$ yr is derived (54).

The above estimates of nebula lifetime appear to be inconsistent with that expected from simple models of viscous accretion disks: Because of viscous dissipation and the accompanying inward transfer of nebula mass toward the central object, the solar nebula is expected to have a lifetime of $\sim 10^5$ yr (39). An even shorter lifetime might be expected for the less massive nebulae of the outer planets from simple scaling considerations. However, two factors may act to extend the lifetimes of these nebulae. First, thermal convection and hence viscosity may diminish as grain opacity is substantially reduced by the accretion of grains into large bodies. Second, there may have been a continual resupply of gas and grains to the nebulae of the outer planets from the surrounding solar nebula as long as the latter was present.

7. SUMMARY

Arriving at definitive answers to such fundamental questions as the origin of the Jovian planets and their satellite systems is an extremely difficult task. Nevertheless, some progress has been made, partly because of the application of modern astrophysical concepts to these problems and partly because of the derivation of key observational constraints.

The determination that the core masses of the outer planets are the same within a factor of several, while their envelope masses differ by about two orders of magnitude, has provided solid evidence in favor of the "core instability" model for their origin. However, it would be premature to say that the matter has been settled in favor of this model. For example, the "gas instability" model more readily allows Jupiter to form before Mars is fully formed and before the accretion of material in the asteroid belt is completed, thereby allowing Jupiter to influence the final phases of formation in these nearby regions of the inner solar system. Hopefully, as more information is gained on the deviations in atmospheric composition of the outer planets from solar elemental abundances (particularly for Uranus and Neptune), additional tests of the two origin models can be devised.

The release of gravitational energy—in the present or past—is the ultimate source of the excess luminosity of Jupiter, Saturn, and Neptune. (Recall that only an upper bound exists on Uranus' excess luminosity.) However, only for Jupiter is there a detailed, uncontroversial quantitative understanding of the relationship between the present excess luminosity and the gravitational energy release. In this case, the present excess is derived from a loss of internal energy that was built up during rapid contractions that characterized phases 1 and 2 and the very earliest portions of phase 3.

This same model falls a factor of 3 shy of reproducing Saturn's excess luminosity. Most likely, Saturn's excess luminosity is derived from the release of gravitational energy produced by the downward-directed segregation of helium in the planet's interior due to its partial immiscibility in the outer portion of the metallic hydrogen region. An apparent deficiency of He in Saturn's observable atmosphere and, more indirectly, the near alignment of its magnetic field with its rotational axis support this explanation. However, the thermodynamic basis for He immiscibility occurring at plausible interior temperatures has been questioned.

Finally, the difference in the excess luminosities of Uranus and Neptune can be attributed to the larger solar input to Uranus' atmosphere acting to "choke off" interior cooling. A simple cooling model for these two planets leads to results in approximate accord with observations for initially "hot" interiors. But *upward* gravitational migration of heavy core material into their envelopes and the attendant more rapid cooling of these planets are not excluded by current observations.

On only some relatively general and qualitative matters is there a fairly unambiguous understanding of the origin of the satellites and ring material of the Jovian-type planets. Presumably, the regular satellites and ring material formed within nebula disks that developed from the outermost layers of the planets as they underwent a large contraction during phase 2. Condensation of nebular gases and the accretion of this material with itself and with preexisting grains led to the formation of a small number of large objects—the satellites. Incomplete accretion and/or later disaggregation of small moons led to the production of rings near most of the giant planets. Decreasing temperatures with increasing distance from the primary led to the preferential condensation of more volatile gases at greater distances; thus, for example, water ice was more available for incorporation into the outer large moons of Jupiter. Also, cooler temperatures are expected for the nebulae of the less massive planets. Hence, ices other than water ice were available for incorporation into some of the satellites of Saturn, but they may not have been available for those of Jupiter.

Detailed modeling of the structure and evolution of the nebulae of the outer planets is in its infancy. Such basic questions as the source of nebula heating—viscous dissipation vs. planetary luminosity—and the lifetimes of these nebulae remain to be answered.

The irregular satellites of the outer planets may have formed within the solar nebula and have been captured by the outer planets when they passed through their extended envelopes and suffered gas drag close to the end of phase 1. The subsequent hydrodynamical collapse or rapid contraction of the planets removed gas from the vicinity of these objects, thus allowing them to remain satellites. If this hypothesis is correct—it is far from proven—irregular satellites may provide a guide as to temperature conditions within the solar nebula, as well as the volume density of 10–100-km-sized planetesimals.

Literature Cited

1. Bodenheimer, P. 1977. *Icarus* 31:356–68
2. Bodenheimer, P., Grossman, A. S., DeCampli, W. M., Marcy, G., Pollack, J. B. 1980. *Icarus* 41:293–308
3. Bodenheimer, P., Pollack, J. B. 1984. *Icarus*. In press
4. Brown, R. H., Clark, R. N. 1983. *Bull. Am. Astron. Soc.* 15:856
5. Deleted in proof
6. Cameron, A. G. W. 1978. *The Moon and the Planets* 18:5–40
7. Cameron, A. G. W., DeCampli, W. M., Bodenheimer, P. 1982. *Icarus* 49:298–312
8. Canuto, V. M., Goldman, I., Hubickyj, O. 1984. *Phys. Rev. Lett.* In press
9. Deleted in proof
10. Cassen, P., Moosman, A. 1981. *Icarus* 48:353–76
11. Columbo, G., Franklin, F. A. 1971. *Icarus* 15:186–91
12. Cruikshank, D. P. 1979. *Rev. Geophys. Space Phys.* 17:165–76
13. Cruikshank, D. P., Brown, R. H., Clark, R. N. 1983. *Bull. Am. Astron. Soc.* 15:857
14. Cuzzi, J. N., Pollack, J. B., Summers, A. L. 1980. *Icarus* 44:683–705
15. Degewij, J., Cruikshank, D. P., Hartmann, W. K. 1980. *Icarus* 44:541–47
16. Elliot, J. L., Mink, D. J., Elias, J. H., Baron, R. L., Dunham, E., et al. 1981. *Nature* 294:526–29
17. Elliot, J. L., Nicholson, P. D. 1984. In *Planetary Rings*. Tucson: Univ. Ariz. Press. In press
18. Fink, U., Larson, H. P., Gautier, T. N. III, Treffers, R. R. 1976. *Ap. J. Lett.* 207:63–68
19. Gautier, D., Owen, T. 1983. *Nature* 304:691–94
20. Deleted in proof
21. Goldreich, P., Ward, W. 1973. *Ap. J.* 183:1051–61
22. Graboske, H. C. Jr., Pollack, J. B., Grossman, A. S., Olness, R. J. 1975. *Ap. J.* 199:265–81
23. Grossman, A. S., Pollack, J. B., Reynolds, R. T., Summers, A. L., Graboske, H. C. 1980. *Icarus* 42:358–72
24. Gulkis, S., Olsen, E. T., Klein, M. J., Thompson, T. J. 1983. *Science* 221:453–54
25. Hanel, R., Conrath, B., Flasar, F. M., Kunde, V., Maguire, W., et al. 1981. *Science* 212:192–220
26. Hanel, R. A., Conrath, B. J., Kunde, V. G., Pearl, J. C., Pirraglia, T. A. 1983. *Icarus* 53:262–85
27. Heppenheimer, T. A. 1975. *Icarus* 24:172–80
28. Heppenheimer, T. A., Porco, C. 1977. *Icarus* 30:385–401
29. Holberg, J. B., Forrester, W. T., Lissauer, L. T. 1982. *Nature* 297:115–20
30. Hubbard, W. B. 1980. *Rev. Geophys. Space Phys.* 18:1–9
31. Hubbard, W. B., MacFarlane, T. T. 1980. *J. Geophys. Res.* 85:225–34
32. Jewitt, D. C. 1982. In *Satellites of Jupiter*, ed. D. Morrison, pp. 44–64. Tucson: Univ. Ariz. Press. 972 pp.
33. Kuiper, G. P. 1951. *Proc. Natl. Acad. Sci. USA* 37:717–21
34. Deleted in proof
35. Lambert, D. 1978. *MNRAS* 182:249–72
36. Lewis, J. S. 1972. *Earth Planet. Sci. Lett.* 15:286
37. Lewis, J. S. 1972. *Icarus* 16:241–52
38. Lewis, J. S., Prinn, R. G. 1980. *Ap. J.* 238:357–64

39. Lin, D. N. C. 1981. *Ap. J.* 246:972–84
40. Lin, D. N. C., Papaloizou, J. 1979. *MNRAS* 186:799–812
41. Lin, D. N. C., Papaloizou, J. 1980. *MNRAS* 191:37–48
42. Lunine, J. I., Stevenson, D. J. 1982. *Icarus* 52:14–39
43. MacFarlane, J. L., Hubbard, W. B. 1984. *Ap. J.* In press
44. Deleted in proof
45. Mizuno, H. 1980. *Prog. Theor. Phys.* 64:544–57
46. Morrison, D. 1982. In *Satellites of Jupiter*, ed. D. Morrison, pp. 3–43. Tucson: Univ. Ariz. Press. 972 pp.
47. Morrison, D., Cruikshank, D. P., Burns, J. A. 1977. In *Planetary Satellites*, ed. J. A. Burns, pp. 3–17. Tucson: Univ. Ariz. Press. 598 pp.
48. Perri, F., Cameron, A. G. W. 1974. *Icarus* 22:416–25
49. Podolak, M., Reynolds, R. T. 1981. *Icarus* 46:40–50
50. Podolak, M., Reynolds, R. T. 1984. *Icarus*. In press
51. Pollack, J. B., Burns, J. A., Tauber, M. E. 1979. *Icarus* 37:587–671
52. Pollack, J. B., Consolmagno, G. 1984. In *Saturn*, ed. T. Gehrels. Tucson: Univ. Ariz. Press. In press
53. Pollack, J. B., Fanale, F. 1982. In *Satellites of Jupiter*, ed. D. Morrison, pp. 872–910. Tucson: Univ. Ariz. Press. 972 pp.
54. Pollack, J. B., Grossman, A. S., Moore, R., Graboske, H. C. Jr. 1976. *Icarus* 29:35–48
55. Pollack, J. B., Grossman, A. S., Moore, R., Graboske, H. C. 1977. *Icarus* 30:111–28
56. Pollack, J. B., Reynolds, R. T. 1974. *Icarus* 21:248–53
57. Pollack, J. B., Yung, Y. L. 1980. *Ann. Rev. Earth Planet. Sci.* 8:425–87
58. Prinn, R. G., Fegley, B. Jr. 1981. *Ap. J.* 249:308–17
59. Safronov, V. S., Ruskol, E. L. 1982. *Icarus* 49:284–96
60. Salpeter, E. E. 1973. *Ap. J. Lett.* 181:L83–86
61. See, T. J. J. 1910. *Researches on the Evolution of the Stellar Systems: The Capture Theory of Cosmical Evolution*, Vol. 2. Lynn, Mass: R. P. Nichols & Sons
62. Slattery, W. L., DeCampli, W. M., Cameron, A. G. W. 1980. *The Moon and the Planets* 23:381–90
63. Deleted in proof
64. Smith, B. A., Soderblom, L., Batson, R., Bridges, P., Inge, J., et al. 1982. *Science* 215:504–37
65. Smith, E. J., Davis, L. Jr., Jones, D. E. 1976. In *Jupiter*, ed. T. Gehrels, pp. 788–829. Tucson: Univ. Ariz. Press. 1254 pp.
66. Smith, E. J., Davis, L. Jr., Jones, D. E., Coleman, D. J. Jr., Colburn, D. S., et al. 1980. *J. Geophys. Res.* 85:5655–74
67. Smoluchowski, R. 1967. *Nature* 215:691–95
68. Stevenson, D. J. 1980. *Science* 208:746–48
69. Stevenson, D. J. 1982. *Planet. Space Sci.* 30:755–64
70. Stevenson, D. J., Salpeter, E. E. 1977. *Ap. J. Suppl.* 35:221–37
71. Stone, E. C., Miner, E. D. 1982. *Science* 215:499–504
72. Trafton, L. 1974. *Ap. J.* 193:477–80
73. Ward, W. R. 1983. *Bull. Am. Astron. Soc.* 15:811

Ann. Rev. Astron. Astrophys. 1984. 22 : 425–44

THE ORIGIN OF ULTRA-HIGH-ENERGY COSMIC RAYS

A. M. Hillas

Physics Department, University of Leeds, Leeds LS2 9JT, England

1. WHY BOTHER WITH ULTRA-HIGH-ENERGY COSMIC RAYS?

Protons, nuclei, and electrons in the energy range 10^9–10^{12} eV account for most of the energy of the cosmic-ray flux and have attracted most attention. Plausible models have been proposed for their acceleration (popularly in supernovae), of their entanglement by galactic magnetic fields, and of their eventual escape from the Galaxy ("leaky box" models). However, the energy spectrum of cosmic rays extends to $\sim 10^{20}$ eV (and *smoothly* to 10^{19} eV). This poses a challenge to these models, because (*a*) the ultra-high energies of these particles rule out most of the accelerating mechanisms that have been discussed, and (*b*) such particles cannot be retained in the disk of our Galaxy by its magnetic fields. Regrettably, the observations also cast doubt on the importance of the most popular shock-wave process for accelerating particles to ultrarelativistic energies.

In most respects, a very firm body of data on these ultra-high-energy cosmic rays now exists, and so it may be hoped that more attention will be paid to its interpretation. However, there are many problems to be solved.

SIZE OF ACCELERATING REGION   The Larmor radius of a relativistic particle of charge Ze in a magnetic field $B_{\mu G}$ (strictly the component of B normal to the particle's velocity) is $r_{\mathrm{L}} = 1.08\, E_{15}/ZB_{\mu G}$ pc, where E_{15} is the particle's energy in units of 10^{15} eV, and $B_{\mu G}$ is in microgauss. Clearly, in gradual modes of acceleration, where the particle makes many irregular loops in the field while gaining energy, the size L of the essential part of the accelerating region containing the field must be much greater than $2r_{\mathrm{L}} \sim 2E_{15}/B_{\mu G}$. In

425

0066–4146/84/0915–0425$02.00

fact, a characteristic velocity βc of scattering centers is of vital importance, and it turns out (Section 3) that L has to be larger than $2r_L/\beta$, so

$$B_{\mu G}L_{pc} > 2E_{15}/Z\beta, \qquad\qquad 1.$$

where L_{pc} is in parsecs. This limitation arises also in one-shot acceleration schemes, where an emf $\sim LvB/c$ (cgs) arises from the motion of a conductor (speed $v = \beta c$) in a magnetic field and may be partly available for particle acceleration (L may be the diameter of a rotating neutron star, for instance).

In Figure 1 are plotted many sites where particle acceleration may occur, with sizes ranging from kilometers to megaparsecs. Sites lying below the diagonal line fail to satisfy condition (1.), even for $\beta = 1$, for 10^{20} eV protons (the dashed line refers to 10^{20} eV iron nuclei): for more reasonable plasma velocities in the range $c > v > 1000$ km s^{-1}, the line will lie even higher, somewhere within the stippled band. Clearly, very few sites remain as possibilities: either one wants highly condensed objects with huge B or enormously extended objects. In either case, very high speeds are required. Among the excluded sites are supernova remnant envelopes.

ARRIVAL DIRECTIONS   Many particles having energy $> 3 \times 10^{19}$ eV have been reported, and many of these arrive from directions very far from the galactic plane (30), as is shown in Figure 2, which depicts a section through the Galaxy. If these particles have been deflected from sources within the active regions of our Galaxy, we require a magnetic field of somewhat

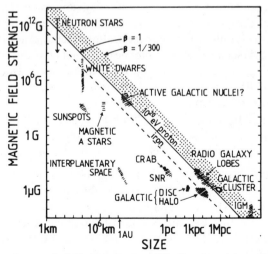

Figure 1   Size and magnetic field strength of possible sites of particle acceleration. Objects below the diagonal line cannot accelerate protons to 10^{20} eV.

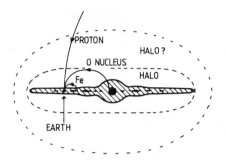

Figure 2   Size of the trajectories of 7×10^{19} eV cosmic-ray nuclei in relation to the Galaxy in a 2-μG magnetic field (assumed to be almost uniform).

implausibly constant direction over a large volume. This field could hardly be stronger than 2 μG (rather like that in the local galactic disk); in Figure 2, the typical energy of the sample is taken to be 7×10^{19} eV, and the trajectories of protons and oxygen and iron nuclei in such a field (normal to the diagram) are shown. Protons would clearly originate outside our Galaxy, and the arrival direction points roughly from the Virgo cluster of galaxies, 15–20 Mpc away (though Southern Hemisphere observers may not be much impressed by this remark). Only if the particles were all more highly charged nuclei and if the magnetic halo of our Galaxy extended to several kiloparsec (as shown) could the particles originate in our Galaxy. The evidence suggests that at least some of the particles are protons (97), but their identification is not easy. Such an identification is of critical importance. If they were to turn out to be entirely highly charged nuclei, we might consider whether young pulsarlike objects could possibly be sources; otherwise, we have to look outside the Galaxy.

2. OBSERVATIONAL DATA

Particles of energy $> 10^{15}$ eV are detected through the extensive cascades of secondary particles that they generate in the atmosphere ("extensive air showers"). These are observed by large arrays of particle detectors on the ground. The largest of these arrays viewing the northern sky have been at Volcano Ranch (USA), Haverah Park (England), and Yakutsk (USSR); another array, Chacaltaya (Bolivia), operates near the equator. The southern sky has as usual been somewhat neglected, but the observers at Sydney (Australia) had the largest exposure of all at the highest energies. In the effort to gather statistics on 10^{20} eV particles, the global exposure to date is ~ 500 km^2 yr.

2.1 *Energy Spectrum and Composition of High-Energy Particles*

ENERGY SPECTRUM   Figure 3 shows how the particle flux at very high energies continues the spectrum that has been well explored below 10^{12} eV. The differential flux $J(E)$ (particles m^{-2} s^{-1} sr^{-1} GeV^{-1}) is plotted as a function of energy E, with J multiplied by $E^{2.5}$ (E in GeV), thus disguising a very steeply falling spectrum. Particle species are not distinguished, as only the total energy given to the shower can be estimated. A few experiments spanning a broad energy range have been selected, and a critical assessment of the data is not attempted. (Some other data suggest that the flux may be slightly higher than shown near 10^{15} eV.) The low Sydney flux does not indicate a real difference in the southern sky, but reflects instead their use of the number of muons (N_μ) to measure shower size. Conversion from N_μ to E is strongly sensitive to the poorly known charge-retention features of high-energy pion interactions; other evidence on E/N_μ from Yakutsk (37) would about double the Sydney energies (see the upper dotted curve in Figure 3). The derivation of E from shower size is usually not a trivial matter, but

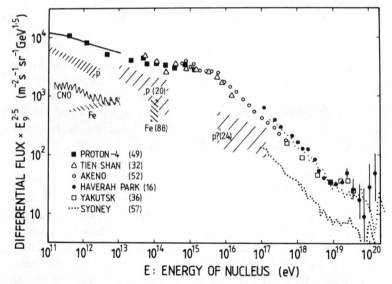

Figure 3   Energy spectrum of cosmic-ray particles from a few selected experiments (references given on diagram). A second version of the Sydney spectrum is also shown, with energies doubled. Shaded bands refer to particular nuclear species or groups. [The Akeno report (52) used two alternative energy conversion factors: the data plotted here are interpolated between these, guided by a shower model.]

three different methods—used at Volcano Ranch, Haverah Park, and Yakutsk—have been shown (17) to agree to within $\sim 20\%$ when applied to individual showers. (The error may nevertheless turn out to be somewhat larger than this.)

The energy spectrum presents the following main features. After J has varied as $E^{-\gamma}$ for many decades, it falls somewhat more steeply above 5×10^{15} eV—the well-known "knee" of the spectrum—and then continues smoothly to 10^{19} eV, beyond which Volcano Ranch, Haverah Park, and Sydney (to a lesser extent) find that it falls less rapidly. A hypothesis blessed by long tradition is that the more-energetic particles above the knee are able to leak out of the Galaxy more rapidly (assuming their sources to be within it). Probably, particles with a galactic origin end near 10^{19} eV (as Figure 2 might suggest), and above that a different source is active (most plausibly in the nearby supercluster of galaxies). However, the maximum energy of cosmic rays is disputed, as a reanalysis of the Yakutsk showers [internal report (36) and informal discussions] indicates an end to the spectrum near 5×10^{19} eV, beyond which they have no data points in Figure 3. The reason for this disagreement with the other three experiments is not at present understood. If the Yakutsk results are vindicated, they could alter considerably our view of cosmic rays, as they might demonstrate the predicted effect of long exposure to interactions with the primeval 2.7 K radiation (48,100), expected if particles have traveled > 50 Mpc.

NATURE OF THE PRIMARY PARTICLES   The fine detail of nuclear composition obtained below 10^{11} eV is not available at much higher energies. According to the "leakage" interpretation of the knee in the spectrum, the spectrum of each nucleus should exhibit a "knee" at the same magnetic rigidity (i.e. at $E_{knee} \propto Z$). Protons would fall away first, iron nuclei last. Some factor appears to complicate this picture, however, as only a single step is apparent in the spectrum; and although many authors claim indirect evidence exists that very heavy nuclei predominate in the flux just above 10^{15} eV, the evidence is far from convincing, and such a dominance is not yet evident in the few top-of-atmosphere observations (21, 88). At 10^{17}–10^{19} eV, the fluctuations in the depth of maximum development of showers indicate that there must be an appreciable proportion of protons among the incoming particles (24, 97); and the few showers of energies $> 10^{19}$ eV measured in detail look no different from 10^{18} eV showers. This has an important bearing on trajectories (Figure 2); and the presence of protons raises severe difficulties in acceleration models (Section 3). Charged dust particles have been dismissed as possible primaries in this energy range (69), as the depth of shower development clearly puts the particle mass $\ll 10^4$ amu.

2.2 *Observed Anisotropy of High-Energy Cosmic Rays*

IS THERE REALLY A SIGNIFICANT ANISOTROPY AT HIGH ENERGIES? This
question is persistently raised by critics. Around 10^{13} eV, there is a clear but
very small (0.06%) pattern of intensity variation (see, for example, 1)—but is
this all? Until recently, the statistics at high energies have only justified the
simplest measure of anisotropy [the first Fourier component (24-hr sidereal
period) of the intensity variation as the sky passes overhead], and the
reported amplitude A in many energy ranges has never been far above
noise. The best test of whether these amplitudes are random noise is to
compare the phases (time of maximum) reported in various energy ranges
by Haverah Park (30, 41, 72, 73) and Yakutsk (42), which have both viewed
the northern sky with good statistics. These are plotted in Figure 4, together
with data from Chacaltaya (3)—although these have only ~60% overlap in
the declination range—and from two less recent experiments near 10^{16} eV
(29, 34). Apart from a region near 10^{19} eV where the direction is unclear,
they do agree: the phases are not random, and there is a real anisotropy.
For $E < 10^{14}$ eV the phase is ~1 h; this phase is altered above the knee and
shows a new pattern above 5×10^{17} eV. Finally, above 10^{19} eV, there is a
rapid change.

The magnitude A of the anisotropy provides a crude test of the "leakage"
interpretation of the knee in the spectrum. If sources in the Galaxy produce
$Q(E)$ ($\propto E^{-\gamma}$, say) cosmic rays of energy E per unit time, and if these move
around for an average time $T(E)$ before escaping, then the number of

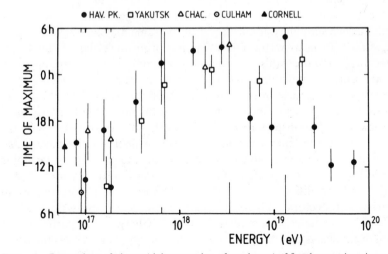

Figure 4   Comparison of phases (right ascension of maximum) of first harmonic anisotropy
reported by different experiments.

particles in the Galactic box will be $N(E) = Q(E)T(E)$, and the spectrum falls off more if leakage becomes more rapid. If t is the average escape time under rectilinear propagation, an observer at a typical position in the box would see an anisotropy $A = kt/T$ very roughly: thus $A \propto 1/T$, and

$$T(E) = N(E)/Q(E) = E^{\gamma}N(E) \propto 1/A. \qquad\qquad 2.$$

In Figure 5, the observed values of A (1, 31, 41, 42, 70, 72, 73; with noise subtracted in quadrature, and values corrected for solar motion below 10^{15} eV) are plotted inversely and compared with the curve $E^{2.47}J(E)$ (i.e. we postulate a production spectrum $E^{-2.47}$ arbitrarily). Of course, there may be local eddies in the flow of cosmic rays, and we have plotted only one component A of a vector flow, but the observations are at least consistent with the model. Even the low-A point at 3×10^{14} eV may be associated with a proton "knee" (56). Most remarkably, the model suggests a power-law spectrum extending to $\sim 10^{19}$ eV (if the final decade were considered to be extragalactic). For comparison, the source spectrum deduced below 10^{11} eV from other data is $\sim E^{-2.4}$. If, instead, the knee is due to features of the accelerating region, the anisotropy would have to reflect very local, rather than global, flow patterns.

ARRIVAL DIRECTIONS IN RELATION TO GALACTIC COORDINATES   The Haverah Park group have found that in the energy range 2×10^{17}–10^{19} eV there is an intensity gradient in galactic latitude (5), with a deficit of flux from the north; this result has now been fully confirmed at Yakutsk (42). At 4×10^{18} eV, for example, the gradient is 0.2% per degree of latitude.

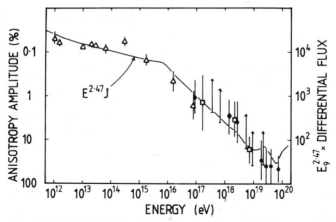

Figure 5 Amplitude of first harmonic as a measure of residence time: variation in A compared with variation in flux (cf. Equation 2).

Above 10^{19} eV, the pattern of arrival directions changes rapidly with energy. Haverah Park finds that above $\sim 3 \times 10^{19}$ eV there is a large excess of particles from high northern galactic latitudes, and Watson (98) considers the most likely explanation for this to be a large contribution from the Virgo supercluster. The Sydney group, though, find no north/south (N/S) anisotropy above 2×10^{19} eV in the part of the sky they see (58), and 8 out of 10 particles above $\sim 8 \times 10^{19}$ eV (doubling their quoted energies again) are within 30° of the galactic plane. Krasilnikov et al. (63) believe, in fact, that the strong northern excess occurs in a limited energy region, and that taken as a whole, the pattern bears a strong relationship to galactic features; thus, we must be seeing trajectories in a large-scale galactic field. A global view of the sky is needed to decide between these viewpoints.

2.3 Specific Identified Sources of Cosmic Rays

A new field is opening up here, with the direct identification of one or two specific sources of very energetic particles. Dzikowski et al. at Lodz (39, 40) reported an excess of air showers detected from the general direction of the Crab Nebula at $> 10^{15}$ eV (about 10^3 times the energy of previously known gamma-ray sources); however, the Akeno and Haverah Park experiments have failed to see such an excess [Hayashida et al. (53) and Lambert et al. (68) appear to set a limit a factor of ~ 50 lower], although the Fly's Eye optical detector may have seen it (15). Thus the present position is unclear. However, Samorski & Stamm at Kiel (84) discovered that the X-ray binary Cygnus X-3 emits 10^{16} eV particles, presumably gamma-rays, which are detected at one point of the 4.8-hr binary period. This has since been confirmed (74): the "pulse" is narrow (duty cycle 2%) and occurs at a phase 0.25 after the X-ray minimum. Near 10^{12} eV, pulses have been reported at a similar phase (90), and also about 0.4 later. The differential spectrum seems to extend as $\sim E^{-2}$ from lower energies but falls off sharply above 10^{16} eV. Presumably a compact object (neutron star? black hole?) accelerates electrons or protons to energies above 10^{16} eV, and the 4.8-hr period is attributed either to interactions with the surroundings of a companion (89, 96) or to precession of a relativistic jet (50). This identified source is of the greatest importance and is a challenge to theorists, as it is hard enough to see how 10^{12} eV electrons or photons get out of a pulsar magnetosphere (see, for example, 4).

3. ACCELERATION MECHANISMS

Attention focuses here on mechanisms for achieving energies of 10^{20} eV (since this is seen by several observers) or 10^{19} eV (approximately the end of that part of the spectrum that extends back smoothly to 10^{15} eV). All the proposed acceleration processes are problematical, however, and new ideas are needed.

STATISTICAL OR DIRECT ACCELERATION?   Figure 1 focused attention on a few plausible acceleration sites, but the question remains as to what mechanism is at work.

1. Particles may gain energy gradually by numerous encounters with regions of changing (moving) magnetic field; such processes are variants of Fermi's mechanism (43). Their advantage is that the energy is spread over many decades, and in the shock-wave variant (7, 8, 12, 64) the spectrum very convincingly emerges as $\sim E^{-2}$. Their disadvantages are that they are slow, and that it is hard to keep up with energy losses at the highest energies.
2. Particles may be accelerated directly to high energy by an extended electric field (e.g. emf arising in rapidly rotating magnetized conductors, such as neutron stars or supermassive objects). Such a mechanism has the advantage of being fast, but it suffers from the circumstance that the acceleration occurs in an environment of very high energy density, where new opportunities for energy loss exist. In addition, the complexity of such an analysis is daunting: and it is usually not obvious how to get a power-law spectrum to emerge.

3.1 Problems Associated with Statistical Acceleration

FERMI ACCELERATION   Cavallo (22) has summarized many of these problems [though Greisen (47) had raised the basic issues before]. Following the same path, we find that the acceptable sites for acceleration are very few indeed. The results here largely carry over to the shock acceleration models.

Statistical acceleration is characterized by an effectively continuous gain of energy with an acceleration time t_A (i.e. $dE/dt = E/t_A$) competing with a possibility that the particle escapes from the system, with mean escape time t_E. Particles emerge with an energy spectrum $J \propto E^{-\gamma}$, where $\gamma = 1 + t_A/t_E$, so we need $t_A \sim t_E$. Apart from the problem that t_A probably rises and t_E falls with increasing E, energy gain will stop if any steady energy loss sets in that cancels the gain. The energy gain may occur by reflection of charged particles from distinct magnetized "clouds" moving randomly with velocity βc in a very inhomogeneous medium, as in the models of supernova remnants proposed by Scott & Chevalier (85) and others; in such a case (43), we have $t_A^{-1} = 2c\beta^2/\lambda$, where λ is the particle's mean free path for scattering from clouds (assuming the motion approximates to a random walk). In a less inhomogeneous medium, particles can be scattered by Alfvén and similar waves moving with speed $\sim v_A$ in the plasma, and the acceleration rate (based on 66) is $t_A^{-1} \sim 3c\beta^2/2\lambda$, where $\beta = v_A/c$. Scattering is now effected by field perturbations of wavelength $\sim 2\pi r_L$, where r_L is the Larmor radius of the particle in the overall field B. In somewhat turbulent conditions, we may assume $\lambda \sim 4$–$25\ r_L$; i.e. $\lambda = \eta r_L$, where (say) $\eta \sim 10$. (The rough lower limit to λ arises because much of the field energy will lie in nonresonant long wavelengths.)

In a strong magnetic field of B gauss, synchrotron radiation causes an energy loss with a time scale $t_S = 1.4/E_{20}B^2$ yr for protons, where E_{20} is the particle energy in 10^{20} eV units. [The value of t_S is increased by a factor of $(A/Z)^4$ for a heavy nucleus.] In a weak field, interactions with low-energy photons provide a more serious energy loss. As a minimum, there are always the primeval 2.7 K photons, and for protons the energy loss time scale t_P is $\sim 7 \times 10^8$ yr at 10^{20} eV (with t_P falling rapidly as E rises) and $\sim 5 \times 10^9$ yr at 10^{19} eV, although t_P will be much reduced near a luminous source. Figure 6 shows the range of Alfvén speeds βc that are required to make $t_A < t_{loss}$ so that an energy of 10^{20} or 10^{19} eV can be reached: the shaded regions are excluded by one or the other of the two losses. If the ratio $\eta = \lambda/r_L$ is reduced from 10 to 4, the effect on the boundary of the allowed

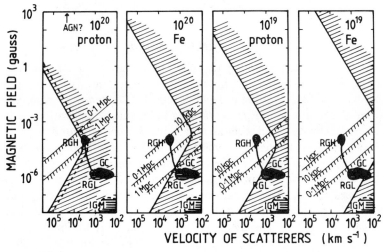

Figure 6 Combinations of magnetic field strength and velocity of scattering centers that allow Fermi acceleration to reach 10^{20} or 10^{19} eV for protons or for iron nuclei. Only the unshaded triangular region at the left in each diagram beats synchrotron losses (above the upper leg of the triangle) and photoreactions (below the lower leg). Any candidates must also lie above the diagonal line appropriate to their radius or diffusive escape will be too rapid. Positions of galactic clusters (GC), radio galaxy lobes (RGL), radio galaxy hotspots (RGH), and the intergalactic medium (IGM) are indicated.

area is shown by the dashed line in the first diagram. Also, the accelerating region must be large enough to ensure that the escape time t_E is not much less than t_A. The diffusive escape time from a sphere of radius R is $\sim 1.5R^2/c\lambda$: this limit is also indicated on Figure 6, although a highly organized field could alter t_E. Among the candidate sources from Figure 1, clusters of galaxies, such as the Virgo or Coma clusters, may have $B \sim 2 \times 10^{-6}$ G (35, 82), $v_A \sim 1000$ km s^{-1} ($\sim$ random motion of the galaxies), and $R \sim 0.5$ Mpc. The lobes of very large radio galaxies (and also the local radio galaxies, Virgo A and Cen A) are probably little different. Hotspots in the very active lobes (such as Cyg A), where $B \sim 10^{-4}$ G, might be the only locations that approach our requirements. In the general intergalactic medium, one can only guess at conditions, but if v_A is a few hundred kilometers per second, as in galactic motions, and $B \sim 3 \times 10^{-8}$ G (or less), then $t_A \gg t_{Hubble}$. For iron nuclei, the losses are smaller, and some of the illustrated sources may be large enough to attain energies $> 10^{19}$ eV. Turning to the more compact (high-B) sources of Figure 1, we can see from Figure 6 that for any speed of scatterers ($< c$), the magnetic field B is < 1 G for 10^{20} eV protons (50 G at 10^{19} eV); thus, this defeats Fermi acceleration at accretion disks around active galactic nuclei (proposed in 59), with the possible exception of acceleration of iron nuclei (but there are other photons around!). Furthermore, if the diffusive escape time $t_E > \frac{1}{2}t_A$, then this requires that $R_{pc}B > 0.5E_{20}/\beta$ (times $\eta/10$), and thus the total magnetic energy over a spherical volume is $> 3.10^{54}E_{20}^5\beta^{-5}A^{-4}$ $(\eta/10)^4$ erg (where A refers to heavy nuclei); this is more than $100/\beta^5$ times the energy available from collapse to a neutron star if $E_{20} = 1.5$.

ACCELERATION BY SHOCK WAVES   Many readers will be familiar with this process, so only a brief introduction is given here. A collisionless shock front, propagating at a speed $v_S = \beta c$ through a plasma carrying a magnetic field, can reflect particles with a large gain in energy if the orientation of the field is just right; but practically all investigators have concentrated on "parallel shocks" (magnetic field parallel to gas flow), in which case the most important process for very-high-energy particles is closely related to Fermi acceleration. Behind the shock, the gas advances with speed $v_2 = \beta_2 c$ ($= \frac{3}{4}v_S$ for a strong shock—very high v_S—if gas pressure dominates), carrying with it magnetic irregularities that again act as scattering centers. These scatterers advance upon the motionless scattering centers in the unshocked gas (this time we neglect their individual motions), and particles crossing the shock front encounter oncoming scatterers, on average gaining energy $4E\beta_2/3$ per back-and-forth crossing. As it diffuses around, a particle makes on average $1/4(\beta - \beta_2)$ cycles of crossing before eventually being left behind by the front (8), and the statistics of crossings yield an energy spectrum E^{-2}

for strong shocks ($\gamma > 2$ as the shock weakens). (For an enthusiastic review, see 6; see also 38, 95.) The initial applications of shock-wave theories were greeted with enthusiasm, since supernovae certainly produce strong shocks that seemed almost inevitably to lead to about the right intensity and spectrum of cosmic rays generated in the Galaxy [Krymsky & Petukhov (65), Blandford & Ostriker (13)].

There have been doubts, however, about the spectrum generated (e.g. 46): most seriously, the spectrum generated in this way by supernova shocks would terminate much too soon. Lagage & Cesarsky (67) have made a detailed study of the limitations imposed by the size and duration of the shock and find that the known scattering mechanisms would yield E_{max} a few times 10^{13} eV for protons, above which the spectrum would fall very steeply. They find, too, that an ensemble of shocks does not extend this limit much. How do shocks improve on normal Fermi acceleration in other sites? Clearly β is greater ($v_S > v_A$), and for various reasons (8, 38) turbulence should ensure a short λ, but that is all. [Axford (6) suggested that shocks might speed up on entering the galactic halo, but this is unlikely to be so (67).] Lagage & Cesarsky find an effective acceleration rate of $t_A^{-1} = c\beta^2/(\lambda_1 + 4\lambda_2) \sim c\beta^2/2\lambda_1$, where λ_1, λ_2 refer to upstream and downstream scattering mean free paths—this is very similar in form to Fermi acceleration, except that β has to be $\sqrt{3}$ times as large for the same effect, so virtually the same diagram as Figure 6 applies in this case too. Shock waves do not rescue the galactic cluster site from the effect of energy losses, since $v_S < 3000$ km s^{-1} for bow shocks of galaxies; nor, by a larger margin, does the intergalactic medium in general appear to be a possible location for the origin of these high-energy particles. The diffusive motion needs a very large space L on either side of the shock: $L \sim \lambda/\beta$, a value of, say, 300 mean free paths or $\gg 300 r_L$. (If $v_S = 1000$ km s^{-1}, a particle makes 4×10^5 scatterings in gaining one decade of energy: is this really the most efficient process?)

From Figure 6, it appears that acceleration should be far easier if $v_S \sim c$, and although this would give a notably flatter spectrum (77), the effect has not been fully explored. The fast-moving "knots" in radio galaxies may be unusually effective shock waves (87) if they are large enough. But the radio galaxies in our vicinity are not so active. So, iron nuclei might possibly be accelerated to 10^{19} eV, but not protons.

3.2 Direct Acceleration

ELECTRICAL GENERATORS   We now consider the high-B candidate sources of Figure 1. If a rotating neutron star has a surface field $\sim 10^{12}$ G, a radius $r \sim 10$ km, and a rotational frequency $\omega/2\pi \sim 30$ s^{-1} (like the Crab pulsar), a circuit connected between pole and equator would see an emf $\sim \omega B r^2/c$ (cgs units) $\sim 10^{18}$ V for an aligned or oblique dipole. If we assume, as is customary, that the part of the magnetosphere corotating within the light cylinder will fill with conducting plasma maintaining $\mathbf{E} \cdot \mathbf{B} = 0$ [although this has been questioned (2, 76)], the emf measured now from pole to the last open field line is reduced by the factor $r\omega/c$. (Note that $\mathbf{E}$ refers to electric field, E to particle energy.) Just where the main potential drop occurs in the external space, which contains current sources (pair production), is still unclear (4). This position affects acceleration of whatever nuclei are around, since the radiative energy loss $-dE/dx = 2\gamma^4 Z^2 e^2/3R_C^2$ (cgs) has to be overcome, where the particle of charge Ze and Lorentz factor γ follows a path of radius R_C. Berezinsky (10) suggests that the electrical potential $V = B\omega^2 r^3/4c^2$ becomes available for acceleration at a distance R very far from the pulsar, perhaps near the light cylinder, so that R_C may be large ($\sim R$); the Crab pulsar could then accelerate iron nuclei to 10^{18} eV (protons to 5×10^{16} eV). His expression varies weakly with ω. This model for a cosmic-ray source would imply that the most energetic nuclei were highly charged, unless there is a mechanism by which the paths could be even less curved by the field. Ferrari & Trussoni (43a) found that even the unrealistic vacuum fields around an oblique rotating dipole (emf $\sim 10^{20}$ V) resulted in virtually the same maximum energies as quoted above. Although it is not obvious that a power-law spectrum is likely, it may just possibly result from pulsar evolution. (According to observation, pulsars do generate electrons with power-law spectra.) It is very challenging to see that Cygnus X-3 accelerates particles to at least 10^{16} eV : if these are electrons, protons might reach a higher energy.

Lovelace (75) offered as another unipolar inductor the rotating accretion disk around a $10^8 M_\odot$ black hole in the center of a radio galaxy ; such a disk could draw in magnetic flux with the gas to give a magnetic field $B \sim 10^4$ G parallel to the rotation axis, generate 10^{19} V, and might fire out a beam of protons axially (if $\mathbf{B} \cdot \omega < 0$). More recently, Rees et al. (81) have proposed that in a galactic nucleus, a $10^8 M_\odot$ black hole may be spun up to near-maximum permissible angular momentum by accretion, after which rotational energy up to $0.29Mc^2$ may in principle be extracted by electrodynamic torques [Blandford & Znajek (14); see also 79]. They consider a magnetic field $B \sim 10^4$ G entering the hole's horizon: an emf $V = 10^{19}(B/10^4\,\text{G})(M_H/10^8 M_\odot)^2$ volt is then available at maximum angular momentum (79). These later papers, however, do not suggest that 10^{19} eV protons will automatically be produced, for although efficient power extraction may be possible (79), there will be a dense positron-electron plasma generated by thermal gamma rays that will modify the electric field and also interact with very energetic protons. An important question again is, How far away does the main potential drop occur? Colgate (26) made a crude estimate of energy losses due to known photon fluxes from active galactic nuclear regions, and even in Cen A (luminosity 10^{-3} of Colgate's example), after correcting for a numerical error, the energy loss time scale is $\ll 1$ yr over a wide energy range; thus the electric fields would have to occur well away from the nucleus, where there were fewer photons. Otherwise, acceleration would be defeated above $\sim 10^{16}$ eV. But the environment of a galactic nucleus is likely to be more complex than imagined, and G. Lake & R. E. Pudritz (preprint) argue that its huge inductance makes the electric circuit outside the spinning nucleus unstable, with disruption of currents inducing large transient fields. They suggest that this should show up in unsteady production of high-energy gamma rays (and neutrinos?).

Potentials of 10^{21} V appear also in Fischhoff's electrostatic cosmology (44); but interactions with photons over 10^9 yr remain a problem.

MORE ABOUT PULSARS   Gunn & Ostriker's proposal (51) that protons and nuclei could reach 10^{21} eV by riding in the outgoing strong wave field beyond the light cylinder of a pulsar has not been seriously considered since it was noted (61) that even a small contamination of plasma in the waves would destroy the necessary phase locking. However, few other schemes could produce the whole range of cosmic-ray energies in the same source, and any modification that made this process available again would have great attractions. Pulsar remnant disks might offer another way to subvert unproductive magnetospheres and regain very high voltages (76).

HYDRODYNAMIC SHOCKS   The value of E_{max} seems much too small in Colgate & Johnson's scheme (27), where a shock wave emanating from the core of a collapsing supernova eventually attains ultrarelativistic speed in the steep density gradient of the stellar atmosphere, accelerating a small fraction of the local matter to cosmic-ray energies directly in the radiation-dominated shock. Colgate (25) has defended this scheme against arguments questioning whether the shock can form (see 86), but he still needs to find a way to propagate the shock beyond the point where $E \sim 3 \times 10^{14}$ eV per nucleon. In addition, evidence indicates that the relativistic particles in supernovae largely gain their energy after the initial outburst (e.g. 23, 85).

3.3 *Final Comments*

None of the schemes discussed above has really exploited first-order Fermi acceleration, which might be much faster than the other statistical processes. In a sense, shock waves involve first-order acceleration at the front itself, but almost all the scattering merely produces diffusion within the gas: one would like to find approaching mirrors with a higher reflection coefficient! M. E. Pesses (private communication) suggests that oblique shocks may play an important role.

4. PROPAGATION OF COSMIC RAYS

Great uncertainties about the configuration of the magnetic field both in the outer regions of our Galaxy and in metagalactic space have discouraged attempts to interpret the arrival directions in detail. The improved body of cosmic-ray data may now encourage further analysis. Since the present situation cannot be described profitably in the short space available, only the barest indication of some recent work is given here.

4.1 *Propagation from Extragalactic Sources*

LOSSES DUE TO PHOTOREACTIONS   Soon after Greisen (48) and Zatsepin & Kuzmin (100) pointed out the serious energy losses that protons must suffer as a result of interactions with the 2.7 K primeval radiation, and the consequent depletion of the flux expected above $\sim 4 \times 10^{19}$ eV, it was generally accepted that 10^{20} eV particles cannot come from sources distributed uniformly throughout the Universe. Instead, the Local super-cluster (at 15–20 Mpc) must make a disproportionately strong contribution to any extragalactic component (99), regardless of whether protons or mixed nuclei made up the flux (55, 80, 91, 94). Even then, the flux would have to drop off a little above 10^{20} eV. We may indeed live near a local density node in a cellular pattern of voids in the Universe (101). If photon interactions have not resulted in flux losses, neither should they have removed very heavy nuclei from the flux; and since most potential sources are likely to contain synthesized elements, we do not expect just a proton flux. However, if the revised Yakutsk spectrum eventually prevails, the whole question may be reopened.

MOTION OF THE PARTICLES   As mentioned above, the (by no means exclusive) tendency of particles above 3×10^{19} eV to arrive from high northern galactic latitudes also drew attention to the Virgo supercluster region. Giler et al. (45) have proposed modeling the motion through intergalactic space with a diffusion, the coefficient D rising with E. Then a rising anisotropy above 10^{19} eV could be expected, while D might be too small to allow space to fill with old particles from farther off. At a distance r from a source, the anisotropy amplitude is $A \sim \lambda/r$, so if $A \sim 0.6$ at $\sim 6 \times 10^{19}$ eV, the mean free path is $\lambda \sim 10$ Mpc if Virgo dominates the flux. The spectrum can flatten somewhat in the 10^{19} eV region in such models, but this depends on an interplay of several factors.

4.2 *Propagation from Galactic Sources*

What if the sources lie within the Galaxy? Are the observed cosmic rays too isotropic for this? Kiraly et al. (62) found that for energies up to 10^{17} eV, the anisotropy was consistent with galactic sources; and a measurement at 10^{17} eV that worried them has since been retracted (28). Figure 5 is certainly favorable. Such an origin for the cosmic rays requires a galactic halo extending many kiloparsecs (Figure 2), but an ordered field of ~ 2 μG extending beyond 10 kpc is quite compatible with radio observations (33, 78), as B^2 must fall off more slowly than synchrotron emission. An ordered field confined to the disk, however, would guide a large concentration of cosmic rays above 10^{17} eV into directions along the galactic magnetic field lines (60, 92), contrary to observation; but Berezinsky & Mikhailov (11) find that a small B component normal to the disk destroys this pattern. They also find that the observed magnitude of the anisotropy is very reasonable for particles originating in the Galaxy (up to 10^{19} eV), even without the requirement of a very extensive halo. The striking tendency toward a southern excess below 10^{19} eV has been interpreted (56) as indicating a local density gradient of cosmic rays in this energy range ($\sim 15\%$ kpc^{-1}) increasing toward the Orion region. (Alternatively, we may live in a hole in the magnetic field, asymmetrically situated north of the central plane of the Galaxy.) Indeed, these energy-dependent anisotropies indicative of density variations on a kiloparsec scale are much more favorable to a galactic origin.

5. CONCLUSIONS

Where do cosmic rays originate?

Are they all extragalactic? Burbidge (18, 19) has pointed out the possibilities of such schemes. At low energies we really only have weak gamma-ray evidence against an extragalactic origin (83), and an earlier conclusion that cosmic-ray protons were less numerous in the outer parts of the Galaxy has now been overturned by observations from COS-B (54). There is, however, a consistent galactic picture, and near 10^{18} eV the 10–20% N/S asymmetries just alluded to speak against a very distant origin.

Are they all Galactic? This would seem possible only if (a) further evidence at 10^{19} eV and above becomes consistent with a composition that is principally carbon and heavier, and (b) there is the necessary very large halo with ordered fields > 1 μG. (One might use the COS-B result to argue for the existence of a very large halo.) The irregular shape of the spectrum near 10^{19} eV—which is not really just a different slope—would probably then relate to large-scale looping of trajectories. The directions of the most energetic Sydney showers favor such an interpretation. A direct mode of acceleration by pulsars would probably be called for.

Do extragalactic particles take over for energies larger than 10^{19} eV? This is at present the most likely hypothesis [supported also in an earlier review by Berezinsky (9)], being more probable than the previous one because of Haverah Park's directions and the fluctuation measurements (though at somewhat lower energy) that favor protons as a large component. However, Virgo is not a particularly impressive source region.

Present Galactic acceleration models have difficulty in getting particles beyond 10^{15} eV, but Cygnus X-3 illustrates the shortcomings of such models (and this source points to the importance of direct rather than statistical acceleration). Do the principal (supernova?) accelerators really give out immediately above the knee of the spectrum, beyond which ($\sim 3 \times 10^{16}$ eV) a new proton-rich flux takes over, as is strongly advocated by some workers (e.g. 71, 93)? Most remarkably, experiments do not show a departure from a smooth spectrum here (there must be a very neat join!), and the smoothness of the spectrum from 10^{16} eV to near 10^{19} eV presents a great challenge to those who build accelerators in the sky. At any rate, this part of the spectrum cannot be a "universal" flux: this would be highly isotropic. The smooth one-bend spectrum and the clear anisotropies below 10^{19} eV are particularly interesting aspects of the present data.

When the methods of deducing composition near 10^{15} eV have been made consistent, we may be able to reduce the number of possibilities considerably. Meanwhile, it is very tantalizing that although progress in extending the upper limit of the spectrum has been very slow, it has shown that a global mapping of the sky just beyond the energy range now studied could make the Galactic versus Virgo choice clear.

ACKNOWLEDGEMENTS

While preparing this review, I have particularly benefited from discussions with F. D. Kahn, J. Linsley, and especially A. A. Watson.

Literature Cited

1. Alexeenko, V. V., Chudakov, A. E., Gulieva, E. N., Sborshikov, V. G. 1981. *Proc. Int. Conf. Cosmic Rays, 17th, Paris* 2:146–49
2. Alfvén, H. 1978. *Astrophys. Space Sci.* 54:279–92
3. Anda, R., Aguirre, C., Tachi, K., Kakimoto, F., Inoue, N., et al. 1981. *Proc. Int. Conf. Cosmic Rays, 17th, Paris* 2:164–67
4. Arons, J. 1981. In *Origin of Cosmic Rays*, ed. G. Setti, G. Spada, A. W. Wolfendale, pp. 175–204. Dordrecht: Reidel
5. Astley, S. M., Cunningham, G., Lloyd-Evans, J., Reid, R. J. O., Watson, A. A.

1981. *Proc. Int. Conf. Cosmic Rays, 17th, Paris* 2:156–59
6. Axford, W. I. 1981. *Proc. Int. Conf. Cosmic Rays, 17th, Paris* 12:155–203
7. Axford, W. I., Leer, E., Skadron, G. 1977. *Proc. Int. Conf. Cosmic Rays, 15th, Plovdiv* 11:132–37
8. Bell, A. R. 1978. *MNRAS* 182:147–56
9. Berezinsky, V. S. 1977. *Proc. Int. Conf. Cosmic Rays, 15th, Plovdiv* 10:84–107
10. Berezinsky, V. S. 1983. *Proc. Int. Conf. Cosmic Rays, 18th, Bangalore* 2:275–78
11. Berezinsky, V. S., Mikhailov, A. A. 1983. *Proc. Int. Conf. Cosmic Rays, 18th, Bangalore* 2:174–77

12. Blandford, R. D., Ostriker, J. P. 1978. *Ap. J. Lett.* 221 : L29–32
13. Blandford, R. D., Ostriker, J. P. 1980. *Ap. J.* 237 : 793–808
14. Blandford, R. D., Znajek, R. L. 1977. *MNRAS* 179 : 433–56
15. Boone, J., Cady, R., Cassiday, G. L., Elbert, J. W., Loh, E. C., et al. 1983. *Proc. Cosmic Ray Workshop, Univ. Utah*, ed. T. K. Gaisser, pp. 268–88. Newark, Del : Bartol Res. Found.
16. Bower, A. J., Cunningham, G., England, C. D., Lloyd-Evans, J., Reid, R. J. O., et al. 1981. *Proc. Int. Conf. Cosmic Rays, 17th, Paris* 9 : 166–69
17. Bower, A. J., Cunningham, G., Linsley, J., Reid, R. J. O., Watson, A. A. 1983. *J. Phys. G* 9 : L53–58
18. Brecher, K., Burbidge, G. R. 1972. *Ap. J.* 174 : 253–91
19. Burbidge, G. R. 1975. In *Origin of Cosmic Rays*, ed. J. W. Osborne, A. W. Wolfendale, pp. 25–36. Dordrecht : Reidel
20. Burnett, T. H., et al. (the JACEE collaboration). 1982. *Workshop Very High Energy Cosmic Ray Interact., Univ. Pa., Philadelphia*, ed. M. L. Cherry, K. Lande, R. I. Steinberg, pp. 221–38
21. Burnett, T. H., et al. (the JACEE collaboration). 1983. *Proc. Int. Conf. Cosmic Rays, 18th, Bangalore* 2 : 105
22. Cavallo, G. 1978. *Astron. Astrophys.* 65 : 415–19
23. Cavallo, G. 1982. *Astron. Astrophys.* 111 : 368–70
24. Chantler, M. P., Craig, M. A. B., McComb, T. J. L., Orford, K. J., Turver, K. E. 1983. *J. Phys. G* 9 : L27–31
25. Colgate, S. A. 1975. In *Origin of Cosmic Rays*, ed. J. W. Osborne, A. W. Wolfendale, pp. 425–66. Dordrecht : Reidel
26. Colgate, S. A. 1983. *Proc. Int. Conf. Cosmic Rays, 18th, Bangalore* 2 : 230–33
27. Colgate, S. A., Johnson, M. H. 1960. *Phys. Rev. Lett.* 5 : 235–38
28. Coy, R. N., Lloyd-Evans, J., Patel, M., Reid, R. J. O., Watson, A. A. 1981. *Proc. Int. Conf. Cosmic Rays, 17th, Paris* 9 : 183–86
29. Cranshaw, T. E., Galbraith, W. 1957. *Philos. Mag.* 2 : 804–10
30. Cunningham, G., Lloyd-Evans, J., Reid, R. J. O., Watson, A. A. 1983. *Proc. Int. Conf. Cosmic Rays, 18th, Bangalore* 2 : 157–60
31. Cutler, D. J., Bergeson, H. E., Davis, J. F., Groom, D. E. 1981. *Ap. J.* 248 : 1166–78
32. Danilova, T. V., Kabanova, N. V., Nesterova, N. M., Nikolskaya, N. M., Nikolsky, S. I., et al. 1977. *Proc. Int. Conf. Cosmic Rays, 15th, Plovdiv* 8 : 129–32
33. de Bruyn, A. G., Hummel, E. 1979. *Astron. Astrophys.* 73 : 196–201
34. Delvaille, J. 1962. PhD thesis. Cornell Univ., Ithaca, N.Y.
35. Dennison, B. 1980. *Ap. J. Lett.* 239 : L93–96
36. Diminstein, O. S., Efimov, N. N., Pravdin, M. I. 1982. *Bull. Naushno-Tech. Inf. Yakutsk* 9 : 537–91
37. Diminstein, O. S., Efimov, N. N., Efremov, N. N., Glushkov, A. V., Kaganov, L. I., et al. 1983. *Proc. Int. Conf. Cosmic Rays, 18th, Bangalore* 6 : 118–21
38. Drury, L. O'C. 1983. *Rep. Prog. Phys.* 46 : 973–1027
39. Dzikowski, T., Gawin, J., Grochalska, B., Wdowczyk, J. 1981. *Philos. Trans. R. Soc. London Ser. A* 301 : 641–43
40. Dzikowski, T., Gawin, J., Grochalska, B., Korejwo, J., Wdowczyk, J. 1983. *Proc. Int. Conf. Cosmic Rays, 18th, Bangalore* 2 : 132–35
41. Edge, D. M., Pollock, A. M. T., Reid, R. J. O., Watson, A. A., Wilson, J. G. 1978. *J. Phys. G* 4 : 133–57
42. Efimov, N. N., Mikhailov, A. A., Pravdin, M. I. 1983. *Proc. Int. Conf. Cosmic Rays, 18th, Bangalore* 2 : 149–52
43. Fermi, E. 1949. *Phys. Rev.* 75 : 1169–74
43a. Ferrari, A., Trussoni, E. 1974. *Astron. Astrophys.* 36 : 267–72
44. Fischhoff, E. 1981. *Proc. Int. Conf. Cosmic Rays, 17th, Paris* 9 : 254–57
45. Giler, M., Wdowczyk, J., Wolfendale, A. W. 1980. *J. Phys. G* 6 : 1561–73
46. Ginzburg, V. L., Ptuskin, V. S. 1981. *Proc. Int. Conf. Cosmic Rays, 17th, Paris* 2 : 336–39
47. Greisen, K. 1966. *Proc. Int. Conf. Cosmic Rays, 9th, London* 2 : 609–15
48. Greisen, K. 1966. *Phys. Rev. Lett.* 16 : 748–50
49. Grigorov, N. L., Gubin, Yu. V., Rapoport, I. D., Savenko, I. A., Yakovlev, B. N., et al. 1971. *Proc. Int. Conf. Cosmic Rays, 12th, Hobart* 5 : 1746–59
50. Grindlay, J. E. 1982. *Proc. Int. Workshop Very High Energy Gamma Ray Astron.*, Ootacamund, ed. G. T. Ramana Murthy, T. C. Weekes, p. 178. Bombay : Tata Inst.
51. Gunn, J. E., Ostriker, J. P. 1969. *Phys. Rev. Lett.* 22 : 728–31
52. Hara, T., Hayashida, N., Honda, M., Ishikawa, F., Kamata, K., et al. 1984. *Proc. Int. Conf. Cosmic Rays, 18th, Bangalore, 1983*, Pap. OG4–16. In press

53. Hayashida, N., Ishikawa, F., Kamata, K., Kifune, T., Nagano, M., Tan, Y. H. 1981. *Proc. Int. Conf. Cosmic Rays, 17th, Paris* 9:9–12
54. Hermsen, W. 1984. *COSPAR/IAU Symp. High Energy Astrophys. Cosmol., Pamporovo, Rojen, 1983,* ed. G. F. Bignami. Dordrecht: Reidel. In press
55. Hillas, A. M. 1975. *Proc. Int. Conf. Cosmic Rays, 14th, Munich* 2:717–22
56. Hillas, A. M. 1983. In *Composition and Origin of Cosmic Rays,* ed. M. M. Shapiro, pp. 125–48. Dordrecht: Reidel
57. Horton, L., McCusker, C. B. A., Peak, L. S., Ulrichs, J., Winn, M. M. 1983. *Proc. Int. Conf. Cosmic Rays, 18th, Bangalore* 2:128–30
58. Horton, L., McCusker, C. B. A., Peak, L. S., Ulrichs, J., Winn, M. M. 1983. *Proc. Int. Conf. Cosmic Rays, 18th, Bangalore* 2:153–56
59. Kafatos, M., Shapiro, M. M., Silberberg, R. 1981. *Comments Astrophys.* 9:179–98
60. Karakula, S., Osborne, J. L., Roberts, E., Tkaczyk, W. 1972. *J. Phys. A* 5:904–15
61. Kegel, W. H. 1971. *Astron. Astrophys.* 12:452–55
62. Kiraly, P., Kota, J., Osborne, J. L., Stapley, N. R., Wolfendale, A. W. 1979. *Riv. Nuovo Cimento (Ser. 3)* 2(7):1–46
63. Krasilnikov, D. D., Ivanov, A. A., Kolosov, V. A., Krasilnikov, A. D., Makarov, K. N., et al. 1983. *Proc. Int. Conf. Cosmic Rays, 18th, Bangalore* 2:145–48
64. Krymsky, G. F. 1977. *Dokl. Akad. Nauk. SSSR* 234:1306–8
65. Krymsky, G. F., Petukhov, S. I. 1979. *Bull. Acad. Sci. USSR, Phys. Ser. (Engl. Transl.)* 43:14–16
66. Kulsrud, R. M. 1979. In *Particle Acceleration Mechanisms in Astrophysics,* ed. J. Arons, C. McKee, C. Max, pp. 13–25. New York: Am. Inst. Phys.
67. Lagage, P. O., Cesarsky, C. J. 1983. *Astron. Astrophys.* 125:249–57
68. Lambert, A., Lloyd-Evans, J., Watson, A. A. 1984. *Proc. Int. Conf. Cosmic Rays, 18th, Bangalore, 1983,* Pap. OG4–28. In press
69. Linsley, J. 1981. *Proc. Int. Conf. Cosmic Rays, 17th, Paris* 2:141–44
70. Linsley, J., Watson, A. A. 1977. *Proc. Int. Conf. Cosmic Rays, 15th, Plovdiv* 12:203–8
71. Linsley, J., Watson, A. A. 1981. *Phys. Rev. Lett.* 46:459–63
72. Lloyd-Evans, J. 1982. PhD thesis. Univ. Leeds, Engl.
73. Lloyd-Evans, J., Pollock, A. M. T.,

Watson, A. A. 1979. *Proc. Int. Conf. Cosmic Rays, 16th, Kyoto* 13:130–33
74. Lloyd-Evans, J., Coy, R. N., Lambert, A., Lapikens, J., Patel, M., et al. 1983. *Nature* 305:784–87
75. Lovelace, R. V. E. 1976. *Nature* 262:649–52
76. Michel, F. C., Dessler, A. J. 1981. *Proc. Int. Conf. Cosmic Rays, 17th, Paris* 2:340–43
77. Peacock, J. A. 1981. *MNRAS* 196:135–52
78. Phillipps, S., Kearsey, S., Osborne, J. L., Haslam, C. G. T., Stoffel, H. 1981. *Astron. Astrophys.* 103:405–14
79. Phinney, E. S. 1984. *Proc. Torino Workshop Astrophys. Jets, 1982.* Dordrecht: Reidel. In press
80. Puget, J. L., Stecker, F. W., Bredekamp, J. H. 1976. *Ap. J.* 205:638–54
81. Rees, M. J., Begelman, M. C., Blandford, R. D., Phinney, E. S. 1982. *Nature* 295:17–21
82. Roland, J. 1981. *Astron. Astrophys.* 93:407–10
83. Said, S. S., Wolfendale, A. W., Giler, M., Wdowczyk, J. 1982. *J. Phys. G* 8:383–91
84. Samorski, M., Stamm, W. 1983. *Ap. J. Lett.* 268:L17–21
85. Scott, J. S., Chevalier, R. A. 1975. *Ap. J. Lett.* 197:L5–8
86. Shapiro, P. R. 1979. In *Particle Acceleration Mechanisms in Astrophys,* ed. J. Arons, C. McKee, C. Max, pp. 295–318. New York: Am. Inst. Phys.
87. Smith, M. D., Norman, C. A. 1980. *Astron. Astrophys.* 81:282–87
88. Sood, R. K. 1983. *Nature* 301:44–46
89. Stepanian, A. A. 1981. *Proc. Int. Conf. Cosmic Rays, 17th, Paris* 1:50–53
90. Stepanian, A. A. 1982. *Proc. Int. Workshop Very High Energy Gamma Ray Astron., Ootacamund,* ed. G. T. Ramana Murthy, T. C. Weekes, pp. 43–63. Bombay: Tata Inst.
91. Strong, A. W., Wdowczyk, J., Wolfendale, A. W. 1974. *J. Phys. A* 7:1489–96
92. Thielheim, K. O., Langhoff, W. 1968. *J. Phys. A* 1:694–703
93. Thornton, G., Clay, R. W. 1979. *Phys. Rev. Lett.* 43:1622–25
94. Tkaczyk, W., Wdowczyk, J., Wolfendale, A. W. 1975. *J. Phys. A* 8:1518–29
95. Toptyghin, I. N. 1980. *Space Sci. Rev.* 26:157–213
96. Vestrand, W. T., Eichler, D. 1982. *Ap. J.* 261:251–58
97. Walker, R., Watson, A. A. 1982. *J. Phys. G* 8:1131–40

98. Watson, A. A. 1981. *Cosmology and Particles: 16th Rencontre de Moriond,* ed. J. Audouze et al., pp. 49–67. Dreux, Fr: Ed. Front.
99. Wolfendale, A. W. 1974. *Philos. Trans. R. Soc. London Ser. A* 277:429–42
100. Zatsepin, G. T., Kuzmin, V. A. 1966. *Sov. Phys. JETP Lett. (Engl. Transl.)* 4:78–80
101. Zeldovich, Ya. B., Einasto, J., Shandarin, S. F. 1982. *Nature* 300:407–13

Ann. Rev. Astron. Astrophys. 1984. 22 : 445–70

THE INFLUENCE OF ENVIRONMENT ON THE H I CONTENT OF GALAXIES

Martha P. Haynes

National Astronomy and Ionosphere Center[1] and Astronomy Department, Cornell University, Ithaca, New York 14853

Riccardo Giovanelli

National Astronomy and Ionosphere Center, Arecibo Observatory, Arecibo, Puerto Rico 00613

Guido L. Chincarini

Department of Physics and Astronomy, University of Oklahoma, Norman, Oklahoma 73069, and European Southern Observatory, Garching bei München, West Germany

1. INTRODUCTION

Morphological classification is a fundamental tool for the identification of activity in galaxies. Even the most casual perusal of the Arp (4) atlas will provide convincing evidence that not all galaxies live quiescent existences. Furthermore, the conclusion immediately arises that it is the presence of neighboring galaxies that somehow leads to the morphological peculiarities.

At least as early as 1931, Hubble & Humason (67) noticed that the distribution of morphological types among galaxies differs between the central parts of the rich clusters and the field; such morphological segregation has been quantified more recently by several authors (21, 32,

[1] The National Astronomy and Ionosphere Center is operated by Cornell University under contract with the National Science Foundation.

0066–4146/84/0915–0445$02.00

82). A basic question raised by these studies concerns the formation of S0 galaxies: Did their formation as a separate Hubble class occur at early epochs, or are they instead the remnants of recently stripped spirals? The understanding of galaxy morphology is related to the very general problem of galaxy formation and evolution and how a galaxy, during its birth and evolution, is affected by its surroundings.

Since its detection in 1951, the 21-cm line of neutral hydrogen has offered clues to the nature of quite diverse galaxian attributes, from the obvious interstellar gas content to the more subtle mass-to-light ratios and the determination of the Hubble constant. As a tracer of potential star formation, the neutral hydrogen content is also an indirect probe of the evolutionary processes at work within a galaxy today. Here, we discuss one aspect of "galaxian sociology": how a galaxy's interstellar gas, as revealed by its neutral hydrogen, is influenced by its environment during its evolutionary history.

The observed morphological segregation implies that somehow the interstellar gas is removed from the galaxies in clusters, thereby reducing the star formation rate so that the galaxies appear as early-type. An alternative possibility is that galaxies of different morphological types are created as such in regions of differing mass density, with ellipticals and S0s preferentially formed in higher density regions, so that the differentiation among types is forced in early stages of evolution. While not attempting to rule out this scenario, we present the current evidence that testifies to the viability of at least some contribution of ongoing gas removal mechanisms, the result of interactions between galaxies and their neighbors as well as with the intergalactic medium within which they are immersed.

Evidence that galaxies are aware of their environment comes from several sources, such as the head-tail radio source morphology, the increased frequency of the [O II] line in cluster galaxies, and the large H I appendages seen extending from galaxies in pairs and small groups. There are a variety of mechanisms that may contribute to the depletion of the interstellar H I gas within a galaxy. Perhaps the most popular external models proposed are galaxy-galaxy collisions, tidal interactions, ram pressure sweeping, and evaporation. Internal ongoing mechanisms may include galactic winds or simply star formation with insufficient gas replenishment. The neutral hydrogen content and its distribution can be used not only to compare galaxies within differing intergalactic environments but also to estimate the relative efficiency of the several proposed gas removal mechanisms. In Section 2, we briefly review the current state of 21-cm line measures of the H I content of galaxies and the expected distribution of the neutral gas. Section 3 examines the environment of small aggregates of galaxies in which tidal encounters among neighbors can remove a substantial portion of a galaxy's interstellar gas, dramatically altering its appearance and subsequent evolution. In Section 4, the evidence for H I deficiency in clusters is presented, along with a discussion of the contributions of the various gas-sweeping mechanisms.

2. THE H I CONTENT OF NORMAL GALAXIES

Normalcy is obviously a relative matter. A major stumbling block in the study of the comparative H I contents of galaxies occurs in defining a proper standard of measurement. In this section, we briefly review how the H I content can be determined with a minimum of selection effects in order to optimize the comparison of different samples of galaxies and to determine environmental influences. For more general summaries of the H I properties of galaxies, the reader is referred to other reviews (89, 93, 112).

2.1 *The Definition of H I Content*

"H I content" is a loosely defined term used to gauge the abundance of the interstellar gas within a galaxy. Strictly speaking, it should be a synonym for the H I mass M_H, a parameter that does not reflect the total amount of gas in interstellar space (a large fraction of which is in the form of molecular hydrogen or of a diffuse, ionized component). H I may actually account for only about one half of a galaxy's interstellar gas (and that fraction may be strongly type dependent), but its flux and distribution, which can be determined with relative ease, are frequently used to diagnose traumas in the recent past. Anomalies in these parameters are usually assumed to be symptomatic of comparable anomalies in the total gas content; it should be kept in perspective, however, that atomic, ionized, and molecular components are distributed differently in a galaxy's gravitational potential well, and they are vulnerable in different degrees to the action of external forces.

In the literature, the measure of H I content has been associated with a variety of observational and derived quantities, most frequently the H I mass M_H, the hydrogen mass to luminosity ratio M_H/L, or the H I surface density M_H/D^2, where D is the linear diameter. The simple quantity M_H appears at first glance to be the most attractive, but as a distance-dependent quantity, it is severely affected by uncertainties in the Hubble constant. Moreover, more massive galaxies also have, on the average, greater H I masses, although the fractional H I masses may not vary. Use of the distance-independent ratios M_H/L or M_H/D^2 has been more popular, but not without controversy.

Historically, the first comparison between field and cluster galaxies was made by Davies & Lewis (28) for galaxies in the Virgo cluster. Their results were immediately questioned (17) as being caused by a luminosity dependence on the ratio M_H/L, in the sense that more luminous galaxies, which characterize the cluster sample but not the field, have intrinsically lower M_H/L values simply because of their higher luminosities. This residual dependence between M_H/L and L has been much debated since the publication of Davies & Lewis' paper (16, 55, 58, 98). As we discuss later, the improved statistics in the Virgo data have removed any doubt of the cluster's relative deficiency, but the Malmquist effect (whereby more distant samples are populated by more luminous galaxies) is a serious problem for comparison of galaxies in differing ranges of absolute magnitude.

Further concerns arise in applying the measure of M_H/L to the fainter galaxies ($m > +14.0$) because of uncertainties in the measured magnitudes. Photoelectric magnitudes for the large number of galaxies being surveyed are unavailable. For the fainter galaxies, the only source of magnitudes is the *Catalog of Galaxies and Clusters of Galaxies* (CGCG ; 119). Conversion to standard photometric systems, such as that of Holmberg (66), for objects for which direct measurements on those standard systems are not available has been a common practice in the literature (12, 30, 84). However, while these conversions may produce reliable results for the larger, brighter galaxies for which there is substantial overlap between the observational sample and the standard sample, the extrapolations necessary for application to smaller, fainter galaxies are very insecure. When properly corrected for systematic biases, especially the anomalies in the first volume, CGCG magnitudes are actually quite reliable, involving random errors on the order of 0.25 mag or less (45).

Another serious problem with the use of M_H/L is the necessity for good morphological classification, as this ratio is highly type dependent. As a practical matter, classification is difficult for distant small angular diameter galaxies, especially where large-scale, high dynamic range plate material is not available. Specifically, classification using the Palomar Sky Survey material is very uncertain for galaxies more distant than $cz > 6000$ or with diameters smaller than one arcmin. A prime example of preliminary misclassification is provided by NGC 5902, once thought to be a gas-rich lenticular but shown later to be an Sb galaxy with a low surface brightness disk (39).

One can correct for the residual dependence of M_H/L on luminosity by replacing luminosity as the second variable by the surface magnitude defined as $SM = m + 5 \log a$, where m is the apparent magnitude corrected for galactic extinction and internal absorption, and a is the optical major diameter (47). Because of the way L is folded into SM, the relation between M_H/L and SM nearly reduces to one between M_H and D alone.

Several authors have noticed the near-independence of the hybrid H I surface density M_H/D^2 of spiral disks on morphological type and the constancy of this parameter when averaged over the disk (22, 63, 98). Apparently, the diameter of a spiral disk is a much more important diagnostic parameter for the H I mass than are the luminosity and morphological type. Both M_H and D are disk properties, while L refers to the combined contributions of both the disk and the relatively gas-free spheroidal bulge. The increasing importance of the bulge component in earlier morphological types reduces the values of M_H/L. A question yet to be answered is whether the morphological dependence still remains if a disk luminosity is used in the computation of the H I mass-to-luminosity ratio.

It is interesting to note the case of the low surface brightness spirals studied by Romanishin et al. (91), a category of relatively isolated objects. When their H I content is judged by the average M_H/L ratio applicable for all late spirals, they appear overabundant by more than a factor of 2 (91, 105). If gauged in terms of their sizes, however, they have perfectly normal H I masses when compared with other isolated galaxies.

It should be underscored that M_H/D^2 is a hybrid quantity, as D is an optical diameter. For isolated galaxies, Hewitt et al. (63) have shown that the H I extent is well correlated with the optical diameter: with near-independence on morphological type, they find 70% of the H I mass to be confined within 1.2 times the blue diameter as taken from the *Uppsala General Catalog of Galaxies* (81a). However, M_H/D^2 is not perfectly invariant. A small residual dependence on D can be detected, in the sense that larger disks have slightly lower M_H/D^2. The correction of such residuals introduces a distance-scale-dependent bias, which is fortunately very small. The parameter M_H/D^2 may provide the most sensitive estimate of the normalcy of a galaxy's H I content.

2.2 The Distribution of H I Gas

If we naïvely assume that the H I distribution expected for a normal galaxy resembles an axisymmetric, flattened disk and that its velocity field likewise shows no asymmetries, we may well conclude, after an examination of the collection of H I maps of galaxies, that there are no normal galaxies. Many, if not all, of the well-studied galaxies show deviations from circular symmetry and evidence for noncircular motions (9, 10). Detailed maps show the H I concentrated in the regions of ongoing star formation, following spiral arms ridges. The large-scale deviations from axial symmetry, warps, and even lopsidedness may be coincident in the H I, continuum, and optical light distributions. Central depressions in the H I distribution such as that seen in the Milky Way seem to be more pronounced in high-luminosity early-type systems, which possess large bulges. Peculiarities in the H I distribution occur even in galaxies that are relatively isolated in space (73).

Of particular importance to the question of environmental influence is the extent of the H I distribution, which can be an indicator of the effectiveness of external gas removal mechanisms expected to favor the outer, less tightly bound portions of the disk. Since isophotal radii are very uncertain when H I extents are measured with poor resolution, H I mapping studies of galaxies made with filled-aperture telescopes have favored the estimate of H I extent in terms of effective radii (38, 44, 63), although synthesis studies that can produce isophotal radii are accumulating (10, 93, 112a, 114). In both cases, the present data suggest that H I disks are generally modest in size, not extending past one or two Holmberg radii (20, 63). Very large H I envelopes do surround a number of systems, however (68a, 93). The better-behaved envelopes are usually found around late-type spirals and Magellanic-type irregulars, while the H I distribution in lenticulars is frequently concentrated in an annulus whose radius lies at the edge of the optical extent (80, 113). As is discussed in Section 3, H I appendages extend from the disks of galaxies believed to be undergoing tidal interactions. In a number of other cases, particularly that of NGC 628 (19, 100), the extended H I emission cannot be understood in terms of a recent encounter with a visible neighbor.

3. GROUPS OF GALAXIES

Some of the most exotic specimens of neutral hydrogen distributions are found where galaxies seem to come together. In fact, neutral hydrogen proves to be an even better tracer of tidal disruption than does starlight. In this section, we examine the occurrence of neutral hydrogen streams, the interactive nature of galaxies in groups, and the ultimate fate of the gas, stars, and galaxies that partake in such encounters.

3.1 *The Group Environment*

Groups of galaxies come in a wide spectrum of sizes and membership; basically, a group is often defined as a density enhancement of about 10^2 to 10^3 with respect to its surroundings. Groups can be further categorized as loose or compact, with the compact structures possibly only temporarily unbound subcondensations existing within more widespread structures (91a). In comparison to the richer clusters, groups have a higher proportion of spiral galaxies; only a very few of the groups in de Vaucouleurs' (29) list are dominated by galaxies earlier than Sa. Velocity dispersions within groups are on the order of a few hundred kilometers per second, and in a few interesting instances, such as the IC 698 group (118), they are much smaller. X-ray emission has been found in prominent galaxies in two nearby loose groups, NGC 3607 (8) and NGC 5846 (7), with X-ray luminosities that follow an extrapolation of the relationship between X-ray luminosity and velocity dispersion found for clusters. Both of these groups, however, are distinctive among other nearby loose groups in having higher velocity dispersion and a larger representation of early-type objects. In general, the density of the intergalactic gas in groups is low, and the major influence of group membership is simply the presence of nearby neighbors.

In both clusters and groups, the increased density of the galaxy population enhances the likelihood of a tidal interaction among members. The number of encounters is expected to be highest in clusters, where the space density is large. But the potential for actual damage done to an individual galaxy, the "disruption damage" as discussed by Cottrell (26), depends on the degree of penetration, the duration of the encounter, and the sense of the orbital and rotational motions. In the impulsive approximation, the disruption damage goes as $1/v_r$, where v_r is the relative velocity of the galaxies involved. Significant perturbations of the disk occur only when the relative velocities are close to the parabolic (108). Tidal encounters exert strong influences on group galaxies that stray too close to one another. However, because the relative velocities expected in clusters are much higher than in groups, a typical collision in a rich cluster will render only about one tenth the damage as one with the same impact parameter and mass ratio that takes place in a small group.

3.2 Tidal Streams in Nearby Groups

Within the Local Group, several examples testify to the disruption of the outer H I disks of galaxies. The bridge between the Magellanic Clouds (65) and the Magellanic Stream (78) is likely the remnant of a recent close passage of the Clouds past the Milky Way; the high-velocity H I clouds seen on the galactic periphery may also be debris from that encounter (42). Moreover, the outer warps of the Milky Way (64), M31 (81), and M33 (90) may be induced rather than generic.

Perhaps the best cases of morphological disruption are the H I appendages, which have been mapped in detail because of their large angular size and favorable viewing geometry. The presently identified systems with intergalactic H I are listed in Table 1. Among the best known of the two-dozen streams are the M81/M82/NGC 3077 system (3, 111), the Leo triplet NGC 3623/3627/3628 (shown in Figure 1; 60, 92), the "Antennae" NGC 4038/4039 (110), and the NGC 4631/4656 system (115). The detailed mapping of the H I distribution permits not only the numerical simulation of the tidal disruption of the sky distribution, but also the tracing of the velocity field. Computer simulations of hypothetical encounters have been developed by numerous authors, especially Toomre & Toomre (108), whose model nicely reproduced the details of the optical images of the "Mice" NGC 4676 and the "Antennae." The observed velocity fields merge with the parent galaxies in numerical models of several encounters. Even with relatively simplistic simulations that ignore the subtler effects of magnetic fields, induced shocks, and self-gravitation within arms, the gross characteristics can be reproduced, although the finer details as seen in such streams as the tail of NGC 3628 (60) elude such models. It thus seems clear that tidal encounters can remove a substantial fraction, even as much as 50%, of the interstellar H I from a perturbed galaxy.

Most of the H I streams that have been well mapped emanate from galaxies showing optical peculiarities that first aroused the interest of observers. In many cases, the neutral hydrogen appendages coincide with optical plumes. Recent high-resolution observations show that the optical and H I tails of NGC 4747 are offset by 11° in position angle (116); the relationship of the stars and gas seen in the plumes is obviously complex. The detailed mapping of a significant number of interacting systems that do not necessarily show optical plumes or wisps has been restricted by the difficulty in achieving the required sensitivity and angular resolution. Several surveys of binary galaxies (62) and loose groups (54) made with the Arecibo telescope have shown that the phenomenon is common, although by no means ubiquitous. Undoubtedly, some of our understanding of the

Table 1 Examples of intergalactic H I

System	References
Galaxy—Magellanic Clouds	65, 78
Stephan's quintet	1
M51	59, P. Appleton and R. Davies, priv. comm.
M81/M82/NGC 3077	3, 111
M95/M96	95
NGC 678/680	54
NGC 1510/1512	52
NGC 3165/3166/3169	54
NGC 3395/3396	62
NGC 3623/3627/3628	60, 92
NGC 4038/4039	110
NGC 4485/4490	113a
NGC 4631/4656	115
NGC 4725/4747	53, 116
NGC 7241	R. Giovanelli and M. Haynes, unpublished
NGC 7448/7463/7464/7465	54
II Zw 70–71	6a
UGC 6922/6956	2

frequency of interactions is encumbered by the severe selection restrictions that pervade recognition of tidal activity, including favorable viewing geometry, the presence of substantial interstellar gas in at least one of the affected systems, and the relatively short time frame over which the tidal remnant is likely to be prominently separate from the galactic disk but yet not too dispersed. Prediction of the occurrence of tidal activity in any one system is hampered by the absence of three-dimensional perspective in the true distance between galaxies, in their orbital parameters and directions of motion, and in the relative sense of rotation and orbital motion. A vivid example occurs in the Leo triplet, shown in Figure 1: the third member of the triplet, NGC 3623, appears unperturbed, while the disruption of NGC 3627 and NGC 3628 is in marked contrast because of the different orientations of spin and orbit.

While the evidence in support of a tidal encounter origin is compelling in many appendages, the chaotic and distinct nature of some extended H I distributions makes their interpretation still uncertain. The case for

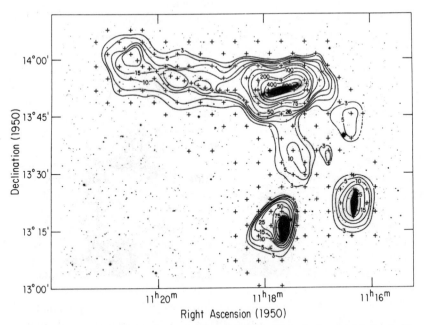

Figure 1   Contours of H I column density (in units of $\int T_A \, dV$) in the Leo triplet of galaxies (clockwise from top: NGC 3628, NGC 3623, NGC 3627), superimposed on an enlargement of the Palomar Sky Survey. A faint optical plume coincides with the H I appendage to the northeast of NGC 3628 (60).

discrete, massive ($\geq 10^8 \, M_\odot$) H I clouds (79) is unconvincing (35, 61). At the same time, however, the high-velocity gas near M51 (59, P. Appleton and R. Davies, private communication), the complex H I distribution in Stephan's quintet (1), and the recently discovered intergalactic "cloud" in the Leo group (95) do not fit straightforward tidal models involving recent encounters between two neighboring galaxies. It is clear that a single two-dimensional snapshot and a few radial velocities are not enough to convict. In these cases, the puzzle and the frustration of our limited view remain.

3.3 *The Fate of Tidal Debris*

By disturbing and perhaps removing a substantial portion of a galaxy's interstellar gas, a tidal encounter can wreak substantial havoc on a victim's evolution. Peculiar morphological characteristics likely resulting from the dramatic tidal events in which the galaxy is participating appear in numerous systems with H I streams, notably NGC 3077 and M82 (111), NGC 4747 (53, 116), and M51 (59, P. Appleton and R. Davies, private communication). In addition to the general tidal disruption, the gas can either be removed from the galaxy's disk or caused to fall toward the center of the galaxy by a loss of angular momentum. Mass transfer from one galaxy to another is likely if the orbital sense coincides with the spin orientation, prolonging the activity of the tidal acceleration. In the latter circumstance, gas that was raised to high-z extents will fall back to the nucleus, perhaps inducing a burst of star formation or other activity. Cottrell (26) suggests that the Irr II galaxies are the end products of collisions between galaxies in groups, and in fact, a significant fraction of the galaxies with H I streams are classified as such.

The potential of mass transfer is of particular importance for the observed H I distribution in early-type galaxies. In many of the elliptical and lenticular galaxies with detectable H I masses, there is a marked discrepancy between the gas distribution and dynamics and those of the stars (71, 85). The incompatibility of the angular momentum vectors between the stellar and gaseous components argues in favor of accretion of the gas from the outside (102). However, direct evidence for the accretion of substantial amounts of H I gas contributed by a late-type galaxy to an earlier-type companion is rare, e.g. NGC 678/680 (54), NGC 1510/1512 (52), and NGC 4026 (2). More often, there is no neighbor with sufficient proximity to explain the mass transfer. Moreover, there may be no relationship between the H I mass of ellipticals and their environment at all (31). At present, there is no satisfactory explanation for either the absence of significant interstellar gas in many early-type galaxies or the presence of H I in others.

4. CLUSTERS OF GALAXIES

Cluster cores are the sites where perhaps the most severe environment-related alterations to the evolution of galaxies occur. The question has been frequently asked, To what extent is the morphological type segregation, observed to be dependent on the galaxian density, the result of these alterations? In broad terms, this question is reviewed by Dressler (33) in this volume. Here, we consider the evidence that has accumulated from H I observations in favor of the effectiveness of mechanisms that dramatically affect the structure of spiral disks.

4.1 *Comparison Samples and Definition of H I Deficiency*

As a rule, the parameters that describe the optical appearance of a galaxy are used to estimate its expected H I content, with the assumption that it does not deviate from the standards of normalcy set by some control sample. The ratio between that expectation and the actual observed value of a given galaxy's H I content is labeled the "deficiency factor" or the "H I deficiency." It is usually expressed as a logarithmic quantity, positive for H I-deficient galaxies. The reliability of this parameter, which depends not only on the quality of the observed quantities but also on the choice of the control sample and of the criteria utilized to estimate the H I content, has been questioned (11). When rigorous criteria are applied, however, it yields quite satisfactory results (46). Earlier cluster work used as comparison samples those of normal galaxies compiled by Roberts (89) and by Balkowski (6). These samples include mainly nearby galaxies, which are sometimes members of groups or pairs, a circumstance frequently seen to favor disruptive phenomena (as discussed in Section 3). They are also vulnerable to the appearance of Malmquist-type biases in the analysis of distant clusters (17, 43). Larger compilations of H I properties of individual galaxies now exist (18, 69), but the largest comparison sample, rigorously selected to include objects affected as little as possible by interactions with neighbors or violent environments, consists of approximately 300 isolated galaxies observed at Arecibo and Green Bank (56, 58, 63). Of course, most cluster galaxy samples include objects found in different regimes of local density, thus providing a measure of built-in relative comparison.

The various observers have adopted different criteria for the definition of a galaxy's H I content, as illustrated in Section 2.1. Although the most frequently used quantity is $\log(M_H/L)$, it is operationally the least accurate indicator of H I content. For the sample of isolated galaxies, the rms dispersions of $\log(M_H/L)$ values around mean estimates are high within all morphological types, ranging between 0.3 and 0.4. By comparison, in commensurable units, that scatter is reduced to about 0.21 (when averaged over all types) if one analyzes H I content in terms of the hybrid surface density M_H/D^2, as outlined in Section 2.1. An added advantage of the second approach is its almost negligible dependence on morphological type; because the estimate of type, especially in faint small angular diameter galaxies, is accompanied by a large margin of error, the minimization of the possible effect of misclassification on the inferences of H I deficiency is a welcome result.

4.2 *Cluster Samples*

Inhomogeneity in the observed cluster samples, as well as differences in the data quality attained and the analytical techniques adopted by the various groups of workers, makes comparison of the H I properties of different clusters difficult. Following the relatively strict selection criteria adopted by a recent comparative study (42a, 46), we list in Table 2 a summary of properties of clusters for which extensive H I observations have been made. It should be pointed out that observations of large samples in the 21-cm line are currently constrained within recession velocities smaller than 10–15,000 km s^{-1}. For each cluster in Table 2, we list its Abell radius r_A (column 2), its systemic velocity and line-of-sight velocity dispersion (columns 3 and 4), the number of observed galaxies in the cluster sample (column 5), the number of these galaxies that are projected within $1r_A$ of the cluster center (column 6), the H I-deficient fraction (column 7), the X-ray luminosity in the 2–6 keV range (column 8), and the references to published material (column 9). Some of the listed quantities require further explanation. Only galaxies of type Sa or later are included; nondetections are used only if the upper limit for the H I flux sets the H I deficiency limits higher than 0.3; and H I deficiencies are all measured by the same criterion, namely, by using the diameter as the diagnostic variable, as described in Sections 2.1 and 4.1. The deficient fraction is defined as the ratio between the number of galaxies within $1r_A$ with H I deficiency in excess of 0.3 and the total number of galaxies observed within the same radius. The Hydra cluster (A1060) sample does not conform with the criteria of sample membership listed above; we have tentatively estimated a deficient fraction from the M_H/L values given in (88), which we list with the caveat of the inhomogeneity in this entry. Because the Virgo cluster has received more detailed attention, we devote Section 4.3 to a discussion of the related results concerning it.

A clear pattern of H I deficiency in the central region is observed in at least 5 clusters: Virgo (see Section 4.3 for a broader discussion), Coma (A1656; 24, 106, 107), A1367 (24, 25, 106, 107), A262 (47), and A2147 (43, 46, 96). A marginal case can be made for A2151 (43, 46, 96) and Hydra (88),

Table 2 Clusters with extensive H I observations

Cluster	r_A (°)	v (km s^{-1})	Δv (km s^{-1})	n_{tot}	n_1	H I def. fraction	L_X (erg s^{-1})	References
A262	1.75	5068	452	85	27	0.57	43.2	47
Cancer	1.92	4607	317	40	13	0.23	<42.6	46, 96
A1060	2.39	3708	676	17	12	—	43.2	88
A1367	1.40	6370	813	43	19	0.47	43.5	24, 25, 46, 106, 107
Virgo	5.00	1026	673	133	51	0.62	42.9	22, 28, 40, 41, 44, 114
A1656	1.29	6950	905	56	17	0.82	44.2	24, 46, 106, 107
Z74−23	1.52	5840	412	32	17	0.00	<42.8	23, 106
A2147	0.83	10867	1189	34	14	0.64	44.1	43, 46, 96
A2151	0.83	11055	920	40	20	0.33	43.6	43, 46, 96
Pegasus	2.20	3990	616	62	16	0.07	<42.5	13, 46, 87

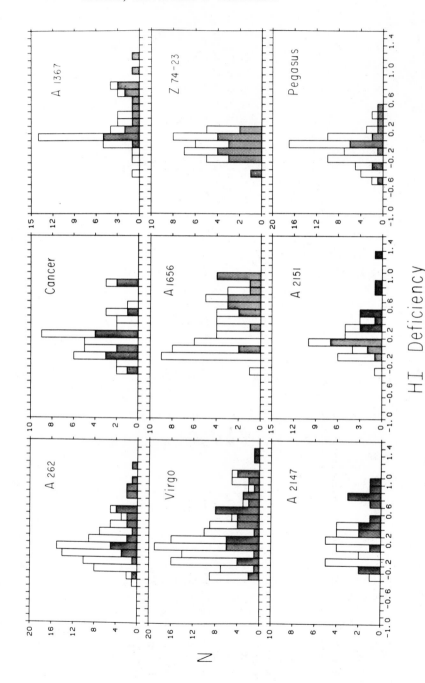

Figure 2   Histograms of H I deficiency in nine clusters. The upper envelope represents all galaxies in each cluster sample; the shaded one, those projected within one Abell radius from the center (42a).

while little or no deficiency is discernible in Cancer (46, 96), Pegasus (13, 46, 87) or Z74 − 23 (23, 106). Figure 2 (from 42a) summarizes the results for nine of the clusters listed in Table 2. In each case, the shaded blocks of each histogram correspond to the galaxies projected within $1r_A$ of the cluster center, while the unshaded portions correspond to the total sample, as selected for Table 2. The deficiency end of the histograms tends to be strongly dominated by galaxies that lie close to the cluster center. In Figure 3 (also from 42a), differences in the radial extent of the deficiency pattern are illustrated for three well-studied clusters. In the case of Coma (A1656), deficient galaxies appear to prevail out to $1.5r_A$, while in A262 and A2147

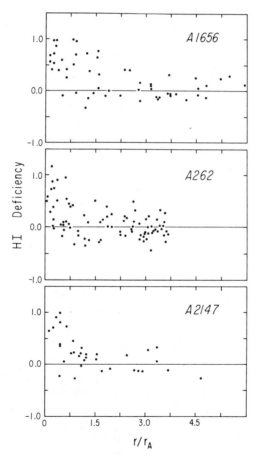

Figure 3   H I deficiency as a function of radial distance in three clusters. Nondetections are plotted at the nominal lower limit of the inferred H I deficiency. Radial distance is scaled by the Abell radius r_A of each cluster.

(which are characterized by smaller deficient fractions) H I deficiency extends to less than $1r_A$. While the high deficient fraction of Coma is partly due to this difference in extent, Coma has also a very high proportion of galaxies with very high deficiencies, a detail not fully appreciated in Figures 2 and 3 because nondetections have been plotted at their nominal limit values.

4.3 The Virgo Cluster

Because of its proximity, the Virgo cluster has been the object of special attention. For comparison, note that the *faintest* galaxies observable in, say, the Hercules cluster would appear at an apparent magnitude of about 12 in Virgo. The measurements of H I content are complete, for the galaxies projected within a 5 or 6° radius from the center, to a magnitude of 13.5, and they are rapidly nearing completion to a magnitude of about 15.0. In addition to offering a much larger sample size, Virgo also lends itself to studies of the spatial variations of H I properties of early-type galaxies, which are very difficult to detect even outside clusters at larger distances because of their low H I content. Finally, and most importantly, the large angular sizes of the Virgo galaxies can be easily resolved not only by synthesis instruments but also by the Arecibo telescope, allowing the study of the variations in the extent of the gaseous disks, the outer parts of which may be most vulnerable to hostile environments.

The first reported result of H I deficiency in Virgo cluster galaxies appeared in 1973 (28). Based on the observations of 25 galaxies, an average deficiency of Virgo galaxies with respect to field spirals of a factor of 1.6 was inferred. As discussed in Section 2.1, this result was subject to skeptical questioning for several years; the results of larger samples, however, did confirm the thrust of the initial finding of Davies & Lewis. Chamaraux et al. (22) extended the analysis to a sample of 56 Virgo galaxies and obtained an average deficiency of 2.2 ± 0.3 with respect to their comparison sample, a result confirmed by studies based on samples observed at Arecibo (40, 44). When H I deficiency is analyzed using the optical diameters as a diagnostic parameter, it is necessary to appraise the possible impact of systematic differences in the sizes of optical disks, within and outside of clusters (see Section 2.1). It appears that Virgo cluster spirals are on the average smaller than those in the field (83); this effect, which is not confirmed in other clusters such as Hercules, may produce underestimates by as much as a factor of 1.6 in the average H I deficiency of Virgo galaxies; thus the average H I deficiency may be as high as a factor of 4 (44). The sky distribution of H I-deficient galaxies appears to slightly favor the region of the cluster around M87 and M86, rather than the southern concentration around

NGC 4472; a recent analysis based on a deep redshift survey of the Virgo cluster (68) suggests that the "clump" around M87 may constitute a separate dynamical unit with higher velocity dispersion and mass-to-light ratio than the cluster as a whole; by contrast, the clump around NGC 4472 has a lower velocity dispersion (489 vs. 761 km s^{-1} for the M87 clump) and much weaker X-ray emission.

The depressed H I content of galaxies projected within the cluster core is also detectable in Virgo if the analysis is restricted to early-type systems; although the scatter in the values of the H I content for these galaxies, which is larger than for later spiral systems, makes the determination of an H I deficiency uncertain, the differences in the detection rates inside and outside the core are quite significant. Figure 4 (after 41) illustrates this effect separately for S0 and S0a–Sa galaxies; the associated detection rates for S0 galaxies are 4/39 inside a 6° radius and 21/49 outside, while the corresponding figures for S0a–Sa galaxies are, respectively, 7/19 and 32/38 (according to 41).

H I sizes of spiral disks in Virgo have been measured both with the Arecibo antenna and synthesis telescopes. The single-dish measurements indicate that the extent of the disks is smaller within the cluster core than outside by about a factor of 1.5 (40, 44); the reduced size of H I disks is also well correlated with H I deficiency, suggesting that the mechanism(s) that causes the H I deficiency does so by depleting preferentially the outer regions of the disks. This result has been confirmed by VLA observations (112a), as is illustrated in Figure 5, although Westerbork results still cast doubt on the exact factor by which cluster H I disks appear to be shrunk (114).

4.4 Gas Removal Mechanisms

In 1951, Spitzer & Baade (103) proposed that the morphological type segregation seen in clusters could be the result of environment-related mechanisms; collisions between galaxies, made frequent by the high densities and high velocity dispersions in cluster cores, would lead to removal of gas from spiral disks. As discussed in Section 3.1, tidal disruption is highest when the relative velocity of the two galaxies is smaller, a circumstance that makes galaxies in clusters less vulnerable to this process than those in small groups. For rapid encounters to be effective in producing stripping, impact parameters must be small, i.e. disks must collide. Chamaraux et al. (22) estimated that the probability that a bright spiral in Virgo undergoes such an encounter in a Hubble time is only about 0.03. In a cluster like Coma, the higher value of the velocity dispersion, which would increase the chance of encounters, is balanced by the reduced

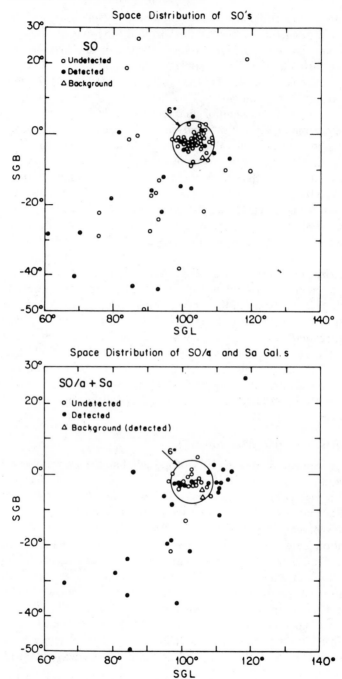

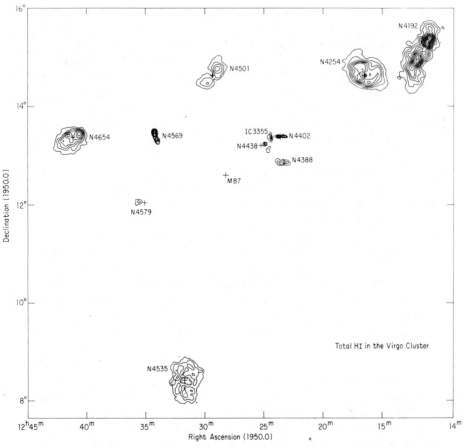

Figure 5 Synthesis H I maps of some of the spirals in the Virgo cluster core. Notice the sizes of the galaxies closer to the center, much smaller than those in the periphery (112a).

spiral fraction; thus there is no reason to expect an enhanced rate of spiral-spiral collisions, at least at the present time. Several authors have considered the integrated effect of encounters that occur when the constraints in the impact parameter are relaxed (34, 86); serious disruption occurs only in the outer regions of disks ($r > 15$ kpc), and stripping of large fractions of the interstellar gas is improbable.

Figure 4 Spatial distribution, in supergalactic coordinates, of early-type galaxies in the Virgo area, coded by detection statistics. The circle has a radius of $6°$ (41).

The intracluster medium (ICM) observed in the X-ray domain in many clusters (36) may play a more important role for galactic structure. The ICM is tightly coupled with the galaxian component of the cluster; it may accrete onto massive galaxies or remove the interstellar gas of galaxies that venture into cluster cores, either by means of the ram pressure generated by their motion relative to the ICM or by evaporation induced by heat conduction into the cooler interstellar gas. In the first instance, gas removal occurs if the ram pressure overcomes the restoring gravitational force per unit area that acts on the disk gas, i.e. if $\rho_{icm}v_\perp^2 > 2\pi G\sigma_*(r)\sigma_g(r)$, where ρ_{icm} is the ICM density, $v_\perp$ the component of the velocity of the galaxy (with respect to the ICM) perpendicular to the disk of the galaxy, G the gravitational constant, and $\sigma_*(r)$ and $\sigma_g(r)$ respectively the surface density of collapsed matter and that of gas in the disk, at a distance r from the galactic center (51). Numerical simulations (48, 75, 97) have shown that ram pressure can be an effective mechanism of gas removal; however, these results should be considered with caution, as they are based on simplified models that neglect the effect of viscosity, cooling instabilities, and thermal conduction. Hydrodynamic calculations that include these effects suggest a more complex picture (97), with the possibility of star formation induced near the center of the galaxy, a prediction that may find observational support in UGC 6697 (11). Clearly, the degree of clumpiness of the interstellar medium will have a major effect on the outcome of the interaction (101). The effects of heat conduction and consequent evaporation of the interstellar gas have been investigated, most notably by Cowie & Songaila (27). The evaporative mass loss rate is a sensitive function of the ICM temperature and, somewhat less sensitive, of the ICM density. For a spiral disk of radius 15 kpc and thickness 200 pc, the evaporative mass loss rate goes as $T_8^{2.5}$ provided that $n_3 > 3T_8^2$, where n_3 and T_8 are the ICM density and temperature in units of 10^{-3} cm^{-3} and 10^8 K, respectively (except when radiation from the interface becomes important, for $n_3 > 10^{2.2}T_8^2$, in which case material condenses onto the galaxy). For $n_3 < 3T_8^2$, heat flux saturation occurs, and the mass loss rate is proportional to $(n_3T_8^2)^{0.6}$ [see Figure 1 in (27)]. Evaporation could also be an effective means of gas removal from galaxies; a major uncertainty lies with the topology of magnetic fields, which may inhibit conduction.

Certainly the gas content of a galaxy does not remain unaltered in the absence of an active environment. Star formation occurs, and gas is restored to the disk via stellar mass loss; and supernovae may power galactic winds (77), a process that has been invoked particularly for elliptical galaxies and that appears to be effective as a gas removal mechanism only at early stages of galaxy evolution (48). The rates of these processes are uncertain; to what extent they are affected by an active environment is even more difficult to say. While the evidence for their relevance grows, the interpretation of the details remains highly speculative.

4.5 *Discussion*

Because graphic evidence that suggests ongoing interaction between galaxies and their environment has been collected for a variety of individual objects (37, 72, 72a, 76, 99), there has been a tendency toward interpreting the H I deficiency as the result of presently active mechanisms, rather than the remnant effect of earlier, now exhausted periods in the evolution of clusters. There has not been, however, a general consensus. As documented by Dressler (32), it is difficult to account for all the aspects of the morphological type dependence on local density purely in terms of secular stripping mechanisms. Hence, conditions at some early phase must be responsible for most of the differentiations: while bulges collapse rapidly, disks may do so slowly (50) and continue to accrete gas from external reservoirs; the tidal stripping of those reservoirs, at early stages of cluster formation, would arrest the growth of disks (74), thus explaining the preponderance of lenticular systems. Could the "arrested growth" hypothesis account for the observed H I deficiencies? This view has been favored by some (11, 70).

One obvious direction to search for corroborating evidence of the H I deficiency measurements is in the colors of the H I-poor cluster galaxies. In fact, colors of Virgo spirals are found to be redder than those of spirals in the field (70, 104), and differences have been reported also for other clusters (14, 43). However, in Virgo, H I-deficient spirals exhibit a large difference in their $(U - V)$ colors with respect to nondeficient spirals (41, 44), a result that suggests a relatively sudden and recent arrest of star formation in the first group and that is at odds with the arrested growth hypothesis. Disk $H\alpha + [N\ II]$ line emission also reveals a striking difference between Virgo and field galaxies (70); on the other hand, nuclear $H\alpha$ emission does not appear to be significantly depressed in Virgo spirals (104), suggesting that the cluster environment has preferentially affected the outer regions of the galaxies. This result is in qualitative agreement with the observed reduction of H I disks discussed in Section 4.3. The morphological appearance of the spiral pattern has been invoked to define a class of anemic spirals (109), suggested to represent a transitional phase of a spiral galaxy when star-forming activity has ceased. Although the morphological classification involves a measure of subjectiveness, anemic spirals, and the more extreme cases of "smooth-arm" spirals (117), have been shown to be very H I-deficient objects.

The pattern of H I deficiency is not equally as pronounced in all clusters. Some authors (15) have reviewed the results, concluding that evidence for a global sweeping mechanism at work is present only in Coma. Others have found that the magnitude of the effect depends on cluster properties, most notably on the cluster's X-ray luminosity (46) (as can be reconstructed from Table 1). Strong X-ray emitters, like Coma and A2147, present the highest fraction of deficient spiral galaxies. The comparison between A2147 and A2151 is of particular interest: while the space density of spirals is higher in A2151, and the velocity dispersions are comparable, a circumstance that would favor a higher collision rate and therefore more effective collisional stripping in A2151, the H I deficiency is more pronounced in A2147, which is dominated by a cD galaxy and a healthy ICM.

Another clue could be provided by comparisons of H I deficiency with a galaxy's velocity relative to the ICM and with the cluster's X-ray temperature, as they may permit distinctions between the efficiency of ram pressure and evaporative processes. In analyzing the velocities of individual galaxies, the lack of knowledge of a galaxy's exact location and velocity vector introduces a great deal of uncertainty. A correlation supportive of ram pressure stripping at work has been obtained for the Virgo cluster, although no confirmation in other clusters has been found (44, 106). Data on X-ray temperatures of the ICM are available for only five of the clusters well studied in H I, and the paucity of the statistics allows little speculation; it is interesting to note, however, that A2147 has the lowest of cluster X-ray temperatures, yet its H I deficiency is quite pronounced.

The various pieces of circumstantial evidence tend to support a picture of ICM-induced gas removal from galaxies in clusters. The large fraction of H I-poor galaxies in several clusters suggests that the efficiency of the gas removal processes is higher now than it was a few billion years ago (46), a conclusion also reached by Gisler (48) on theoretical grounds. Schemes based on evolutionary models that may distinguish a lenticular galaxy that originated from a stripped spiral from one whose morphology originated in earlier evolutionary stages have not yet been developed. However, there appears to be an enhanced fraction of S0s, and a matching depression of the fraction of spirals, at any given value of the local density of galaxies in clusters that are strong X-ray emitters (32). Furthermore, de Freitas et al. (28a) have found that the axial ratios of S0 galaxies tend to be higher in regions of high galaxian density than in the field; they identify the flat cluster S0s as swept spirals. These statistical results suggest that morphological evolution, secularly imposed by environmental factors, is reinforcing the characteristics of the segregation that may have been established *ab initio* (46).

5. CONCLUDING REMARKS

The 21-cm line of neutral hydrogen has been used since its discovery in 1951 as a probe of the interstellar matter within our own and other galaxies and, both directly and indirectly, of the intergalactic gas that surrounds the galaxy distribution. Investigations of the impact of the environment on the H I content of galaxies have shown that under the right conditions of intergalactic density, external influences upon galaxy evolution can be dramatic.

In groups, where relative velocities are small, tidal damage is frequent, and because the outer regions of the disks are most susceptible, disruptive encounters may be best delineated by the H I distribution. The formation of bridges and tails provides spectacular displays in which the appendages extend several galaxy diameters. H I masses in excess of $10^8 \, M_\odot$ may be freed from a parent galaxy; by falling back onto the disk of a parent or neighbor, such gas may then fuel bursts of activity. Still, in a number of the systems in which gas is detected far outside the disk, explanations in terms of recent tidal events are inadequate; are such gas clouds the remnants of the galaxy formation era or of not too recent disruptive encounters? While rare, some galaxies are endowed with extraordinarily extended H I envelopes; are these at the extreme of a normal distribution, or do they belong to a special category? And if the external gas reservoirs required by the continuous accretion models of slow spiral evolution (50) are only swept away by tides after encounters, why do the disks of isolated galaxies not all have extended gas distributions?

In clusters, the tidal damage is reduced by the much higher values of the relative velocities during close encounters. However, stripping of the gaseous component of spirals occurs in clusters, with consequences far more drastic than those of the tidal encounters in groups. Stripping appears to be associated primarily with the presence of a pervasive ICM, which acts on the interstellar gas via ram pressure or conductive processes. But is such galaxy–intergalactic medium interaction too efficient? If so, then why do we see spirals in clusters at all?

The analysis of H I deficiency in cluster galaxies is a delicate business demanding both optical and 21-cm data of high quality. The number of clusters studied in the 21-cm line is still relatively small, and the identification of statistical trends only tentative. Aperture synthesis observations with long integration times will be necessary in order to extend the sample of observed clusters beyond the redshift limits currently accessible to filled-aperture instruments. Much work remains to be done in the Virgo cluster. Further investigation is necessary to confirm the correlation between H I deficiency, the velocity relative to the cluster as a whole, and the density of the intergalactic gas. Observations of integrated CO content will provide an indication of the fate of the clumpier molecular gas.

Perhaps the most perplexing issues relate to the ultimate fate of the gas-swept spirals: What do they look like after a few billion years? Can we somehow distinguish between those lenticulars whose type was determined at the time of cluster collapse and those perhaps similar in appearance that originated from recently stripped spirals? In other words, can we tell the genetic attributes from the sociological?

Literature Cited

1. Allen, R. J., Sullivan, W. T. 1980. *Astron. Astrophys.* 84:181–90
2. Appelton, P. N. 1983. *MNRAS* 203:533–44
3. Appleton, P. N., Davies, R. D., Stephenson, R. J. 1981. *MNRAS* 195:327–52
4. Arp, H. A. 1966. *Ap. J. Suppl.* 14:1–20
5. Athanassoula, E., ed. 1983. *Internal Kinematics and Dynamics of Galaxies.* Boston: Reidel. 432 pp.
6. Balkowski, C. 1973. *Astron. Astrophys.* 34:43–55
6a. Balkowski, C., Chamaraux, P., Weliachew, L. 1978. *Astron. Astrophys.* 69:263–70
7. Biermann, P., Kronberg, P. P. 1983. *Ap. J. Lett.* 268:L69–74
8. Biermann, P., Kronberg, P. P., Madore, B. F. 1982. *Ap. J. Lett.* 256:L37–40
9. Bosma, A. 1981. *Astron. J.* 86:1791–1824
10. Bosma, A. 1981. *Astron. J.* 86:1825–46
11. Bothun, G. D. 1982. *Ap. J. Suppl.* 50:39–54
12. Bothun, G. D., Schommer, R. A. 1982. *Ap. J. Lett.* 225:L23–28
13. Bothun, G. D., Schommer, R. A., Sullivan, W. T. 1982. *Astron. J.* 87:725–30
14. Bothun, G. D., Schommer, R. A., Sullivan, W. T. 1982. *Astron. J.* 87:731–38
15. Bothun, G. D., Schommer, R. A., Sullivan, W. T. 1984. *Astron. J.* 89:466–74
16. Bottinelli, L. 1982. See Ref. 57, pp. 5–8
17. Bottinelli, L., Gouguenheim, L. 1974. *Astron. Astrophys.* 36:461–62
18. Bottinelli, L., Gouguenheim, L., Paturel, G. 1982. *Astron. Astrophys. Suppl.* 47:171–92
19. Briggs, F. H. 1982. *Ap. J.* 259:544–58
20. Briggs, F. H., Wolfe, A. M., Krumm, N.,

Salpeter, E. E. 1980. *Ap. J.* 238:510–23
21. Butcher, H., Oemler, A. 1978. *Ap. J.* 226:559–65
22. Chamaraux, P., Balkowski, C., Gerard, E. 1980. *Astron. Astrophys.* 83:38–51
23. Chincarini, G. L., Giovanelli, R., Haynes, M. P. 1979. *Astron. J.* 84:1500–10
24. Chincarini, G. L., Giovanelli, R., Haynes, M. P. 1983. *Ap. J.* 269:13–28
25. Chincarini, G. L., Giovanelli, R., Haynes, M. P., Fontanelli, P. 1983. *Ap. J.* 267:511–14
26. Cottrell, G. A. 1978. *MNRAS* 184:259–64
27. Cowie, L. L., Songaila, A. 1977. *Nature* 266:501–3
28. Davies, R. D., Lewis, B. M. 1973. *MNRAS* 165:231–44
28a. de Freitas, J. A., de Souza, R. E., Arakaki, L. 1983. *Astron. J.* 88:1435–41
29. de Vaucouleurs, G. A. 1975. See Ref. 94, pp. 557–600
30. Dickel, J. R., Rood, H. J. 1976. *Ap. J.* 223:391–409
31. Dressel, L. L., Bania, T. M., O'Connell, R. W. 1982. *Ap. J.* 259:55–66
32. Dressler, A. 1980. *Ap. J.* 236:351–65
33. Dressler, A. 1984. *Ann. Rev. Astron. Astrophys.* 22:185–222
34. Farouki, R., Shapiro, S. L. 1981. *Ap. J.* 243:32–41
35. Fisher, J. R., Tully, R. B. 1981. *Ap. J. Lett.* 243:L23–26
36. Forman, W., Jones, C. 1982. *Ann. Rev. Astron. Astrophys.* 20:547–85
37. Forman, W., Schwarz, J., Jones, C., Liller, W., Fabian, A. C. 1979. *Ap. J. Lett.* 234:L27–31
38. Fouque, P. 1983. *Astron. Astrophys.* 122:273–81
39. Gallagher, J. S. 1979. *Astron. J.* 84:1281–92

40. Giovanardi, C., Helou, G., Salpeter, E. E., Krumm, N. 1983. *Ap. J.* 267: 35–51
41. Giovanardi, C., Krumm, N., Salpeter, E. E. 1983. *Astron. J.* 88: 1719–35
42. Giovanelli, R. 1980. *Astron. J.* 86: 1468–79
42a. Giovanelli, R. 1984. See Ref. 49. In press
43. Giovanelli, R., Chincarini, G. L., Haynes, M. P. 1981. *Ap. J.* 247: 383–402
44. Giovanelli, R., Haynes, M. P. 1983. *Astron. J.* 88: 881–908
45. Giovanelli, R., Haynes, M. P. 1984. *Astron. J.* 89: 1–4
46. Giovanelli, R., Haynes, M. P. 1984. *Ap. J.* In press
47. Giovanelli, R., Haynes, M. P., Chincarini, G. L. 1982. *Ap. J.* 262: 442–50
48. Gisler, G. R. 1976. *Astron. Astrophys.* 51: 137–50
49. Giuricin, G., Mardirossian, F., Mezzetti, M., eds. 1984. *Clusters and Groups of Galaxies.* Boston: Reidel. In press
50. Gunn, J. E. 1982. In *Astrophysical Cosmology, Pontif. Acad. Sci. Scr. Varia* 48: 233–76
51. Gunn, J. E., Gott, J. R. 1972. *Ap. J.* 176: 1–19
52. Hawarden, T. G., van Woerden, H., Mebold, U., Goss, W. M., Peterson, B. A. 1979. *Astron. Astrophys.* 76: 230–39
53. Haynes, M. P. 1979. *Astron. J.* 84: 1830–36
54. Haynes, M. P. 1981. *Astron. J.* 86: 1126–54
55. Haynes, M. P. 1982. See Ref. 57, pp. 9–12
56. Haynes, M. P., Giovanelli, R. 1980. *Ap. J. Lett.* 240: L87–91
57. Haynes, M. P., Giovanelli, R., eds. 1982. *The Comparative H I Content of Normal Galaxies.* Green Bank, W.Va: NRAO Publ. 124 pp.
58. Haynes, M. P., Giovanelli, R. 1984. *Astron. J.* In press
59. Haynes, M. P., Giovanelli, R., Burkhead, M. S. 1978. *Astron. J.* 83: 938–45
60. Haynes, M. P., Giovanelli, R., Roberts, M. S. 1979. *Ap. J.* 229: 83–90
61. Haynes, M. P., Roberts, M. S. 1979. *Ap. J.* 227: 767–75
62. Helou, G., Salpeter, E. E., Terzian, Y. 1982. *Astron. J.* 87: 1443–64
63. Hewitt, J. N., Haynes, M. P., Giovanelli, R. 1983. *Astron. J.* 88: 272–95
64. Henderson, A. P. 1979. In *The Large Scale Characteristics of the Galaxy,* ed. W. B. Burton, pp. 493–500. Boston: Reidel. 611 pp.
65. Hindman, J. V., Kerr, F. J., McGee, R. X. 1963. *Aust. J. Phys.* 16: 570–83
66. Holmberg, E. 1958. *Medd. Lunds Astron. Obs. Ser. II, No. 136.* 102 pp.
67. Hubble, E., Humason, M. 1931. *Ap. J.* 74: 43–80
68. Huchra, J. P. 1984. See Ref. 49. In press
68a. Huchtmeier, W. K., Richter, O.-G. 1982. *Astron. Astrophys.* 109: 331–35
69. Huchtmeier, W. K., Richter, O.-G., Bohnenstengel, H.-D., Hauschildt, M. 1984. *Astron. Astrophys.* In press
70. Kennicutt, R. C. 1983. *Astron. J.* 88: 483–88
71. Knapp, G. R. 1983. See Ref. 5, pp. 297–304
72. Kotanyi, C., Ekers, R. D. 1983. *Astron. Astrophys.* 122: 267–72
72a. Kotanyi, C., van Gorkom, J. H., Ekers, R. D. 1983. *Ap. J. Lett.* 273: L7–9
73. Krumm, N., Shane, W. W. 1982. *Astron. Astrophys.* 116: 237–47
74. Larson, R. B., Tinsley, B. M., Caldwell, C. N. 1980. *Ap. J.* 237: 692–707
75. Lea, S. M., de Young, D. S. 1976. *Ap. J.* 210: 647–65
76. Longmore, A. J., Hawarden, T. G., Cannon, R. D., Allen, D. A., Mebold, U., et al. 1979. *MNRAS* 188: 285–96
77. Mathews, W. G., Baker, J. C. 1971. *Ap. J.* 170: 241–60
78. Mathewson, D. S., Cleary, M. N., Murray, J. D. 1974. *Ap. J.* 190: 291–96
79. Mathewson, D. S., Cleary, M. N., Murray, J. D. 1975. *Ap. J. Lett.* 195: L97–100
80. Mebold, U., Goss, W. M., van Woerden, H., Hawarden, T. G., Siegman, B. 1979. *Astron. Astrophys.* 74: 100–7
81. Newton, K., Emerson, D. T. 1977. *MNRAS* 181: 573–90
81a. Nilson, P. 1973. *Uppsala General Catalog of Galaxies, Uppsala Astron. Obs. Ann.* 6. 456 pp.
82. Oemler, A. 1974. *Ap. J.* 194: 1–20
83. Peterson, B. M., Strom, S. E., Strom, K. M. 1979. *Astron. J.* 84: 735–43
84. Peterson, S. D. 1979. *Ap. J. Suppl.* 40: 527–76
85. Raimond, E., Faber, S. M., Gallagher, J. S., Knapp, G. R. 1981. *Ap. J.* 246: 708–21
86. Richstone, D. 1976. *Ap. J.* 204: 642–48
87. Richter, O.-G., Huchtmeier, W. K. 1982. *Astron. Astrophys.* 100: 155–65
88. Richter, O.-G., Huchtmeier, W. K. 1983. *Astron. Astrophys.* 125: 187–92
89. Roberts, M. S. 1975. See Ref. 94, pp. 309–57
90. Rogstad, D. H., Wright, M. C. H., Lockhart, I. A. 1976. *Ap. J.* 204: 703–16
91. Romanishin, W., Krumm, N., Salpeter, E. E., Knapp, G., Strom, K. M., Strom, S. E. 1982. *Ap. J.* 263: 94–100

470 HAYNES, GIOVANELLI & CHINCARINI

91a. Rose, J. A. 1977. *Ap. J.* 211:311–18
92. Rots, A. H. 1978. *Astron. J.* 83:219–23
93. Sancisi, R. 1981. In *Structure and Evolution of Normal Galaxies*, ed. S. M. Fall, D. Lynden-Bell, pp. 149–68. Cambridge: Univ. Cambridge Press. 272 pp.
94. Sandage, A., Sandage, M., Kristian, J., eds. 1975. *Galaxies and the Universe.* Chicago: Univ. Chicago Press. 818 pp.
95. Schneider, S. E., Helou, G., Salpeter, E. E., Terzian, Y. 1983. *Ap. J. Lett.* 273: L1–6
96. Schommer, R. A., Sullivan, W. T., Bothun, G. D. 1981. *Astron. J.* 86:943–52
97. Shaviv, G., Salpeter, E. E. 1982. *Astron. Astrophys.* 110:300–15
98. Shostak, G. S. 1978. *Astron. Astrophys.* 68:321–41
99. Shostak, G. S., Hummel, E., Shaver, P. A., van der Hulst, J. M., van der Kruit, P. C. 1982. *Astron. Astrophys.* 115:293–307
100. Shostak, G. S., van der Kruit, P. C. 1984. *Astron. Astrophys.* In press
101. Silk, J. 1978. *Ap. J.* 220:390–400
102. Silk, J., Norman, C. A. 1979. *Ap. J.* 234:86–99
103. Spitzer, L., Baade, W. 1951. *Ap. J.* 113:413–28
104. Stauffer, J. R. 1983. *Ap. J.* 264:14–23
105. Strom, S. E. 1982. See Ref. 57, pp. 13–17
106. Sullivan, W. T., Bothun, G. D., Bates,

B., Schommer, R. A. 1981. *Astron. J.* 86:919–42
107. Sullivan, W. T., Johnson, P. E. 1979. *Ap. J.* 225:751–55
108. Toomre, A., Toomre, J. 1972. *Ap. J.* 178:623–66
109. van den Bergh, S. 1976. *Ap. J.* 206:883–87
110. van der Hulst, J. M. 1979. *Astron. Astrophys.* 71:131–40
111. van der Hulst, J. M. 1979. *Astron. Astrophys.* 75:97–111
112. van der Kruit, P. C., Allen, R. J. 1978. *Ann. Rev. Astron. Astrophys.* 16:103–39
112a. van Gorkom, J., Balkowski, C., Kotanyi, C. 1984. See Ref. 49. In press
113. van Woerden, H., van Driel, W., Schwarz, U. 1983. See Ref. 5, pp. 99–104
113a. Viallefond, F., Allen, R. J., de Boer, J. A. 1980. *Astron. Astrophys.* 82:207–220
114. Warmels, J. 1984. See Ref. 49. In press
115. Weliachew, L., Sancisi, R., Guelin, M. 1978. *Astron. Astrophys.* 65:37–45
116. Wevers, B., Appleton, P. N., Davies, R. D., Hart, L. 1984. *MNRAS.* In press
117. Wilkerson, M. S. 1980. *Ap. J. Lett.* 240:L115–19
118. Williams, B. A. 1983. *Ap. J.* 271:461–70
119. Zwicky, F., Herzog, E., Karpowicz, M., Kowal, C. T., Wild, P. 1960–68. *Catalog of Galaxies and Clusters of Galaxies.* Pasadena: Calif. Inst. Technol. Press. 6 vols.

Ann. Rev. Astron. Astrophys. 1984. 22:471–506

BLACK HOLE MODELS FOR ACTIVE GALACTIC NUCLEI

Martin J. Rees

Institute of Astronomy, Madingley Road, Cambridge CB3 0HA, England

1. INTRODUCTION

It is now 20 years since active galactic nuclei (AGNs) became widely acknowledged as an important astrophysical phenomenon (33, 109). Over the entire subsequent period, one of the few statements to command general agreement has been that the power supply is primarily gravitational: the whole bestiary of models involving dense star clusters, supermassive stars, or black holes at least have this feature in common. Systems dependent on gravitational energy have something else in common: they all undergo an inexorable runaway as the central potential well gets deeper and deeper. According to conventional physics, the almost inevitable endpoint of any dense star cluster or supermassive star will be the collapse of a large fraction of its total mass to a black hole. This is the "bottom line" of Figure 1. Such arguments suggest that massive black holes should exist in the nuclei of all galaxies that have ever experienced a violently active phase. Furthermore, physical processes involving black holes offer a more efficient power supply than any of the "precursor" objects depicted in Figure 1. So massive black holes may not merely be the defunct remnants of violent activity; they may also participate in its most spectacular manifestations.

Considerations such as these have shifted the emphasis of theoretical work away from dense star clusters and supermassive stars and have motivated fuller (or at least less perfunctory) investigations of how black holes might generate the power in quasars, radio galaxies, and related objects. All of the evolutionary tracks in Figure 1 deserve more study: none can be dismissed as irrelevant to the AGN phenomenon. The present review is nevertheless focused on black hole models. Moreover, its scope is even more restricted: I am primarily concerned here with what goes on close to the black hole—in the region where the gravitational potential is not merely

471

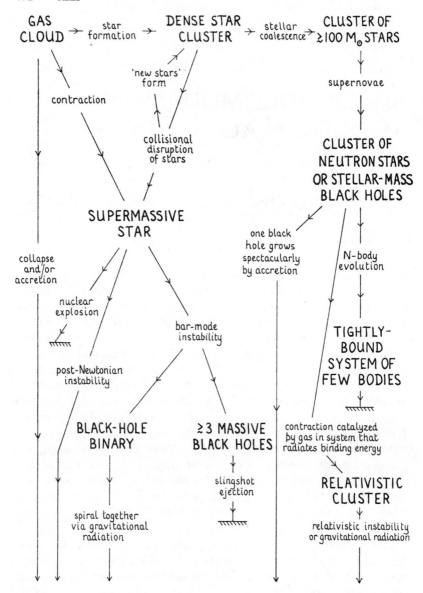

massive black hole

Figure 1   Schematic diagram [reproduced from Rees (106)] showing possible routes for runaway evolution in active galactic nuclei.

"(1/r)," but where intrinsically relativistic features can also be significant. Although this is where the power output is concentrated, many conspicuous manifestations of AGNs—the emission lines, the radio components, etc.—involve some reprocessing of this energy on larger scales. For this reason (and also because of space limitations), little is said here about phenomenology: I merely discuss some physical processes and simple idealized models that have been advanced as ingredients of AGNs.

Two obvious generic features of active galactic nuclei are (a) the production of continuum emission, which in some cases at least must be nonthermal (probably synchrotron); and (b) the expulsion of energy in two oppositely directed beams. The activity is manifested on many scales—up to several megaparsecs in the case of the giant radio sources. It is a tenable hypothesis, however—and one implicitly adopted here—that the central prime mover is qualitatively similar in all of the most highly active nuclei, and that the wide differences observed reflect "environmental" factors on larger scales (where the primary energy output can be reprocessed) and perhaps orientation effects as well.

2. INFERENCES INSENSITIVE TO DETAILED MODEL

2.1 Some Fiducial Numbers

Before focusing on specific properties of black holes, it is interesting to consider some general features of compact ultraluminous sources. Certain order-of-magnitude quantities are involved in any model.

A central mass M has a gravitational radius

$$r_g = \frac{GM}{c^2} = 1.5 \times 10^{13} \, M_8 \text{ cm,} \qquad 1.$$

where M_8 is the mass in units of $10^8 \, M_\odot$. The characteristic minimum time scale for variability is

$$r_g/c \simeq 500 \, M_8 \text{ s.} \qquad 2.$$

A characteristic luminosity is the "Eddington limit," at which radiation pressure on free electrons balances gravity:

$$L_E = \frac{4\pi G M m_p c}{\sigma_T} \simeq 1.3 \times 10^{46} \, M_8 \text{ erg s}^{-1}. \qquad 3.$$

Related to this is another time scale (112):

$$t_E = \frac{\sigma_T c}{4\pi G m_p} \simeq 4 \times 10^8 \text{ yr.} \qquad 4.$$

This is the time it would take an object to radiate its entire rest mass if its luminosity were L_E. The characteristic blackbody temperature if luminosity L_E is emitted from radius r_g is

$$T_E \simeq 5 \times 10^5 \, M_8^{-1/4}. \qquad\qquad 5.$$

We can further define a characteristic magnetic field, whose energy density is comparable with that of the radiation. Its value is

$$B_E \simeq 4 \times 10^4 \, M_8^{-1/2} \text{ G}. \qquad\qquad 6.$$

The expected field strengths induced by accretion flows can be of this order. The corresponding cyclotron frequency is

$$\nu_{cE} \simeq 10^{11} \, M_8^{-1/2} \text{ Hz}. \qquad\qquad 7.$$

The Compton cooling time scale for a relativistic electron of Lorentz factor γ_e (equivalent to the synchrotron lifetime in the field B_E) is

$$t_{cE} \simeq (m_e/m_p)\gamma_e^{-1}(r_g/c) \simeq 0.3 \, \gamma_e^{-1} \, M_8 \text{ s}. \qquad\qquad 8.$$

The photon density n_γ within the source volume is $\sim (L/r^2 c)/\langle h\nu \rangle$. If a luminosity fL_E emerges in photons with $h\nu \approx m_e c^2$, which can interact (with a cross section $\sim \sigma_T$) to produce electron-positron pairs (40), then these photons will interact before escaping if

$$f > (m_e/m_p)(L/L_E)^{-1}(r/r_g). \qquad\qquad 9.$$

Several inferences now follow about the radiation processes, given only the assumption that a primary flux with $L \simeq L_E$ is generated within radii a few times r_g:

1. Thermal radiation from *optically thick* material would be in the far-ultraviolet or soft X-ray region; if, however, thermal gas in the region were hot enough to emit X-rays, reabsorption would be unimportant.
2. If the bulk of the luminosity L were synchrotron radiation in a field $B \simeq B_E$ (Equation 6), then the self-absorption turnover would be (99)

$$\nu_{sE} = 2 \times 10^{14} \, M_8^{-5/14} \qquad\qquad 10.$$

(i.e. typically in the infrared). No significant *radio* emission can come directly from $r \simeq r_g$ unless some coherent process operates at $\nu \simeq \nu_{cE}$. Synchrotron emission at $\sim \nu_{sE}$ would require electrons with $\gamma_e \simeq 40 \, M_8^{1/14}$.

3. The synchrotron or inverse Compton lifetimes of relativistic electrons is $\ll (r_g/c)$ under these conditions, so in any model involving such mechanisms, the radiating particles must be injected or repeatedly reaccelerated at many sites distributed through the source volume.

4. If a substantial fraction of the radiation were generated as gamma rays with energies $\gtrsim 1$ Mev, then electron-positron pairs would inevitably be produced.

This last point is less familiar than the previous three, and so it may merit some elaboration. Photons with energies above 0.5 Mev will experience an optical depth to pair production that exceeds unity whenever (Lf/r) exceeds a value equivalent to $\sim 5 \times 10^{29}$ erg s^{-1} cm^{-1}. Moreover, the annihilation rate constant for these pairs is $\sim \sigma_{\mathrm{T}} c$ if they are subrelativistic, and smaller by $\sim \gamma_{\mathrm{e}}^2$ if they are ultrarelativistic (104). This has the important consequence that a compact source that produces gamma rays (either thermally or nonthermally) at a steady rate satisfying (9.) will shroud itself within an optically thick "false photosphere" of electron-positron pairs, which scatters and Comptonizes all lower-energy photons (58).

2.2   Processes in Ultrahot Thermal Plasma

The only quantities entering into the above discussion have been essentially those involving the electromagnetic energy densities. We now consider the physical conditions in plasma near a collapsed object. If thermal plasma can radiate efficiently enough, it can cool (even at $r \simeq r_{\mathrm{g}}$) to the relatively modest temperature T_{E} (Equation 5). However, two-body cooling processes can be inefficient at low densities; for this reason, and also because the energy available in the relativistically deep potential well may amount to 100 Mev ion^{-1}, the plasmas in AGNs may get hotter than those familiarly encountered elsewhere (even by astrophysicists).

At ion temperatures up to, say, $kT_{\mathrm{i}} = 100$ Mev the ions are of course nonrelativistic, but the thermal electrons may be relativistic. The main distinctive effects arise because the time scale for establishing electron-ion equipartition via two-body processes, or even for setting up a Maxwellian distribution among the electrons themselves, may exceed the time scale for radiative cooling via the same two-body effects. Moreover, other cooling processes may hold the electron temperature to $\lesssim 1$ Mev even if the ions are much hotter. Detailed discussions of these various time scales are given by Gould (54–56) and Stepney (121).

COMPTONIZATION   If photons of energy $h\nu$ are scattered by electrons with temperature T_{e} such that $kT_{\mathrm{e}} \gg h\nu$, then there is a systematic mean gain (67, 125) in photon energy of $(\delta\nu/\nu) \simeq (kT_{\mathrm{e}}/m_{\mathrm{e}}c^2)$ until, after many scatterings, a Wien law is established. If soft photons are injected in an optically thick $(\tau_{\mathrm{T}} > 1)$ source, then the emergent spectrum depends essentially on the parameter $y = \tau_{\mathrm{T}}^2(kT_{\mathrm{e}}/m_{\mathrm{e}}c^2)$: if $y \ll 1$, nothing much happens; if $y \gg 1$, a Wien law is set up; but in the intermediate case when $y \simeq 1$, the emergent spectrum has an approximate power-law form. When $kT \simeq m_{\mathrm{e}}c^2$, the

Table 1   Main production/annihilation processes for electron energies >0.5 Mev

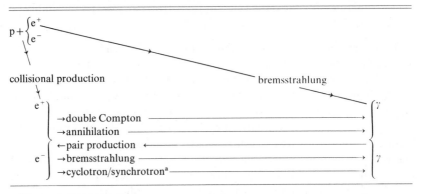

ᵃ If magnetic field is present.

energy change in each scattering is too large for a diffusion approximation to be valid, and Monte Carlo methods are needed (57).

PAIR PRODUCTION EFFECTS   When the electron energies on the tail of the Maxwellian distribution exceed a threshold of 0.5 Mev, collisional processes can create not only gamma rays but also e^+-e^- pairs. These pairs then themselves contribute to the cooling and opacity; the physical conditions must therefore be computed self-consistently, with pairs taken into account (22). Discussions have been given by several authors (22, 40, 69, 70, 132).

The main production/annihilation processes are summarized in Table 1. Further high-energy processes can operate above 50 Mev (46). The fullest discussions of thermal balance in relativistic plasma that take pairs into account are due to Lightman and collaborators (7, 69, 70) and Svensson (122–124). There is a maximum possible equilibrium temperature, of order 10 Mev; but if the heat input is raised beyond a certain value, the increment in pair density is so great that the temperature falls again toward 1 Mev. Note that to extend the usual cooling function $\Lambda(T_e)$ into the temperature range where pair production is important, one must specify the column density $n_i r$ of the source as a second parameter (n_i being the ion density). When $n_i r \ll 1$, the dominant pair production is via e-p collisions; but for sources of higher column density, relation (9.) may be fulfilled, and more pairs come from $\gamma + \gamma$ encounters.

2.3   Cyclotron/Synchrotron and Inverse Compton Cooling

Suppose that the magnetic energy is q times the rest mass density of the plasma: we might expect $q \lesssim kT_i/m_p c^2$ for accretion flows. The ratio of the cyclotron cooling time (neglecting reabsorption) to the bremsstrahlung time for a subrelativistic electron is $\alpha_f(m_e/m_p)q^{-1}(kT_e/m_ec^2)^{-1/2}$, which is $\ll 1$ for a plasma with $kT_i \gtrsim 1$ Mev with an equipartition field; for ultrarelativistic electrons the dominance of synchrotron losses over bremsstrahlung is even greater. Analogously, Compton losses can be very important: indeed, in any source where Thomson scattering on electrons (or positrons) yields $\tau_T > 1$, the requirement that the Compton y-parameter be $\lesssim 1$ implies that the electrons or positrons must be mostly subrelativistic.

The conventional distinction between thermal and nonthermal particles becomes somewhat blurred in these contexts where two-body coupling processes cannot necessarily maintain a Maxwellian distribution. Various acceleration mechanisms (relativistic shocks, reconnection, etc.) may, moreover, boost some small fraction of the particles to high γ: such mechanisms operate in many contexts in high-energy astrophysics and should be even more efficient in an environment where the bulk velocities and Alfvén speeds are both $\sim c$. These particles would then emit synchrotron or inverse Compton radiation. Such acceleration would be "impulsive," in the sense that its time scale is $\ll r_g/c$. The accelerating force would be eE, where E ($\lesssim B$) is the electric field "felt" by the charge. There is then a characteristic peak energy attainable by such processes (39), namely that for which the radiative drag due to synchrotron and inverse Compton emission equals eB. For $B = B_E$ (Equation 6), this yields $\gamma_{drag} = 4 \times 10^5\ M_8^{1/4}$. For acceleration along straight fieldlines, synchrotron losses are evaded, and the terminal energy could be $\sim B_E r_g$ (corresponding to $\gamma_e = 3 \times 10^{14}\ M_8^{1/2}$) if linear acceleration operated over the whole scale of the source. Such limiting energies have emerged from specific studies of accretion disk electrodynamics (34, 72). However, inverse Compton losses cannot be evaded in this way, and they would set a limit not much greater than γ_{drag}. (Individual *ions*, not subject to radiative losses, could in principle get more energetic than electrons.) The parameter γ_{drag} scales as $B^{-1/2}$, and electrons with this energy emit synchrotron photons with $h\nu \simeq \alpha_f^{-1}m_ec^2$ (i.e. 60 Mev) (58, 99). Inverse Compton radiation from the same electrons could of course have photon energies right up to $\gamma_{drag}m_ec^2$. There is thus no reason why a (power-law?) spectrum should not extend up to the gamma-ray band.

3. RADIAL ACCRETION FLOWS

The plasma around black holes will be in some dynamical state—participating in an accretion flow, or perhaps in a wind or jet. Realistically, it would probably be very inhomogeneous: a "snapshot" might reveal many dense filaments at $T \simeq T_E$, embedded in ultrahot thermal plasma filling most of the volume, as well as localized sites where ultrarelativistic electrons are being accelerated. But it is a basic prerequisite for such modeling to know how the various cooling and microphysical time scales compare with the dynamical time at a radius r ($\gtrsim r_g$). The latter can be written as

$$t_{inflow} = \alpha^{-1}(r_g/c)(r/r_g)^{3/2}.$$ 11.

The parameter α, equal to one for free-fall, is introduced explicitly at this stage because the numbers all scale straightforwardly to cases (with $\alpha < 1$) where the inflow is impeded by rotation or by pressure gradients. (In deriving these characteristic numbers, we approximate the flow as spherically symmetric: although this is roughly true for thick tori, further geometrical factors obviously enter for thin disks.)

If accretion with efficiency ε provides the power, the value of $\dot{M}$ needed to supply a luminosity L can be written as $\dot{m}\dot{M}_E = (L/L_E)\varepsilon^{-1}$, where $\dot{M}_E = L_E/c^2$. The particle density at radius r corresponding to an inflow rate $\dot{m}$ is

$$n \simeq 10^{11}\dot{m}\alpha^{-1}M_8^{-1}(r/r_g)^{-3/2} \text{ cm}^{-3}.$$ 12.

Another quantity of interest is the Thomson optical depth at radius r, which is

$$\tau_T \simeq \dot{m}\alpha^{-1}(r/r_g)^{-1/2}.$$ 13.

The "trapping radius," within which an accretion flow would advect photons inward faster than they could diffuse outward [i.e. within which $\tau_T > (c/v_{inflow})$] is

$$r_{trap} = \dot{m}r_g.$$ 14.

Note that this depends only on $\dot{m}$ and not on α.

In Figure 2 are shown the ratios of various physically important time scales to t_{inflow} for a radial free-fall with $\dot{m} = 1$, calculated on the assumption that the ions at each radius are at the virial temperature [i.e. $kT_i = m_pc^2(r/r_g)^{-1}$]. This assumption is self-consistent because bremsstrahlung cooling and electron-ion coupling are indeed ineffective for $\dot{m}\alpha^{-2} = 1$. If the magnetic field is close to equipartition, synchrotron cooling is effective for the electrons (except insofar as it is inhibited by self-absorption); Comptonization is important whenever (kT_e/m_ec^2) max$[\tau_T, \tau_T^2] > 1$. This diagram helps us to understand the detailed results derived for various specific cases.

3.1   Spherical Accretion

The specific angular momentum of accreted material is likely to control the flow pattern, especially when close to the hole. Nevertheless, it is worthwhile to start off with the simpler case of spherically symmetric accretion. Some of the quantities derived in this section (for relative time scales, etc.) can, moreover, be straightforwardly scaled to cases where inflow occurs at some fraction α of the free-fall speed.

If the inflow is laminar, then the only energy available for radiation is that derived from PdV work; therefore, any smooth inflow at high Mach

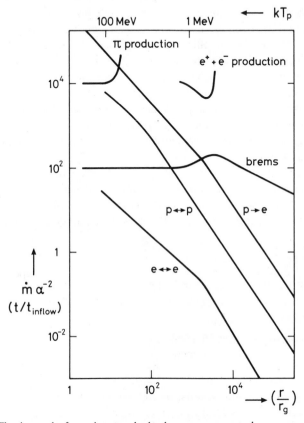

Figure 2   The time scales for various two-body plasma processes are here compared with the inflow time scale for an accretion flow. Processes shown are the self-equilibration time for electrons (e-e) and protons (p-p) (the latter includes nuclear as well as Coulomb effects at > 10 Mev); the time scale for transferring the proton thermal energy to the electrons (p → e); the bremsstrahlung cooling time for the electrons; and the effects of $e^+ + e^-$ and π production. The proton temperature is taken as the virial temperature $[kT = m_p c^2 (r/r_g)^{-1}]$, but the electron temperature is assumed to vary as $r^{-1/2}$ when the electrons are relativistic. The diagram shows that for free-fall accretion at the "critical" rate ($\dot{m} = 1$), two-body cooling processes are inefficient, and the electrons and protons are not thermally coupled by Coulomb interactions when $kT > 1$ Mev. The ratio of the various time scales to t_{inflow} scales as $\dot{m}^{-1}$; for inflow at α times the free-fall speed, the ratio scales as α^2 (for given $\dot{m}$). Data on cross sections are from Stepney (121).

number is certain to be inefficient irrespective of the radiation mechanism. Higher efficiency is possible if the Mach number is maintained at a value of order unity, or if there is internal dissipation (83). However, the fact that the bremsstrahlung cross section is only $\sim \alpha_f \sigma_T$ means that this mechanism alone can never be operative on the free-fall time unless $\dot{m} \gg 1$, in which case (from Equation 14) most of the radiation is swallowed by the hole. Several authors have discussed the important effects of Comptonization. If the only photons are those from bremsstrahlung, then merely a logarithmic factor is gained in the radiative efficiency. However, if the magnetic field is comparable with the value corresponding to full equipartition with bulk kinetic energy, then photons emitted at harmonics of the cyclotron frequency can be Comptonized up to energies such that $h\nu \simeq kT$. The most detailed work on this problem is that of Maraschi and collaborators (42, 80): the calculated spectrum is a power law of slope ~ -1 extending upward from the cyclotron/synchrotron self-absorption turnover to the gamma-ray band.

When a high luminosity L emerges from $r \simeq r_g$, Compton heating or cooling of material at larger r can create important feedback on the flow (45, 92). If the central source emits power $L(\nu)\, d\nu$ at frequencies between ν and $\nu + d\nu$, then Compton processes tend to establish an electron temperature such that

$$kT_e \simeq \tfrac{1}{4}h\langle \nu L(\nu)\rangle/L. \qquad 15.$$

(This formula strictly applies only if $h\nu < m_e c^2$ for all the radiation, and if induced processes can be neglected.) The time scale for this temperature to be established is

$$t_{\text{Comp}} \simeq (m_e/m_p)(L/L_E)^{-1}(r/r_g)^2(r_g/c). \qquad 16.$$

If $t_{\text{Comp}}(r) < t_{\text{inflow}}(r)$, and if no other heating or cooling processes come into play, the consequences depend on whether $kT_e \gtrsim kT_{\text{virial}} = m_p c^2 (r/r_g)^{-1}$. If $T_e < T_{\text{virial}}$, then the inflow must be supersonic, with the pressure support unimportant. Conversely, if there is a range of r where $t_{\text{Comp}} < t_{\text{inflow}}$ but $T_e > T_{\text{virial}}$, steady inflow is impossible: if the flow were constrained to remain spherically symmetrical, "limit cycle" behavior would develop; but in more general geometry, inflow in some directions could coexist with outflow in others (18, 19).

A characteristic feature of the region where $kT_i \gg m_e c^2$ is that the electron-ion coupling time is so long that equality of the electron and ion temperatures is not guaranteed. For low $\dot{m}$, the collisional mean free paths for each species may exceed r (see Figure 2), though even a very weak magnetic field would suffice to make the inflow fluidlike. However, if there were no such field at all, then each electron or ion could orbit the hole many times between collisions (a situation resembling stellar dynamics around a massive central object): the net inflow velocity would be $\ll c(r/r_g)^{-1/2}$, and the density (and hence the radiative efficiency) would be higher than for the fluidlike free-fall solution with the same value of $\dot{M}$ (85).

Material infalling toward a collapsed object obviously eventually encounters the relativistic domain (51). It is therefore necessary to take note of what general relativity tells us about black holes; this is done in the next section.

4. BLACK HOLES ACCORDING TO GENERAL RELATIVITY

The physics of dense star clusters and of supermassive objects are complex and poorly understood. In contrast, the final state of such systems—if gravitational collapse indeed occurs—is comparatively simple, at least if we accept general relativity. According to the so-called no-hair theorems, the endpoint of a gravitational collapse, however messy and asymmetrical it may have been, is a standardized black hole characterized by just two parameters—mass and spin—and described exactly by the Kerr metric. If the collapse occurred in a violent or sudden way, it would take several dynamical time scales for the hole to settle down; during that period, gravitational waves would be emitted. But the final state would still be the Kerr solution, unless the material left behind constituted a strong perturbation. [The perturbation due to the infalling material in steady accretion flows is a negligible perturbation by a wide margin of order $(r_g/c)/t_E$.]

The expected spin of the hole—an important influence on its observable manifestations—depends on the route by which it formed (see Figure 1). A precursor spinning fast enough to be significantly flattened by rotational effects when its radius was $\gg r_g$ would probably have more specific angular momentum than the critical value GM/c. A massive black hole that forms "in one go" is thus likely to have been fed with as much spin as it can accept and to end up near the top of the range of angular momentum permitted by the Kerr metric; the same is true for holes that grow by gradual accretion of infalling galactic gas (11) (though the expectation is less clear if they grow by tidal disruption of stars). We should therefore take full cognizance of the distinctive properties of spinning black holes.

4.1 The Kerr Metric

The Kerr metric changes its character, and the event horizon disappears, if the specific angular momentum $J = J_{max} = GM/c$. The so-called cosmic censorship hypothesis would then require that holes always form with

$J < J_{\max}$. The Kerr solution then has a critical radius called the static limit, within which particles must corotate with the hole, though they can still escape. This arises because the frame-dragging is so strong that even light cones necessarily point in the ϕ direction. This critical surface, with equatorial radius

$$r_{\text{stat}} = r_{\text{g}}\{1 + [1 + (J/J_{\max})^2 \cos^2 \theta]^{1/2}\}, \qquad 17.$$

is not the event horizon itself; the latter occurs at a smaller radius. The region between the event horizon ($r = r_{\text{EH}}$) and the static limit is called the "ergosphere," because one can in principle extract energy from it via a process first proposed by Penrose (96): a particle entering the ergosphere can split in two in such a way that one fragment falls into the hole, but the other leaves the ergosphere with more energy than the original particle. The extra energy comes from the hole itself. A Kerr hole can be considered to have two kinds of mass-energy: a fraction associated with its spin, which can be extracted via the Penrose process, and an "irreducible" mass (14, 41). The fraction that can in principle be extracted is

$$1 - 2^{-1/2}\{1 + [1 - (J/J_{\max})^2]^{1/2}\}^{1/2}, \qquad 18.$$

which is 29% for a maximally rotating hole. The above limit is an instance of a general theorem in black hole physics, according to which the area of the event horizon (a quantity analogous to entropy) can never decrease: a Kerr hole has smaller surface area than a Schwarzschild hole of the same mass. There have been various attempts to incorporate Penrose-style energy extraction into a realistic astrophysical model (64, 100). Those mechanisms that involve particle collisions or scattering operate only for a special subset of trajectories (14), and they would be swamped by accompanying processes. However, a process involving *electromagnetic* effects—the Blandford-Znajek mechanism (29)—seems more promising (and is discussed further below and in Section 5).

ORBITS   The binding energy per unit mass for a circular orbit of radius r around a Schwarzschild hole (with $J = 0$) is

$$c^2\left\{1 - \left[\frac{r - 2r_{\text{g}}}{(r^2 - 3rr_{\text{g}})^{1/2}}\right]\right\}. \qquad 19.$$

For $r \gg r_{\text{g}}$ this reduces to $GM/2r$, which is just the Newtonian binding energy. However, the binding energy has a maximum of $0.057c$ for an orbit at $r_{\min} = 6r_{\text{g}}$, with angular momentum $\mathcal{L}_{\min} = 2\sqrt{3}r_{\text{g}}c$. Circular orbits closer in than this have more angular momentum and are less tightly bound (as for orbits in classical theory when the effective force law is $\propto r^{-n}$, with $n > 3$): the orbits have zero binding energy for $r = 4r_{\text{g}}$ (with corresponding angular momentum $\mathscr{L}_0 = 4r_g c$); and for $r = 3r_g$, the expression (19.) goes to infinity, which implies that photons can move in circular orbits at this radius. In the Kerr metric, the behavior of orbits depends on their orientation with respect to the hole and on whether they are corotating or counterrotating (14). For corotating equatorial orbits, the innermost stable orbit moves inward (as compared with the Schwarzschild case); it becomes more tightly bound, with a smaller $\mathscr{L}_{min}$. For $(J/J_{max}) > 0.94$, r_{min} actually lies within the ergosphere. As $J \to J_{max}$ the stable corotating orbits extend inward toward $r = r_g$, and their binding energy approaches $(1 - 3^{-1/2})c = 0.42c$. These numbers determine the maximum theoretical efficiency of accretion disks.

4.2   Three Astrophysically Important Relativistic Effects

THE MINIMUM ANGULAR MOMENTUM   An important inference from the above is that there are no stationary bound orbits whose angular momentum is less than a definite threshold value: particles whose angular momentum is too low plunge directly into the hole. This qualitative feature of the orbits means that *no stationary axisymmetric flow pattern* can extend too close to the rotation axis of a black hole (even well away from the equatorial plane)—no such constraint arises for flows around an object with a "hard" surface. Many authors have suggested that the resultant "funnels" play a role in the initial bifurcation and collimation of jets.

LENSE-THIRRING PRECESSION   An orbit around a spinning (Kerr) hole that does not lie in the equatorial plane precesses around the hole's spin axis with an angular velocity [discussed by Bardeen & Petterson (13)] of

$$\omega_{BP} \sim 2(r/r_g)^{-3}(c/r_g)(J/J_{max}).   20.$$

This precession has a time scale longer than the orbital period by a factor of $\sim (r/r_g)^{3/2}(J/J_{max})^{-1}$. However, if material spirals slowly inward (at a rate controlled by viscosity) in a time much exceeding the orbital time, then the effects of this precession can mount up. The important consequence follows that the flow pattern near a black hole, within the radius where $2\pi/\omega_{BP}$ is less than the inflow time, can be axisymmetric with respect to the hole irrespective of the infalling material's original angular momentum vector. The Lense-Thirring precession, an inherently relativistic effect, thus guarantees that a wide class of flow patterns near black holes will be axisymmetric—an important simplification of the problem.

ELECTROMAGNETIC PROPERTIES OF BLACK HOLES   Interactions of black holes with magnetic fields imposed on their surroundings can have important astrophysical effects. When a hole forms from collapsing magnetized material, the magnetic field outside the horizon decays ("redshifts away") on the collapse time scale r_g/c. But if, for instance, an external electric field were applied to a Schwarzschild hole, then after transients had decayed, a modified field distribution would be established where the electric field appeared to cross the horizon normally. The event horizon (or "surface") of the hole thus behaves in some respects like a conductor (47, 76, 134). It does not have *perfect* conductivity, however: if it did, electromagnetic flux would never be able to penetrate the horizon. Comparing the decay time scale for transients around a black hole (r_g/c) with the time scale ($r_g^2/4\pi\sigma$) appropriate to a sphere of radius r_g and conductivity σ, we can associate a surface resistivity of 377 Ω with the horizon. This analogy can be put on a more rigorous basis (134), and the "resistance" of a black hole is found to be $Z_H \simeq 100$ ohms. More generally, a *Kerr* black hole behaves like a *spinning* conductor. A simple discussion (98) valid for $J \ll J_{max}$ shows that a hole embedded in a uniform magnetic field B_0 would acquire a quadrupole distribution of electric charge

$$-\Omega^H \frac{r_{EH}}{8\pi c} B_0(3\cos^2\theta - 1),\qquad\qquad 21.$$

where $\Omega^H = (J/J_{max})c/2r_{EH}$ is the effective angular velocity of the hole. The corresponding poloidal electric field in a nonrotating frame is

$$E_\theta = -\frac{2\Omega^H r_{EH}}{c} B_0 \sin 2\theta.\qquad\qquad 22.$$

Just as in a classical "unipolar inductor," power can be extracted by allowing a current flow between a spinning hole's equator and poles. The maximum electric potential drop is $\sim B_0 r_g(J/J_{max})$, where B_0 is the imposed field. This can be very large, as it is when a similar argument is applied to spinning magnetized neutron stars in conventional models for pulsars.

For the fiducial field strength B_E (Equation 6), this emf is

$$\sim m_e c^2 r_g(v_{cE}/c)^{-1}(J/J_{max}) \simeq 3 \times 10^{15} M_8^{1/2}(J/J_{max})m_e c^2.\qquad 23.$$

A single test charge introduced into this electromagnetic field will extract from the hole an energy of this order. However, the magnetosphere is unlikely just to contain a few "test charges"; indeed, the bare minimum charge density needed to modify the imposed field is

$$3 \times 10^{-4} M_8^{-3/2}\ \text{cm}^{-3}\qquad\qquad 24.$$

(cf. Equation 12), and pair production generates far more charges than this (see Section 5). Just as in pulsars (8), a realistic magnetospheric current system and plasma distribution, though very hard to calculate, is likely to "short-out" the electric field. A relevant parameter is then Ω^F, the angular velocity of the field lines at large distance from the hole. This is related to the ratio of the effective resistance Z_∞ to the resistance of the hole Z_H:

$$(\Omega^H - \Omega^F)/\Omega^F = Z_\infty/Z_H. \qquad\qquad 25.$$

In the charge-starved limit, corresponding to infinite resistance at infinity, $\Omega^F = 0$. The "matched" case when $Z_\infty = Z_F (\Omega^F = \frac{1}{2}\Omega^H)$ corresponds to the maximum power extraction for a given B_0. This power is of order

$$B_0^2 r_g^2 (J/J_{max})^2 c. \qquad\qquad 26.$$

The efficiency in this case is lower than when $Z_\infty \to \infty$ (zero power), in the sense that half of the power is dissipated in the hole, and raises its irreducible mass; nevertheless, 9.2% of the rest energy could be extracted while slowing down a hole that started off with $J = J_{max}$.

Electromagnetic extraction of energy from black holes seems a realistic and important possibility. Its astrophysical context is discussed in Section 5.

SUMMARY   The results of this section can be summarized by saying that three distinctively relativistic features of black holes are important in models for galactic nuclei:

1. There is a definite lower limit to the angular momentum of any stably orbiting material.
2. The Lense-Thirring precession enforces axisymmetry on any inward-spiraling flow pattern near the hole; consequently, any directed outflow initiated in the relativistic domain will be aligned with the hole's spin axis and will squirt in a constant direction (irrespective of the provenance of the infalling gas), except insofar as precession or accretion processes can reorient the hole's spin (105).
3. A rotating hole's latent spin energy can be tapped by externally applied magnetic fields; this can provide a power source far exceeding that from the accretion process itself.

5. ACCRETION FLOWS WITH ANGULAR MOMENTUM

5.1 Origin of Infalling Matter

The accreted material could fall in from the body of the galaxy (gas expelled from ordinary stars via stellar winds and supernovae); it could even come from intergalactic clouds captured by the galaxy. [Relevant here is the evidence that galaxies are more likely to be active if they are interacting with a neighbor (10, 43), and that quasars may be in interacting galaxies (62).] Alternatively, the gas supply may originate in the central parts of the galaxy: e.g. (*a*) debris from stars tidally disrupted by the hole (60, 61); (*b*) debris from stellar collisions in a compact star cluster around the hole (52); or (*c*) a positive feedback process whereby stars are induced to lose mass (and thereby provide further fuel) by irradiation from a luminous central source (82).

The accretion flow pattern depends on the angular momentum of the infalling gas: if this is large and has a steady orientation, then an accretion disk may extend out to very large values of (r/r_g); but the Lense-Thirring effect renders the flow pattern near the hole (where the power is primarily released) insensitive to conditions at large r, provided only that the matter has enough angular momentum to prevent it from falling directly into the hole. Accretion disks have been reviewed by Pringle (101) in a general astronomical context; I summarize here some new developments insofar as they may relate to massive holes in galactic nuclei.

5.2 Thin Disks

The simplest hypothesis is that the central object is being fueled steadily via an accretion disk (35, 73, 117). The standard thin disk model assumes that the gas at each radius is in a nearly Keplerian orbit. Slow radial infall occurs as viscosity transfers angular momentum outward. Energy dissipated by the viscous stress is radiated locally at a rate three times the local rate at which gravitational energy is liberated ($GM\dot{M}\, dr/r$ between r and $r+dr$). The factor of 3 arises because viscous stresses transport energy as well as angular momentum outward. This local imbalance is globally rectified in the innermost region of the disk, where the local release of binding energy exceeds the dissipation. For thin disks, slow inflow can be maintained down to the innermost stable orbit; the efficiency then equals the fractional binding energy for this orbit.

A disk has a scale height h normal to the orbital plane such that $(h/r) \simeq c_s/v_{\text{virial}}$, where c_s is the internal sound speed, and is "thin" if this is $\ll 1$. One can write

$$(h/r)^2 \simeq (kT_{\text{gas}}/m_p c^2)(1 + p_{\text{rad}}/p_{\text{gas}})(r/r_g). \qquad 27.$$

In this expression, T_{gas} is the gas temperature in the plane of symmetry (which could significantly exceed the surface temperature if the optical depth were very large); the quantity on the right-hand side is essentially the ratio of thermal and gravitational energies. Generally, the vertical support is provided by gas pressure at large r and for low accretion rates (116). Disks with high $\dot{M}$ are strongly radiation dominated in their inner regions: this is more true when the central hole is supermassive than for a stellar-mass hole because [for a given L/L_E, and thus a given (h/r)] the gas pressure per particle, proportional to T_{gas} (cf. Equation 5), scales as $M^{-1/4}$.

The very simplest models for such disks predict a thermal spectrum typically peaking in the ultraviolet (cf. Equation 5); they thus cannot in themselves account for the very broadband radiation from galactic nuclei. But the major uncertainties in the theory of these disks are the interlinked questions of viscosity and magnetic fields. These fields, amplified by shearing motions (49) and possibly by turbulence-driven dynamo action (102, 103), probably provide the main viscosity. Only crude estimates can be made of the resultant α-parameter. Moreover, it is unclear whether the magnetic stresses build up to a fixed fraction of the *total* pressure or only of the *gas* pressure. The argument for the latter view (44, 110, 111) is that large-amplitude density contrasts can be induced as soon as magnetic stresses become competitive with gas pressure, and buoyancy effects then elevate the flux into the disk's "corona," impeding further amplification. This can happen, however, only if the radiation is able to diffuse relative to the gas: in the limit of very large optical depths, the field could be amplified by differential rotation on time scales much shorter than those on which density inhomogeneities could develop. Gas and radiation would then act like a single composite fluid, and only the total pressure would be relevant. The answer to this somewhat confusing (though well-posed) theoretical question makes a big numerical difference to the inward drift time scale; more importantly, it determines whether such a disk would be unstable to the "visco-thermal" instability (101).

Magnetic fields may also have a big effect on the radiation spectrum emerging from a realistic thin disk. Energy transported by magnetic buoyancy into a hot corona could dominate the (approximately blackbody) radiation from the dense part of the disk. Magnetic flares in the corona may accelerate relativistic electrons that radiate nonthermally.

Blandford (24) has emphasized that there is no obvious ultimate repository from the angular momentum of disks in galactic nuclei (whereas the companion star and the orbit serve this role for binary star systems). If the magnetic field were sufficiently well ordered, a coronal wind (rather than outward transfer via viscosity within the disk itself) could be the main sink for the angular momentum of accreted material (23, 26). An alternative resolution of the problem, suggested by Ostriker (91), is that the angular momentum is transferred via dynamical friction to a star cluster in which the disk is embedded.

Most of the recent theoretical work on thin disk structure is aimed primarily at understanding cataclysmic variables, X-ray binaries, etc., but it is relevant also in the galactic nucleus context. In all disks, the thermal balance of the outer parts is likely to be controlled by irradiation (causing photoionization, Compton heating, etc.) from the central region. Even where such disks exist, they could be embedded in hotter quasi-spherical structures. There may thus be no clear demarcation in the real world between thin disks and the toroidal structures to which we next turn.

5.3   General Structure of Tori or Thick Disks

Disks become geometrically thick, with $h \simeq r$, if the internal pressure builds up so that $c_s \simeq (GM/r)^{1/2}$. This can happen either because radiation pressure becomes competitive with gravity or because the material is unable to radiate the energy dissipated by viscous friction, which then remains as internal energy. Before discussing the (very different) internal physical conditions in these two kinds of tori, let us consider their general equilibrium structure.

In thick disks, radial pressure gradients cannot be ignored; the angular velocity is therefore not Keplerian and becomes (within certain constraints) a free parameter. Uncertainty about the viscosity is a major stumbling block. This uncertainty is not crucial to many qualitative features of thin disks (e.g. their overall energetics). However, in thick disks one must deal explicitly with shear stresses in two directions. The stresses determine the distributions both of angular momentum and enthalpy, and therefore the shape of the isobars inside the disk; internal circulation patterns may be important for energy transport. There is always a pressure maximum at $r = r_{max}$ in the equatorial plane. Outside r_{max}, the angular velocity is sub-Keplerian, but for $r < r_{max}$ it is faster than Keplerian. Such structures around Kerr holes were investigated by Bardeen (12) and by Fishbone & Moncrief (50; see also 36, 37). Recent work, from a more astrophysical viewpoint, has been spearheaded by Abramowicz and colleagues (1–3, 63, 65, 93, 129). They have exploited an important simplifying feature: the shape of a torus depends only on its *surface distribution* of angular momentum. If the angular velocity $\Omega(\mathcal{L})$ is given as a function of angular momentum $\mathcal{L}$, then the surface binding energy U is given implicitly by

$$dU/U = \Omega \, d\mathcal{L}/(c^2 - \mathcal{L}\Omega^2). \qquad 28.$$

A simple special case is that for which $\mathcal{L}$ is the same everywhere. The binding energy is then constant over the whole surface of the torus; there is thus, for each value of $\mathcal{L}$, a family of such tori, parametrized by the surface binding energy U. As U tends to zero, the tori "puff up," and the part of the surface close to the rotation axis acquires a paraboloidal shape. The gravitational field is essentially Newtonian throughout most of the volume, but relativistic effects come in near the hole if $\mathcal{L} \simeq \mathcal{L}_{min}$, the angular momentum of the smallest stable orbit. For $\mathcal{L}$ in the range $\mathcal{L}_{min} < \mathcal{L} < \mathcal{L}_0$, special significance attaches to the torus for which U exactly equals the binding energy of the (unstable) orbit of angular momentum $\mathcal{L}$. There is then a cusplike inner edge, across which material can spill over into the hole (just as material leaves a star that just fills its Roche lobe in a binary system). This particular relation between U and $\mathscr{L}$ would approximately prevail at the inner edge of any torus where quasi-steady accretion is going on (see Figure 3 and caption).

More generally, one can consider (99) tori where Ω goes as some power of $\mathscr{L}$. Such tori exist in all cases where the increase of angular momentum with Ω is slower than Keplerian. The funnels tend to be conical rather than paraboloidal if the rotation law is nearer to Keplerian; they extend closer to $r = r_g$ when the black hole is rapidly rotating.

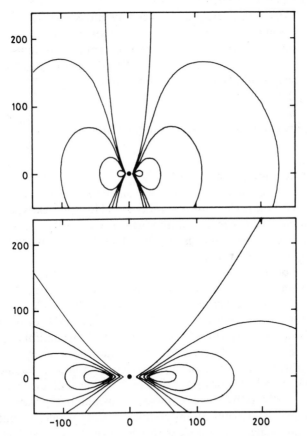

Figure 3 This diagram shows the shape of isobars for tori around a nonrotating (Schwarzschild) hole. The upper picture shows the case $\mathscr{L}$ = constant; in the lower picture, the angular momentum law is $\Omega \propto \mathscr{L}^{-4}$ (i.e. less different from Keplerian), and the funnels are less narrow. For a given rotation law, narrower funnels (extending inward to smaller r) are possible if the hole is rapidly rotating ($J \simeq J_{max}$). The units of length are r_g [from Phinney (99)].

Accretion flows where high internal pressures guarantee $h \simeq r$ [from (27.)] could resemble such tori if the viscosity parameter were low enough that the flow was essentially circular, and provided also that the configuration were stable (though there is frankly no firm basis for confidence in either of these requirements).

A generic feature of accretion tori is that they are less efficient—in the sense that they liberate less energy per gram of infalling matter—than thin disks. The efficiency is given by the binding energy of the material at the cusp; this depends on the angular momentum profile (via Equation 28), but for an $\mathscr{L}$ = constant torus of outer radius r_0, it is $(r_0/r_g)^{-1}$, which implies very low efficiency for large tori.

In any torus with $r_0 \gg r_g$ and a strongly sub-Keplerian rotation law, rotation is unimportant (gravity being essentially balanced by pressure gradients, and the isobars almost spherical) except near the funnel along the rotation axis. To avoid convective instability, the density must fall off with radius at least as steeply as the isentropic laws

$$n \propto r^{-3} \qquad\qquad 29.$$

for $\gamma = 4/3$ (e.g. radiation pressure support), and

$$n \propto r^{-3/2} \qquad\qquad 30.$$

for $\gamma = 5/3$ (e.g. ion pressure support).

The two very different cases of radiation-supported and ion-supported tori may incorporate elements of a valid model for some classes of galactic nuclei. I discuss them here in turn, and then (in Section 6) I consider another question: whether the "funnels" in such flow patterns are important in collimating the outflowing jet material.

The foregoing discussion begs the question of whether these tori are stable and whether stability requirements narrow down the possible forms for $\Omega(\mathscr{L})$. Local instabilities can arise from unfavorable entropy and angular momentum gradients (66, 115). These presumably evolve to create marginally stable convection zones, as in a star. Dynamically important magnetic fields may induce further instabilities. Moreover, tori may be seriously threatened by nonaxisymmetric instabilities. Papaloizou & Pringle (94) recently demonstrated that an $\mathscr{L}$ = constant toroidal configuration marginally stable to axisymmetric instabilities possesses global, nonaxisymmetric dynamical instabilities, which would operate on a dynamic time scale. It is not clear to what extent more general angular momentum distributions are similarly vulnerable, but it may turn out that funnel regions where pressure gradients are balanced by centrifugal effects rather than by gravity are *never* dynamically stable.

5.4   *Radiation-Supported Tori*

A thick structure can be supported by radiation pressure only if it radiates at $L \simeq L_E$. Indeed, in any configuration supported in this way, not only the *total* luminosity but its *distribution over the surface* is determined by the form of the isobars. Tori with long narrow funnels have the property that their total luminosity can exceed L_E by a logarithmic factor (118). More interestingly, most of this radiation escapes along the funnel, where centrifugal effects make the "surface gravity" (and hence the leakage of radiation) much larger than over the rest of the surface. If accretion powers such a torus, then $\dot{m} \times$ (efficiency) $\gtrsim 10$.

If the outer parts are sufficiently slowly rotating that (29.), or a still steeper law, approximately holds, the characteristic Thomson optical depth must depend on radius r at least as steeply as

$$\tau_T(r) \propto r^{-2}. \qquad\qquad 31.$$

This in turn implies that the torus cannot remain optically thick (in the sense that $\tau_T > 1$) out to $r \gg r_g$ unless the viscosity parameter α at $r \simeq r_g$ is very low indeed. (This has been thought by some to be an implausible feature of such models. However, one could argue contrariwise that these objects resemble stars, in which the persistence of differential rotation certainly implies an exceedingly low effective α. Pursuing this analogy further suggests that large-scale circulation effects may play as big a role in energy transport as radiative diffusion does.)

If LTE prevails in such a torus, then the temperature at radius r, at locations well away from the rotation axis, is

$$T(r) \simeq [\tau_T(r_g)]^{1/4} T_E (r/r_g)^{-1} \qquad\qquad 32.$$

(cf. Equation 5). The condition for LTE [i.e. that photons can be thermalized within their diffusion time scale $\tau_T(r)(r/c)$] is more stringent than $\tau_T > 1$. Indeed, even at the pressure maximum ($r \simeq r_g$), the requirement is

$$\dot{m}\alpha^{-1} \simeq \tau_T(r_g) > 2 \times 10^3 M_8^{1/17}, \qquad\qquad 33.$$

and radiation pressure dominates gas pressure by a factor of $\sim 10^6 [\tau_T(r_g)]^{-1/4} M_8^{1/4}$—much larger than ever occurs in stellar structure. If $\tau_T(r_g)$ is even larger than (33.), so that LTE prevails out to $r \gg r_g$, the hole may be sufficiently well smothered that all the radiation effectively emerges from a photosphere, in appearance rather like an O or B star (24).

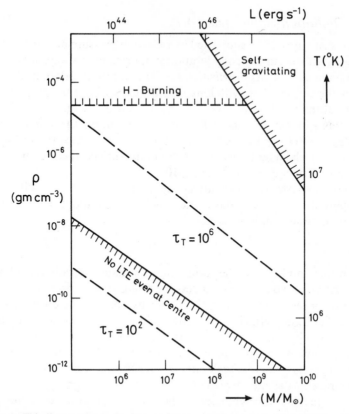

Figure 4   This diagram shows physical conditions near the pressure maximum of an optically thick radiation-supported torus around a hole of mass M. For a given M, the free parameter is the density (or Thomson optical depth τ_T), which scales as $\dot{m}\alpha^{-1}$. Rather high values of this parameter are needed in order to achieve LTE, even in the inner parts of the torus where the pressure is maximal (and where the temperature would then have the value given on the right-hand scale); if LTE is to extend out nearer the surface, then the torus may be so dense and massive near the center that nuclear burning and self-gravitation become significant. (In the approximate treatment given here, the pressure maximum is taken to occur at $r \simeq r_g$; in fact it is always at a larger radius than this, by an amount related inversely to the hole's angular momentum parameter.)

5.5   *Ion-Supported Tori*

We have seen that for spherically symmetric inflow, the cooling time scale— and even the electron-ion coupling time—can be longer than the free-fall time; the same conditions can prevail even for inflow with angular momentum, provided that $\dot{m}$ is low enough. As compared with Figure 2, all that is changed is that the inflow time is $\alpha^{-1}t_{\text{free-fall}}$ and the characteristic density for a given $\dot{m}$ is higher by α^{-1}. The condition for electron-ion coupling to be ineffective in the inner parts of a torus (cf. Figure 2) is

$$\dot{m}\alpha^{-2} < 50. \qquad\qquad 34.$$

When (34.) holds, the ions can remain at the virial temperature even if synchrotron and Compton processes permit the electrons to cool, and the disk swells up into a torus. The dominant viscosity is likely to be magnetic. Estimates of magnetic viscosity are very uncertain; Eardley & Lightman (49) suggest that α falls in the range 0.01–1.0. However, there is no reason why the magnetic α should fall as $\dot{m}$ is reduced, so (34.) should definitely be fulfilled for sufficiently low accretion rates.

An accretion flow where $\dot{m}$ is small, and where (furthermore) the radiative efficiency is low, may seem a doubly unpromising model for any powerful galactic nucleus. However, such a torus around a spinning black hole offers an environment where the Blandford-Znajek (29) process could operate (108). Even though it may not radiate much directly, the torus can then serve as a catalyst for tapping the hole's latent spin energy. Three conditions are necessary:

1. *Magnetic fields threading the hole must be maintained by an external current system.* The requisite flux could have been advected in by slow accretion; even if the field within the torus were tangled, it would nevertheless be well ordered in the magnetosphere. The torus would be a good enough conductor to maintain surface currents in the funnel walls, which could confine such a field within the hole's magnetosphere. The only obvious upper limit to the field is set by the requirement that its total energy should not exceed the gravitational binding energy of the torus. (An equivalent statement is that B should not exceed $\dot{m}^{1/2}\alpha^{-1/2}B_E$.)

2. *There must be a current flowing into the hole.* Although an ion-supported torus radiates very little, it emits some bremsstrahlung gamma rays. Some of these will interact in the funnel to produce a cascade (31) of electron-positron pairs (99, 108), yielding more than enough charge density to "complete the circuit" and carry the necessary current—enough, indeed, to make the magnetosphere essentially charge-neutral, in the sense that $(n^+ + n^-) \gg |(n^+ - n^-)|$, so that relativistic MHD can be applied.

3. *The proper "impedance match" must be achieved between the hole and the external resistance.* Phinney (99) has explored the physics of the relativistic wind, whose source is the pair plasma created in the magnetosphere and that flows both outward along the funnel and into the hole. By considering the location of the critical points, he finds consistent wind solutions where Ω^F is as large as 0.2 Ω^H. This corresponds (cf. Equation 25) to 60% of the maximum power extraction (for a given B-field). Although

some energy is dissipated in the hole, this would still permit a few percent of the hole's rest mass energy to be transformed into a mixture of Poynting flux and a relativistic electron-positron outflow.

The Blandford-Znajek process could operate even if the field threading the hole were anchored to a thin disk, but a thick ion-supported torus provides an attractive model for strong radio galaxies because it could initiate collimated outflow (see the discussion in Section 6). The possibility of such tori depends, however, on the assumption that Coulomb scattering alone couples electrons to ions. This raises the question of whether some collective process might, realistically, be more efficient—if so, the electrons could drain energy from the ions and the torus would deflate. There are bound to be shearing motions, owing to differential rotation, which generate local pressure anisotropies in the plasma. There are certainly instabilities that isotropize the ion plasma, as well as instabilities that isotropize the electron plasma. The key question—which still seems open— is whether these two isotropization processes act almost independently, or whether they can transfer energy from ions to electrons.

[Although electromagnetic extraction of energy is especially important for ion-supported tori (objects where the accretion process is inevitably inefficient), this process could also augment the power generated within a radiation-supported torus. There is in principle no limit to the power that could be extracted from a spinning hole embedded in a dense and strongly magnetized cloud, provided that this power can escape preferentially along the rotation axis without disrupting the cloud. These optically thick radiation-driven jets (21), discussed primarily in the different context of SS 433, could occur in quasars. If the cloud were not sufficiently flattened to permit the excess energy to escape in preferential directions, material would be blown from the cloud, reducing its central pressure: this condition would persist until the total (accretion plus electromagnetic) power fell to L_E, but only a fraction came from accretion.]

6. JET FORMATION

Directed outflow is a ubiquitous feature of active galactic nuclei, and it is also seen in some small-scale prototypes of AGNs in our own Galaxy (e.g. SS 433). This is in itself evidence that a spherically symmetric model cannot be entirely realistic. For a full review of theories of jet propagation, with special relevance to radio galaxies, the reader is referred to Begelman et al. (17). The direct evidence for jets pertains exclusively to scales much larger than the primary power source. The scales probed by VLBI are typically a few parsecs ($\gtrsim 10^4$ r_g for plausible central masses); the only evidence for smaller-scale beaming comes from indirect arguments about the physics of optically violent variables (OVVs), or "blazars" (6, 87, 88). There are theoretical reasons for postulating that the relativistic outflow is initiated on scales of order r_g, but there are really no grounds for believing that a *narrow* collimation angle is established until the jets get out to VLBI scales or beyond: indeed, conditions in the medium $\lesssim 1$ pc from the central source cannot readily provide the kind of pressure-confined "nozzles" (27) that could best collimate them (107).

The radiation from the jets—the emission detected by VLBI and other radio techniques, as well as the emission in other wave bands from (for instance) the M87 jet—is presumably synchrotron radiation from electrons accelerated in situ. Plainly, any high-γ *random* motions produced at $r \simeq r_g$ would have been eliminated by radiative and adiabatic losses before the jet got out to 1 pc. In the superluminal sources, there is direct evidence for bulk relativistic outflow ($\gamma_b \gtrsim 5$). We do not know whether this outflow involves ordinary matter, electron-positron plasma, or even Poynting flux, and various authors have suggested schemes involving each of these options.

Any disk structure near a black hole provides a pair of preferred directions along the rotation axis; moreover, within the Lense-Thirring effect's domain of influence, this axis is maintained steady by the hole's gyroscopic effect. Magnetically driven winds from tori or from thin disks (23, 26) could generate outflowing jets with the attractive attribute of a self-confining toroidal field.

The evacuated vortices along the axes of thick accretion tori, which can be very narrow for an angular momentum distribution close to $\mathscr{L} = $ constant, suggest themselves as possible preexisting channels for directed outflow. The most widely discussed version of this idea, first proposed by Lynden-Bell (74), utilizes radiation pressure. A simple order-of-magnitude argument shows that a test particle (electron plus ion) released from rest outside a source with $r \simeq r_g$ and $(L - L_E)/L_E \gtrsim 1$ would attain a relativistic speed; a radiation-supported torus whose vortex has cone angle θ emits within this cone a greatly enhanced luminosity $\sim \theta^{-2} L_E$ per unit solid angle, which suggests that this photon beam might impart high Lorentz factors to any matter in its path.

Detailed study reveals flaws in this superficially attractive idea (4, 5, 90, 119). The main problem is that the radiation field within a long, narrow funnel is almost isotropic: there may indeed be a super-Eddington outward flux along it, but the radiation density far exceeds (flux/c) because of scattering, or absorption and reemission, by the walls. Consequently, a test electron travels *sub*relativistically along the funnel, at a speed such that the radiation appears nearly isotropic in its moving frame. The radiation flux only becomes well collimated by the time the particle escapes from the funnel, at $r = r_0$. Even for the (probably unstable) $\mathscr{L} = $ constant tori, r_0 is at least $\theta^{-2}r_g$; and out there the dilution (because r is now $\gg r_g$) cancels out the θ factor gained from the beaming. The net result is that γ-values of only ~ 2 can be reached for an electron-ion plasma, and maybe up to ~ 5 for electron-positron plasma. A second difficulty is that the Thomson depth along the funnel would become > 1, vitiating the test-particle approach adopted in the calculations, if the particles were numerous enough to carry a substantial fraction of L. [However, in the limit of very large optical depths, where radiation and matter can be treated as a single fluid, radiation pressure around a supercritical central source—a "cauldron" (21)—could efficiently generate a jet of ordinary matter with high γ_b.]

Quite apart from these theoretical difficulties, models involving radiation-supported tori cannot be relevant to the objects where the most spectacular jets are seen (radio galaxies, M87, etc.). We have *upper* limits to the thermal luminosity from these AGNs; we also have *lower* limits to the energies involved in producing large-scale radio structure and, hence, to the masses involved. Combining these limits precludes there being any object emitting a thermal luminosity L_E (the level of isotropic emission that would be an inevitable concomitant of a radiation-supported torus with a narrow funnel).

An ion-supported torus maintained by accretion with low $\dot{M}$ can provide funnels along the rotation axis, just as a radiation-supported torus can. The expelled material would then be an electromagnetically driven wind of electron-positron plasma (99, 108). The rest mass energy of the pairs could be $\ll L/c^2$—indeed, most of the outflow could be in Poynting flux rather than being carried by the pairs themselves—making high beam Lorentz factors γ_b no problem. An energy flux of this kind could readily be converted into relativistic particles at large distances from its point of origin and is thus an attractive model for radio sources.

Two factors constrain the content and the Lorentz factor of jets emerging from scales of $\sim r_g$ (99, 107). First, an e^+-e^- jet that started off with too high a particle density would suffer annihilation before moving one scale height: this means that an energy flux L_E in pair kinetic energy, rather than in Poynting flux, is impossible unless γ_b is high. [The particle flux is then less for a given L; furthermore, the time scale available for annihilation, measured in the moving frame, is only $\gamma_b^{-1}(r/c)$.] But radiation drag effects give a second countervailing constraint that precludes particle jets with very high values of γ_b. Radiation pressure provides an acceleration only if it comes from the backward direction after transforming into the moving frame (97). If radiation comes from a source of finite size r_s, then the acceleration at a distance r would always saturate for $\gamma_b \simeq (r/r_s)$, no matter how high the luminosity of the source. Moreover, in a realistic model for a galactic nucleus, some fraction of the luminosity is scattered or reemitted on scales out to ~ 1 pc. This quasi-isotropic flux exerts a Compton drag force on any beam, and it is particularly serious for e^+-e^- beams, which have the least inertia relative to their scattering cross section.

The interaction of jets with the material at ~ 1 pc in AGNs is an interesting topic that has only recently been seriously discussed (86). Possibly, the beams generally deposit their energy in the emission-line region, and only in especially favorable cases does the jet material get collimated sufficiently to penetrate beyond $r \simeq 1$ pc.

7.   SOME COMMENTS ON PHENOMENOLOGY

7.1   The Continuum Spectrum

The only direct clue to physical conditions in the central region (i.e. within a radius of, say, $100r_g$) is the rather featureless continuum luminosity; spectral lines originate farther out. The models we have discussed can radiate either thermally or nonthermally: indeed, one of the hardest things to estimate is what fraction of the power dissipated via viscous friction in a realistic flow pattern would go directly into ultrarelativistic particles (via shocks, magnetic reconnection, etc.) rather than being shared among all the particles. Unfortunately, observations are little help in discriminating between various continuum radiation mechanisms: a smooth spectrum could be produced equally well by several alternative mechanisms. For instance (99), there are at least four ways of getting a spectrum with $L(v) \propto v^{-1/2}$:

1. Thermal processes can mimic a power law if the spatial properties of the emitting medium vary in a suitable way (84). This particular slope arises if we consider bremsstrahlung from a spherical distribution of gas with density $n \propto r^{-3/2}$ (corresponding to free-fall) and $T \propto T_{\text{virial}} \propto r^{-1}$.
2. Relativistic particles may be steadily injected with a high Lorentz factor and then lose energy by synchrotron or inverse Compton emission before escaping, yielding nonthermal radiation with $L(v) \propto v^{-1/2}$.
3. Relativistic particles could be accelerated with an E^{-2} differential spectrum. An (over)simple theory of shock acceleration (9, 48) actually yields this law for a compression factor of 4 (the value expected for a strong nonrelativistic shock).
4. Comptonization of injected soft photons yields a power law, which would have the particular value $-\frac{1}{2}$ for a value of the parameter $y = (kT/m_e c^2)\tau_T^2$ that is slightly geometry dependent but typically close to unity.

It is true that theoretical arguments can rule out some of these emission processes in some particular instances: for example, bremsstrahlung can never generate a high luminosity ($L \simeq L_E$) without τ_T being so large that Comptonization reshapes the spectrum (71). These examples of mechanisms, any or all of which could be occurring within a single source, nevertheless highlight the necessity of other indicators (such as polarization or spectral breaks) for discriminating between them.

Obviously the values of M and $\dot{M}$ are crucial in determining the properties of an accreting hole; the angular momentum parameter (J/J_{max}) is also important. We conclude further, and somewhat less trivially, that it is the value of $\dot{m} = \dot{M}/\dot{M}_E$ that determines the nature of the inflow. The value of M itself only enters explicitly (and with weak fractional powers) when reabsorption effects are important. This means that there is a genuine physical similarity, not merely a crude resemblance, between active galactic nuclei and the stellar-scale phenomena (X-ray binaries, etc.) observed within our own Galaxy.

While it is perhaps foolhardy to put forward any fully comprehensive unified scheme for the various kinds of AGNs, there have been several proposals to relate particular categories of objects, or particular features in their spectra, to specific mechanisms.

7.2   Ordinary QSOs

Most QSOs are radio-quiet and are neither violently variable nor highly polarized. The main bolometric luminosity, in the near-ultraviolet, could come from the photosphere of a radiation-supported torus around a (10^7–10^8) $M_\odot$ hole. Blandford (24) has suggested that the characteristic surface temperature is determined by He recombination, which changes the mean molecular weight. An isentropic torus of the type discussed in Section 5.3 would need to have a very high central density (and a correspondingly low value of α) in order to be sufficiently optically thick to thermalize radiation out at the putative photosphere—indeed, its central pressure and temperature might have to be so high that nuclear energy released via hydrogen-burning (16) dominates accretion-generated power (see Figure 4).

Even if one accepts that there is something special about a photospheric temperature $T = 20,000$ K, the configuration need not resemble a stable torus. A more tentative and less controversial conjecture would simply be that typical QSOs are objects where the central hole is smothered by plasma clouds at distances (10^2–10^3) r_g, which are dense enough to be close to LTE [but which are not necessarily supported quasi-statically by an $n \propto r^{-3}$ density distribution (cf. Equation 29) at smaller r]. Such a hypothesis would suffice to explain the "UV bump" in quasar spectra (78, 79). The filaments emitting the broad spectral lines would lie outside this photosphere. Realistically, one expects an additional nonthermal component due to shocks and/or magnetic flaring (by analogy with O star photospheres, except that in AGNs the escape velocity, and probably also the characteristic Alfvén speed, would be very much higher). The X rays could be attributed to this component, since in such a model no radiation would escape directly from $r \simeq r_g$.

7.3   Radio Galaxies

In radio galaxies, the direct radiative output from the nucleus is typically $\sim 10^{42}$ erg s^{-1}, less than the inferred output of the beams that fuel the extended radio components. The energy carried by the beams in Cygnus A exceeds the central luminosity by a factor of ~ 10. These objects must therefore channel *most* of their power output into directed kinetic energy. Moreover, the mass involved in producing the large-scale radio structure must be large—certainly $> 10^7\, M_\odot$. The thermal output from these AGNs is therefore $\lesssim 10^{-3}\, L_E$, implying that they cannot involve radiation-supported tori; nor can radiation pressure be important for accelerating the jet material. Such considerations suggest that strong radio sources may involve massive spinning black holes onto which matter is accreting very slowly (maybe $10^{-3}\, M_\odot$ yr^{-1}) to maintain an ion-supported torus, so that the holes' energy is now being tapped electromagnetically and being transformed into directed relativistic outflow (108).

7.4   Radio Quasars and Optically Violent Variables (OVVs)

Data on OVVs (also known as "blazars") have been reviewed by Angel & Stockman (6; see also 87, 88). For the extreme members of the class, such as OJ 287 and AO 0235 + 164, the case for beaming seems compelling. The less luminous objects might also be beamed, but they could alternatively involve unbeamed synchrotron emission from $r \lesssim 10\, r_g$. More evidence on the hard X-ray spectrum of such objects would help to decide between these options. If gamma rays were emitted and (9.) were fulfilled, the resultant "false photosphere" of electron-positron pairs would scatter the optical photons and destroy any intrinsic high polarization (58). One would then be disposed to invoke relativistic beaming, which would increase the intrinsic source sizes compatible with the observed variability and reduce the luminosity in the moving frame; this would mean that (9.) was no longer fulfilled, and gamma rays could escape without being transformed into pairs.

7.5   Hard X-Ray and Gamma-Ray Sources

Boldt & Leiter (30, 68) have proposed a scheme whereby the output in gamma rays relative to X rays increases as $\dot{M}$ decreases. Low-redshift objects are postulated to have low $\dot{M}$ and to emit gamma rays; their high-z counterparts, however, are fueled at a higher rate, and they yield most of the

X-ray background without contributing proportionately to the gamma-ray background.

According to White et al. (128), the characteristic X-ray spectrum of active nuclei depends on whether their primary luminosity in hard photons is $\gtrsim 10^{-2} L_E$. For a source size of $\sim 10 r_g$, this determines whether or not a pair photosphere is produced (9.). In sources with high L/r, where a pair photosphere *is* produced, the emergent Comptonized spectrum is softer. A small-scale analogue of this phenomenon may be the galactic compact source Cygnus X-1, which undergoes transitions between "high" and "low" states, the spectrum being softer for the former. The fact that many AGNs emit variable X rays with a flat spectrum (energy index 0.6; 89, 131) suggests that e^+-e^- production is inevitable, and that the effects of pairs on dynamics (99) and radiative transfer (130) need further attention.

8.   DEMOGRAPHY OF AGNs

Even if AGNs are precursors on the route toward black hole formation (cf. Figure 1) rather than structures associated with black holes that have already formed, it seems hard to escape the conclusion that massive black holes must exist in profusion as remnants of past activity; they would be inconspicuous unless infall onto them recommenced and generated a renewed phase of accretion-powered output or catalyzed the extraction of latent spin energy.

Estimates of the masses and numbers of "dead" AGNs are bedeviled by uncertainty about how long individual active objects live and the evolu-tionary properties (i.e. the z-dependence) of the AGN population. As regards the latter, see, for instance, (81, 114, 127) for recent reviews of optical data, and (95, 126) for radio studies. It has long been known that the evolution is *strong*, amounting to a factor of up to 1000 in comoving density for the strongest sources; the evolution is *differential*, being less steep for lower-luminosity objects of all kinds. It is now feasible to refine these statements, though it is still premature to be extremely precise about the redshift dependence of the multivariate function $f(L_{rad}, L_{opt}, L_X)$. And we are still a long way from having much astrophysical understanding of *why* the luminosity function evolves in this way. Anyway, at the epoch $z = 2$, the population of strong sources declined on a time scale $t_{Ev} \simeq 2 \times 10^9 h_{100}$ yr; this is of course an *upper limit* to the "half-life" of a particular source, since there may be many generations of objects within the period t_{Ev}.

Soltan (120) has given an argument that bypasses the uncertainty in AGN lifetimes but nevertheless yields useful constraints on the masses involved in such phenomena and the kinds of galaxies in which they can reside. The overall energy budget for AGNs is dominated by QSOs (most of which are "radio quiet"); they contribute an integrated background luminosity amounting to

$$\sim 3000 M_\odot c^2 \text{ Mpc}^{-3}. \tag{35.}$$

This estimate involves an uncertain bolometric correction; the main measured contribution to $\nu L(\nu)$ is typically at $\sim 10^{15}$ Hz (i.e. in the ultraviolet). Although some individual quasars could emit more power in (for instance) the far-ultraviolet, hard X-rays, or gamma rays, we know enough about the isotropic background in these bands to be sure that such emission cannot permit a huge bolometric correction for the typical quasar. The main contribution to (35.) comes from quasars with $19 < B < 21$ (corresponding to bolometric luminosities of 10^{45}–$10^{46} h_{100}^{-2}$ erg s^{-1} if they are typically at $z \simeq 2$); the counts flatten off at fainter magnitudes, so (35.) is unlikely to be a severe underestimate. The energy output from radio galaxies and other manifestations of active nuclei is much smaller than that from optically selected quasars; we are therefore probably justified in using (35.) in the discussion that follows.

The individual remnant masses can be estimated as

$$8 \times 10^7 h_{100}^{-3} \varepsilon^{-1}(t_Q/t_{Ev}) \, M_\odot, \tag{36.}$$

where ε is the overall efficiency with which rest mass is converted into electromagnetic radiation over a typical quasar's active lifetime t_Q (defined as the time for which the magnitude is $M < -24$). If quasars were associated with all "bright" galaxies ($M < -21.3$), whose space density is known, the mean hole mass would be $\sim 2 \times 10^6 h_{100}^{-3} \varepsilon^{-1} M_\odot$. If only a small fraction of galaxies had ever harbored active nuclei, the masses (and lifetimes) of each would be correspondingly increased.

The above discussion is important if one wishes to relate nuclear activity to galactic morphology. However, it is also relevant as a discriminant between different models (99). If individual quasars are as long lived as is compatible with the cosmological evolution of the quasar population [i.e. $t_Q = t_{Ev}$ in (36.)], their remnant masses must be as large as $\sim 10^9 h_{100}^{-3}(\varepsilon/0.1) M_\odot$ (the present space density of remnants being only about that of radio sources with $P_{178} > 10^{23} h_{100}^{-2}$ W Hz^{-1} sr^{-1}). But the luminosity of a "typical" quasar ($M \simeq -25.5$ for $h_{100} = \frac{1}{2}$) corresponds to the Eddington limit for a mass of "only" $\sim 10^8 M_\odot$; so if quasars resemble radiation-supported tori, then they must have lifetimes such that $(t_Q/t_{Ev}) \ll 1$, and individual quasars must be "switched off" by something internal to the particular system, rather than being influenced by any change in the overall cosmic environment (which could occur only on time scales $\sim t_{Ev}$) (38, 77).

Because of space limitations, I do not speculate here about how different forms of AGNs might be interrelated. Another contentious issue is the role of *beaming* (25, 28, 32, 113), which has been advocated to explain compact radio sources (and perhaps also extreme optical outbursts in OVVs). There is no reason to invoke it, however, for the typical radio-quiet quasar, and in any case it cannot affect the estimation of (35.). Suffice it to say, as emphasized by Phinney (99), that tentative "demographic" studies of AGNs imply the following:

1. If strong radio sources involve ion-supported tori around holes of mass $\gtrsim 10^9$ $M_\odot$, they could be the "reactivated" remnants of long-lived quasars (with $t_Q = t_{Ev}$).
2. If quasars radiate at the Eddington limit, then their lifetimes would be $\sim 4 \times 10^7 h(\varepsilon/0.1)$ yr (cf. Equation 4). Even if their efficiency ε is high, they must be short lived compared with t_{Ev}. The space density of remnants cannot exceed $10^{-3} h_{100}^2 (\varepsilon/0.1)^{-1}$ Mpc^{-3} unless their luminosities are highly "super-Eddington."
3. If Seyfert galaxies are quasar remnants, they cannot be modeled by radiation-supported tori.

9.   SOME CONCLUDING COMMENTS

The foregoing sections have discussed some physical processes and some idealized models that are relevant to the general phenomenon of active galactic nuclei. It is, however, depressingly evident how tenuous are the links between these models and the actual observations. Partly, this is because the subject is just beginning; but it is also partly because the observations relate only very indirectly to the primary energy source—they may instead tell us about secondary reprocessing that has occurred on much larger scales.

Exhortations and hopes for the future can be summarized in three categories:

1. On the purely theoretical level, even the simple "toy" models discussed here need further investigation—they involve effects in Kerr geometry, collective processes and radiative transfer in pair-dominated plasma, and acceleration of high gamma particles, none of which are yet well explored or understood. We need to clarify the stability of the various axisymmetric configurations: this should narrow down the embarrassing freedom we now have in specifying the angular momentum and the enthalpy distribution in tori. Large-scale computer simulations could be crucial here.

Computer simulations should permit us also to relax the assumption of stationarity (59), which has been implicit in most work on accretion flows. It may be more realistic to envisage that the feeding process and the subsequent viscous redistribution of angular momentum and drainage into the hole are sporadic. There is, after all, observational evidence for variability on all time scales. Three-dimensional gas-dynamical codes could also check whether the Lense-Thirring effect does indeed align the flow in the way simple arguments suggest. A further valuable computational development will be the advent of MHD codes able to treat electromagnetic processes around black holes, as well as the initiation (and possible magnetic confinement) of relativistic jets.

Detailed computations would also be worthwhile on other classes of relativistic systems relevant to the evolutionary tracks in Figure 1 (20). In particular, supermassive stars with realistic differential rotation should be investigated. For a suitable angular momentum distribution, these could acquire a high gravitational binding energy [cf. the massive disks (15, 75) that have been treated analytically]. Redistribution of angular momentum within such objects would be likely to cause their inner regions to collapse, leaving a massive torus around the resultant black hole. If too massive, this would be subject to gravitational instability and could fragment. Otherwise, it would evolve on a Kelvin or viscous time scale (whichever was shorter). Such models also remind us that evolution need not be restricted to the slow time scales of order t_E (Equation 4), but that rare "hypernovae" may occur.

2. The "peripheral fuzz" at $r \gg r_g$ in the emission-line region and the radio structures involves physics that is less extreme and more familiar than that in the central relativistic domain. However, it is here that one is perhaps more pessimistic about theoretical progress. This is because in the central region, even though the physics may be exotic, we have a relatively "clean" problem : axisymmetric flow in a calculable gravitational field. On the other hand, in the large-scale sources, environmental effects are plainly crucial : progress will be slow, for the same reasons that weather prediction is difficult.

The subject has proceeded in a highly compartmentalized way : the central engine, the emission-line region, the radio jets, etc., are modeled somewhat disjointly. To a certain extent this is inevitable—after all, the relevant scales may differ by many powers of ten. As data proliferate on source morphology, it no longer seems premature to develop more comprehensive models, nor to understand the relation of AGNs to their parent galaxies : If we compare spirals and ellipticals, are the central masses different? Is the fueling different? What other environmental influences determine the kind of AGN that is observed? And do stellar-mass compact objects within our own Galaxy offer many clues to the mechanisms of AGNs?

3. It is perhaps salutary (especially for relativists) to remain aware that

Einstein's theory is empirically validated only in the weak-field limit. An extra motive for studying the central region is therefore to seek a diagnostic (by refining our models for galactic nuclei) that could test strong-field general relativity and check whether the space-time around a rotating black hole is indeed described by the Kerr metric.

Ginzburg (53) has recently remarked on how surprisingly *slowly* most sciences develop. Concentrated activity over a short time-base may give the illusion that progress is fast, but the advance of science—particularly where data are sparse—displays a slowly rising trend, with large-amplitude "sawtooth" fluctuations superposed on it as fashions come and go. There has been progress toward a consensus, in that some bizarre ideas that could be seriously discussed a decade ago have been generally discarded. But if we compare present ideas with the most insightful proposals advanced when quasars were first discovered 20 years ago (such proposals being selected, of course, with benefit of hindsight), progress indeed seems meager. It is especially instructive to read Zeldovich & Novikov's (1964) paper entitled "The Mass of Quasi-Stellar Objects" (133). In this paper, on the basis of early data on 3C 273, they conjectured the following: (a) Radiation pressure perhaps balances gravity, so the central mass is $\sim 10^8 \, M_\odot$. (b) For a likely efficiency of 10%, the accretion rate would be 3 $M_\odot$ yr^{-1}. (c) The radiation would come from an effective "photosphere" at a radius $\sim 2 \times 10^{15}$ cm (i.e. $\gg r_g$), outside of which line opacity would cause radiation to drive a wind. (d) The accretion may be self-regulatory, with a characteristic time scale of ~ 3 yr. These suggestions accord with the ideas that remain popular today, and we cannot yet make many firmly based statements that are more specific.

ACKNOWLEDGMENTS

I am grateful for discussions and collaboration with many colleagues, especially M. C. Begelman, R. D. Blandford, A. C. Fabian, and E. S. Phinney.

Literature Cited

1. Abramowicz, M., Calvani, M., Nobili, L. 1980. *Ap. J.* 242:772
2. Abramowicz, M., Jaroszynski, M., Sikora, M. 1978. *Astron. Astrophys.* 63:221
3. Abramowicz, M. A., Lasota, J. P. 1980. *Acta Astron.* 30:35
4. Abramowicz, M., Piran, T. 1980. *Ap. J. Lett.* 241:L7
5. Abramowicz, M. A., Sharp, N. A. 1983. *Astrophys. Space Sci.* 96:431
6. Angel, R., Stockman, H. S. 1980. *Ann. Rev. Astron. Astrophys.* 18:321
7. Araki, S., Lightman, A. P. 1983. *Ap. J.* 269:49
8. Arons, J. 1979. *Space Sci. Rev.* 24:437
9. Axford, W. I. 1981. *Proc. Tex. Symp., 10th, Ann. NY Acad. Sci.* 375:297
10. Balick, B., Heckman, T. M. 1982. *Ann. Rev. Astron. Astrophys.* 20:431
11. Bardeen, J. M. 1970. *Nature* 226:64
12. Bardeen, J. M. 1973. In *Les Astres Occlus*, ed. C. DeWitt, B. S. DeWitt, p. 219. New York: Gordon & Breach

13. Bardeen, J. M., Petterson, J. A. 1975. *Ap. J. Lett.* 195:L65
14. Bardeen, J., Press, W., Teukolsky, S. 1972. *Ap. J.* 178:347
15. Bardeen, J. M., Wagoner, R. V. 1971. *Ap. J.* 167:359
16. Begelman, M. C. 1984. *Proc. IAU Symp. 112, Very-Long-Baseline Interferometry,* ed. R. Fanti, K. Kellermann, G. Setti. In press
17. Begelman, M. C., Blandford, R. D., Rees, M. J. 1984. *Rev. Mod. Phys.* In press
18. Begelman, M. C., McKee, C. F. 1983. *Ap. J.* 271:89
19. Begelman, M. C., McKee, C. F., Shields, G. A. 1983. *Ap. J.* 271:70
20. Begelman, M. C., Rees, M. J. 1978. *MNRAS* 185:847
21. Begelman, M. C., Rees, M. J. 1984. *MNRAS* 206:209
22. Bistnovatyii-Kogan, G. S., Zeldovich, Y. B., Sunyaev, R. A. 1971. *Sov. Astron. AJ* 15:17
23. Blandford, R. D. 1976. *MNRAS* 176:465
24. Blandford, R. D. 1984. In *Numerical Astrophysics,* ed. J. Centrella, R. Bowers, J. LeBlanc. In press
25. Blandford, R. D., Konigl, A. 1979. *Ap. J.* 232:34
26. Blandford, R. D., Payne, D. G. 1982. *MNRAS* 199:883
27. Blandford, R. D., Rees, M. J. 1974. *MNRAS* 169:395
28. Blandford, R. D., Rees, M. J. 1978. *Pittsburgh Conf. BL Lac Obj.,* ed. A. M. Wolfe, p. 328. Pittsburgh: Univ. Pittsburgh Press
29. Blandford, R. D., Znajek, R. L. 1977. *MNRAS* 179:433
30. Boldt, E., Leiter, D. 1984. *Ap. J.* 276:427
31. Bonometto, S., Rees, M. J. 1971. *MNRAS* 152:21
32. Browne, I. W. A., Orr, M. J. L., Davis, R. J., Foley, A., Muxlow, T. W. B., Thomasson, P. 1982. *MNRAS* 198:673
33. Burbidge, G. R., Burbidge, E. M., Sandage, A. 1963. *Rev. Mod. Phys.* 35:947
34. Burns, M. L., Lovelace, R. V. E. 1982. *Ap. J.* 262:87
35. Callahan, P. S. 1977. *Astron. Astrophys.* 59:127
36. Carter, B. 1979. In *Active Galactic Nuclei,* ed. C. Hazard, S. Mitton, p. 273. Cambridge: Cambridge Univ. Press
37. Carter, B. 1979. In *General Relativity,* ed. S. W. Hawking, W. Israel, p. 294. Cambridge: Cambridge Univ. Press
38. Cavaliere, A., Giallongo, E., Messina, A., Vagentti, F. 1983. *Ap. J.* 269:57
39. Cavaliere, A., Morrison, P. 1980. *Ap. J. Lett.* 238:L63
40. Cavallo, G., Rees, M. J. 1978. *MNRAS* 183:359
41. Christodoulou, D. 1970. *Phys. Rev. Lett.* 25:1596
42. Colpi, M., Maraschi, L., Treves, A. 1984. *Ap. J.* In press
43. Condon, J. J., Condon, M. A., Mitchell, K. J., Usher, P. D. 1980. *Ap. J.* 242:486
44. Coroniti, F. V. 1981. *Ap. J.* 244:587
45. Cowie, L. L., Ostriker, J. P., Stark, A. A. 1978. *Ap. J.* 226:1041
46. Dahlbacka, G. H., Chapline, G. F., Weaver, T. A. 1974. *Nature* 250:37
47. Damour, T. 1975. *Ann. NY Acad. Sci.* 262:113
48. Drury, L. O'C. 1983. *Rep. Prog. Phys.* 46:973
49. Eardley, D. M., Lightman, A. P. 1976. *Ap. J.* 200:187
50. Fishbone, L. G., Moncrief, V. 1976. *Ap. J.* 207:962
51. Flammang, R. A. 1982. *MNRAS* 199:833
52. Frank, J. 1979. *MNRAS* 187:883
53. Ginzburg, V. L. 1983. *Waynflete Lectures on Physics.* Oxford: Pergamon
54. Gould, R. J. 1981. *Phys. Fluids* 24:102
55. Gould, R. J. 1982. *Ap. J.* 254:755
56. Gould, R. J. 1982. *Ap. J.* 263:879
57. Guilbert, P. W. 1981. *MNRAS* 197:451
58. Guilbert, P. W., Fabian, A. C., Rees, M. J. 1983. *MNRAS* 205:593
59. Hawley, J., Smarr, L. L. 1984. In *Numerical Astrophysics,* ed. J. Centrella, R. Bowers, J. LeBlanc. In press
60. Hills, J. G. 1975. *Nature* 254:295
61. Hills, J. G. 1977. *MNRAS* 182:517
62. Hutchings, J. B., Campbell, B. 1983. *Nature* 303:584
63. Jaroszynski, M., Abramowicz, M. A., Paczynski, B. 1980. *Acta Astron.* 30:1
64. Kafatos, M., Leiter, D. 1979. *Ap. J.* 229:46
65. Kozlowski, M., Jaroszynski, M., Abramowicz, M. A. 1978. *Astron. Astrophys.* 63:209
66. Kandrup, H. 1982. *Ap. J.* 235:691
67. Katz, J. I. 1976. *Ap. J.* 206:910
68. Leiter, D., Boldt, E. 1982. *Ap. J.* 260:1
69. Lightman, A. P. 1982. *Ap. J.* 253:842
70. Lightman, A. P., Band, D. L. 1981. *Ap. J.* 251:713
71. Lightman, A. P., Giacconi, R., Tananbaum, H. 1978. *Ap. J.* 224:375
72. Lovelace, R. V. E. 1976. *Nature* 262:649
73. Lynden-Bell, D. 1969. *Nature* 223:690
74. Lynden-Bell, D. 1978. *Phys. Scr.* 17:185

75. Lynden-Bell, D., Pineault, S. 1978. *MNRAS* 185:679
76. Macdonald, D., Thorne, K. S. 1982. *MNRAS* 198:345
77. MacMillan, S. L. W., Lightman, A. P., Cohn, H. 1981. *Ap. J.* 251:436
78. Malkan, M. A. 1983. *Ap. J.* 268:582
79. Malkan, M. A., Sargent, W. L. W. 1982. *Ap. J.* 254:22
80. Maraschi, L., Roasio, R., Treves, A. 1982. *Ap. J.* 253:312
81. Marshall, H. L., Avni, Y., Tananbaum, H., Zamorani, G. 1983. *Ap. J.* 269:35
82. Matthews, W. G. 1983. *Ap. J.* 272:390
83. Meszaros, P. 1975. *Astron. Astrophys.* 49:59
84. Meszaros, P. 1983. *Ap. J. Lett.* 274:L13
85. Meszaros, P., Ostriker, J. P. 1984. *Ap. J.* In press
86. Miley, G., Norman, C. 1984. *Ap. J.* Submitted for publication
87. Moore, R. L., Stockman, H. S. 1981. *Ap. J.* 243:60
88. Moore, R. L., Stockman, H. S. 1984. *Ap. J.* In press
89. Mushotzky, R. F. 1984. *Proc. Tex. Symp. Relativ. Astrophys., 11th.* In press
90. Narayan, R., Nityananda, R., Wiita, P. 1983. *MNRAS* 205:1163
91. Ostriker, J. P. 1983. *Ap. J.* 273:99
92. Ostriker, J. P., McCray, R., Weaver, R., Yahil, A. 1976. *Ap. J. Lett.* 208:L61
93. Paczynski, B., Abramowicz, M. A. 1982. *Ap. J.* 253:897
94. Papaloizou, J. C. B., Pringle, J. E. 1984. *MNRAS.* In press
95. Peacock, J. A., Gull, S. F. 1981. *MNRAS* 196:611
96. Penrose, R. 1969. *Riv. Nuovo Cimento* 1:252
97. Phinney, E. S. 1982. *MNRAS* 198:1109
98. Phinney, E. S. 1983. In *Astrophysical Jets,* ed. A. Ferrari, A. G. Pacholczyk, p. 201. Dordrecht: Reidel
99. Phinney, E. S. 1983. PhD thesis. Univ. Cambridge, Engl.
100. Piran, T., Shaham, J. 1977. *Ap. J.* 214:268
101. Pringle, J. E. 1981. *Ann. Rev. Astron. Astrophys.* 19:137
102. Pudritz, R. 1981. *MNRAS* 195:881
103. Pudritz, R., Fahlman, G. C. 1982. *MNRAS* 198:689
104. Ramaty, R., Meszaros, P. 1981. *Ap. J.* 250:384
105. Rees, M. J. 1978. *Nature* 275:516
106. Rees, M. J. 1978. *Observatory* 98:210
107. Rees, M. J. 1984. *Proc. IAU Symp. 112,* Very-Long-Baseline Interferometry, ed. R. Fanti, K. Kellermann, G. Setti. In press
108. Rees, M. J., Begelman, M. C., Blandford, R. D., Phinney, E. S. 1982. *Nature* 295:17
109. Robinson, I., Schild, A., Schucking, E., eds. 1964. *Proc. Tex. Conf. Relativ. Astrophys., 1st.* Chicago: Univ. Chicago Press
110. Rosner, R. 1984. Preprint
111. Sakimoto, P. J., Coroniti, F. V. 1981. *Ap. J.* 247:19
112. Salpeter, E. E. 1964. *Ap. J.* 140:796
113. Scheuer, P. A. G., Readhead, A. C. S. 1979. *Nature* 277:182
114. Schmidt, M., Green, R. F. 1983. *Ap. J.* 269:352
115. Seguin, F. G. 1975. *Ap. J.* 197:745
116. Shakura, N. I., Sunyaev, R. A. 1973. *Astron. Astrophys.* 24:337
117. Shakura, N. I., Sunyaev, R. A. 1976. *MNRAS* 175:613
118. Sikora, M. 1981. *MNRAS* 196:257
119. Sikora, M., Wilson, D. B. 1981. *MNRAS* 197:529
120. Soltan, A. 1982. *MNRAS* 200:115
121. Stepney, S. 1983. *MNRAS* 202:467
122. Svensson, R. 1982. *Ap. J.* 258:131
123. Svensson, R. 1982. *Ap. J.* 258:335
124. Svensson, R. 1984. *MNRAS.* In press
125. Takahara, F., Tsuruta, S. 1982. *Prog. Theor. Phys.* 67:485
126. Van der Laan, H., Windhorst, R. A. 1983. In *Astrophysical Cosmology,* ed. H. Bruck, G. Coyne, M. Longair, p. 349. Vatican City: Vatican Publ.
127. Veron, P. 1983. *Proc. Liège Symp. Quasars Gravitational Lenses.* Liège: Obs. Publ.
128. White, N. E., Fabian, A. C., Mushotzky, R. F. 1984. *Astron. Astrophys.* In press
129. Witta, P. 1982. *Ap. J.* 256:666
130. Wilson, D. B. 1982. *MNRAS* 200:881
131. Worrall, D. M., Marshall, F. E. 1984. *Ap. J.* 276:434
132. Zdiarski, A. A. 1982. *Astron. Astrophys.* 110:L7
133. Zeldovich, Y. B., Novikov, I. D. 1964. *Dokl. Acad. Nauk. SSSR* 158:811
134. Znajek, R. L. 1977. *MNRAS* 179:457

Ann. Rev. Astron. Astrophys. 1984. 22 : 507–36

OBSERVATIONS OF SS 433

Bruce Margon

Department of Astronomy, University of Washington, Seattle, Washington 98195

1. INTRODUCTION

There are few stellar objects that have gone from complete obscurity to a state of intensive study as rapidly as Stephenson-Sanduleak 433. Clark (23) has pointed out that there were 2 published papers on the object in 1978, 28 in 1979, 73 in 1980, and 122 in 1981! Under these circumstances, it is already impossible to attempt a comprehensive observational and theoretical review of this most curious stellar system. In this paper I therefore set more modest goals: a very brief review of basic observations at each wavelength and somewhat more detailed coverage of recent observations, theoretical progress, and problems posed by SS 433. Some recent (but, unfortunately, already outdated) reviews of observational (106–109) and theoretical (126, 137, 157) problems related to SS 433 have appeared elsewhere, as have several anecdotal and historical accounts (23, 104).

SS 433 was first catalogued in an objective prism survey (166) as one of many emission-line objects with very strong Hα. The prominent radio emission from the star was independently discovered (24), but initially without an optical identification, as was also the case for the X-ray emission from the object (40, 155). The realization that all these sources of radiation were one and the same object was reached independently by several workers in 1978 (25, 151), with Clark & Murdin (25), in particular, drawing attention to the triple coincidence of the radio/optical/X-ray sources and the possible role of the extended surrounding supernova remnant, W50. They also presented the first moderate-resolution spectral data that revealed the extraordinary complexity of SS 433.

The optical spectrum of SS 433, in addition to very strong, broad Balmer and He I emission lines, contains a number of prominent broad emission features at unfamiliar wavelengths (99, 101). These latter emissions are due to Doppler-shifted Balmer and He I lines (95, 102), one set with a very large

507

but consistent redshift and the other with a huge blueshift. The Doppler-shifted features are seen to change in wavelength and drift through the spectrum on a time scale of days as the velocity of the two shifted systems changes smoothly (99, 101, 102). The variable Doppler shifts achieve impressive magnitudes: a maximum of about 50,000 km s^{-1} in the redshift, and a mimimum of 30,000 km s^{-1} in the blueshifted features. This change in wavelength proves to be periodic, with a time scale of about 164 days (102), although the underlying clock is somewhat imperfect (5). The mean velocity of the sum of the red- and blueshift systems remains approximately constant at a value of about 12,000 km s^{-1} throughout the cycle, although the true systemic velocity of the object, as measured from the Balmer and He I "rest wavelength" emissions, for example, is only about 70 km s^{-1} (30).

Two of the earliest theoretical papers on the system remain the cornerstone of the most widely discussed model for SS 433. Both Fabian & Rees (39) and Milgrom (124) proposed collimated, oppositely ejected jets as a possible explanation of the simultaneous redshifts and blueshifts; one of a variety of scenarios proposed by the latter author also suggested that the observed velocity modulation could be achieved by a periodic rotation of the jet axis. Abell & Margon (1), who had the benefit of a large number of observations of Doppler shift values, were able to considerably elaborate the Milgrom model and provide solutions for the free parameter values that are essentially identical to those in use today. These authors made the advance prediction that the "moving" emission-line systems would cross through each other, with the redshift system acquiring a blueshift and vice versa, for part of the 164-day cycle; this effect was indeed observed shortly thereafter (8, 105).

The "kinematic model," as it has come to be called, hypothesizes that matter is ejected in two opposing jets that are collimated and oppositely aligned to within a few degrees. The ejection velocity, a free parameter in the solution, proves to be 0.26c and is remarkably stable (127). The axis of these jets rotates with a period of ~ 164 days; the central axis of the rotation cone is inclined by 79° from the line of sight, and the half-angle of this cone is 20°. (The two inclination angles are in fact degenerate in the model, but radio observations described in Section 3.2 resolve this ambiguity.) The time variable aspect of the jets with respect to the observer explains the observed modulation of the redshifts and blueshifts. The near-relativistic ejection velocity of the jets creates a substantial and directly observable transverse (second-order) Doppler shift, a unique situation in astronomy. The Lorentz factor for the observed $v = 0.26c$ corresponds to $\gamma = 1.035$, and this "time dilation" redshift is thus present in the velocity data as a zero point, regardless of the geometric phase of the jet axis; this explains the 12,000 km s^{-1} symmetry value for the Doppler shifts. Both radio (49, 65,

66) and X-ray (156, 175) imaging observations provide strong support for the jet model.

The model leaves the cause of the periodic change in the direction of the jet axis unspecified, although simple rotation can almost surely be rejected, because the inferred kinetic energy in the jets grossly exceeds the store of rotational energy in a compact object with such a long period (1, 75). Most workers (e.g. 75, 170) immediately turned to precession as a more likely candidate for the 164-day clock. The need for a mechanism to drive the precession, as well as the requirement for a source of matter for accretion to supply the jets and produce the X-ray emission, made the hypothesis of a compact star in a close binary system clearly attractive. The discovery of a 13-day periodicity in the radial velocity of the "stationary" SS 433 emission-line system (30, 32) provided elegant confirmation of this picture. The need to exert the precession torque on a reasonably large object, the desire for a simple symmetry plane to explain the excellent opposed alignment of the jets, the knowledge that substantial mass transfer must occur, and again the analogy to X-ray binary systems lead naturally to the concept that a substantial accretion disk must be present; we describe below a variety of observations that confirm this.

The kinematic model is exactly that: a convenient and very simple device for understanding and predicting in advance the kinematics of the ejected jets. However, it leaves unspecified almost all of the interesting physical aspects of the system, e.g. the nature of the underlying star or stars, the mechanisms that eject, collimate, and precess the beams and create the Doppler-shifted radiation, and certainly the evolutionary sequence that has led to this most peculiar behavior. The more detailed optical, radio, and X-ray observations described below attempt to elucidate these issues.

1.1   Basic Data

For convenience we provide here a summary of a variety of basic data on SS 433, also known as V1343 Aquilae (82). Some of this information is relatively uncontroversial in the sense that it is derived from model-independent measurements; in other cases (e.g. distance) the properties are conclusions from specific models that are described in more detail below. The optical object is located at (36, 73) $\alpha(1950) = 19^h09^m21.282^s \pm 0.003^s$, $\delta(1950) = +04°53'54.04'' \pm 0.05''$, epoch of 1980. The corresponding galactic coordinates are $l = 39.7°$, $b = -2.2°$. Finding charts at various scales may be found in [51, 86 (object 16, but coordinates in error), 93, 104, 136]. Nearby photoelectric comparison sequences are given in (51, 93). The optical position coincides to within measurement uncertainties with the position of the compact radio source associated with the object. There is as yet no optically detectable proper motion (101). The distance to SS 433, as deduced from radio observations of the time variable jet structure (65) as well as from 21-cm absorption studies (172), is 5 kpc; this figure is probably uncertain by no more than 10%.

Typical optical magnitudes and colors for the object are $V = 14.2$, $(B - V) = 2.1$, and $(U - B) = 0.6$ (18, 101, 129), although it is known to exhibit extensive periodic and aperiodic variability of amplitude up to 1 mag. The interstellar extinction is somewhat uncertain but known to be severe $(A_V \sim 8$ mag). Typical radio flux densities (summed over the central point and extended sources) are 1 Jy at 11 and 20 cm, 0.5 Jy at 3.7 and 6 cm, and 0.2 Jy at 2 cm (65), although again chaotic variability is common. The observed X-ray flux is 5 μJy in the 2–10 keV range (155), once again with variability.

We adopt for the binary orbital period 13.087 ± 0.003 days (4, 79); limits on possible period changes are quoted in (4, 51). The typical precession period hovers near 164 days (102, 105) with marked instability (5); for brevity, we refer to it as 164 days throughout.

2.   OPTICAL AND INFRARED OBSERVATIONS

One of the curious features of SS 433 is that despite its unusual properties, it is relatively bright in the optical region of the spectrum and may thus be studied intensively with small instruments. Indeed, a large fraction of the initial spectroscopy that revealed the Doppler-shifted spectral lines, perhaps still the chief outstanding peculiarity of the system, was obtained (101) with a 60-cm telescope! There are therefore numerous observations of SS 433 in the optical and infrared (using almost every conceivable technique) already in the literature. The following sections select for discussion either basic properties or recent developments, and are therefore not comprehensive.

2.1   Doppler-Shifted Spectral Lines

The majority of spectroscopic observations of the "moving" emission lines, as they are often called, have been reported by two groups [one at Asiago (20, 21, 99, 100), and one consisting of the author and colleagues (1, 5, 79, 101, 102, 105, 110, 127)], although a variety of important independent investigations have also been reported (e.g. 8, 95, 129). The lines observed are inevitably Balmer and He I lines, although with suitable infrared instrumentation, Paschen and Brackett shifted lines are also reported (3, 102, 118, 169). There are no reports of He II lines, although $\lambda4686$ is prominent in the "rest" (nonmoving) emission-line system. The absence of these features is clearly a crucial constraint on the excitation of the beams, and more quantitative limits would be desirable; predictions (10) are that

He II may well be exceedingly weak. There are no reports of shifted lines from heavy elements, but the strength of the observed shifted He I features is entirely consistent with the cosmic H/He ratio, and the lack of heavy element lines thus far is almost surely a sensitivity problem. The shifted lines are known to be unpolarized (95).

The breadth of the features and their sometimes-irregular profile may make them annoyingly difficult to detect on photographic plates, almost independent of spectral resolution, but they have proven especially well suited to study by the modern generation of spectrophotometric detectors. The individual lines are typically several thousand kilometers per second wide at the base, and their profiles range from extremely complex (101, 105, 129) to astonishingly Gaussian (110), although, as expected, all H and He lines in the same jet have the same velocity structure at a given time. At times the red- and blueshifted features show mirror-image velocity structures (101, 102), while at other times the two systems have unrelated structures. The profiles may change on a time scale short compared with one day.

The intensity of the shifted lines is also highly variable on similar time scales. A typical strength for the shifted lines may be 30% of the analogous unshifted features, but on occasion the lines have been seen to decrease in intensity by a factor of 10 in 24 hours, rendering them completely undetectable (110). Equally rapid reappearances have been seen. These profile symmetry and disappearance/reappearance episodes are synchronized in the red- and blueshifted emission-line systems to high accuracy; limits on time delays between the beams are <1 day, implying that the two radiating regions are separated by <100 AU. This is comparable to the length of each individual (optically radiating) jet inferred from brightness temperature considerations (34). Although the shifted features may be absent for several days, there is no reported evidence for any extended (months) periods when these lines are missing.

Somewhat independent of these daily intensity variations is the question of how the "moving" lines actually move. Several workers have described the process in terms of "shadows" or "bullets" (58, 105, 129), discussing discrete episodes when individual lines seem to grow and then fade, to be replaced by a neighbor that performs similarly. In other series of spectra this effect is not seen. Because we never observe the same physical patch of gas for an extended time scale, the distinctions between "bullet" or "picket-fence" descriptions of the line motion are chiefly morphological and semantic, rather than kinematic.

The simple "kinematic model" (1) provides a straightforward framework to both interpret and predict the gross behavior of the Doppler-shifted spectral features in SS 433. As of this writing, it may be applied to a data base consisting of almost 500 individual nights of observation of the Doppler shifts, spanning a 2000-day period in the interval 1978–83. Figure 1 displays these data, together with the best-fit kinematic model. Doppler shifts in the kinematic model may be computed from the formula

$$(1+z) = \gamma(\pm v \sin \theta \sin i \cos \psi \pm v \cos \theta \cos i + 1). \qquad 1.$$

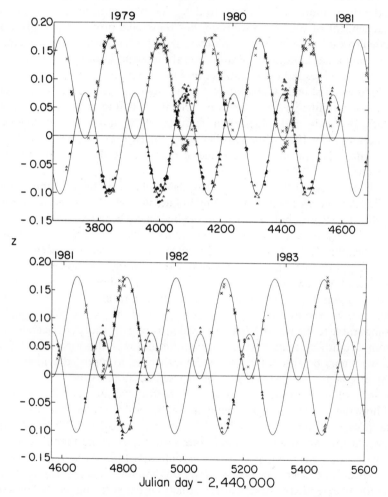

Figure 1   Doppler shifts of SS 433 on 450 nights in the period 1978–83. The majority of these data were obtained by the author and colleagues, supplemented by sources cited in (105). The solid curve is a least-squares best fit to the simple "kinematic model" (1). The free parameter values and their associated 1σ uncertainties (notation as in 105) for this fit are $v/c = 0.2601 \pm 0.0014$, $\theta = 19.80° \pm 0.18°$, $i = 78.82° \pm 0.11°$, $t_0 = $ JD 2,443,562.27 ± 0.39, $P = 162.532 \pm 0.062$ days.

Here, γ is the Lorentz factor,

$$\gamma = (1-v^2)^{-1/2},$$ 2.

$$\psi(t) = \psi_0 + 2\pi(t-t_0)/P_0,$$ 3.

and

$$\psi_0 = 2\pi\alpha = \arccos(-\cot i \cot \theta).$$ 4.

The notation here is as in (1, 105), and the "best-fit" values of the parameters are given in the caption of Figure 1. It should be noted that a typesetting error of precisely 10,000 Julian Days (JD) in the best-fit epoch quoted in (105) has propagated into a number of papers by other authors. We provide in the caption formal uncertainty estimates for the free parameter values as derived from the least-squares fitting procedure, but these must be regarded with caution because in at least some cases they are small compared with the known systematic deficiencies of the simple kinematic model (discussed below).

2.2   Deviations from the Kinematic Model

It is clear from inspection of Figure 1 that in certain respects the kinematic model is remarkably successful: it provides a reasonable explanation for an observed velocity variation of full amplitude 80,000 km s^{-1}! A variety of more sophisticated analyses show that in some sense the SS 433 system is elegantly (some might say frustratingly) simple. For example, the spread of velocities in the jets is remarkably small [as yet undetectable in the data (127)], and the system is sufficiently compact that differential light travel times are not seen (110, 132, 184), contrary to some early claims (28). Nevertheless, at least three separate types of significant deviations from the simple model are currently evident, and each provides important physical insights into the system.

Superposed on the 164-day periodic motions of the moving spectral lines are at least two shorter periods (21, 79, 100, 132). The dominant short-term period is 6.28 days, at an amplitude of 5–10% of the major (164-day) variation. The interpretation of the short-period variation (79) in a model-independent way proves to be straightforward. The 13-day orbital period of the binary companion is not negligibly short compared with the 164-day precession, so there is a time-variable torque on the accretion disk. Therefore, we expect a resulting periodic modulation of this torque at half the synodic period, and if the jets are constrained to follow the disk normal, this modulation should then be observable in the Doppler shifts. This is in fact the case. Some more general comments on these disk "nodding" motions have also been made (94, 115), and inferences on the disk structure and precession mechanism available from these data are discussed in Section 6.2. The 80,000 km s^{-1} jet velocity amplifies the nodding motion to an amplitude that permits rapid, accurate spectroscopic observation. As the frequency of this 6.28-day clock is conveniently high, these nodding motions provide a ready mechanism for observationally limiting any orbital period change in the system, which is ultimately expected because of the mass ejection (4, 51). A 5.8-day periodic component is also present in the observed nodding motion (79).

The second of the three deviations from the simple model is evident to the eye through inspection of Figure 1. There are extended (weeks to months) periods of time when the best-fit simple model runs systematically "early" or "late" from the data, suggesting that either the model precession period is incorrect and/or changing, or that the underlying clock is somehow more complex. A phase residual analysis of the currently available data (5) indicates that period derivatives of both signs are present at a statistically significant level, i.e. discrete sections of observations may be found that illustrate either increasing or decreasing periods. There is as yet little physical motivation other than mischief to suggest that these derivative fluctuations are themselves periodic, although this cannot yet be excluded. The permitted periods are of order 4 yr, and thus probing them will require very extended observations.

The fact that these two period derivatives are observed presents both physical and practical problems. Among the latter is that a lifetime estimator traditional for use in radio pulsars, $(1/P)(dP/dt)$, is clearly inapplicable to SS 433. The early announcement of a 40-yr lifetime for the system by Collins & Newsom (27) was due to an (in hindsight premature) analysis of a limited duration portion of the total observations (29), which unfortunately proved to lie in a short stretch of particularly large negative period derivative. Such a short lifetime is obviously not only anti-Copernican, but also contradicts the minimum lifetime of thousands of years needed to explain the current size of W50 (see Section 5). If one performs the identical calculation with currently available data and naively insists upon the best-fit constant period derivative, a characteristic lifetime of several thousand years (compatible with the size of W50) does in fact emerge, but this result must be regarded with great suspicion; past experience clearly shows that the inference may be merely a function of when the calculation is made. A second practical difficulty introduced by the complex period fluctuation is that it becomes awkward to predict values of the Doppler shifts more than 1–2 months in advance to an accuracy of better than many thousands of kilometers per second; this complicates the planning of observational programs.

Anderson et al. (5) point out that the period fluctuations may well be stochastic, and that a simple physical analog has been available for ten years—namely, the instability of the 35-day precession period in Her X-1 (45), which is of approximately the same relative amplitude as that seen in SS 433. Pending further data on the exact behavior of the instability, it seems likely that the simple, 5-free-parameter kinematic model will in fact be the most effective advance predictor of Doppler shifts. This conclusion may also be reached by inspection of Figure 1, which shows that there are very extended periods of time, both at the beginning and also the end of the 2000-day observation period, where the simple model is an excellent fit; this vividly illustrates that there is as yet no evidence for a large, systematic period change.

The third class of deviations from the simple model is also visible to the eye in Figure 1. It consists of departures from the predicted ephemeris of up to several thousand kilometers per second, of apparently random sign, on a time scale of one to a few days. This behavior has been termed "jitter" and studied by several authors (78, 106; S. Radford & B. Margon, in preparation). Interestingly, the amplitude of the ephemeris residuals is unchanged before and after subtraction of the 6.3- and 5.8-day periodic components, and thus the jitter is apparently unrelated to the disk "nodding" motion. Furthermore, the amplitude of the jitter has no obvious phase dependence in the 164-day period. This is significant, as it eliminates random perturbations of such parameters as the beam velocity as a candidate explanation; such deviations would, because of the high system inclination, have their apparent amplitude sharply modulated by the 164-day period, even if random in the frame of the star. The most likely explanation seems to be stochastic deviations in the beam-pointing direction on a time scale of days. A characteristic deviation of the pointing angle (106) is 0.04 rad, which is comparable to the inferred opening angle of the beam. Katz & Piran (78) consider some constraints that this mechanism places on the angular momentum of components of the SS 433 system.

2.3 *The Stationary Spectrum and "Normal" Star*

The set of emission and absorption lines that do not share the large periodic Doppler shifts are here for convenience called the "stationary spectrum." These lines are superposed on an extremely red continuum, the slope of which is presumably due to interstellar extinction. The stationary spectrum is dominated by very intense, broad Balmer and He I emission lines; Hα emission can exceed 500 Å equivalent width, and it is, of course, responsible for the inclusion of the object in the SS catalog. Most of the references cited for spectroscopy of the moving lines in Section 2.1 also discuss the stationary emission spectrum. At least the Balmer emission lines are highly variable in equivalent width and probably also absolute intensity, both periodically and aperiodically (4). In common with many galactic X-ray sources and cataclysmic variables, the spectrum also inevitably displays He II $\lambda4686$ and $\lambda10,124$ emission (although not nearly as strong as in the AM Her-type objects) and the C III/N III $\lambda4640$–50 emission blend. In X-ray sources, this anomalously strong blend has been attributed to the Bowen fluorescence mechanism, triggered by the wavelength coincidence of a copious supply of He II Lα $\lambda304$ ionizing photons with O III resonance lines (119); in the specific case of HZ Her/Her X-1, this is known to be the case (112). The O I $\lambda8446$ emission line is also prominent in SS 433 (99, 102), and its great strength again suggests a preferential emission process, possibly Lβ fluorescence (57).

If we neglect for the moment interstellar features, the spectrum of SS 433 is remarkably deficient in absorption lines. The He I and Balmer emission lines often show weak but significant P-Cygni absorption wings, the former somewhat more frequently than the latter, although the great intensity and breadth of the Balmer emission may often mask the true absorption. The origin of this absorption in the system is unclear. Amusingly, the only other prominent stellar absorption features of which this author is aware are of very low excitation, viz. Fe II $\lambda5169$ and O I $\lambda7773$. These two features are highly variable in intensity, but they can occasionally be very strong (see Figure 1 of 102). Weaker Fe II lines in the same multiplet are also reported (30). Such absorptions are often seen in shell stars and A–F supergiants. A phase dependence of Fe II absorption equivalent width with the 164-day period was suspected by some investigators (33, 105) and is confirmed strongly in unpublished data of the author and colleagues. The sense of the variation is that maximum absorption strength is reached when the moving emission lines are near equal Doppler shift, precisely the time when one expects to see the accretion disk edge-on. Thus these narrow lines either originate from cool matter in the disk or reflect the photometric variation of the disk.

Quite apart from the unprecedented "moving" lines, the total impression presented by the stationary absorption plus emission spectrum of SS 433 is unique. The broad emission is reminiscent of a variety of systems dominated by accretion disks (e.g. cataclysmic variables, low-mass X-ray binaries), but it also shares features in common with certain early-type stars undergoing copious mass loss (e.g. Wolf-Rayet stars). It is certainly not obvious which *if any* components of the spectrum—emission, absorption, and continuum—are dominated by the "normal" (i.e. noncompact) member of the system.

The radial velocity variations of the stationary emission and absorption have been studied in detail by Crampton and colleagues (30, 32). The first of these two papers concentrated on the H emission, leading to a discovery of the 13-day binary nature of the system through H radial velocity variations and an estimate of the mass function of 0.5 $M_\odot$. As the large majority (if not all) of bright galactic X-ray sources contain neutron stars of $\sim 1 M_\odot$, these data naturally led to many models of SS 433 having systems of mass ratio about unity, containing a low-mass, relatively normal object as the unseen secondary. The second work (32) complicated the situation considerably by demonstrating that the He II λ4686 emission illustrates periodic variations of the same period but substantially different amplitude *and phase* than the Balmer emission. The phase offset of 0.27 reported between the H and He II emission demonstrates that at least one of these emission regions is not simply associated with either of the two stars in the system. Crampton & Hutchings (32) hypothesize that the He II emission originates from the immediate vicinity of the compact X-ray source, not only because of similar situations in other X-ray binaries, but because the orbital motion of the H emission occasionally became undetectable in their data. As we see below, the combination of photometry and the disk "nodding" motions provides elegant verification of this hypothesis. The mass function implied if one accepts the λ4686 variations as truly indicative of the system orbit is 11 $M_\odot$ (32), a value greatly different from that previously inferred from the Balmer lines, and one that necessitates a massive, early-type companion, independent of the issue of whether such a star is actually visible in the spectrum.

2.4   *Interstellar Lines and Extinction*

Clearly relevant to the question of the nature of the underlying star is a quantitative estimate of the interstellar extinction to SS 433. Simple eye inspection of photographs of the region show extinction in the area to be large and patchy; thus, methods that determine the reddening very near to or coincident with the star are clearly preferred, although given the obscure nature of the entire system, as many independent estimates as possible would be comforting. Prominent interstellar absorption lines and bands appear in the spectrum and confirm that most or all of the extreme red color of the continuum must be due to extinction. Among the strong features reported are the 4430, 5778/80, and 6284 Å diffuse interstellar bands, as well as Ca H & K and Na D (101, 129); undoubtedly, numerous weak features are also present. Indeed, given the combination of very large intrinsic optical luminosity and the high reddening that we infer for the object, one could argue that SS 433 is an excellent background star for studies of interstellar lines and bands!

The correlation of extinction with the equivalent width of the interstellar features is one of the methods used to estimate the reddening of SS 433 (25, 101, 129), although these relations are known to have large scatter; thus, this method is probably of limited reliability. Decrements in the emission spectrum of optical filaments of W50 (83, 130) provide another approximation, but those filaments bright enough for spectroscopy have rather large angular offset from SS 433, which introduces further uncertainty. Similar difficulties are potentially shared by studies of extinction of stars near the field (30). The low-energy turnover in the X-ray spectrum carries information on the reddening (113, 152), but it is convolved with the uncertain intrinsic spectral shape and the possible circumstellar X-ray opacity. Infrared spectrophotometry, which enables a comparison of Brackett to Balmer emission-line ratios (118, 169), should be reasonably accurate, unless the well-known complexities that affect this technique in QSO spectra are also applicable here. Perhaps the most sophisticated approach to determination of the extinction also unfortunately has the most free parameters, namely a fitting of the continuum slope with the assumption of various underlying spectral types, reddening laws, reddening magnitudes, and variable amounts of extraneous (e.g. accretion disk) light of variable spectral slope. However, Murdin et al. (129) and especially Cherepashchuk et al. (18) argue persuasively that $A_V \sim 8$ mag, and I recommend adoption of this value. Most of the (individually weaker) determination methods discussed above are also compatible with this figure. If the continuum at some binary phases is indeed at least partially due to the normal star, one infers a spectral type of late O to early B, with uncertainty unfortunately too large to aid in the important measurement of the mass of the compact star. Depending upon the uncertain bolometric correction, the absolute magnitude of SS 433 is probably more luminous than -7, perhaps substantially so. Thus, SS 433 is one of the most intrinsically luminous stars in the Galaxy; of course, the kinetic energy of the ejected jets may well exceed this radiated optical luminosity. A further corollary of these extinction estimates, however uncertain, is that SS 433 is probably not detectable in the ultraviolet, even with the sensitivity of *Space Telescope*–class instruments.

The continuum of SS 433 is linearly polarized by a few percent (122), presumably (at least in part) because of interstellar extinction. However, McLean & Tapia (120) have reported that this polarization is time variable on at least a scale of weeks, implying that some significant fraction of the polarization is in fact intrinsic to the system.

2.5   *Optical and Infrared Photometry*

Extensive photometry, both broad- and narrowband, has been reported for SS 433 by several groups. Broadband work has been pursued primarily by consortia at Oregon/Israel/Japan (63, 80, 81, 91, 116, 167) and the USSR (16, 18, 50–52); narrowband work has been reported by the author and colleagues (4). The object is found to be variable on virtually all time scales, from hours through years; however, searches for very rapid optical variability (0.1 ms to 1 s) have proven negative (90). The majority of the detected variability is not merely due to the passage of the moving spectral lines through filter bandpasses, but rather it is known to be the result of genuine continuum variations.

Both of the spectroscopic periods, 13 and 164 days, are also seen clearly in the photometry. The 13-day variation is "double peaked," with a full amplitude of ~ 0.7 mag in B and V. The primary minimum agrees with the spectroscopic phase of maximum positive radial velocity of the stationary Balmer emission lines. Furthermore, the zero phase of the 6.3-day "nodding motion" of the Doppler-shifted spectral lines (79) is offset from this primary minimum by ~ 0.25 orbital phase, the identical phase offset found by Crampton & Hutchings (32) between the Balmer and He II emission-line velocities. The accuracy of agreement of this peculiar phase offset as measured by such completely disjoint techniques is impressive. Because the only plausible interpretation of the nodding motion requires that it originate in the vicinity of the compact star, any remaining ambiguity about which radial velocity curve (Balmer or He II) represents the true orbital motion is surely removed, and the 11 $M_\odot$ mass function (32) seems irrevocably established. A reasonable interpretation of the light curve (e.g. 4, 16, 92) invokes mutual eclipses of the massive, early-type normal star and the extended accretion disk.

Infrared photometry of the system has been reported by several investigators (15, 26, 46, 47, 182). Although the 13-day variation is not obvious in these data, Cherepashchuk et al. (18) claim that a phasing of the data of Catchpole et al. (15), using a more accurate binary period than available to those authors, reveals a periodic orbital modulation at infrared wavelengths.

The 164-day variation is now also evident in photometry, even though only ~ 10 cycles have been observed (4, 63). Historical photographic photometry showing variability is also available (50, 179), but it is unclear how to properly phase these data because of the known complex instability of the 164-day clock, inferred from the moving-line Doppler shifts (5). The 164-day variation is again double peaked, with full amplitude of ~ 0.5 mag in V. The phasing of the variations is nicely compatible with a model that attributes most of this modulation to light from an extended accretion disk; primary maximum coincides with the maximum projected surface area of the disk, as inferred from the observed jet orientation.

3. RADIO OBSERVATIONS

SS 433 has been intensively observed at radio wavelengths by a large variety of techniques, including single-dish studies at many frequencies and VLBI. However, studies made with the Very Large Array (VLA) are surely among the most intriguing. The radio properties of SS 433 are uncannily well suited to observation by the VLA; this is especially curious, as the design of the instrument predates the recognition of the importance of the object! Could the designers realistically have hoped for the later discovery of an object with ~ 1 Jy of total intensity, whose spatial structure, interesting on a scale of 1 arcsec and thus suitable for mapping, would obligingly change on a time scale of weeks?

3.1   *The Central Source*

The radio source associated with SS 433 was probably first observed ten years prior to recognition of its unusual properties, a situation not dissimilar to that of the optical object. The source 4C 04.66 (53) is almost surely SS 433 despite the substantial positional discrepancy, which was presumably due to confusion in this crowded region. The positional coincidence between the optical object and the well-studied radio source is now known to be at least 0″.2 (65, 73).

Observations of the central source have been reported by numerous observers from 160 MHz to at least 22 GHz; a typical spectral index at the higher frequencies is −0.6, common for nonthermal sources (64, 154). A spectrum obtained simultaneously over a very broad range has been given by Seaquist et al. (153). Limits exist at higher (88) and lower (153) frequencies; the latter report shows that the spectrum flattens and probably turns over at frequencies less than 300 MHz. A variety of mechanisms intrinsic to the source can explain the flattening.

The intensity of the central source is highly and erratically variable, at least by a factor of four, at virtually all wavelengths (62, 71, 72, 143, 151) on time scales as rapid as one day, and occasionally hours (151, 154). There is evidence from VLBI observations (44) that, at least at 10.65 GHz, a lower limit to the angular size of the central radio-emitting region is about 5 mas (~ 20 AU at the distance of SS 433), so we probably should not expect more rapid variability. Extensive observations have been analyzed in (71, 72) in a search for analogs to the 164- and 13-day optical cycles of SS 433, and no such periodicities are found. Simultaneous radio and optical observations (19, 131) yield no convincing evidence of aperiodic but correlated behavior between the two wavebands, although the generally chaotic variations on all time scales at both wavelengths, together with the possibility of time delays between the radio and optical variations, make interpretation of this type of data exceedingly difficult. Apparently the coupling between radio and optical variability is weak or perhaps nonexistent. Simultaneous X-ray and radio observations occurred on two occasions (152); evidence for correlated variability was seen in one observation but not the other, again leaving the status of activity correlations at these two wavebands unclear.

3.2 The Extended Components

Shortly after intensive radio observations began, it was realized that a significant fraction of the flux from SS 433 is extended on spatial scales of a few arcseconds, and that the morphology of this extension is time variable (48, 49, 164). The extended emission is heavily polarized, up to 20% (65, 71), which, together with the nonthermal spectral index, certainly points toward highly relativistic synchrotron emission as the mechanism. A remarkable series of intensive observations on a variety of spatial scales, using the VLA and VLBI techniques, have elucidated the nature of the structural variability. Very high spatial resolution (10–100 mas) work has been reported primarily by two collaborations [(145–148) and (133, 134, 174)]. The VLA data of ~1″ resolution have been extensively discussed by Hjellming & Johnston (64–66). These efforts are nicely complementary, as the VLBI work probes the "central engine," where the radio-emitting particles are produced. The VLA data, on the other hand, are an excellent probe of the kinematics of the jets, because they survey a volume sufficiently large that the 164-day precession of the beams should have observable effects on the trajectories of the radio-emitting material. Furthermore, the highly multiplexed nature of VLA observing lends itself to very frequent observations of this rapidly changing system, interleaved with unrelated programs, quite analogous to the optical spectroscopic programs.

The observed morphology changes markedly on a time scale of days, giving the impression of a "corkscrew" pattern projected onto the plane of the sky. If one attributes these changes to ballistically coasting matter ejected from the twin jets, a natural algebraic framework for analysis of the morphology emerges, i.e. the radio equivalent of the optical kinematic model. The results of this analysis (e.g. 65, 66) are remarkable: the free parameter values derived agree to high precision with those inferred quite independently from the optical data analyses. This then represents a most elegant verification that the twin-jet model is in fact the correct interpretation of the optical "moving" spectral lines. This is by no means a redundant function, because the spectroscopic observations are at least formally compatible with a variety of other configurations in addition to the twin jet. In the radio data, however, we directly view the relativistic blobs coasting outward, rather than infer this motion indirectly, as was previously necessary. Furthermore, the radio observations permit derivation of other key system parameters not available from the optical analyses. Because the radio-emitting region probed by the VLA observations has characteristic dimensions 100 times greater than the optically radiating portions of the jets, light travel delays become important, and consistent kinematic solutions are impossible without their inclusion. The "corkscrew" patterns, in fact, track the equivalent of proper motions of the coasting blobs, but because the linear velocity of the material in the jet is also known, an elegant derivation of the distance to SS 433 becomes available. The resulting value of 5 kpc is almost surely more accurate than any optically derivable values. The degeneracy of the system inclination angles inherent in the optical kinematic model is broken by the radio solution; the 79° angle must be the system inclination. Even the sense of the jet rotation—clockwise (left-handed)—emerges from the solution.

The optical observations will still serve a distinct role in studies of the jet kinematics on time scales of days (e.g. the "nodding" motions), simply because the radio-emitting volume is too large to conveniently probe such rapid changes. However, there seems little reason why an extension of the current radio data will not yield information on the instability of the 164-day period, completely analogous to and independent of the optical data.

4. X- AND GAMMA-RAY OBSERVATIONS

X-ray observations of the objects we now know as SS 433 and W50 began almost a decade ago, although the most recent data from the Einstein Observatory have been extraordinarily useful in elucidating the nature of the high-energy emission. The first reported X-ray observations of this rather crowded region of the X-ray sky were obtained by *Ariel V* in 1974–75 and reported by Seward et al. (155). These authors suggested the identification of the rather poorly located source with W50, and, noting intensity variability, they presciently commented that the source was a candidate for identification with a compact stellar remnant. The X-ray source also appears in the 4U catalog (40). Neither the *Ariel* nor the *Uhuru* observations would suggest anything unusual about the object as compared with the several dozen brighter X-ray sources in the plane of the Galaxy. A typical observed flux for the source is about 1×10^{-10} erg cm^{-2} s^{-1} in the 2–10 keV band, but at least the compact source is known to vary. Little correction is required for interstellar photoelectric absorption in this bandpass, so we may confidently estimate the X-ray luminosity as a few times 10^{35} erg s^{-1}; this value is trivially small compared with the inferred optical luminosity, and of course also with the kinetic energy of the material in the jets.

Using observations made with the *HEAO-1 A-2* experiment, Marshall et al. (113) analyzed the X-ray spectrum of the source in the 2–30 keV range. The data lack spatial resolution, and so they must apply to the sum of compact and diffuse components of the source. These workers could not distinguish between thermal and power-law spectra, but both models require a very prominent (580-eV equivalent width) emission line at 6.8 keV.

This line, a result of a variety of unresolved transitions of highly ionized iron, is common in compact galactic X-ray sources. The observed equivalent width suggests a thermal plasma of temperature ~ 14 keV, with large uncertainty. A spectrum from *Ariel VI* shows a similar emission feature (141). Grindlay et al. (61) place very stringent upper limits on the presence of Si emission near 1.9 keV, which is expected to be very strong for thermal sources with kT between 0.35 and 1.3 keV; once again, this result suggests that thermal interpretations of the X-ray emission must invoke substantially hotter plasmas.

The most interesting contributions by *Einstein* on this object are doubtless on the spatial structure of SS 433. The first X-ray imaging observations (156) showed that although 90% of the soft (1–3 keV) X-ray flux is coincident with the optical object, the remaining 10% is contained in two extended (30') jets closely aligned with the major axis of W50. As Seward et al. (156) point out, the inferences from this one image are multiple: the association of SS 433 and W50 is irrevocably confirmed, the evidence for the existence of the ejected jets is further strengthened, and a minimum age of several thousand years is established for the ejection phenomenon. Analysis of further imaging data (175) shows that the spectrum of the X-ray lobes differs markedly from the central source, not a surprising result given that this emission must reflect the interaction of the ejected jets with the ambient medium.

As has been appreciated since the initial observations (155), the X-ray intensity of SS 433 is highly variable on a number of time scales. Significant daily variations have been observed (152), which may on some, but not all, occasions be correlated with radio variability. A search for variability synchronized with the 13-day orbital period has been reported (61), with inconclusive results: variability at this frequency is not statistically significant, but the data when phased with the optically measured binary period do agree in phase with the broadband photometry. Given the number of different phases of interest in the SS 433 orbital period, the importance of this agreement is unclear. The *Einstein* data are of insufficient quantity to search for X-ray analogs to the 164-day period (61); a tentative discussion of this effect in the very extended *Ariel V* observations is given in (141). There is no variability observed on short (minutes to seconds) time scales (61), in contrast to many, but not all, compact galactic X-ray sources. Either the nature of the compact object in SS 433 is different than the bright X-ray variables, or, perhaps more likely, our view in X rays does not extend completely into the central, compact volume.

A recent and particularly fascinating observation of high-energy emission from SS 433 is the report (87) of one or possibly two intense gamma-ray emission lines near 1 MeV. If the radiation is assumed isotropic, the inferred gamma-ray luminosity is 2×10^{37} erg s^{-1}, a significant fraction of the radiation budget of the object. The observed wavelengths of the gamma-ray features do not correspond to any common laboratory transitions, but they have been attributed (87) to a nuclear transition of ^{24}Mg at the Doppler shift of the jets appropriate for the time of the gamma-ray observations. There are several difficulties with this interpretation: the inferred kinetic energy in Mg alone approaches 10^{52} erg during the source lifetime; for cosmic abundances, much stronger nuclear transitions of C and O should appear in the spectrum but are not observed; and the observed features are narrow compared with the expected motion of the Doppler-shifted lines during the 10-day period of the observations. Furthermore, irrespective of the possibility of exotic heavy element abundances in SS 433, proton bombardment of ^{24}Mg must simultaneously produce very strong lines of Na and Ne, but these lines are absent in the reported spectrum (135). Given the esoteric nature of any explanation for the gamma-ray feature, it is clear that observational confirmation is badly needed. A preliminary result of an independent experiment (98) does not confirm the existence of the lines at the previously reported intensity.

5.   W50

Because SS 433 and the diffuse nebula W50 are so intimately linked, both from the point of view of the energy budget of the nebula and the evolution of the stellar system, we would be remiss not to at least briefly summarize some of the properties of the nebula. This very bright (150 Jy at 178 MHz), extended radio structure has long been studied (180) and classified as a supernova remnant (e.g. 22, 67); it has a variety of synonyms, e.g. CTB 69, M43, LMH 40, D 121, MSH 19+01, HC 28. Its largest dimension is about 45', corresponding to 65 pc at the distance of SS 433. The structure of the elliptically shaped nebula has been delineated in increasing detail at both low and high frequency (24, 37, 42, 173). It is one of a small but distinct class of filled-shell nebulae sometimes dubbed "plerions" (177, 178) [from the Greek *pleres* ("full")], typified by the Crab Nebula. The strong, variable radio point source (24, 42, 143, 151) corresponding to the stellar object SS 433 is remarkably centrally located in the elliptical remnant, leading naturally to the suggestion (25, 143) of an association.

A second flat-spectrum, variable, unresolved point source of about the same 1.4-GHz flux is also embedded within W50, 30' NE of SS 433 (143). If one accepts the now virtually indisputable association of SS 433 and W50, then this second object, 1910+052, must surely be a chance superposition on the remnant; optical studies of the field (111) indeed show no unusual objects, and H I absorption measurements (172) are consistent with an extragalactic origin. This source stands as proof that unusual but irrelevant point radio sources will surely be found projected against supernova remnants if one pursues searches to sufficiently faint flux levels; some additional cases are discussed in Section 7.

To complete the confusing radio picture in this area, an H II region, S74, is superposed on the NW portion of W50. Optical studies (30, 31, 161) indicate that this is almost certainly a foreground, unrelated object. An extended and prominent molecular cloud, mapped in CO, also overlaps W50 (68), but it is also quite possibly in the foreground.

Optical emission coincident with W50 is visible on the Palomar Observatory Sky Survey red print, but it was noticed only after independent discovery of visible filaments on new plate material (171, 183). Moderate-resolution spectroscopy of the filaments has been reported in (83, 130, 162, 183). The spectrum is best observed in the red because of the high extinction, which also of course affects SS 433. The large observed ratio of [S II] to Hα is characteristic of the optical filaments in supernova remnants. A very high Balmer decrement again shows the effects of reddening, but the W50 structure is of such large angular extent that there must surely be differential extinction across the remnant, especially given its low galactic latitude; thus, extinction estimates for W50 derived from line ratios are unfortunately probably not directly applicable to SS 433.

Certainly the most interesting property of the spectrum of the filaments is the very great strength of [N II] relative to Hα. Among other observed remnants, only Puppis A seems to share this anomaly to this extent. There is as yet no general agreement as to whether this presumed abundance anomaly reflects conditions in the local interstellar medium or is more specifically related to SS 433. A very high resolution (15 km s^{-1}) study of [N II] at a number of different locations (117) shows complex velocity structure across the remnant, perhaps as could have been expected, and also hints that the sense of rotation of the jets is opposite to that inferred from the radio model (7, 66).

Several authors were quick to point out that the enormous kinetic energy output of the ejected jets of SS 433 is potentially capable of influencing the structure and evolution of W50 (10, 83, 85, 183). The fact that W50 does not obey the standard radio surface brightness correlation with remnant diameter (22) is a hint that such an interaction might be important. However, the excellent observed alignment of the major axis position angle of W50 with that of the precession cone axis inferred from the optical kinematic model, and the directly observed symmetry axes of the extended radio and X-ray emission, show that this proposed interaction must certainly be true; only the matter of degree is in question. The most extreme position on this question might be to wonder if W50 is in fact even a supernova remnant (85); depending on the poorly known lifetime of the SS 433 phenomenon, the time-integrated kinetic energy output of the beam may be comparable to that of a supernova, and thus the energetic need for the explosion event to inflate the nebula is removed.

Because the major axis alignment of W50 with the SS 433 jets indicates that at least a minimal interaction of the jets with the ambient medium is already visible, one can constrain the kinetic energy and mass flow rate through the jets. The uncertainties in the extinction to the source and in the precise physical conditions in the optically emitting regions of the jets make this more model-independent approach particularly attractive. The results of most of these calculations (e.g. 10, 34, 83) indicate that the kinetic energy of the flow must be at least 10^{39} erg s^{-1}; some estimates are more than a factor of 100 above this.

6.   SOME MAJOR UNCERTAINTIES

The above sections hopefully demonstrate that we have a reasonably good macroscopic understanding of certain facets of SS 433, especially considering that intensive observations have been conducted only for a few years, equivalent to less than a dozen precession periods of the system. In particular, the success of the kinematic model in explaining the gross optical and radio properties, and the elegant agreement between system parameters inferred independently from the radio and optical data analyses, seem to put the twin-jet model on a firm footing. On the other hand, the list of vital areas where we have imperfect microscopic understanding of processes in SS 433 is still large, and it seems appropriate to touch upon some of them here. As there have been more than 100 theoretical papers published on SS 433 in the past few years, I again offer the disclaimer that the reference citations in this section are necessarily incomplete.

6.1   *Jet Energetics, Collimation, and Acceleration*

The kinetic luminosity in the jets is perhaps the single most impressive statistic of the SS 433 system: it quite possibly exceeds the radiative luminosity of any star in the Galaxy. On general grounds, the most appealing mechanisms to supply this considerable demand are either rotation of or accretion onto a compact star. As there are no observations suggesting the presence of rapid rotation in the compact object of this system, most workers have concentrated on accretion as the logical energy source. Supercritical accretion (i.e. beyond the Eddington limit) has been actively discussed in connection with SS 433 (10, 14, 69, 70, 75, 96, 121, 163, 170) for at least two reasons: (a) unless the compact object is extremely massive ($\geq 10\ M_\odot$) or the jet radiation mechanism sufficiently exotic that the mass flow rate (and thus the kinetic energy requirement) has been badly overestimated, the Eddington limit must be exceeded to power the jets; (b) all models for SS 433 are permitted the luxury of at least one very different parameter from other compact binary systems, simply because the object exhibits such unique behavior! It is noteworthy that several years before the discovery of the jets in SS 433, Shakura & Sunyaev (158) considered supercritical accretion onto compact objects and concluded that collimated ejection might result (see especially their Figure 9).

The remarkably good collimation of the jets, as evidenced by the relative narrowness of the moving spectral lines, is another major theoretical problem. Because the phase difference between the 164-day cycles of the photometric variations and the moving lines provides positive observational evidence that the jets are transverse to and follow the cyclically changing orientation of the accretion disk, and because the disk provides a natural symmetry plane, it seems reasonable to assume that the disk plays an intimate role in the collimation process (34, 75, 163). Similar arguments have been made in connection with the collimation problem in active galactic nuclei (2, 97). Other mechanisms are certainly available as well: Begelman & Rees (9) suggest that the collimation occurs in nozzles near the magnetopause of a magnetized neutron star, and Eichler (38) points out that the ambient medium may also play a role in collimation.

Yet another challenge is an explanation of the jet acceleration mechanism. The observed degree of isotropy of jet velocities (127) is impressive. Milgrom (125) has elegantly pointed out an observational clue that is either a key to this process or else a cruel hoax of nature: the observed velocity of the jets, $0.26c$, is (to within a few percent) the value needed to Doppler shift the hydrogen Lyman continuum limit to the wavelength of Lα, including the special relativistic correction. The implication is that acceleration is provided by radiative absorption into the sub-Lyman continuum of photons from a central, luminous source whose intrinsic spectrum is blanketed at the Lyman limit, presumably by cool gas. The acceleration can therefore bring the jet velocity up to the value where the longest available continuum photon at 912 Å can no longer "line-lock" with Lα, and the jet velocity then stabilizes at that value. Some difficulties with this process include the considerable ultraviolet luminosity required from the central continuum source, as well as the problem of avoiding complete ionization of the gas in the jet, which necessitates the hypothesis of severely clumpy geometry (160). Hybrid acceleration mechanisms are, of course, also possible, in which this line-locking mechanism plays only the secondary role of stabilizing the velocity once it is achieved by some different process.

6.2   The Disk and Precession Mechanism

Recent observations have provided considerable insight into the structure of the accretion disk, an issue intimately linked to the details of the mechanism that precesses this disk. The amplitude and shape of the 164-day photometric component of the light curve indicate that the disk is remarkably thick (4, 12); a disk half-thickness to radius ratio of 2/3 has been suggested. This conclusion is supported by an entirely independent observation, namely the existence of the 6.3-day nodding motion (79). The disk must be extremely viscous to rapidly transmit the nodding from the outside, where significant torque is applied, to the interior, where the jets originate and are seen to respond to the motion (76). If, to ensure stability, one demands that the disk internal pressure equal or exceed the large inferred viscous stresses, then a thick disk of about these same proportions is also implied. This thick disk, perhaps so extreme in proportions that the term "loaf" might be more applicable, fits nicely with the type thought theoretically to be useful in collimating the outflow, as discussed above. An attempt to constrain the temperature of the disk through multiple free parameter fitting of a decomposition of the optical spectrum has been reported in (18). Probably more certain are inferences on the disk linear dimensions, derived (4) simply from the amplitude of the observed 164-day photometric variations and plausible emissivity limits: the disk must be quite extended, with dimensions of order 10^{12} cm (i.e. comparable to the separation of the two stars). Again, the existence of the nodding motion provides independent confirmation of the extent of the disk, because the torque can be significant only on a rather large structure.

Considerable effort has been devoted to understanding the cause of the 164-day precession of the disk in SS 433. This is not a new problem in the context of compact stars in binary systems; for example, the 35-day X-ray modulation of Hercules X-1 is widely attributed to disk precession, and a variety of other X-ray sources are now known to have long-term periods probably also due to disk precession (89, 138). The difficulties in precise modeling of the precession mechanism for all of these systems (e.g. can there be a single, stable precession period in a differentially rotating disk?) are certainly present in SS 433 as well. Early, innovative suggestions (114, 144) that the SS 433 precession mechanism might instead be the relativistic Lense-Thirring effect ("dragging of frames") can probably now be ruled out, because such schemes require a highly compact disk, contrary to the inferences from the photometry and the nodding motions discussed above.

Additional model discriminants available from photometry and spectroscopy are discussed in (77). The remaining candidate mechanisms most often considered are classical, driven precession, where the normal companion exerts a torque on the disk (74), and "slaved" precession (142, 159), where a short residence time for matter in the disk permits it to follow the precessional motion of a misaligned companion star. These mechanisms have been discussed in the context of SS 433 in some detail by a variety of authors (35, 69, 75, 76, 170, 181), with the slaved mechanism most often suggested. Katz et al. (79) claim that the observed amplitude of the nodding motion strongly favors slaved over driven precession as the dominant process. Sadly, the details of how and why a fluid star precesses remain to be conquered!

6.3   The Nature of the Compact Star

The near-relativistic ejection velocity of the jets, as well as the X-ray emission from the system, certainly suggests the presence of a collapsed object, either a neutron star or black hole, as a companion to the early-type star in the system. Despite the intensive studies of SS 433 described above, however, the choice between these two alternatives is still (at least in my view) unclear. Traditional tests to distinguish between these objects include the detection of sustained periodicities, as in neutron star pulses, or a mass measurement for the collapsed object (6). While no rapid variations of any sort have yet been reported in SS 433 at any wavelength, this is of course inconclusive, as such behavior might be ultimately observed later. Because the system eclipses and is at least a single-lined spectroscopic binary, in theory the optically derived mass function in conjunction with light curve synthesis techniques can produce a mass estimate for the compact star; by now-traditional arguments, results exceeding a threshold of $\sim 3\ M_\odot$ may be indicative of a black hole. Such modeling has been attempted by an increasing number of authors (e.g. 17, 92), but the problems are the same as those that have plagued light curve synthesis efforts for decades: the results are of uncertain uniqueness, and/or the magnitude of the errors is difficult to evaluate. This is not meant to denigrate this technique; indeed, it may well prove to be the only reasonable method ultimately available to ascertain the nature of the unseen star. It is certainly exciting to note that at least some of these efforts do result in inferences of very massive compact objects; Leibowitz et al. (92), for example, find a mass $> 4.3\ M_\odot$.

If one had confidence that the collimation mechanism of the jets was well understood, an additional piece of evidence on the nature of the compact star would then be available. For example, the nozzle collimation process discussed in (9) probably requires a neutron star; but the thick disk scenarios discussed may require black holes. However, it has been argued

(34) that general relativistic effects are unimportant in the collimation process.

Most galactic X-ray sources are now known to contain neutron stars rather than black holes, so an argument of simplicity might opt for a neutron star in SS 433 until proven otherwise. On the other hand, it is obvious that if one intuitively expects stellar black holes to have any curious or interesting observational consequences, then SS 433 is a far better black hole candidate than Cygnus X-1!

6.4 *The Extragalactic Analogy*

One of the most interesting implications of the twin-jet model for SS 433 has been appreciated since its inception and was in fact a primary motivation (39) for this line of interpretation: collimated, relativistic outflow of material is thought to be a basic, recurring phenomenon in a variety of active galactic nuclei, including radio galaxies and quasi-stellar objects (123). If there is in fact similarity in the basic physical processes between SS 433 and these extragalactic cases, this would be extraordinarily fortunate. We can study the microscopic details of the jet ejection, collimation, and precession process in the former object through repeated observations of a few minutes duration with telescopes as small as 60 cm or with the VLA, and the time scales for observable changes in the SS 433 jets are probably $\sim 10^5$ times more rapid than the extragalactic cases. Furthermore, we have in SS 433 a precise, unambiguous measurement of the ejection velocity, a basic datum normally lacking in the extragalactic cases. Recent observations of expanding radio structure in Cygnus X-3 (41) may represent a second such galactic jet velocity measurement.

Rees and collaborators (139, 140) have presented arguments for the physical significance of this scaling argument, implying that the analogy between SS 433 and extragalactic jets is indeed more than a morphological coincidence. Meanwhile, recent VLA and VLBI data continue to increase our confidence that precession is indeed the clock in at least some extragalactic jets (11, 54–56, 128).

7. ANALOGOUS GALACTIC OBJECTS

Very few, if any, astronomical objects are truly unique, so it seems quite appropriate to conclude this review by touching upon the question of stellar objects analogous to SS 433. The fact that the distance to SS 433 is many kiloparsecs rather than tens or hundreds of parsecs suggests immediately that the galactic population of these sources, if there is in fact more than one, consists of only a handful of objects. Thus, if one or two analogs are "hidden" by extinction on the far side of the disk, it is not impossible that we already know of the only observable example of the SS 433 phenomenon in the Galaxy.

An initial suggestion of an "SS 433 class," a group of compact, variable radio sources positioned inside of supernova remnants, was made by Ryle et al. (143), who listed nine objects, including SS 433 and Circinus X-1, as putative members. Although there is little question that Cir X-1 is indeed a binary system with a compact member resulting from a supernova (13, and references therein), this system, despite extensive observations, shows no observational evidence for the chief peculiarity of SS 433, viz. collimated, relativistic jets. The remaining seven objects in the list have been studied with increasing detail, and evidence is mounting not only that none are SS 433 analogs, but quite possibly that none are even galactic. Two of the sources are now definitely known to be chance superpositions of extragalactic objects onto foreground galactic supernova remnants: CL4, a quasi-stellar object of $z = 3.18$ (111, 168), and $0125 + 628$, a radio galaxy of $z = 0.018$ (84, 165). For three additional sources in the Ryle et al. (143) list (including $1910 + 052$, the second radio source embedded in W50 quite near to SS 433), recent 21-cm absorption measurements indicate distances in conflict with estimates for the respective supernova remnants (172), again arguing for chance superpositions upon the remnants.

Occasional odd, isolated objects have also been suggested as SS 433 analogs. The pathological "Red Rectangle," the bipolar nebula surrounding HD 44179, was one such suggestion (176), based upon its collimated, symmetric morphology as well as its unusual optical spectrum; further spectroscopy (149) and radio observations (43), however, lend no support to this idea. The pulsing X-ray source $2259 + 586$, centered in the supernova remnant G109.1 − 1.0, has also been suggested as being closely related (59, 60), but again further optical (103) and X-ray (61) observations indicate that this is an interesting but probably quite different object.

Depending upon the (somewhat uncertain) estimates of the interstellar extinction of SS 433, it seems apparent that this object is of comparable optical absolute magnitude to the most luminous known stars. Thus an analogous object should be easily observable in the Magellanic Clouds, perhaps (in a curious repetition of history) already catalogued but unrecognized! If located in M31, SS 433 would be within the spectroscopic reach of large telescopes, *if* one knew where to search. If the frequency of the phenomenon scales simply with galaxian mass, one might expect a few currently active examples in M31 but none in the Clouds. However, at least the Large Magellanic Cloud is relatively overabundant in early-type stellar phenomena, so it might still be a logical host for one or two such objects. Because the X-ray luminosity of SS 433 is not spectacular compared with several dozen other known galactic objects of varied nature, a reasonable speculation is that attention might be called to an extragalactic version through either the radio emission or morphology, or of course the peculiar optical emission lines.

It seems a not overly stringent requirement that a true analog to SS 433 should exhibit direct observational evidence for the most unique and interesting feature in the prototype object: collimated, near-relativistic expulsion of opposed, precessing jets. At the time of this writing, no such contender has arisen.

ACKNOWLEDGMENTS

I am indebted to S. F. Anderson, R. A. Downes, S. A. Grandi, and J. I. Katz for a collaboration on observations and interpretation of SS 433 that has spanned five years. D. Clark is responsible for my original interest in this object. The unselfish and repeated contributions of observing time by a variety of University of California astronomers are largely responsible for the intensive record of spectroscopic observations evident in Figure 1: among this group are L. H. Aller, E. M. Burbidge, H. C. Ford, C. D. Keyes, A. T. Koski, M. Plavec, A. Shafter, H. E. Smith, H. Spinrad, J. Stauffer, and R. P. S. Stone. The literature survey for this work was conducted by H. Preston. I acknowledge the financial support of the NSF, NASA, and the Alfred P. Sloan Foundation during stages of much of the work reported here.

Literature Cited

1. Abell, G. O., Margon, B. 1979. *Nature* 279:701–3
2. Abramowicz, M. A., Piran, T. 1980. *Ap. J. Lett.* 241:L7–11
3. Allen, D. A. 1979. *Nature* 281:284–85
4. Anderson, S. F., Margon, B., Grandi, S. A. 1983. *Ap. J.* 269:605–12
5. Anderson, S. F., Margon, B., Grandi, S. A. 1983. *Ap. J.* 273:697–701
6. Bahcall, J. N. 1978. *Ann. Rev. Astron. Astrophys.* 16:241–64
7. Barker, B. M., Byrd, G. G. 1981. *Ap. J. Lett.* 245:L67–69
8. Bedogni, R., Braccesi, A., Marano, B., Messina, A. 1980. *Astron. Astrophys.* 84:L4–7
9. Begelman, M. C., Rees, M. J. 1984. *MNRAS* 206:209–20
10. Begelman, M. C., Sarazin, C. L., Hatchett, S. P., McKee, C. F., Arons, J. 1980. *Ap. J.* 238:722–30
11. Biretta, J. A., Cohen, M. H., Unwin, S. C., Pauliny-Toth, I. I. K. 1983. *Nature* 306:42–44

12. Bochkarev, N. G., Karitskaya, E. A., Kurochkin, N. E., Cherepashchuk, A. M. 1980. *Astron. Circ. USSR No. 1147*
13. Bradt, H. V. D., McClintock, J. E. 1983. *Ann. Rev. Astron. Astrophys.* 21:13–66
14. Calvani, M., Nobili, L. 1981. *Astrophys. Space Sci.* 79:387–95
15. Catchpole, R. M., Glass, I. S., Carter, B. S., Roberts, G. 1981. *Nature* 291:392–94
16. Cherepaschuk, A. M. 1981. *MNRAS* 194:761–69
17. Cherepashchuk, A. M. 1981. *Sov. Astron. Lett.* 7:111–12
18. Cherepashchuk, A. M., Aslanov, A. A., Kornilov, V. G. 1982. *Sov. Astron. Lett.* 26:697–702
19. Ciatti, F., Mammano, A., Bartolini, C., Guarnieri, A., Piccioni, A., et al. 1981. *Astron. Astrophys.* 95:177–83. Erratum in *Astron. Astrophys.* 100:330
20. Ciatti, F., Mammano, A., Vittone, A. 1981. *Astron. Astrophys.* 94:251–58
21. Ciatti, F., Mammano, A., Iijima, T.,

Vittone, A. 1983. *Astron. Astrophys. Suppl.* 52:443–53

22. Clark, D. H., Caswell, J. L. 1976. *MNRAS* 174:267–305

23. Clark, D. H. 1984. *The Quest for SS 433.* New York: Viking. In press

24. Clark, D. H., Green, A. J., Caswell, J. L. 1975. *Aust. J. Phys. Astrophys. Suppl.* 37:75–86

25. Clark, D. H., Murdin, P. 1978. *Nature* 276:45–46

26. Clark, T. A., Milone, E. F. 1981. *Publ. Astron. Soc. Pac.* 93:338–43

27. Collins, G. W. II, Newsom, G. H. 1980. *IAU Circ. No. 3547*

28. Collins, G. W. II, Newsom, G. H. 1981. *Vistas Astron.* 25:169–72

29. Collins, G. W. II, Newsom, G. H. 1982. *Astrophys. Space Sci.* 81:199–208

30. Crampton, D., Cowley, A. P., Hutchings, J. B. 1980. *Ap. J. Lett.* 235:L131–35

31. Crampton, D., Georgelin, Y. M., Georgelin, Y. P. 1978. *Astron. Astrophys.* 66:1–11

32. Crampton, D., Hutchings, J. B. 1981. *Ap. J.* 251:604–10

33. Crampton, D., Hutchings, J. B. 1981. *Vistas Astron.* 25:13–21

34. Davidson, K., McCray, R. 1980. *Ap. J.* 241:1082–89

35. DeCampli, W. M. 1980. *Ap. J.* 242:306–18

36. deVegt, Chr., Gehlich, U. K. 1979. *Astron. Astrophys.* 79:L16–17

37. Downes, A. J. B., Pauls, T., Salter, C. J. 1981. *Astron. Astrophys.* 103:277–87

38. Eichler, D. 1983. *Ap. J.* 272:48–53

39. Fabian, A. C., Rees, M. J. 1979. *MNRAS* 187:13p–16

40. Forman, W., Jones, C., Cominsky, L., Julien, P., Murray, S., et al. 1978. *Ap. J. Suppl.* 38:357–412

41. Geldzahler, B. J., Johnston, K. J., Spencer, J. H., Klepczynski, W. J., Josties, F. J., et al. 1983. *Ap. J. Lett.* 273:L65–69

42. Geldzahler, B. J., Pauls, T., Salter, C. J. 1980. *Astron. Astrophys.* 84:237–44

43. Geldzahler, B. J., Cohen, N. L. 1983. *Publ. Astron. Soc. Pac.* 95:489–90

44. Geldzahler, B. J., Downes, A. J. B., Shaffer, D. B. 1981. *Astron. Astrophys.* 98:205–6

45. Giacconi, R., Gursky, H., Kellogg, E., Levinson, R., Schreier, E., et al. 1973. *Ap. J.* 184:227–36

46. Giles, A. B., King, A. R., Cooke, B. A., McHardy, I. M., Lawrence, A. 1979. *Nature* 281:282–83

47. Giles, A. B., King, A. R., Jameson, R. F., Sherrington, M. R., Hough, J. H., et al. 1980. *Nature* 286:689–91

48. Gilmore, W., Seaquist, E. R. 1980. *Astron. J.* 85:1486–95

49. Gilmore, W. S., Seaquist, E. R., Stocke, J. T., Crane, P. C. 1981. *Astron. J.* 86:864–70

50. Gladyshev, S. A., Kurochkin, N. E., Novikov, I. D., Cherepashchuk, A. M. 1979. *Astron. Circ. USSR No. 1086*

51. Gladyshev, S. A., Goranskij, V. P., Kurochkin, N. E., Cherepashchuk, A. M. 1980. *Astron. Circ. USSR No. 1145*

52. Gladyshev, S. A. 1981. *Sov. Astron. Lett.* 7:330–33

53. Gower, J. F. R., Scott, P. F., Wills, D. 1967. *Mem. RAS* 71:49–144

54. Gower, A. C., Hutchings, J. B. 1982. *Ap. J. Lett.* 253:L1–5

55. Gower, A. C., Hutchings, J. B. 1982. *Ap. J. Lett.* 258:L63–66

56. Gower, A. C., Gregory, P. C., Hutchings, J. B., Unruh, W. G. 1982. *Ap. J.* 262:478–96

57. Grandi, S. A. 1975. *Ap. J.* 196:465–72

58. Grandi, S. A., Stone, R. P. S. 1982. *Publ. Astron. Soc. Pac.* 94:80–86

59. Gregory, P. C., Fahlman, G. G. 1980. *Nature* 287:805–6

60. Gregory, P. C., Fahlman, G. G. 1981. *Vistas Astron.* 25:119–25

61. Grindlay, J. E., Band, D., Seward, F., Leahy, D., Weisskopf, M. G., Marshall, F. E. 1984. *Ap. J.* 277:286–95

62. Heeschen, D. S., Hammond, S. E. 1980. *Ap. J. Lett.* 235:L129–30

63. Henson, G. D., Kemp, J. C., Barbour, M. S., Kraus, D. J., Leibowitz, E. M., Mazeh, T. 1983. *Ap. J.* 275:247–50

64. Hjellming, R. M., Johnston, K. J. 1982. In *Extragalactic Radio Sources, IAU Symp. No. 97,* ed. D. S. Heeschen, C. M. Wade, pp. 197–204. Dordrecht: Reidel

65. Hjellming, R. M., Johnston, K. J. 1981. *Nature* 290:100–7

66. Hjellming, R. M., Johnston, K. J. 1981. *Ap. J. Lett.* 246:L141–45

67. Holden, D. J., Caswell, J. R. 1969. *MNRAS* 143:407–35

68. Huang, Y. L., Dame, T. M., Thaddeus, P. 1983. *Ap. J.* 272:609–14

69. Hut, P., van den Heuvel, E. P. J. 1981. *Astron. Astrophys.* 94:327–32

70. Jaroszynski, M., Abramowicz, M. A., Paczynski, B. 1980. *Acta. Astron.* 30:1–34

71. Johnston, K. J., Santini, N. J., Spencer, J. H., Klepczynski, G. H., Kaplan, G. H., et al. 1981. *Astron. J.* 86:1377–83

72. Johnston, K. J., Geldzahler, B. J., Spencer, J. H., Waltman, E. B., Klepczynski, W. J., et al. 1984. *Astron. J.* 89:509–14

73. Kaplan, G. H., Kallarakal, V. V., Harrington, R. S., Johnston, K. J.,

Spencer, J. H. 1980. *Astron. J.* 85:64–65

74. Katz, J. I. 1973. *Nature Phys. Sci.* 246:87–89
75. Katz, J. I. 1980. *Ap. J. Lett.* 236:L127–30
76. Katz, J. I. 1980. *Astrophys. Lett.* 20:135–36
77. Katz, J. I. 1981. *Astron. Astrophys.* 95: L15–17
78. Katz, J. I., Piran, T. 1982. *Astrophys. Lett.* 23:11–15
79. Katz, J. I., Anderson, S. F., Margon, B., Grandi, S. A. 1982. *Ap. J.* 260:780–93
80. Kemp, J. C., Barbour, M. S., Arbabi, M., Leibowitz, E. M., Mazeh, T. 1980. *Ap. J. Lett.* 238:L133–38
81. Kemp, J. C., Barbour, M. S., Kemp, G. N., Hagood, D. M. 1981. *Vistas Astron.* 25:31–43
82. Kholopov, P. N., Samus', N. N., Kukarkina, N. P., Medvedeva, G. I., Perova, N. B. 1981. *Inf. Bull. Var. Stars No. 1921*
83. Kirshner, R. P., Chevalier, R. A. 1980. *Ap. J. Lett.* 242:L77–81
84. Kirshner, R. P., Chevalier, R. A. 1978. *Nature* 276:480
85. Königl, A. 1983. *MNRAS* 205:471–85
86. Krumenaker, L. E. 1975. *Publ. Astron. Soc. Pac.* 87:185–87
87. Lamb, R. C., Ling, J. C., Mahoney, W. A., Riegler, G. R., Wheaton, W. A., Jacobson, A. S. 1983. *Nature* 305:37–39
88. Landau, R., Epstein, E. E., Rather, J. D. G. 1980. *Astron. J.* 85:363–67
89. Lang, F. L., Levine, A. M., Bautz, M., Hauskins, S., Howe, S., et al. 1981. *Ap. J. Lett.* 246:L21–25
90. Lebedev, V. S., Pimonov, A. A. 1981. *Sov. Astron. Lett.* 7:333–35
91. Leibowitz, E. M., Mazeh, T., Mendelson, H., Kemp, J. C., Barbour, M. S., et al. 1984. *MNRAS* 206:751–65
92. Leibowitz, E. M., Mazeh, T., Mendelson, H. 1984. *Nature* 307:341–42.
93. Leibowitz, E. M., Mendelson, H. 1982. *Publ. Astron. Soc. Pac.* 94:977–78
94. Levine, A. M., Jernigan, J. G. 1982. *Ap. J.* 262:294–300
95. Liebert, J., Angel, J. R. P., Hege, E. K., Martin, P. G., Blair, W. P. 1979. *Nature* 279:384–87
96. Lipunov, V. M., Shakura, N. I. 1982. *Sov. Astron. AJ* 26:386–89
97. Lynden-Bell, D. 1978. *Phys. Scr.* 17:185–91
98. MacCallum, C. J., Leventhal, M., Stang, P. D. 1983. *Bull. Am. Astron. Soc.* 15:939
99. Mammano, A., Ciatti, F., Vittone, A. 1980. *Astron. Astrophys.* 85:14–19

100. Mammano, A., Margoni, R., Ciatti, F., Christiani, S. 1983. *Astron. Astrophys.* 119:153–59
101. Margon, B., Ford, H. C., Katz, J. I., Kwitter, K. B., Ulrich, R. K., et al. 1979. *Ap. J. Lett.* 230:L41–45
102. Margon, B., Ford, H. C., Grandi, S. A., Stone, R. P. S. 1979. *Ap. J. Lett.* 233:L63–68
103. Margon, B., Anderson, S. F. 1983. *Astrophys. Lett.* 23:211–15
104. Margon, B. 1980. *Sci. Am.* 243:54–65
105. Margon, B., Grandi, S. A., Downes, R. A. 1980. *Ap. J.* 241:306–15
106. Margon, B. 1981. *Ann. NY Acad. Sci.* 375:403–13
107. Margon, B. 1982. *Science* 215:247–52
108. Margon, B. 1982. In *Galactic X-Ray Sources*, ed. P. W. Sanford, P. Laskarides, J. Salton, pp. 417–35. Chichester, Engl: Wiley
109. Margon, B. 1983. In *Accretion Driven Stellar X-Ray Sources*, ed. W. H. G. Lewin, E. P. J. van den Heuvel, pp. 287–301. Cambridge: Cambridge Univ. Press
110. Margon, B., Anderson, S. A., Aller, L. H., Downes, R. A., Keyes, C. D. 1984. *Ap. J.* 281:313–17
111. Margon, B., Downes, R. A., Gunn, J. E. 1981. *Ap. J. Lett.* 249:L1–4
112. Margon, B., Cohen, J. G. 1978. *Ap. J. Lett.* 222:L33–36
113. Marshall, F. E., Swank, J. H., Boldt, E. A., Holt, S. S., Serlemitsos, P. J. 1979. *Ap. J. Lett.* 230:L145–48
114. Martin, P. G., Rees, M. J. 1979. *MNRAS* 189:19p–22
115. Matese, J. J., Whitmire, D. P. 1982. *Astron. Astrophys.* 106:L9–11
116. Mazeh, T., Leibowitz, E. M., Lahav, O. 1981. *Astrophys. Lett.* 22:185–91
117. Mazeh, T., Aguilar, L. A., Treffers, R. R., Königl, A., Sparke, L. S. 1983. *Ap. J.* 265:235–38
118. McAlary, C. W., McLaren, R. A. 1980. *Ap. J.* 240:853–58
119. McClintock, J. E., Canizares, C. R., Tarter, C. B. 1975. *Ap. J.* 198:641–52
120. McLean, I. S., Tapia, S. 1980. *Nature* 287:703–5
121. Meier, D. L. 1982. *Ap. J.* 256:706–16
122. Michalsky, J. J., Stokes, G. M., Szkody, P., Larson, N. R. 1980. *Publ. Astron. Soc. Pac.* 92:654–56
123. Miley, G. 1980. *Ann. Rev. Astron. Astrophys.* 18:165–218
124. Milgrom, M. 1979. *Astron. Astrophys.* 76:L3–6
125. Milgrom, M. 1979. *Astron. Astrophys.* 78:L9–12

126. Milgrom, M. 1981. *Vistas Astron.* 25: 141–51
127. Milgrom, M., Anderson, S. F., Margon, B. 1982. *Ap. J.* 256: 222–26
128. Moore, R. L., Readhead, A. C. S., Baath, L. 1983. *Nature* 306: 44–46
129. Murdin, P., Clark, D. H., Martin, P. G. 1980. *MNRAS* 193: 135–51
130. Murdin, P., Clark, D. H. 1980. *MNRAS* 190: 65p–68
131. Neizvestnyi, S. I., Pustil'nik, S. A., Efremov, V. G. 1980. *Sov. Astron. Lett.* 6: 368–71
132. Newsom, G. H., Collins, G. W. II. 1981. *Astron. J.* 86: 1250–58
133. Niell, A. E., Lockhart, T. G., Preston, R. A. 1981. *Ap. J.* 250: 248–53
134. Niell, A. E., Lockhart, T. G., Preston, R. A., Backer, D. C. 1982. In *Extragalactic Radio Sources, IAU Symp. No. 97*, ed. D. S. Heeschen, C. M. Wade, pp. 207–8. Dordrecht: Reidel
135. Norman, E. B., Bodansky, D. 1984. *Nature* 308: 212
136. Overbye, D. 1979. *Sky Telesc.* 58: 510–16
137. Petterson, J. A. 1981. *Adv. Space Res.* 1: 49–61
138. Priedhorsky, W. C., Terrell, J. 1984. *Ap. J.* In press
139. Rees, M. J. 1982. In *Extragalactic Radio Sources, IAU Symp. No. 97*, ed. D. S. Heeschen, C. M. Wade, pp. 211–22. Dordrecht: Reidel
140. Rees, M. J., Begelman, M. C., Blandford, R. D. 1981. *Ann. NY Acad. Sci.* 375: 254–86
141. Ricketts, M. J., Hall, R., Page, C. G., Pounds, K. A., Sims, M. R. 1981. *Vistas Astron.* 25: 71–74
142. Roberts, W. J. 1974. *Ap. J.* 187: 575–84
143. Ryle, M., Caswell, J. L., Hine, G., Shakeshaft, J. 1978. *Nature* 276: 571–73
144. Sarazin, C. L., Begelman, M. C., Hatchett, S. P. 1980. *Ap. J. Lett.* 238: L129–32
145. Schilizzi, R. T., Fejes, I., Romney, J. D., Miley, G. K., Spencer, R. E., Johnston, K. J. 1982. In *Extragalactic Radio Sources, IAU Symp. No. 97*. Heeschen, C. M. Wade, pp. 205–6. Dordrecht: Reidel
146. Schilizzi, R. T., Norman, C. A., van Breugel, W., Hummel, E. 1979. *Astron. Astrophys.* 79: L26–27
147. Schilizzi, R. T., Miley, G. K., Romney, J. D., Spencer, R. E. 1981. *Nature* 290: 318–20
148. Schilizzi, R. T., Romney, J. D., Spencer, R. E. 1983. In *VLBI and Compact Radio Sources, IAU Symp. No. 110*, pp. 16–24. Dordrecht: Reidel
149. Schmidt, G. D., Cohen, M., Margon, B. 1980. *Ap. J. Lett.* 239: L133–38
150. Deleted in proof
151. Seaquist, E. R., Garrison, R. F., Gregory, P. C., Taylor, A. R., Crane, P. C. 1979. *Astron. J.* 84: 1037–41
152. Seaquist, E. R., Gilmore, W. S., Johnston, K. J., Grindlay, J. E. 1982. *Ap. J.* 260: 220–32
153. Seaquist, E. R., Gilmore, W., Nelson, G. J., Payten, W. J., Slee, O. B. 1980. *Ap. J. Lett.* 241: L77–81
154. Seaquist, E. R. 1981. *Vistas Astron.* 25: 79–85
155. Seward, F. D., Page, C. G., Turner, M. J. L., Pounds, K. A. 1976. *MNRAS* 175: 39p–46
156. Seward, F., Grindlay, J., Seaquist, E., Gilmore, W. 1980. *Nature* 287: 806–8
157. Shaham, J. 1981. *Vistas Astron.* 25: 217–33
158. Shakura, N. I., Sunyaev, R. A. 1973. *Astron. Astrophys.* 24: 337–55
159. Shakura, N. I. 1973. *Sov. Astron. AJ* 16: 756–62
160. Shapiro, P. R., Milgrom, M., Rees, M. J. 1982. In *Extragalactic Radio Sources, IAU Symp. 97*, ed. D. S. Heeschen, C. M. Wade, pp. 209–10. Dordrecht: Reidel
161. Sherwood, W. A. 1974. *Publ. R. Obs. Edinburgh* 9: 85–116
162. Shuder, J. M., Hatfield, B. F., Cohen, R. D. 1980. *Publ. Astron. Soc. Pac.* 92: 259–61
163. Sikora, M., Wilson, D. B. 1981. *MNRAS* 197: 529–41
164. Spencer, R. E. 1979. *Nature* 282: 483–84
165. Spinrad, H., Stauffer, J., Harlan, E. 1979. *Publ. Astron. Soc. Pac.* 91: 619–23
166. Stephenson, C. B., Sanduleak, N. 1977. *Ap. J. Suppl.* 33: 459–69
167. Takagishi, K., Jugaku, J., Eiraku, M., Matsuoka, M. 1981. *ISAS Res. Note 143*
168. Tapia, S., Turnshek, D. A. 1982. *Bull. Am. Astron. Soc.* 14: 577
169. Thompson, R. I., Rieke, G. H., Tokunaga, A. T., Lebofsky, M. J. 1979. *Ap. J. Lett.* 234: L135–38
170. van den Heuvel, E. P. J., Ostriker, J. P., Petterson, J. A. 1980. *Astron. Astrophys.* 81: L7–10
171. van den Bergh, S. 1980. *Ap. J. Lett.* 236: L23
172. van Gorkom, J. H., Goss, W. M., Seaquist, E. R., Gilmore, W. S. 1982. *MNRAS* 198: 757–65
173. Velusamy, T., Kundu, M. R. 1974. *Astron. Astrophys.* 32: 375–90
174. Walker, R. C., Readhead, A. C. S., Seielstad, G. A., Preston, R. A., Niell, A. E., et al. 1981. *Ap. J.* 243: 589–96

175. Watson, M. G., Willingdale, R., Grindlay, J. E., Seward, F. D. 1983. *Ap. J.* 273:688–96
176. Webster, A. 1979. *MNRAS* 189:33p–36
177. Weiler, K. W. 1983. *Observatory* 103:85–106
178. Weiler, K. W., Panagia, N. 1980. *Astron. Astrophys.* 90:269–82
179. Wenzel, W. 1980. *Mitt. Veränderl. Sterne* 8:141–43
180. Westerhout, G. 1958. *Bull. Astron. Inst. Neth.* 14:215–60
181. Whitmire, D. P., Matese, J. J. 1980. *MNRAS* 193:707–12
182. Wynn-Williams, C. G., Becklin, E. E. 1979. *Nature* 282:810–11
183. Zealey, W. J., Dopita, M. A., Malin, D. F. 1980. *MNRAS* 192:731–43
184. Zhou, Y. Y., Li, Q. B., Borner, G. 1983. *Astrophys. Space Sci.* 92:161–64

Ann. Rev. Astron. Astrophys. 1984. 22 : 537–92
Copyright © 1984 by Annual Reviews Inc. All rights reserved

NEUTRON STARS IN INTERACTING BINARY SYSTEMS

Paul C. Joss and Saul A. Rappaport

Department of Physics, Center for Space Research, and Center for Theoretical Physics, Massachusetts Institute of Technology, Cambridge, Massachusetts 02139

1. INTRODUCTION

The early history of the development of our understanding of neutron stars has been described by others, and we summarize it only briefly here. Following an early speculation by Baade & Zwicky (1934) soon after the discovery of the neutron in 1932, Oppenheimer & Volkoff (1939) presented a classic analysis of the theoretical viability of neutron stars, based on the general theory of relativity and the nuclear physics that was then known. A number of further theoretical studies were published during the next three decades, but it seemed that neutron stars might be virtually unobservable and thus remain little more than a theorist's curiosity piece.

In a prophetic paper, Pacini (1967) noted that a rotating neutron star with a magnetic field whose axis was misaligned with the rotation axis should emit intense magnetic dipole radiation. Almost immediately thereafter (but quite independently), Hewish et al. (1968a,b) announced the discovery of radio pulsars. For a brief period, several theoretical models for the pulsar phenomenon were in contention. However, the discovery of the pulsar NP 0532 in the Crab Nebula (Staelin & Reifenstein 1968, Comella et al. 1969), followed by a notably straightforward set of observations and theoretical arguments, soon clinched the case for neutron stars (Gold 1969a).[1] (For a review of these remarkable developments, see Gold 1969b.)

---

[1] Recent theoretical efforts [see Baym & Pethick (1979) for references] have pointed increasingly to the possibility that neutron-rich matter undergoes a transition to a fluid composed of "free" (but strongly interacting) quarks at densities not far in excess of nuclear-matter densities ($\sim 2.5 \times 10^{14}$ g cm^{-3}). If this is the case, then much of the matter within a

0066–4146/84/0915–0537$02.00

An enormous quantity of data concerning isolated radio pulsars has been accumulated since the discovery of these objects [see Manchester & Taylor (1977) for a review]. However, these data have revealed disappointingly little about the fundamental properties of the underlying neutron stars. The highly complex features of the radio pulses appear, instead, to depend most directly upon the properties of plasmas (i.e. the neutron-star atmosphere and surface layers) under extreme physical conditions and upon the magnetic field configuration on the surface of the neutron star. Only limited information concerning neutron-star interiors has been gleaned from such pulse-timing data as the pulsar spin-down rates and pulse period "glitches" (Ruderman 1972, Manchester & Taylor 1977, Sieber & Wielebinski 1981, and references therein), and relatively little can be learned about neutron-star masses or radii.

Since 1971, three other types of astrophysical systems containing neutron stars have been discovered: binary X-ray pulsars, X-ray burst sources, and binary radio pulsars. In all three cases, the neutron stars are members of close binary stellar systems, and this membership indirectly provides much more powerful handles on the intrinsic properties of the neutron stars themselves. In the first two classes of sources, a large amount of information is available because the systems are interacting (i.e. matter is being transferred from the companion star to the neutron star). This review addresses the phenomenology of systems in these two classes and the neutron stars that they contain, and the constraints on the properties of neutron stars that may be inferred from observations of these systems. (For reviews of the binary radio pulsars, see Taylor 1981 and Blandford & De Campli 1981.)

The first of the binary X-ray pulsars was discovered in 1971 with the *Uhuru* satellite (Giacconi et al. 1971, Schreier et al. 1972a, Tananbaum et al. 1972) and with a balloon-borne instrument (Lewin et al. 1971). A total of ~ 21 such objects are now known (see Figure 1), with pulse periods ranging from less than 0.1 s to nearly 10^3 s. It was almost immediately recognized that the source of energy to drive the X-ray pulsations was accretion of matter onto the magnetic polar caps of a rotating neutron star. The accreted matter is transferred to the neutron star from a relatively normal binary companion star. Thus the energy source for X-ray pulsars is very different from that of the radio pulsars, which extract their energy from the rotational kinetic energy of the neutron stars (although a few radio pulsars,

---

"neutron" star may actually be composed of quark matter, and such objects might be more accurately called "quark stars." However, many of the observable macroscopic properties of quark stars (masses, radii, moments of inertia, and so forth) need not be different from those of neutron stars (see, for example, Fechner & Joss 1978). For simplicity, in this review we refer to all such objects as "neutron stars."

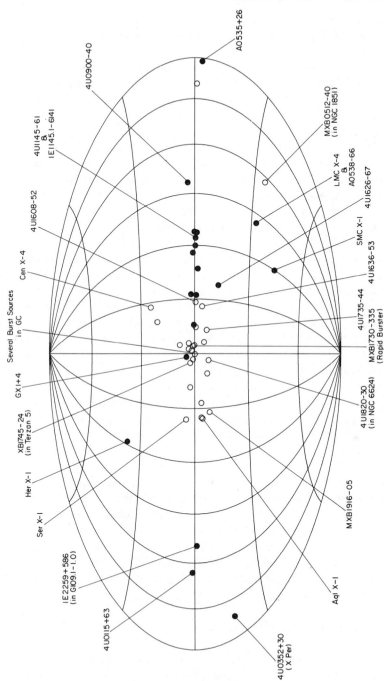

Figure 1 Sky map, in galactic coordinates, of 21 binary X-ray pulsars (●) and 27 X-ray burst sources (○) for which reasonably accurate positional determinations are available (adapted from Lewin & Joss 1981). Some of the more important members of each class are identified. Some burst sources are located in globular clusters, as indicated. The tendency of the X-ray pulsars to be distributed along the galactic equator and the concentration of X-ray burst sources toward the galactic center are both apparent.

such as the Crab pulsar, also emit X-ray pulsations). The Doppler delays of the pulsations resulting from the orbital motion of the neutron star about its binary stellar companion, together with properties of the companion star that are observable in optical light (such as Doppler shifts in spectral features resulting from orbital motion), permit important constraints to be placed on the properties of the binary system. In particular, useful estimates of the mass of the neutron star may be obtained.

In 1975, X-ray bursts were discovered by Grindlay et al. (1976) and Belian et al. (1976). Following the initial discovery, much of the observational work was carried out with the *SAS-3* X-ray astronomy satellite under the leadership of W. Lewin and G. Clark and, more recently, with the *Hakucho* satellite under the leadership of M. Oda. At this writing, about 35 X-ray burst sources have been identified (see Figure 1). Typical individual bursts are very energetic, with $\sim 10^{39}$ ergs of energy radiated in only a few seconds or tens of seconds; the intervals between bursts from a given source are sometimes quite regular but never strictly periodic, and they typically range from hours to a day or so. There is now persuasive evidence that these sources, too, are neutron stars in close binary systems. The neutron star must again be accreting matter from its companion star, but in this case the surface magnetic field of the neutron star is probably relatively weak ($\ll 10^{12}$ G). This weak field is unable to direct the flow of the accreting matter, so that X-ray pulsations are absent, but the lack of funneling of the accretion flow also alters the properties of the neutron-star surface layers in such a way that the freshly accreted matter may undergo strong thermonuclear flashes. It is these flashes that result in the emission of X-ray bursts. The observed properties of the bursts, in turn, may provide constraints on the fundamental properties of the underlying neutron star, largely through the effects of gravitational redshift upon the properties of radiation emitted from the neutron-star surface.

In Section 2 of this review, we describe the observational properties and theoretical implications of binary X-ray pulsars. Several outstanding puzzles in the interpretation of these systems are also discussed. In Section 3, we describe the X-ray burst sources and discuss their interpretation as thermonuclear flashes on accreting neutron stars. The current status of our empirical knowledge of these systems and the neutron stars that they contain, and the prospects for future developments in this field, are briefly summarized in Section 4.

2. BINARY X-RAY PULSARS

2.1 *Pulse Profiles and Spectra*

The pulse periods of the known binary X-ray pulsars range over four decades in period, from 69 ms to 835 s (see Figure 2). Within the limited

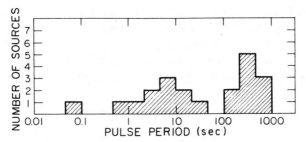

Figure 2 The distribution of pulse periods for the known binary X-ray pulsars as of September 1983.

statistics of this small sample, the pulse period distribution is consistent with a constant number per logarithmic interval in period, with a slight excess for periods near ~ 300 s. Observational selection effects undoubtedly affect the discovery rate for periods below ~ 1 s and above ~ 1000 s. Nevertheless, this distribution probably reflects the evolution of a typical X-ray pulsar, which may undergo enormous changes in pulse period during its lifetime, from very short ($P \ll 1$ s) to very long (~ 1000 s) and back to a final low value (~ 1 s) (see, for example, van den Heuvel 1977; other references given in Section 2.2).

Sample pulse profiles for 14 of the ~ 21 known binary X-ray pulsars[2] are presented in Figure 3. The profiles are characterized by (a) large "duty cycles" of $\gtrsim 50\%$ [compared with the $\sim 3\%$ duty cycles typical of radio pulsars (Manchester & Taylor 1977)], (b) modulation factors that range from ~ 20 to $\sim 90\%$, (c) a range from symmetric to highly asymmetric shapes, (d) *no* obvious trend in pulse morphology as a function of pulse period, (e) *no* obvious correlation of pulse morphology with X-ray

---

[2] In this review, we restrict our attention to sources that are widely believed to be neutron stars undergoing accretion in binary stellar systems. We do not discuss other sources that have been observed to emit periodic X-ray pulsations. In particular, we exclude from further consideration such sources as the Crab pulsar NP 0532 [which is thought to be an isolated neutron star whose radiated energy is derived from its rotational kinetic energy (see Manchester & Taylor 1977)], the 5 March 1979 gamma-ray transient [whose underlying character is unknown, but which shows no evidence of binary membership (Mazets et al. 1979)], and sources, including H2252−035 (White & Marshall 1981), that are thought to be accreting degenerate dwarfs rather than neutron stars. We also exclude from our discussion sources, such as 4U1700−37, that have many properties in common with binary X-ray pulsars but are not observed to emit periodic pulsations, so that the identification of the accreting object as a neutron star is not yet demonstrated.

After the manuscript for this article was completed, two additional binary X-ray pulsars were discovered: 4U1907+09 [$P \simeq 438$ s, $P_{orb} \simeq 8.4$ days, $a_x \sin i \simeq 105$ lt-s, $e \simeq 0.2$ (Tanaka 1983)] and V0332+53 [$P \simeq 4.38$ s, $P_{orb} \simeq 34$ days, $a_x \sin i \simeq 55$ lt-s, $e \simeq 0.35$ (Stella & White 1983, White et al. 1984)].

luminosity L_x over a range of $\sim 10^5$ in L_x, and (f) a slight but significant positive correlation between pulse period and binary orbital period (van den Heuvel & Rappaport 1982). For several sources (e.g. A0535 + 26 and 4U0900 − 40) the pulse profiles vary radically with X-ray energy, while for others (e.g. Cen X-3 and 4U0352 + 30) the basic profile is retained over a

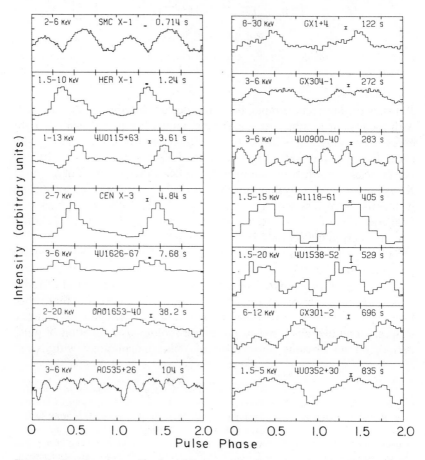

Figure 3 Sample pulse profiles for 14 X-ray pulsars (from Rappaport & Joss 1981). The profiles for Her X-1, 4U0115 + 63. Cen X-3, OAO1653 − 40, A1118 − 61, and 4U1538 − 52 are from Joss et al. (1978a), Johnston et al. (1978), Ulmer (1976), White & Pravdo (1979), Ives et al. (1975), and Becker et al. (1977), respectively; the remaining profiles are from unpublished *SAS-3* data. (For additional references, see Table 1 and Rappaport & Joss 1977a.) In each case, the data are folded modulo the pulse period and plotted against pulse phase for two complete cycles. The approximate pulse periods and X-ray energy intervals are indicated for each pulsar. Nonsource background counting rates have been subtracted. A typical ± 1σ error bar, derived from photon counting statistics, is indicated for each profile.

wide range of energies. There is some tendency for the more complex dependences of the pulse profiles on energy to occur among the longer period pulsars.

Representative X-ray pulsar spectra from six sources are shown in Figure 4 [see Holt & McCray (1982) and White et al. (1983) for extensive reviews of the subject]. Though these spectra differ in detail, the spectra of X-ray pulsars have the following general features: (a) broadband emission not dominated by sharp spectral features, (b) power mostly emitted in the range 2–20 keV, (c) rapid falloff in flux for energies above ~ 20 keV, with the falloff often less abrupt for systems with lower X-ray luminosities, (d) no simple description in terms of a blackbody, thermal bremsstrahlung, or power-law shape, and (e) an iron K-shell emission feature near 7 keV in many, but not all, cases (see, for example, Becker et al. 1978, White & Pravdo 1979, White et al. 1980).

Cyclotron lines   Two X-ray pulsars, Her X-1 and 4U0115+63, exhibit X-ray spectral features that have been widely interpreted as cyclotron radiation from the vicinity of the neutron-star surface (Trümper 1983, and references therein). A feature in the spectrum of Her X-1 may result from either absorption at ~ 38 keV or emission at ~ 53 keV (Trümper et al. 1977, Trümper 1983). Correcting for the effects of gravitational redshift, one finds that the observed energy of the fundamental cyclotron resonance is

$$E_c = \frac{\hbar\omega_B}{1+z_s} \simeq 10\left(\frac{B}{10^{12}\,G}\right)\left(\frac{1+z_s}{1.2}\right)^{-1} \text{keV}, \qquad 1.$$

where B is the magnetic field strength at the accreting pole of the neutron star, $\omega_B = eB/mc$ is the electron cyclotron frequency, and $z_s = (1 - 2GM/Rc^2)^{-1/2} - 1$ is the gravitational redshift, relative to a distant observer, from the surface of a neutron star with mass M and radius R. Thus, a cyclotron resonance producing the observed spectral feature would correspond to a field strength of ~ 3.8 or $\sim 5.3 \times 10^{12}$ G if the feature results from absorption or emission, respectively. Similarly, Wheaton et al. (1979) reported a feature in the spectrum of 4U0115+63 at ~ 20 keV; further detailed studies by White et al. (1983) indicated features at 11.5 and 23 keV, which varied from absorption lines at some pulse phases to emission lines at other phases. These features, if interpreted as the fundamental and first harmonic of the cyclotron resonance, correspond to a magnetic field strength of $\sim 1.2 \times 10^{12}$ G. If, in general, such spectral features have been correctly associated with cyclotron radiation, then their study should become an accurate probe of the surface magnetic fields of neutron stars. A number of theoretical investigations involving the production of cyclotron radiation and the transfer of this radiation through the hot, dense, highly

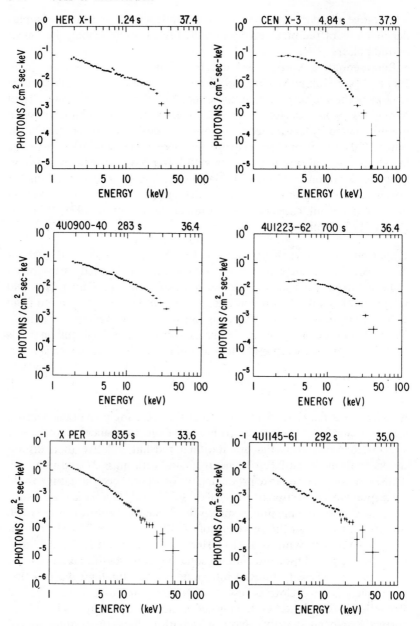

Figure 4 Sample X-ray spectra, averaged over pulse phase, for six X-ray pulsars (from White et al. 1983). The spectra typically show broadband emission devoid of prominent features (aside from iron K-shell emission near 7 keV in many cases) and a rapid falloff in flux for X-ray energies above ~20 keV (see text).

magnetized plasma near the neutron-star surface have yielded spectra with features similar to those observed (Mészàros et al 1980, and references therein, Yahel 1980, and references therein, Nagel 1981a,b). However, as is discussed below, these calculations have not been carried out in a self-consistent manner, and therefore the interpretation of the observed features as cyclotron lines is not yet completely settled.

In fact, we now lack a complete theoretical understanding of the complex pulse shapes or their dependence on X-ray energy. It is widely accepted, however, that the pulsations result from an X-ray beam pattern that is misaligned with the rotation axis of an accreting, magnetized neutron star and viewed from different directions as the star rotates (Pringle & Rees 1972, Davidson & Ostriker 1973, Lamb et al. 1973, Baan & Treves 1973). The X-ray luminosity is generated by the release of the gravitational potential energy of the accreting matter, which is in turn supplied by a close binary stellar companion. The X-ray beam pattern is determined by the external magnetic field of the neutron star, the resultant funneling of the accretion flow, and the complex radiative transfer processes for X-rays propagating from the vicinity of the neutron-star surface through regions of accreting, magnetized plasma.

The infalling matter may be decelerated onto the neutron star and its energy thermalized in one or more of the following ways:

1. Deceleration by radiation pressure (a "photon cushion"). In this picture (Davidson 1973, Inoue 1975, Basko & Sunyaev 1976, Wang & Frank 1981), an essentially blackbody radiation field, with a temperature corresponding to the soft X-ray range, is present within the accretion column owing to its high optical depth. The incoming electrons are decelerated by interactions with the photons, and this deceleration is then transmitted electrostatically to the infalling ions, whose kinetic energy (~ 100 MeV nucleon^{-1}) is thereby thermalized.

2. Collisions beneath the neutron-star surface. Here, the infalling matter is confined to beams that strike the magnetic polar caps of the neutron star at nearly the free-fall velocity (Lamb et al. 1973, Basko & Sunyaev 1975, Pavlov & Yakovlev 1976, Kirk & Galloway 1981, Mészàros et al. 1983). The beams consist essentially of free particles, and their kinetic energy is not thermalized until after they reach the surface of the neutron star. The infalling ions are then stopped by binary collisions (nuclear or Coulomb) within a short distance beneath the neutron-star surface.

3. Passage through a shock followed by settling. In this picture, the incoming matter is assumed to undergo a strong shock above the surface of the neutron star (Inoue 1975, Shapiro & Salpeter 1975, Basko & Sunyaev 1976, Wang & Frank 1981, Langer & Rappaport 1982). At the shock, a large fraction of the kinetic energy of the incoming matter is thermalized.

The matter in the postshock region then radiates and cools as it settles onto the neutron-star surface. Only recently have the effects of the strong magnetic field upon the emission processes and the heat exchange between electrons and ions been incorporated into such a model (Langer & Rappaport 1982).

The success of any such ideas depends ultimately on their ability to describe the emergent X-ray spectra and beam patterns. At present, only crude X-ray spectra have been calculated for several simplified models (Davidson 1973, Inoue 1975, Pavlov & Shibanov 1979, and references therein, Kanno 1980, Mészàros et al. 1980, and references therein, Yahel 1980, and references therein, Nagel 1981a,b, Langer & Rappaport 1982, Mészàros et al. 1983). We do not even know how the energy characterizing the observed X-ray spectra (~ 10 keV) arises. There are at least two characteristic energies associated with the accretion process that are about the correct magnitude. One is the characteristic cyclotron line energy (Equation 1). The other is the thermalization energy E_T characteristic of a hot polar cap of area A (measured locally) that is radiating luminosity L_x at temperature T (as measured by a distant observer):

$$E_T \simeq kT \simeq k\left(\frac{L_x}{\sigma A}\right)^{1/4}(1+z_s)^{-1/2}$$

$$\simeq 9\left(\frac{L_x}{10^{38}\text{ erg s}^{-1}}\right)^{1/4}\left(\frac{A}{10^{10}\text{ cm}^2}\right)^{-1/4}\left(\frac{1+z_s}{1.2}\right)^{-1/2}\text{ keV.} \qquad 2.$$

In any case, the complex frequency and directional dependence of photon scattering, spatial variations in the magnetic field, and thermal Doppler broadening effects may help to distort the spectra into the shapes that we observe (see, for example, Langer & Rappaport 1982).

Early attempts to derive theoretical pulse profiles (see, for example, Bisnovatyi-Kogan 1973, Lamb et al. 1973, Gnedin & Sunyaev 1973, Basko & Sunyaev 1975, Tsuruta 1975, Wang 1975) utilized highly simplifying approximations for the actual physical processes. These efforts suggest emission patterns ranging from pencil beams directed along the magnetic axes of the neutron star to fan beams perpendicular to these axes. More recently, a few workers (Yahel 1980, and references therein, Pravdo & Bussard 1981) have used Monte Carlo techniques to simulate the generation of the pulse profiles and X-ray spectra. These calculations have taken into account many of the important scattering processes for X-rays propagating through the accretion column. However, such calculations have not yet been carried out in a self-consistent manner, in that the combined effects of the radiation upon the plasma and the emission of radiation by the plasma have not been fully considered (but see Mészàros et al. 1983).

Further studies of the general accretion flow problem, in which the emission and transfer of radiation in the intense magnetic field are coupled to the mass flow, are now needed. Some progress is currently being made in this area (R. Klein, J. Arons & S. Lea, in preparation). The similarities among the pulse profiles and energy spectra for X-ray pulsars with a very wide range of pulse periods and luminosities provide hope that a basic understanding of the accretion processes may in fact be attainable in the near future.

2.2   Pulse Period Variations

The pulse periods for a number of X-ray pulsars have been sufficiently well measured over the past decade to provide important information regarding the torques exerted on the neutron stars by the accreting material. Nearly all of the available pulse period measurements (as of June 1981) for the eight best-measured X-ray pulsars are shown in Figure 5. When we examine the pulse period histories of these eight sources, it is apparent that the "spin-up" trend first noted in Her X-1 and Cen X-3 (Giacconi 1974, Gursky & Schreier 1975, Schreier & Fabbiano 1976) is very prominent in at least five of them.

The trend in most of the X-ray pulsars toward a secular decrease in pulse period can be understood in terms of torques exerted by the matter accreting onto the neutron star. These torques can be readily calculated for the case where the matter has roughly circular Keplerian velocities at the magnetopause of the neutron star, as would be the case if the accretion is mediated by a disk. Such calculations (Pringle & Rees 1972, Lamb et al. 1973, Ghosh & Lamb 1979, and references therein) show that the rate of change $\dot{P}$ of the intrinsic pulse period P is related to the X-ray luminosity and the physical properties of the neutron star:

$$\frac{\dot{P}}{P} \simeq -3 \times 10^{-5} \left(\frac{\xi v_r}{v_{ff}}\right)^{1/7} \left(\frac{M}{M_\odot}\right)^{-10/7} \left(\frac{R}{10 \text{ km}}\right)^{6/7} \left(\frac{R_g}{10 \text{ km}}\right)^{-2}$$

$$\times \left(\frac{\mu}{10^{30} \text{ G cm}^3}\right)^{2/7} \left(\frac{L_x}{10^{37} \text{ erg s}^{-1}}\right)^{6/7} \left(\frac{P}{1 \text{ s}}\right) \text{ yr}^{-1}. \quad 3.$$

Here, ξ is the fractional solid angle subtended at the neutron star by the infalling matter at the magnetopause; v_r/v_{ff} is the ratio of the average radial infall velocity of a particle to its free-fall velocity just outside the magnetopause; M, R, R_g, and μ are the mass, radius, radius of gyration, and magnetic dipole moment of the neutron star, respectively; and L_x is the

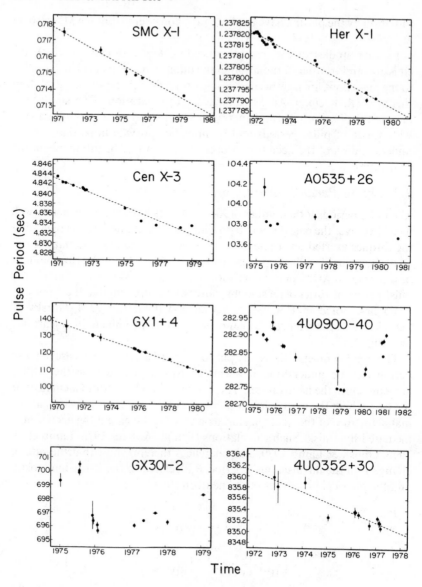

Figure 5   Pulse period histories for eight X-ray pulsars (from Rappaport & Joss 1983). The heavy dots are individual measurements of pulse period; the vertical bars represent the 1σ uncertainties in the period determinations. The data are from the *Uhuru, Copernicus, Ariel 5, SAS-3, OSO-8,* and *HEAO-1* satellites, the Apollo-Soyuz Test Project, and a balloon observation and sounding rocket observation. In general, the original references are quoted elsewhere in the text (for additional references, see Rappaport & Joss 1977a,b, White et al. 1977, Lamb et al. 1978, and Nagase 1981). The dashed lines are linear fits to the data points.

accretion-driven luminosity. The quantity $(\xi v_r/v_{ff})^{1/7}$ is not expected to differ greatly from unity (see, for example, Lamb et al. 1973). The overall minus sign in Equation 3 is explicitly for the case where the sense of the orbital angular momentum in the accreting matter is the same as that of the rotation of the neutron star.

For simplicity, we can rewrite Equation 3 as

$$\frac{\dot{P}}{P} = -3 \times 10^{-5} f\left(\frac{P}{1 \text{ s}}\right)\left(\frac{L_x}{10^{37} \text{ erg s}^{-1}}\right)^{6/7} \qquad 4.$$

(Rappaport & Joss 1977b), where the dimensionless function f is expected to be of order unity for a neutron star and contains parameters that are not yet measurable for most or all of the X-ray pulsars.

Following Rappaport & Joss (1977b), we have estimated the spin-up rate for each of 13 X-ray pulsars in the following way. For the 5 sources in Figure 5 that show a clear secular trend toward decreasing pulse period (4U0900 −40 is excluded on the basis of observations by Nagase et al. 1981), we have adopted the average value of $\dot{P}$ (i.e. the dashed lines in Figure 5). In the cases of the remaining 3 sources shown in Figure 5, we have adopted the average value of $\dot{P}$ during the longest interval with generally decreasing pulse period as a representative measure of the accretion torques during times when Equations 3 and 4 are most likely to be valid (see Elsner & Lamb 1976, Ghosh & Lamb 1979). For 4U0115+63, we have used the value of $\dot{P}$ measured during its 1978 outburst (Rappaport et al. 1978). For two other sources, LMC X-4 and 4U1145−61, essentially only upper limits exist for $\dot{P}$ (Kelley et al. 1983a, White et al. 1980). Finally, for 4U1626−67 and 2S1417 −62, values of $\dot{P}$ are available (Rappaport et al. 1977, Kelley et al. 1981), but the distance to the source, and hence the X-ray luminosity, is not known; we have here assumed a value of 10 kpc as a reasonable upper limit to the source distances.

The empirical data relating the spin-up rate and the quantity $PL_x^{6/7}$ (see Equation 4) for these 13 X-ray pulsars are shown in Figure 6 [see Ghosh & Lamb (1979) for an alternative method of presenting these data]. The solid line is the relation expected from Equation 4 if the X-ray stars are neutron stars with commonly accepted values of mass, radius, radius of gyration, and magnetic moment as indicated in Equation 3 (i.e. if $f = 1$). The dashed line, lying ~ 2 orders of magnitude below the data points, is the expected relation for accreting degenerate dwarfs. The general agreement between the solid line and the empirical data (see also Mason 1977) provides a compelling quantitative argument that the X-ray pulsars are, in fact, accreting neutron stars.

The preceding discussion is based on the premise that the accreting matter has sufficient angular momentum to enter a circular Keplerian orbit outside the magnetosphere of the neutron star. This situation is expected if the accreting matter is lost from the companion star at relatively low velocity through the inner saddle point of the critical potential lobe (see Section 2.3.1); in fact, if the net specific angular momentum of the accreting matter is sufficiently large, the accretion will be mediated by a large disk that surrounds the neutron star. If, on the other hand, matter is accreted from a high-velocity stellar wind via the Bondi-Hoyle (1944) mechanism (Davidson & Ostriker 1973), then the net specific angular momentum of the captured matter is likely to be too small to result in the formation of an accretion disk (Shapiro & Lightman 1976, Wang 1981). In this case, the accretion torques exerted on the neutron star will be determined by the

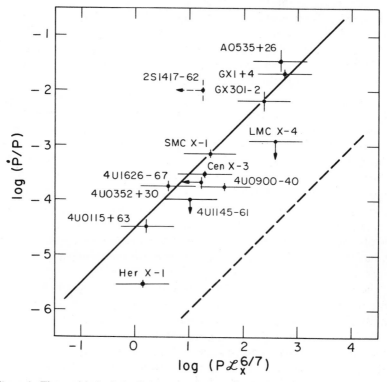

Figure 6   The empirical relation between the fractional rate of change of pulse period, $\dot{P}/P$, and the parameter (pulse period × luminosity$^{6/7}$) for 13 X-ray pulsars (adapted from Rappaport & Joss 1977b and Rappaport & Joss 1981). The units of $\dot{P}/P$, P, and L are yr^{-1}, s, and 10^{37} erg s^{-1}, respectively. The solid line is the relation expected from Equation 4 if the X-ray stars are neutron stars with commonly accepted values of mass, radius, radius of gyration, and magnetic moment. The dashed line is the expected relation for $\sim 1\,M_\odot$ degenerate dwarfs. See Elsner & Lamb (1976), Kelley et al. (1983a), and Kelley et al. (1981) for discussions of the special cases of Her X-1, LMC X-4, and 2S1417−62, respectively.

properties of the stellar wind and the binary orbit, and they will be nearly independent of the properties of the neutron-star magnetosphere (Ghosh & Lamb 1979, Arons & Lea 1980, Wang 1981). In fact, the sense of the accreted angular momentum may even be reversed over that given in Equation 4 (see Wang 1981). The dependence of the spin-up (or spin-down) time scale on pulse period and X-ray luminosity for the case of wind capture will still be qualitatively similar to that resulting from disk accretion, but the dependence on the binary system parameters will differ (Ghosh & Lamb 1979, Arons & Lea 1980, Wang 1981). At present, the observed pulse period variations do not permit a conclusive comparison between these two accretion modes (see Arons & Lea 1980). However, even if wind accretion prevails in some of the X-ray pulsars that we have considered, the conclusion that the underlying object is a neutron star (see Figure 6) would be unaffected.

We note that even for the five sources in Figure 5 that exhibit a clear "spin-up" trend, the change in pulse period is not always monotonic on short time scales (see, for example, Giacconi 1974, Fabbiano & Schreier 1977, Ögelman et al. 1977, Darbo et al. 1981). As noted above, for three of the sources in Figure 5 [GX301 − 2, 4U0900 − 40, and A0535 + 26 (Kelley et al. 1980, Nagase et al. 1981, Li et al. 1979)] there are long intervals where the pulse period is nearly constant or even increasing with time. Observations of 4U0900 − 40 with the *Hakucho* satellite (Nagase 1981) indicate that after its pulse period had decreased for at least four years (1975.0–1979.0), the period consistently increased during the following two years; as a result, the pulse period as of March 1981 was substantially the same as it had been six years earlier.

It is possible that A0535 + 26 becomes a "fast pulsar" (Elsner & Lamb 1976) when in the low state between transient outbursts and thereby undergoes periods of spin-down (Li et al. 1979, Elsner et al. 1980, Ziolkowski 1980). The pulse period behavior of GX301 − 2 and 4U0900 − 40 might be explained by occasional reversals in the sense of rotation of the accreted matter at the magnetopause (see the above discussion and Wang 1981). (For further discussions of such effects, see Ghosh et al. 1977, Ghosh & Lamb 1978, 1979, Davies et al. 1979, and references therein.) It is suggestive in this regard that the three sources in Figure 5 that do not exhibit a consistent spin-up trend are most likely to be undergoing accretion from a stellar wind whose velocity is large compared with the orbital velocity (E. P. J. van den Heuvel and H. Henrichs, private communication), rather than being fed by Roche lobe overflow (in which case the angular momentum of the accreting matter should always retain the sense of the orbital angular momentum, as described above). Finally, we note that (a) the low upper limit on $\dot{P}/P$ for LMC X-4 may be related to the fact that it apparently exhibits X-ray pulsations only during brief flaring episodes (Kelley et al. 1983a); and (b) the very large value of $\dot{P}/P$ for 2S1417 − 62 should be regarded as tentative, since we do not yet have secure limits on the Doppler effects due to its binary orbital motion.

Detailed studies of pulse arrival times can also potentially yield information concerning the internal structure of neutron stars (Lamb et al. 1978, Boynton & Deeter 1979, Boynton 1981). In particular, for a simple two-component model of neutron-star structure (Baym et al. 1969), there is a crust whose rotational angular frequency is presumably the same as the frequency of the X-ray pulsations, and a superfluid component of the interior that may rotate at a different rate. Fourier analysis of the pulse phase residuals, following subtraction of the known orbital Doppler effects and secular spin-up (or spin-down) trends, can yield the characteristic time scale on which the crust is dynamically coupled to the superfluid, as well as the relative moments of inertia of the two components. Boynton & Deeter (1979) and Boynton (1981) found that the accretion torque noise in Her X-1 is apparently "white" for frequencies between ~ 0.003 and ~ 1 day^{-1}, with no evidence for a "feature" due to crust-superfluid coupling. They showed that if the crust-superfluid coupling time is in the range of ~ 2–30 days, then the ratio of the moment of inertia of the crust to that of the entire star is probably greater than ~ 0.2. This same type of analysis is also presently being applied to extensive *HEAO-1* data from 4U0900 − 40 (P. E. Boynton and F. K. Lamb, private communication).

See Henrichs (1983) for a more complete review of the theory and phenomenology of pulse period variations in X-ray pulsars.

2.3   *Orbital Determinations and the Evaluation of Binary System Parameters*

2.3.1   ASSUMPTIONS AND TECHNIQUES   Measurements of the pulse arrival times from a number of binary X-ray pulsars have been used very successfully to determine the orbits of these systems. As is discussed below, the measured binary orbital parameters constitute the key ingredients that allow a determination of many of the physically and astrophysically interesting properties of these systems (Schreier et al. 1972a, Tananbaum et al. 1972).

The method of determining orbits from the Doppler delays of X-ray pulse arrival times (Schreier et al. 1972a) is analogous to classical optical measurements of Doppler shifts in spectral lines. For a perfect clock emitting pulses at uniform intervals and moving with constant velocity, a plot of pulse arrival time as a function of pulse number will reveal a simple linear relation. If the intrinsic rate of the clock increases with time, as is the case for a neutron star that is spinning up (see Section 2.2), the same type of plot yields a curved line. If, in addition, the clock is in a Keplerian orbit, then periodic Doppler delays in the arrival times, due to the time-of-flight of the pulses across the orbit, will be superposed. For the case where the curvature due to orbital motion is much greater than that due to changes in the intrinsic pulse period (or where the measurement interval contains a number of orbital cycles), the orbit can be determined by subtracting a low-order polynomial (in time) from the arrival-time plot. Complications arise when the reverse of the above condition obtains, and in such circumstances the orbits are difficult to measure. This is the case for slow X-ray pulsars (with their concomitantly large values of $\dot{P}$) in long-period orbits (see, for example, Li et al. 1979).

An example of an orbital determination from a measurement of the Doppler delays of X-ray pulses is shown in Figure 7. The Doppler delays in the arrival times of pulsations from SMC X-1 (Primini et al. 1977), from which the expected arrival times for a constant pulse period have been

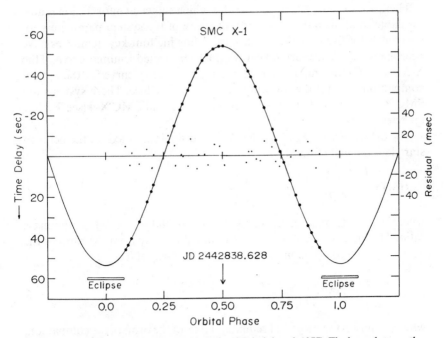

Figure 7   Doppler delay data for SMC X-1 (from Primini et al. 1977). The large dots are the measured Doppler delays in the pulse arrival times from SMC X-1 as a function of the $3^{d}.892$ orbital phase, with the scale indicated on the left-hand side of the figure. The curve represents the best-fit circular orbit. Small dots indicate the residual differences between the measured delays and the best-fit orbit, with the scale indicated on the right-hand side of the figure. The rms scatter in the residuals is ~ 10 ms.

subtracted, are shown for one orbital cycle. The resultant data points can be fitted extremely well by a simple sinusoidal function, which indicates that the orbit is very nearly circular (eccentricity $<7 \times 10^{-4}$). The projected semimajor axis of the orbit of the neutron star can be read directly from the figure ($a_x \sin i \simeq 53.5$ lt-s, where a_x is the semimajor axis and i is the inclination of the orbit to the plane of the sky). At present, the main limitation on the accuracy of this orbital determination is the statistical precision with which the pulse arrival times can be measured. Even now, however, the orbit of SMC X-1 is the most precisely determined binary stellar orbit in an extragalactic system.

All the measured orbits of binary X-ray pulsars [with the exceptions of 4U1626 − 67, which is discussed in Section 2.3.2, and the newly discovered pulsar 2S1553 − 54 (Kelley et al. 1983c)] are shown to scale in Figure 8, in order of increasing size of the semimajor axis (see also Table 1). Nominal values for the masses and radii of the companion stars are indicated in the figure; the derivation of these parameters is described below.

There are six binary X-ray pulsars for which sufficient information is now available to allow a determination of many of the system parameters. In particular, for five of these systems, the following four key quantities have been measured: (a) the orbital period, (b) the projected semimajor axis of the X-ray star, (c) the amplitude of the Doppler velocity curve for the optical companion, and (d) the duration of the X-ray eclipse. These systems are SMC X-1, Cen X-3, 4U0900 − 40, 4U1538 − 52, and LMC X-4 (see Table 1 for references).

The orbital period P_{orb} and the projected semimajor axis of the neutron star yield the mass function:

$$f(M) = \frac{4\pi^2 (a_x \sin i)^3}{GP_{orb}^2} = \frac{M_c \sin^3 i}{(1+q)^2}, \qquad 5.$$

where M_c is the mass of the companion star and $q \equiv M/M_c$ is the mass ratio. The latter quantity can be determined directly from the ratio of the velocity of the companion star to that of the X-ray star:

$$q = \frac{a_c \sin i}{a_x \sin i} = \frac{K_c P_{orb}\sqrt{1-e^2}}{2\pi a_x \sin i}, \qquad 6.$$

where $a_c \sin i$ is the projected semimajor axis of the orbit of the companion, K_c is the semiamplitude of the optical Doppler velocity curve, and e is the orbital eccentricity.

The final ingredient for determining the system parameters is the orbital inclination angle. This can be computed approximately with the aid of a simple model. If we replace the companion star by a sphere of radius R_c

whose volume equals the actual volume of the star, we obtain the relation

$$R_c \simeq a[\cos^2 i + \sin^2 i \sin^2 \theta_e]^{1/2}, \qquad\qquad 7.$$

where $a \equiv a_x + a_c$ is the separation of the centers of mass of the two stars, and the eclipse half-angle θ_e is π times the fraction of the orbital period that the X-ray star is in eclipse behind its companion. This approximation is surprisingly accurate, even for stars that suffer appreciable tidal or

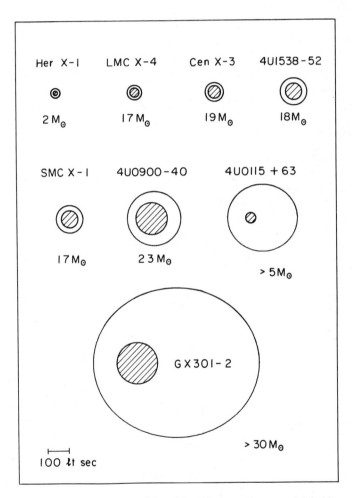

Figure 8 Schematic sketch, to scale, of the orbits and companion stars of eight binary X-ray pulsars (adapted from Rappaport & Joss 1981). The approximate mass of the companion star is indicated below each orbit. See text for a discussion of the derivation of the orbital parameters, stellar masses, and stellar radii.

Table 1 Binary X-ray pulsars[a]

Source	Optical counterpart	Pulse period (s)	$-\dot{P}/P$[b] (yr^{-1})	Orbital period (days)	$a_x \sin i$ (lt-s)	$f(M)$ ($M_\odot$)	Orbital eccentricity	K_c (km s^{-1})	Eclipse half-angle (degrees)	References[c]
A0538−66	d	0.069	—	16.66	—	—	>0.4	—	—	WH78d, JO79, SK80, CH81, PA81, SK82
SMC X-1	Sk 160	0.714	7.1×10^{-4}	3.892	53.46 (3)	10.8	<0.0007	19 (2)	26.5-29	SC72b, LU76, PR76, HU77a, PR77
Her X-1	HZ Her	1.24	2.9×10^{-6}	1.700	13.1831 (3)	0.85	<0.0003	20.2 (3.5)	24.4-24.7	TA72, MI76, SC76, FE77, JO80, DE81
1E2259+586	d	3.49	—	<0.08	—	—	—	—	—	GR80, FA81, FA82, MI82
4U0115+63	d	3.61	3.2×10^{-5}	24.31	140.13 (10)	5.00	0.3402 (2)	—	0	CO78, JO78a, RA78, HU81a
Cen X-3	V779 Cen	4.84	2.8×10^{-4}	2.087	39.792 (5)	15.5	0.0008 (1)	24 (6)	35-40	GI71, SC72a, PO75, PO76, FA77, HU79
4U1626−67	KZ TrA	7.68	1.9×10^{-4}	0.0288	<0.04	$<8 \times 10^{-5}$	—	280 (76)	0	RA77, MC77a, IL78, JO78b, PR79, LI80, MI81
2S1553−54	—	9.26	—	30.7 (2.8)	165 (30)	5.1 (2.9)	<0.07	—	—	KE83C
LMC X-4	d	13.5	$<1.2 \times 10^{-3}$	1.408	30 (5)	15 (8)	<0.2	37.9 (2.4)	25.5-33	CH77, HU78a, LI78, WH78a, KE83a
2S1417−62	—	17.6	9×10^{-3}(?)	>15	>25	—	—	—	—	GR81, KE81, LE82
OAO1653−40	—	38.2	5.4×10^{-3}	—	—	—	—	—	—	PO78, WH79, PA80c

A0535+26	HDE 245770	104	3.5×10^{-2}	111 (?)	—	—	—	≤ 20	—	RO75, ST76, HU78b, LI79, NA82, PR83
GX1+4	V2116 Oph	122	2.1×10^{-2}	>15	>60	—	—	—	—	LE71, BE76, WH76a, DO81
GX304−1	d	272	—	132 (?)	—	—	—	—	—	MC77b, HU77b, PA80b, PR83
4U0900−40	HD 77581	283	1.7×10^{-4}	8.965	113.0 (8)	19.3	0.092 (5)	21.8 (1.2)	31–37	UL72, FO73, MC76, RA76, VA77, WA77, RA80
4U1145−61	HEN 715	292	$<10^{-4}$	187 (?)	>100	—	—	—	—	WH78c, WH80, HU81b, WA81, PR83
1E1145.1−6141	d	297	$<10^{-4}$	>12	>50	—	—	—	—	WH78c, LA80, WH80, HU81b
A1118−61	HEN 3−640	405	—	—	—	—	—	—	—	IV75, JA81
4U1538−52	QV Nor	529	$<2 \times 10^{-3}$	3.730	55.2 (3.7)	13	—	33 (7)	25–33	BE77, DA77a, DA77b, CR78, PA78
GX301−2	WRA 977	696	7×10^{-3}	41.4	367 (3)	31	0.47 (1)	—	0	VI73, WH76a, WH78b, KE80, PA80a, WA82
4U0352+30	X Per	835	1.8×10^{-4}	580 (?)	—	—	—	<20	—	BR72, MO74, WH76b, HU77c, WH77

[a] The quoted uncertainties are approximate 1σ confidence limits and are given in parentheses [e.g. 140.13(10) ≡ 140.13 ±0.10]. Upper and lower limits represent approximate 95% confidence limits. When no uncertainty is indicated, the numerical value is measured to an accuracy greater than the given number of digits. For the eclipse half-angle, the range of values reasonably commensurate with the observational data is given.

[b] Measured during the longest interval of sustained spin-up (see text for details).

[c] References are given in abbreviated format (e.g. White et al 1976 ≡ WH76).

[d] The optical counterpart has been identified but has not yet been assigned a catalog name.

rotational distortion (see, for example, Avni 1976, Rappaport 1982). Moreover, the size and shape of the critical potential lobe can be computed from the mass ratio and the rotation rate of the companion star. (The critical potential lobe represents the largest volume that a dynamically stable companion star can occupy before it starts to lose matter through the inner saddle point of the effective gravitational potential between the two stars.) The effective radius R_L of the critical potential lobe has been found (Plavec 1968, Avni 1976) to be reasonably well fitted by the expression

$$R_L \simeq a[A + B \log q + C \log^2 q], \qquad 8.$$

where A, B, and C are constants that depend on the ratio Ω of the rotational frequency of the companion star to the orbital frequency. [We assume, for simplicity, that the rotational and orbital angular momentum vectors are parallel (cf. Avni & Schiller 1982).] We have calculated R_L/a as a function of Ω and q and devised the following fitting formulae:

$$A \simeq 0.398 - 0.026 \, \Omega^2 + 0.004 \, \Omega^3,$$
$$B \simeq -0.264 + 0.052 \, \Omega^2 - 0.015 \, \Omega^3, \qquad 9.$$
$$C \simeq -0.023 - 0.005 \, \Omega^2.$$

These expressions give R_L to within an accuracy of $\sim 2\%$ over the range $0 \leq \Omega \leq 2$ and $0.02 \leq q \leq 1$ (see also Plavec 1968, Avni 1976, Bahcall 1978b, Rappaport 1982). If we define the radius of the companion as some fraction β of the radius of the critical potential lobe,[3] then Equations 7 and 8 may be combined to yield an expression for the inclination angle:

$$\sin i \simeq [1 - \beta^2 (R_L/a)^2]^{1/2}/\cos \theta_e. \qquad 10.$$

We have ascertained that the use of Equations 8–10 over a wide range of values of q, θ_e, and Ω appropriate to X-ray binaries yields inclination angles with typical errors of only ~ 1–$2°$. For the case of 4U0900$-$40, we modified Equations 7–10 to take into account the appreciable eccentricity of the orbit (see Avni 1976).

From the above relations, we see that if K_c, $a_x \sin i$, and θ_e are measured, and if we make reasonable assumptions about the rotation rate of the companion star and the fraction of its critical potential lobe that it occupies, then all of the system parameters mentioned above may be determined. The amplitude of the observed ellipsoidal light variations in several of these systems indicates that the companion stars nearly fill their critical potential

---

[3] Note that our definition of β uses the effective radii of the critical potential lobe and the companion star, and does not refer to distances along the line joining the two stars.

lobes (i.e. $\beta \gtrsim 0.9$; see, for example, Avni & Bahcall 1975a,b). It can then be argued that tidal dissipation should force these systems into approximately synchronous rotation (i.e. $\Omega \simeq 1$; see Zahn 1975, 1977, Lecar et al. 1976). However, in view of the ongoing evolution of such binaries under the influence of mass exchange within the system and mass and angular momentum losses from the system, it is uncertain whether synchronous rotation can always be enforced (see, for example, Kelley et al. 1983b). Measurements of the rotational velocities of the companion stars in several of the X-ray binaries (see, for example, Conti 1978, Crampton et al. 1978, Hutchings et al. 1978a), while lacking high precision, do yield values consistent with $0 \lesssim \Omega \lesssim 1.5$. (For other discussions of these issues, see van den Heuvel 1975, Avni & Bahcall 1976, Rappaport & Joss 1977c, Avni 1978, Bahcall 1978a,b, Conti 1978, Petterson 1978, Avni & Schiller 1982).

Values for the system parameters are easily found from Equations 5–10 by inserting values for K_c, $a_x \sin i$, and θ_e, and reasonable guesses for β and Ω. We have evaluated the system parameters and their uncertainties from these equations by means of a Monte Carlo error propagation technique (Rappaport et al. 1980). In 2×10^4 trial evaluations, $a_x \sin i$ and K_c were chosen randomly with respect to Gaussian distributions with the appropriate mean values and widths, while the values of θ_e were chosen randomly and uniformly in the appropriate range, in order to reflect the experimental uncertainties. To simulate the theoretical and observational uncertainties in the values of β and Ω, we chose values of β randomly and uniformly over the range 0.9–1.0 and Ω randomly and uniformly between 0 and 1.5.

2.3.2 PROPERTIES OF THE BINARY SYSTEMS AND COMPANION STARS   The ranges of companion-star masses and radii for SMC X-1, Cen X-3, 4U0900 −40, and 4U1538−52, as determined by our Monte Carlo error propagation technique, are shown in Figure 9 as contours of constant probability and are also listed in Table 2. The most probable parameter values (see Table 2) are in good agreement with previously reported values (see Table 1 for references). The outer contour in Figure 9, which contains at least 95% of the Monte Carlo events, represents reasonably secure error limits for the masses and radii.

In the case of the Her X-1/HZ Her system, a reliable optical Doppler velocity curve is not yet available; however, extensive studies of optical pulsations (Middleditch & Nelson 1976), which are evidently produced by the reprocessing of X-ray pulsations that strike the surface of the companion star, have provided additional information needed to estimate the system parameters (Middleditch & Nelson 1976, Bahcall & Chester 1977, Chester 1978). We have determined the most probable Her X-1 system parameters and their uncertainties by use of a modified version of our Monte Carlo analysis code. This analysis incorporates the measurements of P_{orb}, $a_x \sin i$, θ_e, and the Doppler velocity of the optical pulsations, and it utilizes a geometric model for the reprocessing of X-ray pulsations into optical pulsations that was developed by Middleditch & Nelson (1976) and Bahcall & Chester (1977; in particular, their Equation 3). For Her X-1 only, we restricted β to the range $0.95 \leq \beta \leq 1.0$ (see, for example, Bahcall & Chester 1977). The results of this analysis are given in Table 2.

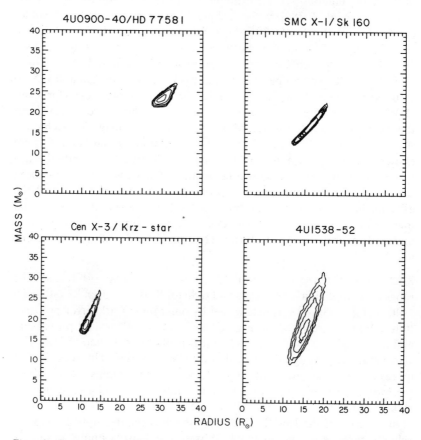

Figure 9 Computer-generated error contours of the mass and radius of four companion stars in binary X-ray pulsar systems (from Rappaport & Joss 1983). The input data are the orbital period, the projected semimajor axis of the neutron-star orbit, the X-ray eclipse duration, and the amplitude of the optical Doppler velocity curve (see text). The error contours are derived from a Monte Carlo analysis (Rappaport et al. 1980; see text) and are generated directly from the raw output of the analysis. The various contours represent arbitrary confidence levels; however, the outer contour contains ∼95% of the Monte Carlo events.

Table 2 Derived binary system parameters[a]

	Companion mass (M_c)	Companion radius (R_c)	Neutron-Star mass (M)	Inclination angle (i) (degrees)
SMC X-1	$17.0 \pm {}^{4.5}_{4.0}$	16.5 ± 4	$1.05 \pm {}^{0.40}_{0.30}$	57–77
Cen X-3	$19.0 \pm {}^{6.0}_{2.5}$	$12.2 \pm {}^{2.8}_{2.2}$	$1.07 \pm {}^{0.63}_{0.57}$	>63
4U0900−40	$23.0 \pm {}^{3.5}_{1.5}$	$31.0 \pm {}^{4}_{3}$	$1.85 \pm {}^{0.35}_{0.30}$	>73
4U1538−52	$18.5 \pm {}^{12}_{6.5}$	$16.0 \pm {}^{5}_{4}$	$1.87 \pm {}^{1.33}_{0.87}$	>60
LMC X-4	$19.0 \pm {}^{35}_{13}$	$9.0 \pm {}^{4.5}_{3.5}$	$1.70 \pm {}^{1.90}_{1.00}$	>58
Her X-1	$2.35 \pm {}^{0.20}_{0.40}$	$4.05 \pm {}^{0.25}_{0.40}$	$1.45 \pm {}^{0.35}_{0.40}$	>80
PSR 1913+16	—	—	1.42 ± 0.12[b]	41–51

[a] All dimensioned quantities are in solar units; the quoted uncertainties are 95% confidence limits.
[b] From Taylor & Weisberg (1982); the mass of the companion star is also measured, but is not included in the table because its nature is very different from the other companion stars (it is probably another neutron star).

Highly compact binaries   The X-ray pulsars discussed thus far typically have massive companion stars, with M_c of the order of 20 $M_\odot$ and R_c of the order of 20 $R_\odot$. Even for the relatively small Her X-1 system, we have $M_c \simeq 2.3 M_\odot$ and $R_c \simeq 4 R_\odot$. However, at least one X-ray pulsar, 4U1626 −67, has been found to be in a highly compact binary stellar system. The source has been identified with a faint blue star ($V \simeq 18.7$) by McClintock et al. (1977a); the inferred absolute magnitude of McClintock's star (KZ TrA) is $M_v \gtrsim 3$. This is consistent with a system containing a late-type dwarf companion with negligible intrinsic optical luminosity, wherein most of the light results from reprocessed X-radiation.

Severe constraints have been placed on the size of the orbit of the X-ray star. The lack of variable Doppler delays in the pulse arrival times has yielded limits on $a_x \sin i$ of ~ 0.1–0.2 lt-s for orbital periods in the range of 10 min $\leq P_{orb} \leq 20$ days and, in particular, $a_x \sin i \lesssim 0.04$ lt-s for $P_{orb} \simeq 1$ h (Rappaport et al. 1977, Joss et al. 1978b, Pravdo et al. 1979, Li et al. 1980, Middleditch et al. 1981). Based on these observations, Joss et al. (1978b), Li et al. (1980), and Rappaport (1982) suggested that the system was highly compact, with an orbital separation of ~ 1 lt-s and a companion mass of $\sim 0.1 M_\odot$ (see Figure 10).

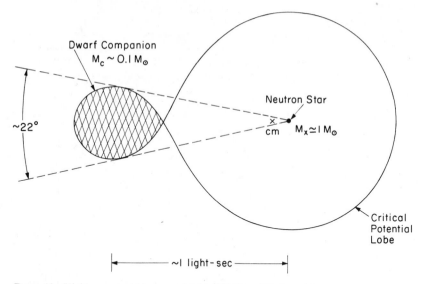

Figure 10   Highly compact binary model for 4U1626−67 (adapted from Joss & Rappaport 1979). This model may also apply to many of the galactic bulge X-ray sources, including the X-ray burst sources (see Section 3.2). The figure is drawn to scale for the indicated illustrative values for the masses of the neutron star and companion star (M_x and M_c, respectively).

Optical photometric pulsations at the X-ray pulse period were discovered by Ilovaisky et al. (1978). Observations by Middleditch et al. (1981) subsequently revealed the presence of two optical pulse periods, one at the frequency of the X-ray pulsations and the other downshifted in frequency by ∼0.4 mHz. The downshifted peak apparently corresponds to X-radiation reprocessed on or near the surface of the companion star; the difference between the two pulse periods can be understood as the difference between the sidereal and synodic rotation periods of the neutron star as seen from the companion, provided that the orbital period is ∼2491 s (Middleditch et al. 1981). Middleditch et al. derived the following system parameters: $M = 1.8$ ($+2.9, -1.3$) $M_\odot$, $M_c < 0.5$ $M_\odot$, $a_c \sin i = 0.36$ (± 0.10) lt-s, $a = 1.14$ (± 0.40) lt-s, and $i = 18°$ ($+18°, -7°$).

Mass transfer in the binary system could be driven by the decay of the orbit due to gravitational radiation (see, for example, Paczyński 1967, Faulkner 1971, Rappaport et al. 1982). The system may have evolved from a cataclysmic variable, wherein the matter accreting onto the degenerate dwarf eventually drives it over the Chandrasekhar limiting mass, forming a neutron star. This evolutionary scenario has been discussed by Gursky (1976), van den Heuvel (1977), and Joss & Rappaport (1979).

2.3.3   NEUTRON-STAR MASSES   The measured neutron-star masses for SMC X-1, Cen X-3, 4U0900−40, 4U1538−52, LMC X-4, and Her X-1, derived from the Monte Carlo error propagation technique discussed in Section 2.3.1, are shown in Figure 11 (cf. Joss & Rappaport 1976, Rappaport & Joss 1977c, Avni 1977, Bahcall 1978b, Rappaport & Joss 1981, 1983). The most probable masses (indicated by the filled circles) and the corresponding uncertainties are obtained directly from the Monte Carlo probability distributions. The shorter error bars shown in Figure 11 represent the uncertainties in the neutron-star masses for the less conservative assumptions that $\beta = \Omega = 1$ (Roche geometry). Also included in the figure is the mass of the binary radio pulsar PSR 1913 + 16 (see Section 1),

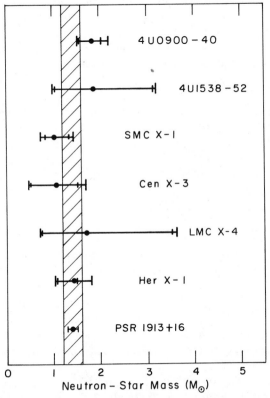

Figure 11 Empirical knowledge of neutron-star masses (updated from Rappaport & Joss 1983). Six of the masses are derived from observations of binary X-ray pulsars (see Section 2.3.3). PSR 1913 + 16 is a binary radio pulsar (Taylor & Weisberg 1982, and references therein) and is added for completeness. The most probable value for the mass of each neutron star is indicated by the filled circle. For the X-ray binary systems, an inner set of error limits is also shown, corresponding to the less conservative assumptions that $\beta = 1$ and $\Omega = 1$ (see Section 2.3.1). The hatched region represents the range of neutron-star masses (1.2–1.6 $M_\odot$) that might be expected on the basis of current theoretical scenarios for neutron-star formation (see Section 2.3.3).

which is taken from Taylor & Weisberg (1982) and is added to our sample for completeness. These seven cases represent the only reliable mass measurements for objects known to be neutron stars.

The masses given in Figure 11 are all consistent, within the uncertainties, with a mass of 1.4 ± 0.2 $M_\odot$ (the shaded region in Figure 11). This is the range of masses that might be expected for neutron stars if they are formed in the collapse of the degenerate cores of highly evolved stars (Arnett & Schramm 1973, Iben 1974, and references therein) or the collapse of accreting degenerate dwarfs in close binary stellar systems (see, for example, Canal & Schatzman 1976, Nomoto 1981, and references therein).

Furthermore, all of the measured neutron-star masses are consistent with neutron-star and quark-star models, in the context of the general theory of relativity, based on conventional many-body nuclear and high-energy physics (Arnett & Bowers 1977, Baym & Pethick 1979, and references therein). Maximum allowed masses for nonextreme equations of state lie between ~ 1.4 and ~ 2.7 $M_\odot$. Thus, the presently available observational mass estimates for neutron stars are marginally sufficient to constrain the equation of state of matter at very high densities. In particular, these mass determinations demonstrate that the high-density equation of state cannot be very much softer than the current predictions of nuclear and high-energy physics, provided that general relativity is the correct theoretical framework for the calculation of neutron-star models.

3.   X-RAY BURST SOURCES

3.1   Observational Properties

The observed properties of X-ray burst sources have been extensively reviewed by Lewin & Joss (1981, 1983). The salient features of these sources include burst rise times of ~ 1 s, decay time scales of ~ 3–100 s, peak luminosities of $\sim 10^{38}$ erg s^{-1}, and total emitted energies of $\sim 10^{39}$ ergs per burst (see Figure 12). The spectra of X-ray bursts can generally be well fitted by blackbody emission from a surface with a peak temperature of $\sim 3 \times 10^7$ K and a roughly constant scale size that corresponds, for a spherical surface, to a radius of ~ 10 km (if general relativistic effects are neglected; see Section 3.3.4). The intervals between bursts from a given source may be regular or erratic and are typically in the range of $\sim 10^4$–10^5 s; many sources undergo burst-inactive phases that can last for weeks or months. Most burst sources are also sources of persistent X-ray emission, and the ratio of average persistent luminosity to time-averaged burst luminosity is typically $\sim 10^2$ during burst-active phases. The properties of the "rapid burster," MXB 1730 − 335, are different from those of all other known burst sources and are discussed separately in Section 3.4.2.

The distribution of burst sources on the celestial sphere (Figure 1) is

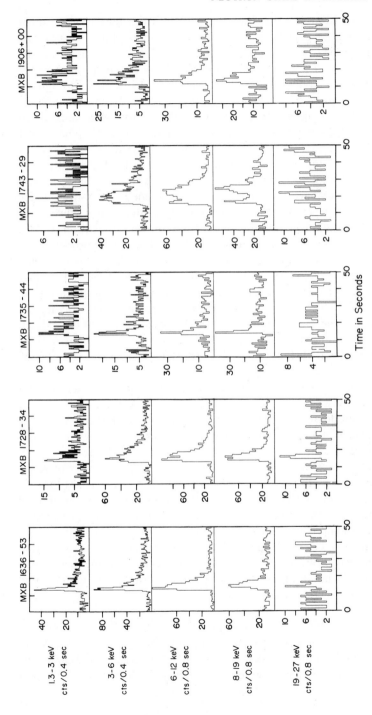

Figure 12 Profiles of typical X-ray bursts from five different sources in five X-ray energy channels (from Lewin & Joss 1977). Note that the gradual decay (known as the burst "tail") persists longer at low X-ray energies than at high energies, indicating cooling of the burst emission region.

strongly reminiscent of stellar Population II, with a strong concentration in the direction of the galactic center. In fact, a number of burst sources have been found to lie in the direction of globular clusters (see Figure 1). This situation contrasts sharply with that of the X-ray pulsars (also shown in Figure 1), which appear to be more uniformly distributed through the galactic disk and whose optical companion stars are often quite massive (see Section 2.3.2), suggesting association with a young stellar population.

The optical counterparts of the burst sources, where they have been identified at all, are found to be faint, blue objects with spectra that are dominated by emission lines [Canizares et al. 1979; see Bradt & McClintock (1983) for additional references]. Thus, the burst sources appear to be members of the larger class of galactic bulge X-ray sources, whose optical and persistent X-ray emissions share the same properties; the only clearly distinguishing feature of the burst sources is the existence of the bursts themselves (see Lewin & Joss 1981, 1983). The optical properties are consistent with the idea that all of these sources are close binary stellar systems containing collapsed objects, such as neutron stars, and intrinsically faint, low-mass companion stars (see Section 3.2). In this picture, most of the optical light results from reprocessing of some of the X-ray flux by matter within the binary system. If this idea is correct, then the galactic bulge sources in general, and the burst sources in particular, may be generically related to the binary X-ray pulsar 4U1626−67 (see Section 2.3.2).

Thermonuclear flashes in the surface layers of an accreting neutron star were first proposed by Woosley & Taam (1976) and Maraschi & Cavaliere (1977) as a model for X-ray burst sources. During the last few years, the preponderance of accumulating evidence has come to strongly favor this model; as is documented below, calculations of such flashes have been remarkably successful in accounting for the general properties of X-ray bursts. Moreover, the theoretical work to date strongly suggests that the characteristics of bursts should be capable of imparting substantial constraints upon the properties (e.g. masses, radii, and internal temperatures) of the underlying neutron stars. In the remainder of Section 3, we concentrate on the theoretical investigations of thermonuclear flashes that have been carried out to date and on the possibilities that the thermonuclear flash model holds for advancing our understanding of the fundamental properties of neutron stars.

3.2   *Burst Sources as Binary Stellar Systems*

The faintness of the optical counterparts of most galactic bulge X-ray sources (including burst sources) rules out giant, supergiant, and early-type main sequence stellar companions (see Lewin & Joss 1981, 1983, and references therein). It therefore seems likely that many of the companion stars, with perhaps only a few exceptions, are low-mass ($\lesssim 0.5\ M_\odot$) main sequence dwarfs or degenerate dwarfs (Joss & Rappaport 1979). It is, moreover, possible that such a star could transfer sufficient mass to the collapsed star only if it nearly fills its critical potential lobe, in which case the orbital separation would be $\lesssim 10^{11}$ cm and the orbital period $\lesssim 0.3$ days (see Figure 10). For such highly compact binary systems, the mass transfer could be effected through the decay of the orbit due to gravitational radiation (see Section 2.3.2), possibly augmented by magnetic braking (Verbunt & Zwaan 1981, Rappaport et al. 1983, Taam 1983), a self-excited stellar wind (Faulkner 1974, Basko et al. 1977), and/or the evolution of the companion (which could be the evolved remnant of a more massive star; see Faulkner 1974).

The properties of such systems can be reconciled with the observational characteristics of most of the galactic bulge sources (Joss & Rappaport 1979). In particular, the faintness of the optical counterparts is a natural consequence of the intrinsic faintness of the companion star; in fact, most of the very blue light that is seen results from reprocessing of X-radiation within the system, rather than from the intrinsic luminosity of the companion. This idea gains support from the discovery that at least two transient X-ray sources, Aql X-1 and Cen X-4, emit bursts (Koyama et al. 1981, Matsuoka et al. 1980) and have optical counterparts with properties typical of those of galactic bulge sources [see Lewin & Joss (1981, 1983) for references] when they are in a state of high X-ray emission; when these sources go into quiescence, the spectra of their counterparts develop an ordinary late-type stellar component (Thorstensen et al. 1978, van Paradijs et al. 1980).

The general lack of observed X-ray eclipses is a bit harder to understand. The discovery (Walter et al. 1982, White & Swank 1982) of 50-min periodic absorption events in the persistent X-ray flux from the burst source 4U1915 -05 is extremely important, in that these events almost certainly reflect the orbital periodicity of an underlying low-mass binary. However, these events cannot be true eclipses by the companion star, since they do not recur on every orbital cycle and have other properties that are incommensurate with true eclipses. Although the small size of the companion compared with the dimensions of the binary systems renders the probability of observing true eclipses in any one system rather small ($\sim 20\%$; Joss & Rappaport 1979), the a priori probability is quite high ($\gtrsim 99\%$) for having detected true eclipses in at least one of the twenty or so galactic bulge sources for which adequate observations are available. As first suggested by Milgrom (1978), the companion may be largely shielded from the X-radiation by an accretion disk surrounding the collapsed star; such shielding may be responsible for the lack of eclipsing behavior and the general lack of optical photometric variability at the orbital period.

If the surface magnetic field of a neutron star has largely decayed away, as is probable for neutron stars in systems with ages of $\gtrsim 10^9$ yr (Ruderman 1972, and references therein, Flowers & Ruderman 1977), then the accretion disk may extend downward to the surface of the neutron star (Milgrom 1978). When this occurs, up to $\sim 1/2$ the gravitational potential energy may be released in the disk rather than on the surface of the neutron star. The relatively large X-radiating surface area of the inner disk plus neutron star may then account for the "soft" X-ray spectra of the galactic bulge sources compared with those of the X-ray pulsars, whose magnetic fields are evidently strong enough to disrupt an accretion disk well above the neutron-star surface and funnel the accretion flow onto the magnetic polar caps of the neutron star (see Sections 2.1 and 2.2).

This picture gains further support from the discovery of optical bursts coincident with some X-ray bursts (Grindlay et al. 1978, Hackwell et al. 1979, McClintock et al. 1979, Pedersen et al. 1979). The optical bursts are delayed and spread in time relative to the X-ray bursts with which they are associated, in a manner quantitatively consistent with the idea that the optical light results from the reprocessing of the X-radiation in an accretion disk whose dimensions are appropriate to the low-mass binary hypothesis (Pedersen et al. 1982a,b). Coherent studies of the optical and X-ray properties of burst sources should, in the future, provide a powerful means of probing the properties of the binary system, the evolutionary status of the companion star, and the character of the accretion process (see, for example, Swank et al. 1984).

3.3   *The Thermonuclear Flash Model*

3.3.1   THE PHYSICAL PICTURE AND DIMENSIONAL ARGUMENTS   Consider a neutron star undergoing accretion from a binary stellar companion. The freshly accreted matter will be rich in hydrogen and/or helium. However, at depths greater than $\sim 10^4$ cm beneath the surface of the neutron star, the density is sufficiently high that nuclear statistical equilibrium will be swiftly achieved; the predominant nuclei will have maximal binding energies, with atomic weights of ~ 60. Hence, the accreting matter must pass through a series of nuclear burning shells as it is gradually compressed by the accretion of still more material. If the core of the neutron star is sufficiently hot or the accretion rate is sufficiently high, the temperature in the surface layers will be high enough that the burning will proceed via thermonuclear reactions, rather than electron capture or pycnonuclear reactions (which are driven by high densities rather than high temperatures). A sketch of the resultant structure of the neutron-star surface layers is given in Figure 13.

It was first realized by Hansen & Van Horn (1975) that these burning shells will tend to be unstable to thermal runaway. The instability is related to the "thin-shell instability" in red giants, which was discovered by Schwarzschild & Härm (1965). The existence and strength of the instability are related to the strong temperature dependence of the thermonuclear reaction rates and the partial degeneracy of the burning material. Discussions of this type of instability, in the context of neutron-star envelopes, have been given by Barranco et al. (1980) and Paczyński (1983a).

Hydrogen is fused into helium in the outermost burning shell of the neutron star. However, the reaction rate of the p-p chains is insufficiently temperature sensitive to produce a thermal runaway in this shell. Most of the hydrogen burning actually occurs via the CNO cycle, which is modified at the prevailing high temperatures and whose rate reaches a saturated value for temperatures in excess of $\sim 7 \times 10^7$ K. This saturation effect results from the appreciable lifetimes ($\sim 10^{2-3}$ s) of the beta-unstable nuclei ^{13}N, ^{14}O, ^{15}O, and ^{17}F that participate in the cycle. Thus, hydrogen-burning runaways are usually unable to release large amounts of energy on the time scale of an X-ray burst (Joss 1977, Lamb & Lamb 1978) (see, however, the discussion of interacting hydrogen-helium shells in Section 3.3.2). The next shell inward is the helium-burning shell, which should generate rapid and energetic thermonuclear flashes over a wide range of conditions. It is unlikely that there will be any other significant burning

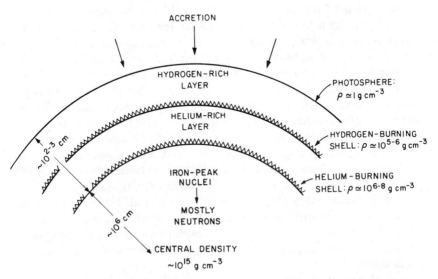

Figure 13   Schematic sketch of the surface layers of an accreting neutron star (adapted from Joss 1979a).

shells, as numerical calculations (see Section 3.3.2) indicate that nearly all matter will usually be processed into massive nuclear species within the hydrogen- and helium-burning shells. Dimensional analysis (Joss 1977, Lamb & Lamb 1978) indicates that helium-burning flashes should have the following properties: (a) They should occur after the accumulation of $\lesssim 10^{21}$ g of fuel and release total energies of $\lesssim 10^{39}$ ergs per flash. (b) For accretion rates comparable with those observed in X-ray pulsars ($\lesssim 10^{17}$ g s^{-1}), the time interval between flashes should be $\sim 10^4$ s, very roughly. (c) The transport of energy through the surface layers should result in the emission of bursts of electromagnetic radiation from the neutron-star photosphere with rise times of ~ 0.1 s, peak luminosities near the Eddington limit ($\sim 10^{38}$ erg s^{-1}), decay time scales of ~ 10 s, and peak effective temperatures of $\sim 3 \times 10^7$ K (if a full 10^{39} ergs of energy is indeed released in a single flash).

3.3.2   NUMERICAL MODELS   The above estimates, though crude, suggest that thermonuclear flashes on accreting neutron stars could account for the observed properties of most X-ray burst sources. With this encouragement, detailed numerical computations of the evolution of the surface layers of an accreting neutron star have been carried out (Joss 1978, Taam & Picklum 1979, Joss & Li 1980, Taam 1980, 1981a, 1982, Ayasli & Joss 1982, Hanawa & Sugimoto 1982, Starrfield et al. 1982, Wallace et al. 1982). The properties of the flashes obtained in these calculations were in good agreement with those expected from dimensional analysis. These calculations also indicated that (a) a full $\sim 10^{21}$ g of matter (or even more, under some circumstances) typically accumulates on the neutron-star surface before each flash, (b) a flash usually consumes much, if not all, of the available nuclear fuel and probably synthesizes mostly iron-peak elements, and (c) most of the energy of a flash is transported to the photosphere and lost as X-radiation, rather than carried inward to heat the interior of the star. These properties of the flashes had not been discerned prior to the performance of detailed evolutionary computations, at least partly because they depend on the highly nonlinear characteristics of the flash growth and decay. Much progress has also been achieved by semianalytic calculations (e.g. those by Taam & Picklum 1978, Czerny & Jaroszyński 1979, Barranco et al. 1980, Érgma & Tutukov 1980, Hōshi 1980, Fujimoto et al. 1981, Taam 1981b, Hanawa & Fujimoto 1982, Paczyński 1983a), which have augmented the available evolutionary computations by exploring a wide range of physical conditions and investigating various aspects of the physics of neutron-star surface layers.

One of the central results of detailed theoretical studies was the discovery that hydrogen burning plays a major role in neutron-star thermonuclear flashes (Taam & Picklum 1978, 1979). The saturation of the CNO cycle by the non-negligible lifetimes of the beta-unstable nuclei that participate in the cycle limits the hydrogen-burning rates to such an extent that at moderately high accretion rates ($\gtrsim 1 \times 10^{16}$ g s^{-1}), the available hydrogen cannot be consumed before helium-burning conditions are achieved. Under these circumstances, the hydrogen- and helium-burning shells overlap. (If the accreting material has Population II chemical abundances, then the lack of CNO seed nuclei results in overlapping shells at even more modest accretion rates.) Although the hydrogen is usually unable to produce significant thermonuclear flashes (see Section 3.3.1), its presence during a helium-burning flash can, via proton-capture reactions, increase by a factor of several the total energy released in the flash. Alpha-capture reactions onto the CNO nuclei, whose relative abundances are heavily modified by their earlier participation in hydrogen-burning reactions, can also play an important role in the flash (Taam & Picklum 1979, Taam 1980, 1981a, Wallace & Woosley 1981, Ayasli & Joss 1982, Starrfield et al. 1982).

If temperatures in excess of $\sim 1 \times 10^9$ K are achieved, multiple proton captures and positron decays cause the intermediate-mass nuclei to leave the range of atomic masses where the CNO cycle is operative and result in the synthesis of exotic, proton-rich nuclei (Kudryashov & Érgma 1980, Wallace & Woosley 1981, Ayasli & Joss 1982, T. Hanawa, D. Sugimoto, and M. Hashimoto, preprint). At still higher temperatures ($\gtrsim 2 \times 10^9$ K), (α, p) reactions compete with proton captures onto these nuclei. Such reactions consume helium nuclei and release protons, resulting in a complex nuclear reaction network in which hydrogen burning, helium burning, and heavy-element nucleosynthesis all occur simultaneously. Unfortunately, the synthesized nuclei should be buried permanently beneath the surface of the neutron star and remain invisible to external observers. The results of some numerical calculations of neutron-star thermonuclear flashes that include these fusion processes are shown in Figure 14.

The role of the mass accretion rate ($\dot{m}$), the internal thermal structure of the neutron star, and the neutron-star magnetic field in determining the behavior of the nuclear burning shells has received considerable attention. The shells are unstable to thermonuclear flashes over a wide range of accretion rates, although the properties of such flashes vary substantially with varying $\dot{m}$ (see, for example, Figure 14). At a given accretion rate, the core of the neutron star will tend toward a state of thermal equilibrium (Lamb & Lamb 1978), wherein the heat flow into the core from the surface layers during thermonuclear flashes (when the surface temperatures are relatively high) is just balanced by the neutrino emission from the core and by the heat transferred radiatively and conductively from the core between

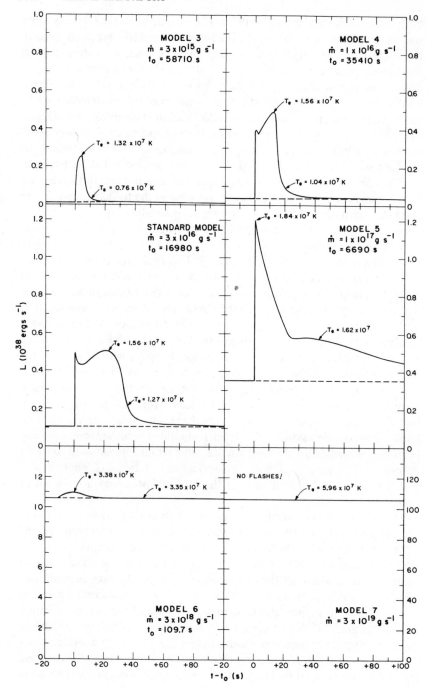

flashes. For moderate accretion rates ($\sim 10^{16}$–10^{17} g s^{-1}), the equilibrium state is reached after $\sim 10^5$ yr of steady accretion and corresponds to core temperatures of ~ 1–2×10^8 K (Ayasli & Joss 1982, Nomoto & Tsuruta 1982, Fujimoto et al. 1984).

If the neutron star has a sufficiently strong magnetic field to funnel the accretion flow onto the magnetic polar caps, the effective accretion rate in the polar cap regions (for a fixed total accretion rate) will be enhanced by a factor of $\sim 10^3$. For $\dot{m} \gtrsim 10^{15}$ g s^{-1}, the high effective accretion rate in the polar cap regions should stabilize the nuclear burning shells against thermonuclear flashes (Joss 1978, Taam & Picklum 1978, Joss & Li 1980, Ayasli & Joss 1982). Even in the absence of funneling effects, magnetic fields in excess of $\sim 10^{12}$ G will significantly reduce the radiative and conductive opacities in the neutron-star surface layers (Lodenquai et al. 1974, Canuto 1970), thereby changing the properties of the nuclear burning shells (Joss & Li 1980, Ayasli & Joss 1982).

Of course, the behavior of the neutron-star surface layers also depends on the neutron-star mass and radius. However, it appears that aside from the general relativistic effects discussed in Section 3.3.4, the observable properties of the thermonuclear flashes are not particularly sensitive to M and R (see, for example, Ayasli & Joss 1982).

Figure 14 The behavior of the surface luminosity L as a function of time t following a thermonuclear flash in the standard model and five additional models calculated by Ayasli & Joss (1982). All significant general relativistic corrections to the equations of stellar structure and evolution were taken into account in these calculations, and the numerical values of all quantities are those that would be measured by a distant observer. In each case, the neutron star is assumed to be nonrotating and unmagnetized and to have a mass of 1.41 $M_\odot$, a radius of 6.57 km, and a core temperature (as measured by an observer on the neutron-star surface) of 1.5×10^8 K; the accreting matter is assumed to be hydrogen rich and to accrete spherically onto the neutron-star surface. The time t_0 is the interval from the onset of accretion (at time $t = 0$) to the start of the flash, the effective temperature T_e is indicated at a few points, and the dashed lines indicate the level of persistent accretion-driven luminosity in each case. The accretion rate is successively higher in models 3 and 4, the standard model, and models 5, 6, and 7, as indicated. In model 3, $\dot{m}$ is sufficiently low that the hydrogen at the base of the freshly accreted matter is consumed by steady hydrogen-burning prior to any flash, and a pure helium flash results. In both model 4 and the standard model, hydrogen is still present at the base of the freshly accreted matter when helium ignites; the resultant X-ray bursts are double-peaked, with the first luminosity maximum reflecting the helium flash and the second maximum reflecting the subsequent rapid burning of hydrogen due to proton-capture reactions involving the moderately heavy nuclei that are largely produced by the helium flash. In model 5, the flash is very strong and generates sufficiently high temperatures to swiftly process the moderately heavy elements into iron-peak elements, so that the hydrogen is unable to burn very rapidly and the second luminosity maximum is replaced by an extended plateau. As $\dot{m}$ continues to be increased, in models 6 and 7, the flashes become weaker and disappear, since the surface layers become sufficiently hot to consume the nuclear fuel virtually as fast as it accumulates (see also Paczyński 1983a).

Nearly all of the calculations of neutron-star thermonuclear flashes to date have relied on the assumption of spherical symmetry. In fact, however, small but significant violations of spherical symmetry might result from the residual angular momentum of the accreting matter or from the effects of a relatively weak ($\ll 10^{12}$ G) surface magnetic field. Such violations could be the key to understanding some of the observed complexities in the structure and recurrence patterns of X-ray bursts (see Lewin & Joss 1981, 1983). For example, a thermonuclear flash that ignites on one portion of the neutron-star surface may propagate around the star, in a pattern that varies from flash to flash and from one burst source to another (Joss 1978, Fryxell & Woosley 1982b, Shara 1982). A thorough investigation of such possibilities will require two- or three-dimensional numerical computations (see, for example, Fryxell & Woosley 1982a), which will be much more difficult than the calculation of spherically symmetric models.

3.3.3 DYNAMICAL PHENOMENA   The possibility of dynamical phenomena following a thermonuclear flash has recently been investigated by several authors. The available nuclear energy per unit mass ($\sim 10^{-2}$–10^{-3} c^2) is always small compared with the gravitational binding energy per unit mass of the neutron-star surface layers ($\sim 10^{-1}$ c^2), so that the bulk of these layers should remain in hydrostatic equilibrium (Joss 1978). However, dynamical effects may still be generated in a small fractional mass of the surface layers that receive a disproportionate share of the released energy. It seems unlikely that shock waves in the outer surface layers can produce important effects (see, for example, Starrfield et al. 1982, Wallace et al. 1982), but high radiative luminosities near the neutron-star surface may result in dynamical mass motions of great observational significance.

The Eddington limit is defined to be

$$L_{\text{ed}} = \frac{4\pi cGM}{\kappa_T(1+z_s)} \simeq 1.22 \times 10^{38}\left(\frac{M}{M_\odot}\right)\left(\frac{1+X}{1.7}\right)^{-1}\left(\frac{1+z_s}{1.2}\right)^{-1} \quad \text{erg s}^{-1}, \quad 11.$$

where $\kappa_T \simeq 0.20\,(1+X)$ cm^2 g^{-1} is the Thomson scattering opacity at the neutron-star surface, X is the fractional abundance (by mass) of hydrogen at the surface, and $(1+z_s)^{-1}$ is a general relativistic correction factor (z_s is defined following Equation 1). When the radiative luminosity, as measured by a distant observer, equals the Eddington limit (and the radiative opacity equals κ_T), the upward radiation pressure just balances the downward gravitational force. Luminosities substantially in excess of the Eddington limit should result in the ejection of large quantities of mass from the surface layers by the action of radiation pressure, but, as noted above, there is insufficient energy to permit such ejection (Joss 1977, Ruderman 1981). If the Eddington limit is even slightly exceeded, the net upward force may cause the outermost surface layers to expand appreciably (Wallace et al. 1982, Taam 1982, Hanawa & Sugimoto 1982, Paczyński 1983b). In fact, if the luminosity remains super-Eddington for a time much longer than the dynamical time scale at the neutron-star surface ($\sim 10^{-4}$ s), a quasi-static wind may become established, resulting in the elevation of the photosphere to a level well above the original neutron-star surface (Wallace et al. 1982, Ebisuzaki et al. 1983, Kato 1983, Melia & Joss 1984a; see Figure 15). It seems likely that such photospheric expansion in response to flashes that involve large energy releases and concomitantly high luminosities will provide an explanation for apparent variations in the size of the X-ray-emitting regions during the course of some X-ray bursts. In such instances, the radius of the (presumably spherical) emitting surface is seen to increase from ~ 10 km to a value several times larger near the peak of the burst, and it is then observed to decrease back to a value near 10 km during the burst decay (Swank et al. 1977, Grindlay et al. 1980, Hoffman et al. 1980, and references therein). The conjecture that such events result from high radiation pressure at the neutron-star photosphere is further strengthened by the observation [see Figure 5 in Oda & Tanaka (1981) or Figure 6 in Oda (1982)] that radius variations are detected primarily among the most luminous bursts from a given source.

3.3.4   GENERAL RELATIVISTIC EFFECTS AND NEUTRON-STAR PARAMETERS   General relativistic corrections have an important influence on the observed properties of X-ray bursts. The potential importance of general relativity can be seen by inspection of the parameter

$$\frac{2GM}{Rc^2} \simeq 0.60 \left(\frac{M}{1.4\,M_\odot}\right)\left(\frac{R}{7\,\text{km}}\right)^{-1}. \tag{12.}$$

The left-hand side of Equation 12 is just the ratio of the Schwarzschild radius of the neutron star ($2GM/c^2$) to its actual radius. For the indicated values of M and R, which have been used in many of the actual model calculations of thermonuclear flashes, it is evident that this parameter is not very much smaller than unity, so that general relativistic effects should be substantial. A recent report by Inoue (1983) of redshifted atomic absorption features in the spectra of some X-ray bursts seems to provide direct evidence for the existence of general relativistic effects in these sources; such a measure of gravitational redshift may prove to be of the utmost importance in advancing our understanding of neutron stars, the influence of general relativity upon the X-ray burst phenomenon, and gravitation theory itself.

Some general relativistic effects seem to be inextricably entwined in the corrections to the full set of stellar structure and evolution equations (see Ayasli & Joss 1982). However, other effects can be more readily interpreted.

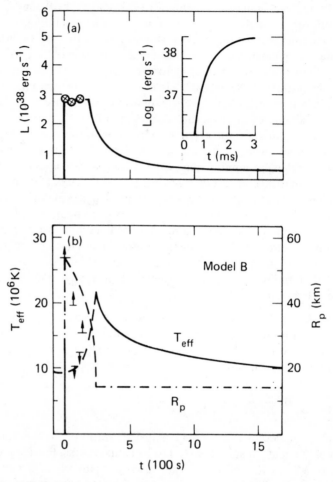

Figure 15 (a) Behavior of the surface luminosity L as a function of time t following a deep thermonuclear flash (involving ∼ 1.4 × 10²³ g of helium fuel) in model B calculated by Wallace et al. (1982). A stellar wind developed when the luminosity approached the Eddington limit, resulting in an increase in the photospheric radius and a decrease in the effective temperature; the model was more finely zoned at three times during this phase, and the resulting luminosities are plotted as data points. The inset shows the initial luminosity increase in more detail. (b) Behavior of the effective temperature T_{eff} (solid curve) and photospheric radius R_p (dot-dashed curve) for the same event. Upper limits to T_{eff} and lower limits to R_p are shown as data points during the wind phase; the dashed lines during that time indicate the qualitative behavior of these quantities. The observable properties of this event strongly resemble those of the fast X-ray transients (see Section 3.4.3 and Figure 19).

In particular, the reductions in the peak X-ray luminosity and effective temperature, as well as the increase in the burst duration and recurrence time, for the general relativistic model calculation displayed in Figure 16 (relative to the Newtonian calculation also shown) are all at least partly due to the direct effects of gravitational redshift upon the radiation emitted from the neutron-star surface.

The general relativistic corrections to observable properties of the emitted radiation, combined with considerations of theoretically expected burst properties, can be used, at least in principle, to place constraints on the masses and radii of the underlying neutron stars (Goldman 1979, van Paradijs 1979, Hōshi 1981, Marshall 1982). The arguments presented in Section 3.3.3 indicate that the peak luminosity in a burst should not substantially exceed the Eddington limit. A simple Newtonian calculation, based on the peak color temperatures of observed bursts and the

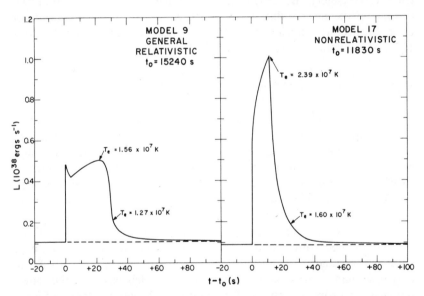

Figure 16   Same as Figure 14 for two further models calculated by Ayasli & Joss (1982). The notation is the same as in Figure 14. The parameter values assumed for model 9 are identical to those of the standard model, except that the neutron-star core temperature is taken to be 2×10^8 K. Model 17 assumes identical parameter values to those of model 9, but all general relativistic corrections to the equations of stellar structure and evolution have been suppressed, and all parameters have Newtonian definitions. The longer X-ray burst duration and recurrence time (t_0), the lower peak X-ray luminosity and effective temperature, the higher level of persistent accretion-driven luminosity, and the more complex burst shape in model 9 compared with model 17 are all the result of gravitational redshift and other general relativistic effects (see Section 3.3.4).

assumptions that (a) the peak burst luminosity is near the Eddington limit and (b) the emitted spectrum is that of a blackbody (i.e. the color temperature equals the effective temperature), yields neutron-star radii of ~ 10 km (van Paradijs 1978, Oda & Tanaka 1981, Oda 1982; for additional references, see Lewin & Joss 1981, 1983). However, as shown below, this argument is severely complicated by general relativistic effects and by deviations of the emitted spectrum from that of a simple blackbody.

Combining Equation 11 for the Eddington limit with the general relativistic expression for the luminosity of a spherical blackbody surface of temperature T_e (as measured by a distant observer) and radius R, one obtains

$$\frac{L_x}{L_{ed}} = \frac{\kappa_T \sigma T_e^4 R^2}{cGM} (1 + z_s)^3 \qquad \left(R \geq \frac{3GM}{c^2} \right), \qquad 13.$$

where σ is the Stefan-Boltzmann constant. For $R < 3GM/c^2$, the recapture of some of the emitted flux by the neutron star, due to the curvature of photon trajectories, must also be taken into account (see also van Paradijs 1979); the resultant expression is

$$\frac{L_x}{L_{ed}} = \frac{27\kappa_T \sigma T_e^4 GM}{c^5} (1 + z_s) \qquad \left(R < \frac{3GM}{c^2} \right). \qquad 14.$$

Loci of M and R for several values of T_e, as described by Equations 13 and 14, are displayed in Figure 17. If the peak luminosity during a burst is no greater than the Eddington limit, and if the emitted spectrum at the burst peak is that of a blackbody at a given temperature, then the corresponding curve in this figure represents the upper envelope of allowed values of M and R for the underlying neutron star. However, most observed bursts have peak color temperatures in excess of 2×10^7 K, and many have peak values of $\sim 3 \times 10^7$ K (see, for example, Cominsky 1981, Marshall 1982). These measured temperatures are based solely on spectral fits and could not be affected by distance miscalibrations. Peak temperatures of $\sim 2 \times 10^7$ K are only marginally consistent with neutron-star models based on the softest physically plausible equations of state for high-density matter (see Figure 17); appreciably higher temperatures are seemingly inconsistent with any reasonable neutron-star models.

The most likely resolutions of this discrepancy involve violations of the Eddington limit (see Section 3.4.1), despite the theoretical arguments noted above. Deviations of the burst spectra from a simple blackbody shape, as considered by J. H. Swank, D. Eardley, and P. J. Serlemitsos (preprint), van Paradijs (1982), and M. Czerny and M. Sztajno (preprint), probably also play a significant role in this discrepancy. Such deviations may be largely caused by the effects of an optically thick electron-scattering atmosphere and complicated by dynamical motions in the outermost surface layers of the neutron star (see Section 3.3.3); much more work on the character of the emitted X-ray spectra remains to be done. Despite the current problems, a better theoretical understanding of neutron-star atmospheres and general relativistic corrections to the observed properties of X-ray bursts should eventually provide significant constraints on the masses and radii of neutron stars and a probe of the behavior of gravity away from the weak-field limit.

3.4   *Theory Versus Observation*

3.4.1   X-RAY BURST SOURCES   The results of the numerical calculations of neutron-star thermonuclear flashes (see Section 3.3.2 and Figures 14–16) strongly support the conjecture that bursts from most X-ray burst sources result from such flashes. In particular, the typical burst rise times, decay times scales, peak luminosities, total emitted energies, spectral properties, and recurrence intervals (see Lewin & Joss 1981, 1983) are reproduced remarkably well by such calculations. However, there remain a number of difficulties in the interpretation of X-ray burst sources as thermonuclear flashes. In the following paragraphs, we describe several of the most severe problems and discuss how these problems might ultimately be resolved.

1. The peak luminosities in some X-ray bursts appear to exceed the Eddington limit (see Equation 11) by factors of ~ 3–10 for neutron stars of reasonable mass (Grindlay et al. 1980, Inoue et al. 1980, and references therein, Grindlay & Hertz 1981, Hōshi 1981). Moreover, the peak color temperatures of some bursts exceed the maximum value for neutron stars emitting blackbody radiation at or below the Eddington limit and obeying a physically reasonable mass-radius relation; this latter problem may, however, be resolved by the development of improved models for neutron-star atmospheres (see Section 3.3.4). In principle, the Eddington limit could be exceeded if the assumption of spherical symmetry were violated (as would be the case, for example, if only a portion of the neutron-star surface area participated in each flash). However, in the absence of beaming of the emitted radiation, a breakdown of spherical symmetry is unlikely to help this problem (Marshall 1982); in fact, the reduction in surface area and concomitant reduction in total burst energy would tend to lower the peak luminosities even further. It is unlikely that the suppression of radiative opacities by a magnetic field (Joss & Li 1980, and references therein) will solve this problem, since magnetic fields that are sufficiently strong ($\gtrsim 10^{12}$ G) to reduce the opacity should also funnel the accretion flow onto the magnetic polar caps of the neutron star, thereby suppressing thermonuclear flashes at moderate accretion rates (Joss & Li 1980; see Section 3.3.2).

Special relativistic effects that reduce the electron-scattering opacity at high temperatures ($\gtrsim 10^8$ K) are significant below the neutron-star surface (van Paradijs 1981, Hanawa & Sugimoto 1982, Paczyński 1983b), but these effects should have little influence upon the opacity at the photosphere itself. Super-Eddington luminosities might be achieved by the conversion of the kinetic energy fluence in a neutron-star wind into radiative luminosity when the wind interacts with the accretion flow (Melia & Joss 1984b); further exploration of this idea would be worthwhile.

2. In the context of the thermonuclear flash model, the ratio α of time-averaged persistent X-ray luminosity to time-averaged burst luminosity from a given source should be equal to the ratio of the gravitational energy per unit mass released by accretion onto the neutron star to the nuclear energy per unit mass liberated in flashes:

$$\alpha \simeq \frac{GM}{\eta Rc^2} \simeq 150 \left(\frac{M}{M_\odot}\right)\left(\frac{R}{10 \text{ km}}\right)^{-1}\left(\frac{\eta}{10^{-3}}\right)^{-1}, \qquad 15.$$

where η is the fractional mass of the surface layers converted into energy by nuclear reactions during a flash. General relativistic corrections to the

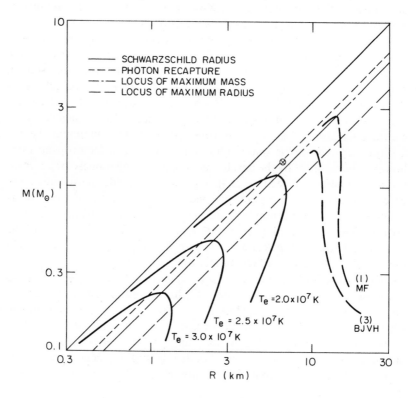

gravitational energy release (see Ayasli & Joss 1982) have been neglected here. For reasonable neutron-star masses and radii and any mixture of hydrogen and helium as the nuclear fuel, one expects that $20 \lesssim \alpha \lesssim 500$.

Most observed burst sources do indeed have observed values of α that fall within this range. Higher values of α are occasionally seen during interludes when bursts from a given source become infrequent or temporarily stop altogether; such interludes might result from episodes when the nuclear burning shells become relatively stable and the bursting phenomenon is intermittently suppressed, and/or from one or more inefficiencies in the nuclear flash process, such as loss of some flash energy to the core of the neutron star and partial burning of the relevant nuclear fuel between flashes (see, for example, Taam 1981a, Ayasli & Joss 1982). More puzzling are several occasions when bursts from a given source were separated by intervals of only ~ 4 to ~ 10 min (Lewin et al. 1976, Murakami et al. 1980; additional references in Lewin & Joss 1981, 1983). In the clearest of these instances, where two bursts from the source XB 1745−24 in the globular cluster Terzan 5 were separated by an interval of ~ 8 min (Oda & Tanaka 1981, Oda 1982), the associated persistent X-ray luminosity was not

---

Figure 17 Constraints on neutron-star parameters derived from general relativistic consider-ations and the theoretical assumptions that (a) the peak luminosity in an X-ray burst is no greater than the Eddington limit and (b) the emitted spectrum is that of a blackbody (cf. van Paradijs 1979, Marshall 1982). The heavy solid curves denote the loci of allowed values of neutron-star mass M and radius R for several values of the peak effective temperature T_e during a burst (as measured by a distant observer), under the assumption that the peak luminosity is equal to the Eddington limit. If the peak luminosity is lower than the Eddington limit, each curve is the upper envelope of permissible values of M and R for the given value of T_e. The heavy dashed curves, labeled (1) and (3), denote, respectively, the mass-radius relations for neutron stars based on the mean-field (MF) equation of state for nuclear matter (Pandharipande & Smith 1975), which is moderately stiff, and the BJVH equation of state (Malone et al. 1975), which is moderately soft. The symbol $\otimes$ denotes the neutron-star mass ($M \simeq 1.4\ M_\odot$) and radius ($R \simeq 6.6$ km) that has been used in many of the actual model calculations of thermonuclear flashes; it lies along the mass-radius relation for a very soft, but physically reasonable, equation of state for nuclear matter (based on the Reid potential for nucleon-nucleon interactions; see Hansen & Van Horn 1975). The solid straight line is the locus of Schwarzschild radii ($R = 2GM/c^2$), above which, according to the general theory of relativity, the star must be a black hole rather than a neutron star; the short-dashed line ($R = 3GM/c^2$) is the locus of parameter values above which the recapture by the neutron star of some of its own emitted photons must be taken into account (see text); the dot-dashed line $[R = (7/2)GM/c^2]$ is the locus of maximum mass along each T_e curve; and the long-dashed line ($R = 5GM/c^2$) is the locus of maximum radius along each T_e curve. Only the softest physically plausible equations of state for extremely dense matter give mass-radius relations consistent with the indicated constraints for $T_e \simeq 2 \times 10^7$ K, and no reasonable equations of state are consistent with higher values of T_e (see text). We thank C. Alcock, who collaborated in the preparation of this figure, for permission to publish it here.

especially high, and α was effectively much less than unity (Oda & Tanaka 1981). Thus, there was insufficient time for nuclear fuel to accumulate onto the neutron star between flashes, so that a storage mechanism that preserves fuel already on the neutron-star surface (Lamb & Lamb 1978) seems to be needed. Violations of spherical symmetry, wherein only a portion of the neutron-star surface participates in each flash, might provide the needed mechanism, but the nearly equal sizes of the adjacent bursts from XB 1745−24 argue against this possibility (Oda & Tanaka 1981, Oda 1982). Another possible mechanism is provided by the incomplete consumption of hydrogen in some calculated flashes (Ayasli & Joss 1982, Woosley 1983, T. Hanawa, D. Sugimoto, and M. Hashimoto, preprint); the residual hydrogen could, if it becomes mixed with heavier nuclei, release sufficient energy for a second burst in rapid succession (though it is unclear whether two nearly identical bursts could be thereby produced).

3. The available numerical calculations of neutron-star thermonuclear flashes do not yet reproduce many of the observed complexities in burst structure and recurrence patterns, which vary from one burst source to another and often vary with time in a given source (see Lewin & Joss 1981, 1983). Indeed, most such calculations have heretofore been carried through only a single flash. It is quite possible that many of these complexities will be better understood when further theoretical complications, such as violations of spherical symmetry (see Section 3.3.2), the thermal inertia of the neutron-star surface layers (Taam 1980, 1981a, Ayasli & Joss 1982), the residual radioactivity of the flashed matter (see, for example, Ayasli & Joss 1982), and the nuclear fuel left unburned by a flash (Ayasli & Joss 1982, Woosley 1983), are fully incorporated into future calculations.

3.4.2   THE RAPID BURSTER   It is important to realize that bursts from the "rapid burster," MXB 1730−335 (see Figure 18 and Lewin & Joss 1981, 1983), are presently unique and almost certainly cannot be the result of thermonuclear flashes. The recurrence intervals between bursts are $\sim 10^1$–10^3 s when the source is active, and on at least one occasion α was less than ~ 0.2 (White et al. 1978a). Hoffman et al. (1978b) have described these bursts as "type II" and those from other sources as "type I." However, Hoffman et al. also found that the rapid burster occasionally emits "special" bursts whose properties much more closely resemble the type I bursts from other sources. Moreover, the ratio of time-averaged luminosity in the type II bursts from the rapid burster to that in its type I bursts is $\sim 10^2$. Hoffman et al. made the intriguing speculation that type I bursts from the rapid burster are the result of thermonuclear flashes on an accreting neutron star, while the type II bursts are the result of an unstable accretion flow onto the same object. A brief review of possible accretion instabilities that might be applicable to the rapid burster is given in Lewin & Joss (1981, 1983).

3.4.3   THE FAST X-RAY TRANSIENTS   The morphology of some of the "fast X-ray transients," which have durations of $\sim 10^2$–10^3 s (see Figure 19), is suggestively similar to that of ordinary type I X-ray bursts (Hoffman et al. 1978a). Such events may be the result of thermonuclear flashes relatively deep within the surface layers of slowly accreting neutron stars (with accretion rates less than $\sim 10^{15}$ g s^{-1}) (Joss 1979b, Érgma & Kudryashov

SAS-3 OBSERVATIONS OF RAPIDLY REPETITIVE
X-RAY BURSTS FROM MXB 1730-335

24-minute snapshots from 8 orbits on March 2/3, 1976

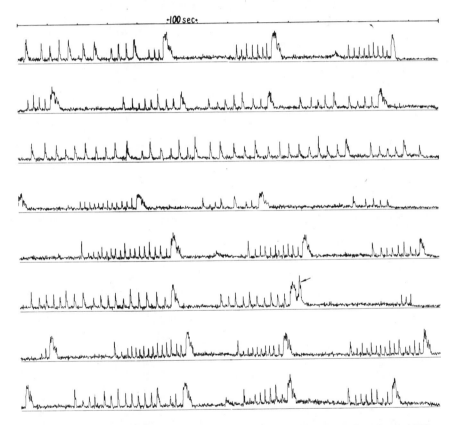

Figure 18   Type II X-ray bursts from the "rapid burster," MXB 1730 – 335 (from Lewin 1977). These are ~ 24-min stretches of data from eight different orbits of the *SAS-3* X-ray astronomy satellite on 1976 March 2–3, when this unique object was discovered (Lewin et al. 1976). Note the strong correlation between the integrated flux in a burst and the duration of the quiescent period that follows. The arrow indicates a type I burst from MXB 1728 – 34.

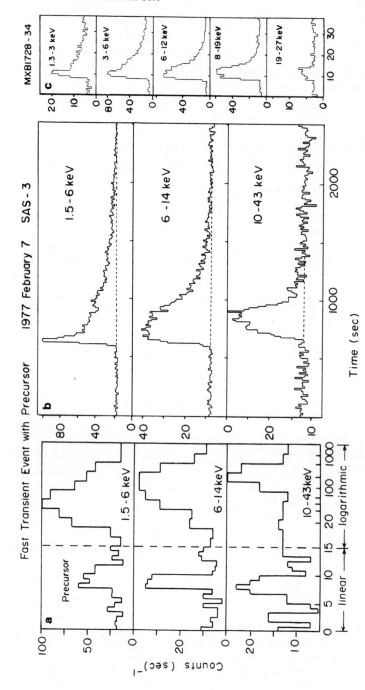

Figure 19 (*a* and *b*) Profiles of the 1977 February 7 fast X-ray transient, which lasted ~ 1500 s. The precursor is not visible in (*b*) because of the summing of the data into coarse time bins. (*c*) Composite profile of type I X-ray bursts from MXB 1728 − 34, shown for comparison. The resemblance between the event shown in (*b*) and the type I X-ray bursts is striking. This figure is from Hoffman et al. (1978a).

1980, Starrfield et al. 1982, Wallace et al. 1982; see Figure 15). If this picture is correct, one would expect outbursts from the fast transients to recur, but only on time scales much longer than those of typical X-ray bursts.

Some fast transients have been observed to have distinct precursors, consisting of relatively brief but intense X-ray emission just before the start of the main burst (Hoffman et al. 1978a; see Figure 19). Lewin et al. (1984) and Tawara et al. (1984) have recently suggested that the interval between the precursor and the main event corresponds to a sharp reduction in the color temperature of the emitted radiation, under conditions of roughly constant bolometric luminosity. The drop in color temperature presumably results from a large increase in the radius of the neutron-star photosphere during the emission of an intense, radiatively driven wind. If this picture is correct, the precursor phenomenon is an extreme example of photospheric variations due to mass loss from the neutron star (see Section 3.3.3); the large magnitude of the variations in photospheric radius are probably related in a way not yet understood to the large total energy of the transient outburst.

3.4.4 GAMMA-RAY BURSTS?   Several authors (Woosley & Taam 1976, Ruderman 1981, Woosley & Wallace 1982, Fryxell & Woosley 1982a, Hameury et al. 1982) have proposed that cosmic gamma-ray bursts may result from thermonuclear flashes on neutron stars. [See Lingenfelter et al. (1982) for reviews of the gamma-ray burst phenomenon and its interpretation.] The fundamental difficulty confronting thermonuclear flash models for gamma-ray bursts is the attainment and maintenance of gamma-ray temperatures at the neutron-star photosphere, which strongly tends to radiate in X rays rather than gamma rays (Joss 1977). The principal mechanism that has been advanced to overcome this problem is the confinement of plasma near the neutron-star surface by strong magnetic fields (Ruderman 1981, Woosley & Wallace 1982). It is presently unclear whether, under detailed scrutiny, this or any other mechanism will ultimately prove to be a viable way to generate gamma-ray bursts from neutron-star thermonuclear flashes.

3.4.5 RELATION TO BINARY X-RAY PULSARS AND THE AGES OF BURST SOURCES   Since X-ray pulsars are also accreting neutron stars, it is puzzling, at first sight, that these objects do not also display bursting behavior. However, at moderate accretion rates, funneling of the accretion flow onto the magnetic polar caps of the neutron star should stabilize the nuclear burning shells of an X-ray pulsar against thermonuclear flashes (see Section 3.3.2). Conversely, those neutron stars whose magnetic fields are too weak to funnel the accreting matter should be precisely those that can undergo thermonuclear flashes. If the magnetic field in such a neutron star was originally as strong as in an X-ray pulsar but has since decayed, then the neutron star must be fairly old (probably older than 10^7 yr; see Ruderman 1972, and references therein, Flowers & Ruderman 1977). The general lack of X-ray eclipses in burst sources, as well as the optical properties of these sources, strongly suggests membership in low-mass binary systems (see Section 3.2) that could be very old. The concentration of X-ray burst sources in the direction of the galactic center and the identification of several of them with globular clusters (see Figure 1) are probably additional manifestations of membership in an older galactic population than the X-ray pulsars.

4.   SUMMARY

During the past decade, the study of binary X-ray pulsars has provided a great stimulus to many areas of astrophysics. Of particular importance are the measurements of the masses of neutron stars (Section 2.3.3) and their relation to stellar evolution, to nuclear and high-energy physics, and to gravitation theory. Detailed information has been obtained on the companion stars (Section 2.3.2), and this should in turn assist the further development of a coherent picture of binary stellar evolution in these systems. Binary X-ray pulsars have also provided a significant new probe into the mass transfer processes in binary stellar systems and an impetus to theoretical investigations of accretion flows, accretion disks, and accretion torques (see Section 2.2). Preliminary efforts have also been made to utilize the X-ray pulsations as probes of the internal structure of both the emitting neutron stars (Section 2.2) and their companion stars.

Our present understanding of X-ray burst sources is somewhat more primitive than that of binary X-ray pulsars. The results of both dimensional analyses (Section 3.3.1) and detailed numerical calculations (Section 3.3.2) provide compelling evidence that the bursts from most cosmic X-ray burst sources result from neutron-star thermonuclear flashes. Moreover, the observed properties of X-ray bursts are capable, at least in principle, of placing important constraints upon the fundamental properties of the underlying neutron stars, especially when general-relativistic effects (Section 3.3.4) are taken into account. In fact, owing to the strong influence of gravitational effects upon the observed properties of bursts, X-ray burst sources may provide us with an astrophysical environment where we can quantitatively study gravitation away from the weak-field limit. However, as documented in Sections 3.3.3, 3.3.4, and 3.4, there remain substantial problems in the interpretation of various observational features of X-ray bursts in terms of neutron-star thermonuclear flashes. Hence, the full utility of the X-ray burst phenomenon as a tool for understanding the physics and astrophysics of neutron stars must await further theoretical developments.

ACKNOWLEDGMENTS

The authors acknowledge numerous helpful discussions with C. Alcock, R. Kelley, and W. Lewin. We thank R. Kelley for carrying out a number of the calculations utilized in this work, and S. Black and S. Makseyn for their able assistance in the preparation of the manuscript. This work was supported in part by the National Science Foundation under grant AST-8217451 and by the National Aeronautics and Space Administration under grants NSG-7643 and NGL-22-009-638 and contract NAS5-24441.

Literature Cited

Arnett, W. D., Bowers, R. L. 1977. *Ap. J. Suppl.* 33:415
Arnett, W. D., Schramm, D. N. 1973. *Ap. J. Lett.* 184:L47
Arons, J., Lea, S. M. 1980. *Ap. J.* 235:1016
Avni, Y. 1976. *Ap. J.* 209:574
Avni, Y. 1977. In *Highlights of Astronomy*, ed. E. A. Müller, 4(1):137. Dordrecht: Reidel
Avni, Y. 1978. In *Physics and Astrophysics of Neutron Stars and Black Holes*, ed. R. Giacconi, R. Ruffini, p. 43. Amsterdam: North-Holland
Avni, Y., Bahcall, J. N. 1975a. *Ap. J.* 197:675
Avni, Y., Bahcall, J. N. 1975b. *Ap. J. Lett.* 202:L131
Avni, Y., Bahcall, J. N. 1976. In *X-Ray Binaries, NASA Spec. Publ.* 389, ed. E. Boldt, Y. Kondo, p. 615. Washington DC: NASA
Avni, Y., Schiller, N. 1982. *Ap. J.* 257:703
Ayasli, S., Joss, P. C. 1982. *Ap. J.* 256:637
Baade, W., Zwicky, F. 1934. *Proc. Natl. Acad. Sci. USA* 20:254
Baan, W. A., Treves, A. 1973. *Astron. Astrophys.* 22:421
Bahcall, J. 1978a. In *Physics and Astrophysics of Neutron Stars and Black Holes*, ed. R. Giacconi, R. Ruffini, p. 63. Amsterdam: North-Holland
Bahcall, J. 1978b. *Ann. Rev. Astron. Astrophys.* 16:241
Bahcall, J. N., Chester, T. J. 1977. *Ap. J. Lett.* 215:L21
Barranco, M., Buchler, J. R., Livio, M. 1980. *Ap. J.* 242:1226
Basko, M. M., Hatchett, S., McCray, R., Sunyaev, R. A. 1977. *Ap. J.* 215:276
Basko, M. M., Sunyaev, R. A. 1975. *Astron. Astrophys.* 42:311
Basko, M. M., Sunyaev, R. A. 1976. *MNRAS* 175:395
Baym, G., Pethick, C. 1979. *Ann. Rev. Astron. Astrophys.* 17:415
Baym, G., Pethick, C., Pines, D., Ruderman, M. 1969. *Nature* 224:872
Becker, R. H., Boldt, E. A., Holt, S. S., Pravdo, S. H., Rothschild, R. E., et al. 1976. *Ap. J. Lett.* 207:L167 (BE76)

Becker, R. H., Rothschild, R. E., Boldt, E. A., Holt, S. S., Pravdo, S. H., et al. 1978. *Ap. J.* 221:912
Becker, R. H., Swank, J. H., Boldt, E. A., Holt, S. S., Pravdo, S. H., et al. 1977. *Ap. J. Lett.* 216:L11 (BE77)
Belian, R. D., Conner, J. P., Evans, W. D. 1976. *Ap. J. Lett.* 206:L135
Bisnovatyi-Kogan, G. S. 1973. *Astron. Zh.* 50:902
Blandford, R. D., De Campli, W. M. 1981. In Sieber & Wielebinski 1981, p. 371
Bondi, H., Hoyle, F. 1944. *MNRAS* 104:273
Boynton, P. E. 1981. In Sieber & Wielebinski 1981, p. 279
Boynton, P. E., Deeter, J. 1979. In *Compact Galactic X-Ray Sources*, ed. F. Lamb, D. Pines, p. 168. Urbana: Phys. Dept., Univ. Ill.
Bradt, H. V. D., McClintock, J. E. 1983. *Ann. Rev. Astron. Astrophys.* 21:13
Brucato, R. J., Kristian, J. 1972. *Ap. J. Lett.* 173:L105 (BR72)
Canal, R., Schatzman, E. 1976. *Astron. Astrophys.* 46:229
Canizares, C. R., McClintock, J. E., Grindlay, J. E. 1979. *Ap. J.* 234:556
Canuto, V. 1970. *Ap. J.* 159:641
Charles, P. A., Booth, L., Densham, R. H., Thorstensen, J. R., Willis, A. J. 1981. *Space Sci. Rev.* 30:423 (CH81)
Chester, T. J. 1978. *Ap. J.* 222:652
Chevalier, C., Ilovaisky, S. A. 1977. *Astron. Astrophys.* 59:L9 (CH77)
Comella, J. M., Craft, H. D. Jr., Lovelace, R. V. E., Sulton, J. M., Tyler, G. L. 1969. *Nature* 221:453
Cominsky, L. R. 1981. PhD thesis. Mass. Inst. Technol., Cambridge
Cominsky, L., Clark, G. W., Li, F., Mayer, W., Rappaport, S. 1978. *Nature* 273:367 (CO78)
Conti, P. S. 1978. *Astron. Astrophys.* 63:225
Crampton, D., Hutchings, J. B., Cowley, A. P. 1978. *Ap. J. Lett.* 225:L63 (CR78)
Czerny, J., Jaroszyński, M. 1979. *Acta Astron.* 30:157
Darbro, W., Ghosh, P., Elsner, R. F.,

Weisskopf, M. C., Sutherland, P. G., Grindlay, J. E. 1981. *Ap. J.* 246:231

Davidson, K. 1973. *Nature Phys. Sci.* 246:1

Davidson, K., Ostriker, J. P. 1973. *Ap. J.* 179:585

Davies, R. E., Fabian, A. C., Pringle, J. E. 1979. *MNRAS* 186:779

Davison, P. J. N. 1977. *MNRAS* 179:35p (DA77a)

Davison, P. J. N., Watson, M. G., Pye, J. P. 1977. *MNRAS* 181:73p (DA77b)

Deeter, J. E., Boynton, P. E., Pravdo, S. H. 1981. *Ap. J.* 247:1003 (DE81)

Doty, J. P., Hoffman, J. A., Lewin, W. H. G. 1981. *Ap. J.* 243:257 (DO81)

Ebisuzaki, T., Hanawa, T., Sugimoto, D. 1983. *Publ. Astron. Soc. Jpn.* 35:17

Elsner, R. F., Ghosh, P., Lamb, F. K. 1980. *Ap. J. Lett.* 241:L155

Elsner, R. F., Lamb, F. K. 1976. *Nature* 262:356

Érgma, É. V., Kudryashov, A. D. 1980. *Astrophys. Lett.* 21:13

Érgma, É. V., Tutukov, A. V. 1980. *Astron. Astrophys.* 84:123

Fabbiano, G., Schreier, E. J. 1977. *Ap. J.* 214:235 (FA77)

Fahlman, G. G., Gregory, P. C. 1981. *Nature* 293:202 (FA81)

Fahlman, G. G., Gregory, P. C. 1982. *IAU Circ. No.* 3701 (FA82)

Faulkner, J. 1971. *Ap. J. Lett.* 170:L99

Faulkner, J. 1974. In *Late Stages of Stellar Evolution, IAU Symp. No.* 66, ed. R. J. Tayler, p. 155. Dordrecht: Reidel

Fechner, W. B., Joss, P. C. 1977. *Ap. J. Lett.* 213:L57 (FE77)

Fechner, W. B., Joss, P. C. 1978. *Nature* 274:347

Flowers, E., Ruderman, M. 1977. *Ap. J.* 215:302

Forman, W., Jones, C., Tananbaum, H., Gursky, H., Kellogg, E., Giacconi, R. 1973. *Ap. J. Lett.* 182:L103 (FO73)

Fryxell, B. A., Woosley, S. E. 1982a. *Ap. J.* 258:733

Fryxell, B. A., Woosley, S. E. 1982b. *Ap. J.* 261:532

Fujimoto, M. Y., Hanawa, T., Iben, I. Jr., Richardson, M. B. 1984. *Ap. J.* 278:813

Fujimoto, M., Hanawa, T., Miyaji, S. 1981. *Ap. J.* 247:267

Ghosh, P., Lamb, F. K. 1978. *Ap. J. Lett.* 223:L83

Ghosh, P., Lamb, F. K. 1979. *Ap. J.* 234:296

Ghosh, P., Lamb, F. K., Pethick, C. J. 1977. *Ap. J.* 217:578

Giacconi, R. 1974. In *Astrophysics and Gravitation, Proc. Int. Solvay Conf. Phys., 16th*, p. 27. Brussels: Univ. Bruxelles

Giacconi, R., Gursky, H., Kellogg, E., Schreier, E., Tananbaum, H. 1971. *Ap. J. Lett.* 167:L67 (GI71)

Gnedin, Yu. N., Sunyaev, R. A. 1973. *Astron. Astrophys.* 25:233

Gold, T. 1969a. *Nature* 221:65

Gold, T. 1969b. In *Pulsating Stars 2, A Nature Reprint*, p. X. London: Macmillan

Goldman, Y. 1979. *Astron. Astrophys.* 78:L15

Gregory, P. C., Fahlman, G. G. 1980. *Nature* 287:805 (GR80)

Grindlay, J. 1981. *IAU Circ. No. 3620* (GR81)

Grindlay, J., Gursky, H., Schnopper, H., Parsignault, D. R., Heise, J., et al. 1976. *Ap. J. Lett.* 205:L127

Grindlay, J. E., Hertz, P. 1981. *Ap. J. Lett.* 247:L17

Grindlay, J. E., Marshall, H., Hertz, P., Soltan, A., Weisskopf, R. F., et al. 1980. *Ap. J. Lett.* 240:L121

Grindlay, J. E., McClintock, J. E., Canizares, C. R., van Paradijs, J., Cominsky, L., et al. 1978. *Nature* 274:567

Gursky, H. 1976. In *Structure and Evolution of Close Binary Systems, IAU Symp. No. 73*, ed. P. Eggleton, S. Mitton, J. Whelan, p. 19. Dordrecht: Reidel

Gursky, H., Schreier, E. 1975. In *Neutron Stars, Black Holes and Binary X-Ray Sources*, ed. H. Gursky, R. Ruffini, p. 175. Dordrecht: Reidel

Hackwell, J. A., Grasdalen, G. L., Gehrz, R. D., van Paradijs, J., Cominsky, L., Lewin, W. H. G. 1979. *Ap. J. Lett.* 233:L115

Hameury, J. M., Bonazzola, S., Heyvaerts, J., Ventura, J. 1982. *Astron. Astrophys.* 111:242

Hanawa, T., Fujimoto, M. Y. 1982. *Publ. Astron. Soc. Jpn.* 34:495

Hanawa, T., Sugimoto, D. 1982. *Publ. Astron. Soc. Jpn.* 34:1

Hansen, C. J., Van Horn, H. M. 1975. *Ap. J.* 195:735

Henrichs, H. 1983. In *Accretion-Driven Stellar X-Ray Sources*, ed. W. H. G. Lewin, E. P. J. van den Heuvel, p. 393. Cambridge: Cambridge Univ. Press

Hewish, A., Bell, S. J., Pilkington, J. D. H., Scott, P. F., Collins, R. A. 1968a. *Nature* 217:709

Hewish, A., Bell, S. J., Pilkington, J. D. H., Scott, P. F., Collins, R. A. 1968b. *Nature* 218:126

Hoffman, J. A., Cominsky, L., Lewin, W. H. G. 1980. *Ap. J. Lett.* 240:L27

Hoffman, J. A., Lewin, W. H. G., Doty, J., Jernigan, J. G., Haney, M., Richardson, J. A. 1978a. *Ap. J. Lett.* 221:L57

Hoffman, J. A., Marshall, H. L., Lewin, W. H. G. 1978b. *Nature* 271:630

Holt, S. S., McCray, R. 1982. *Ann. Rev. Astron. Astrophys.* 20:323

Hōshi, R. 1980. *Prog. Theor. Phys.* 64:820

Hōshi, R. 1981. *Ap. J.* 247:628

Huckle, H. E., Mason, K. O., White, N. E., Sanford, P. W., Maraschi, L., et al. 1977. *MNRAS* 180:21 (HU77b)

Hutchings, J. B. 1977. *MNRAS* 181:619 (HU77c)

Hutchings, J. B., Bernard, J. E., Crampton, D., Cowley, A. P. 1978b. *Ap. J.* 223:530 (HU78b)

Hutchings, J. B., Cowley, A. P., Crampton, D., van Paradijs, J., White, N. E. 1979. *Ap. J.* 229:1079 (HU79)

Hutchings, J. B., Crampton, D. 1981. *Ap. J.* 247:222 (HU81a)

Hutchings, J. B., Crampton, D., Cowley, A. P. 1978a. *Ap. J.* 225:548 (HU78a)

Hutchings, J. B., Crampton, D., Cowley, A. P. 1981. *Astron. J.* 86:871 (HU81b)

Hutchings, J. B., Crampton, D., Cowley, A. P., Osmer, P. S. 1977. *Ap. J.* 217:186 (HU77a)

Iben, I. Jr. 1974. *Ann. Rev. Astron. Astrophys.* 12:215

Ilovaisky, S. A., Motch, Ch., Chevalier, C. 1978. *Astron. Astrophys.* 70:L19 (IL78)

Inoue, H. 1975. *Publ. Astron. Soc. Jpn.* 27:311

Inoue, H. 1983. Talk presented at Workshop High Energy Transients Relat. Phenom., Univ. Calif., Santa Cruz

Inoue, H., Koyama, K., Makishima, K., Matsuoka, M., Murakami, T., et al. 1980. *Nature* 283:358

Ives, J. C., Sanford, P. W., Bell Burnell, S. J. 1975. *Nature* 254:578 (IV75)

Janot-Pacheco, E., Ilovaisky, S. A., Chevalier, C. 1981. *Astron. Astrophys.* 99:274 (JA81)

Johns, M., Koski, A., Canizares, C., McClintock, J. 1978. *IAU Circ. No. 3171* (JO78a)

Johnston, M. D., Bradt, H. V., Doxsey, R. E., Griffiths, R. E., Schwartz, D. A., Schwarz, J. 1979. *Ap. J. Lett.* 230:L11 (JO79)

Johnston, M., Bradt, H., Doxsey, R., Gursky, H., Schwartz, D., Schwarz, J. 1978. *Ap. J. Lett.* 223:L71

Joss, P. C. 1977. *Nature* 270:310

Joss, P. C. 1978. *Ap. J. Lett.* 225:L123

Joss, P. C. 1979a. *Comments Astrophys.* 8:109

Joss, P. C. 1979b. In *Compact Galactic X-Ray Sources*, ed. D. Pines, F. Lamb, p. 89. Urbana: Phys. Dept., Univ. Ill.

Joss, P. C., Avni, Y., Rappaport, S. 1978b. *Ap. J.* 221:645 (JO78b)

Joss, P. C., Fechner, W. B., Forman, W., Jones, C. 1978a. *Ap. J.* 225:994

Joss, P. C., Li, F. K. 1980. *Ap. J.* 238:287

Joss, P. C., Li, F., Nelson, J., Middleditch, J. 1980. *Ap. J.* 235:592 (JO80)

Joss, P. C., Rappaport, S. 1976. *Nature* 264:219

Joss, P. C., Rappaport, S. 1979. *Astron. Astrophys.* 71:217

Kanno, S. 1980. *Publ. Astron. Soc. Jpn.* 32:105

Kato, M. 1983. *Publ. Astron. Soc. Jpn.* 35:33

Kelley, R., Apparao, K., Doxsey, R., Jernigan, G., Naranan, S., Rappaport, S. 1981. *Ap. J.* 243:251 (KE81)

Kelley, R. L., Rappaport, S., Ayasli, S. 1983c. *Ap. J.* 274:765 (KE83c)

Kelley, R. L., Jernigan, J. G., Levine, A., Petro, L. D., Rappaport, S. 1983a. *Ap. J. Lett.* 264:568 (KE83a)

Kelley, R. L., Rappaport, S., Clark, G. W., Petro, L. D. 1983b. *Ap. J.* 268:790

Kelley, R., Rappaport, S., Petre, R. 1980. *Ap. J.* 238:699 (KE80)

Kirk, J. R., Galloway, J. J. 1981. *MNRAS* 195:45p

Kirk, J. G., Trümper, J. 1983. In *Accretion-Driven Stellar X-Ray Sources*, ed. W. H. G. Lewin, E. P. J. van den Heuvel, p. 261. Cambridge: Cambridge Univ. Press

Koyama, K., Inoue, H., Makishima, K., Matsuoka, M., Murakami, T., et al. 1981. *Ap. J. Lett.* 247:L27

Kudryashov, A. D., Érgma, É. V. 1980. *Pis'ma Astron. Zh.* 6:712. Republished in *Sov. Astron. Lett.* 6:375

Lamb, D. Q., Lamb, F. K. 1978. *Ap. J.* 220:291

Lamb, F. K., Pethick, C. J., Pines, D. 1973. *Ap. J.* 184:271

Lamb, F. K., Pines, D., Shaham, J. 1978. *Ap. J.* 224:969

Lamb, R. C., Markert, T. H., Hartman, R. C., Thompson, D. J., Bignami, G. F. 1980. *Ap. J.* 239:651 (LA80)

Langer, S. H., Rappaport, S. 1982. *Ap. J.* 257:733

Lecar, M., Wheeler, J. C., McKee, C. F. 1976. *Ap. J.* 205:556

Levine, A. 1982. Private communication (LE82)

Lewin, W. H. G. 1977. *Ann. NY Acad. Sci.* 302:210

Lewin, W. H. G., Doty, J., Clark, G. W., Rappaport, S. A., Bradt, H. V. D., et al. 1976. *Ap. J. Lett.* 207:L95

Lewin, W. H. G., Joss, P. C. 1977. *Nature* 270:211

Lewin, W. H. G., Joss, P. C. 1981. *Space Sci. Rev.* 28:3

Lewin, W. H. G., Joss, P. C. 1983. In *Accretion-Driven Stellar X-Ray Sources*, ed. W. H. G. Lewin, E. P. J. van den Heuvel, p. 41. Cambridge: Cambridge Univ. Press

Lewin, W. H. G., Ricker, G. R., McClintock, J. E. 1971. *Ap. J. Lett.* 169:L17 (LE71)

Lewin, W. H. G., Vacca, W. D., Basinska, E. M. 1984. *Ap. J. Lett.* 277:L57

Li, F. K., Joss, P. C., McClintock, J. E., Rappaport, S., Wright, E. L. 1980. *Ap. J.* 240:628 (LI80)

Li, F., Rappaport, S., Clark, G. W., Jernigan, J. G. 1979. *Ap. J.* 228:893 (LI79)

Li, F., Rappaport, S., Epstein, A. 1978. *Nature* 271:37 (LI78)

Lingenfelter, R. E., Hudson, H. S., Worrall, D. M., eds. 1982. *AIP Conf. Proc.* 77, *Gamma-Ray Transients and Related Astrophysical Phenomena (La Jolla Institute, 1981)*. New York: Am. Inst. Phys. 500 pp.

Lodenquai, J., Canuto, V., Ruderman, M., Tsuruta, S. 1974. *Ap. J.* 190:141

Lucke, R., Yentis, D., Friedman, H., Fritz, G., Shulman, S. 1976. *Ap. J. Lett.* 206:L25 (LU76)

Malone, R. C., Johnson, M. B., Bethe, H. A. 1975. *Ap. J.* 199:741

Manchester, R. N., Taylor, J. H. 1977. *Pulsars.* San Francisco: Freeman. 281 pp.

Maraschi, L., Cavaliere, A. 1977. In *Highlights in Astronomy*, ed. E. A. Müller, 4(1):127. Dordrecht: Reidel

Marshall, H. L. 1982. *Ap. J.* 260:815

Mason, K. O. 1977. *MNRAS* 178:81p

Matsuoka, M., Inoue, H., Koyama, K., Makishima, K., Murakami, T., et al. 1980. *Ap. J. Lett.* 240:L137

Mazets, E. P., Golenetskii, S. V., Il'inskii, Y. N., Aptekar, R. L., Guryan, Yu. A. 1979. *Nature* 282:587

McClintock, J. E., Canizares, C. R., Bradt, H. V., Doxsey, R. E., Jernigan, J. G., Hiltner, W. A. 1977a. *Nature* 270:320 (MC77a)

McClintock, J. E., Canizares, C. R., van Paradijs, J., Cominsky, L., Li, F. K., Lewin, W. H. G. 1979. *Nature* 279:47

McClintock, J. E., Rappaport, S., Joss, P. C., Bradt, H., Buff, J., et al. 1976. *Ap. J. Lett.* 206:L99 (MC76)

McClintock, J. E., Rappaport, S., Nugent, J. J., Li, F. K. 1977b. *Ap. J. Lett.* 216:L15 (MC77b)

Melia, F., Joss, P. C. 1984a. In *High Energy Transients*, ed. S. Woosley. New York: Am. Inst. Phys. In press

Melia, F., Joss, P. C. 1984b. *Ap. J. Lett.* Submitted for publication

Mészàros, P., Harding, A. K., Kirk, J. G., Galloway, D. J. 1983. *Ap. J. Lett.* 266:L33

Mészàros, P., Nagel, W., Ventura, J. 1980. *Ap. J.* 238:1066

Middleditch, J., Mason, K. O., Nelson, J., White, N. 1981. *Ap. J.* 244:1001 (MI81)

Middleditch, J., Nelson, J. 1976. *Ap. J.* 208:567 (MI76)

Middleditch, J., Pennypacker, C., Burns, S. 1982. *IAU Circ. No. 3701* (MI82)

Milgrom, M. 1978. *Astron. Astrophys.* 67:L25

Mook, D. E., Boley, F. I., Foltz, C. B., Westpfahl, D. 1974. *Publ. Astron. Soc. Pac.* 86:894 (MO74)

Murakami, T., Inoue, H., Koyama, K., Makishima, K., Matsuoka, M., et al. 1980. *Ap. J. Lett.* 240:L143

Nagase, F. 1981. In *X-Ray Astronomy, Proc. ESLAB Symp., 15th*, ed. R. D. Andresen, p. 395. Dordrecht: Reidel

Nagase, F., Hayakawa, S., Kunieda, H., Makino, F., Masai, K., et al. 1981. *Nature* 290:572

Nagase, F., Hayakawa, S., Kunieda, H., Makino, F., Masai, K., et al. 1982. *Ap. J.* 263:814 (NA82)

Nagel, W. 1981a. *Ap. J.* 251:278

Nagel, W. 1981b. *Ap. J.* 251:288

Nomoto, K. 1981. In *Fundamental Problems in the Theory of Stellar Evolution, IAU Symp. No. 93*, ed. D. Sugimoto, D. Q. Lamb, D. N. Schramm, p. 295. Dordrecht: Reidel

Nomoto, K., Tsuruta, S. 1982. In *Accreting Neutron Stars*, ed. W. Brinkmann, J. Trümper, p. 275. Munich: Max-Planck-Inst. Phys. Astrophys.

Oda, M. 1982. In Lingenfelter et al. 1982, p. 319

Oda, M., Tanaka, Y. 1981. *Res. Note No. 150*, Inst. Space Astronaut. Sci., Tokyo

Ögelman, H., Beuermann, K. P., Kanbach, G., Mayer-Hasselwander, H. A., Capozzi, D., et al. 1977. *Astron. Astrophys.* 58:385

Oppenheimer, J. R., Volkoff, G. 1939. *Phys. Rev.* 55:374

Pacini, F. 1967. *Nature* 216:567

Paczyński, B. 1967. *Acta Astron.* 17:287

Paczyński, B. 1983a. *Ap. J.* 264:282

Paczyński, B. 1983b. *Ap. J.* 267:315

Pakull, M., Parmar, A. 1981. *Astron. Astrophys.* 102:L1 (PA81)

Pandharipande, Y. R., Smith, R. A. 1975. *Phys. Lett.* 593:15

Parkes, G. E., Mason, K. O., Murdin, P. G., Culhane, J. L. 1980a. *MNRAS* 191:547 (PA80a)

Parkes, G. E., Murdin, P., Mason, K. O. 1978. *IAU Circ. No. 3184* (PA78)

Parkes, G. E., Murdin, P. G., Mason, K. O. 1980b. *MNRAS* 190:537 (PA80b)

Parmar, A. N., Branduardi-Raymont, G., Pollard, G. S. G., Sanford, P. W., Fabian, A. C., et al. 1980. *MNRAS* 193:49p (PA80c)

Pavlov, G. G., Shibanov, Yu. A. 1979. *Sov. Phys. JETP* 49:741

Pavlov, G. G., Yakovlev, Yu. A. 1976. *Sov. Phys. JETP* 43:389

Pedersen, H., Lub, J., Inoue, H., Koyama, K., Makishima, K., et al. 1982a. *Ap. J.* 263:325

Pedersen, H., Oda, M., Cominsky, L., Doty, J., Jernigan, G., et al. 1979. *IAU Circ. No. 3399*

Pedersen, H., van Paradijs, J., Motch, C., Cominsky, L., Lawrence, A., et al. 1982b. *Ap. J.* 263:340

Petterson, J. A. 1978. *Ap. J.* 224:625

Plavec, M. 1968. *Adv. Astron. Astrophys.* 6:201

Polidan, R. S., Pollard, G. S. G., Sanford, P. W., Locke, M. C. 1978. *Nature* 275:296 (PO78)

Pounds, K. A. 1976. Talk presented at Meet. High Energy Astrophys. Div., Am. Astron. Soc., Cambridge, Mass. (PO76)

Pounds, K. A., Cooke, B. A., Ricketts, M. J., Turner, M. J., Elvis, M. 1975. *MNRAS* 172:473 (PO75)

Pravdo, S. H., Bussard, R. W. 1981. *Ap. J. Lett.* 246:L115

Pravdo, S. H., White, N. E., Boldt, E. A., Holt, S. S., Serlemitsos, P. J., et al. 1979. *Ap. J.* 231:912 (PR79)

Priedhorsky, W. C., Terrell, J. 1983. Preprint (PR83)

Primini, F., Rappaport, S., Joss, P. C. 1977. *Ap. J.* 217:543 (PR77)

Primini, F., Rappaport, S., Joss, P. C., Clark, G. W., Lewin, W., et al. 1976. *Ap. J. Lett.* 210:L71 (PR76)

Pringle, J. E., Rees, M. J. 1972. *Astron. Astrophys.* 21:1

Rappaport, S. 1982. In *Galactic X-Ray Sources*, ed. P. W. Sanford, P. Laskarides, J. Salton, p. 159. Chichester: Wiley

Rappaport, S., Clark, G. W., Cominsky, L., Joss, P. C., Li, F. K. 1978. *Ap. J. Lett.* 224:L1 (RA78)

Rappaport, S., Joss, P. C. 1977a. *Nature* 266:123

Rappaport, S., Joss, P. C. 1977b. *Nature* 266:683

Rappaport, S., Joss, P. C. 1977c. *Ann. NY Acad. Sci.* 302:460

Rappaport, S., Joss, P. C. 1981. In *X-Ray Astronomy with the Einstein Satellite*, ed. R. Giacconi, p. 123. Dordrecht: Reidel

Rappaport, S., Joss, P. C. 1983. In *Accretion-Driven Stellar X-Ray Sources*, eds. W. H. G. Lewin, E. P. J. van den Heuvel, p. 1. Cambridge: Cambridge Univ. Press

Rappaport, S., Joss, P. C., McClintock, J. E. 1976. *Ap. J. Lett.* 206:L103 (RA76)

Rappaport, S., Joss, P. C., Stothers, R. 1980. *Ap. J.* 235:570 (RA80)

Rappaport, S., Joss, P. C., Webbink, R. F. 1982. *Ap. J.* 254:616

Rappaport, S., Markert, T., Li, F. K., Clark, G. W., Jernigan, J. G., McClintock, J. E. 1977. *Ap. J. Lett.* 217:L29 (RA77)

Rappaport, S., Verbunt, F., Joss, P. C. 1983. *Ap. J.* 275:713

Rosenberg, F. D., Eyles, C. J., Skinner, G. K., Willmore, A. P. 1975. *Nature* 256:628 (RO75)

Ruderman, M. 1972. *Ann. Rev. Astron. Astrophys.* 10:427

Ruderman, M. 1981. In *Progress in Particle and Nuclear Physics*, ed. D. Wilkinson,

6:215. Oxford: Pergamon

Schreier, E. J., Fabbiano, G. 1976. In *X-Ray Binaries*, *NASA Spec. Publ. 389*, ed. E. Boldt, Y. Kondo, p. 197. Washington, DC: NASA (SC76)

Schreier, E. J., Giacconi, R., Gursky, H., Kellogg, E., Tananbaum, H. 1972b. *Ap. J. Lett.* 178:L71 (SC72b)

Schreier, E. J., Levinson, R., Gursky, H., Kellogg, E., Tananbaum, H., Giacconi, R. 1972a. *Ap. J. Lett.* 172:L79 (SC72a)

Schwarzschild, M., Härm, R. 1965. *Ap. J.* 142:855

Shapiro, S. L., Lightman, A. P. 1976. *Ap. J.* 204:555

Shapiro, S. L., Salpeter, E. E. 1975. *Ap. J.* 198:671

Shara, M. M. 1982. *Ap. J.* 261:649

Sieber, W., Wielebinski, R., eds. 1981. *Pulsars, IAU Symp. No. 95.* Dordrecht: Reidel. 475 pp.

Skinner, G. K., Bedford, D. K., Elsner, R. F., Leahy, D., Weisskopf, M. C., Grindlay, J. 1982. *Nature* 297:568 (SK82)

Skinner, G. K., Shulman, S., Share, G., Evans, W. D., McNutt, D., et al. 1980. *Ap. J.* 240:619 (SK80)

Staelin, D. H., Reifenstein, E. C. III. 1968. *Science* 162:1481

Starrfield, S., Kenyon, S., Sparks, M., Truran, J. W. 1982. *Ap. J.* 258:683

Stella, L., White, N. E. 1983. *IAU Circ. No. 3902*

Stier, M., Liller, W. 1976. *Ap. J.* 206:257 (ST76)

Swank, J. H., Becker, R. H., Boldt, E. A., Holt, S. S., Pravdo, S. H., Serlemitsos, P. J. 1977. *Ap. J. Lett.* 212:L73

Swank, J. H., Taam, R. E., White, N. E. 1984. *Ap. J.* 277:274

Taam, R. E. 1980. *Ap. J.* 241:358

Taam, R. E. 1981a. *Ap. J.* 247:257

Taam, R. E. 1981b. *Astrophys. Space Sci.* 77:257

Taam, R. E. 1982. *Ap. J.* 258:761

Taam, R. E. 1983. *Ap. J.* 268:361

Taam, R. E., Picklum, R. E. 1978. *Ap. J.* 224:210

Taam, R. E., Picklum, R. E. 1979. *Ap. J.* 233:327

Tanaka, Y. 1983. *IAU Circ. No. 3882*

Tananbaum, H., Gursky, H., Kellogg, E. M., Levinson, R., Schreier, E., Giacconi, R. 1972. *Ap. J. Lett.* 174:L143 (TA72)

Tawara, Y., Hayakawa, S., Hirano, T., Kii, T., Kunieda, H., Nagase, F. 1984. *Ap. J. Lett.* Submitted for publication

Taylor, J. H. 1981. In Sieber & Wielebinski 1981, p. 361

Taylor, J. H., Weisberg, J. M. 1982. *Ap. J.* 253:908

Thorstensen, J., Charles, P., Bowyer, S. 1978. *Ap. J. Lett.* 220:L131

Trümper, J., Pietsch, W., Reppin, C., Sacco, B., Kendziorra, E., Staubert, R. 1977. *Ann. NY Acad. Sci.* 302:538

Tsuruta, S. 1975. *Ann. NY Acad. Sci.* 262:391

Ulmer, M. P. 1976. *Ap. J.* 204:548

Ulmer, M. P., Baity, W. A., Wheaton, W. A., Peterson, L. E. 1972. *Ap. J. Lett.* 178:L121 (UL72)

van den Heuvel, E. P. J. 1975. *Ap. J. Lett.* 198:L109

van den Heuvel, E. P. J. 1977. *Ann. NY Acad. Sci.* 302:14

van den Heuvel, E. P. J., Rappaport, S. 1982. In *Proc. Asian Pac. Reg. Meet. IAU, 2nd*, ed. B. Hidayat. In press

van Paradijs, J. 1978. *Nature* 274:650

van Paradijs, J. 1979. *Ap. J.* 234:609

van Paradijs, J. 1981. *Astron. Astrophys.* 101:174

van Paradijs, J. 1982. *Astron. Astrophys.* 107:51

van Paradijs, J., Verbunt, F., van der Linden, T., Pedersen, H., Wamsteker, W. 1980. *Ap. J. Lett.* 241:L161

van Paradijs, J., Zuiderwijk, E. J., Takens, R. J., Hammerschlag-Hensberge, G., van den Heuvel, E. P. J., de Loore, C. 1977. *Astron. Astrophys. Suppl.* 30:195 (VA77)

Verbunt, F., Zwaan, C. 1981. *Astron. Astrophys.* 100:L7

Vidal, N. V. 1973. *Ap. J. Lett.* 186:L81 (VI73)

Wallace, R. K., Woosley, S. E. 1981. *Ap. J. Suppl.* 45:389

Wallace, R. K., Woosley, S. E., Weaver, T. A. 1982. *Ap. J.* 258:696

Walter, F. M., Bowyer, S., Mason, K. O., Clarke, J. T., Henry, J. P., et al. 1982. *Ap. J. Lett.* 253:L67

Wang, Y.-M. 1975. *Nature* 253:249

Wang, Y.-M. 1981. *Astron. Astrophys.* 102:36

Wang, Y.-M., Frank, J. 1981. *Astron. Astrophys.* 93:255

Watson, M. G., Griffiths, R. E. 1977. *MNRAS* 178:513 (WA77)

Watson, M. G., Warwick, R. S., Corbet, R. H. D. 1982. *MNRAS* 199:915 (WA82)

Watson, M. G., Warwick, R. S., Ricketts, M. J. 1981. *MNRAS* 195:197 (WA81)

Wheaton, W. A., Doty, J. P., Primini, F. A., Cooke, B. A., Dobson, C. A., et al. 1979. *Nature* 282:240

White, N. E. 1978. *Nature* 271:38 (WH78a)

White, N. E., Carpenter, G. F. 1978. *MNRAS* 183:11p (WH78d)

White, N. E., Davelaar, J., Parmar, A. N., Stella, J. 1984. *IAU Circ. No. 3912*

White, N. E., Marshall, F. E. 1981. *Ap. J. Lett.* 249:L25

White, N. E., Mason, K. O., Carpenter, G. F., Skinner, G. K. 1978a. *MNRAS* 184:1p

White, N. E., Mason, K. O., Huckle, H. E., Charles, P. A., Sanford, P. W. 1976a. *Ap. J. Lett.* 209:L119 (WH76a)

White, N. E., Mason, K. O., Sanford, P. W. 1977. *Nature* 267:229 (WH77)

White, N. E., Mason, K. O., Sanford, P. W. 1978b. *MNRAS* 184:67p (WH78b)

White, N. E., Mason, K. O., Sanford, P. W., Murdin, P. 1976b. *MNRAS* 176:201 (WH76b)

White, N. E., Parkes, G. E., Sanford, P. W., Mason, K. O., Murdin, P. G. 1978c. *Nature* 274:665 (WH78c)

White, N. E., Pravdo, S. H. 1979. *Ap. J. Lett.* 233:L121 (WH79)

White, N. E., Pravdo, S. H., Becker, R. H., Boldt, E. A., Holt, S. S., Serlemitsos, P. J. 1980. *Ap. J.* 239:655 (WH80)

White, N. E., Swank, J. H. 1982. *Ap. J. Lett.* 253:L61

White, N. E., Swank, J. H., Holt, S. S. 1983. *Ap. J.* 270:711

Woosley, S. E. 1983. Talk presented at Workshop High Energy Transients Relat. Phenom., Univ. Calif., Santa Cruz

Woosley, S. E., Taam, R. E. 1976. *Nature* 263:101

Woosley, S. E., Wallace, R. K. 1982. *Ap. J.* 258:716

Yahel, R. Z. 1980. *Astron. Astrophys.* 90:26

Zahn, J.-P. 1975. *Astron. Astrophys.* 41:329

Zahn, J.-P. 1977. *Astron. Astrophys.* 57:383

Ziolkowski, J. 1980. In *Close Binary Stars: Observations and Interpretation*, ed. M. J. Plavec, D. M. Popper, R. K. Ulrich, p. 335. Dordrecht: Reidel

Ann. Rev. Astron. Astrophys. 1984. 22 : 593–619

HELIOSEISMOLOGY :
Oscillations as a Diagnostic of the Solar Interior

Franz-Ludwig Deubner

Institut für Astronomie und Astrophysik, Universität Würzburg, Würzburg, Federal Republic of Germany

Douglas Gough

Institute of Astronomy and Department of Applied Mathematics and Theoretical Physics, University of Cambridge, Cambridge, England

1. INTRODUCTION

The detection of a rich spectrum of resonant oscillations of the Sun in the course of the past decade is making it possible to investigate important basic properties of the solar interior. The term "helioseismology" is now widely used to describe this new field of research.

About ten years after the first observation of the "five-minute oscillations" by Leighton and his coworkers (115, 115a) and by Evans & Michard (65), Ulrich (151) presented a sound theoretical description of the phenomenon. A similar, though less complete, theory was advanced by Leibacher & Stein (113). The oscillations are standing acoustic waves, most of which are trapped beneath the photosphere in the upper layers of the convection zone. Their amplitudes are low, so linearized theory is valid. Ulrich, and subsequently Ando & Osaki (1), predicted sequences of eigenfrequencies that depend on the horizontal wavelength of the oscillations. Five years later, the theory was confirmed by the observations of Deubner (44).

At about the same time, new observations with different kinds of instruments (13, 97, 141) revealed the existence of other oscillations at periods (20–60 min, 160 min) quite different from the familiar five-minute range. These findings marked the beginning of a new phase of interest in hydrodynamical phenomena deep in the interior of the Sun.

593

0066–4146/84/0915–0593$02.00

In this article we outline the most recent advances in the diagnostic aspects of the subject. A more detailed account of the early development of helioseismology may be found in a number of other reviews (45, 46, 73, 94, 148). We begin by discussing the conditions that must be satisfied for low-amplitude waves to be able to propagate, to make it apparent where in the Sun the waves are trapped, and to display the aspects of the solar structure that are predominantly responsible for determining the frequencies. Our approach is somewhat different from previous discussions, and thus we hope it will be useful even to readers who are already familiar with the subject. We then summarize briefly the observational techniques and review the major areas where progress is being made, discussing in turn the modes of high, low, and intermediate degree. Interpretations of limb observations are still uncertain, and we have refrained from presenting a detailed critique.

The current main product of helioseismology is a theoretical model of the spherically symmetrical component of the Sun. Oddly enough, that model is quite similar to the predictions of so-called standard evolution theory and suggests that the simplifying approximations in that theory do not appear to introduce severe errors in the basic hydrostatic structure of the star. It also reinforces the difficulties already encountered in explaining the low observed neutrino flux. However, there are systematic discrepancies between the observed solar oscillation frequencies and the eigenfrequencies of the best-fitting theoretical model that have not yet been eliminated. Therefore, one must treat with caution any claim that the model is certainly a close approximation to reality.

In principle, the fine structure of the oscillation spectrum should enable us to set bounds on deviations from spherical symmetry. These can be caused by rotation, large-scale magnetic fields, and convection. Already there are reports in the literature of several preliminary investigations, but few firm conclusions have yet been drawn.

2.   RESONANT CAVITIES IN THE SUN

A linear normal mode of oscillation can be regarded as a standing wave resulting from the interference between oppositely directed propagating waves. The interference pattern has the following mathematical form with respect to spherical polar coordinates (r, θ, ϕ):

$$\xi(r, \theta, \phi, t) = \text{Re}\{\Xi(r)P_l^m (\cos \theta)_{\sin}^{\cos} m\phi e^{i\omega t}\}, \qquad 2.1$$

where ξ is the vertical component of the fluid displacement $\boldsymbol{\xi}$, and P_l^m is the associated Legendre function. For any given *degree* l, there are many possible eigenfunctions Ξ. These are labeled with an integer n, called the *order* of the mode (see Appendix). Broadly speaking, n measures the vertical component of the wave number, and l measures the horizontal component.

To a first approximation the Sun can be considered to be spherically symmetrical. Consequently, all orientations of the coordinate axis are equivalent. The eigenfrequencies ω cannot, therefore, depend on the *azimuthal order m*, for m, which measures the number of nodes on the equator, depends on the choice of coordinates. There are two sequences of modes: acoustic or p modes, for which the principal restoring force is provided by pressure fluctuations; and gravity or g modes, for which the restoring force is buoyancy.

Conditions permitting the propagation of acoustic-gravity waves have been reviewed before (e.g. 41, 114, 148, 149, 155). If the wavelength is much less than the solar radius, the local effects of spherical geometry on the dynamics can be ignored. Then the equations of motion for adiabatic oscillations can be reduced in the manner chosen by Lamb (110), leading to

$$\Psi'' + K^2\Psi = 0, \qquad\qquad 2.2$$

where $\Psi = \rho^{1/2}c^2 \operatorname{div} \xi$, ρ and c are the density and sound speed of the equilibrium state, and a prime denotes differentiation with respect to r. Perturbations to the gravitational potential have been ignored. The vertical component of the local wave number K is given by

$$K^2 = \frac{\omega^2 - \omega_c^2}{c^2} + \frac{l(l+1)}{r^2}\left(\frac{N^2}{\omega^2} - 1\right), \qquad\qquad 2.3$$

where

$$\omega_c^2 = \frac{c^2}{4H^2}(1 - 2H'), \qquad\qquad 2.4$$

H being the density scale height, and

$$N^2 = g\left(\frac{1}{H} - \frac{g}{c^2}\right) \qquad\qquad 2.5$$

is the square of the buoyancy (Brunt-Väisälä) frequency; g is the magnitude of the acceleration due to gravity. Equation 2.2 is a good approximation except near $r = 0$, where curvature effects are important. However, as is evident from the discussion that follows, aside from low-order modes of low degree, only modes with $l \ll n$ penetrate sufficiently close to the center of the Sun for the errors introduced by the approximation to have a significant effect on the solutions.

For the solutions to (2.2) to be wavelike, it is necessary that $K^2 > 0$. It can be seen immediately that for spherically symmetrical modes ($l = 0$), this occurs only when ω exceeds the critical frequency ω_c, which is a generalization of Lamb's (109) acoustical cutoff frequency. For nonradial modes it is convenient to rewrite Equation 2.3 in the form

$$c^2 K^2 = \omega^2 \left(1 - \frac{\omega_+^2}{\omega^2}\right)\left(1 - \frac{\omega_-^2}{\omega^2}\right), \qquad 2.6$$

where

$$\omega_\pm^2 = \tfrac{1}{2}(S_l^2 + \omega_c^2) \pm [\tfrac{1}{4}(S_l^2 + \omega_c^2)^2 - N^2 S_l^2]^{1/2}. \qquad 2.7$$

Here,

$$S_l = [l(l+1)]^{1/2} \frac{c}{r}, \qquad 2.8$$

which is sometimes called the Lamb frequency. It follows from Equation 2.6 that $K^2 > 0$ either when $\omega_\pm^2$ are complex or when $\omega_\pm^2$ are real and (a) $\omega^2 > \omega_-^2$ and $\omega^2 > \omega_+^2$ or (b) $\omega^2 < \omega_-^2$ and $\omega^2 < \omega_+^2$.

The critical frequencies ω_+ and ω_- are plotted in Figure 1 for a standard solar model. The horizontal lines represent modes; they are continuous in the propagating regions and dashed in the evanescent regions where neither of the conditions (a) or (b) is satisfied. It is evident that there are several regions in the Sun within which modes can be trapped.

Eigenfrequencies can easily be estimated for high-order modes, for which K^{-1} is typically much smaller than the range (r_1, r_2) of a cavity within which $K^2 > 0$. Barring accidental resonances, which would permit the mode to have high amplitude in two cavities at once, the condition for a normal mode is essentially that an integral number (n) of half wavelengths fit between r_1 and r_2. Thus

$$\int_{r_1}^{r_2} K \, dr = \omega \int_{r_1}^{r_2} \left[\left(1 - \frac{\omega_+^2}{\omega^2}\right)\left(1 - \frac{\omega_-^2}{\omega^2}\right)\right]^{1/2} \frac{dr}{c} \simeq (n + \varepsilon)\pi, \qquad 2.9$$

where the phase ε accounts for the fact that rather than vanishing at $r = r_1$ and $r = r_2$, the wave amplitude matches onto decaying solutions outside the cavity.

In the deep interior of the Sun, S_l^2 is usually rather greater than ω_c^2 and N^2, and therefore $\omega_+ \approx S_l$ and $\omega_- \approx N$. This permits us to associate conditions (a) with acoustic modes: waves can propagate vertically provided the horizontal phase speed $\omega r/l$ does not exceed c. The vertical wave number K is influenced by buoyancy through the term $1 - N^2/\omega^2$, causing a reduction of the eigenfrequencies below what one would expect from a pure acoustic mode.

An asymptotic expression for ω when $n/l \gg 1$ has been derived by Vandakurov (156) and extended by Tassoul (150):

$$\omega_{n,l} \sim 2\pi(n + \tfrac{1}{2}l + \varepsilon_p)\nu_0 + \delta_{n,l}, \qquad 2.10$$

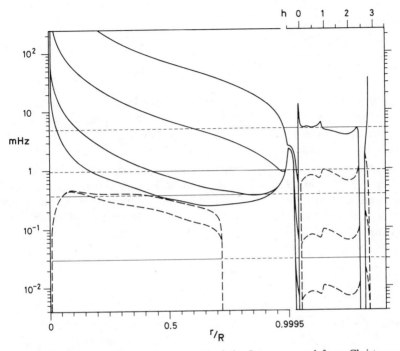

Figure 1 Propagation diagram for a model of the Sun, computed from Christensen-Dalsgaard's (23a) solar Model 1 beneath the photosphere and the temperature–optical depth relation of the Harvard-Smithsonian reference atmosphere (70b) above the photosphere. The corona is not included. Solid curves represent $\omega_+/2\pi$, and dashed curves $\omega_-/2\pi$, in the regions where the critical frequencies $\omega_\pm$ are real. Propagation at any frequency is possible where $\omega_\pm$ are complex. The lower abscissa scale extends to $r/R = 0.9995$; beyond that value, the scale is expanded by a factor of 100, and the scale is indicated on the upper boundary of the diagram: The independent variable is the height h above the photosphere measured in units of $10^{-3} R$. The curves $\omega_\pm$ are for $l = 1, 5, 50$, and 500. In all cases, $\omega_\pm$ are increasing functions of l at fixed r/R, which permits the identification of the curves: In the interior the ω_- curves for $l = 5$, $l = 50$, and $l = 500$ are essentially indistinguishable, as are all four ω_+ curves in the atmosphere, where $\omega_+ \simeq \omega_c$. The thin horizontal lines represent normal modes; they are continuous in zones of propagation and dashed in evanescent regions. The lowest-frequency mode is a high-degree ($l \gtrsim 25$) g mode. Its amplitude is likely to be substantial in either the interior or the atmosphere, but not both. The next mode represents $g_1(l = 1)$, which has the character of a p mode in its outer zone of propagation. The third mode is $p_4(l = 5)$, which is a simple p mode confined to a single region of propagation. The highest-frequency mode is $p_6(l = 500)$; because the evanescent regions are thin, the mode could have a substantial amplitude both in the photosphere and in the chromosphere.

where

$$\delta_{n,l} = -2\pi\nu_0 \frac{\alpha l(l+1)-\beta}{n+\frac{1}{2}l+\varepsilon_{\mathrm{p}}}, \qquad\qquad 2.11$$

$$\nu_0 = \left(2\int_0^R \frac{dr}{c}\right)^{-1}, \qquad\qquad 2.12$$

and R is the radius of the Sun. Here $\varepsilon_{\mathrm{p}} = \frac{1}{2}(\bar{\mu}+\frac{1}{2})$, where $\bar{\mu}$ is an effective polytropic index near $r = r_2$, and α and β are other constants that depend on the equilibrium state. In this limit, $r_1 \sim 0$ and $r_2 \sim R$, and the leading term of (2.10) follows immediately from Equation 2.9, except that l is replaced by $\sqrt{[l(l+1)]}$; an approximation to the remaining term is also given by Equation 2.9, and this is asymptotically correct when $l \gg 1$. When l is small, a more careful treatment that takes into account the curvature terms omitted from Equation 2.2 must be undertaken. Notice that the asymptotic frequencies are roughly uniformly spaced.

As $l/n \to \infty$, Equation 2.9 reduces to

$$\frac{n+\varepsilon}{\omega} \sim \frac{1}{\pi} \int_{r_1}^{r_2} \left(1 - \frac{k^2 c^2 + \omega_c^2}{\omega^2}\right)^{1/2} \frac{dr}{c}, \qquad\qquad 2.13$$

where $k = [l(l+1)]^{1/2}R^{-1} \simeq lR^{-1}$ is the horizontal component of the wave number at the solar surface. The modes are confined near the surface, penetrating to a depth of about $2\Lambda(n+\varepsilon)k^{-1}$, where Λ is a constant of order unity that depends on the manner in which c^2 varies with depth. Approximating the surface layers by a polytrope of index μ yields a constant gradient $G \equiv -dc^2/dr$, provided the adiabatic exponent γ can be assumed constant. Then $\Lambda = 1$, and Equation 2.13 reduces to

$$\omega^2 \sim \frac{2(n+\varepsilon)\gamma}{\mu+1} gk = 2(n+\varepsilon)Gk. \qquad\qquad 2.14$$

The plane-parallel polytrope can actually be treated exactly (e.g. 23, 73, 74) by extending the analyses of both Lamb (110), who considered acoustic waves in a shallow polytropic layer, and Spiegel & Unno (145), who considered convective modes in a deep layer. The result has the same form as (2.14) when $n \gg 1$, with $\varepsilon = \frac{1}{2}\mu$.

Conditions (b) describe g modes. The buoyancy frequency N is the frequency of a fictitious, adiabatically vertically oscillating fluid parcel that does not displace, and remains in pressure equilibrium with, its environment. In practice there is not time to reach perfect pressure balance, and there is an acoustic modification, which increases with increasing $\omega/(kc)$.

Moreover, the environment must be displaced sideways to make way for the parcel, which effectively increases the inertia of the fluid and reduces ω. As $l \to \infty$ at fixed n, the motion becomes nearly vertical and the latter effect vanishes; ω increases to the maximum value of N, but now the mode is severely confined within the small cavity in the vicinity of this maximum (23). Moreover, the decay of the amplitude away from the edges of the propagating regions is rapid, because the opposing influences of motions separated horizontally by half a wavelength λ largely cancel beyond a radial distance of about λ into the evanescent region. Hence, high-degree internal gravity modes have very low amplitude in the photosphere and are therefore difficult to detect (see also 25, 60). As $n \to \infty$ at fixed l, Equation 2.9 yields

$$\omega^{-1} \sim (n+\varepsilon)\pi[l(l+1)]^{1/2}\left(\int_{r_1}^{r_2} \frac{N\,dr}{r}\right)^{-1} \equiv \frac{(n+\varepsilon)P_0}{2\pi[l(l+1)]^{1/2}}, \qquad 2.15$$

indicating that the oscillation periods are asymptotically uniformly spaced. Further details and extensions of the high-order asymptotic relations are given in (111, 150, 156, 164) and the references therein. The results have the form (2.15), with $r_1 = 0$, r_2 equal to the radius of the base of the convection zone, and $\varepsilon = \frac{1}{2}l + \varepsilon_g$ provided $N^2 > 0$ in the core; the value of the constant ε_g depends on the manner in which the convection zone matches onto the radiative interior.

Equation 2.2 does not apply to f modes, whose character is clear when $l \gg 1$. Then they are essentially surface gravity waves, satisfying $\Psi = 0$. The displacement amplitude is proportional to $\exp[-k(R-r)]$, and

$$\omega^2 \simeq gk, \qquad 2.16$$

which is independent of the stratification of the Sun.

In the atmosphere, $\omega_+ \simeq \omega_c$ for the values of k that have been observed (see Figure 2). Modes whose frequencies are comparable with this value have vertical wavelengths that are comparable with the scale of variation of ω_+, and an accurate description of the oscillations requires a more careful treatment of the wave equation (2.2) than has been given here. Nevertheless, the condition $K^2 > 0$ provides a rough guide and indicates that a wave is trapped within the Sun only if its frequency is less than the value of ω_c characteristic of the atmosphere. This frequency corresponds to a period of about 3 min. Notice that the horizontal wavelengths of the modes are rather greater than the wavelengths in the vertical. Therefore, in a single oscillation period, information cannot be communicated over a distance great enough to recognize the horizontal variation of the wave, and the dynamics is locally independent of l. Consequently, the criterion for propagation derived by Christensen-Dalsgaard et al. (24) for radial modes (using the more natural acoustical radius as the independent variable) and the subsequent discussions by Christensen-Dalsgaard & Frandsen (26, 28) are relevant to all the observed nonradial p modes.

Notice also that, in principle, modes such as the highest-frequency mode indicated in Figure 1 can be trapped between the two maxima of ω_+ in the chromosphere. These have been called chromospheric modes. Since the evanescent barriers between the zones of propagation are quite thin, significant penetration can occur, and the modes can have substantial amplitudes simultaneously in more than one zone. Gravity waves can also be trapped in the atmosphere.

Except when both n and l are small, the wave-induced perturbation Φ' to the gravitational potential is small and hardly influences the frequencies. Cowling (40) provided some justification for ignoring Φ', even when l or n is not large, and to do so has subsequently become known as the Cowling approximation. It provides a good qualitative description of high-order monopole and dipole modes and of all modes with $l \geqslant 2$.

The resonant cavities are clearly defined only for modes of very high order or very high degree. Otherwise, the decline in amplitude through the evanescent regions is gentle, and modes can exist simultaneously in more than one cavity. As is evident in Figure 1, a mode can behave like a gravity wave in one part of a star and an acoustic wave in another. It is sometimes convenient to discuss oscillations in these terms (e.g. 155), but it is also useful to be able to assign a precise value to the order n without computing a large part of the spectrum of modes in order to apply the definition suggested in the Appendix. In the Cowling approximation, this is straightforward. Eckart (61) discusses how the phase differences between vertical displacement and pressure fluctuation decrease with height for p modes and increase for g modes, so that n can be computed by first assigning a sign to each zero in Ξ according to the direction of variation of the phase difference, and then counting the zeros algebraically. This criterion was first applied to stellar oscillations by Scuflaire (137) and Osaki (122). It has subsequently been found to work well for oscillations with $l \geqslant 2$ of most stars (including the Sun) when Φ' is correctly taken into account, though no wholly reliable simple criterion valid for low l has yet been found.

3. OBSERVATIONAL METHODS

The basic observations leading to the discovery of the resonant oscillations have all been carried out with conventional solar telescopes equipped with conventional spectrographs. The Doppler effect was used to determine the oscillatory velocities in selected spectral lines (64, 99, 115). Sometimes the central intensity of a line has been used as an indicator of temperature fluctuations associated with the oscillations (117). The sensitivity of the Doppler measurement with this type of instrumentation, using photoelectric detectors, is typically about 10 ms^{-1}/(5 arcsec)^2/s. Recognition of the importance of measuring a substantial fraction of the solar surface has led to the development of various scanning techniques, either in one dimension, with a pointlike scanning aperture (44) or with a long rectangular aperture (128), or in two dimensions, using array detectors (16, 63).

One- or two-dimensional spatial resolution is necessary to distinguish the high-degree oscillation modes. The signal of low-degree modes has been extracted from observations with little or no spatial resolution, without scanning. Some observers have compared the Doppler shift in light from an outer annulus of the solar image with that from either the entire image (105) or a circular portion of it (50); others have worked with integrated light from the entire solar disk (14, 92). The differential method eliminates the velocity of the detector relative to the Sun and also much of the noise introduced by the spectrograph. It is most sensitive to modes of degree somewhat higher than those selected by the whole-disk measurements (33).

To obviate the instability of large spectrographs arising from internal seeing and the impracticality of an absolute wavelength reference, other groups of investigators have developed devices with high spectral resolution using resonance scattering of the sunlight by a suitable alkali vapor (14, 92). The comparatively low quantum efficiency of these detectors is unimportant if the instrument is used to analyze light integrated from much or all of the solar disk. A sensitivity of a fraction of 1 ms^{-1} per individual measurement is then achieved. Such high sensitivity is needed to detect the low-amplitude signals from individual modes of oscillation. The performance of a magneto-optical filter (20), which combines the high wavelength stability of resonant scattering devices with imaging capability, is presently being studied (125).

The new development of a "Fourier tachometer" (asymmetric Michelson interferometer placed behind a broadband prefilter in combination with panoramic detector arrays) takes advantage of high quantum efficiency to obtain stable measurements of the Doppler radial velocity with arbitrary spatial resolution (63).

An active cavity radiometer has recently been flown on the *Solar Maximum Mission* (SMM) to record temporal fluctuations of the solar luminosity (159). With its high stability (better than 10^{-6}), this instrument is capable of recording the extremely low-amplitude brightness oscillations of the integrated solar disk (160).

Observations of the solar limb have provided yet another means of studying pulsations, either by measuring fluctuations of the apparent diameter of the solar disk (21) or by looking locally at changes of the structure of the limb-darkening function in the vicinity of its inflection point (12, 147, 163). H. Hill and his coworkers, who started this branch of helioseismology, have developed a sophisticated numerical procedure to overcome the tremendous difficulties imposed by the various seeing effects of the terrestrial atmosphere on the interpretation of these measurements (19, 98). Recently, another experiment for limb measurements has been put into operation, which relies both on the ability to "freeze" the solar image by rapid scanning and on an observing site with superior seeing (130).

4. HIGH-DEGREE MODES

In an extensive study of the five-minute oscillations using a spatio-temporal Fourier decomposition of Doppler and brightness data, Frazier (69) discovered that power was concentrated in two regions in the k-ω plane, one of which was situated where atmospheric waves are evanescent. This finding was already suggestive of normal modes trapped beneath the photosphere.

It took considerable further effort before the dispersion relation could be measured (44, 128). The observations were designed to isolate sectoral (and nearly sectoral) modes, which resemble plane waves propagating around a narrow equatorial zone. The characteristic parabolic dispersion relation (2.12) was evident for modes with $l \gtrsim 100$ (see Figure 2), and the close agreement of that relation with the eigenfrequencies of a solar model, already computed by Ulrich (151) at Frazier's instigation, convincingly established the nature of the oscillations: they are f modes and p modes of high degree. The oscillations are coherent in space and time over intervals at least as large as the limits attained by the observations (1000 Mm, 0.5 days). It has not been possible to resolve individual modes, so the power spectrum is a series of continuous ridges.

The detection and identification of these modes was soon followed by attempts to search for other high-degree modes. Chromospheric p modes with periods near 240 s and 180 s may have been detected by T. Duvall & J. Harvey (private communication) and Kneer et al. (104), and ridge structure in the k-ω diagram at frequencies below 1 mHz [Brown & Harrison (18)] is suggestive of g modes trapped in the atmosphere. In the last two studies, brightness fluctuations, rather than velocities, were observed. Because modes that are confined to the atmosphere have comparatively little inertia, they are prone to scattering by spatial and temporal atmospheric inhomogeneities. This exacerbates the observational problems resulting from seeing and instrumental noise.

Propagating acoustic waves generated, for example, by convective turbulence are expected in the atmosphere at frequencies above the cutoff frequency. By radiative and shock dissipation, and by coupling to MHD modes, they may contribute substantially to the heating of the lower chromosphere. Although these very high-degree oscillations have been

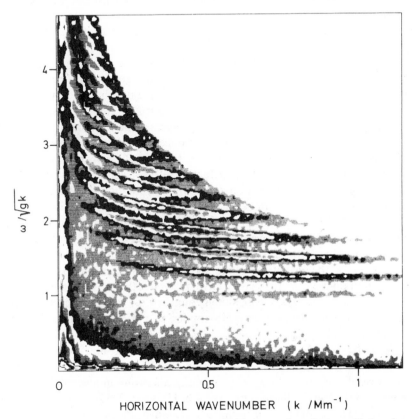

HORIZONTAL WAVENUMBER (k /Mm^{-1})

Figure 2 Two-dimensional power spectrum of high-degree five-minute oscillation data obtained by Deubner (47a). Different shadings signify different levels of power in the k-σ plane, where $\sigma = \omega/\sqrt{(gk)}$. The horizontal ridge of power at $\sigma = 1$ is produced by the f mode. The remaining ridges are produced by p modes. All modes in a ridge have a common order n, and n increases as σ increases at fixed k. At large horizontal wave numbers, σ depends only weakly on k, which indicates that near the surface the variation of sound speed with depth is similar to that of a polytrope. The rise in σ at low k arises partly from the fact that at greater depths c^2 increases more rapidly. There is also a geometrical contribution: Because the Sun is spherical, deeply penetrating waves travel a shorter distance than they would if the solar envelope were plane; they return to the surface more quickly and therefore resonate at a higher frequency than that given by Equation 2.14. The curvature of the ridges is greater at higher order, because at fixed k the depth of penetration is an increasing function of frequency.

studied intensively for many years (42, 88, 112, 116, 119, 120, 136), only recently has their wave character been established more firmly (48, 62, 146). Since they are mainly an atmospheric phenomenon, we do not discuss them further here. However, results obtained from these studies may well be useful for measuring the structure of the Sun's atmosphere, with implications concerning the outer boundary conditions that must be imposed on oscillations in the solar envelope.

Solar Structure

The first inference to have been drawn from the dispersion relation concerned the stratification of the convection zone. The frequencies observed were lower than the theoretical eigenvalues that had previously been computed (1, 151). According to Equation 2.14, this implies that the mean gradient G characterizing the upper layers of the convection zone was too large in the theoretical models. To reduce it would entail lowering the superadiabatic temperature gradient in the upper convective boundary layer, which in turn leads to a reduction of the specific entropy in the adiabatic part of the convection zone; trends in standard solar envelope computations suggested that such a reduction would imply an increase of the depth D of the convection zone and a higher opacity in the radiative interior (73). Numerical envelope computations by Ulrich & Rhodes (153) were consistent with this suggestion, and subsequent analyses of the sensitivity of the eigenfrequencies to other uncertain aspects of solar envelope models (8, 9, 118) seem to have ruled out other possibilities. Thus it appears that $D \simeq 2.0$–2.3 Mm. When coupled with the so-called standard stellar evolution theory, this value implies an initial solar helium abundance $Y_0 \simeq 0.25$.

More recently, Duvall (52) has pointed out that the observed dispersion relation is such that $(n + \varepsilon)/\omega$ is a function of ω/k alone, where ε is constant. His results, together with more recent data from modes of low and intermediate degree obtained by Duvall & Harvey (53, 55), are displayed in Figure 3. This is the law for nondispersive waves in a waveguide of fixed dimensions, and it can be seen to follow from Equation 2.13, provided ω_c^2 does not contribute substantially to the integral. Of course, the dimensions of the solar waveguide are not fixed, but the law is preserved because the depth of penetration of the waves is also a function of ω/k. Gough (81) and Christensen-Dalsgaard et al. (24a) have discussed how Duvall's law can be used to estimate the variation of c with depth. The law also puts limits on the amount of local dispersion in the waves, and hence on the degree of contamination by buoyancy (81); it is the first explicit demonstration that the bulk of the convection zone is close to being adiabatically stratified.

Subphotospheric Velocities

Horizontal flow advects the wave pattern of high-degree five-minute modes at a speed equal to the mean flow speed, weighted by the energy density of the oscillation (74, 87). This shifts the ridge positions in the k-ω power spectrum. The original interest was in measuring the subphotospheric rotation, and early results of Deubner et al. (49) indicated that at some depth the velocity, which was presumed to indicate rotation, exceeded that at the photosphere (74). Subsequent observations by Rhodes et al. (126)

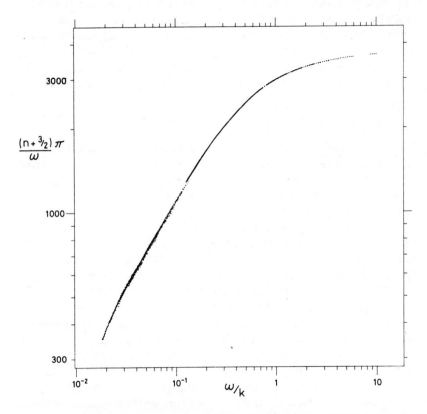

Figure 3 Representation, in the manner of Duvall (52), of 2783 five-minute modes of oscillation reported by Duvall (52) and Duvall & Harvey (53, 55), with degrees l ranging from 1 to 892 [from Christensen-Dalsgaard et al. (24a)]. Frequency ω is measured in s^{-1}, and the photospheric phase speed $v = \omega/k$ in units of Mm s^{-1}. It is evident from the discussion of resonant cavities, that the angular speed v/R of the wave pattern in the photosphere is approximately equal to the angular speed c/r of a sound wave propagating horizontally at the lower boundary $r = r_1$ of the region of propagation.

yielded inconclusive results, possibly because there are significant flow variations associated with giant convective cells, and the positions of these cells were such as to annul the variable component of advection at the time of the more recent measurements. Evidence for temporal variations in the ridge positions that might have been caused by the passage of giant cells across the field of view has since been reported by F. Hill et al. (93, 93a).

5. LOW-DEGREE MODES

The existence of individual low-degree modes was first reported by Severny et al. (141) and Brookes et al. (13). Power in Doppler data was found near 3 mHz, which corresponds to periods near 5 min, as well as in the 40–60 min range. A peculiar mode with a period of 160.01 min, close to one ninth of a terrestrial day, was also reported, and after considerable controversy (e.g. 89, 162) is now generally accepted as being of solar origin (106, 133, 142).

High-Order p Modes

The dense population of detectable frequencies requires high frequency resolution to separate the individual lines in the power spectrum. A partial separation of high-order modes, with periods near 5 min, was first recognized by Claverie et al. (36, 37) from whole-disk Doppler observations with a resonance spectrometer. The separation was possible because the modes fall into groups of odd and even degree that are just sufficiently separated in frequency to be resolvable by a single day's observations; as can be seen from the asymptotic relation (2.10), increasing l by 2 and decreasing n by unity leaves the frequency almost unchanged, since $\delta_{n,l}$ is small for high-order modes. Individual modes were first detected by Grec et al. (90, 91) in a continuous five-day observing run at the South Pole. This too was a whole-disk measurement with a resonance spectrometer. Identification of the degree of the modes was then achieved by comparing the frequency distribution with theory, and the observed amplitudes with the instrumental sensitivity (31). The latter comparison was justified by a theoretical argument that the true low-degree mode amplitudes are independent of l at fixed frequency (33). This argument subsequently found independent observational support for modes with $l \lesssim 150$ (47); typically amplitudes decline with l when $l \gtrsim 150$.

In order to reduce stochastic amplitude variations, Grec et al. (90, 91) averaged their data in the manner illustrated in the lower portion of Figure 4. The correspondence with theory leaves little room to doubt the identification of the degrees l of the modes.

Soon after, the South Pole observations were more-or-less confirmed and apparently surpassed by the University of Birmingham group [Claverie et al. (38, 39)]. A power spectrum of the most recently analyzed data is shown in Figure 5. Pairs of modes with cyclic frequencies separated by $\delta_{n,l}/2\pi$ are clearly visible. The pairs have alternately odd and even degrees and are distributed uniformly with separation $\simeq v_0$.

On an expanded frequency scale, a similar spectrum of earlier data appeared to exhibit a splitting of the degeneracy with respect to the

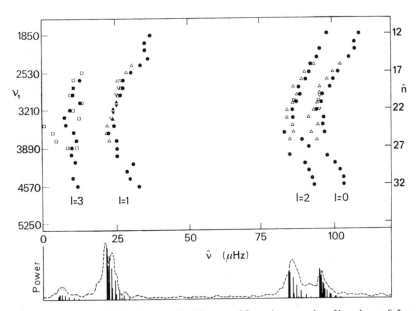

Figure 4   The upper portion is an echelle diagram of five-minute modes of low degree [after Scherrer et al. (134)]. A frequency axis is divided into 136-μHz segments, which are placed in a vertical row; a point in the resulting diagram represents the cyclic frequency v of a mode according to $v = v_1 + \hat{v}$, where v_1 and $\hat{v}$ are the ordinate and abscissa, respectively. The data have been compiled from various sources: ● Grec et al. (91), △ Claverie et al. (38), □ Scherrer et al. (135), ▽ Woodward & Hudson (160). Full symbols indicate coincidence (better than 0.5 μHz) with the South Pole data. The 136-μHz frequency interval is roughly the value of v_0 defined by Equation 2.12. It follows from Equation 2.10 that the almost vertical columns correspond with modes of like degree l, with order n increasing downward in steps of unity. The values of l were inferred by comparing the superposed power spectrum of the observations (91), shown as a dashed curve in the lower part of the figure, with the theoretical predictions of power that are plotted as vertical lines in the manner of Christensen-Dalsgaard (23b). The amplitudes are computed as the product of the expectation of the actual amplitudes and the instrumental sensitivity (33). The orders of the modes are given by $n = \hat{n}$ for $l = 1, 2$ and $n = \hat{n}+1, \hat{n}-1$ for $l = 0$ and 3, respectively. The absolute values of $\hat{n}$ were originally inferred from theory (30). They were subsequently confirmed unambiguously by the observations of Duvall & Harvey (53), which connected these data with the high-degree modes illustrated in Figure 2.

azimuthal order m: radial ($l = 0$) modes were unsplit, dipole ($l = 1$) modes appeared to be split into triplets, and there was evidence that quadrupole modes ($l = 2$) were split into quintuplets. The average splitting was 0.75 μHz, and this was interpreted as being produced by rotation. The observation is not what one would expect, however, because although there are $2l + 1$ values of m, modes with odd $l + m$ should not have been detected (33). Five-minute oscillations have also been found in the SMM-ACRIM intensity data (160, 161). The frequencies are in good agreement with those from Doppler measurements; however, it has not been possible to resolve degeneracy splitting (159a).

Whole-disk measurements of p modes are sensitive only to those modes with $l \lesssim 3$. Detection of five-minute modes with $l = 3, 4$, and 5 was achieved at Stanford University [Scherrer et al. (134, 135)] by subtracting Doppler signals from an inner disk and an outer annulus of the solar image. Octupole modes were identified by comparing frequencies with the South Pole and Birmingham data (see Figure 4), and the remaining values of l were deduced from theory. The results are in fair agreement with the whole-disk data, and they confirm the linear dependence of $\delta_{n,l}$ on $l(l + 1)$ at fixed $n + \frac{1}{2}l$ predicted by asymptotic theory.

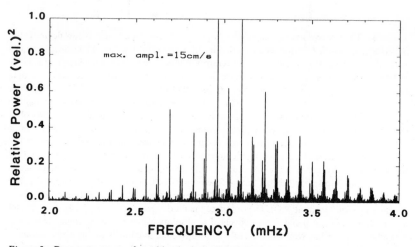

VELOCITY POWER SPECTRUM

1981: 3 months. Tenerife&Hawaii, up to 22 hrs. per day

Figure 5 Power spectrum of combined whole-disk Doppler observations by the University of Birmingham group obtained from Hawaii and Tenerife over a period of 3 months. The separations $\delta_{n,l}$ between modes with like values of $n + \frac{1}{2}l$ are clearly visible.

Gravity Modes

Recently, evidence for low-degree g modes has been found in the Stanford (132) and Birmingham data (157), with periods of 3–5 hr, and in the SMM-ACRIM data (70), with periods of 4–140 hr. The method was to search among the highest peaks in the power spectra for sequences of frequencies corresponding to uniformly spaced periods, as is predicted by the asymptotic formulae (2.15) and (2.16). The high density of peaks (representing true eigenfrequencies, their aliases, and possibly noise) renders identification somewhat uncertain, and indeed the reports from Stanford and SMM, with $P_0 = 38.6$ min, conflict with the Birmingham data, from which it was inferred that $P_0 = 41.2$ min.

Solar Structure

The first diagnostic from low-degree modes to be discussed was the measure v_0 of the mean sound speed. Christensen-Dalsgaard & Gough (31) made a rough estimate of the mean separation between adjacent peaks in the p-mode power spectrum (of 36) and compared it with theoretical separations computed from a sequence of three solar models with different chemical composition. The separation observed was lower than expected. Within the confines of standard stellar evolution theory, the discrepancy suggested that the solar helium abundance might be somewhat low. A similar conclusion was subsequently drawn by Claverie et al. (36) using theoretical estimates extrapolated from the computations of Iben & Mahaffy (100). However, after refining the eigenfrequency computations and making a more detailed comparison between theory and observation, it became evident that precise agreement would not be possible (30); though the measured frequency separations in the 2–4 mHz range were lower than expected, the values of the frequencies themselves were rather high. In a sequence of solar models with varying initial helium abundance Y_0, the frequency scale expands as Y_0 is increased, increasing both the values of the frequencies and their mean separation.

Subsequently, various workers have adjusted parameters in the theory, experimented with the equation of state, or added a mock magnetic field (70a, 121, 127, 138, 139, 140, 144, 152). They have been able to reduce the discrepancy, but they have not been able to remove it.

A diagnostic of considerable interest is the small separation $\Delta_{n,l} = v_{n,l} - v_{n-1,l+2}$; according to (2.10) and (2.11),

$$\Delta_{n,l} \simeq 2(2l+3)\alpha v_0^2 v_{n,l}^{-1}. \qquad 5.1.$$

Noting the sensitivity of this expression to l, one would expect the region in the Sun that influences $\Delta_{n,l}$ to be contained sufficiently close to the center that sound can travel a horizontal wavelength in an oscillation period; for only there can a sound wave experience directly its horizontal variation.

Since $\omega_c \ll S_l$ in the core, it can be seen from Equations 2.6–2.8 that this condition is satisfied only in the region $r \lesssim r_1$. However, there the wave is evanescent. As r decreases below r_1, the influence of the equilibrium state on the wave must diminish. Therefore, one might anticipate $\Delta_{n,l}$ to measure conditions close to r_1.

Unfortunately, the situation is not quite that simple. First, since sound speed, and consequently the wavelength of the modes, is greater near the center, it is more difficult for p modes to resolve small-scale structure as r decreases. Second, it appears that terms in the asymptotic expansion of ω beyond those given in (2.10–2.12), and which include perturbations in the gravitational potential, might have a significant influence (W. Dziembowski & D. Gough, unpublished), so that oscillatory motion in the core is coupled instantaneously to the rest of the star. Nevertheless, the local contribution to $\Delta_{n,l}$ is probably the more important; it depends on the variation of c in the core, and decreases as the chemical inhomogeneity produced by nuclear reactions increases (82).

Theoretical computations by various groups (31, 144, 154) have been compared with the observations of Claverie et al. (38) and Fossat et al. (68) by Gough (79). The difference $\Delta_{n,l}$ varies slowly with n over the 2–4 mHz frequency range ($15 \lesssim n \lesssim 30$); its average, according to the observations, is about 9 and 15 μHz for $l = 0$ and 1, respectively. Standard unmixed solar models require an initial helium abundance Y_0 as great as 0.25 to produce such low values. These models have high neutrino fluxes. They also have relatively deep convection zones, in accord with the implications of the high-degree modes. Any partial mixing in the core is likely to increase $\Delta_{n,l}$ and consequently Y_0 would have to be increased to compensate.

Attempts to estimate Y_0 have been made by fitting by least-squares interpolated p-mode eigenfrequencies of a sequence of solar models to the observed five-minute frequencies. The procedure also determines the order n of the modes. By weighting all frequencies equally, Christensen-Dalsgaard & Gough (32) and Shibahashi et al. (144a) found $Y_0 = 0.27$ and 0.23, respectively. The best-fitting model does not agree with the g-mode data: The periods of g modes of that model corresponding to those reported by Scherrer & Delache (132), when fitted to Equation 2.15, yield $P_0 \approx 34.5$ min (83). To increase this value would require a reduction in the buoyancy frequency. This can be achieved either by mixing the interior (10, 10a) or by reducing Y_0. Either would increase $\Delta_{n,l}$.

6. FIVE-MINUTE MODES OF INTERMEDIATE DEGREE

By projecting Doppler data onto zonal harmonics, Duvall & Harvey (53) have identified five-minute modes with $l \leqslant 140$, distinguishing more than

20 branches of the dispersion relation. Thus they were able to connect the previously acquired low-degree data with the high-degree modes and confirm the values of n that had already been inferred for the low-degree modes by model fitting (30, 32, 144a). These data satisfy the waveguide law (see Figure 3) found by Duvall (53) for high-degree modes. They also share with the low-degree modes the property that the ratio $(v_{n+1,l} - v_{n,l}) : v_{n,l}$ is lower than can be explained by theory. More recent attempts to isolate sectoral harmonics (54) have produced similar results.

A thorough comparison of intermediate-degree data with theory has not yet been made. However, preliminary analyses (34, 55) have revealed that the systematic discrepancy mentioned earlier between theoretical low-degree five-minute eigenfrequencies and observation has a frequency dependence that is independent of l for $l \lesssim 30$; at higher values of l, the dependence of the discrepancy on frequency changes. Recalling that the wave senses l directly only in the vicinity of its lower turning point at $r = r_1$ leads one to conclude that the most deep-seated error in the theoretical model that has a substantial influence on p modes is likely to be located somewhere near the level at which $S_{30} \simeq \omega$. This is near the base of the convection zone.

There are several other programs to measure modes of intermediate degree. Spatial resolution is to be achieved either by sequential scanning of the solar image (15) or by arrays of detectors (53, 63). Groups of modes can be isolated by performing suitable projections of the data (23c). Individual modes cannot be isolated at any one instant, because for this it would be necessary to view the entire surface of the Sun, but there are grounds to anticipate that a subsequent comparison of the temporal information with theory could separate the individual modes in each group (17).

7. LIMB OBSERVATIONS

At about the same time as the first measurement of the dispersion relation for high-degree modes (44) and the discovery of discrete frequencies in spatially unresolved observations (13, 141), H. Hill and his collaborators (97) reported an accidental detection of oscillations in the apparent diameter of the Sun. There followed a considerable dispute over whether or not the origin of these oscillations is solar (e.g. 67, 95). Many of the objections, which related to atmospheric distortions of the solar image, have been circumvented in the most recent observations (12, 130), where local fluctuations in the limb-darkening function were analyzed. Nevertheless, the interpretation of even these data is still in doubt.

Current limb observations are most sensitive to oscillations of degree $l \lesssim 50$ (94). They do not have the spatial and temporal resolution to separate the individual modes, except possibly for frequencies roughly in the range 0.4–0.6 mHz, where the higher-degree modes do not exist. There is reasonable hope (85), however, that with more spatial information, future limb observations might provide valuable information about the modes of low order.

8. PROBLEMS FOR THE IMMEDIATE FUTURE

Our discussion has been concerned mainly with the hydrostatic stratification of the Sun, for this is the aspect of the solar structure about which oscillations have so far yielded the most reliable information. However, it is already apparent that this is merely the first step in a subject that promises to progress significantly in the next decade. There are good reasons to believe that we shall soon put useful bounds on large-scale asymmetries in the structure of the Sun that split the degeneracy with respect to m in the oscillation eigenfrequencies. These will no doubt raise new questions about the internal dynamics of the Sun.

Aside from improvements in our knowledge of the density stratification, an issue with which we are most likely to make imminent progress is the distribution of angular momentum within the Sun. This has an obvious direct bearing on the internal dynamics of stellar spin-down and is likely to shed light on ill-understood questions of wave transport. The ramifications concern material mixing by Ekman-Eddington-Sweet circulation and the influence of the centrifugally induced distortion of the gravitational potential on the precession of planetary orbits (20a, 55a, 77, 96).

Several reports of rotational splitting have been made (39, 43, 54, 96). Taken at face value they all imply that somewhere at depth, rotation is more rapid than at the surface, though by how much is not yet clear. Unless one accepts that the Sun's angular velocity Ω is a rapidly varying function of position or time, the data appear to be contradictory (81). The most extensive data (54) suggest that throughout most of the Sun, Ω is somewhat less than the equatorial angular velocity in the photosphere, and that in a central core (corresponding roughly to where nuclear reactions have raised the molecular weight) Ω is substantially greater than it is elsewhere (55a).

Degeneracy splitting by intense magnetic fields might also enrich the spectrum (49a, 58, 59, 78, 86).

Recent developments in instrumentation for making spatially resolved observations (e.g. 63) will no doubt permit us to improve substantially our measurements of high-degree modes. Thus, we might make coarse measurements of velocity and temperature variations immediately beneath the photosphere, although the extent to which they will be able to resolve any of the theoretical issues about the dynamics of the convection zone is still an open question. There is perhaps more hope for high temporal resolution of modes of low degree. Here, one is hindered less by the problems of seeing through the Earth's atmosphere. Variations in the eigenfrequencies of these modes during the course of the solar cycle are likely to be large enough to be detectable (80), and they may reveal whether the cycle is controlled by a turbulent dynamo in the convection zone or by some more deep-seated process.

It may be necessary to restrict attention to modes of relatively low order, because the lifetimes of the high-order oscillations may be insufficient for accurate eigenfrequencies to be determined. Short lifetimes of high-frequency modes were predicted by theory (71, 75), and now there is also evidence from the observations: The peaks in Figure 5 are substantially broader at high frequencies. It is also evident from Figure 5 that the amplitudes of the lowest-order modes are small, rendering it difficult to extract the signal of the regular oscillations from the background noise.

Almost all the diagnostic studies that have been made have relied on model fitting. Oscillation eigenfrequencies of a solar model defined by certain parameters are computed (this is called the "forward problem"), and the parameters are adjusted to find the model that best fits the observations. This procedure has the advantage of being relatively simple. It also provides useful insight, especially if the parameters have a clear physical significance. However, inverse calculations designed to extract certain aspects of the structure of the Sun directly from the data are likely to increase substantially our diagnostic capabilities. A variety of methods apparently suitable for helioseismology have been developed by geophysicists (6, 7, 103, 123, 124, 131, 158) and used with considerable success to infer the structure of the Earth. The product of the techniques is a theoretical model, together with an estimate of its accuracy. A few inverse calculations have been performed on artificial solar data (34, 74, 83, 84) to assess what information is likely to be acquired from future observations.

Any diagnostic calculation depends on knowing in considerable detail the physics of the oscillations. In the Sun there are many uncertainties, so our problems are exacerbated. In particular, the equation of state is inadequately understood, so we could not calculate the sound speed accurately even if we knew the composition and thermodynamic state of the gas (129). Of course, there may be sufficient information in the oscillation eigenfrequencies to impose important constraints on the possible forms of the equation of state in the dense interior (81), adding yet another role to the Sun as a physics laboratory.

It is not only an analysis of oscillation eigenfrequencies that is likely to provide valuable information about the solar interior. The principal processes that excite waves and control their amplitudes are also important. At present we do not know what these are. Some have maintained that the oscillations are self-excited by modulating the thermal energy flow with an Eddington valve, in much the same way as in the Cepheids and RR Lyrae stars (2–5). In that case amplitudes might be limited by nonlinear coupling to stable modes of oscillation (56, 57), but it is not yet clear whether this process can extract energy fast enough at the low amplitudes observed. Alternatively, the modes might be stable, as a result of their interaction with convection (9, 71). Stochastic fluctuations in the turbulence might then be responsible for forcing the modes to oscillate at low amplitude (72). Amplitude estimates based on this idea are roughly correct for solar five-minute oscillations (75), but other modes may need to be driven differently. What is required now is a careful assessment of all the nonlinear processes affecting the modes, perhaps along the lines proposed by Dolez et al. (51). If successful, it may then be possible to invert the argument to learn about convection.

Spatially unresolved oscillation measurements can be carried out on other stars. Christensen-Dalsgaard & Frandsen (27) have estimated amplitudes under the assumption that the modes are excited stochastically by turbulence, and they have concluded that on the main sequence, F stars are the most suitable candidates for study. As a result, Fossat (66) recently observed α Centauri and reported having detected oscillations. Also, Kurtz (108) appears to have found a rapidly oscillating Ap star to have a spectrum of p modes excited (143), like the Sun. These promising beginnings are likely to stimulate further interest in asteroseismology.

We cannot end this review without mentioning the 160-min oscillation. At first there was considerable doubt about its origin, particularly because its period is one ninth of a terrestrial day (89, 162). But now it seems likely that the oscillation is solar (106, 107, 133). However, its nature is still a mystery. The first plausible suggestion was that it is a low-degree g mode (29), but this then raised the question of why only one of a rich spectrum of modes should predominate. One possibility is that it is excited by a chance resonance between p modes (29, 73). If so, the potential modification of the accuracy of the resonance during the solar cycle might have interesting observable consequences. Isaak (102) has suggested that the oscillation is driven by the gravitational radiation from a close binary star. On the other hand, Childress & Spiegel (22) have argued that the oscillation may not be a linear mode at all, but instead is caused by a soliton propagating around the solar core and driven by the perturbations it produces in the nuclear reactions. More exotic but unproven suggestions have been entertained (11).

After the five-minute modes, this long-period oscillation was the first to be discovered. It may be the last to be explained.

Acknowledgments

We are grateful to G. Isaak and his colleagues for providing us with Figure 5, and to J. Christensen-Dalsgaard both for his help in producing Figure 1 and for his comments on the original manuscript.

APPENDIX

CLASSIFICATION OF STELLAR OSCILLATIONS It is important to define a precise nomenclature in order to avoid ambiguity. Since to a first approximation the equilibrium state of the Sun is spherically symmetrical, linearized vertical displacements can be represented as a superposition of normal modes of the form (2.1). The vertical velocity and the perturbations in all thermodynamical state variables have a similar form.

Under certain idealized conditions (e.g. adiabatic motion and perfect reflection of the waves at the stellar surface), the governing equations of motion and their boundary conditions admit a sequence of discrete eigensolutions for which Ξ and ω are real. These can be arranged in order of increasing ω. The modes are then labeled sequentially with an integer n, which is called the *order* of the mode. In reality the idealized conditions are not met precisely, though they are a good approximation in the Sun for the modes that have been observed. Therefore, the true modes are quite similar to the idealized ones, and n is probably well defined.

There has been some diversity in the naming of l and m; here, we call the degree l of the associated Legendre function the *degree* of the mode, and the order m we call the *azimuthal order* of the mode.

Formally, the frequency is unbounded below (except when $l = 0$), and were it not for the solar atmosphere it would also be unbounded above (see Section 2). Therefore it is not immediately obvious where to choose the origin of n, and again there is some diversity in the literature. It appears to be possible (though it has not yet been proved) to choose n such that as $n \to \pm \infty$ at fixed l, $|n|$ is the number of zeros in Ξ (excluding the zero at $r = 0$ when $l \geq 2$). In the case of the spherically symmetrical modes $(l = 0)$, we label the lowest-frequency (or fundamental) mode with $n = 1$; once again n is the number of zeros in Ξ when n is large, although now the zero at $r = 0$ must be counted.

In simple stellar models, modes with $n > 0$ and $n < 0$ have the characteristics of acoustic and internal gravity waves; they were designated p modes and g modes by Cowling (40). Modes with $n = 0$ are essentially surface gravity waves. They are the fundamental g modes, or f modes, and have the property that when $l \gg 1$ they have no zeros in Ξ.

In this review we have adopted the common practice of ignoring the sign of n and designating the mode with g, f, or p. Then both the g and p sequences start from $n = 1$. A p mode with $n = 2$ and $l = 3$, for example, is designated $p_2(l = 3)$. Spherically symmetric gravity modes cannot exist, so the $l = 0$ modes are p modes. This is partly why the fundamental $l = 0$ mode was labeled with $n = 1$. There is also a mathematical reason for this choice, but we do not discuss that here.

In the case of high-degree g modes, such as the lowest-frequency mode illustrated in Figure 1, the decline in amplitude in the evanescent region immediately beneath the photosphere is very severe (25, 60), and communication between the two regions of propagation is extremely small. Except in cases of accurate resonance between oscillations trapped in the two regions of propagation (which is a mathematical possibility for essentially undamped oscillation modes of idealized solar models, but is unlikely to be relevant in practice), the interior and the atmosphere oscillate independently. Consequently, it is sometimes expedient to classify the interior and atmospheric modes separately (155). Otherwise, a modification to the structure of the atmosphere that has no influence on the interior could change the formal classification of a mode that exists essentially only beneath the convection zone. Notice that in this scheme a theoretical mode that happens to resonate with both the interior and the atmosphere would be classified twice.

In reality, stars are not spherically symmetrical, for they rotate and contain magnetic fields. The oscillation eigenfunctions are no longer precisely of the form (2.1), and, in particular, the degeneracy of the frequency with respect to m is split. Nevertheless, in the Sun the deviations from spherical symmetry are small, and except in certain rare cases of resonance (which we do not discuss here), the modifications to the normal modes are small too. Thus we can identify without ambiguity the corresponding modes of a similar spherically symmetrical model of the Sun, and we adopt the classification of the latter for the solar modes.

Literature Cited[a]

1. Ando, H., Osaki, Y. 1976. *Publ. Astron. Soc. Jpn.* 27:581–603
2. Ando, H., Osaki, Y. 1977. *Publ. Astron. Soc. Jpn.* 29:221–33
3. Antia, H. M., Chitre, S. M. 1984. In CATANIA. In press
4. Antia, H. M., Chitre, S. M., Kale, D. M. 1978. *Sol. Phys.* 56:275–92
5. Antia, H. M., Chitre, S. M., Narasimha, D. 1982. *Sol. Phys.* 77:303–27

[a] We use the following abbreviated notations for some frequently cited conference proceedings:

TUCSON = *Nonradial and Nonlinear Stellar Pulsation*, ed. H. A. Hill, W. A. Dziembowski. Berlin: Springer. 497 pp.
CATANIA = *Oscillations as a Probe of the Sun's Interior*, ed. G. Belvedere, L. Paterno. *Mem. Soc. Astron. Ital.* In press
SNOWMASS = *Solar Seismology from Space*, ed. J. W. Harvey, E. J. Rhodes, Jr., J. Toomre, R. K. Ulrich. Washington DC: NASA. In press

6. Backus, G. E., Gilbert, J. F. 1967. *Geophys. J. R. Astron. Soc.* 13:247–76
7. Backus, G. E., Gilbert, J. F. 1970. *Philos. Trans. R. Soc. London Ser. A* 266:123–92
8. Belvedere, G., Gough, D. O., Paternò, L. 1983. *Sol. Phys.* 82:343–54
9. Berthomieu, G., Cooper, A. J., Gough, D. O., Osaki, Y., Provost, J., Rocca, A. 1980. In TUCSON, pp. 307–12
10. Berthomieu, G., Provost, J., Schatzman, E. 1984. In CATANIA. In press
10a. Berthomieu, G., Provost, J., Schatzman, E. 1984. *Nature* 308:254–57
11. Blinnikov, S. I., Khlopov, M. Yu. 1983. *Sol. Phys.* 82:383–85
12. Bos, R. J., Hill, H. A. 1983. *Sol. Phys.* 82:89–102
13. Brookes, J. R., Isaak, G. R., van der Raay, H. B. 1976. *Nature* 259:92–95
14. Brookes, J. R., Isaak, G. R., van der Raay, H. B. 1978. *MNRAS* 185:1–17
15. Brookes, J. R., Isaak, G. R., van der Raay, H. B. 1981. *Sol. Phys.* 74:503–8

16. Brown, T. M. 1981. In *Solar Instrumentation: What's Next?*, ed. R. B. Dunn, pp. 150–54. Sunspot, N. Mex: Sacramento Peak Obs.
17. Brown, T. M., Christensen-Dalsgaard, J., Mihalas, B. 1984. In SNOWMASS. In press
18. Brown, T. M., Harrison, R. L. 1980. *Ap. J. Lett.* 236:L169–73
19. Brown, T. M., Stebbins, R. T., Hill, H. A. 1978. *Ap. J.* 223:324–38
20. Cacciani, A., Rhodes, E. J. Jr., Ulrich, R. K., Howard, R. 1984. In SNOWMASS. In press
20a. Campbell, L., McDow, J. C., Moffat, J. W., Vincent, D. 1983. *Nature* 305:508–10
21. Caudell, T. P., Knapp, J., Hill, H. A., Logan, J. D. 1980. In TUCSON, pp. 206–18
22. Childress, S., Spiegel, E. A. 1981. In *Variations of the Solar Constant*, ed. S. Sofia, pp. 273–91. *NASA Conf. Publ. 2191*
23. Christensen-Dalsgaard, J. 1980. *MNRAS* 190:765–91
23a. Christensen-Dalsgaard, J. 1982. *MNRAS* 199:735–61
23b. Christensen-Dalsgaard, J. 1983. *Adv. Space Res.* 2:11
23c. Christensen-Dalsgaard, J. 1984. In SNOWMASS. In press
24. Christensen-Dalsgaard, J., Cooper, A. J., Gough, D. O. 1983. *MNRAS* 203:165–79
24a. Christensen-Dalsgaard, J., Duvall, T. L. Jr., Gough, D. O., Harvey, J. W. 1984. *Nature.* In press
25. Christensen-Dalsgaard, J., Dziembowski, W., Gough, D. O. 1980. In TUCSON, pp. 313–41
26. Christensen-Dalsgaard, J., Frandsen, S. 1983. *Sol. Phys.* 82:165–204
27. Christensen-Dalsgaard, J., Frandsen, S. 1983. *Sol. Phys.* 82:469–86
28. Christensen-Dalsgaard, J., Frandsen, S. 1984. In CATANIA. In press
29. Christensen-Dalsgaard, J., Gough, D. O. 1976. *Nature* 259:89–92
30. Christensen-Dalsgaard, J., Gough, D. O. 1980. *Nature* 288:544–47
31. Christensen-Dalsgaard, J., Gough, D. O. 1980. In TUCSON, pp. 184–90
32. Christensen-Dalsgaard, J., Gough, D. O. 1981. *Astron. Astrophys.* 104:173–76
33. Christensen-Dalsgaard, J., Gough, D. O. 1982. *MNRAS* 198:141–71
34. Christensen-Dalsgaard, J., Gough, D. O. 1984. In SNOWMASS. In press
35. Deleted in proof
36. Claverie, A., Isaak, G. R., McLeod, C. P., van der Raay, H. B., Roca Cortez, T. 1979. *Nature* 282:591–94
37. Claverie, A., Isaak, G. R., McLeod, C. P., van der Raay, H. B., Roca Cortez, T. 1980. *Astron. Astrophys. Lett.* 91:L9–10
38. Claverie, A., Isaak, G. R., McLeod, C. P., van der Raay, H. B., Roca Cortez, T. 1981. *Nature* 293:443–45
39. Claverie, A., Isaak, G. R., McLeod, C. P., van der Raay, H. B., Roca Cortez, T. 1982. *Sol. Phys.* 74:51–57
40. Cowling, T. G. 1941. *MNRAS* 101:367–75
41. Cox, J. P. 1980. *Theory of Stellar Pulsation.* Princeton Univ. Press. 380 pp.
42. Cram, L. E. 1978. *Astron. Astrophys.* 70:345–54
43. Delache, P., Scherrer, P. H. 1983. *Nature* 306:651–53
44. Deubner, F.-L. 1975. *Astron. Astrophys.* 44:371–75
45. Deubner, F.-L. 1976. In *The Energy Balance and Hydrodynamics of the Solar Chromosphere and Corona*, ed. R. M. Bonnet, P. Delache, pp. 45–68. Clairmont-Ferrand: G. de Bussac
46. Deubner, F.-L. 1980. In *Highlights of Astronomy*, ed. P. A. Wayman, 5:75–87. Dordrecht: Reidel
47. Deubner, F.-L. 1981. In *The Sun as a Star*, ed. S. Jordan, pp. 65–84. *NASA-SP-450*
47a. Deubner, F.-L. 1983. *Sol. Phys.* 82:103–9
48. Deubner, F.-L., Endler, F., Staiger, J. 1984. In CATANIA. In press
49. Deubner, F.-L., Ulrich, R. K., Rhodes, E. J. Jr. 1979. *Astron. Astrophys.* 72:177–85
49a. Dicke, R. H. 1982. *Nature* 300:693–97
50. Dittmer, P. H. 1977. PhD thesis. Stanford Univ., Stanford, Calif.
51. Dolez, N., Poyet, J.-P., Legait, A. 1984. In CATANIA. In press
52. Duvall, T. L. Jr. 1982. *Nature* 300:242–43
53. Duvall, T. L. Jr., Harvey, J. W. 1983. *Nature* 302:24–27
54. Duvall, T. L. Jr., Harvey, J. W. 1984. *Nature.* In press
55. Duvall, T. L. Jr., Harvey, J. W. 1984. In SNOWMASS. In press
55a. Duvall, T. L. Jr., Dziembowski, W. A., Goode, P. R., Gough, D. O., Harvey, J. W., Leibacher, J. W. 1984. *Nature.* In press
56. Dziembowski, W. A. 1982. *Acta Astron.* 32:147–71
57. Dziembowski, W. A. 1983. *Sol. Phys.* 82:259–66
58. Dziembowski, W. A., Goode, P. R. 1983. *Nature* 305:39–42
59. Dziembowski, W. A., Goode, P. R. 1984. In CATANIA. In press

60. Dziembowski, W. A., Pamjatnykh, A. A. 1978. In *Pleins Feux sur la Physique Solaire*, ed. J. Rösch, pp. 135–40. Paris: CNRS
61. Eckart, C. 1960. *Hydrodynamics of Oceans and Atmospheres*. London: Pergamon
62. Endler, F., Deubner, F.-L. 1983. *Astron. Astrophys.* 121:291–96
63. Evans, J. W. 1981. In *Solar Instrumentation: What's Next?*, ed. R. B. Dunn, pp. 150–54. Sunspot, N. Mex: Sacramento Peak Obs.
64. Evans, J. W., Michard, R. 1962. *Ap. J.* 135:812–21
65. Evans, J. W., Michard, R. 1962. *Ap. J.* 136:493–506
66. Fossat, E. 1984. In CATANIA. In press
67. Fossat, E., Grec, G., Harvey, J. W. 1981. *Astron. Astrophys.* 94:95–99
68. Fossat, E., Grec, G., Pomerantz, M. 1981. *Sol. Phys.* 74:59–63
69. Frazier, E. N. 1968. *Z. Ap.* 68:345–56
70. Fröhlich, C., Delache, P. 1984. In CATANIA. In press
70a. Gabriel, G., Scuflaire, R., Noels, A. 1982. *Astron. Astrophys.* 110:50–53
70b. Gingerich, O., Noyes, R. W., Kalkofen, W., Cuny, Y. 1971. *Sol. Phys.* 18:347–65
71. Goldreich, P., Keeley, D. A. 1977. *Ap. J.* 211:934–42
72. Goldreich, P., Keeley, D. A. 1977. *Ap. J.* 212:243–51
73. Gough, D. O. 1976. In *The Energy Balance and Hydrodynamics of the Solar Chromosphere and Corona*, ed. R. M. Bonnet, P. Delache, pp. 3–36. Clairmont-Ferrand: G. de Bussac
74. Gough, D. O. 1978. *Proc. EPS Workshop Sol. Rotation*, ed. G. Belvedere, L. Paternò, pp. 255–68. Catania: Catania Univ. Press
75. Gough, D. O. 1980. In TUCSON, pp. 273–99
76. Gough, D. O. 1982. *Irish Astron. J.* 15:118–19
77. Gough, D. O. 1982. *Nature* 298:334–39
78. Gough, D. O. 1982. *Nature* 298:350–54
79. Gough, D. O. 1983. *Proc. ESO Workshop Primordial Helium*, ed. P. A. Shaver, D. Kunth, K. Kjar, pp. 117–36. Garching: ESO
80. Gough, D. O. 1983. In *Pulsations in Classical and Cataclysmic Variables*, ed. J. P. Cox, C. J. Hansen, pp. 117–37. Boulder, Colo: JILA
81. Gough, D. O. 1984. In CATANIA. In press
82. Gough, D. O. 1983. *Phys. Bull.* 34:502–7
83. Gough, D. O. 1984. In SNOWMASS. In press
84. Gough, D. O. 1984. *Philos. Trans. R. Soc. London Ser. A.* In press
85. Gough, D. O., Latour, J. 1984. *Astr. Expr.* 1:9–25
86. Gough, D. O., Taylor, P. P. 1984. In CATANIA. In press
87. Gough, D. O., Toomre, J. 1983. *Sol. Phys.* 82:401–10
88. Gouttebroze, P., Leibacher, J. W. 1980. *Ap. J.* 238:1134–51
89. Grec, G., Fossat, E. 1979. *Astron. Astrophys.* 77:351–53
90. Grec, G., Fossat, E., Pomerantz, M. 1980. *Nature* 288:541–44
91. Grec, G., Fossat, E., Pomerantz, M. A. 1983. *Sol. Phys.* 82:55–66
92. Grec, G., Fossat, E., Vernin, J. 1976. *Astron. Astrophys.* 50:221–25
93. Hill, F., Toomre, J., November, L. J. 1983. *Sol. Phys.* 82:411–25
93a. Hill, F., Gough, D. O., Toomre, J. 1984. In CATANIA. In press
94. Hill, H. A. 1978. In *The New Solar Physics*, ed. J. A. Eddy, pp. 135–214. Boulder, Colo: Westview
95. Hill, H. A., Bos, R. J., Caudell, T. P. 1983. *Sol. Phys.* 82:129–38
96. Hill, H. A., Bos, R. J., Goode, P. R. 1982. *Phys. Rev. Lett.* 49:1794–97
97. Hill, H. A., Stebbins, R. T., Brown, T. M. 1976. In *Atomic Masses and Fundamental Constants*, ed. J. H. Sanders, A. H. Wapstra, pp. 622–28. New York: Plenum
98. Hill, H. A., Stebbins, R. T., Oleson, J. R. 1975. *Ap. J.* 200:484–98
99. Howard, R. 1967. *Sol. Phys.* 2:3–33
100. Iben, I. Jr., Mahaffy, J. 1976. *Ap. J. Lett.* 209:L39–43
101. Isaak, G. R. 1982. *Nature* 296:130–31
102. Isaak, G. R. 1984. In CATANIA. In press
103. Jackson, D. D. 1972. *Geophys. J. R. Astron. Soc.* 28:97–109
104. Kneer, F., Newkirk, G. Jr., von Uexküll, M. 1982. *Astron. Astrophys.* 113:129–34
105. Kotov, V. A., Severny, A. B., Tsap, T. T. 1978. *MNRAS* 183:61–87
106. Kotov, V. A., Severny, A. B., Tsap, T. T., Moiseev, I. G., Efanov, V. A., Nesterov, N. S. 1983. *Sol. Phys.* 82:9–19
107. Koutchmy, S., Koutchmy, O., Kotov, V. A. 1980. *Astron. Astrophys.* 90:372–76
108. Kurtz, D. 1983. *MNRAS* 205:11–22
109. Lamb, H. 1908. *Proc. London Math. Soc.* 7:122–41
110. Lamb, H. 1932. *Hydrodynamics*. Univ. Cambridge Press. 6th ed.
111. Ledoux, P., Perdang, J. 1980. *Bull. Soc. Math. Belg.* 32:133–59
112. Leibacher, J. W., Gouttebroze, P., Stein, R. F. 1982. *Ap. J.* 258:393–403
113. Leibacher, J. W., Stein, R. F. 1971. *Astrophys. Lett.* 7:191–92

114. Leibacher, J. W., Stein, R. F. 1981. In *The Sun as a Star*, ed. S. Jordan, pp. 263–88. *NASA SP-450*
115. Leighton, R. B. 1960. *Proc. IAU Symp.* 12:321–25
115a. Leighton, R. B., Noyes, R. W., Simon, G. W. 1962. *Ap. J.* 135:474–99
116. Lites, B. W., Chipman, E. G. 1979. *Ap. J.* 231:570–88
117. Livingston, W., Milkey, R., Slaughter, C. 1977. *Ap. J.* 211:281–87
118. Lubow, S. H., Rhodes, E. J. Jr., Ulrich, R. K. 1980. In TUCSON, pp. 300–6
119. Mein, N. 1977. *Sol. Phys.* 52:283–92
120. Mein, N., Schmieder, B. 1981. *Astron. Astrophys.* 97:310–16
121. Noels, A., Scuflaire, R., Gabriel, M. 1984. *Astron. Astrophys.* 130:389–96
122. Osaki, Y. 1975. *Publ. Astron. Soc. Jpn.* 27:237–58
123. Parker, R. L. 1977. *Ann. Rev. Earth Planet. Sci.* 5:35–65
124. Parker, R. L. 1977. *Rev. Geophys. Space Phys.* 15:446–56
125. Rhodes, E. J. Jr., Cacciani, A., Blamont, J., Tomczyk, S. 1984. In SNOWMASS. In press
126. Rhodes, E. J. Jr., Harvey, J. W., Duvall, T. L. Jr. 1983. *Sol. Phys.* 82:111
127. Rhodes, E. J. Jr., Ulrich, R. K., Brunish, W. M. 1984. In CATANIA. In press
128. Rhodes, E. J. Jr., Ulrich, R. K., Simon, G. W. 1977. *Ap. J.* 218:521–29
129. Rogers, F. 1984. In SNOWMASS. In press
130. Rösch, J., Yerle, R. 1983. *Sol. Phys.* 82:130–50
131. Sabatier, P. C. 1977. *J. Geophys.* 43:115–37
132. Scherrer, P. H., Delache, P. 1984. In CATANIA. In press
133. Scherrer, P. H., Wilcox, J. M. 1983. *Sol. Phys.* 82:37–42
134. Scherrer, P. H., Wilcox, J. M., Christensen-Dalsgaard, J., Gough, D. O. 1982. *Nature* 297:312–13
135. Scherrer, P. H., Wilcox, J. M., Christensen-Dalsgaard, J., Gough, D. O. 1983. *Sol. Phys.* 82:75–87
136. Schmieder, B. 1978. *Sol. Phys.* 57:245–53
137. Scuflaire, R. 1974. *Astron. Astrophys.* 36:107–11
138. Scuflaire, R., Gabriel, M., Noels, A. 1981. *Astron. Astrophys.* 99:39–42
139. Scuflaire, R., Gabriel, M., Noels, A.

1982. *Astron. Astrophys.* 110:50–53
140. Scuflaire, R., Gabriel, M., Noels, A. 1982. *Astron. Astrophys.* 113:219–22
141. Severny, A. B., Kotov, V. A., Tsap, T. T. 1976. *Nature* 259:87–89
142. Severny, A. B., Kotov, V. A., Tsap, T. T. 1984. In CATANIA. In press
143. Shibahashi, H. 1984. In CATANIA. In press
144. Shibahashi, H., Noels, A., Gabriel, M. 1983. *Astron. Astrophys.* 123:283–88
144a. Shibahashi, H., Noels, A., Gabriel, M. 1984. In CATANIA. In press
145. Spiegel, E. A., Unno, W. 1962. *Publ. Astr. Soc. Jpn.* 14:28–32
146. Staiger, J. 1984. Thesis. Univ. Freiburg, Fed. Rep. Germany
147. Stebbins, R., Wilson, C. 1983. *Sol. Phys.* 82:43–54
148. Stein, R. F., Leibacher, J. W. 1974. *Ann. Rev. Astron. Astrophys.* 12:407–35
149. Stein, R. F., Leibacher, J. W. 1981. In *The Sun as a Star*, ed. S. Jordan, pp. 289–300. *NASA SP-450*
150. Tassoul, M. 1980. *Ap. J. Suppl.* 43:469–90
151. Ulrich, R. K. 1970. *Ap. J.* 162:993–1002
152. Ulrich, R. K. 1982. *Ap. J.* 258:404–13
153. Ulrich, R. K., Rhodes, E. J. Jr. 1977. *Ap. J.* 218:521–29
154. Ulrich, R. K., Rhodes, E. J. Jr. 1983. *Ap. J.* 265:551–63
155. Unno, W., Osaki, Y., Ando, H., Shibahashi, H. 1979. *Nonradial Oscillations of Stars.* Univ. Tokyo Press. 323 pp.
156. Vandakurov, Yu. V. 1967. *Astron. Zh.* 44:786
157. van der Raay, H. B., Claverie, A., Isaak, G. R., McLeod, J. M., Roca Cortez, T., et al. 1984. In CATANIA. In press
158. Wiggins, R. A. 1972. *Rev. Geophys. Space Phys.* 10:251–85
159. Willson, R. C., Hudson, H. S. 1981. *Ap. J. Lett.* 244:L185
159a. Woodard, M. 1984. *Nature.* In press
160. Woodard, M., Hudson, H. 1983. *Sol. Phys.* 82:67–73
161. Woodard, M., Hudson, H. 1984. In CATANIA. In press
162. Worden, S. P., Simon, G. W. 1976. *Ap. J. Lett.* 210:L163–66
163. Yerle, R. 1981. *Astron. Astrophys. Lett.* 100:L23–25
164. Zahn, J.-P. 1970. *Astron. Astrophys.* 4:452–61

SUBJECT INDEX

CUMULATIVE INDEXES

CONTRIBUTING AUTHORS, VOLUMES 12–22

629

CHAPTER TITLES, VOLUMES 12–22

ORDER FORM

Annual Reviews Inc.

A NONPROFIT SCIENTIFIC PUBLISHER

4139 EL CAMINO WAY • PALO ALTO, CA 94306-9981 • (415) 493-4400

ers for Annual Reviews Inc. publications may be placed through your bookstore; subscription agent; par-
ating professional societies; or directly from Annual Reviews Inc. by mail or telephone (paid by credit
or purchase order). Prices subject to change without notice.

viduals: Prepayment required in U.S. funds or charged to American Express, MasterCard, or Visa.
itutional Buyers: Please include purchase order.
dents: Special rates are available to qualified students. Refer to Annual Reviews *Prospectus* or contact
ual Reviews Inc. office for information.
fessional Society Members: Members whose professional societies have a contractural arrangement with
ual Reviews may order books through their society at a special discount. Check with your society for
rmation.

ular orders: When ordering current or back volumes, please list the volumes you wish by volume number.
nding orders: (New volume in the series will be sent to you automatically each year upon publication. Can-
ation may be made at any time.) Please indicate volume number to begin standing order.
publication orders: Volumes not yet published will be shipped in month and year indicated.
fornia orders: Add applicable sales tax.
tage paid (4th class bookrate /surface mail) by Annual Reviews Inc.

ANNUAL REVIEWS SERIES		Prices Postpaid per volume USA/elsewhere	Regular Order Please send:	Standing Order Begin with:
			Vol. number	Vol. number
ual Review of ANTHROPOLOGY				
Vols. 1-10	(1972-1981)	$20.00/$21.00		
Vol. 11	(1982)	$22.00/$25.00		
Vols. 12-13	(1983-1984)	$27.00/$30.00		
Vol. 14	(avail. Oct. 1985)	$27.00/$30.00	Vol(s). _____	Vol. _____
ual Review of ASTRONOMY AND ASTROPHYSICS				
Vols. 1-19	(1963-1981)	$20.00/$21.00		
Vol. 20	(1982)	$22.00/$25.00		
Vols. 21-22	(1983-1984)	$44.00/$47.00		
Vol. 23	(avail. Sept. 1985)	$44.00/$47.00	Vol(s). _____	Vol. _____
ual Review of BIOCHEMISTRY				
Vols. 29-34, 36-50	(1960-1965; 1967-1981)	$21.00/$22.00		
Vol. 51	(1982)	$23.00/$26.00		
Vols. 52-53	(1983-1984)	$29.00/$32.00		
Vol. 54	(avail. July 1985)	$29.00/$32.00	Vol(s). _____	Vol. _____
ual Review of BIOPHYSICS				
Vols. 1-10	(1972-1981)	$20.00/$21.00		
Vol. 11	(1982)	$22.00/$25.00		
Vols. 12-13	(1983-1984)	$47.00/$50.00		
Vol. 14	(avail. June 1985)	$47.00/$50.00	Vol(s). _____	Vol. _____
ual Review of CELL BIOLOGY				
Vol. 1	(avail. Nov. 1985)	est. $27.00/$30.00	Vol. _____	Vol. _____
ual Review of EARTH AND PLANETARY SCIENCES				
Vols. 1-9	(1973-1981)	$20.00/$21.00		
Vol. 10	(1982)	$22.00/$25.00		
Vols. 11-12	(1983-1984)	$44.00/$47.00		
Vol. 13	(avail. May 1985)	$44.00/$47.00	Vol(s). _____	Vol. _____
ual Review of ECOLOGY AND SYSTEMATICS				
Vols. 1-12	(1970-1981)	$20.00/$21.00		
Vol. 13	(1982)	$22.00/$25.00		
Vols. 14-15	(1983-1984)	$27.00/$30.00		
Vol. 16	(avail. Nov. 1985)	$27.00/$30.00	Vol(s). _____	Vol. _____

1

		Prices Postpaid per volume USA/elsewhere	Regular Order Please send: Vol. number	Standing O Begin with Vol. numb
Annual Review of ENERGY				
Vols. 1-6	(1976-1981)	$20.00/$21.00		
Vol. 7	(1982)	$22.00/$25.00		
Vols. 8-9	(1983-1984)	$56.00/$59.00		
Vol. 10	(avail. Oct. 1985)	$56.00/$59.00	Vol(s).	Vol.
Annual Review of ENTOMOLOGY				
Vols. 8-16, 18-26	(1963-1971; 1973-1981)	$20.00/$21.00		
Vol. 27	(1982)	$22.00/$25.00		
Vols. 28-29	(1983-1984)	$27.00/$30.00		
Vol. 30	(avail. Jan. 1985)	$27.00/$30.00	Vol(s).	Vol.
Annual Review of FLUID MECHANICS				
Vols. 1-5, 7-13	(1969-1973; 1975-1981)	$20.00/$21.00		
Vol. 14	(1982)	$22.00/$25.00		
Vols. 15-16	(1983-1984)	$28.00/$31.00		
Vol. 17	(avail. Jan. 1985)	$28.00/$31.00	Vol(s).	Vol.
Annual Review of GENETICS				
Vols. 1-15	(1967-1981)	$20.00/$21.00		
Vol. 16	(1982)	$22.00/$25.00		
Vols. 17-18	(1983-1984)	$27.00/$30.00		
Vol. 19	(avail. Dec. 1985)	$27.00/$30.00	Vol(s).	Vol.
Annual Review of IMMUNOLOGY				
Vols. 1-2	(1983-1984)	$27.00/$30.00		
Vol. 3	(avail. April 1985)	$27.00/$30.00	Vol(s).	Vol.
Annual Review of MATERIALS SCIENCE				
Vols. 1-11	(1971-1981)	$20.00/$21.00		
Vol. 12	(1982)	$22.00/$25.00		
Vols. 13-14	(1983-1984)	$64.00/$67.00		
Vol. 15	(avail. Aug. 1985)	$64.00/$67.00	Vol(s).	Vol.
Annual Review of MEDICINE: Selected Topics in the Clinical Sciences				
Vols. 1-3, 5-15	(1950-1952; 1954-1964)	$20.00/$21.00		
Vols. 17-32	(1966-1981)	$20.00/$21.00		
Vol. 33	(1982)	$22.00/$25.00		
Vols. 34-35	(1983-1984)	$27.00/$30.00		
Vol. 36	(avail. April 1985)	$27.00/$30.00	Vol(s).	Vol.
Annual Review of MICROBIOLOGY				
Vols. 17-35	(1963-1981)	$20.00/$21.00		
Vol. 36	(1982)	$22.00/$25.00		
Vols. 37-38	(1983-1984)	$27.00/$30.00		
Vol. 39	(avail. Oct. 1985)	$27.00/$30.00	Vol(s).	Vol.
Annual Review of NEUROSCIENCE				
Vols. 1-4	(1978-1981)	$20.00/$21.00		
Vol. 5	(1982)	$22.00/$25.00		
Vols. 6-7	(1983-1984)	$27.00/$30.00		
Vol. 8	(avail. March 1985)	$27.00/$30.00	Vol(s).	Vol.
Annual Review of NUCLEAR AND PARTICLE SCIENCE				
Vols. 12-31	(1962-1981)	$22.50/$23.50		
Vol. 32	(1982)	$25.00/$28.00		
Vols. 33-34	(1983-1984)	$30.00/$33.00		
Vol. 35	(avail. Dec. 1985)	$30.00/$33.00	Vol(s).	Vol.

2